Advances in Intelligent Systems and Computing

Volume 1328

The series "Advances in Intelligent Systems and Computing" contains publications on theory, applications, and design methods of Intelligent Systems and Intelligent Computing. Virtually all disciplines such as engineering, natural sciences, computer and information science, ICT, economics, business, e-commerce, environment, healthcare, life science are covered. The list of topics spans all the areas of modern intelligent systems and computing such as: computational intelligence, soft computing including neural networks, fuzzy systems, evolutionary computing and the fusion of these paradigms, social intelligence, ambient intelligence, computational neuroscience, artificial life, virtual worlds and society, cognitive science and systems, Perception and Vision, DNA and immune based systems, self-organizing and adaptive systems, e-Learning and teaching, human-centered and human-centric computing, recommender systems, intelligent control, robotics and mechatronics including human-machine teaming, knowledge-based paradigms, learning paradigms, machine ethics, intelligent data analysis, knowledge management, intelligent agents, intelligent decision making and support, intelligent network security, trust management, interactive entertainment, Web intelligence and multimedia.

The publications within "Advances in Intelligent Systems and Computing" are primarily proceedings of important conferences, symposia and congresses. They cover significant recent developments in the field, both of a foundational and applicable character. An important characteristic feature of the series is the short publication time and world-wide distribution. This permits a rapid and broad dissemination of research results.

Indexed by DBLP, EI Compendex, INSPEC, WTI Frankfurt eG, zbMATH, Japanese Science and Technology Agency (JST), SCImago.

All books published in the series are submitted for consideration in Web of Science.

More information about this series at http://www.springer.com/series/11156

Michael E. Auer · Tiia Rüütmann
Editors

Educating Engineers for Future Industrial Revolutions

Proceedings of the 23rd International Conference on Interactive Collaborative Learning (ICL2020), Volume 1

Springer

Editors
Michael E. Auer
Carinthia University of Applied Sciences
Villach, Austria

Tiia Rüütmann
Tallinn University of Technology
Tallinn, Estonia

ISSN 2194-5357 ISSN 2194-5365 (electronic)
Advances in Intelligent Systems and Computing
ISBN 978-3-030-68197-5 ISBN 978-3-030-68198-2 (eBook)
https://doi.org/10.1007/978-3-030-68198-2

This Springer imprint is published by the registered company Springer Nature Switzerland AG
The registered company address is: Gewerbestrasse 11, 6330 Cham, Switzerland

Preface

ICL2020 was the 23rd edition of the International Conference on Interactive Collaborative Learning and the 49th edition of the IGIP International Conference on Engineering Pedagogy.

This interdisciplinary conference aims to focus on the exchange of relevant trends and research results as well as the presentation of practical experiences in interactive collaborative learning and engineering pedagogy.

ICL2020 has been organized by University of Technology Tallinn, Estonia, from September 23 to 25, 2020, as an online event.

This year's theme of the conference was "Educating Engineers for Future Industrial Revolutions."

Again outstanding scientists from around the world accepted the invitation for keynote speeches:

- **Ruth Graham**, Higher Education Consultant, USA. Speech title: ***Reward and Recognition of University Teaching***
- **Tarmo Soomere**, President of Estonian Academy of Science. Speech title: ***Connecting Science, Training, Society and Policy***
- **Hanno Hortsch**, Technische Universität Dresden (TUD), President of IGIP. Speech title: ***The New Prototype Curriculum of IGIP in Engineering Pedagogy***

The following very interesting workshops have been held:

- ***Teaching Environmentally and Sustainability-Conscious Design Projects in Higher Education, using GRANTA EduPack***
  Facilitator: **Vakhitova Tatiana Vadimovna** PhD, ANSYS Granta/Academic Relations Team (UK)
- ***Decentralizing Education Using Blockchain Technology***
  Facilitator: Dr. **Alexander Mikroyannidis**, Knowledge Media Institute, the Open University (UK)
- ***Idea Generation Board Game for Product Development "Create Products"***
  Facilitators: **Erich Scheffl** and **Jürgen Jantschgi**, (Austria)

- ***IGIP Workshop on methodologies to build conceptual questions for assessing important misconceptions in engineering related areas***
  Facilitators: **Teresa Restivo** and **Diana Urbano** (Portugal)

We would like to thank the organizers of the following special sessions:

- ***Games in Engineering Education (GinEE)***
  Chairs
  Matthias C. Utesch, Technical University of Munich, Germany
  Marek Milosz, Faculty of Electrical Engineering and Computer Science, Lublin University of Technology, Poland
- ***Entrepreneurship in Engineering Education 2020" (EiEE'20)***
  Chairs
  Stefan Vorbach, University of Technology Graz, Austria, stefan.vorbach@tu-graz.at
  Jürgen Jantschgi, HTL Wolfsberg, Austria, juergen.jantschgi@htl-wolfsberg.at
- ***Public-Private Partnership in Engineering Education (SYNERGY)***
  Chair
  Svetlana V. Barabanova, Kazan National Research Technological University, Russia
- ***IoT, IIoT and Energy Harvesting in Future of Industrial Revolution (IoT-EHFIR)***
  Chairs
  Doru Ursutiu and **Paul Nicolae Borza**, "Transilvania" University Brasov, Romania

Since its beginning, this conference is devoted to new approaches in learning with a focus to collaborative learning and engineering education. We are currently witnessing a significant transformation in the development of education. There are at least three essential and challenging elements of this transformation process that have to be tackled in education:

- the impact of globalization and digitalization on all areas of human life, and
- the exponential acceleration of the developments in technology as well as of the global markets and the necessity of flexibility and agility in education
- the new generation of students, who are always online and do not know live without Internet

Therefore, the following main themes have been discussed in detail:

- Collaborative Learning
- New Learning Models and Applications
- Project-Based Learning
- Game-Based Education
- Educational Virtual Environments
- Computer-Aided Language Learning (CALL)
- Teaching Best Practices

- Engineering Pedagogy Education
- Public-Private Partnership and Entrepreneurship Education
- Research in Engineering Pedagogy
- Evaluation and Outcomes Assessment
- Internet of Things and Online Laboratories
- IT and Knowledge management in Education
- Real-World Experiences

As submission types have been accepted:

- Full Paper, Short Paper
- Work in Progress, Poster
- Special Sessions
- Round Table Discussions, Workshops, Tutorials

All contributions were subject to a double-blind review. The review process was very competitive. We had to review more than 400 submissions. A team of about 225 reviewers did this terrific job. Our special thanks go to all of them.

Due to the time and conference schedule restrictions, we could finally accept only the best 156 submissions for presentation.

The conference had near 200 online participants from 34 countries from all continents.

Our special thank goes to **Anton Jegorov** and the technicians of University of Technology Tallinn, who made the online conference a reality. We thank **Sebastian Schreiter** for the technical editing of this proceedings.

ICL2021 will be held in Dresden, Germany.

Michael E. Auer
ICL General Chair

Tiia Rüütmann
ICL2020 Chair

Organization

Committees

General Chair

Michael E. Auer	CTI, Frankfurt/Main, Germany

ICL2020 Conference Chair

Tiia Rüütmann	Tallinn University of Technology, Tallinn, Estonia

International Chairs

Samir A. El-Seoud	The British University in Egypt, Africa
Neelakshi Chandrasena Premawardhena	University of Kelaniya, Sri Lanka, Asia
Alexander Kist	University of Southern Queensland, Australia/Oceania
Alaa Ashmawy	American University Dubai, Middle East
David Guralnick	Kaleidoscope Learning New York, USA, North America
Uriel Cukierman	UTN Buenos Aires Argentina, Latin America

Honorary Advisors

Hanno Hortsch (IGIP President)	Technical University Dresden, Germany
Mati Lukas	Tallinn University of Technology, Estonia
Hendrik Voll	Tallinn University of Technology, Estonia
Hans J. Hoyer	IFEES/GEDC General Secretary
Viacheslav Prikhodko	Moscow Technical University, Russia
Krishna Vedula	University of Massachusetts Lowell, USA

Technical Program Chair

Sebastian Schreiter	IAOE France

Workshop and Tutorial Chair

Barbara Kerr	Ottawa University, Canada

Special Sessions Chair

Andreas Pester	The British University in Egypt

Publication Chair

Sebastian Schreiter	IAOE France

Awards Chairs

Tiia Rüütmann	Tallinn University of Technology, Estonia
Teresa Restivo	University of Porto, Portugal

Local Arrangement Chairs

Merili Deemant	Tallinn University of Technology, Estonia
Meeli Semjonov	Tallinn University of Technology, Estonia

Senior Program Committee Members

Samir Abou El-Seoud	The British University in Egypt
George Ioannidis	University of Patras, Greece
Eleonore Lickl	College for Chemical Industry, Vienna, Austria
Andreas Pester	The British University in Egypt
Tatiana Polyakova	Moscow State Technical University, Russia
Doru Ursutiu	University of Brasov, Romania
Axel Zafoschnig	Ministry of Education, Austria

Program Committee Members

Alexander Soloviev	Russia
Christian Guetl	Austria
Christos Bouras	Greece
Cornel Samoila	Romania
Demetrios Sampson	Greece
Despo Ktoridou	Cyprus
Hants Kipper	Estonia

Herwig Rehatschek	Austria
Igor Verner	Israel
Imre Rudas	Hungary
Istvan Simonics	Hungary
Ivana Simonova	Czech Republic
Jürgen Mottok	Germany
Martin Bilek	Czech Republic
Matthias Utesch	Germany
Monica Divitini	Norway
Nael Barakat	USA
Pavel Andres	Czech Republic
Rauno Pirinen	Finland
Santi Caballé	Spain
Teresa Restivo	Portugal
Tiia Rüütmann	Estonia
Yu-Mei Wang	USA

Local Organizing Committee Members

Jakob Kübarsepp	Tallinn University of Technology, Estonia
Fjodor Sergejev	Tallinn University of Technology, Estonia
Kristo Karjust	Tallinn University of Technology, Estonia
Tiia Rüütmann	Tallinn University of Technology, Estonia
Jaak Umborg	Tallinn University of Technology, Estonia
Ants Soon	Tallinn University of Technology, Estonia
Raivo Sell	Tallinn University of Technology, Estonia
Inna Kamenev	Tallinn University of Technology, Estonia
Jaan Tamm	Tallinn University of Technology, Estonia
Merili Deemant	Tallinn University of Technology, Estonia
Meeli Semjonov	Tallinn University of Technology, Estonia
Marika Olander	Tallinn University of Technology, Estonia

Contents

Collaborative Learning

Cross-Disciplinary Remote Course for Mixed Group of Students and Professionals

Oskari Lähdeaho[1(✉)] and Olli-Pekka Hilmola[1,2]

[1] LUT University, Kouvola Unit, Prikaatintie 9, 45100 Kouvola, Finland
oskari.lahdeaho@lut.fi
[2] Tallinn University of Technology, Estonian Maritime Academy,
Kopli 101, 11712 Tallinn, Estonia

Abstract. The vast amount of openly available information enables data-driven study of the surrounding world. The aim of this article is to present dissemination of methods to calculating exhaust gas emission and assessing the effect of weather in dispersion of said emissions that are possible to carry out with data from open data repositories. As the measures to prevent spread of COVID-19 virus has led to a temporary cease of contact learning, this course is implemented remotely online. The course offers a controlled environment to test the foreknowledge of the participants regarding environmental challenges tied in transport, and the effectiveness of the remote course in enhancing the awareness in said issues. Data will be gathered on the success of the course. It is used to enable continuous improvement of the course in terms of quality of the content and successfulness of the remote setting. In addition, the gathered data is used to determine the effectiveness of remote education and the preparedness of scholars to operate under special conditions. Through collaborative effort of the course organizers and attendees, the remote implementation can be successful. Also, the mixed group of organizers, students and professionals enable problem-oriented approach to the topic that is difficult to limit into a specific discipline. The pandemic and succeeding mitigation of COVID-19 spreading showcases the importance of planning education for such special situations. While the resulting circumstances are not pleasant, they serve as a promoter for rigorous planning of educational events.

Keywords: Road traffic emissions · Road weather · International · Cross-disciplinary · Education

1 Introduction

The recent concern on environmental sustainability of human presence on earth has sprung action towards lowering the negative environmental impact of all sectors. One of the major contributing factors to enhanced environmental sustainability of existence is education of this subject matter. Through knowledge will the current and coming generations be able to perform informed decision-making on legislation and regulation as well as in organizations or at individual level. The current targets globally towards reduced emissions are relatively strict when considering the short timeframe they have been scheduled. For example, massive reductions in EU are set to be reached by 2030

M. E. Auer and T. Rüütmann (Eds.): ICL 2020, AISC 1328, pp. 3–12, 2021.
https://doi.org/10.1007/978-3-030-68198-2_1

[1]). Therefore, it is vital for achieving those goals to spread the knowledge on possible emission mitigation methods to existing professionals. Furthermore, the acquainted perception on status quo by those professionals enhances the relevancy of education passed to aspiring professionals, i.e., university students.

As the ongoing COVID-19 pandemic has forced almost all nations to act under special conditions, educational activities must also adapt to these circumstances. Social distancing practices have forced majority of the universities to close their doors, effectively ceasing all contact learning. Thus, remote courses have emerged as the focal part of higher education [2]. Whereas certain courses have been carried out remotely in the past, most of the courses must be implemented such way during 2020. While incumbent view has been that remote learning, e.g., via online, is not as effective as conventional contact learning, this forced "experimental" setting could prove that modern technologies can enable a sound learning environment by essentially removing the spatial and temporal constraints of a physical class. In addition to educational effectiveness, remote learning can significantly enhance the environmental sustainability of education by mitigating, e.g., travel and the need for printed material [3]. Furthermore, in the setting of higher education, concept of environmental sustainability extends from substantial means to reduce negative environmental impact to also cover related spread of knowledge and research [4].

This research focuses on a course with the topic of methods for assessing transport pollution and approaches to reduce the environmental burden. The implementation of this course involves organizations from Finland and Russia, and it is targeted to university students as well as professionals from both countries. The teaching staff consists of scholars and experts involved with the mentioned subject. The course is carried out remotely via online learning environment, due to travel restrictions caused by the pandemic and to allow higher accessibility for participants from different countries, life situations and backgrounds [5–8]. Hence, this research focuses on answering the following questions:

- RQ1: How should the remote course involving different organizations and nationalities be designed?
- RQ2: What are the immediate and possible future benefits of implementing the proposed type of remote learning course?

This article proceeds by reviewing the relevant literature on benefits and challenges associated with remote learning in Sect. 2. Thereafter, the proposed course design is presented in Sect. 3, succeeded by the expected results of the course implementation in Sect. 4. Discussion reflecting the course design decisions with the established literature as well as conclusions for this study with outlook on future steps regarding the remote course implementation are presented in Sect. 5.

2 Remote Collaborative Learning

Da Silva et al. [9] raise concerns regarding student interaction and degree of involvement when implementing a course in a full remote manner. While pedagogy remains as the central element to enable efficient education, in the case of remote learning the level of technology must offer adequate support [10]. Lakhal et al. [6] assign a high cost for technology and connection related issues in remote teaching. Especially, during special circumstances, such as social distancing to mitigate spread of COVID-19, servers for remote meetings are under extended stress, which could quite possibly cause connection issues. In online teaching environment, perceived isolation and lack of interaction could undermine the students' learning capabilities [6]. Glukhikh and Norina [11], as well as Ramsey et al. [12] note that video lectures lack informal aspects of interaction present during a physical class. This in turn could lead to suppressed development of professional communication skills. In addition, video lectures limit the communication through facial expressions and body language [2]. Furthermore, courses implemented online reduce the possibilities for improvised development decisions regarding the upcoming lesson made on the fly by the teacher [10]. However, Raes et al. [13] showcase a study, where the teacher could successfully use their existing methods of communication in a remote setting.

In their research, Lakhal et al. [6] establish conditions for successfully implementing remote learning. These conditions consist of appropriate selection of supporting technologies and tools (e.g., experience shows that sound quality is one of the key factors). Using proper equipment for delivering the education and employing technical support personnel is another requisite. In addition, adapting the selected technologies and tools through testing and practice, recording the online sessions for further use, complementing the online synchronous sessions with asynchronous communication (e.g., message boards, wikipages and email), mixing different educational approaches and strategies, and limiting student population in the remote setting (less than 25 participants is proposed as an optimal). In correspondence to the special learning conditions caused by the COVID-19 pandemic, Bao [2] composes six instructional strategies to promote online learning in a case study about Peking University. Strategies include preparing for unexpected technical difficulties (e.g., with Internet connection), separating the course content to shorter modules (length of 25 min is proposed), and emphasizing the manner in which voice is used due to the lack of other aspects of communication. Also, teaching assistants should be more closely involved to the process, focus of the course should be on the independent learning outside of the online class, and the online and offline studies need to be effectively mixed (e.g., through assignments and additional materials). Raes et al. [13] raise the importance of effectively using quizzes during remote classes to boost students' learning motivation.

Several main points in establishing successful remote learning can be deduced from the findings of the abovementioned scholars. In short, used technical equipment and solutions should be carefully considered, the teaching should be arranged with those tools in mind, the interaction with the participants must be actively established, and asynchronous communication and additional offline materials are required to complement the online sessions. In addition, the role of teacher in higher education should shift from traditional to more represent a facilitator of independent learning for the students [11]. Moreover, in those disciplines where it is appropriate, remote laboratories improve the educational quality of a remote course [14]. While these types of laboratories are being utilized increasingly, some avoid them due to relatively high costs and low interest from potential participants to justify those costs [7].

A successful implementation of remote learning could bring various benefits for the providers of education (e.g., universities) as well as to the students. Offering remote learning enhances the flexibility and accessibility of the studies [5–8]. This in turn could increase attractiveness of the study programs offered by a university [7]. Remote setting also offers a pathway for collaboration between industries and universities, where students can learn from challenges faced by companies in practice [15]. While the current situation forces, the wide emerge of remote learning possibilities, in future the already established remote courses could be repurposed as blended courses (introducing present and remote learners to the same class). Such courses with blended synchronous learning can be utilized to mitigate the shortcomings in interactivity of pure remote learning [16]. In addition, blended learning promotes accessibility of courses by enabling simultaneous learning on-site and in other locations [17], e.g., during travel restrictions.

3 Course Design

The remote course will be offered to students from Finnish and Russian Universities in the area spanning from Southeast Finland to Northwest Russia (St. Petersburg). In addition, professionals from these regions, who are interested in the theme, are welcome to participate. Since the participants form a group of people from different countries, the course will be held in English. Common language is important to promote collaborative learning and interaction throughout the course. As the course is still being piloted, it is not a part of any study program and can be regarded as elective. Despite being implemented for the first time and not being a compulsory course to any study program in university, number of university students and local professionals have already enrolled to participate.

The purpose of the course is to showcase methods for assessing road transport emissions and approaches to reduce the environmental burden of road transport activities. The course offers the participants basic knowledge on key properties and processes in atmosphere. These will be applied to create understanding on the influence of meteorological factors to the spread of emissions in living areas and in the vicinity of highways. In addition, methods and means for measuring anthropogenic pollution and reducing environmental safety risks. Through these methods, the participant will be able to assess the risks of road transport emissions to environment and human health. Moreover, the assessment of environmental impact of road transportation will be learned through real life scenarios, which require use of openly accessible data via internet. The course will introduce the participants to different types of data needed in the environmental assessment and where this data can be gathered openly by anyone.

The remote setting of the course will combine synchronous and asynchronous communication to ensure achievement of the established learning goals. The synchronous part of the course will be delivered in form of remote lectures, that include classical lecturing on the topic but also group work, assignments and related discussion. Asynchronous part of the communication will be established through course message boards, where the participants can communicate with each other or with the course organizers. Furthermore, video lectures will be available on the course website to support the individual learning.

As a prerequisite, the participants of the course are required to study a book on emission calculation and assessment methods written by the organizers. The book will introduce the reader to the topic of this course and offer means to perform environmental impact assessment of road transport on their own. In addition to the book, the course will be supported by additional learning materials which consist of peer-reviewed journal articles, other texts and books, and learning material made by the organizers. The repository for the learning materials on the course website can be seen in Fig. 1. Unfortunately, the main web page for the course (which contains the video lectures and the assignments) is still under construction and cannot be showcased in this article. Throughout the course, at the end of each subtopic, the participants must answer to an online test on the content of the respective subtopic. These are completed in form of quizzes, as recommended by Raes et al. [13].

Fig. 1. Learning material repository on the course website [18].

The composition of the teaching group for this course includes Finnish and Russian specialists with diverse experiences and backgrounds. Therefore, the course is based on combined knowledge in road transportation technology, business and management, economics, sustainability studies, and meteorology of the organizers. This cross-disciplinary setting enables problem-oriented learning from the common topic of road traffic emissions. The course is constructed from seven separate modules (presented in Table 1), each of which contribute on the common goal of solving environmental sustainability issues in road traffic. The modules contain learning of modeling and calculation techniques, emission impact assessment and best practices in emissions mitigation.

Table 1. Learning modules of the proposed remote course.

Topic	Subtopic	Number of classes
Weather, climate and traffic pollution	The influence of meteorological conditions on the spread of anthropogenic pollution	2
	Devices and methods for assessing anthropogenic pollution	4
	Estimating direct air quality emissions of the road transport and air quality dispersion modeling	2
Risks of environmental and transport safety	Road traffic emissions calculation and analysis	2
	The main directions of reducing the negative impact of the road transport on the environment	2
	Using weather information to improve road safety	2
	Road weather forecast as a tool to increase road transport safety	2

As Walder [8] states, assessment is one of the most important aspects in providing education. Through assessment, continuous improvement of education can be achieved. Hence, the proposed course will include feedback gathering from the participants in the form of survey and free form discussion at the end. The feedback will concern three topics: quality of teaching, success of the remote setting and the adequateness of the course material and content. Despite all the educational innovation and employed technology, quality of the teaching remains as cornerstone for successful education [10], which is why it is vital aspect to measure. As setting up remote learning environment requires technological solutions, the success in this area should be monitored also. Those results can be in turn used to evaluate the further needs for technological improvement, and to include only the needed solutions in the mix, as proposed by Lakhal et al. [6]. Lastly, as the theme of the course is subject to constant development through research, best practices and regulatory action, the course content and offered materials need to remain up to date. Therefore, it is important to measure the content via feedback, especially since the mixed group of participants can voice their opinion from the perspective of both theory and practice. This data will also help the organizers to improve their book on emission calculation and environmental impact assessment.

The course will be held for the first time in the autumn of 2020, and quite possibly social distancing and travel restrictions caused by the COVID-19 pandemic are still in place by that time. Hence, the course will be held completely in a remote setting. As uncertainty lingers over the possible start of contact learning at universities, the course might be run remotely for several times. Moreover, international travel could remain restricted even after universities open their doors, which would create a demand for blended learning environments. Such environment could be one where participants gather in their local university and the course is ran remotely between these universities, allowing collaboration face-to-face on site as well as between the remote locations.

4 Expected Results

As mentioned, the course will be held for the first time in the autumn 2020. Hence, the current work has been dedicated on designing and preparing the course according to the documented best practices in the scientific literature. In addition, the long spanning experience and expertise of the members in the organizing group has been consulted extensively in this phase. It is important to define expected results in order to establish measurement and development procedures, and to subsequently enable sound process of continuous improvement. In this section of the article, expected results are drawn based on the competences of the organizing group and the typical challenges in establishing a remote course.

Since the facilities used to create the remote learning setting are up to date and tested regularly, most probably the bottle neck in delivering the education remotely will be in the participant side equipment (computers and their audiovisual capabilities) and in the interface between these (internet connection). Since the course at hand is implemented by multiple organizations from locations with different standards for, e.g., software in education, the remote setting is implemented with using only the most general tools (Microsoft Office package, Skype and similar software) and placing the materials on a website run by these organizations. This approach further limits some of the problems that specific software could create in terms of compatibility. Therefore, the technological set up and performance of the remote setting is expected to be optimal.

As established, the organizing group consists of experts in various fields from both academic and professional backgrounds. The group also possesses vast amount of experience regarding education, be it in the higher education or professional setting. Hence, the quality of teaching can be expected to be of high level. Of course, it is important to regard the challenges that multicultural and cross-disciplinary setting brings to any educational event. The lecturers must be able to deliver their lecture in a language that is common for the group, i.e., English in this case. Furthermore, since the course employs multiple different disciplines, it is highly important to define common terminology and concepts, and make sure that these are used correctly in the intercourse.

The topic of this course is extremely relevant in the contemporary world, which leads to constant research, development, cultural and technological change as well as growing number of observations, conclusions and differing opinions. Thus, courses focused on environmental challenges of modern world must practice continuous improvement to keep up and stay relevant. In other words, the described operational environment requires the course material and topics to be excellent in quality, significance and originality. Therefore, the quality of the educational content seems to be the most critical point for the success of the course. It is difficult to estimate how the course will perform in this aspect for the first time, but the expertise of the organizing group should guarantee adequate course content in the mentioned qualities.

As the course focuses on environmental sustainability of road transport, it is quite fitting to also evaluate the sustainability of the course in terms of transport involved (e.g., required travelling). As pointed out by Roy et al. [3] and Caird and Roy [4], remote courses can significantly reduce the stress on environment created by higher education. Measuring the environmental sustainability of the course at hand also creates an excellent case study for the participants.

5 Concluding Discussions

This article focuses in designing a remote course delivering cross-disciplinary knowledge and education for a mixed group of student and professionals. The work was completed by following the frameworks of Lakhal et al. [6] and Bao [2], and by employing the experience of the scholars in the organizing group. This approach led to identifying three critical points for the success of this remote course. First and foremost, the level of education and pedagogy remains vital in implementing any course. In the case of modern higher education and remote courses, the teacher should act more as a facilitator for independent study, than as a lecturer [11]. The mix of technologies used in the remote setting must be then selected in order to complement the pedagogical competences of the teacher. In addition, it is recommendable to select only the amount of different technologies that is truly needed. Lastly, the provided content within the course should be relevant. The adequateness of the content seems to be the most critical point in success for this course. Therefore, extended attention has been given to the course content, which includes a book on the course topic written by the organizers, video lectures and additional readings. Also, as suggested by Raes et al. [13], the course is complemented with quizzes in between the subtopics to engage the participants more. The communication in this course is carried out both synchronously (during the online classes) and asynchronously (via message boards).

As the course is piloting a setting that is currently not very common, the importance of feedback and research to enable improvement is high. After the first run, lots of room for development will probably remain. The organizers must be ready to improve all the measured aspects of the course. Especially the course content can be expected to be subject to constant improvement due to the nature of the course topic. Therefore, the continuous improvement of the course requires predictive work by the organizers in addition to regarding the course feedback.

If successful, such remote courses as the one described in this article could enhance the flexibility and accessibility of the studies in the involved universities [5–8]. This is important benefit for modern universities, as it could increase the attractiveness towards new students [7] in the world where competition in higher education is dire. In addition, a course which welcomes professional participants also enables universities to more effectively solve societal and business practice challenges [15].

Similarly as described by Pisoni [15], after the current special conditions ease up (e.g., universities open their doors, travel restrictions start to be lifted) remote implementation offers a pathway to continue interorganizational collaboration via synchronous online classes, where the physical class is situated in one of the partner institutes and other participants may join via online. At the same time potential for extending the course to other institutions could be recognized (i.e., appreciation from the voluntary participants). In this case, the course could act as a bridge for extended collaboration between Finnish and Russian higher education institutes. This type of development would be desired, as it enables continuous longitudinal development of this particular course, as well as addition of new courses or even creation of a specific program revolving around road traffic emissions. Nowadays, even the implementation of fully remote or blended university program is not out of the question [12].

References

1. European Commission, A European Strategy for Low-Emission Mobility. https://eur-lex.europa.eu/legal-content/en/TXT/?uri=CELEX:52016DC0501, Accessed 26 May 2020
2. Bao, W.: COVID-19 and online teaching in higher education: a case study of Peking University. Hum. Behav. Emerg. Technol. **2**, 113–115 (2020)
3. Roy, R., Potter, S., Yarrow, K.: Designing low carbon higher education systems: environmental impacts of campus and distance learning systems. Int. J. Sustain. High. Educ. **9**(2), 116–130 (2008)
4. Caird, S., Roy, R.: Sustainable higher education systems. In: Encyclopaedia of Sustainability and Higher Education, p. 11. Springer, Cham (2019)
5. Lightner, C.A., Lightner-Laws, C.A.: A blended model: simultaneously teaching a quantitative course traditionally, online, and remotely. Interact. Learn. Environ. **24**(1), 224–238 (2016)
6. Lakhal, S., Bateman, D., Bédard, J.: Blended synchronous delivery modes in graduate programs: a literature review and how it is implemented in the master teacher program. Collected Essays Learn. Teach. **10**, 47–60 (2017)
7. Katzis, K., Dimopoulos, C., Meletiou-Mavrotheris, M., Lasica, I.E.: Engineering attractiveness in the European educational environment: can distance education approaches make a difference? Educ. Sci. **8**(1), 16 (2018)
8. Walder, A.M.: Pedagogical innovation in Canadian higher education: professors' perspectives on its effects on teaching and learning. Stud. Educ. Eval. **54**, 71–82 (2017)
9. da Silva, M.G., Pereira, L.D., Carrilho, J.A.D., Neto, J., Marcelino, M.J., Mateus, M., Silva Brito, N., Pedrosa, S.: A distance-learning course on indoor environmental comfort in buildings. Int. J. Interact. Mob. Technol. **11**(5), 118–129 (2017)
10. Rehn, N., Maor, D., McConney, A.: Navigating the challenges of delivering secondary school courses by videoconference. Br. J. Edu. Technol. **48**(3), 802–813 (2017)
11. Glukhikh, V.N., Norina, N.V.: Experience of technical disciplines remote training at the St. Petersburg State University of Architecture and Civil Engineering. Educ. Inf. Technol. **21**(5), 1401–1412 (2016)
12. Ramsey, D., Evans, J., Levy, M.: Preserving the seminar experience. J. Polit. Sci. Educ. **12** (3), 256–267 (2016)
13. Raes, A., Vanneste, P., Pieters, M., Windey, I., Van Den Noortgate, W., Depaepe, F.: Learning and instruction in the hybrid virtual classroom: an investigation of students' engagement and the effect of quizzes. Comput. Educ. **143**, 103682 (2019)
14. Pastor, R., Tobarra, L., Robles-Gómez, A., Cano, J., Hammad, B., Al-Zoubi, A., Hernández, R., Castro, M.: Renewable energy remote online laboratories in Jordan universities: tools for training students in Jordan. Renew. Energy **149**, 749–759 (2020)
15. Pisoni, G.: Collaborative learning in a shared course between two Universities. In: ACM International Conference Proceeding Series, pp. 534–537 (2019)
16. Szeto, E., Cheng, A.Y.N.: Towards a framework of interactions in a blended synchronous learning environment: what effects are there on students' social presence experience? Interact. Learn. Environ. **24**(3), 487–503 (2016)

17. Fenn, N.E., Sprunger, T., Gonzalvo, J.D., Isaacs, A.N., Sheehan, A.H., Ramsey, D.C., Beckett, R.D.: Global collaboration to deliver a live pharmacy teaching and learning curriculum. Curr. Pharm. Teach. Learn. **12**, 307–312 (2020)
18. Green InterTraffic, Documents of the project. https://en.greenintertraffic.ru/greenintertraffic/materials/documents-of-the-project.html, Accessed 02 July 2020

Realization of Elite Education at Engineering Higher Education Institution

Olga Khatsrinova, Julia Khatsrinova(✉), and Veronika Bronskaya

Kazan National Reaserch Technological University, Kazan, Russian Federation
khatsrinova12@mail.ru

Abstract. The article reflects the organizational and methodological aspects of teaching elite students in an engineering university. The task of the "Introduction to Engineering Activities" discipline should not be a simple study of the essence, time stages of engineering activity, but a problematic description of the relationship between past achievements and current problems of the present. After studying the course, we revealed a high level of students' interest in engineering activities, the level of readiness for project activities increased. In general, students have a higher importance of educational - cognitive and professional motives.

Keywords: Elite education · Student · Technological University

1 Context

In recent decades, the contradiction between the needs of society in the formation of engineering personnel, ready for active creative activity and quick adaptation to new, often changing conditions and the ability of graduates to carry out algorithmic activities, has been clearly indicated.

The improvement of the engineering education system should be based on the analysis of fundamental reformist shifts in the scientific, technical and socio-economic spheres of activity in the 21st century. Today it is necessary: a systematic representation of the goals and values of engineering activities in the future; taking into account the emerging philosophy of vocational education; taking into account the personal characteristics of a specialist engineer in his own way of entering the engineering culture; installation on self-development and professional creativity; taking into account the connections of academic disciplines of various blocks of the curriculum for the training of engineers, etc.

The engineering education system is designed to create conditions for the evolutionary formation of a new generation of highly educated professionals in the field of engineering, for which the installation of self-development, professional culture and skill, the development of an individual style of activity are priority throughout life.

Note that improving the of will ensure the high competitiveness of the future engineer in the face of intense competition for a job, as well as for the quality of work performed.

M. E. Auer and T. Rüütmann (Eds.): ICL 2020, AISC 1328, pp. 13–29, 2021.
https://doi.org/10.1007/978-3-030-68198-2_2

Every year, educational organizations market a large number of specialists with higher education. At the same time, about 30% of them are not employed in their specialty. One of the reasons for this situation is the inconsistency of the educational services market and labor market requirements for the quality of training. The requirements of employers are focused mainly on the conformity of professional training with the requirements of real professional activity and the desire of specialists to increase professional growth.

Fundamental changes in the economies of developed countries occur mainly due to high technology. There is a revolution in the technical, technological and economic spheres. This can be seen by accelerating the emergence of new products on the market, their production cycles and shortening their life time. Moreover, this acceleration is not 1–2%, but an order of magnitude higher in 3–4 years. The life time of a product on the market begins to be comparable to the time it was created. For example, if a product is in demand throughout the year and then its popularity falls due to the appearance of another similar product with higher consumer qualities, it is often necessary to spend from six months to a year to develop and introduce it into production. Rapid changes in production based on the latest information technologies are also a motive in improving the quality of training.

At the same time, the needs of training engineering personnel put on the agenda the study of the mechanisms of self-development of a specialist, since the development and education of an individual does not end when a diploma of higher education is received, but perhaps it is just beginning. In this regard, the university is faced with the task of setting the future specialist on self-realization, self-improvement, and achievement of the heights of professionalism.

The engineering education system is designed to create conditions for the evolutionary formation of a new generation of highly educated professionals in the field of engineering, for which the installation of self-development, professional culture and skill, the development of an individual style of activity is a priority throughout life. A modern engineer is not only a good production worker, but a specialist who understands the economic, environmental, social and other problems of society [1]. He should be distinguished by scientific and technical erudition, a desire for the constant development of his professional interests, a critical approach to finding constructive solutions to problems, and the ability to work with people. In any field, a true engineer must act independently, proactively and creatively.

High school responds to the challenges of the time by finding gifted people to study at the university and developing their creative abilities. To do this, flexible competency-based educational programs are created, a transition to an individually-oriented model of the organization of the educational process and the project model is carried out.

Innovative engineering education is developing all over the world, aimed not only at the formation of fundamental knowledge and skills, but also at special competencies focused on how to put them into practice when creating new competitive equipment and technologies.

Therefore, the university must decide for itself the question of what will be the priority focus of attention, expenditure of time and energy. To cut talent or tow middle peasants? How to solve this problem, which is one of the significant components of the quality of education? How to organize the educational process so that everyone would be equally interested?

The article reflects the organizational and methodological aspects of preparing elite students in an engineering university, which are determined by the need to improve the quality of training for future specialists focused on the creation of innovative technologies, work in multidisciplinary projects.

The study of this issue meets the needs of employers in the training of specialists who are ready to quickly and productively solve various levels of complexity, professional tasks, implement innovations in their activities. The training of such an engineer can be carried out not only in higher education, but also in additional professional education.

The new requirements for vocational education are most fully and promptly transmitted by the system of continuing education, which becomes the main link between the interests of the individual, society and the state, various levels of vocational education and economic and social sectors, the requirements of employers and the requests of direct consumers of educational services. Additional professional education is designed to meet the requirements of the changing social, economic and cultural situation in the country. Being one of the rapidly developing forms of continuing professional education, it is becoming one of the strategic priorities of state educational policy. In modern conditions, additional professional education is constructive, bearing the main burden of adapting university graduates to fulfill new functional responsibilities. This system is more flexible and effective, turning into a stage of professional education. In contrast to higher education, which creates a base of fundamental knowledge, skills and professional competencies of a specialist, additional professional education provides an opportunity for a superstructure. New skills and knowledge that are in demand on the labor market are added to existing skills and knowledge. This makes the specialist more competitive. It is necessary to draw a lot of attention to the early vocational guidance of future students, starting from high school and continuing in undergraduate studies.

In Kazan National Research Technological University, for the preparation of successful students, the Tehnolider School of Continuing Professional Education was created [2]. Students of the school are trained according to the curriculum, which contains specialized training modules that ensure fundamental, innovative, professional orientation and creativity of the learning process. At the final stage of training, students should present a scientific development or project to potential customers who are ready to introduce them into production. About 200 students are selected annually to study at the school, which makes up about 10% of the freshmen contingent who have gone to budget places to study in engineering areas and specialties. Unfortunately, by the 3rd year there are about half of them left. The screening is due to the high requirements for the participants of the program and the difficulties of its development, which only the most talented, purposeful and strong- willed students can overcome. Those who do not cope with the curriculum return to the trajectory of training according to the basic curriculum. However, a "reverse movement" also occurs when successful "standard" students of the 2nd year of study move on to the trajectory of elite engineering education. Thus, according to statistics, this program is mastered by no more than 2.5% of students admitted annually to the 1st year to study in engineering. However, it is these graduates who are able to make a significant contribution to solving the problems of the new industrialization and technological development of the country. The system of elite technical education at the Kazan National Research University will be improved taking into account the experience gained from the beginning of its implementation in 2018, as well as the best practices in the functioning of similar systems in domestic and foreign universities. It is focused on training leaders in the engineering profession in priority areas in the field of engineering and technology for the resource-efficient economy of the country.

2 Purpose or Goal

In modern conditions, the competitiveness and viability of universities is determined by their ability to adapt to changes in the external and internal environment, the desire for a dynamic and flexible response to current and future needs of the market, society and the state. These tasks actualize the search, development, definition of priority areas for the development of universities in the future. Therefore, the significance of the problem of training elite specialists in an engineering university is due to the new requirements of modern society for graduates.

Elite engineering education is the training of future leaders in the engineering profession, ready for innovation and entrepreneurship, able to discern the challenges of modern society, knowledgeable in breakthrough areas of science and technology, as well as modern engineering methods and means, able to systematically, critically and creatively think in a dynamic a changing world and with the skills that allow them to organize a team and lead a project. Social responsibility and organization are important qualities of future engineering leaders.

This necessitates the identification of talented students and the organization of their individual training in the chosen specialty, which will allow the graduate to be competitive and in demand on the labor market. The main goal of elite education is to prepare for innovation.

Students who have received such an education must have a high level of development of competencies, personal and professional qualities. And here the ability to predict and formulate problems, to identify ways to solve them is of great importance. The purpose of training at the school is to prepare future specialists for successful work in innovation and inventive spheres, to participate in the development and implementation of projects, and to form a successful professional career.

At present, the Tehnolider School of Continuing Professional Education in KNIRU is developing and improving to increase the international competitiveness of the university and its graduates, including using the experience and best practices of training elite specialists in leading domestic and foreign universities.

Those who study at school acquire unique competencies in the following areas:

1) foresight (forming a picture of the future, identifying problems and paradoxes, synthesis of understanding of needs and opportunities, imagination, creative thinking, development of concepts for engineering solutions);

2) semantic analysis of the context (comprehension of the surrounding world, understanding the context of the problem situation, creating a mental map of conditions, assessing the natural and social situation, awareness of customer requests and enterprise requirements);
3) implementation of concepts (movement from abstractions to innovations and inventions, their implementation, building and managing organizations, project planning and management, decision evaluation);
4) practical engineering knowledge (understanding the essence of engineering problems, analysis and synthesis, knowledge of research methods, task setting skills.
5) interpersonal interaction and effective communication (building relationships within the organization and between organizations, the ability to hear others, understand and accept their point of view, convincingly argue and defend their position, negotiate).

Elite engineering education is being developed taking into account the peculiarities of engineering activity in a post-industrial society based on knowledge, taking into account the peculiarities of the advanced requirements for training personnel.

Engineering in the modern world is becoming more integrated, integrated and innovative. Integrated engineering activity, being complex and multi-component, covers the solution of a wide range of various issues in the planning, design, production and application of technical objects, systems and technological processes. To prepare specialists for such an activity, the leading universities of the world are implementing the CDIO concept developed at the Massachusetts Institute of Technology in the United States with the participation of leading technical universities in Sweden - the Royal Institute of Technology (Kungliga Tekniska högskolan, KTH), Chalmers University of Technology1.

The concept of CDIO (Conceive - Design - Implement - Operate) is based on the principle of preparing graduates - bachelors for complex engineering activities, that is, those capable of "Think - Design - Produce - Apply" [3].

The main thing in CDIO is practice-oriented training, during which students independently develop new products or technologies, conduct design work and think out strategies for introducing know-how into real production. Along with the CDIO Syllabus, which sets the requirements for learning outcomes, 12 CDIO Standards have been developed that establish the requirements for educational programs in technical areas and specialties of universities. CDIO standards define the philosophy of programs for preparing graduates for integrated engineering activities (CDIO Standard 1), set requirements for learning outcomes (CDIO Syllabus) and curriculum development (CDIO Standards 2, 3, 4), the educational environment of the university (CDIO Standards 5.6), teaching methods (CDIO Standards 7.8), teachers (CDIO Standards 9.10) and methods for assessing student learning outcomes and the program as a whole (CDIO Standards 11, 12).

Already in the first year of study, students should be shown the relationship of the proposed teaching material with their future engineering activities, the prospects for the technical, technological, economic and social development of society. It is especially important to focus on practice-oriented teaching of disciplines.

In order for students to learn to apply the acquired engineering knowledge in engineering, to understand their importance for practical work, a certain integration of disciplines is required, which implies both “fundamentalization of the professional” and “specialization of the fundamental”.

Below are the attributes of the future leaders of the engineering profession, to which the system of elite engineering education at KNRTU is oriented.

1. Fundamental education is provided by in-depth training of students in the field of natural, fundamental sciences, economics and a foreign language. As a result, a stablenatural-scientific worldview is formed, horizons are expanded, systemic thinking and logic are developed, engineering is positioned in the economic context of society development, international experience is gained, world information resources are made available through English proficiency, the foundations for continuous self-education and self-improvement in throughout life.
2. Professionalism is achieved by active research, inventive and design activities of students. As a result, their professional creative culture is being formed, an interdisciplinary approach to the problem is being developed, the ability to pose and solve priority engineering problems, practical experience is gained in modeling, experimenting, solving inventive problems, the technology for performing individual and group projects, including in the international environment, is being mastered, academic and professional mobility skills.
3. Innovation is acquired through the development of critical thinking and initiative of students, the analysis of modern problems and values, the study of trends in the formation of society’s needs in new equipment and technologies, and the development of ways to satisfy them. As a result, students have a vision of the prospects for improving engineering activities, develop skills in studying the market of modern technical products, assess their consumer properties, actualize motivation to create new competitive equipment and technologies that provide a new social and economic effect.
4. Entrepreneurship skills are formed in the course of students’ practical activities in organizing educational and real production of innovative products. As a result, skills are acquired in the field of marketing and business planning, the ability to analyze the demand for products, assess the technical and technological feasibility of projects, the economic aspects of production and the financial viability of projects, manage intellectual property, and develop project and production management.
5. Leadership is formed when students gain experience in managing a team of developers of new technical and technological solutions. As a result, students master the skills of foresight, they are ready to pose problems, generate ideas and concrete proposals for solving them, the ability to plan team work, share responsibility and authority among team members, develop the art of communication and building relationships in the team, and professional ethics is developed, perseverance in achieving the goal and responsibility for the results of the team.

The motivation of students is formed in the learning process, as students understand that they have the opportunity to obtain foresight knowledge, work with famous scientists, participate in various projects, the opportunity to implement their development in real production, create their own start - up.

The training of "innovative engineers" capable of introducing new technological solutions, managing large-scale technical projects requires the generation of new types of programs that will form the graduates' competence in system engineering, which is distinguished by a holistic approach to the perception of engineering problems, understanding the importance of taking into account the entire life cycle of an engineering product, development creative thinking, teamwork abilities in custom-made developing breakthrough technological areas of engineering teams.

By choosing a specific set of modules in the educational program, one can achieve the preparation of a given orientation.

Since this is additional training for students, little time is allocated for it (6 h a week), the task is to focus on the formation of a superdisciplinary understanding of the essence of engineering activity and the direction of its development. The compensatory function of these classes is designed to actualize the main problems of the content of academic disciplines by identifying and detailing the search for solutions to problems. In the first semester of study, the discipline "Introduction to Engineering" (Standard 4 CDIO) is studied, aimed at stimulating interest and increasing students' motivation for engineering, encouraging them to engage in engineering. Discipline creates a general idea of engineering practice. This presentation includes a wide range of tasks and responsibilities of the engineer, as well as the application of disciplinary knowledge to solve these problems. The course provides for the acquisition of personal and interpersonal skills, knowledge and skills that are necessary at the initial stage of development of the program (Standards 6, 8 CDIO). During classes, students individually and in teams carry out simple tasks on the analysis and synthesis of machines, learn to think outside the box. After studying this discipline, students know the types and tasks of professional activity, are able to apply methods of rational organization of mental work, understand the importance of engineering work, are aware of the social responsibility of an engineer, and are actively involved in solving problems of engineering.

Therefore, the task of the "Introduction to Engineering Activities" discipline should not be a simple study of the nature, content of the time stages of engineering activity, but a problematic description of the relationship between the achievements of the past and current problems of the present. The purpose of the discipline is to develop students' interest in engineering, increase motivation to form the skills and abilities described in CDIO Syllabus. Discipline is implemented during one academic semester and includes the theoretical and practical part - team creative projects.

Teaching this discipline should orient and engage students in functioning in a changing engineering activity, identify possible ways of orientation in this activity and build a career growth strategy.

In the domestic pedagogical system, the main type of training is predominantly supportive training, which is aimed at reproducing the existing technology, methods and ways of implementing professional activities. And innovative training should prepare a "person acting". Any of his activities is formed in stages: from awareness of the problem to the search for information; evaluation options; and further executive, control, corrective stages. Therefore, the involvement of students from the first year of study in the design process under the guidance of teachers significantly contributes to solving the problems of elite engineering education. The student must determine for himself the further direction of his educational, scientific and research activities.

3 Approach

Issues of elite training at an engineering university are reflected in the studies of many Russian scientists [4]. The elitism of training, its significance, is ensured by careful selection of students according to the level of knowledge and the development of motivation, the small composition of student groups and the individualization of the learning process. Elite education is the highest quality education. Experience in training elite specialists in Russia is represented by many leading engineering universities. Extensive foreign experience has also been gained. An analysis of the existing experience of training such specialists led to the conclusion about the identity of the organization of training [5]. He showed that their appearance is due to the need of the economy for highly professional technical personnel capable of solving non- standard large-scale tasks and navigating in the modern information space. In the context of the transition from a raw materials economy to an innovative one, the totality of traditional knowledge and skills becomes insufficient, since it is necessary in the engineer's work to comprehensively assess the problem and find its solutions embodied in projects, developments, and research activities. To do this, the future engineer needs to form a set of general and professional competencies that will effectively solve innovative problems, navigate modern trends and be competitive in the labor market.

The engineering elite is trained in universities according to special programs, identifying the most talented students, developing their natural abilities and actively involving them in practical research activities. You can identify the distinctive features in the priorities of the formation of competencies. The emphasis of foreign universities is on the humanitarian component of engineering education and the development of universal (personal) competencies, the emphasis on Russian ones is on natural-scientific, mathematical and technological training.

Innovations in engineering ensure the development of production. It is known that an education system that is not related to production cannot train specialists for practical work.

To create and introduce innovations, specialists with special competencies to identify and solve new problems, including those with leadership qualities, are required. The number of such specialists is no more than 3–5% of the entire engineering corps, but they are absolutely necessary for the production of new ideas and ways to solve problems. The training of such a level of specialists justifies itself. Also, training is based on a student-centered approach, which is implemented during all types of classes, which allows for the individualization and differentiation of training. The system-activity approach is characterized by orientations on practical activities. The pedagogy of cooperation allows us to introduce students and teachers as equal participants in educational activities.

Based on the foregoing, it is possible to formulate the principles of organization of the educational process in the module "Introduction to Engineering" in the ideology of CDIO.

The principle of the context of engineering education is sustained in the selection of the content of lectures, practical exercises, design tasks and integrated tasks.

The principle of practice-oriented education is superimposed on the principle of context with an emphasis on the activities of enterprises that are directly employers.

The principle "from idea to product" is realized through the implementation of research, practical tasks.

The principle of project training through the use of pedagogical design technologies.

The principle of active learning is manifested in the selection of interactive pedagogical technologies that work on the development of competencies.

The principle of integration is leading, since all the others are based on integration at different levels. This conclusion can be reached based on the historical traditions of integration processes in education: the unity of integration and differentiation as a way of self-organization of education; anthropocentrism, determining the position of the student and teacher in the integrated educational system; cultural diversity characterizing the attitude of education to its cultural environment.

Summarizing all the principles and approaches is a synergistic approach. We consider the use of the synergetic approach appropriate, since synergetics today is regarded as an interdisciplinary direction in modern science, devoted to the study of the general laws of the dynamics of complex nonequilibrium systems, and can provide general guidelines for the scientific search in complex social systems, which include the educational system. We believe that the implementation of the synergistic approach in the teacher's activity is manifested in updating the content, methods and forms of teaching, taking into account factors such as openness, self-organization, self-development, non-linear thinking, management and self-management. In the synergetic approach to teaching, the personal importance of the student occupies an important place, which is a pillar for its development. The most important condition for the manifestation of the positive synergistic effect is an understanding of the common tasks and goals of the team.

Such training significantly increases the interactivity of the educational process, stimulates the manifestation of students' creative abilities. Based on the data of theoretical and methodological approaches, technologies and principles, as well as the peculiarities of preparing students at national research universities, we came to the conclusion that the goal of studying at the Technolider school should be to create conditions aimed at creating the ability to master ready-made solutions; determine the conditions for innovation; willingness to work in a team; own the necessary knowledge; ability to highlight a problem; formulate a task; design; analyze the technical level of the facility; ability to work with information in computer networks.

If earlier the doubling of the volume of new knowledge took place in about a century, humanity and the education system managed to adapt to the new flow of information, today the situation has changed. In the modern world, the doubling of knowledge, for example, in physics occurs during the life of one generation. First, the actual material quickly becomes obsolete. Secondly, after graduation, a university graduate falls into another professional environment, where the knowledge that he received at the university and spent a lot of time on them simply may not be useful to him. But the general laws of nature, fundamental knowledge never become obsolete. They also develop, but they enable a person to quickly switch from one technology to another, because all technologies are based on knowledge of fundamental science and understanding of nature. This postulate needs to be conveyed to students so that they recognize the importance of fundamental knowledge and are able to further use them in teaching special disciplines. It is also the subject of the Introduction to Engineering.

Therefore, the objectives of teaching this discipline are quite multifunctional, they are a structural component of the entire training process.

Thus, the conceptual provisions of the CDIO technology, which is a comprehensive poly-paradigmatic approach to organizing the learning process of a new generation of engineers and the formation of an innovative educational environment in engineering universities, are implemented at the basis of training at the Technolider school.

The content of elite training programs developed taking into account the requirements of a modern engineer confirms the thesis that an elite technical education is able to provide a student with the necessary cluster of competencies that will work in the future [6]. This can be ensured by the orientation of the educational process on the acquisition by students along with theoretical knowledge of a wide range of practical skills. At the same time, the graduate should become a mobile engineer, able to initiate a new one, quickly rebuild in accordance with the tasks set, think professionally, implement bold engineering solutions and be responsible for the results of his activities.

4 Results

An important place in the formation of professional competence, and, therefore, in the content of training a future specialist, is given to the development of skills to apply theoretical knowledge in solving practical problems, for example, in the field of comparative analysis of technologies, quality of products and others. It is this content that the Introduction to Engineering Activity discipline has, which creates a general idea of engineering practice, considers a wide range of tasks and responsibilities of an engineer, as well as the application of disciplinary knowledge to solve these problems.

The discipline "Introduction to Engineering" is built on the basis of the systemic interconnection of CDIO standards 1, 2, 4, 5, 7, 8, 11. It includes topics substantiating the implementation of standards 1 and 4: "Chemical technology in the ideology of CDIO"; "Culture of intellectual activity as a condition of self-development"; "The place of chemistry in society, in the worldview, the stages of the formation of the chemical industry"; "Features of engineering and the role of an engineer in the modern world." The technological map of the organization of mastering the discipline includes discussions, round tables, work in small groups (Standard 8). An innovative approach to assessment is prescribed in the module's work program through the selection of monitoring activities, in accordance with the verified learning outcomes in the form of essays, tests, presentations, project defense, speeches at a scientific conference, portfolio (Standard 11).

Students should also be willing to study and use scientific and technical information, domestic and foreign experience on research topics; the ability to apply basic natural science, mathematical and special knowledge necessary for engineering activities; solving engineering problems, conducting theoretical and experimental research, including searching and studying the necessary scientific and technical information, to observe labor safety rules and environmental protection requirements in their proposals, to work individually and as a member of a group. Students should demonstrate knowledge of the social aspects of integrated engineering, develop the ability to self-study throughout life, to continuous self-improvement the engineering profession. The use of active forms of learning will mobilize the internal potential of students and, in a game situation, simulate a solution to the problems of practical activity. The methods and techniques mastered in practical exercises should be fixed in the course of independent work.

The content of the discipline includes not only consideration of the most significant engineering objects for a given period, but also consideration of the use of these solutions in modern society. In the classroom, practical tasks are used, for the solution of which it is necessary to apply knowledge from different subject areas and acquired by students in practice.

It is advisable to conduct briefing and entrance control at the first practical lesson using the business card "Business card" method: "Hello, what are you doing?" Purpose of the game: acquaintance and the formation of competence in choosing the direction of research activity. Students tell us what brought them to the university, what they are interested in, what problems they started to solve, what successes have been achieved and what they want to get in the future? Many talk about participating in the "Nobel Hopes of KNRTU" contest, studying at the summer chemical school "Orbital" at KNRTU, where schoolchildren from all over Russia and foreign countries come to spend summer vacations, who decided to devote their activities to the natural sciences. Thus, many students of the Technolider school are not random people, but those who have already proved themselves in the study of chemical disciplines, who purposefully chose both the university and their future specialty. We have studied the motivation of students (Figs. 1, 2 and 3).

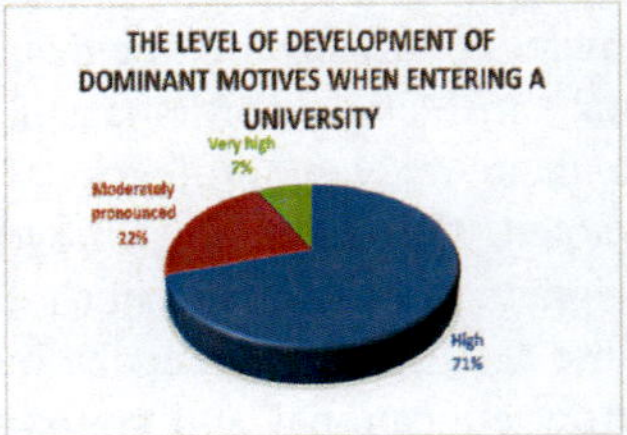

Fig. 1. Diagram of the level of development of dominant motives for entering a university.

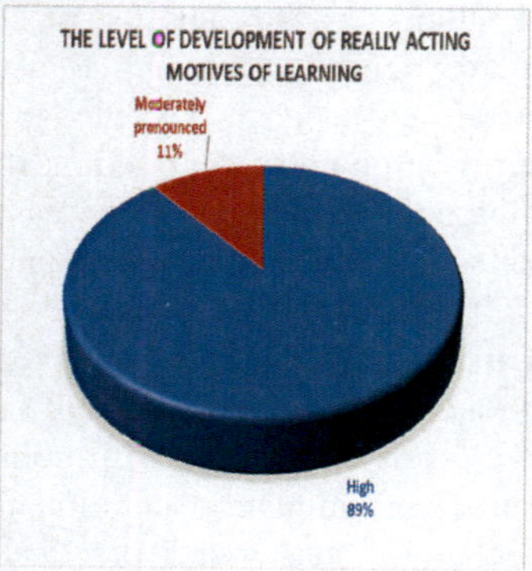

Fig. 2. A diagram of the level of development of real-life motives for learning.

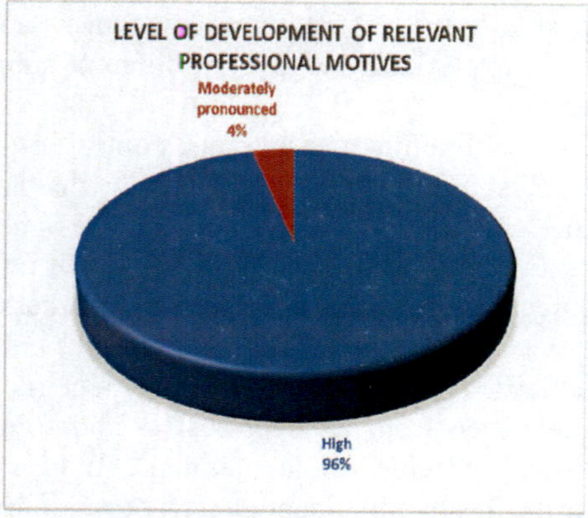

Fig. 3. Chart of development level of relevant professional motives

We determined the dominant motives for entering a university: interest in the profession; the desire to get higher education and the actual motives of the doctrine: the desire to use the acquired knowledge in your life, the desire to acquire deep and lasting knowledge, receive intellectual satisfaction, participate in competitions and olympiads in the subjects studied. The main relevant professional motives of students of the Technolider school [6] are noted: the desire to achieve social recognition, respect; desire to continue studies in graduate school (master's program); Build a successful professional career. During the test, results were obtained that characterize the low level of development of narrow educational and cognitive motives of first-year students, such as free admission, family traditions, and the desire of parents; advice of friends, acquaintances; prestige, authority of the university and faculty; desire to obtain a diploma of higher education. The levels of irrelevant professional motives of first-year students are established: the desire to achieve the approval of others, to have a guarantee of stability; the desire to get a well-paid job, to work in government or in private organizations; get a leadership position.

Thus, students are motivated by sustainable motives for obtaining a profession and building a future career. Therefore, the task of the teacher is to consolidate this motivation and direct the development of their cognitive activity. Students analyze the topics of lectures for the preparation of materials in the first lesson. For example, topics may be as follows. What is the difference between the picture of the world of medieval man and the picture of the world of man of the Renaissance? What are the achievements of science and technology in the Renaissance? What are the results of the scientific revolution of the seventeenth century? What is the significance of the industrial revolution of the late XVIII - mid XIX centuries? What are the main scientific and technical achievements in the era of the New Age? What is the role of the scientific and technological revolution of the twentieth century in the history of science and technology? There is a keynote speaker and co-rapporteur who adds material that is missing for the full disclosure of the topic. The final word is given to the teacher, who makes a synthesis of all the material on the basis of which students independently try to formulate conclusions.

As knowledge accumulates, an independent written work is carried out - an essay. One of the essays is based on a curious document from the beginning of 1481, in which the thirty-year-old Leonardo da Vinci (1452–1519) offers his services to the ruler of Milan Lodovico Sforza and where he characterizes the diversity of knowledge of the engineer. And then comes the listing of engineering solutions owned by the great Renaissance scientist. Students describe one solution and analyze it from a modern point of view. The following is a proprietary solution to an engineering problem. Mutual assessments are formed during the discussion at the round table or colloquium, which is characterized by a more clear formalized nature of the scenario. Students identify the strengths and weaknesses of the submitted works, there is a mutual assessment.

The consolidation of acquired knowledge-skills-possessions is carried out in the form of interactive interaction. Oral survey-training is carried out in the form of the game "I know…", where the name of the game is variable. Say, "I know that the drawing, according to G. Monge, is "the language of the engineer", and the problem is further revealed that he is also the language of communication with performers: technicians, craftsmen, workers. "I know about the contribution of M.V. Lomonosov's development of science…" and his role in the development of chemical research is revealed. Students themselves choose to discuss the achievement of a particular scientist, study his methods of solving an engineering problem and the ability to transfer his methods of solution to modern reality.

Questions to be solved in practical classes: "How did scientists solve the following problems before?": "In the event of a fire alarm, how to safely leave this audience?", "How to dispose of waste?", "How to heat a room without losses?" other. As an independent work, we propose the task "Select the actual problem, the engineering and technical task that you consider relevant, in the solution of which you would like to participate."

The result of solving the problem should be of cognitive and professional significance for students [7]. These tasks form the basis of the teaching methodology of the discipline. Thus, when studying the course, real problem situations are considered that allow students to participate in the development of a group solution to the problem.

At the final stage of training, it is necessary to present a project, the theme of which is selected independently by a group of students and represents a way to solve a professional problem based on a design methodology. When designing a model of a training lesson, the teacher should simulate the learning situation, questions should be formulated for the organization of educational reflection:

1. "What is the task of the project?" (a question of the analytical genre urging the student to reproduce the existing problem).
2. "What did not work out for you?" (the question is aimed at finding students a place of difficulty, error). If students cannot build their version from the current situation, then they can turn to a teacher for help. It is important to emphasize that in such a situation the problem of "lack of interest in students to study" disappears.

Project protection takes the form of presenting the results of each group to members of other teams. Students reveal the field of study and the possible expected effect. All present can ask questions and take part in the discussion.

For example, the following projects received a positive assessment for their effectiveness.

Social project of an environmental focus "without paper". Project goal: to popularize the recycling of used waste paper among KNITU students by conducting promotions in educational buildings and active educational activities in social networks. Students expect to establish active cooperation with business owners on mutually beneficial conditions. The significance of this project for students studying is clear.

The "synthetic oil" project, the idea of which is to create synthetic oil with a formula from which it will be possible to produce plastic. This plastic will decompose better and faster. As a result, environmental damage will not be caused.

The project "folding printer" is aimed at solving the urgent problem of students - the ability to instantly print the necessary document.

Positive emotions aroused the work of "3D toy souvenir".

Only one project received a negative rating. This is a "nanoshack" project that does not need to be washed, ironed, etc. Students did not like that she would have to be worn for a long time and she would become unfashionable.

A final survey of students was conducted to identify the level of competency formation. Suggested answers in the questionnaire: yes, no, partially. The table shows the competencies that, according to students, they have been formed.

Competency ranks

No p/p	Competencies	Answers,%		
		Yes	No	Partially
1	Ability to master ready-made solutions	98	2	
2	Ability to determine innovation conditions	59	11	30
3	Willingness to work in a team	35	25	40
4	Knowledge required	67	11	22
5	Ability to highlight a problem	61	9	30
6		62	8	30
7	The ability to formulate a task	56	5	39
8	Ability to design	79	7	14
9	Ability to analyze the technical level of an object	34	40	26
10	Ability to work with information in computer networks	33	45	22

We conducted a test questionnaire of satisfaction with the educational activities of L.V. Mishchenko to identify satisfaction with student learning activities and methods of studying the motivation of learning Fig. 4. 70.8% of students study well, the student is satisfied with communication with classmates, teachers, confident in their future professional relevance, his requests do not exceed household and budget reality, in 20.8% the educational activity is within normal limits, but not gives you the opportunity to realize all your abilities,some dissatisfaction arises only in certain areas of educational and professional activity and 8.3% of educational activities are not going well enough, the student has a number of communicative difficulties.

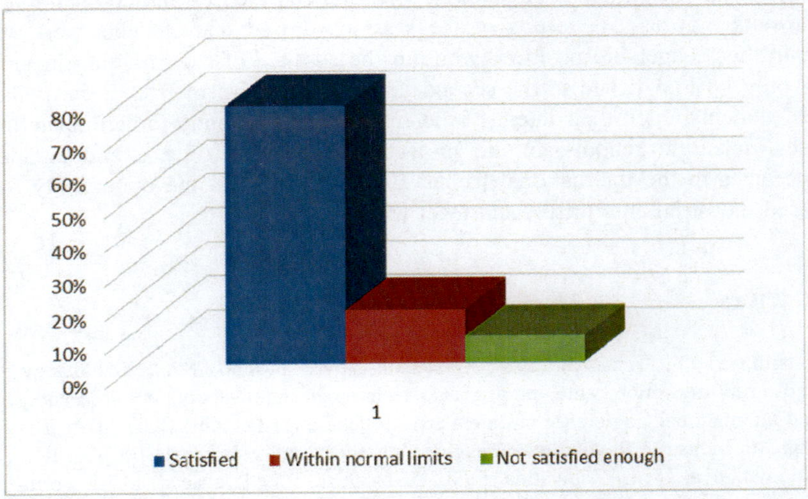

Fig. 4. Survey results.

5 Conclusions

Education in the school of continuing education "Technolider" is aimed at building students' competencies for innovative engineering activities. At the same time, students retain a high positive motivation for the engineering profession and for the learning process as a whole. This is focused on all academic disciplines, including the discipline "Introduction to Engineering." The complex of professionally-oriented tasks used in its content performs an important function. It involves an independent search for information, determining the problem and how to solve it. The knowledge gained during the study of the discipline "Introduction to Engineering" can improve the understanding of special disciplines studied in the future, competently approach the solution of scientific and technical problems, and develop creative abilities. After studying this discipline, students know the types and tasks of modern engineering, are able to apply the methods of rational organization of mental work, understand the importance of engineering work, are aware of the social responsibility of an engineer, and are actively involved in solving creative problems.

The advantages of the discipline include: 1) strengthening the creative nature of educational activities (the ability to creatively and unconventionally solve future professional problems, quickly navigate in large volumes of information; 2) the acquisition of knowledge and orientation in the field of engineering; 3) the formation of motivation for a future profession; 4) interprofessional communication.

In addition, after studying the course, we revealed a high level of students' interest in engineering (86%). Preparedness for project activities is being formed - 36%. In general, students have a higher importance of educational - cognitive and professional motives, which can be attributed to intrinsic motivation: awareness of the value of creative discovery - 56%; increasing student satisfaction with the quality of education - 75%; annual participation of students in scientific conferences and competitions with high results - 56%. The results of the session showed that students were able to maintain their leadership position, while in the absence of such training in previous years, only 68% switched to the second year with positive marks. In the polls conducted, students noted their interest in learning, the opportunity to participate in conferences, scientific schools, in the project to popularize science and engineering competencies in the international format Engineering Slam, the opportunity to visit advanced industrial enterprises, and meet with famous scientists.

References

1. Khatsrinova, O., Tarasova, E.N., Ovsienko, L.V., Yushko, S.V., Galikhanov, M.F.: Engineering education: elite training at a technological university. In: Proceeding of the 22nd International Conference on Interactive Collaborative Learning (ICL 2019). The Impact of the 4th Industrial Revolution on engineering education, vol. 1, pp. 366–376. ICL (2019)
2. Solodovnikova, O.M., Zamyatina, O.M., Mozgaleva, P.I., Lychaeva, M.V.: Formation of competencies of an elite technical specialist. Prof. Educ. Russia Abroad **3**(11), 65–71 (2013)

3. Yuryevna, K.O., Tarasova, E.M., Ovsienko, L.V., Yushko, S.V., Galikhanov, M., Julia, K.: Engineering education: elite training at a technological university. In: Proceeding of the 22nd International Conference on Interactive Collaborative Learning (ICL 2019). The Impact of the 4th Industrial Revolution on Engineering Education, vol. 1, pp. 366–376 (2019)
4. Vorobyova, I.M.: Strengthening the role of engineering education and the practical component of educational programs in a technical university. Young Sci. **11**(91), 1304–1307 (2015). https://moluch.ru/archive/91/19565/, Accessed 02 June 2020
5. Chubik, P.S., Mogilnitsky, S.B.: The system of elite training of engineers in TPU. Qual. Educ. **10**, 22–28 (2012)
6. Chubik, P.S., Chuchalin, A.I., Soloviev, M.A., Zamyatina, O.M.: Training of elite specialists in the field of engineering and technology. Issues Educ. **2**, 200 (2013)
7. Bulgakov, Yu.V.: The role of professional orientation in a technical university. In: Innovations in Science: Sat. Art. by Mater. International Scientific-Practical Conference, vol. 6, no. 55, pp. 73–77. SibAK, Novosibirsk (2016)

Innovative Common Study Block Framework for Joined Collaborative Curriculums Development

Eduard Shevtshenko[1,2](✉), Kati Nõuakas[2], Lea Murumaa[2], Oliver Kallas[2], and Tatjana Karaulova[1]

[1] Tallinn University of Technology, Ehitajate tee 5, 19086 Tallinn, Estonia
{Eduard.sevtsenko,Tatjana.karaulova}@taltech.ee
[2] TTK University of Applied Sciences, Pärnu mnt 62, 10135 Tallinn, Estonia
{knoukas,Lea.Murumaa,oliver.kallas}@tktk.ee

Abstract. The field of logistics is constantly changing. These changes must also reach the classroom as quickly as possible to ensure high-quality teaching and maintain the reputation of the Institute of Logistics alumni as valued professionals in the labour market. Employers expect to have a graduate-level employee that can be directly assigned to work tasks. However, in the provision of higher education the community faced with a situation where the university can instead create a basement on which the employer begins to build a specialist based on company-specific need. This gap between the qualifications of graduates and company expectations fostering extensive cooperation between institutes and entrepreneurs. Companies are gaining experience in dealing with international higher education institutions. The purpose of this article is to give an overview of the carried out research resulted in Innovative Common Study Block Framework development and introduce how the framework has been applied to improve the form and content of teaching in the format of collaborative projects.

The methods of teaching and study objects are created to enable the graduated students to possess competencies expected by future employers also have to be adjusted to correspond to rapid changes. The new generation entering the labour market expects curriculums and courses to be created and adapted to their requirements. The heterogenous studentship expects to be involved in the teaching-learning process and suitable for them studying environment. The teaching process is moving more and more towards using a combination of activities held out both in physical and e-environment.

1 Introduction

Previous research identified that universities mostly focus on developing and assessing the knowledge competencies of future engineers, while employers need skills that allow university graduates solving actual engineering problems immediately upon graduation [1]. Another major challenge is the need for further integration of cross-sectoral skills into the teaching and learning process [2].

M. E. Auer and T. Rüütmann (Eds.): ICL 2020, AISC 1328, pp. 30–41, 2021.
https://doi.org/10.1007/978-3-030-68198-2_3

There is the gap that existing between the needs of the regions and education response in the area of the Career Guidance at secondary schools and the introduction of youth with professions demanded in selected industries [1].

Skills needs and skills mismatches are the key current issues for education and human capital development in many European countries. Policymakers, education providers and social partners claim that significant efforts are needed in evoking the interest of young people in choosing vocational pathways of a career in the different sectors of the economy [4]. Mendoza et al. stated that higher educational institutions started activities related to Circular Economy [5]. Klein et al. state that public sector received very little attention in CE research [6]. Ronkova and Dimitrov recognise that study syllabi and curricula always need improvement to correspond to the new requirement of future engineers [7]. Wojcik et al. have developed the web application for the formation of the knowledge component of the planned learning, to increase the cognitive abilities of students [8].

The target of the current research is to introduce the innovative framework for common study blocks development, that can be applied for bachelor and master curriculums in different universities. The potential users of the framework would be teaching and research staff working with supply chain subjects. The goal of the framework is developing teaching staff professional competence and provide the ready modules for curriculum development in international universities either.

To apply on practice created framework research group has initiated the UniLog project. The UniLog project aims to validate an innovative framework for the development of common learning blocks that can be applied to undergraduate and graduate programs at different universities. Potential users of the framework would be supply chain teaching and research staff in partner universities. The project aims to develop the professional competence of teachers and to create ready modules for the development of curricula at international universities. The opportunity to connect with businesses allows authors to identify the gap between current curricula and today's business needs. This research aims to optimise labour market opportunities by increasing the number of strategic business specialists in the Baltic Sea countries. The authors have prepared study and training materials for university teachers in the form of a selected 1 ECTS module.

Within the UniLog project, the Institute of Logistics has been developing the logistics curriculum and creating the content of the subject blocks required for the subjects. The project team includes representatives from Sweden (KTH Royal Institute of Technology), Finland (LAB University of Applied Sciences), Latvia (Riga Technical University) and Estonia. In the early stages of the project, an overview of logistics training opportunities and companies' expectations for training in the field was prepared.

2 Research of Companies Requirements

In previous research, authors have determined the general principles of how to connect guidance and secondary school students' competencies development related to specific industries [3]. Authors have collected information on employers needs corresponding to their staff competencies in design, technology and economics areas and elaborated recommendations for career counsellor in-service training program development. The project team has used the questionnaire for data collection [4]. There are more than 2000 metalworking and machine-building companies, employing more than 35000 employees, but the average percentage of dropouts in Estonian higher schools was 20% during the last three years. The sector needs more engineers whose skills and knowledge are suitable for starting work after graduating, for developing the business that creates more significant added value [9].

An important task is to encourage young people to choose a job in the fields of mathematics, science or technology because high demand for graduates in those fields is significant for further innovation and growth of the industries. Adequate training of qualified specialists is possible only in close cooperation between schools, universities and enterprises. Educational programs of the universities should correspond to the needs of the industrial companies developed in a particular region. Therefore, there is a need for constant monitoring of the competencies of the participants in education and production systems [10, 11].

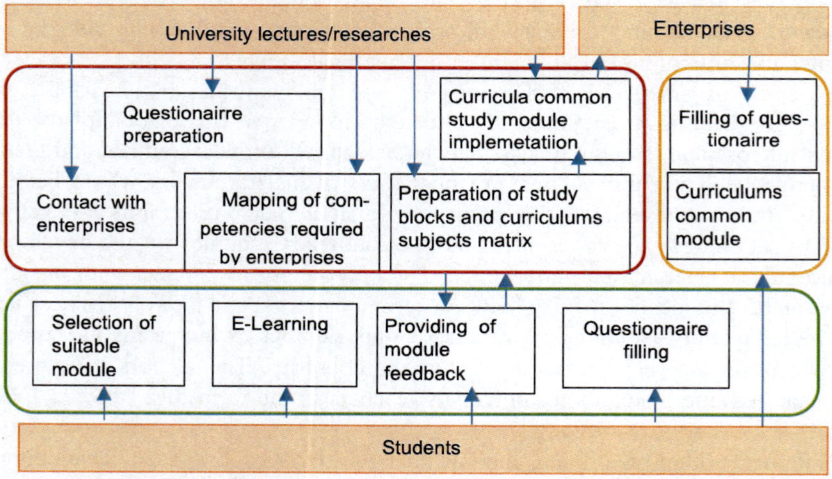

Fig. 1. Functions of companies, lecturers and students in the development of study modules

To address the issues described above authors have created an innovative framework for the development of common study blocks that can be applied to the bachelor's and master's degree programs at different universities, see Fig. 1. Potential users of the framework would be logistics, as well as management and administration teachers, companies and students. To design the study block, the lecturer prepares a questionnaire and introduces it to partner companies that are interested in university students. Using the questionnaire, the lecturer assesses the competencies required for the company and analyses how they are covered in the existing curricula. Then, in cooperation with the partner universities, new additional study blocks are developed to cover the necessary competencies. The completed modules are offered to students either within the existing subjects or as electives, with particular attention to development of the critical thinking and reflection skills [12]. After completing the modules, student feedback is analysed, appropriate corrections are made, and the study blocks are ready for use in bachelor's and master's degree programs. After implementation in the curriculum the authors asked students to fill in the questionnaire to validate that the required skills are acquired. After the students are employed, the company representatives were asked to confirm that the students can use the received knowledge from common modules on practice.

3 Validation of Innovative Framework for Joined Collaborative Curriculums Development During the UniLog Project

Not only in Estonia but also in many other European countries, the critical issues in education and human capital development are skills and skills mismatches. Policy-makers, education providers and social partners argue that significant efforts are needed to highlight the interest of young people in choosing career opportunities in different sectors of the economy [4]. Therefore, it is essential to cooperate with partners outside Estonia as well. For example, opportunities have been found for cooperation with universities in neighbouring countries in various projects, because just like the Institute of Logistics at TTK University of Applied Sciences, universities in other countries face similar problems and challenges, such as student motivation and timely graduation. Here, the authors of the article rely on the results of cooperation projects and communication with partner universities.

The project aims to develop the professional competence of teachers and to create ready-made modules for the development of curricula in international universities. The opportunity to interact with companies allows authors to identify the gap between current curricula and the modern needs of companies. This research aims to better link curricula with the expected competencies of the labour market by increasing the strategic number of business specialists in the Baltic Sea States. Study blocks that take into account the needs of the labour market enable students to acquire knowledge and practical skills that are an advantage when starting a job. The authors have prepared study and training materials for university teachers in the form of a selected 1 ECTS module. During UniLog project implementation, the Institute of Logistics has been engaged in the logistics curriculum development and the creation of the content of the topic blocks necessary for the subjects. The project team includes representatives from Sweden (KTH Royal Institute of Technology), Finland (LAB University of Applied Sciences), Latvia (Riga Technical University) and Estonia. At the beginning of the project, the project team investigated the study opportunities in the field of logistics. After that, the expectations of companies to the study outputs in this field were compiled for all countries.

The study prepared within the project is entitled "Improvement needs in the Central Baltic logistics competence and the related applied higher educations". The study identified answers to the following questions: what are the key indicators for the future that companies in the Baltic Sea Region need; what are the differences and similarities in logistics competencies in Sweden, Finland, Latvia and Estonia; what kind of logistics education is offered in the study countries and what in the partner institutions; what are the differences and similarities in higher education in the field of logistics in the countries of the Baltic Sea Region and the project partner institutions; what gaps exist between the need for future competencies and the higher education currently offered.

The subsequent is a summary table to follow up on the results of the survey conducted in by the partner countries' universities (see Table 1).

Table 1. Participating enterprises by country of residence and industry [11].

Country	Companies interviewed						
	Automotive	Pharmacy	Food industry	Other industries	Transport and warehousing	Retail	Others
Sweden	3	2	2	3	2	1	2
Finland	1		1	1	3	4	5
Latvia			1	3	11		
Estonia		2	2	2	8		

Thus, companies from as many different fields as possible were included in the study, but the main focus remained on companies related to transport and warehousing. Representatives from each country conducted interviews with at least ten different companies from different fields of activities (see Table 1). The results of the interviews are presented (see Table 2), i.e. the main points that were assessed as the necessary knowledge that should be available in higher education and the most significant differences between employers' expectations and employees' skills. The results presented in Table 2 were the main basis for the development and testing of the new learning study blocks. First, the curricula of four higher education institutions partners were reviewed and compared with the information obtained from interviews with employers regarding the expectations of entrepreneurs. The missing subjects were then selected, within which it would be necessary to implement the developed and improved learning outcomes and the content of the subjects. Within the framework of the project, study blocks with a volume of 1 ECTS were created. Blocks are integrated at both the bachelor and master's level, to eliminate the existing gap between employee expectation and curriculum actual output. The knowledge achieved are validated after the students passed the subjects in the form of tests and exams. Companies will be asked to fill in the questionnaire on regular basis in future, to receive the employee's feedback on modified curriculums.

Table 2. Knowledge required and major differences between employers' expectations and employees' skills [13].

Country	Skills	In general	Operative level	Managerial level
Sweden	Key skills required	Holistic overview, understanding of the financial situation, sustainability and digitalisation, engagement	Sourcing, packaging, sustainability [14] (spec.packaging), cost understanding, a combination of theory and application	Multi-criteria decision, leadership, long term vision
	Gaps in skills and employer expectations	Not rated	To understand the overall situation and how all impact on each other, problem-solving ability, to extract the right data, lean thinking in logistics, a feedback loop	To understand the overall situation and how all impact on each other, more significant gap than on operational level, problem-solving ability, cost of poor quality, feedback and keeping goal focus
Finland	Key skills required	The "Big Picture", business principles and finances, customer service, basics of logistics operations, processes, project work, data analysis, language skills, business understanding and customer focus, process understanding	Understanding the entire process, able to work in all functionalities (eg. in warehouse), work and occupational safety legislation, mental fitness, šelf-governance, problem-solving, innovation	Managing people, Interpersonal skills, Entrepreneurial skills, Challenges of international logistics
	Gaps in skills and employer expectations	Practical realities of the work, "Big Picture", numbers and business finance, the interface of storage and transport, IT, Project management, customer focus, interpersonal skills	Labour law, occupational safety law, rights, responsibilities and obligations of employees and employers, IT skills	Managing people interpersonal skills
Latvia	Key skills required	Understanding of logistics process, comprehension of the relevant legislation, to be able to make decisions independently, react quickly, communicate efficiently, to take responsibility, foreign language skills, Computer skills, ability to collect/analyse data, understanding of how effective use of resources	Understanding the organisation of cargo flows and experience with cargo transportation documents; planning and organisation, ability to concentrate for a long time; customer-focused, problem-solving skills, teamwork/collaboration, foreign (English) language knowledge, knowledge and experience with industry-specific equipment/machinery use	Ability to apply professional knowledge (i.e. acquired education) in practice (distribution logistics, supply chain management, critical/analytical thinking, risk and quality management, legislation); comprehensive knowledge of the industry ("big-picture"), understanding of business management processes, management/leadership skills, process development skills

(*continued*)

Table 2. (*continued*)

Country	Skills	In general	Operative level	Managerial level
	Skills required Gaps in skills and employer expectations	Not rated	Experience working whit Industry equipment, technical knowledge (IT skills), specific education, ability to read, understand cargo documentation	Ability to apply theoretical knowledge to practise, lack of basics economics knowledge, fear of risk, lack of appropriate education in logistics
Estonia	Key skills required	Basics of logistics	"Helicopter view", understanding of the entire supply chain, knowledge of business model suppliers and key clients, knowledge of new technologies, leading and management skills, motivation, right KPIs	Client and supplier need, lean process and efficiency, new technologies
	Skills required Gaps in skills and employer expectations	Basic economic knowledge, understanding of cost structure and efficiency	Problem-solving approach, afraid of new technologies and solutions, statistics and prognoses explanation, the gap between theory and practical work	Different new generation approach

The working language of the project is English. That is why the created study blocks and their descriptions are presented in English. In the piloting process, it was up to each partner to decide whether to translate the study blocks into their language of instruction or to use the English version. The following are the study blocks created for the undergraduate logistics curriculum:

1. Foundation of industry 4.0 and CPS in logistics
2. Concepts of continuous process improvement
3. Logistics improvement methods and tools
4. Lean management
5. Future trends in SC
6. E-commerce changing last mile
7. Risk management in SC
8. ERP
9. Customer value creation in SC
10. Transportation means and technologies
11. TMS
12. Reverse logistics
13. Sustainable and intelligent transport systems
14. E-services

In addition to bachelor's studies, partner universities in Sweden and Latvia also offer master's level logistics studies. As master's studies are also an essential part of the provision of education in the field of logistics, the master's study blocks were also selected and created, which are mentioned as follows:

15. Lean and green logistics concepts
16. Sustainability models for logistics
17. Simulation models for production logistics
18. Supply chain transparency and the role on interoperable systems
19. Material handling-how technology can support
20. Lean and optimisation models-how to consider different criteria
21. Technology introduction in logistics-risks and opportunities
22. Advanced strategic SCM
23. Advanced business logistics
24. Transport regulations
25. Logistics regulations
26. KPI and logistics quality management
27. SC simulation
28. SCOR model
29. FIATA regulations
30. Basics of Logistics Information and Communication Systems

The content of the created study blocks was discussed in bilateral meetings, and almost all the designed blocks were piloted or are still being piloted in different partner universities. The piloting process is still ongoing, but as a preliminary result, students' assessment of the subject modules piloted in the autumn semester 2019/2020 at TTK University of Applied Sciences can be presented. The results of the student feedback are shown in the table below (see Table 3), where the first number means the number of positive replies and second the number of negative replies.

According to the student's feedback, the developed study blocks primarily develop and enrich study activities. The main disadvantage was the non-compliance of the content of the study block with the volume, i.e. in the case of some blocks, the students estimated that the volume of the material was more significant than the 1 ECTS received for it.

The international view and the experience of various higher education institutions have created the basis on which high-quality and meaningful subject modules have been created. The activities of the UniLog project will last until the end of June 2020, after which the lecturers will be able to pilot the already independently produced study blocks in teaching and, if necessary, adjust them based on their experience and opinions. The UniLog project has given the Institute of Logistics essential cooperation, and the curriculum for transport and logistics has also been developed in accordance with international experience. Next step is the development of a common study module for three curriculums of TTK University of Applied Sciences: Transport and Logistics (L), Purchasing and Procurement (P), and Manufacturing Management (M). Authors have

Table 3. Student feedback on piloting study modules.

Course	Correspondence of the content to the aim of the course subject		Module integrated into the course		Clarity of content		Compliance with volume 1 ECTS		Teaching methods suitability	
	Yes	No	Yes	No	Yes	No	Yes	No	Yes	No
Transportations mean and tchnologies	15/1		16		14/2		12/4		16	
Transport management system	10/2		10/2		12		12		11/1	
Reverse logistics	8		8		8		7/1		7/1	
Sustainable and intelligent transport systems	16/3		17/2		19		13/6		19	
Basics of logistics information and communication systems	11		11		10/1		9/2		9/2	
E-services	10		10		10		9/1		10	

Curriculums / Study blocks	L1	L2	L3	L4	L5	L6	L7	L8	P1	P2	P3	P4	P5	P6	P7	P8	P9	P10	P11	P12	P13	M1	M2	M3	M4	M5	M6	M7	M8
1						x																							
2												x											x						
3																													
4																													
5					x										x									x					
6						x																							
7					x										x									x					
8						x			x																	x			
9					x																			x					
10			x													x													
11																													
12								x												x				x					
13							x											x							x				
14		x							x																			x	
15					x										x									x					
16																													
17																	x												x
18											x											x							
19				x									x														x		
20												x											x						
21							x											x							x				
22																			x										
23	x																				x								
24																													
25																													
26																													
27														x													x		
28														x										x					
29							x											x							x				
30									x																		x		

Fig. 2. Matrix of common study blocks mapping with related curriculums subjects: (L1–L8 Logistics and Transport Curriculum subjects; P1–P13 Purchasing and Procurement Curriculum Subjects; M1–M8 Manufacturing Management Curriculum subjects)

developed the mapping matrix, which enables to cover the existing curriculums subjects by suitable developed study blocks, see Fig. 2. Advantage of solution that lecturer of one study module can be learned simultaneously in different subjects of existing curriculums. For this purpose, the timetable of the corresponding subjects should be synchronised.

4 Summary

In this research, the authors created an innovative framework for developing common learning blocks and validated within the UniLog project. The potential users of the framework would be teaching and research staff working with supply chain subjects. The framework allows universities to take into account the competence needs of companies and to build 1EAP subject blocks on their basis. Subject blocks are integrated with existing curriculum subjects at the undergraduate and graduate levels. Study blocks created after completion of the UniLog project are available to all universities. The authors plan to use the framework established to apply for new projects and are considering the possibility of creating a master's program.

The integration of common module into curriculums creates an opportunity to improve the teaching process in terms of both quality and cost-effectiveness. The authors of the article are developing cooperation and integration of common module into curricula in the fields of transport and logistics, production management and purchasing and procurement management at TTK University of Applied Sciences. In the case of these curricula, it has also been possible to apply the study blocks created to develop common study module and piloted in the international project UniLog. The common study module provides an opportunity to exchange lecturers between different institutes within the school, as well as to involve foreign lecturers from partner universities. 1 ECTS size study blocks can be applied as parts of different curricula subjects, and at the same time, guest lecturers can be involved within the subjects. The involvement of visiting and foreign lecturers enriches the knowledge passed on during the teaching. There is a significant variety of views that are also valued by employers. Interviews with employers and cooperation on a daily basis have revealed the skills that the business world appreciates. One of them is the ability to see the so-called big picture, which can be developed precisely as interdisciplinary cooperation. Today, logistics is no longer limited to the topic of freight transport and warehousing; it is necessary to see connections with other areas when acquiring a speciality. The three curricula mentioned in the article are interrelated, and today the integration of these curricula has been implemented. Integration ensures the flexibility of teaching, which provides, among other things, the joint education of electives, for example. Also, the use of new learning methods, such as the implementation of most e-learning. When implementing e-learning, it is possible to involve foreign visiting lecturers from other parts of the world conveniently and cost-effectively. At the moment, there is a cooperation agreement with the lecturers of the partner universities participating in the UniLog project to conduct e-learning lessons. Still, it is always possible to expand the current group of lecturers. The participation of students in the lectures of foreign lecturers provides an opportunity to develop language skills, which are also considered extremely necessary by employers. Today, both Russian and English language skills are required to work in the field studied in Estonia, and any additional language skills provide a competitive advantage for working internationally. Integration in teaching must be primarily results-oriented, which provides graduates with benefits in the labour market.

References

1. Shevtshenko, E., Karaulova, T., Igavens, M., Strods, G., Tandzegolskienė, I., Tūtlys, V., Seyed, T., Kuts, V.: Dissemination of engineering education at schools and its adjustment to needs of enterprises. In: Katalinic, B. (ed.) Proceedings of the 28th DAAAM International Symposium, Zadar, Croatia, 08–11 November 2017, pp. 44–53. DAAAM International, Vienna (2017)
2. 21st Century Skills for Students and Teachers. https://ainamulyana.blogspot.com/2017/06/21st-century-skillsfor-students-and.html, Accessed 18 Sept 2019

3. Shevtshenko, E., Karaulova, T., Igavens, M., Strods, G., Tandzegolskiene, I., Tutlys, V., Mahmood, K.: Innovative methods of engineering education popularization at schools. Proc. Estonian Acad. Sci. **68**(4), 356–363 (2019)
4. Tandzegolskienė, I., Tūtlys, V., Jurgilė, V., Strods, G., Igavens, M., Shevtshenko, E.: Attracting youth to the occupations in the food industry, agriculture and engineering: issues for policy and practice. In: Proceedings of the 11th annual International Conference of Education, Research and Innovation, Seville, Spain, 12–14 November 2018, ICERI 2018 Proceedings, pp. 1637–1643 (2018)
5. Mendoza, J.M.F., Gallego-Schmid, A., Azapagic, A.: Building the business case for implementation of circular economy in higher education institutions. J. Cleaner Prod. **220**, 553–567 (2019)
6. Klein, N., Ramos, B.T., Deutz, P.: Circular economy practices and strategies in public sector organisation: an integrative review. Sustainability. **12**(10), 1–24 (2020). https://doi.org/10.3390/su12104181
7. Ronkova, V., Dimitrov, Y.: Education in engineering – challenges and contributions.. In: Katalinic, B. (ed.) Proceedings of the 30th DAAAM International Symposium, Zadar, Croatia, 23–26 October 2019, pp. 717–723. DAAAM International, Vienna (2019)
8. Wojcik, W., Kubekov, B., Naumenko, V.: Project: Competency based approach and the ontological model of knowledge representation of the planned learning. J. Electron. Telecommun. **65**(1), 45–49 (2019)
9. Summary of proposals of the OSKA expert panel on the manufacturing of metal products, machinery and equipment (2016). https://oska.kutsekoda.ee/wp-content/uploads/2016/05/MMT_ENG.pdf, Accessed 01 Oct 2017
10. Shevtshenko, E., Karaulova, T., Igavens, M., Strods, G., Tandzegolskienė, I., Tūtlys, V., Seyed, T., Kuts, V.: Implementing interdisciplinarity in career guidance for secondary school students in forestry and wood, metal and machinery, agriculture and food sectors of industry. Holistic Learn. **3**, 53–60 (2018)
11. Strods, G., Igavens, M., Shevtshenko, E., Karaulova, T., Tūtlys, V., Tandzegolskienė, I.: Implementing interdisciplinarity in career guidance: guidebook for career counselors. TalTech, Rezekne (2019)
12. Rüütmann, T.: Development of critical thinking and reflection. In: International Conference on Interactive Collaborative Learning, pp. 895–906 (2018)
13. Sallinen, N., Baalsrud Hauge, J., Birze, M., Kempe, I., Hintsov, T.: Improvement needs in the Central Baltic logistics competence and the related applied higher education: Part I. Requirements and needs analysis. Unilog WP2 (2018)
14. Shevtshenko, E., Bashkite, V., Maleki, M., Wang, Y.: Sustainable design of material handling equipment: a win-win approach for manufacturers and customers. Mechanika **18** (5), 561–568 (2012). https://doi.org/10.5755/j01.mech.18.5.2703

Optimizing Experimental Science Learning Outcomes Through the Inquiry Based Method and Team Making Using a Sociometric Software Tool

Charilaos Tsihouridis[1(✉)], Nick Petrou[2], Marianthi Batsila[3], and Denis Vavougios[1]

[1] University of Thessaly, Volos, Greece
{hatsihour, dvavou}@uth.gr
[2] Secondary Education, Larissa, Greece
PETRounick@gmail.com
[3] Directorate of Secondary Education, Ministry of Education, Larissa, Greece
marbatsila@gmail.com

Abstract. The present study investigates the interactivity between members of secondary education student groups based on three team making processes, the interpersonal relationships formed using them and their effectiveness in the students' learning outcomes in science concepts. More specifically, the research aimed to investigate students' opinion on three different methods used for team making, comparing groups formed based on "Group Dynamics" to groups formed based on students' preferences and groups based on student random selection by the teacher. What is more, it aims to investigate which method they would like to form groups with and which method they think has helped them be more effective in their virtual laboratory electricity concept tasks. For the purposes of the study an intervention took place with 67 randomly selected Senior High School learners. The study employed a mixed method approach (qualitative and quantitative) using the balanced test D.I.R.E.C.T. to assess students' learning outcomes, observation and note-taking, students' self and peer-assessment, three student focus groups discussion and twelve students' personal interviews to assess students' social skills and opinion on the effectiveness of experimental teaching and sociometric techniques. The findings indicate positive results on students' cognitive level, whereas the learners who used Group Dynmics displayed better cognitive results than the rest of the groups.

Keywords: Science teaching · Inquiry method · Cooperative learning · Sociogram

1 Introduction

Today, education offers teachers a vast number of tools and methods to help them differentiate their practices in order to achieve the desired goals in science teaching, among which **experimental lab teaching**. The importance of laboratory practice in science at all levels of education and its pedagogical value in teaching physics has been

M. E. Auer and T. Rüütmann (Eds.): ICL 2020, AISC 1328, pp. 42–53, 2021.
https://doi.org/10.1007/978-3-030-68198-2_4

known for years [1, 2]. It has been found that experimental teaching has positive effects in learning as it leads to analytical and scientific thinking with positive outcomes [2–4]. Experiments are considered one of the most important educational tools and strategies in the classroom, especially for the teaching of difficult or abstract concepts [5]. Experiments may be real, virtual or both. It has been found that each one of them have their contribution to science teaching with pros and cons. Virtual experiments can be implemented with digital technologies, virtual tools and relevant software. Research reveals that virtual experiments display many advantages [3]. Among them we discern their ability to be conducted without fear of accidents or mistakes, while less time is needed for their implementation as opposed to real ones.

Experiments can be implemented in teams, based on the inquiry based approach which favors the practice of students' social and collaborative skills. The use of the inquiry based method, or else "collaborative teaching and learning" for experimental science concepts, where students work in groups/teams, has been found to enhance learning [6]. It has been shown that collaborative learning promotes self-efficacy and cooperation, combines knowledge and practice and offers students the opportunity to participate actively, to express their opinion freely in the group and enhance their social skills [7]. Collaborative learning also promotes authentic thinking, constructive dialogues, opportunities for learners to engage in discussions, the formulation of questions and queries and above all collective work by helping learners combine knowledge and experiences [6]. Teams can be formed by the teacher (randomly or based on certain criteria), or by learners (based on their preferences). The aforementioned methods display a number of difficulties [8]. For instance, for groups formed by learners themselves, there may cases with students that are not easily adapted in the group and therefore are not accepted [9]. Additionally, there may be groups formed only with very weak students, thus, possibly not showing any progress [10]. Similarly, if a group is formed by good students only, there may an antagonism among them, with each student seeking top place for himself/herself [10]. In the cases where groups are formed by the teacher, there may be conflicts between members (if a student or more are not accepted by the peers), resulting into lack of interaction, communication and cooperation [11]. In cases where there is a mixture of both good and bad students, there is a danger of "good" students undertaking the heavy work load compared to the weaker students [12].

Whatever the case, it is an unquestionable fact that, for a team to function well, it should display student responsibility, positive attitude, feedback and interaction among all members, collaborative skills and role taking [9, 11]. However, this is not always the case. Contemporary schools display a considerable diversity of the student population. This diversity of today's school classes is usually accompanied by learning and behavioral problems, which make the educational process all the more difficult. Quantitative and/or qualitative mapping of student heterogeneity could prove useful in the prevention, diagnosis and addressing of such problems. Sociometry is a method which uses tests (i.e. Group Dynamics) to systematically describe the type, quantity and quality of interpersonal relationships and interactions that occur between members of a social group in a classroom [13]. Sociometric tools can be used to draw useful personality traits, thus, providing teachers with a comprehensive picture of the students' social relationships and interactions [14]. Data collection is done by formulating positive and negative questions, through which preference or rejection is expressed for one or some members of the group. However, although students' replies remain anonymous for the class, and only the teacher knows them, questions should be carefully set to avoid a sense of feeling rejected by the others. Suggestions are to ask questions such as: "Which student/s would like to ride a bike with/go shopping with/watch a movie with" and so on, instead of "Which student/s you like or you don't like" [15].

2 Rationale of the Research

Experiments play a dominant role in physics and in our attempt to understand the processes of natural phenomena that happen in the world. Experiments can be implemented with virtual or real objects. Researchers argue that the use of virtual experiments in science teaching display significant pedagogical value. Science concepts experiments can be implemented through the inquiry based approach which is considered an effective method to train learners into becoming "scientists" themselves. According to researchers, this method can be a challenge for learners and the learning of science concepts as it provides them with opportunities to ask questions, find answers, take initiatives, give explanations, and provide evidence for the topics under investigation and study.

Team making is the technique used in inquiry based experimental teaching to maximize the learning outcomes through the interaction of team members. However, teams are not always effective [16]. Today the student population is diversified and this diversity may cause behavioral issues which may impede the successful team making process, student collaborative work in experimental teaching and consequently the learning outcomes [17]. Thus, for experiments to be effective, interaction between all members of groups should take place. Otherwise, if a member or more in the group do not communicate or respond to the group obligations, if there are conflicts between members or dislikes among them, the group may not display the desirable learning results [18]. To this end, team making seems crucial for experimental science teaching and learning. Groups can be formed in various ways such as a) based on teachers' selection, b) on students' choices or c) on sociometric tools, such as "Group Dynamics". Based on the above and in order to draw conclusions on the interactivity that takes place based on the aforementioned methods, the interpersonal relationships that the students develop in the classroom working in groups, formed based on these methods, and the effectiveness of these team making methods in experimental science concepts teaching processes, based on the inquiry based approach, it was decided to do this research.

3 Research Methodology

3.1 Research Purpose and Research Questions

The research was conducted in two phases and was implemented following a mixed method approach (qualitative and quantitative) aiming to investigate the interactivity that takes place based on the target team making processes and their effectiveness in the students' learning outcomes in science concepts. Due to the bulk of data received and the limited space provided here, only the first phase of the research (pilot phase) will be presented in this paper. Therefore, the purposes of the pilot phase of the research, presented here, is to investigate the interactivity between members based on the target team making processes, the interpersonal relationships formed using them and their effectiveness in the students' learning outcomes in science concepts, as seen by the learners. More specifically, the research aimed to investigate students' opinion on three different methods used for team making, which method they would like to form groups with and which method they think has helped them be more effective in their tasks. This part of research was conducted following a qualitative approach and the research questions are as follows: a) What is learners' opinion on the three team making techniques used to teach the electric circuit concepts based on the inquiry experimental based method? b) Which method, of the three, according to the learners' opinion, is preferable to form groups with? c) Which group method, according to the learners, is more effective towards their science concepts learning outcomes? For the next part of the research, which will be described in a subsequent phase, students continued their teaching interventions taking a pre and post-D.I.R.E.C.T test (Determining and Interpreting Resistive Electric Circuits Concepts Test) Version 1.0. [19] to assess their performance in electric circuits. The aim was to verify students' focus groups and interviews results, based on which, team making based on Group Dynamics was preferable and more effective in their performance.

3.2 Research Participants

A number of sixty seven randomly selected students took part in the research. They all came from three different Secondary Education classes of the same randomly selected school with twenty one, twenty two and twenty four learners per class (67 total – three groups/classes and 21, 22 and 24 students per group/class). The participant students attended a city Senior High School and they are sixteen to seventeen years of age. Before the interventions, permission was asked and received from the school principal, the students' teacher and their parents. Before they gave their voluntary consent to conduct the research, they were all explained its purpose and were reassured of their anonymity as well as the possibility to withdraw anytime they wished or felt uncomfortable. They were also explained that, if they wished, they could have a copy of their focus group discussions content, a copy of the students' interviews content and/or a copy of the research outcomes, upon the end of the interventions.

3.3 Research Method and Research Tools

In this pilot research phase, presented here, the research tools employed to assess students' opinion about the process, social skills, cooperation/communication as well as metacognitive skills were a short self-assessment and a peer assessment questionnaire of ten different questions each. Student self-assessment is an assessment process, "during which students reflect on the quality of their work, judge the degree to which it reflects explicitly stated goals or criteria, and revise accordingly" [20]. Peer assessment is described as "a strategy that involves students in providing their opinion on the cooperation and contribution of their team members when they work in collaborative learning activities" [21]. To shed more light into the research issues three focus groups (15 students total - 5 students per class/focus group) as well as another twelve short personal interviews (12 students total - 4 students per class) were also conducted. The duration of the focus groups ranged between thirty two to thirty nine minutes whereas the interviews had duration between sixteen to twenty one minutes. Additionally, observations and notetaking by the class teacher and one researcher were also used as data collection tools. However, in the second phase of the research, which, due to limited space here, will be presented in a next step, the test D.I.R.E.C.T. (Determining and Interpreting Resistive Electric Circuits Concepts Test) Version 1.0. Was also used as a pre and post-test research tool to assess the effectiveness of the teaching interventions based on the three different team making processes. D.I.R.E.C.T. is a tool that has been translated and used by many researchers around the globe and was stabilized for the purposes of this study.

3.4 Research Process

For the purposes of the whole research (pilot and second phase), three teaching interventions (of 3 h each-total 9 h of teaching interventions) were designed and took place in the three randomly selected classes. All teaching interventions were implemented by one of the researchers for a number of reasons: a) it is very important that teaching interventions can be realized by the same teacher-researcher and to learners that had not been taught by him/her before (Roussos and Tsaousis 2011). In this way, two factors are significantly reduced: a) the different level of communicability of every teacher towards his/her learners for any school subject and which depends on different particularities of their personalities and the communication that have developed with the students and b) the actual class teacher might have been influenced by having taught the participant groups prior to the research itself, and might have developed a certain relationship with the learners.The reduction of these factors leads to the elimination of the focus on teaching style, and enhances the emphasis on the effectiveness of the teaching aims which are the learning process and learners' outcomes, thus, contributing to the credibility of the research process.

As already explained, the research was realized in two phases. In the first pilot phase, presented here, all three groups were taught the same material (electric circuits concepts) for the first three hours of the interventions (out of total 9 hourly interventions). Every hour, each class worked in groups, formed differently. In this phase, the aim was to detect the interactions between the groups and the dynamics of the groups.

The following table depicts the interventions of the first three hours of teaching and the methods of team making per class and per intervention in the pilot phase of the research (Table 1):

Table 1. Formation of groups per class and intervention of first pilot phase

Groups	1st Intervention	2nd Intervention	3rd intervention
A1	Randomly formed groups	Groups formed based on sociometric tool	Groups based on students' preferences
A2	Groups formed based on sociometric tool	Groups based on students' preferences	Randomly formed groups
A3	Groups based on students' preferences	Randomly formed groups	Groups formed based on sociometric tool

In the second phase, which is not presented here, each class was taught another six hours of electric circuit concepts. The aim was to verify the results of the pilot phase, based on which, and according to the learners, Group Dynamics had a better impact on their interactions and performance. However, as already explained, due to the bulk of the data and the limited space, it was decided hereby to present only the results of the pilot first phase, whereas the second phase results will follow in a subsequent step.

More analytically, all the steps, throughout the research were the following:

1st phase (1 h): Students were explained the research and in a form of a game, they were given two questions to answer (Which three classmates would you like to go to a party with? Which three classmates would you not like to go to a party with) in order to form the teams based on Group Dynamics, which the researcher would process on his own, later on. Students also took a D.I.R.E.C.T pre-test.

2nd phase (9 h): (Pilot three hours – second phase another six hours). In this phase each class first formed their groups. For every hour of intervention each group was taught based on a different team making process (Table 1 above). All classes used the software "Edison 4" to work on the same material of electric circuits, followed by appropriately designed worksheets. Throughout the interventions, the class teacher and one or the researchers were observant, taking notes in relation to the students' interactions, cooperation and communication.

3rd phase (1 h): Students took a post D.I.R.E.C.T test.

4th phase (1 h): Students took a self-assessment and peer-assessment questionnaire to state their opinion on the process. Additionally, twelve short personal interviews with randomly selected learners were also implemented regarding their opinion on the team making methods and the process.

5th phase (1 h): Three focus group discussions with five randomly selected learners per class (but not the same that took part in the interviews) also took place.

4 Data Analysis Methods

For the analysis of the focus groups and personal interviews the "content analysis" method was employed. After the transcription, the repeated listening and reading of the focus groups and interviews discussions content, the most significant parts that linked to the research questions were isolated and recorded and the data analysis units/key words were determined. The key words were "interesting", "innovative", "motivating", "different", "helpful", "effective", "communication", "competence", "cooperation", "collaboration" "interactions", "relationships" that were included in the participants' answers, in relation to the aims of the research and research questions. Due to the large body of information, all data were classified in groups, based on which the content analysis took place. The results of the students' self-assessment and peer assessment that were answered with replying with a tick in a blank space were also qualitatively analyzed with the same method. The pre and post D.I.R.E.C.T tests results were analyzed with the SPSS statistical tool.

5 Results

D.I.R.E.C.T. Test Results

Due to the bulk of the data received, and the limited space of this paper, the full analysis of the D.I.R.E.C.T. tests and students' self-assessment questionnaires - related to their opinion on the effectiveness of the interventions on their learning, are not presented hereby. However, based on the results received from the D.I.R.E.C.T tests and the self-assessment questionnaires it can be argued that students who worked based on the Group Dynamics team making method outperformed the students that worked forming groups with the other two methods. The results of the tests comply with the answers of the students' focus groups, personal interviews, peer assessment presented below.

Students' Peer Assessment Results

The questions of the peer assessment tasks included questions about the extent to which the members of the teams 1. cooperated with one another, 2. participated equally in the activities, 3. listened carefully to one another, 4. supported the work done with their views, 5. discussed with one another to find a solution to the problems, 6. completed the work assigned to them, 7. respected one another, 8. helped one another, 9. contributed equally to the activities, and 10. had a good time working in their team. In relation to the groups formed by the teacher, the results revealed that the majority of the learners had problems to support the discussions with their views, to complete the tasks, to help or contribute to the assignments. With this in mind, the students in the focus groups were asked to provide possible reasons for these answers. According to their answers, lack of bonds among the team members and the participation of less popular students in the group created a passive atmosphere which led to indifference for work or overall cooperation.

Regarding the groups formed by the learners themselves, the findings showed that the students did not listen actively to one another or did not contribute equally in the team. When asked in the focus groups why that might be, students explained that some members in their teams were more dynamic or demanding than others, there was little patience to wait for students with "slower pace" to finish a task, there was grumbling and complaints and therefore, communication had leaks or gaps. As regards the groups formed with the Group Dynamics process, the majority of the learners noted that they were satisfied with the cooperation and the interactions that took place within the groups, where there seemed to be equal contribution, trust and acceptance by the team members. The final results showed a percentage of 59% satisfaction for the formation of Group Dynamic teams which is considered very encouraging compared to the other two methods (groups formed by the teacher 19% and groups formed by the learners 29%).

Students' Focus Groups Results

The focus groups results revealed a number of issues in relation to the research questions. According to the answers, students mentioned a number of roles, taken within groups and with all group formations. Almost all roles [9] were referred to in their answers, discussed however in a different way. Students were asked to refer to these roles that defined their interactions in the groups. Thus, for the work done with groups based on the teacher's decision or their own preference, they talked about the student that "does not exist" or "we never talk to", most probably referring to the neglected type of learner, they talked about the classmate that "few like", "others don't listen to", is "bad", a "nerd" or "boring", probably referring to the rejected type of learner in a group, they discussed about the learner who "everyone wants to approach", that "we all want to ask", that is "nice" or "smart" probably referring to the popular type of student within a group. Some types of learners were not very clear about which role they had in the group and therefore it was decided to categorize them as controversial or average [9]. These were students that "are lazy" (but this does not necessarily mean they may not be good students), "negative" (but nobody can accuse them for not being hard workers perhaps), "leave it all to others" (but nobody can tell whether these students have such an attitude because they are lazy, indifferent or over confident).

They admitted that having the teacher form groups for them "was not a good idea" and that they don't like to "be left aside" when it comes to who they can best work. Some students insisted that "they know better who to keep company with" when they form groups based on their preferences but others contradicted to this belief and argued that "this is not true" reminding them that "[name] and [name] argued many times for silly things" and that they both wanted to "show off" and "made a mess out of it", probably referring to bad task results. When asked what they thought of the interactions that took place in groups formed by the teacher or themselves, they admitted that their cooperation was average when the groups were their own decision but much worse when they were the teacher's. They argued that it was not easy to work very effectively in some of these groups as some learners were not "present really" and the work of the whole team was delayed.

When asked to expand further on this they explained that some students in those groups looked like they meant to work but they did not contribute really, some others dealt with other tasks during the group work, some students did not know one another very well, so they were shy and some other learners were very annoying and caused trouble within the groups. As learners admitted, for someone "being obnoxious", or "Mr know-it-all" is one of the things in a group no one can easily "take or like". When asked what they think of their performance within groups formed as aforementioned they replied that this was not "what they would have liked", "they were not very happy about it", "it did not go very well", "it could have been better", "of 'not of a high' level", or "blurred". When asked to expand further on their answers and how they knew this they replied that "it just didn't work right", "the other method [group dynamics selection] was better", they had "a complicated communication", or "it didn't satisfy" them as much after all.

When asked which method of group formation they would prefer to work with they answered the method with Group Dynamics. According to the learners their interactions in such groups were more effective for a number of reasons. As they argued things "ran rather smoothly", "rolled well" and they "fit well in the team". In our questions as to what they implied by that they explained that "no one had to insist or complain", "communication was good", "there was no fuss or arguments", "everyone knew what to do", and time was "organized well" by everyone. The students explained that they liked the fact that they all worked equally and that there didn't seem to be any conflicts or "negative attitudes". They argued that the latter is very important to them, especially for the good students, who need mutual cooperation and understanding to reach a higher level of performance. It was surprising to hear that even mediocre learners were satisfied with this "serenity" during work, which allowed them to focus on the tasks. As they argued, they felt that this could very well be the cause they thought they had done a lot better at their tasks than when they worked in groups formed differently and caused trouble and noise.

Students admitted that when members work well together it is better because then they are not distracted, they do not have to worry about "hiding their feelings" or "wanting to burst" and that "time runs fast" without boredom, anxiety or impatience. They also explained that the lesson becomes enjoyable, conflicts are rare, tasks seem easier due to effective collaboration, and a sense of happiness is quite evident. There were also comments that made us wonder about the level of the students they derived. Specifically, learners admitted that it was easier for them to overcome difficulties in the task, that they were not afraid to ask questions in the fear of being ridiculed by others, that they like being in school, that they did not have to experience anger or rude classmates. Based on these answers, we questioned ourselves whether these learners had a low self-esteem previously [before being in Group Dynamics formed teams) due to low performance and that within Group Dynamics teams they found they could work better, trust themselves and others more and adapt in the team more easily. However, there were also a couple of learners that were very quiet and passive. According to them, they worked the same in all three different type of groups and "did what they were told".

Students' Interviews Results

Students' personal interviews were held in order to allow learners talk freely without having to worry about "revealing" things in front of others or exposing themselves in front of their friends/colleagues. Thus, when learners were asked their opinion about the formation of the groups based on the teacher's decision they replied that this is what happens many times in class, that they don't always like it, that the teacher does it without asking their opinion and that he "has his own reasons". They explained that they didn't like very much this kind of group formation as they cannot "feel safe", they "cannot be with mates", they don't have support from friends in the tasks or that they are forced to work when they don't want to do it. They also explained that they did not think this method helped them understand the tasks very well, as they were shy to ask questions, or they had to do a lot for others that did minimum or nothing at all, that they did not feel comfortable with some classmates or that they were obliged to be with learners they "never hang around with". In relation to the groups made based on their preferences they explained that this "did not work much" all the time because either they were friends and kept talking, wasting valuable time to finish the tasks, there were difficulties nobody in the group could address effectively as they were all "not very good students" and some wanted "to do their own thing" and did not cooperate.

Regarding the Group Dynamics method of forming groups they expressed their preference and explained that there was lack of tension, there was understanding and communication, they did not feel they were "used" by others and cooperated well for the tasks. The majority felt the interactions that took place were mutual and on equal basis, they collaborated for the tasks with everyone contributing in their own way. A learner explained that his psychology was a lot better this time whereas two other students admitted they felt relieved they didn't have to work with naughty students.

However, there were a couple of learners that said that they did not mind which method their group is formed and that either way they "will do what they have to do because they are used to it". Another student said she did not see any difference and that she thought that it is best when the teacher selects the team members because he knows best and because he has his method. The majority however admitted that they felt their performance was much better when they worked in the group that was formed based on Group Dynamics and that they enjoyed the "professionalism" of their team and the "seriousness" of their actions.

6 Discussion

In this research we have investigated learners' opinion on three different modes of group formation (based on the teacher, on students' preferences and on sociometry), looking into the interactivity features of the team making processes, the interpersonal relationships formed and their effectiveness in the learning outcomes, as seen by learners. Based on these results it can be said that the use of sociometry gives another dimension, the evaluation of the ability of students' social adjustment and behavior. In order for the teacher to "decode" the inter-relation signs of the classroom, it is important to explore the interpersonal relationships among students, which can affect their psychology as well as their performance in the classroom. Sociometry seems to be able to develop these social relations between members of a group. This seems crucial as the classroom is not just a gathering of individuals but a dynamic social team, which means that both the composition and the quality of the relationships cultivated between its members play an important role and have a decisive influence on their education. Furthermore, learning does not only occur due to the family background and relationships but also in other settings like in the school with peers who form the social system of the school or of each class [14]. This fact is extremely important for the classroom, because the structure of its members continues to be the same for many years throughout the academic studies. Thus, the establishment of negative relationships may harm those students who, for example, would be forced to endure for years the feeling of disapproval or even isolation. In addition, it is now certain that the success of learning takes place more successfully within a framework of smooth interpersonal relationships as well as positive environmental influences. It is therefore extremely important for the teacher to be aware of the possible existence of individual groups within the whole class, their structure and the relationships between them. So, once the teacher has applied the sociogram of his school class, he gets the opportunity to fulfill some of his students' wishes, to study and discover causes that cause isolation, inhibitions, repulsions, conflicts and controversies, as well as to solve problems and conflicts that are presented among his students in order to ensure the basic emotional conditions for learning and socialization.

In conclusion, based on the research participant answers, the use of the Group Dynamics team making process, used to teach experimental science - virtual labs on electric circuits - enhanced the learning outcomes of science teaching, ensured a positive attitude of each group towards the lesson, increased the responsibility of the groups, created a good competition between them and facilitated the participation of students. Additionally, the research results revealed that the groups that used "Group Dynamics" for their team making had better cognitive results than those forming their teams in a different way. In relation to the learners' social and cooperation skills as well as metacognitive skills, it can be said that the students that formed their teams following the sociometric team making method knew already from the beginning of the intervention what they needed and how to do it. They focused more on communication, paid more attention to the lesson, fewer students were rejected, there was no student exclusion, there was more homogeneity in the groups, each student's role was defined from the beginning and their responsibilities could be easily explained within the group. Additionally, any obstacles were overcome and the groups worked really well communicating and interacting with one another effectively. In short, these groups outperformed the other two groups.

The results have implications for educators who may explore the possibility of using the method to have better results in students' collaboration and help learners progress through their work.

References

1. Marshall, J.A., Young, E.S.: Pre-service teacher's theory development in physical and simulated environments. J. Res. Sci. Teach. **43**(9), 907–937 (2006)
2. Hofstein, A., Lunetta, V.N.: The laboratory in science education: foundations for the Twenty-First Century. Sci. Educ. **88**(1), 28–54 (2004)

3. Tsihouridis, C., Vavougios, D., Ioannidis, G.S.: The timeless controversy between virtual and real laboratories in science education - "And the winner is…". In: International Conference of Interactive Collaborative Learning and Engineering Pedagogy, (ICL), Kos, Greece, 25–28 September2018 (2018)
4. Chinn, C.A., Malhotra, B.A.: Epistemologically authentic inquiry in schools: a theoretical framework for evaluating inquiry tasks. Sci. Educ. **86**(2), 175–218 (2002)
5. Baser, M., Durmus, S.: The effectiveness of computer supported versus real laboratory inquiry learning environments on the understanding of direct current electricity among pre-service elementary school teachers. Eurasia J. Math. Sci. Technol. Educ. **6**(1), 47–61 (2010)
6. Felder, R.M., Brent, R.: Cooperative learning. In: Mabrook, P.A. (ed.) Active Learning: Models from the Analytical Sciences. ACS Symposim Series, vol. 970, pp. 34–53. American Chemical Society, Washington (2007)
7. Kvarnstrom, S.: Difficulties in collaboration: a critical incident study of interprofessional health care teamwork. J. Interprof. Care **22**(2), 191–203 (2008)
8. Roberts, T.S., McInnerney, J.M.: Seven problems of online group learning (and their solutions). Educ. Technol. Soc. **10**(4), 257–268 (2007)
9. Coie, J.D., Dodge, K.A., Coppotelli, H.: Dimensions and types of social status: a cross-age perspective. Dev. Psychol. **18**(4), 557–570 (1982). https://doi.org/10.1037/0012-1649.18.4.557
10. Estevez-Lopez, E., et al.: Aggressive and non-aggressive rejected students: an analysis of their differences. Psychol. Schools **43**(3), 387–400 (2006)
11. Cadwallader, T.W.: Sociometry reconsidered: the social context of peer rejection in childhood. Int. J. Action Methods **53**(3), 99–118 (2000)
12. Kalkan, M., Epli-Koç, H.: Perceived social support from friends as determinant of loneliness in a sample of primary school. US–China Educ. Rev. **8**(4), 547–551 (2011)
13. Garcia-Magarino, I., Lombas, A.S., Plaza, I., Medrano, C.: ABS-SOCI: an agent-based simulator of student sociograms. Appl. Sci. MPDI **7**(11), 1126 (2017)
14. Drahota, A., Dewey, A.: The sociogram: a useful tool in the analysis of focus groups. Nurs. Res. **57**, 293–297 (2008)
15. Zafiri, I.: Research for the improvement of the relationships and the interactions between students of a school classroom with the use of a digital game for educational purposes, Master Thesis, Kapodistrian University, Athens (2016)
16. Contandriopoulos, D., Larouche, C., Breton, M., Brousselle, A.: A sociogram is worth a thousand words: proposing a method for the visual analysis of narrative data. Qual. Res. **8** (1), 70–87 (2018)
17. Sobieski, C., Dell'Angelo, T.: Sociograms as a tool for teaching and learning: 18 discoveries from a teacher research study. Educ. Forum **80**(4), 417–429 (2016)
18. Leung, B.P., Silberling, J.: Using sociograms to identify social status in the classroom. Calif. School Psychol. **11**, 57–61 (2006)
19. Engelhardt, P.V.: Examining Students' Understanding of Electrical Circuits Through Multiple-Choice Testing and Interviews. Thesis, North Carolina State University (1977)
20. Andrade, H.L., Valtcheva, A.: Promoting learning and achievement through self-assessment. Theory Pract. **48**, 12–19 (2009)
21. Noonan, B., Duncan, R.: Peer and self-assessment in high school. Pract. Assess. Res. Eval. **10**(17), 1–8 (2005)

The Lessons of Forced Distance Learning: Software Engineering Approach in the Gap of Generations of Educational Software

Ekaterina Beresneva, Mariia Gordenko, Olga Maksimenkova(✉), and Alexey Neznanov

National Research University Higher School of Economics, Moscow, Russia
{eberesneva, mgordenko, omaksimenkova, aneznanov}@hse.ru

Abstract. Quite recently, considerable attention has been paid to distance learning due to mass schools and universities closing because of coronavirus disease. This fact brought to light not only strengths but also weaknesses of existing educational software and use cases. This paper is motivated by the questions which are rinsed by thoroughly adopted educational software and the deep gap between educational methodologies and up-to-date distributed information infrastructure capabilities. The lack of an engineering approach in the deployment of distance learning tools leads to both technological and methodological problems. We present exploratory analysis of data gathered from authority web sources. The significant ideological differences between educational software generations are discussed with special attention to non-cloud-based and cloud-based collaborative technologies and corresponding platforms. The authors conclude that the power of integration based on industry-wide interoperability standards is useful to solve current problems in distance education related to software.

Keywords: Distance learning · Integration · Interoperability · Educational software · Software engineering

1 Introduction

This year the World faces COVID-19 as a powerful driver of distance learning which created a unique situation for the real load testing of educational software of different kind. It became possible because educators all over the World started rapidly the force transition to online and distance learning.

As a result, the questions of technology and instructional design for distance learning become the cutting edge for lots of people from educational stuff to parents. These people run into difficulties due to the unplanned manner of this transition and great amount of unusual troubles with hard and software. It is well known that the education in general is a traditional sphere of great social responsibility regarding safety and security as of paramount importance.

Naturally, the shock to global education from COVID-19 is just seems something new and quite similar to World global conflicts of XX century which seriously affected and shifted education. Thus, Conner and Bohan [1] reported on great secondary

M. E. Auer and T. Rüütmann (Eds.): ICL 2020, AISC 1328, pp. 54–65, 2021.
https://doi.org/10.1007/978-3-030-68198-2_5

educational curriculum reconstruction during the Second World War. And, Jaworski [2] comprehensively studied change in women's education in that period. On the one hand, it is good for society that such situations are very rare, but on the other hand that makes them unique enough and the solutions during rapid transformation should be fast, relevant the time, and easy to operate.

From this perspective, the educational society potentially can find help in the fields which focus on software and its use cases, such as Software Engineering. A short review reveal that software engineers have a number of modern software technologies and instruments used in distant working process that are expected to be applied in online learning activity [3, 4]. This work pays attention to the one of software engineering fields – Educational technology (EdTech or EduTech), which refers to fundamental issues of learning, teaching and social organization that makes use of technology [5].

This paper aims both to detect the troubles in the current state of distance learning infrastructure by investigating guidelines and recommendations from different educational governance and to expose their inadequacy from the software engineering point of view. Thus, the context of this paper lays between education and software engineering and it relies on transdisciplinary approach.

2 Overview

To make the context of the paper clear this section introduces educational generations and explains the significant ideological differences between them. It also overviews the impact of these generations on current education during a global pandemic.

2.1 Historical Overview of Educational Software Generations

For decades, the global educational community faced four paradigms of educational software which sequentially arise because of scaling.

The first generation is associated with the growth of personal computers (PCs). In the early 1980s the era of widespread personal computers began beneath the headline "You've just run out of excuses for not owning a personal computer": devices were priced less than 1000$ and people could afford to buy and use them at home. It changed the field of software in general with specific implications for educational software: the cost of copying and distribution of a software product significantly decreased, so the first standalone educational tools came to the market.

The second generation is characterized by Sun Microsystems motto "The Network Is the Computer." From a software engineering point of view that shift was marked by junction of PCs over the network. Since client equipment gradually became cheaper and more compact, mobile devices appeared. As a result, the emergence of the network and the spread of mobile devices lead the network to be mobile. The development of handheld devices and wireless network allowed applying new ways of learning. For example, United States, Australia and the United Kingdom became the pioneers of the 1:1 computing initiative [6]. Later One Laptop Per Child (OLPC) program was introduced in developing countries [7]. These ideas were aimed at mass introduction of personal computers in educational practice in order to supply each student with computing device having Internet connection. It should be noted that not only PCs were used for educational purposes but also laptop computers (in the 1990s), handhelds (in the early 2000s) that were later transformed into smartphones or tablet computers (in the late 2000s). The net result is that one-to-one computing environments changed and improved classroom dynamics owing to computation and network communication capabilities that augment face-to-face interactions.

Later, with the advancement in technologies, network bandwidth increased significantly resulting in expansion of geography coverage. Increased number of users incremented their PCs load but even though the volume of information storage vastly multiplied, it still was limited. Users had a demand for the absence of these restrictions, as well as the no need for modernization and maintenance of their own hardware infrastructure. And they also wanted to receive durable services with a high level of availability and low risks of inoperability. Thus, cloud computing [4] gained its popularity as a result of convergence of telecommunication technologies and the convergence of computing and data storage based on virtualization, together with the advent of new server hardware and management of distributed systems.

Cloud computing technology started its ascendant in the late 1990s and provided an incredible cost reduction because of rapid elasticity and resource pooling: optimal use of software and hardware were achieved, hence increasing the efficiency and effectiveness of the available resources [4]. Other advantages that accrue from cloud computing are convenience and improved accessibility – due to transfer of computations to cloud server, the devices are not obliged to download and deploy any third-party software, and they have to provide only an access to Internet [8]. So, these devices became client terminals receiving data computed on a server side in a remote cloud data center [9]. These facts allowed educational software tools to become even more affordable and applicable to online learning process.

Classical learning management systems (LMS) and learning spaces was superseded by cloud-based ones, that have additional advantages owing to heterogeneity. Thus, more specific educational tools that need integration with LMS for adequate administration and methods of data exchange, commonly known as *omnichannel communication*, for comfortable usage became necessary. These tools should be based on modern interoperability standards, for example, from IMS Global (LTI and QTI APIP).

As a result, and converged computing results, known as *cloud, edge, hybrid, fog computing*, coupled with the *omnichannel communication* and ability to provide automatic context change forms *the third generation,* which the most modern educational software belong to.

And the transition to *the fourth generation* will take place when all learning tools transform to intelligent and adaptive systems powered by artificial intelligence (AI). Squirrel, ALEKS, Smart Sparrow, IBM Tutor are examples of products sustaining the fourth paradigm shifting.

2.2 COVID-19 Impact on Usage of Educational Software Generations

Traditionally, users from the education area are not the early adopters of modern generations of educational software. Nowadays, most of the educational software belongs to the second generation and there are only plans to transit to the next ones. Of course, there are educators who intensively use modern technologies before the outbreak, so they are not up against with the force distant transition. However, there are learning organizations which have not yet fully utilized the tools of the current generation in the learning process. And they are facing significant uncertainty.

The first issue is connected with the fact that educators can easily notice that the second generation has an extensive user base, advanced environment and numerous developments. Consequently, it provokes a continuing usage of second-paradigm technologies. So, the innovator's dilemma arises, described by Clayton M. Christensen [10], but not reflexed by the educational community. The book introduces a "disruptive innovation" concept implicating that old products become uncompetitive because of the parameters, used in an earlier comparison, ceases to be relevant. It means that COVID-19 impact on educational technology generations empower innovator's dilemma; and a decision to stay on second generation of educational software is equal to defeat.

The next issue is associated with the practicability of switching to the fourth generation. Some people postpone the transition waiting for the future paradigm since the "fifth" one can be potentially around the corner. And in this regard, an undesired period of any transition rejection may begin. However, the outbreak pushes educational organizations to make this force transition, and here the gap of generations of learning software is emerged.

3 The World Distance Learning Infrastructure Readiness to COVID-19

To support the main idea of the paper it is crucial to understand the current with educational software all other the World. Thus, this chapter provides the overview of the main recommendations from World organizations and individual countries about transition to distance learning [11, 12]. In total, more than 250 various sources were analysed, including governments' and international organizations' websites, local school platforms and reports about distance learning shifting. The overview seems representative enough because there are over 90% of the world's student population have been affected [13] by the April, 1.

The response of international educational organizations and communities to force transition to distance learning due to COVID-19 is analysed. In Table 1 it is shown that these recommendations sum up the existing open knowledge hubs, well-known MOOC platforms and supporting tools and software. However, it should be mentioned that despite the importance of resources presented, they can cause bewilderment among educators. It can be noticed that these tools cannot be used to fully construct educational process but only to complement it. It means that listed software can interfere the educational process with inept use.

According to the International Monetary Fund classification, there are advanced and developing countries. Some experience of advanced countries which is summarized in Table 2 demonstrates that almost all advanced countries created or refined their national educational platforms with video-lessons, assessments, videoconference abilities. However, the rapid introduction of these systems has shown that they cannot fully and completely satisfy all the needs of the educational process, so the countries have published extra lists of supportive resources and tools. From Table 3 which contains findings about local guidelines of developing countries we can conclude that the situation is the same as in advanced, but there are the technical problems with internet or device accessibility. Consequently, some of the governments of these countries decided to translate lessons on public TV.

Table 1. The reaction of international and educational organizations

Organization	Suggestions/Tips	Lessons/webinars/courses
The Commonwealth of Learning [14]	The organization provides the list of learning resources separated by educational levels (Khan Academy, MIT OpenCourseWare and etc.) [15]	The guides about blended learning and pedagogical innovations are published
The World Bank	The WorldBank collected resources and platforms in one guide to support remote learning. The guide summarizes information about national learning platforms and other platform and software with separation by needs (assessment, file manager, LMS, training, video conference system) [11]	On website the guide about methods and methodologies of distance learning can be found
UNESCO	UNESCO provides the list of educational applications, platforms including resources providing psychosocial support, digital learning management systems, systems with strong offline functionality, MOOC platforms, self-directed learning content, mobile-reading applications and tools for teachers to create the digital learning content [12]	UNESCO provides support, holds webinars and provides a platform for the exchange of educators‘experience

Table 2. The reaction of advanced countries

Organization	Suggestions/Tips
Canada	The Canada provides for free the big collections of learning tools, databases and etc. for each province both in English and French [16]. For example, Manitoba Ministry of Education gave access to “Manitoba_My learning at home” platform with the educational resources and everyday learning activities [17]. Quebec states also provided own platform “Quebec Open school” with educational support for everyone from 9.00 am to 8.00 pm from Monday to Friday [18]
Germany	The wide range of supportive tools were provided for educators and students in Germany. There are secured online platforms for teachers, supportive guides for teachers, databases, lists and overviews on learning tools and websites with materials for both students and educators [16]

(*continued*)

Table 2. (*continued*)

Organization	Suggestions/Tips
Great Britain	The UK government publish the list of useful online recourses and tools for home education, including such subject-specific recourses as math, English, science and etc. [19]
France	With support of Ministry of National Education, a 52-min Maison Lumni daily program for 8–12 years old students were launched The "My class at home" is a national platform, where students can select interactive recourses, do some tasks, and meet with teachers and classmates
Japan	The Ministry of Education, Culture, Sports, Science and Technology (MEXT) developed centralized website, where tips and all useful information for educators are collected The project "Future classroom" shows EdTech events and platforms for support remote educational process [20]. On MEXT website the platform for support e-learning by age, level and subject was developed. Its platform the lessons and materials for self-education are presented [21]
Singapore	There is a Ministry e-learning portal "Singapore Student Learning Space", where helpful materials for students and teachers are available [22]
South Korea	The start of spring semester on 2020 was delayed from March to April Korean Education Broadcasting System (EBS) contains a lot of multimedia content [23]. Korea Education & Research Information Service (KERIS) is an aggregator of different platforms, also provides digital textbook services [24]

Table 3. The reaction of developing countries

Organization	Suggestions/Tips
Angola	The Ministry of Education and the Public Television of Angola shows online lessons on TV [18]
Iran	Daily TV program for educational purposes for all learning grades was launched. The Ministry of Education developed the app "SHAD" to assist learning [18]
Iraq	Lessons for all educational level are stored on YouTube-channel [25]
Kazakhstan	The educational website "Kundelik" with tools and materials for remote learning are the main platform [26]
Russia Federation	The Ministry of Education of Russia support "Russian e-school" platform, where lessons and assessment for school' learners are available [27]
Saudi Arabia	The Ministry of Education developed national e-learning platform "Vschool" and official learning portal "Ien National e-portal" [28, 29]
Tanzania	The Ministry of Education in Tanzania with supportive of the Tanzania Institute for Education launched a series of educational TV-films for learning due to COVID-19 closures [18]

In addition to the list of recommendations, there are restrictions on usage of some tools, for example, popular among the educators video-communication platform *Zoom* was banned by governments, companies and educational organizations by various reasons including cybersecurity and politics, see the most interesting cases in Table 4. The full list of all restricted countries and regions can be found on *Zoom* official website [30]. This restriction also affected the educational process because teachers and children had already managed to adapt to it, and the prohibitions by states and educational organizations on its use forced them to switch to new tools.

Surely, the mentioned guidelines help lots of teachers and learners, but thoroughly absent users' support led to several deep problems. Learners do not understand how to use educational resources and get confused in different tools. Moreover, parents worry about children' health, as prolonged use of gadgets can adversely affect vision and the musculoskeletal system. Students cannot always concentrate on the lesson because there are many other interesting and entertaining content. Students are very tired of gadgets and are not able to perceive a lot of information at once [37, 38]. For example, Russian Federation, Ministry of Education published recommendation to exclude the usage of smartphones for distance learning. It is accompanied by the impact on the student by complex unfavourable factors: electromagnetic radiation, the small size of characters and images, the inability to comply with a rational working posture, strong tension of the muscles of the neck and shoulder girdle and others. Moreover, the duration of continuous usage a computer with a liquid crystal monitor in the lessons ranges from 20 to 35 min, depending on the class of the student. The total duration of such work is from 1 to 3 lessons per day. Such measures are taken to prevent the negative impact of electronic devices on the student's healthy [39].

Table 4. The list of prohibitions

Initiator	Reason	Comment
Taiwan	Security reason/Politics reason	Taiwan was the first country, which banned "Zoom" for public institutions, because some traffic routed through China, which does not recognize Taiwan independence [31]
Rwanda	Security reason/Politics reason	Rwanda Information Society Authority (RISA) advised against using "Zoom" to public institutions because it routes traffic through China although it was said that traffic remains in the country where the call was made. There are also vulnerabilities that allow third parties to enter the conference and broadcast indecent content [32]
Singapore	Security reason	On lesson two unknown men add lewd comments [33]
Google	Security reason/Competitor reason(?)	Google bans using "Zoom" for employees on their personal gadgets due to many security breaches from "Zoom" [34]. Also, Google has own videoconferencing system "Hangouts Meet"
SpaceX	Security reason	Security issues and "Zoombombing" were the main reasons for ban [35]
New York City	Security reason	Zoom was banned to using in schools [36]

The teacher burden increased because of ongoing communication with children and parents. Teachers are concerned that students do not receive individual consultations and do not always understand terms, especially mathematical. Learning creative subjects and art in the current distance format is almost impossible. The quality of education has decreased due to the lack of control over students [40]. Of course, most of the problem is related to the technologies, which used to distance learning. As far as the authors can conclude, recommendations from governments and professional educator organizations are limited with lists of software and distance learning methods without connections among them [3, 41]. Surprisingly, in spite of the development of modern learning management systems, most teachers control the learning process via instant messengers (e.g. WhatsApp, Telegram) or send the necessary information to students by e-mail [40].

Due to the abundance of educational resources, both teachers and students are confused in their accounts. Educators are often confused in the list of recommendations and do not understand what tools they should use. Not always competent use of existing platforms leads to variety entry points and the lack of integration of heterogeneous systems due to mixing products of different generations. Political, economic, competitive reasons also influence on educators and educational process.

It should me mentioned that all types of educational software experienced the sudden stress testing induced by the announced lockdown. It is observed that the educational systems had serious technical crashes and malfunctions caused by scaling problems [42, 43].

4 The Software Engineering View

Here we discuss the problems identified in the previous section from the software engineering point of view. It is important to emphasize that the generation gap was already predicted and recognized by system engineers. Software engineers succeed in such recognition and are currently developing "fourth-paradigm" solutions based on global network infrastructure and cognitive (AI) platforms.

The essential point is that coronavirus decease did not make great problems for software engineers because of their readiness. IT industry is get used to agile management methods, which brought on approaches for organizing virtual teamwork and state-of-the-art software substituted for tools of previous generation. Large-scale development of cloud computing technology prompted to the breakout of heterogeneity platforms and microservices. So, the solutions, constructed on their basis, have such features as rapid transformation, easy operation, and very fast and scalable deployment. So, actual generation of corporate information systems (CIS) not only reacts much faster to the COVID-circumstances than widely adopted educational software but also supplement functionality with specific education features (see examples of Zoom, Teams, etc.).

It is notorious that due to the COVID-19 many employees started working from home. Those organizations which utilized old technology stack and not used highly scalable cloud-based software, had more demand in anykey-men, support, and cybersecurity specialists. Most interestingly, 15% of cybersecurity staff said that they did not have the tools and resources to provide security for remote working [44]. At the same time, 34% said they have only a temporary solution. However, almost half of cybersecurity specialists was transferred to the IT support for helping other employers in a new environment [44]. Thus, unfortunately, that decision is largely responsible for increase of the number of cyber-attacks due to the COVID-19. In a contrast, there were organizations which systematically supported early adoption initiatives; and they did not face problems with attack and vulnerabilities protection. So, the force transition did not put inconvenience to their working process.

In addition, there is a software engineering education sphere which requires a constant peg of knowledge of emerging professional standards to successfully train learners under the current labor market realities. There are not only special communities and conferences (for example, CSEE&T) to discuss education in this area, but also declared bodies of knowledge covering working process of IT specialists. For instance, SEBoK v.2.1 and SWEBoK v.3 are the key knowledge sources of systems and software engineering. Another authority guide EITBoK contains a vast amount of material in specific areas of enterprise IT practice. ProdBoK and PMBoK v.6 with its Software Extension guides discuss the effective application of knowledge in professional practice of product and project managers, respectively. And DAMA DMBoK2 is a body of knowledge which presents a review of the best accepted practices and alternatives data management approaches.

Thus, we can conclude that educational organizations are not the pioneers in this technological evolution. In addition, the outbreak proves that those who was not put cloud-based tools and agile methodologies into usual educational/working practice were unable to painlessly adapt to new conditions. So, the educators are counseled to be in the track of the software engineers' experience.

5 Lessons Learned

This section summarizes main findings obtained through previously classified and systemized data and introduces lessons learned. As a result, we can see:

1. Proof that the current state of telecommunication infrastructure as a very good basis for scaling cloud-based solutions (with many examples of CISs scaling up to billions of users);
2. Triumph of modern CISs in conjunction with open educational resources (OER);
3. Backwardness of many official recommendations from various worldwide and national wide organizations caused by the heterogeneity of use cases of educational processes and supporting software, which is the essential cause of confusions and uncertainty of the end users (e.g. teachers, tutors, learners, instructional administrators).

We posit underestimation of:

1. The rate of technological advancement (some recommend using previous generation of technologies that do not support diverse educational use cases);
2. The significance of integration and interoperability;

3. The significance of platform-based nature of actual CISs (some recommend tools without underlying platform and corresponding authorization and integration consideration);
4. "Engineering literacy" as meaningful addition to so called "digital literacy";
5. Pragmatic approach to information security.

The findings of this study indicate the following inferences:

1. The changes are not "so disruptive" from the technological point of view.
2. The burden of accidental complexity in addition to essential complexity of education process is high and it's the consequence of "educational debt" by analogy with "technical debt" [45].

Overall, the emerged gap of generations of educational software is a very straightforward corollary from the history of five decades. However, fortunately, both software and system engineering experienced great technological breakthrough over this period. Thus, educators can rely on their practice and use time-tested technological solutions adopted to learning.

6 Conclusion

The sudden COVID-19 pandemic became a significant push to distant and online learning this year. Understanding the growing role of technical pros and cons in this type of education, the authors exploded the role of software engineering methodology in total distance learning deployment. The principle conclusion that can be drawn is that the main user roles of distance learning software are in want of personally configured sets of products that supply their needs. The above outcomes show that the lack of an engineering approach in the deployment of distance learning leads to problems and serves as an inhibitor of this process.

Summarizing, our findings demonstrate a strong effect of interoperability standards on usage of distance learning software especially in collaborative educational practices. For today the World educational society have already had software platforms with supported API, but we still have a long way to go to mutually integrated educational systems, and the authors see in this the widest field for the future work.

Acknowledgement. The article was prepared within the framework of the Basic Research Program at the National Research University Higher School of Economics (HSE) and supported within the framework of a subsidy by the Russian Academic Excellence Project "5–100".

References

1. Conner, C., Bohan, C.: The Second World War's impact on the progressive educational movement: assessing its role. J. Soc. Stud. Res. **38**(2), 91–102 (2014)
2. Jaworski, T.: "You're in the army now:" the impact of world war II on women's education, work, and family. J. Econ. Hist. **74**(1), 169–195 (2014)

3. Tatnall, A.: Encyclopedia of Education and Information Technologies. Springer International Publishing (2020)
4. Penuel, W.R.: Implementation and effects of one-to-one computing initiatives: a research synthesis. J. Res. Technol. Educ. **38**(3), 329–348 (2006)
5. OLPC, Inc.: One Laptop per Child. http://one.laptop.org. Accessed 01 April 2020
6. National Institute of Standards and Technology: The NIST Definition of Cloud Computing. http://nvlpubs.nist.gov/nistpubs/Legacy/SP/nistspecialpublication800-145.pdf. Accessed 01 April 2020
7. Almajalid, R.: A Survey on the Adoption of Cloud Computing in Education Sector. arXiv preprint arXiv:1706.01136 (2017)
8. Lee, G.: Data center evolution – mainframes to the cloud. In: Cloud Networking, Morgan Kaufmann, pp. 11–35 (2014)
9. Christensen, C.M.: The Innovator's Dilemma: When New Technologies Cause Great Firms to Fail. Harvard Business Review Press, US (1997)
10. COVID-19 Educational Disruption and Response, UNESCO. http://en.unesco.org/covid19/educationresponse. Accessed 03 May 2020
11. World Bank Group: Remote Learning, Distance Education and Online Learning During the COVID19 Pandemic: A Resource List by the World Bank's Edtech Team (English), 26 March 2020. http://documents.worldbank.org/curated/en/964121585254860581/pdf/Remote-Learning-Distance-Education-and-Online-Learning-During-the-COVID19-Pandemic-A-Resource-List-by-the-World-Banks-Edtech-Team.pdf. Accessed 01 April 2020
12. UNESCO: Distance learning solutions. http://en.unesco.org/covid19/educationresponse/solutions. Accessed 03 May 2020
13. Commonwealth of Learning. http://www.col.org. Accessed 03 May 2020
14. Keeping the doors of learning open COVID-19, Commonwealth of Learning. http://www.col.org/resources/keeping-doors-learning-open-covid-19#B.Basicguides. Accessed 03 May 2020
15. UNESCO: National learning platforms and tools. http://en.unesco.org/covid19/educationresponse. Accessed 03 May 2020
16. My Learning at Home, Manitoba. http://www.edu.gov.mb.ca/k12/mylearning/index.html. Accessed 03 May 2020
17. Quebec Open school, Gouvernement du Québec. http://ecoleouverte.ca/en/. Accessed 03 May 2020
18. DfE Online Education Resources, UK Department of Education, 7 April 2020. http://www.gov.uk/government/publications/coronavirus-covid-19-online-education-resources/coronavirus-covid-19-list-of-online-education-resources-for-home-education. Accessed 03 May 2020
19. Future classroom. http://www.learning-innovation.go.jp/covid_19/. Accessed 03 May 2020
20. MEXT. http://www.mext.go.jp/a_menu/ikusei/gakusyushien/index_00001.htm. Accessed 03 May 2020
21. Singapore Student Learning Space, Ministry of Education, Singapore. http://vle.learning.moe.edu.sg/. Accessed 03 May 2020
22. Educational Broadcasting system (EBS). http://www.ebs.co.kr/main%25252523this. Accessed 03 May 2020
23. KERIS. http://www.keris.or.kr/eng/main.do. Accessed 03 May 2020
24. Iraq educational channel on YouTube. http://www.youtube.com/channel/UC_hXlmP8BLXHcO__66C_Aqg/featured. Accessed 03 April 2020
25. Kundelik. http://portal.kundelik.kz. Accessed 03 May 2020
26. Russian e-school. http://resh.edu.ru. Accessed 03 May 2020
27. Vschool. http://vschool.sa. Accessed 03 May 2020

28. Ien National e-portal. http://ien.edu.sa/Home/Dashbord. Accessed 03 May 2020
29. Restricted countries or regions. http://support.zoom.us/hc/en-us/articles/203806119-Restricted-countries-or-regions. Accessed 03 May 2020
30. Zoom banned by Taiwan's government over China security fears. http://www.bbc.com/news/technology-52200507. Accessed 03 May 2020
31. Rwanda warns against the use of Zoom over security concerns. http://www.newtimes.co.rw/technology/rwanda-warns-against-use-zoom-over-security-concerns. Accessed 03 May 2020
32. Coronavirus: Teachers in Singapore stop using Zoom after 'lewd' incidents. http://www.bbc.com/news/world-asia-52240251. Accessed 03 May 2020
33. Google Told Its Workers That They Can't Use Zoom On Their Laptops Anymore. http://www.buzzfeednews.com/article/pranavdixit/google-bans-zoom. Accessed 03 May 2020
34. Elon Musk's SpaceX Bans Zoom After Security And Privacy Warnings. http://www.forbes.com/sites/kateoflahertyuk/2020/04/02/elon-musks-spacex-bans-zoom-after—security-and-privacy-warnings/#66420e44775a. Accessed 03 May 2020
35. Whittaker, Z.: New York City bans Zoom in schools, citing security concerns. http://techcrunch.com/2020/04/05/zoom-new-york-city-schools/. Accessed 03 May 2020
36. UNESCO: Adverse consequences of school closures. http://en.unesco.org/covid19/educationresponse/consequences. Accessed 03 May 2020
37. Motiejūnaitė-Schulmeister, A., Crosier, D.: How is Covid-19 affecting schools in Europe?. http://eacea.ec.europa.eu/national-policies/eurydice/content/how-covid-19-affecting-schools-europe_en. Accessed 03 May 2020
38. Minprosveshcheniya rekomenduet ne ispol'zovat' mobil'niki pri organizacii distancionnogo obucheniya shkol'nikov. http://futurerussia.gov.ru/nacionalnye-proekty/minprosvesenia-rekomenduet-ne-ispolzovat-mobilniki-pri-organizacii-distancionnogo-obucenia-skolnikov. Accessed 03 May 2020
39. National Survey Tracks Impact of Coronavirus on Schools: 10 Key Findings. http://www.edweek.org/ew/articles/2020/04/10/national-survey-tracks-impact-of-coronavirus-on.html?cmp=eml-enl-eu-news1&M=59369645&U=1938472&UUID=673de9615ec0303ccbae2ae5793fd7be&fbclid=IwAR1gSHigmOVa-OKomz-Ztucap39wrmCWghSkccQVH82tNouoVjMI5-qEpZI. Accessed 03 May 2020
40. UNICEF: COVID-19: More than 95% of children are out of school in Latin America and the Caribbean, 23 March 2020. http://www.unicef.org/press-releases/covid-19-more-95-cent-children-are-out-school-latin-america-and-caribbean. Accessed 03 May 2020
41. Bourque, P., Fairly, R. (eds.): IEEE Computer Society, SWEBOK v3.0. Guide to the Software Engineering Body of Knowledge, p. 335. IEEE (2014)
42. Blackboard Has Worst LMS Problem in Years, Taking K-12 District Offline for Days. http://philonedtech.com/blackboard-has-worst-lms-problem-in-years-taking-k-12-district-offline-for-days/. Accessed 03 May 2020
43. The remote-learning response to COVID-19 is remarkable. It also highlights a problem. http://bigthink.com/Charles-Koch-Foundation/teacher-education. Accessed 03 May 2020
44. Cybersecurity staff are being transferred to IT support. That's adding to the risk of data breaches. http://www.zdnet.com/article/cybersecurity-staff-are-being-transferred-to-it-support-thats-adding-to-the-risk-of-data-breaches/. Accessed 03 May 2020
45. Brooks, F.: The Mythical Man-Month: Essays on Software Engineering, Addison-Wesley Professional (1995)

The EPUM Platform: A Novel Collaboration Paradigm

Christos Mettouris(✉), Evangelia Vanezi, Theodora Kosti, Alexandros Yeratziotis, Nadia Charalambous, and George A. Papadopoulos

University of Cyprus, 1 University Avenue, Nicosia, Cyprus
{mettour, evanez01, tcosti02, yeratziotis.alexandros, george}@cs.ucy.ac.cy, charalambous.nadia@ucy.ac.cy

Abstract. In this paper we present an innovative collaborative online platform developed in the context of the EU funded project EPUM (Emerging Perspectives on Urban Morphology: Researching and Learning through multiple practices). The platform attempts to eliminate research and institutional barriers in educational cultures through the development and use of digital resources, structured under Collaborative Learning Activities, a novel concept proposed in this work. CLAs offer an innovative way for collaboration in the education system, by making available resources accessible, not only to those enrolled in higher education programmes, but to anyone wanting to access training regardless of their geographical location, educational culture or ability to travel.

Keywords: Collaborative Learning Activities · Blended learning · EPUM platform · ICT tools

1 Introduction

A variety of approaches on understanding urban form[1], both theoretical and operational, have been developed in the past decades to respond to a variety of forms of socio-spatial patterns and increasing social, economic, and political fragmentation in contemporary cities, strongly related to their urban form. However, these are characterized by specific disciplinary and geographical trends and have seen the emergence of separate schools of thought. Each approach tends to be associated with a main research centre or with certain individual researchers and, despite some exceptions, they have traditionally been applied in isolation. The teaching of urban form analysis in higher education institutions across Europe is also addressing contemporary cities' issues from often isolated perspectives based on the aforementioned different schools of thought, either reflecting specific national educational trends or opting for a globalized approach cutting the knots with local specificities. There is still a lack of learning spaces which foster a multidisciplinary thinking about contemporary cities' issues and which enable

[1] The physical patterns, layouts, and structures that constitute an urban center.

M. E. Auer and T. Rüütmann (Eds.): ICL 2020, AISC 1328, pp. 66–78, 2021.
https://doi.org/10.1007/978-3-030-68198-2_6

the participation of all relevant stakeholders in the debate about contemporary cities' problems and solutions.

EPUM identified the need to establish a network linking the different approaches, developing learning platforms that foster the exchange of knowledge, providing opportunities for contact between members and encouraging the dissemination of findings. The coming together of researchers, educators and learners from different geographical areas and disciplines will provide the basis for a multidisciplinary exploration and the opportunity to establish common theoretical foundations for the growing number of urban form studies in many parts of the world. It will provide the means to engage all stakeholders currently within introverted disciplinary, institutional and geographical boundaries, in a fruitful discussion through a collaborative open learning curriculum (OLC). The mode of learning which proved to be suitable for such an OLC is one that facilitates both face-to-face activities, so as to allow institutions to work independently, with on-line activities which enable the synchronous or asynchronous collaboration and learning across institutional barriers offering a blended learning system.

The EPUM platform aims at supporting a blended learning approach, breaking down research and institutional barriers in educational cultures through the development and use of digital resources, structured under specific activities in various thematic areas proposed by both professors and students. The activities are represented by "learning environments", referred to as Collaborative Learning Activities (CLAs). CLAs offer an innovative way for collaboration in the education system, by making available resources which are accessible, not only to those enrolled in higher education programmes, but to anyone wanting to access training regardless of their geographical location, educational culture or ability to travel. The innovative framework and tools proposed comprise a visual index to students' work, providing them with the capability to upload data files to assignments, incorporating their peers' feedback and review, as well as tutors' feedback to students for their work, and the ability for discussion around any of the topics or works. Furthermore, the add-on tools provide the capability to visualize a network of activity interactions and present it in a way that it is appealing and understandable to different stakeholders, both registered and non-registered users.

In Sect. 2 we present related work in terms of collaborative blended learning approaches in urban form studies, as well as in collaborative ICT platforms. Section 3 describes the methodology used, while Sect. 4 presents the technical implementation of the EPUM platform. Section 5 describes the evaluation of the EPUM platform with end-users, Sect. 6 discusses the evaluation results and concludes the paper with future work.

2 Related Work

2.1 Collaborative Blended Learning Approaches in Urban Form Studies

Blended learning, which generally speaking refers to a learning environment which combines face-to-face instruction with computer-mediated instruction, has gained much popularity in higher education in the past years. However, a number of studies [1,

2] suggest that blended learning implementations are most often used for the purposes of efficiency and supplementation, with only a low number fully exploiting the potential of this mode of learning to enhance the learning experience and initiate collaborative activities, particularly in the field of urban form studies. Even though we can find examples of intertwining a specific learning style with blended-learning [3] there still is not much investigation either about the relationships between both or the potential to foster collaborations.

As Garrison and Vaughan [4] point out, the value of blended-learning transcends the mere application of ICT for teaching and learning, "recombining concepts that were previously considered contradictory, such as collaborative-reflection and asynchronous community" [4]. Other components which might become blended also include learners, learning styles, academic programs, subject-matters, disciplines, and institutional frameworks [5]. This possibility to combine learning activities which can be carried out at different times and in different places (on-line, in the classroom) combined in interaction with other learning resources, requires specific pedagogic methodologies which take advantage of their collaborative potential and point to the creation of alternative learning environments. Such a learning environment was suggested by Punie [6] to describe ICT-enabled educational spaces which transcend existing limits, physical, conceptual and institutional. Punie highlighted the potential of such environments to place students at the centre of the learning, enabling the personalization of learning as well as social interaction at different scales (from learning individuals and communities to learning cities and regions), while being flexible enough to integrate various learning styles, teachers' skills, and curriculums, gradually becoming informal platforms to share expertise and knowledge across organizations. In the OIKONET project, the term learning space to pursue goals which are in line with those described by Punie has been initiated in the field of housing studies [5].

The EPUM project, identifying the lack of such collaborative, blended learning environments in the field of urban form studies, explored the potential to link different approaches to the study of urban form in different parts of the world, through the development of learning platforms that aim at fostering the exchange of knowledge cutting across institutional and geographical boundaries.

2.2 Collaborative ICT Platforms

In the context of Computer Supported Collaborative Learning (CSCL), learning takes place through social interaction by using computers. CSCL software systems are collaborative learning environments that utilize technology to facilitate user (teacher to learner and learner to learner) interaction and communication, as well as learning coordination [7, 8]. An effective collaborative learning environment is one that effectively and efficiently supports knowledge sharing in a formed learning group [7]. A number of literature works have contributed towards the development of such learning environments.

In [7] the authors propose using semantic web technologies to build a software tool for knowledge sharing through the usage and management of multimedia annotations in CSCL. Four annotation categories are supported: definitions (descriptions and explanations), comments, questions, and associations (e.g., links to other resources).

Users are able to annotate web content and link documents and other resources to their annotations. The experimental evaluation showed that students doing collaborative group reading using this system achieved on average much higher scores than students doing group reading using other methods. In addition, the study showed that students in the first group have used collaborative multimedia annotations that contributed to enhanced knowledge sharing.

In case of multidisciplinary learning groups, the authors in [9] proposed an algorithm for composing optimal learning groups in situations where people have different domain backgrounds. The algorithm is integrated in an ontology-based e-learning system that creates self-built educating communities: trainees that participate in the education process gain points through achievements and ultimately become trainers. User profile information is explicitly acquired by having users filling forms. Based on this profile, users are assigned to (or are recommended) learning groups by maximizing the diversity within a group and minimizing the diversity between groups.

In the subject of exchanging data between learner groups, the authors in [10] propose an XML-based procedure using web-services for CSCL data exchange. The data to be transferred from one learner group to another include Moodle forum discussions, online chats and votes. This allows learner groups to not having to start their discussions from scratch, without any reference of other groups' discussion. The positives of having other learner groups' data available is that the preceding discussions could effectively be used as scaffolding information to participate in the discussions [10]. In addition, the preceding discussions and the information shared provide adequate cognitive workload for the learner to be able to participate in the discussion [10]. A negative point is that the process of actively collecting information and discussion itself is a collaborative learning process in which the user does not participate.

On another dimension, regarding open source available platforms, Moodle's[2] Assignment activity provides a space where students submit assignments in different formats and teachers are able to provide feedback, comments or grades for them. However, it does not provide the opportunity for students to provide feedback on fellow-students assignments, to discuss about the topic, or to upload and access helpful material. These tasks can be related to separate Moodle activities such as forum, and lesson. Nevertheless, the material from all activities will not be packed together as in the CLAs concept, even if they are under the same course, as they may be mixed up with different other topics and material. In like manner, MOOC platforms incorporate collaborative tools that can be independently added to promote community-based learning, likewise the separate Moodle activities.

The EPUM platform facilitates collaboration by offering collaborative spaces, the CLAs. CLAs, among other, make resources widely accessible, promote and share students' work while also enabling their peer review, as well as facilitate discussions and interactions. Each EPUM CLA contains information around a specific topic, a space for assignments submission, a discussion about the specific tasks, and additionally allows both students and teachers to provide feedback, all into the same collaboration space. Each CLA is a place meant to enhance and promote complete

[2] https://moodle.org.

collaboration in the context of a specific topic, between the participants in a specific task. CLAs are described in the following sections.

3 Methodology

The first phase of the project focused on the development of a collaborative open learning curriculum through a pedagogic model which can facilitate a flexible interaction between courses included in the academic programs of the participating institutions in relation to different approaches to the analysis of the urban form and collaborative learning activities. The pedagogic model proposed, regards the implementation of a blended learning environment which enables various learning activities and learning tasks to be executed across institutions. The methodology of aligned learning and teaching [11] proposed by Biggs along with Bloom's [12] approach of taxonomies inform the development of the pedagogical model implemented in the project, aiming at blending different components such as subject-matters, delivery formats, learners from different institutions and levels, learner styles as well as physical, institutional and disciplinary barriers.

The open learning curriculum includes both theoretical and practical learning material and focuses on (a) aligning complementary approaches to develop a comprehensive analysis of urban form and social phenomena and (b) combining on-line (digital platform) and on-site (courses and seminars taking place at the participating institutions) Collaborative Learning Activities, CLAs. Such a pedagogic model enables the development of an open learning and teaching space that cuts across institutional and research boundaries. Building on the precedent of OIKONET, the key (to the learning process) is to intertwine the activities that can be carried out within the programme at each institution with the collaborative tasks amongst the institutions, either synchro-nously or asynchronously [5]. The sequences of tasks (or assignments) can evolve in an open-ended manner and include both online collaborative activities and tasks (through the digital platform) and face-to-face activities and tasks (intensive programme workshops).

The blended learning approach adopted and supported by the digital platform, proved extremely important for the implementation of this project resulting in the creation of a number of Collaborative Learning Activities among partners throughout Europe, facilitating a community of inquiry which is constituted above and beyond institutional and physical barriers, at the same time allowing multiple levels and types of instruction to be adopted. In that sense, it provided the adequate conditions for the implementation of a free and open dialogue, critical debate, negotiation and agreement between different, and often isolated urban form approaches in the participating institutions. Such context therefore, offered an open educational practice which helped partners to share through their teaching, freely and openly, ideas, knowledge, tools, approaches and materials used in urban form studies. At the same time, it enabled participating institutions to keep their own academic program, structure and curriculum; in other words, it enabled the participants to work independently and collaboratively. The small-scale learning activities facilitated by the project enabled the gradual establishment of a network of relationships among the courses, students and topics facilitating the study and exploration of Urban Morphology; exposed students and teachers to other approaches across geographical areas and prepared students and teachers for large-scale activities involving both online pre-workshop activities and onsite intensive workshops with target groups and stakeholders.

4 Technical Implementation

The EPUM digital platform functionality was broken down into components, that were developed as plugins on top of a WordPress [13] Content Management System (CMS) instance, the basis for the development. The CLA management plugin constitutes the EPUM platform's innovative functionality. Additionally, a few existing plugins were used, e.g. the Ultimate member plugin was used for user management functionality, including a set of GDPR requirements. Additional plugins were developed to enhance the platform's GDPR compliance, as it handles the personal data of both teachers and students.

In Fig. 1, the architecture of the EPUM platform is presented. Users can access the platform through the web with their devices. Four plugin components are connected to the core WordPress CMS: the Ultimate Member Plugin, the CLA Plugin that includes a set of functionalities related to collaborative activities, and two plugins developed in the GDPR context.

Figure 2 presents the functionality of the platform that is available to users, either visitors or logged-in users. The CLAs Calendar, some static information, and the Data Processing Privacy Policy are made available to all visitors of the platform. On the other hand, in regards to maintaining and managing user registrations and accounts a User Management Mechanism is embedded. Registration and login are available to all

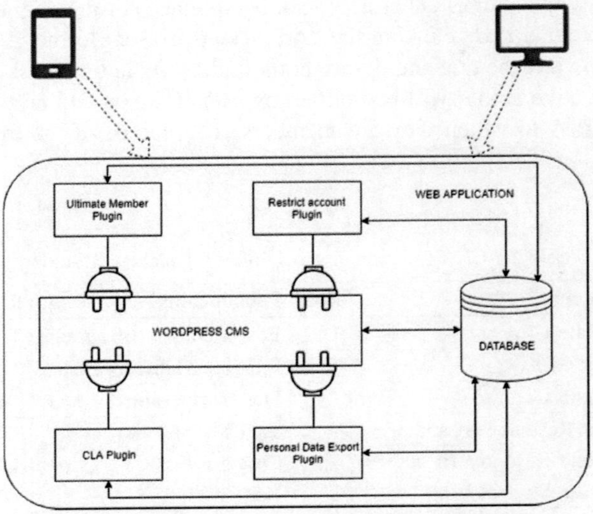

Fig. 1. The EPUM platform architecture

visitors. Login will only succeed upon providing correct, existing credentials. Registration needs to be accepted by a platform admin. Finally, an additional social mechanism is incorporated, i.e. a discussion forum for logged-in users.

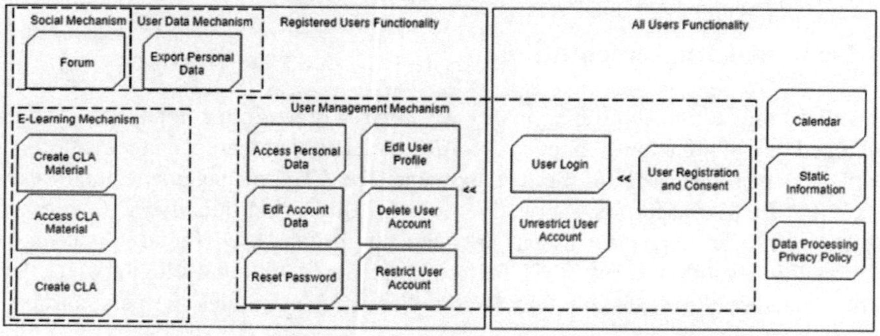

Fig. 2. EPUM platform functionality overview

Collaborative Learning Activities. CLAs constitute the main functionality of the platform. Each CLA is conceptualised as a distinct space within the platform, with its own title and description, an author, content, and member users. CLA functionality is accessible through the top bar menu of the EPUM platform, providing different user interfaces, different capabilities, and different information to users depending on their role: teacher or student.

Table 1 displays the functionalities available to each role. CLAs can only be created by teachers by filling-in the respective form (Fig. 3). The author is granted editing access to the specific CLAs resources, discussions and memberships. CLA details can be edited only by the author. Other users can be assigned (or unassigned) membership to a CLA by its author, by clicking the "Add/Remove users to the CLA" button. All users are able to view the title and description of all CLAs and request access to any of them. The respective author will be notified by e-mail. Users can also create an event for a specific CLA to which they are members. The author of the event can edit or delete it.

Table 1. CLA functionality per role (T:Teacher, S:Student)

Functionality	Role	Functionality	Role
Create CLA	T	Edit Members (if author)	T/S
Edit CLA	T	Preview Members List	T/S
Create Resource	T	Start Discussion	T/S
Edit Resource (if author)	T	Start Peer Review	T/S
Delete Resource (if author)	T	Create Event	T/S
Create Student Work	S	Edit Event	T/S
Edit Student Work	S	Delete Event	T/S
Delete Student Work	S		

An innovation of the CLAs is their ability to provide a set of resources to their members. Resources are created by teachers and are perceived as learning resources to the students and other teachers, as well as informative material to courses, events, etc. They are available to all members of the CLA. Students are not allowed to upload any resources, but are instead enabled to upload their Student Work.

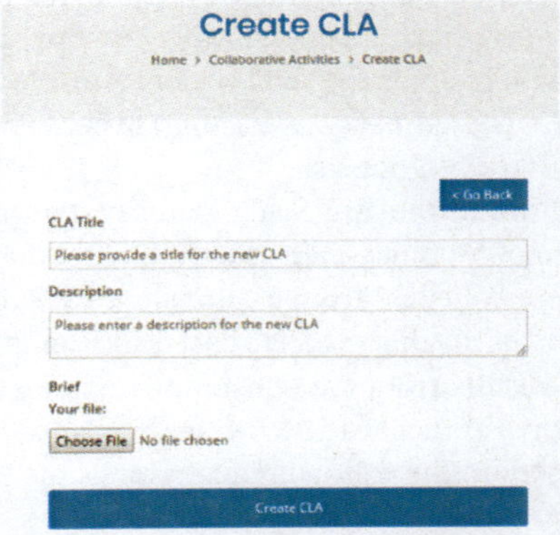

Fig. 3. Create a new CLA

Both teachers and students may view all resources and student works of the CLAs they are members of. This includes the option to download any files, if available. CLA members are able to take part in discussions regarding the CLA, as well as conduct peer reviews on student works by selecting a student work and posting their comment at the corresponding section below. Reviews may be conducted by both students and teachers.

5 Evaluation

In order to develop an innovative, usable and effective platform that could bring to life the promised benefits of the envisioned CLAs within a virtual environment, it was imperative to pilot-test the methodology of how a CLA is experienced within the platform.

A Big Scale Activity (workshop) was organized within the project, where students and tutors interacted on collaborative activities. During the pre- and post-workshop phases, participants had collaborated and interacted through the platform. During the workshop, collaborative activities were conducted, which included both onsite physical collaborative activities but also on-line interaction on the platform.

The workshop was organized in Nicosia, Cyprus, and the main goal was to effectively combine different morphological approaches – historico-geographical approach, process-typological approach, space syntax and relational approach – in the analysis of a physical form and determine the main challenges that form faces today. Drawing on the results, the combined approaches were ultimately applied, not only in the analysis, but also in the design of one particular area in the city of Nicosia. It should be noted that tutors had applied each of their morpho-logical approaches in isolation.

Features incorporated into the design of the platform, such as support for collaboration, offering accessible and available resources, offering a visual index to students' work, capability to upload data files to assignments, incorporating peers' feedback and review, as well as tutors' feedback to students, support for discussion/debate around any of the topics or work, and capability to visualize a network of activity interactions and present it in a way that it is appealing and understandable to different stakeholders, both registered and non-registered users, were thus experienced, exhibited and evaluated in the Big Scale Activity of Nicosia.

Following the completion of the Big Scale Activity, the participating students and tutors completed an online 5-point scale questionnaire that assessed their overall experience. Responses were collected from 7 tutors and 18 students, coming from the 5 institutional partners involved in this project, based in London, Nicosia, Porto, Rome and Wien. Moreover, the results from the questionnaire helped in refining the learning activities, processes and environment of EPUM. Responses were entered anonymously into a database and analysed. In the following paragraphs the key results are presented, which indicate the value of CLAs and the platform.

– Questionnaire item: **The whole teaching experience was positive/good (Fig. 4).**

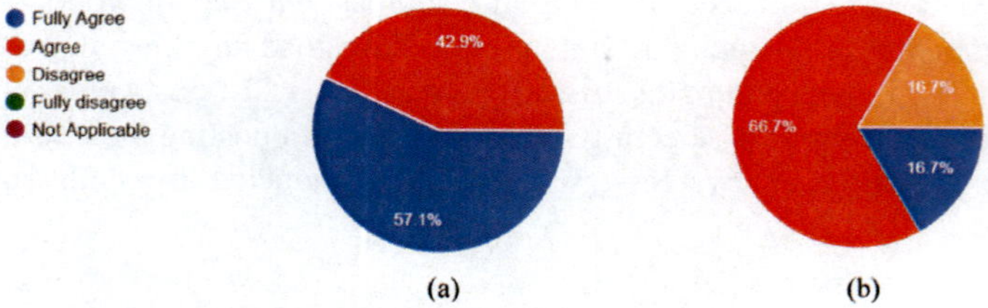

Fig. 4. (a) Tutors (b) Students

From the responses of the tutors, it was established that they were positive towards the collaborative teaching experience which was blended in nature and consisted of both on-line and on-site collaborative activities (see Fig. 4a where 7 out of 7 responded Fully Agree and Agree). From the responses of the students, results were similarly positive (see Fig. 4b where 15 out of 18 responded Fully Agree and Agree). There were 3 students however, that responded Disagree, without offering any additional comments. Besides these 3 students, holistically, the experience was very good for tutors and students.

– Questionnaire item: **The materials provided by all tutors to the students (reference literature, maps) were useful for the workshop implementation.**

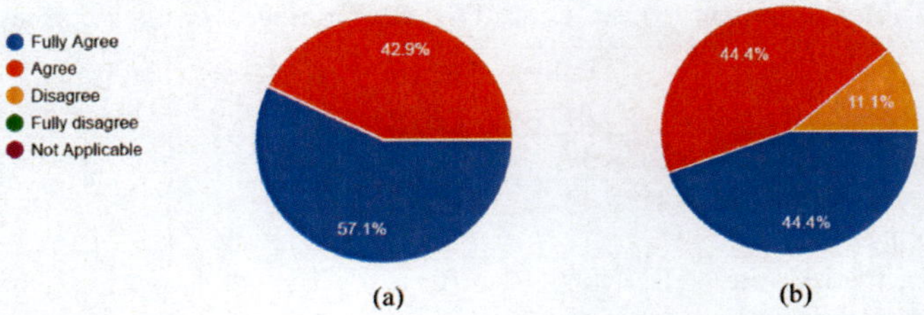

Fig. 5. (a) Tutors (b) Students

From the responses of the tutors, it was established that the materials were useful and thus important for the collaboration activity (see Fig. 5a where 7 out of 7 responded Fully Agree and Agree). From the responses of the students, results were similarly positive (see Fig. 5b where 16 out of 18 responded Fully Agree and Agree). Two students responded Disagree, without offering any additional comments. Holistically, the materials were useful, and deemed important in the implementation of such collaborative activities. Resources being available online are therefore critical.

– Questionnaire item: **Interacting with other students/teachers (e.g., in the Facebook group "EPUM Intensive Workshop Nicosia 2019") has been a good experience.**

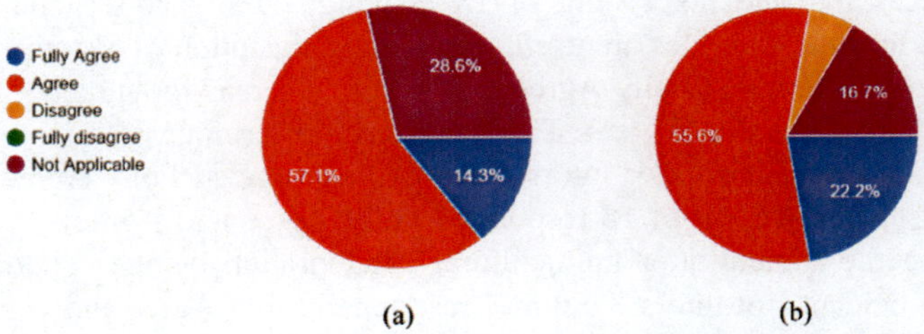

Fig. 6. (a) Tutors (b) Students

From the responses of the tutors, it was established that the interaction between students and/or tutors in online environments, such as FB and the platform itself was positive (see Fig. 6a where 5 out of 7 responded Fully Agree and Agree and 2 out of 7 responded Not Applicable). From the responses of the students, results were similarly positive (see Fig. 6b where 14 out of 18 responded Fully Agree and Agree). There was one student however, that responded Disagree, and three responded Not Applicable. A couple mentioned that they used/would have preferred to use WhatsApp instead of FB for example.

– Questionnaire item: **Please respond to the following questions regarding the workshop.**

Table 2. Student (S) and tutor (T) views on aspects of the workshop

		Fully Agree	Agree	Disagree	Fully Disagree	Not Applicable
The studio activity was well aligned with the workshop theme	S	3	10	5		
	T	4	2	1		
The tutoring activities were useful for the work in the Studio activity	S	4	10	3		1
	T	4	1	2		
The Interim Critique sessions were useful	S	1	15		1	1
	T	4	2			1
Working collaboratively with students/teachers from other countries has been a good experience	S	8	9		1	
	T	4	3			
I was able to communicate effectively with the other members of my group	S	3	11	2	2	
	T	3	4			

What can be summarized from Table 2 is that there is strong support in aligning a studio activity with a workshop theme (for tutors 6 out of 7 responded Fully Agree and Agree, whereas for students 13 out of 18 responded Fully Agree and Agree) and ensuring that tutoring activities are useful toward the execution of a studio activity (for tutors 5 out of 7 responded Fully Agree and Agree, whereas for students 14 out of 18 responded Fully Agree and Agree). The importance of critiques sessions was evident for both students and tutors (for tutors 6 out of 7 responded Fully Agree and Agree, whereas for students 16 out of 18 responded Fully Agree and Agree).

Regarding the student-tutor multicultural collaboration being a good experience, this was also evident (for tutors 7 out of 7 responded Fully Agree and Agree, whereas for students 17 out of 18 responded Fully Agree and Agree). Moreover, effective communication was achieved to a large extent in the online and onsite group collaboration activities (for tutors 6 out of 7 responded Fully Agree and Agree, whereas for students 14 out of 18 responded Fully Agree and Agree).

6 Conclusions and Future Work

The preliminary results of the survey showcase the potential of the EPUM platform, and demonstrate the usefulness and benefits of CLAs on students' learning experience. Based on the aforementioned results of the evaluation, the design of the platform adheres to the CLA requirements of participants as follows:

- Large size related resources (e.g. studio activity information, maps) for a specific CLA can be added to by tutors and shared with its members (tutors and students).
- Critique and peer review on students work submissions by both fellow students and tutors are supported through comments.
- Collaboration and discussion are supported through the specific CLA discussion area.

The CLAs methodology implements a collaboration approach, incorporating all stages relating to a specific task, topic or assignment. All participating organisations play an active role in the design of CLAs in order to identify the educational needs for the effective delivery of training on each different approach of urban form analysis. It can include an archive of key resources, material, discussion, instructions, assignment submission and feedback from both students and teachers. This approach will eventually create and formulate an online community of practice [14] where the active membership of learners and teachers will facilitate an educational social praxis. In that sense, learning "involves co-construction and co-evolution of knowledge" ([15] p.179) among partners and often isolated schools of thought in the study of urban form. Future work includes conducting a large-scale evaluation of the capabilities of the EPUM platform with students and professors from many different Universities from many countries.

Acknowledgment. The current publication is created within the project Emerging Perspectives on Urban Morphology: Researching and Learning through multiple practices (EPUM). The project is funded by the European Union's Erasmus + Program (2014–2020) under Grant Agreement 2017-1-CY01-KA203026745. The content of this publication represents the views of the authors only and is their sole responsibility. The European Commission does not accept any responsibility for use that may be made of the information it contains.

References

1. Driscoll, M.: Blended learning: Let's get beyond the hype. Retrieved August 30, 2010, from E-learning (2002). http://elearningmag.com/ltimagazine
2. Hofmann, J.: Why blended learning hasn't (yet) fulfilled its promises: answers to those questions that keep you up at night. In: Bonk, C.J., Graham, C.R. (eds.) Handbook of Blended Learning: Global Perspectives, Local Designs, pp. 27–40. Pfeiffer, San Francisco (2006)
3. Donnelly, R.: Harmonizing technology with interaction in blended problem-based learning. Comput. Educ. **54**(2), 350–359 (2010)
4. Randy, G.D., Vaughan, N.D.: Blended Learning in Higher Education: Framework, Principles, and guidelines. John Wiley & Sons, Hoboken (2008)
5. Madrazo, L., Sentieri, C., Charalambous, N.: Applying a blended learning methodology to the study of housing. In: Rodrigues Couceiro da Costa, M.J., Roseta, F., Pestana Lages, J., Couceiro da Costa, S., (Eds.) "Architectural Research Addressing Societal Challenges", CRC Press, Taylor and Francis Group, vol. II, pp. 1051–1058 (2017)
6. Punie, Y.: Learning spaces: an ICT-enabled model of future learning in the Knowledge-based Society. Eur. J. Educ. **42**(2), 185–199 (2007)

7. Yang, S.J.H., Zhang, J., Su, A.Y.S., Tsai, J.J.P.: A collaborative multimedia annotation tool for enhancing knowledge sharing in CSCL. Interact. Learn. Environ. **19**(1), 45–62 (2011). https://doi.org/10.1080/10494820.2011.528881
8. Zurita, G., Nussbaum, M.: Computer supported collaborative learning using wirelessly interconnected handheld computers. Comput. Educ. **42**, 289–314 (2004)
9. Dascalu, M.-I., Bodea, C.-N., Lytras, M., Ordoñez de Pablos, P., Burlacu, A.: Improving e-learning communities through optimal composition of multidisciplinary learning groups. Comput. Hum. Behav. **30**, 362–371 (2014). https://doi.org/10.1016/j.chb.2013.01.022. ISSN 0747-5632
10. Tamura, Y., Sumi, K., Yamamuro, T., Maejima, M.: CSCL data structurization and inter-LMS sharing with use of web services. In: Lovrek, I., Howlett, R.J., Jain, L.C. (eds) Knowledge-Based Intelligent Information and Engineering Systems. KES 2008. Lecture Notes in Computer Science, vol. 5179. Springer, Berlin, Heidelberg (2008)
11. Biggs, J., Tang, C.: Teaching for quality learning at university third edition teaching for quality learning at university. High. Educ. **9**, 165–203 (2007)
12. Bloom, B.S.: Taxonomy of Educational Objectives. David McKay Company Inc. (Ed.), New York (1956–1964)
13. WordPress. wordpress.com. Accessed 21 Feb 2020
14. Lave, J., Wenger, E.: Situated Learning: Legitimate Peripheral Participation. Cambridge University Press, Cambridge (1991)
15. Banerjee, I.: From classrooms to learning landscapes: new socio-spatial imaginaries of learning and learning spaces. In: Tognagchi, C., Knierbein, S. (eds.) Public Space and Relational Perspectives, New Challenges for Architecture and Planning, pp. 167–182. Routledge, New York (2016)

The MopBot Cleaning Robot – An EPS@ISEP 2020 Project

Corina Tuluc[1], Frederique Verberne[1], Szymon Lasota[1], Tomás de Almeida[1], Benedita Malheiro[1,2], Jorge Justo[1], Cristina Ribeiro[1,3], Manuel F. Silva[1,2(✉)], Paulo Ferreira[1], and Pedro Guedes[1,2]

[1] ISEP/PPorto – School of Engineering, Polytechnic of Porto, Porto, Portugal
mss@isep.ipp.pt
[2] INESC TEC, Porto, Portugal
[3] INEB – Instituto de Engenharia Biomédica, Porto, Portugal
epsatisep@gmail.com
http://www.eps2020-wiki2.dee.isep.ipp.pt

Abstract. Waste is one of the biggest problems on Earth today. In the spring of 2020, a team of students enrolled in the European Project Semester at Instituto Superior de Engenharia decided to contribute with the design of an ethically and sustainability-oriented autonomous cleaning robot named MopBot. The project started with the research on similar solutions, ethics, marketing and sustainability to define a concept and create a functional, ethical and sustainability driven design, including the complete control system. Finally, given the undergoing pandemic, the operation of the MopBot was simulated using CoppeliaSim. MopBot is a medium-sized vacuum cleaner, with two vertical brushes, intended to clean autonomously large areas inside buildings such as shopping malls or corridors. It is shipped with a sustainable packaging solution which can be re-purposed as a disposal box for electrical components.

Keywords: Collaborative learning · European Project Semester · Sustainability

1 Introduction

The European Project Semester (EPS) is a one semester capstone programme created by Arvind Andersen in 1995 with engineering students in mind [2]. It proposes a student-centred project-based learning framework with emphasis on multicultural multidisciplinary teamwork to develop scientific, technical and interpersonal skills. Since then, EPS has been embraced by a network of 19 European universities, including, since the academic year of 2010/2011, the Instituto Superior de Engenharia do Porto (ISEP) [4].

In the spring of 2020, a team composed of four students (Corina Tuluc, a Telecommunications student from Romania, Frederique Verberne, a Civil Engineering student from Netherlands, Szymon Lasota, a Business and Technology

M. E. Auer and T. Rüütmann (Eds.): ICL 2020, AISC 1328, pp. 79–90, 2021.
https://doi.org/10.1007/978-3-030-68198-2_7

student from Poland, and Tomás de Almeida, a Mechanical Engineering student from Portugal) chose to create an autonomous litter collecting robot. The goal of this robot is to support the waste collection and management of large commercial or services facilities. This challenge matched the interests of the Team as whole and allowed members to individually contribute with their diverse knowledge.

Rubbish and waste products are one of the biggest problems on Earth. Nowadays, rubbish management, from collection to recycling, has become one of the biggest challenges for municipalities around the globe. Specifically, waste removal is a labour intensive ineffective task. The main objective of the Team was to create a sustainable waste collecting robot to clean the inside of large buildings in an efficient safe way with limited human interaction. The challenge is then to design an innovative functional product that contributes to solve one of the biggest challenges in today's world.

This paper, which reports the work of this EPS@ISEP Team, is structured in four additional sections. The Background section presents related projects as well as research on marketing, sustainability and ethical aspects of the project. Next, the Design and Development section describes the architecture of the prototype. Then, the Tests and Results section details the planned functional tests. Finally, the Conclusion section draws the conclusions and identifies future directions.

2 Background

The background research of the project covers existing solutions and the related ethics, marketing and sustainability studies.

2.1 Related Solutions

The industry of autonomous vacuum cleaners is not new. The first fully developed robot of this kind was the Trilobite by the Swedish Electrolux [6] in 1997 (Fig. 1a). Trilobite weighted 5 kg, had a storage capacity of 1.2 L, a velocity of 1.4 m/h and cleaned up to 28 m^2/h. Although it was not a sales success, many other companies started to develop cleaning robots.

This was the case of the American company iRobot, which has created the most popular autonomous vacuum cleaner of our times: Roomba (Fig. 1b). Roomba is equipped with sophisticated sensors to detect obstacles and find alternative paths, allowing it to autonomously clean a relatively small horizontal area. It uses one vertical brush, two horizontal brushes and a vacuum system. With a recommended operation time of 1.5 h, a storage capacity of approximately 0.5 L, a velocity of 1.7 km/h and a weight of 4 kg, Roomba can clean 185 m^2/h [8]. However, these high-end robots, due to their reduced dimensions, are intended to clean small horizontal areas like apartments and offices.

For larger spaces, like streets, public squares, industries or large offices it is necessary to design large powerful robots. There are two robots in this segment that stand out from the competition: the ENWAY Autonomous Sweeping and DustClean. The ENWAY Autonomous Sweeping, shown in Fig. 1c, was developed in Germany with European Union (EU) funds [3]. It has almost the dimensions of a Smart car and weights around 850 kg. It relies on a powerful vacuum system and two front vertical brushes for the cleaning, offers a storage capacity of approximately 150 L and can operate non-stop for 6 h, reaching a maximum speed of 8 km/h and sweeping 9000 m^2/h. Along with the software, ENWAY relies on a combination of Laser Imaging, Detection, and Ranging sensor (LIDAR), cameras, Radio Detection And Ranging (RADAR), Fast Network Simulation Setup (FNSS), and wheel odometry to navigate. This robot can be programmed to follow human workers with the aim of making collection safer and more efficient [5].

DustClean is a ROBOTECH product also funded by the EU. As ENWAY, it relies on two front vertical brushes and a vacuum system for the cleaning. DustClean can operate autonomously and safely using preloaded environment information, such as the map of the area, and on-board sensors. Specifically, the robot is able to follow a working path planned autonomously or defined by a user, avoiding obstacles during navigation [10]. DustClean weights 150 kg and has a storage capacity of 37 L. Due to the 10 h battery autonomy and the selected wheel motors, it can reach a velocity of 3 km/h and clean an area of 200 m^2/h. The design of the DustClean is shown in Fig. 1d.

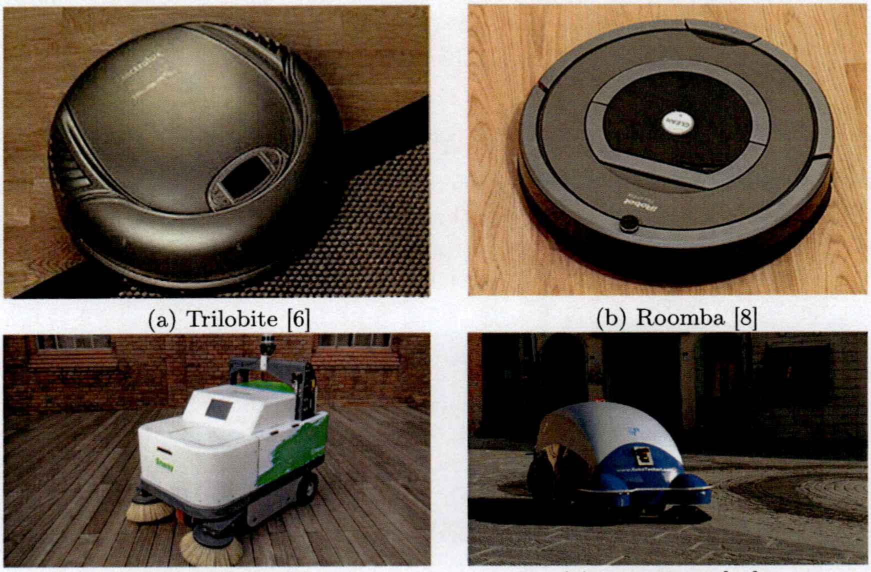

(a) Trilobite [6] (b) Roomba [8]

(c) ENWAY [5] (d) DustClean [10]

Fig. 1. Cleaning robots

2.2 Ethics

When designing a new product or service, it is essential to consider the involved ethical dimensions. Ethics has a big, overarching meaning involving, in this context, engineering, marketing and environmental components.

- Engineering ethics is related with the decisions engineers make when designing technological solutions for real problems. As technology influences more and more human life, the greater is the need to consider engineering ethics. The decision-making process must be sustained in the following fundamental canons: (*i*) engineers shall hold paramount the safety, health, and welfare of the public; (*ii*) engineers shall perform services only in the areas of their competence; (*iii*) engineers shall issue public statements only in an objective and truthful manner; (*iv*) engineers shall act for each employer or client as faithful agents or trustees; and (*v*) engineers shall avoid deceptive acts [1].
- Marketing ethics means that the marketing of product or service must be informative and inspiring, but never misleading, so that the customer can make and informed decision [7]. Ethics in marketing and sales is fair marketing. As long as the marketing of a brand is truthful and fair, the product or service will be sold in an ethical way.
- Environmental ethics considers the ethical relationships between humanity and non-human world [9]. The goal is to ensure the well-being of humans in relation to nature and sustainability. This can be achieved by complying with sustainability standards, ensuring parsimonious usage of resources as well as a positive impact of the product or service on the planet. Companies following these environmental ethics principles benefit from a strong brand image and attract employees and costumers sharing the same values.

2.3 Marketing

Having in mind that marketing is more about selling the benefits rather the product itself, the Team researched the service robot, cleaning robot and professional cleaning robot markets. The final decision was to concentrate on the Business-to-Business segment and, specifically, on large commercial facilities like shopping malls. In this segment, the benefits of an autonomous cleaning robot are low labour costs and high cleaning performance. The Team applied the 4C marketing mix, considering Consumer, Convenience, Cost and Communication, to specify product features matching client satisfaction such as operation time on a single battery charging, cleaned area, dimensions and autonomy of the robot. To build brand awareness, the robot was named "MopBot", a matching logo designed and the "Cleans more than You think" promotional claim was adopted to increase customer awareness and focus on the main benefit. The Team became committed to differentiate MopBot from other autonomous cleaning robots.

2.4 Sustainability

Sustainable design and development is essential to protect all life forms and promote well-being in the planet. In this context, it includes the analysis of the economic, environmental and social sustainability as well as the life cycle of the proposed solution.

- Environmental sustainability is concerned with building the most efficient cleaning robot with the lowest possible environmental impact. This means, for example, using recyclable materials or operating with reduced power consumption. The goal is to be part of the solution, not the problem.
- Economical sustainability aims to balance sustainability and cost. The goal is to choose sustainable yet affordable materials and components. The reuse of good condition second-hand materials saves money and preserves the planet.
- Social sustainability mitigates environmental impacts with negative effects on current and future generations. The proposed solution needs to satisfy societal needs without jeopardising the health and quality of living of mankind.
- Life cycle analysis manages the environmental impact of a product from design, to production, operation and disposal. To reduce this impact, manufacturing should be done by local Portuguese industry, using a robotised production line. The manufacturing process should be employee-friendly, sustainable and minimise waste. A report on the efficiency, sustainability and required actions for continuous improvement should be produced every three months. The product should be packed disassembled in a cardboard box for protection and transportation. The expected operation costs are related with electricity (charging the batteries) and maintenance (scheduled or unscheduled substitution of components). The MopBot company would offer a 3-year guarantee together with a discount disposal programme to encourage customers to return the device for recycling at the end of its life.

3 Proposed Solution

The proposed solution consists of a cleaning autonomous robot called MopBot. This robot was designed for large indoor areas such as the halls and corridors found in shopping malls or universities. For this reason, MopBot needs to balance garbage storage capacity with amenable physical dimensions to avoid tables, chairs or bins that normally populate these spaces.

During the design process, essential ethical concerns were taken into consideration. From engineering ethics, MopBot focuses on the safety of the product and health of the users. The safety is guaranteed with extra safety sensors. In the case this project creates a spin-off, the marketing of the MopBot will be fair, i.e., no misleading information will be provided. Environmental ethics and sustainability had a strong impact in the packaging solution and on the MopBot polystyrene chassis, which are fully recyclable.

MopBot is 585 mm long, 300 mm wide, 300 mm high, weights 20 kg, navigates at a velocity of 2.52 km/h, operates for 2 h without recharging and cleans up to 755 m^2/h. Although the robot has sensors to avoid obstacles and stop, the most efficient operation period is during the night.

3.1 Concept

To efficiently use power and clean the floor, MopBot uses a vacuum system and two vertical front brushes. The vacuum system consists of a fan attached to the air vents on top of the tank (Fig. 2a), which is operated by a high-speed direct current (DC) motor. The fan creates an air flow that exits through the vents, sucking trash and dust into the tank. Figure 2b displays the brushes placed in the front of the chassis. Each brush is connected to a DC motor through a shaft and rotates from the outside to the inside. The shafts are housed in a purpose-built shaft support showed in Fig. 2c. This part, which lodges and holds a bearing on the top, acts as a radial support for the shaft on the internal face. In the cylindrical part of the shaft support there is another bearing held by two elastic rings on the top and bottom. These supports prevent the brushes from bending when they hit an obstacle. Figure 2d shows the brushes assembly layout.

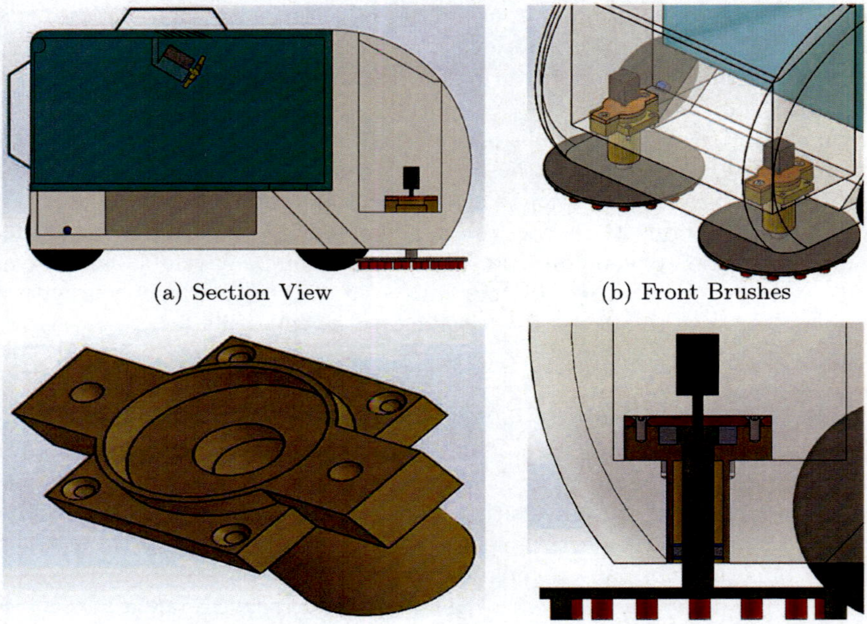

(a) Section View (b) Front Brushes

(c) Shaft Suport (d) Brushes assembly layout

Fig. 2. MopBot key components

Brush Shafts Since the steel shafts are subject to torsion, the safety coefficients to yield stress and to fatigue stress were calculated. Equation 1 returns the minimum radius R_m of the shaft where T is the torsion moment, S_y the safety coefficient to yield stress ($S_y = 2$) and τ the yield shear stress of steel ($\tau_y = 180\,\text{MPa}$). Since $R_m = 2.1\,\text{mm}$, the shaft radius $R \geq 2.1\,\text{mm}$.

$$R_m = \sqrt[3]{\frac{2 \times T \times S_y}{\pi \times \tau}} \qquad (\text{mm}) \tag{1}$$

Given the selected shaft radius $R = 7.5\,\text{mm}$, the DC motor power $P = 2.1\,\text{W}$, the brush angular velocity $\omega = 13.6\,\text{rad/s}$ and the polar moment of inertia $J = 4967.6\,\text{mm}^4$, Eq. 2 determines the maximum shear stress $\tau_a = 0.24\,\text{MPa}$.

$$\tau_a = \frac{P \times 1000}{\omega} \times \frac{R}{J} \qquad \text{(MPa)} \tag{2}$$

The resulting safety coefficient to yield stress $S_y = 774$ is given by Eq. 3 where τ is the yield shear stress of steel.

$$S_y = \frac{\tau}{\tau_a} \tag{3}$$

To calculate the safety coefficient to fatigue stress, first the fatigue stress is determined using Eq. 4, where σ_{f0} represents the fatigue limit stress ($\sigma_{f0} =$ 204 MPa) and k_x the fatigue correction factors. These factors take into account the: superficial finish of the shaft k_a ($k_a = 0.82$ for machined shafts); diameter of the shaft and applied torsion moment k_b ($k_b = 0.91$ in this case); reliability of the selected shafts to the applied effort k_c ($k_c = 1$ in this case); temperature k_d ($k_d = 1$ for air temperature); notches in the shaft k_e ($k_e = 0.7018$ in this case); and unspecified remedial effects k_f ($k_f = 1$ for no additional effects).

$$\sigma_f = k_a \times k_b \times k_c \times k_d \times k_e \times k_f \times \sigma_{f0} \tag{4}$$

Equation 5 determines the safety coefficient to fatigue stress S_f of the shaft where $\sigma_f = 107.34$ MPa is the fatigue stress and $\sigma_a = 0.12$ MPa the amplitude stress caused by a variation of the torsion moment on the shaft between 0 N mm to 154.3 N mm. The result was $S_f = 532$.

$$S_f = \frac{\sigma f \times 0.58}{\sigma_a} \tag{5}$$

The conclusion is that, based on the obtained yield and fatigue stress safety coefficients, $S_y = 774$ and $S_f = 532$, the shafts are extremely robust.

Control System. The MopBot is controlled by an Arduino Uno, which reads the sensor values and commands the motors, to navigate autonomously. It includes a Sound Navigation and Ranging (SONAR) sensor to detect when the tank is full and a laser-ranging VL53L0X sensor to detect obstacles. The wheels are actuated by two DC motors with the help of L298N bridge. The vacuum system and the two vertical brushes are operated by three additional DC motors. It also contains a voltage divider to detect when the battery is low and alert visually the operator. For safety measures, the Mopbot has a rotating warning light and, if necessary, an emergency button to switch off.

3.2 Design

The design considered simultaneously functionality and attractiveness. MopBot was designed to have as few parts as possible for simplicity (maintenance), weight (power saving) and cost related reasons. Figure 3 presents an overview of the 3D model. The polystyrene chassis, showed in Fig. 3c, was designed to accommodate all electrical components. A small door in the front of the chassis provides access

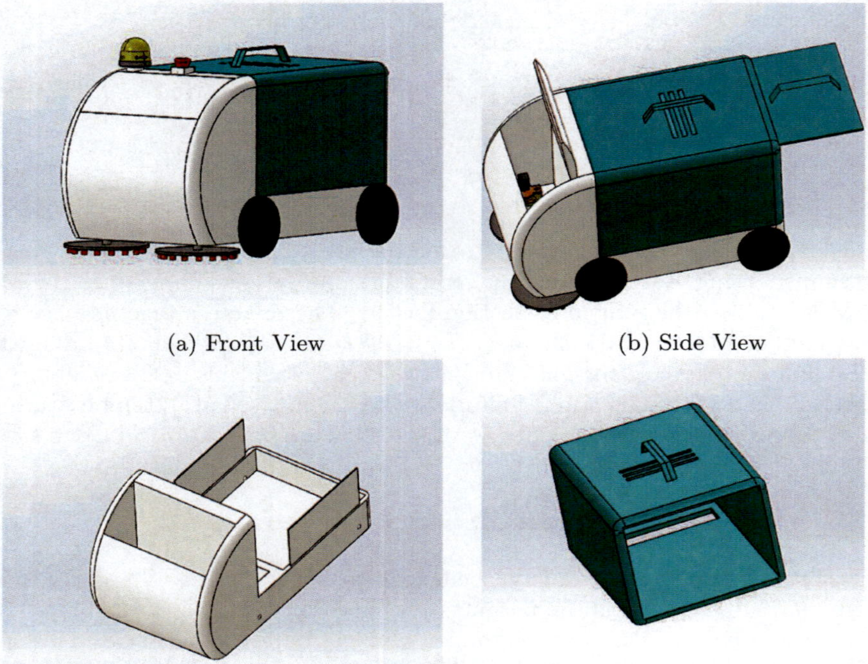

(a) Front View (b) Side View

(c) Chassis (d) Tank

Fig. 3. MopBot design

to the brush system, including the obstacle detection sensors (Fig. 3a). The tank, located in the back, is secured by the shape of the chassis. The sucked trash passes through a dedicated opening in the chassis to the tank. The battery compartment is located in the lowest part of the chassis, below the tank, to lower the centre of gravity and, thus, increase stability. To facilitate turning, MopBot is rear-wheel drive. The driving motors are positioned near and at the same level as the batteries. Once sucked, the trash accumulates inside the tank. The vacuum system is located on the top of the tank, aligned with the trash opening on the chassis, to optimise suction. When the level sensors detect full capacity, a human operator removes the tank through the back door (Fig. 3b) and empties the tank (Fig. 3d) into a bag or bin.

Additionally, the Team designed a sustainable packaging solution. Specifically, sustainable packaging needs to be: (*i*) functional – to guarantee product protection; (*ii*) cost effective – worth to choose less expensive solutions; and (*iii*) supporting long-term human and ecological health – to spread good practices within society. Figure 4 shows the designed packaging solution. The decision was to use 5-layers of BC-wavy cardboard because it is organic, ethical, sustainable and reusable (Fig. 4a). As a "second-life", the packaging can be transformed into a disposal box for discharged batteries or unused electrical equipment (Fig. 4b).

Used batteries and accumulators are a source of valuable recyclable materials and their proper collection allows to neutralise toxic heavy metals as well as to recover raw materials for the production of new batteries, saving the extraction energy. The disposal box can be placed in the malls and corridors that MopBot cleans. Collected equipment will be returned to appropriate services.

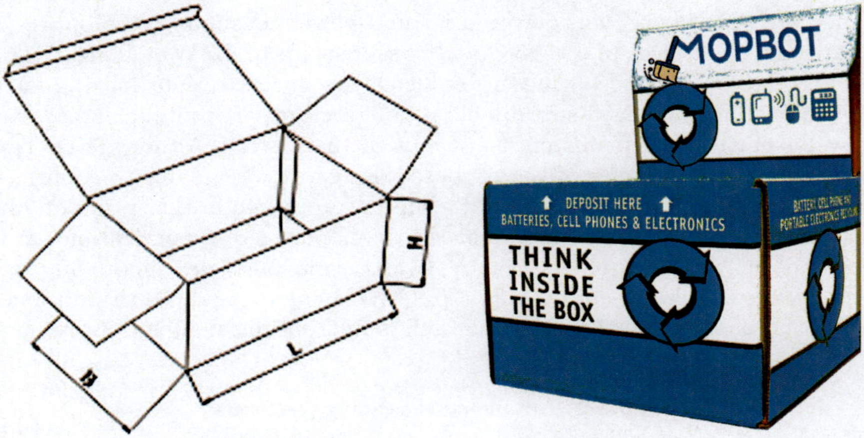

(a) Package (b) Disposal box

Fig. 4. MopBot packaging

3.3 Simulation

The simulation was performed with CoppeliaSim simulator and the Lua programming language. Figure 5 displays two simulation scenarios where MopBot follows a pre-defined path (black line): (*i*) in a hall (Fig. 5a); and (*ii*) in a corridor (Fig. 5b).

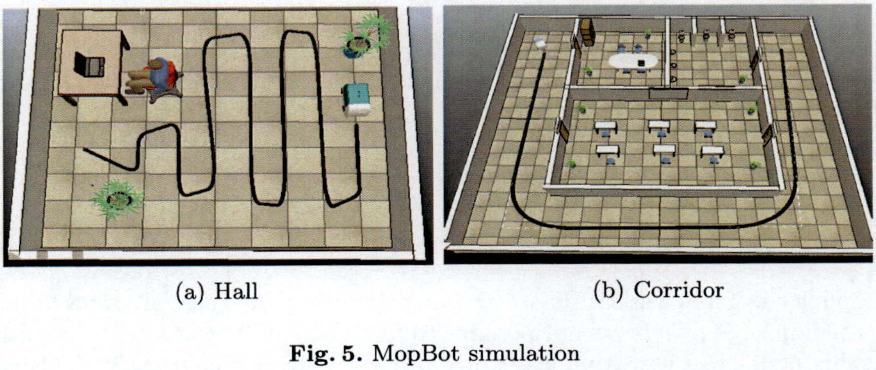

(a) Hall (b) Corridor

Fig. 5. MopBot simulation

4 Discussion

Table 1 displays the weight, storage capacity, velocity, hourly cleaned area and maximum uninterrupted operation time of Trilobite, Roomba, ENWAY, Dust-Clean and MopBot. From this comparison, Mopbot stands out as a medium size platform targeted for an unexplored market segment. Specifically, it presents the highest storage to weight and velocity to weight ratios and the second largest cleaned area per hour. Considering price and power consumption, MopBot has an expected shelf price of 194.55 € and consumes up to 9.7 W h. This suggests that MopBot is likely to be chosen by businesses or institutions looking for an autonomous medium-sized sustainable robot to clean large interior areas. Naturally, its efficiency depends on the layout of these areas. Although the Team is confident on the virtues of the proposed design, it was unable to confirm its performance given the impossibility to build a prototype in the spring of 2020. Finally, the addition of a camera and the development of a notification service would allow: (*i*) the visual detection of litter and obstacles, improving path planning, the appearance of the robot and saving energy; and (*ii*) the automatic notification of the operator when the tank is full, optimising operation time.

Table 1. Autonomous cleaning platforms

Solution	Weight (kg)	Capacity (L)	Velocity (km/h)	Area (m^2/h)	Δt (h)
Trilobite	5	1.2	1.44	28	1.0
Roomba	4	0.5	1.7	185	1.5
ENWAY	850	150	8.00	9000	6.0
DustClean	150	37	3.00	200	10.0
MopBot	20	25	2.52	756	2.0

5 Conclusion

The goal embraced in February of 2020 by the Team was to develop an eco-friendly cleaning robot with a constrained budget. This originally challenging task became harder with the unexpected need to implement remote multicultural multidisciplinary teamwork overnight. Nevertheless, the Team managed to design and simulate an autonomous vacuum cleaner that rivals with competitors.

5.1 Project Outcomes

Although the Team was satisfied with designed prototype, there are aspects that can be improved. As it was not possible to physically build the robot due to the ongoing pandemic, it was impossible to test the correct operation of the brush and wheel motors, namely, verify if the motors have enough power or if there is enough ground friction. Such issues could be solved by choosing alternative motors or adopting tracks instead of wheels. MopBot should also be marketed with different versions and prices, depending on the power autonomy, starting with a 2 h-autonomy basic version for 194.55 €.

5.2 Personal Outcomes

As a multicultural and multidisciplinary project-based learning framework, EPS aims to prepare engineering undergraduates to fulfil future professional challenges. The Team members provide below testimonials regarding their own learning experience as EPS@ISEP students during the spring of 2020.

- "The project was very challenging for all of us. Having to return to Romania and work from home due to the COVID-19 outbreak, made it more difficult than it already was. I am grateful for being part of EPS because I learned how to communicate better, gained more information about different cultures and how to work in a diverse team. I also discovered that I really like sustainability and I want to learn more about this topic in the future. Overall it was an interesting experience that I will definitely recommend." – Corina.
- "The European Project Semester has been very interactive and educational. During our stay in Porto, we were able to get to know each other personally. After the pandemic outbreak in March, we sadly had to leave Porto. This increased the communication problems which already existed due to our diverse cultural and scientific backgrounds. I definitely learned how to communicate better within our group, in person and remotely. We also had to deal with an internal problem during the project, which ended with a member leaving at our suggestion. For me, as project leader, it was difficult to do this as harmlessly and professionally as possible. This was definitely something I struggled with to learn that soft but clear communication is the best way. Overall, I think the group did a good job and is proud of the result. Since everything went a little different due to the COVID-19 outbreak, I think this project is even more an accomplishment." – Frederique
- "Thinking about the whole project, I identify many opportunities to develop hard skills and (maybe even more crucial) soft skills. Working in a team is a vital part of EPS and, thanks to the initial team-building activities, we were able to work together in a most efficient way. Personally, the most interesting topic was project management: the application of the Scrum methodology was really amazing. I think it somehow "saved" our teamwork despite the quarantine and loosing members. Considering that these problems were also opportunities, I think that Team did a great job and, if it was up to me, I would stay in Porto for one more EPS semester." – Szymon
- "Without doubt, this project was very challenging for the Team members. The team-building activities in the first week were very good given the different cultural and scientific backgrounds of the members. The Team was

progressing well, until the pandemic forced the switch from proximate to distance teamwork, increasing greatly the level of difficulty. This change required members to stay highly motivated and focused on the project. Despite these difficulties, the Team managed to design and simulate a medium-sized autonomous vacuum cleaner. EPS, therefore, enabled us to learn how to work as a team in an international remote set-up. We also improved our proficiency in English since it is the official communication language of EPS. However, the soft skills, were what the members felt they had developed most, including how to solve conflicts." – Tomás.

Acknowledgements. This work was partially financed by National Funds through the FCT – Fundação para a Ciência e a Tecnologia (Portuguese Foundation for Science and Technology) as part of project UID/EEA/50014/2019.

References

1. Accreditation Board for Engineering and Technology. Code of Ethics of Engineers (1977). https://engineering.purdue.edu/MSE/academics/undergraduate/ethics.pdf. Accessed May 2020
2. Andersen, A.: The European project semester: a useful teaching method in engineering education. In: de Campos, L.C., Dirani, E.A.T., Manrique, A.L., van Hattum-Janssen, N. (eds.) Project Approaches to Learning in Engineering Education: The Practice of Teamwork, pp. 15–28. SensePublishers, Rotterdam (2012)
3. Bauhof. Start-up: Enway entwickelt autonome kehrmaschinen und abfallsammelfahrzeuge (2017). https://www.bauhof-online.de/d/start-up-enway-entwickelt-autonome-kehrmaschinen-und-abfallsammelfahrzeuge/. Accessed May 2020
4. Duarte, A.J., Malheiro, B., Arnó, E., Perat, I., Silva, M.F., Fuentes-Durá, P., Guedes, P., Ferreira, P.: Engineering education for sustainable development: the European project semester approach. IEEE Trans. Educ. **63**(2), 108–117 (2020)
5. ENWAY. Autonome Kehrmaschinen für Industrielle Umfelder. https://enway.ai/en/. Accessed May 2020
6. Hanlon, M.: Electrolux trilobite robotic vacuum cleaner v2.0 (2015). https://newatlas.com/electrolux-trilobite-robotic-vacuum-cleaner-v20/2829/. Accessed May 2020
7. Hunt, S.D., Vitell, S.: A general theory of marketing ethics. J. Macromarketing **6**(1), 5–16 (1986)
8. iRobot. Roomba®980 robot vacuum (2020). https://store.irobot.com/default/roomba-vacuuming-robot-vacuum-irobot-roomba-980/R980020.html. Accessed May 2020
9. Nash, R.F.: The Rights of Nature: A History of Environmental Ethics History of American Thought and Culture. University of Wisconsin Press, Madison (1989)
10. ROBOTECH. Dustclean - robot sweeper. https://www.robotechsrl.com/dustclean-en-robot-sweeper/. Accessed May 2020

Case Study in Experiential Learning - From Chaos to Order: Sensemaking with the Interactive Timeline Tool in Architecture and Civil Engineering Studies

Nele Nutt[1], Sirle Salmistu[1](✉), Cassi Meitl[2], and Katrin Karu[3]

[1] Tartu College, School of Engineering, Tallinn University of Technology, Tartu, Estonia
sirle.salmistu@taltech.ee

[2] School of Planning, Design and Construction, Michigan State University, East Lansing, MI, USA

[3] School of Educational Sciences, Tallinn University, Tallinn, Estonia

Abstract. Many students have a difficult time memorizing and retaining facts that seem in-consequential to their everyday lives. Experiential learning techniques can make learning more interactive and meaningful, and thereby easier for students to comprehend and retain the material long after the course has ended. This paper presents a longitudinal case study of an interactive teaching method developed for the history of architecture courses over several years (2012–2019), which are compulsory for civil engineering, architecture and landscape architecture students. The professions related to the field of architecture are creative in nature, therefore learning methods based on experience and visual memory are very suitable to teach within these professions.

Keywords: Experiential learning · Digital learning module · Built environment and architecture · Knowledge management

1 Introduction

Traditional teaching approaches, which are based on remembering multiple facts about a subject area have been employed in Estonian schools (from primary to high school) for a long time. This inevitably leads to compartmentalizing subject areas and memorizing facts just long enough to take the test. Consequently, many students have a difficult time memorizing and retaining facts that seem inconsequential to their everyday lives or understanding how they fit into a larger body of knowledge. This "learn by rote" method does little to help students to understand the significance of the material, how it relates to other subject areas and how it links to concurrent activities.

The need to change the learning process to a more learner-centred approach has been stated in the Estonian Lifelong Learning Strategy 2020, which guides the development of education for different stages. University teachers are expected to use new innovative and interactive teaching methods (Haridus- ja Teadusministeerium, n. d.) and the teaching should be focused on students' experiences in order to encourage

M. E. Auer and T. Rüütmann (Eds.): ICL 2020, AISC 1328, pp. 91–102, 2021.
https://doi.org/10.1007/978-3-030-68198-2_8

their active and critical thinking, and finding solutions to problems and issues (Fink 2003). According to Uiboleht (2019), research demonstrates that university studies should involve elements of learner-centred approaches in teaching, such as an interesting presentation (lecture), real-life examples, a variety of teaching methods, pleasant learning environment (atmosphere), and authentic and challenging assignments as well as teaching methods that provide structure, guidance and support for the learning process.

Unfortunately, the teaching-centred (or content-centred) approach of learning is still existent in college setting as noticed by the authors over time. Students frequently are unable to retain material for a longer period of time and to make sufficient connections or even connections at all between the facts between subjects, fields, and disciplines. Students often demonstrate a lack of comprehensive knowledge when there are too many chaotic facts. Furthermore, they tend to forget the knowledge gained during previous courses relatively quickly. Research has shown that students learn better through personal experience, by discussing with each other and explaining concepts to others. One possibility to encourage this is to use experiential learning methods.

In addition, visual oriented fields such as architecture and design tend to attract students that are more sensory-oriented rather than fact/data conscious. In this case, visual versus textual materials play a key role in teaching and learning the material. In traditional teaching methods, data and facts are described in a textual rather than a visual manner, which leads to students not being able to make visual connections and losing the gained information rather quickly. The professions related to the field of architecture, including landscape architecture and civil engineering are creative in nature, therefore learning methods based on experiences and visual memory are very suitable to teaching these professions.

At the university level, our teaching goals are to support students to study and retain the selected information, encourage their creative and critical thinking and to teach them how to study effectively through experiential learning. Consequently, we have modified the traditional coursework by using new tools and techniques (incl. interactive and digital models) to attract and motivate the students to actively participate in the learning process. Our aim is to use an experiential learning approach to motivate and support the students for meaningful learning. Thus, we pursue to create interactive and digital tools that have many positive aspects. Firstly, they are attractive to the students and secondly they are helpful tools to study, organize and self-manage because the process is easily repeated. Everyone is able to access the tool, whether at home or at school. For teachers it is easy and helpful to provide consistency in the process from year to year and to improve it constantly.

We have noted that quite often the facts are chaotic and messy in students' heads. They are not able to manage the huge amount of data or to create connections inside or across disciplines. Thus, the hypothesis of this study was that integrated experiential learning methods develop students' critical thinking, analyzing and memorizing skills to acquire the most important data and to make logical connections within and across the subject fields.

Consequently, the purpose of this study was to use a more efficient teaching-learning method and to develop a tool to facilitate students' learning process. The aims were to create an instrument that helps to make sense of a large amount of facts, create meaningful connections within and across the subjects, and eventually, also provide a digital, attractive and contemporary tool for a new generation of students, which advances e-learning.

2 Theoretical Framework

Higher education is transitioning to an increased emphasis on student learning, and moving away from the traditional focus on lecturer teaching (Ashwin et al. 2015; Barr and Tagg 1995). In traditional learning, the teacher may provide facts and figures, and then expect the student to regurgitate the information without analyzing or comprehending it. By creating a meaningful learning experience, the teacher provides the learner with a more effective and long-lasting form of education (Beard and Wil-son 2006). A focus on student learning requires higher education teachers to deal with teaching innovations and the pressure to use new media (Bachmann 2018). Learner-centered pedagogical approaches include active learning, experiential learning, problem based learning, community based learning, and service learning. These approaches are expected to influence the whole student (Camp 1996) by providing a venue to expand their professional practice knowledge and skills, develop life-long learning skills, and enhance interpersonal skills (Crawford et al. 2011).

With the learner-centered approach, Mary Huba and Jann Freed (2000) recommend using relevant, real world, problems to develop critical thinking, problem solving, and the range of incidental skills needed to transition into professional practice:

"Solving ill-defined [real-world] problems require judgment, planning, the use of strategies, and the implementation of previously learned skill repertoires. Addressing ill-defined problems helps develop inquiry skills as students become researchers, seeking out and evaluating new information in their discipline, integrating it with what is known, organizing it for presentation, and having the opportunity to talk about it with others. In order to prepare students for life beyond graduation, learner-centered teaching should focus on solving ill-defined problems that require the integration of many skills and abilities at once." (p. 203).

Learning and acquiring experience are closely intertwined. Experience is the process to all forms of learning, as it incorporates new and significant ideas into a broader conceptual framework. Learning incorporates forms of doing, such as drawing, reading, writing, listening, and talking to express and receive symbolic information that is used to represent our world (Beard 2010). Beard and Wilson (2006) define experiential learning as "the sense-making process of active engagement between the inner world of the person and the outer world of the environment" (p. 2). With thoughts, feelings, and physical activity, experiential learning involves the whole per-son and the whole environment, both internal and external. Learning occurs with the interaction between the self and the environment (Beard and Wilson 2006). For the professional disciplines of planning and design, the educational foundation is built on an experiential learning by doing process (Wagner and Gansemer-Topf 2005). Experiential immersion is advocated as critical for students to learn the complex disciplinary components of the profession and how to integrate their skills in practice (Educating LAs 1998).

Experiential learning is a learning process in which the learner is central to the teaching and is actively involved in its implementation. The most well-known experiential learning model is based on Kolb`s theory (1984), which explains learning as a cyclical process, which means that first, learners are involved fully in new experiences, second, they must have the time to be able to reflect on their experience from different perspectives and get feedback from others. Third, learners must be able to form and reform, and process their ideas, take ownership of them and integrate their new ideas into logical theories. Then their understanding enables them to make decisions, solve problems, assess implications, and apply solutions to new situations (Wadhwa 2006).

The experiential learning cycle may be applied in only one single class or the entire course. Learning promotes curiosity, critical thinking and understanding, the experience must be examined and reflected on in the light of theory. Experiential learning has the potential to make students more enthusiastic, engaged, and give them a deeper, more nuanced understanding of the subject matter (Atherton and Moore 2016) as depending on the learning situation and the nature of assignment students can be either in the role of an observer, actor, analyzer or participant. A key aspect of learning is a student's beliefs about themselves, particularly their views about their ability to learn (Kolb and Yeganeh 2011). Therefore, teachers should choose those learning activities that encourage students to trust their experiences and the learning process, and reassess and monitor knowledge. Hulaikah and colleagues (2020) have listed the activities for each step on experiential learning (Table 1).

Table 1. Activities of experiential learning.

Steps	Activity
Concrete experience	Laboratories, Observations, Primary Text Reading, Simulations/Games, Field Work, Trigger Films, Readings, Problem Sets, Examples, Case Studies, Living Case Studies, Internships, Job Shadowing
Reflective observation	Logs, Journal, Discussions, Brainstorming, Thought Questions, Rhetorical Questions
Abstract conceptualization	Lecture, Papers, Model Building, Project, Analogies
Active experimentation	Simulations, Case Studies, Laboratory, Fieldwork, Projects, Homework

Experiential learning activities challenge students to overcome obstacles and problems by including physical and intellectual activities to test mental and physical endurance (Beard and Wilson 2006). As the use of senses increase in an activity, the student becomes more involved; therefore creating a memorable learning experience that can have the greatest effect on future thought and behavior. Students also need confidence, great minds, and resilience to survive the problems. A key consideration is the extent to which a learning activity is real or perceived as real. Although a learning activity might be experienced as only a game or recreational activity, the behaviors such as communication, team work, and leadership exhibited by participants during the activity are very real (Beard 2010). Experiential learning can increase student learning by providing more visual information, specifically in the field of architecture and design. The visualization can remove cognitive barriers which may prevent students from fully engaging with the challenging course material. It can be argued that architecture history courses impose a significant cognitive load on the students. The cognitive load theory proposed by Chandler and Sweller (1991, as cited in Snyder 2003, p. 32) suggests that "learning can be limited by the amount and type of material that can be processed in working memory during instruction". Activity based learning has the potential to mitigate problems created by high cognitive load courses by making the subject more accessible for the students and increasing their motivation to learn.

Experiential learning can also provide opportunities to integrate subject material, thus increasing the relevance of the subject for the students (Snyder 2003). A timeline displays facts in a linear fashion by correlating information in chronological order. As a teaching method, the timeline tool increases the relevance of the historical study of architecture to landscape design or other fields, thus encouraging students to perceive history as more valuable. The integration of architectural history and landscape design through a problem-based activity builds a connection to the student's goal of pursuing a career in landscape architecture or architecture. They are more likely to offer solutions and become socialized into the profession (Snyder 2003).

Learning experiences combine the modern use of technology to encourage experiential technology in the modern classroom. This general term describes the virtual experience for students provided by a broad range of hardware and software combinations (Snyder 2003). Experiential technology can enhance classroom activities by increasing student's interest in and engagement with the material because the technology is new and useful, which leads to greater learning and retention (Snyder 2003). The future of experiential exploration with digital technology in architecture is the digital reconstruction and simulation of the built environment, and the replication of complete spaces, and open environments. The models of extant spaces would represent an advance over two-dimensional static images (Snyder 2003). Consequently, combining experiential learning and teaching with the technology makes it very "real" for students.

3 Case Study

3.1 Approach

This section describes the experiential learning method and the interactive learning tool (i.e., Interactive Timeline Tool) that was developed and used in the courses of Landscape Architecture History and Architecture History at the Tallinn University of Technology, and applied in practice from 2012 until 2019. History of Architecture is a compulsory course for civil engineering, architecture and landscape architecture students. Landscape Architecture History course is compulsory only to landscape architecture students. Usually the students participating in the course are 2nd year Bachelor level students. The content used for creating the timeline tool is based on the most essential knowledge and learning outcomes of the particular history courses. Observations, interviews with students and student feedback were used throughout the courses as methods of data collection for this study to test the effectiveness of the tool and to identify changes in learning processes and outcomes. Participation of students in this study was volo· untary. Finally, qualitative analysis were carried out and improvements were planned for the next time.

3.2 Developing the Interactive Timeline Tool

First, the manual version of the interactive and collaborative learning tool was provided to students. The major reason why we decided to change the study format and use the experiential method was the fact that during several years we have pre-tested the landscape architecture students who started the course with a creation of historical timeline for architecture and landscape architecture (seen as simultaneous processes) whereas they have completed the course of architecture history with an exam in the previous semester. We were curious to find out their basic level of knowledge as a starting point.

In the first class of the course we carry out a workshop of memorizing the key factors from the previous course. The task is to create a historic timeline with given architecture styles or eras, and place the key-objects of architecture (buildings and constructions) in the historical and chronological timeline based on their characteristics and properties. The assignment required students to critically think and apply previous knowledge, collaborate within groups and interactively work with historic timelines, which display facts in a linear fashion.

Later on, the same task is repeated with the landscape architecture key examples (carried out only with landscape architecture students). The difference is that the students have studied history of architecture before, but not landscape architecture. Therefore, for the landscape architecture part their performance is not evaluated for correctness. We consistently find that very few students are able to correctly place the key objects (i.e., well known architectural pieces) on the right place in the timeline. They are not able to assess the landscape architecture pieces and apply their understanding of general architectural styles or eras. If they don't recognize the object, or if they don't know or memorize the fact, they are not able to continue the task in an effective way. We are convinced that there is lack of critical thinking in general way and interrelating systems.

The experiential learning activity occurs in several stages during the course. Firstly, it is dealt with in the first class by the workshop described above. This is for testing students' current knowledge and understanding from previous courses. Then the method is viewed and discussed in the beginning of every lecture in order to position the topic of the lecture and provide orientation. Repeating and turning back to the timeline tool helps with progress during the studies. Finally, the method is used again in the end of the course to study for the exam as a self-testing tool, which reinforces the knowledge gained through the lectures.

During the first seminar (workshop), the students, who are working in groups, are provided with a roll of paper (ca 1 per 5 m) for constructing the appropriate layout for the timeline, scissors, tape, markers, printed pictures with characteristic architectural artefacts (to be distributed in all time periods), and the task of creating a chronological timeline from ancient times to the present. The students must note all significant time periods within the western European space of culture, relying only on their memories and understandings from previous courses. They are not allowed to use external facilities, like the internet or study materials. Only when the students agree to a timeline of events, notes with factual data are distributed, and time is allotted to make corrections (see Fig. 1).

Fig. 1. Students in action with putting the pictures in sequence on the timeline.

Interestingly, all students are able to locate architectural objects from the early times (e.g., the Egyptian pyramids) on the right spot in timeline, but they fail to sequence correctly objects from the later periods of time. Although the students are able to distinguish the Roman, Gothic, Baroque etc. architectural styles/periods, they do not succeed to order them properly. This was especially apparent with recent and contemporary periods and styles, which emerged relatively fast in the short period of time.

The next step is to provide students with characteristic pictures and photographs of objects, designs and events. Their task is to assign the pictures to the timeline by recognizing the style unique to a particular era. After the pictures are assigned, factual information is provided and students are once again given time to make corrections. This activity results in the timeline of architecture history.

The next step is to repeat the procedure with the history of landscape architecture (not relevant to students other than landscape architecture). The major difference is that the students have not been exposed to the material earlier through lectures and seminars. They must draw on clues from other courses to follow patterns, characteristics and styles to accomplish this task. As students make educated guesses on which example of the landscape architectural object matches with the example of architecture, connections between the disciplines begin to occur.

When this experiential learning activity was originally introduced, instructors assumed that the students would be able to complete the timeline tasks with little effort. Experience however, has shown that quite frequently students do not remember data and facts learned in the prior semester. The activity helps students to make connections and remember important information, therefore creating a system for themselves. Students are able to learn from their mistakes.

Later on, the timeline tool is followed and amended throughout the whole coursework. Most of the pictures used in the timeline appear again during subsequent lectures and the reappearance helps to reinforce the knowledge. This is reflected by the discussions when they are able to make clear arguments and can create connections within the facts, as well as derive from the context. Students are encouraged to return to their work and make additions to the timeline with new pictures and important key-factors in history as new information is introduced. When combined, the architecture and landscape architecture timelines provide the student with a more comprehensive understanding of the developments and connections between the two disciplines. Students can easily identify the periodic changes and differences in architecture and landscape design.

3.3 Digital Improvement

With regard to integrated learning, today's educators are faced with the millennium generation - those students who were born after 1992. These individuals thrive in a world with computer technology, cell phones, and nearly instantaneous communication. Millennium generation students are receptive to integrated learning with the characteristics of multitasking, social decision-making, and tackling "real-life" issues (Tucker 2006; Tyler 2007). The integrated curriculum is congruent with their social approach to problem solving.

As Estonia is rather well developed in the IT field, e-learning courses are becoming more and more popular among students and teachers. The availability of resources and funds in order to create digital materials and e-learning tools has significant importance. Since the previously discussed experiential activity with paper and scissors demonstrates success in student retention of material, the instructors have also implemented a digital model to appeal to the millennium generation students. The digital version of the timeline activity was sponsored by the European Social Fund, BeST program, which supported the creation of e-learning tools (objects).

The advantage of the digital version compared to the manual version is the fact that it's repeatable multiple times, it's accessible as a self-evaluation tool and provides the opportunities to study and prepare for the exam at the appropriate time and place for the students. In addition, it's trendy (contemporary) as nowadays a majority of students work digitally regardless of time and place.

After using the manual/print version for a few years we developed the method into a digital module (i.e., digital tool) and we added it to the course resources. Our intention was to use the advantages of the digital format. Thus, the manual version of the method of experiential learning was used for designing the digital module (digital equivalent). The first e-learning object (i.e., digital module, digital tool) provides the possibility to study the logical structure of the general sequence of different historic time periods and styles (see Fig. 2). The center of the interactive page is a timeline with a period ranging from the 2nd century BC to modern times. The screen is divided in two parts; the upper part shows the logistics of periods and duration in history of architecture and the lower part shows the same phenomena in landscape architecture. In the column to the right there are two lists of time periods and styles (in Estonian) in chronological order.

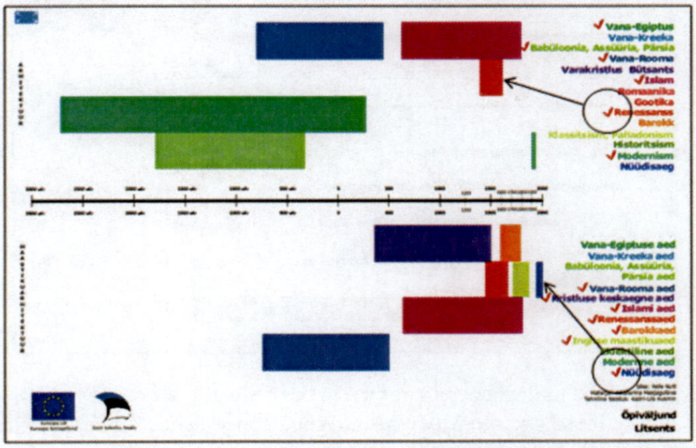

Fig. 2. Screenshot of the tool to learn historic periods and styles of architecture (above) and landscape architecture (below). By clicking on the name of the period/style (on the right), a check-mark appears and the colored ribbon of the selected period appears in the layout of the timeline. It displays simultaneous events in history showing overlapping and duration.

This module provides a visual tool to compare two disciplines and study their historical layout. Similar keywords of the periods (and relevant "time-ribbons") are marked with the same colour. By selecting the name of the period, it appears in the timeline. This visual provides the opportunity to understand what happened in the same time in architecture and landscape architecture history.

The digital version is easily used during the entire course and can be managed by the students more efficiently and effectively. The digital model works generally the same way as the manual version; first the timeline is created and then the pictures are dragged to the correct time periods. Much like a virtual game, this version incorporates assessment by eliminating incorrect answers from the timeline. Users are allowed two attempts before the timeline clears and the user must begin again.

The second e-learning objective (i.e., module/tool) aims to create an overall picture of the entire system by exploring the content of the picture and dragging them to the correct place in the timeline (see Fig. 3). The logic of the layout is similar to the earlier figure. In the center of the interactive page is a timeline with a period ranging from the 2nd century BC to the present. The upper part is for architecture and the lower for landscape architecture. In this module, the periods/styles are fixed in place on the timeline. On the edges of the page there are pictures with characteristic architectural and landscape architectural objects. The goal is to select the picture and place it to the correct period/style on the timeline. Students can identify the object by name and title by first clicking on the picture.

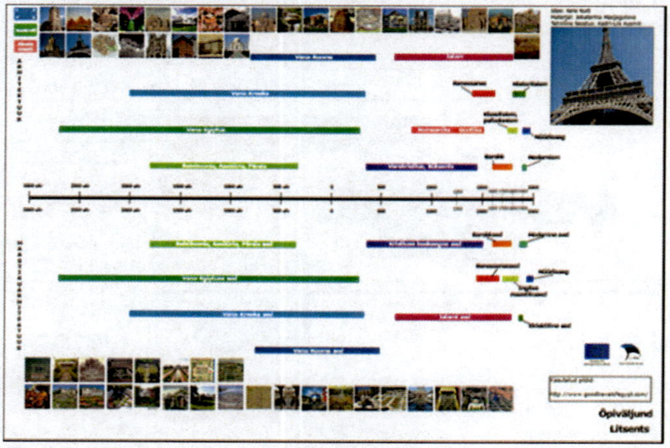

Fig. 3. Screenshot of the tool where pictures are to be dragged in the corresponding period/style in the timeline, e.g., Eiffel Tower to Modernism. When completed and after clicking the control button, all the pictures that were placed incorrectly return to their starting position. All pictures that were placed correctly remain in the timeline. Incorrect pictures can be dragged again. The aim is to repeat the process until all the pictures are placed in the corresponding periods/styles.

4 Discussion and Conclusions

The traditional teaching methods that require memorizing large amounts of facts have been employed for a long time in Estonian schools. We have identified that students are not able to make sufficient connections between subjects, fields, and disciplines based on this teaching method. Students studying in the field of architecture are particularly affected. They have a higher capability of visual memory and are more likely to excel in the classroom through experiential learning provided by the interactive timeline tool. This method (tool) contains activities for each step of experiential learning as described by Hulaikah and colleagues (2020). Concrete experience is created by the playful quiz (game), reflective observation is accomplished by the discussions and brainstorming, and abstract conceptualization is carried out through the model building activity (i.e., building the timeline). Finally, the active experimentation step is performed by simulations, case studies, and homework (e.g., using online timeline tool to prepare for the exam). The timeline approach encourages the students to critically think, analyze and remember the most important data and make logical connections between the architecture history and other fields, and even in larger scale (e.g., the history of town planning, engineering and technology, and general/political history).

A survey among the students at the end of the course revealed that the method of timeline as the central guideline throughout the coursework was very helpful tool to make sense in the course material and enabled to make connections within the whole system. The main aim of the survey was a self-reflection when using the timeline method. The majority of respondents pointed out that the timeline tool is effective at making connections between facts and events, and it helps to memorize the most important information when information is represented visually (on pictures) rather than textually. Interestingly, students of different areas of study had similar attitudes and reflection to the method.

Our findings have convinced us that the classroom studies with experiential and collaborative learning methods will help students to succeed in their studies and eventually in their careers. The result of the method (which could be used during several courses) would be settled in their knowledge – they remember more, acquire the ability to derive facts from context, the ability to think and act in an interdisciplinary wayacross topics.

As demonstrated, the aspect of interdisciplinary is the major advantage of the timeline tool. It allows students to create meaningful connections, see the relations between several subjects, and identify parallel developments over time. Another advantage is that the method helps to manage large amounts of facts by creating connections and logical sequences. Digital models and digital courses are easily used with the possibility to repeat and improve in the future after they are created.

After using this experiential learning method for four to five years, the results showed more positive outcomes than expected each year. A greater number of students (subjective evaluation by the authors) were able to reach the expected outcomes in a more efficient way. Also, as a result, the average grade of students increased. However, this may not represent adequate impact of the developed learning tool as the student level varies from year to year. Nevertheless, providing a simple means of linking information to a timeline to visualize concurrent events or a particular sequencing of information can be an enormous help to students.

Experiences from using this tool over years have demonstrated that the activity is fun and exciting and motivates the student to work. Frequently, the initial reactions from students is surprising. They are puzzled with the idea of working on the floor, cutting, taping and playing with the pictures. Quite often we hear the comments that they feel like they are in kindergarten since they expect university studies to be rigid and serious, not fun and playful. Soon the students seem to enjoy testing themselves and studying on the floor. There is sometimes a healthy competition between the groups to finish first. We have seen a lot of commitment and even confusion about how a simple task becomes serious work and how collective action contributes to learning. The success and effectiveness of the timeline method has demonstrated that experiential and collaborative learning methods should be further developed and incorporated in (theoretical) architecture and other relevant courses, such as Art History, Town Planning History and Urbanism, and relevant engineering studies (e.g., history of materials and constructions). Although technology is changing dramatically over years and there are areas of improvement to make this online tool more user-friendly, the main idea and usefulness of it is still relevant.

And as a final note, the ability to make facts and data understandable and relevant to a larger audience, brings increased value to our work. This method encourages the use of other simple, interactive and fun ways to teach and learn.

References

Ashwin, P., et al.: Reflective teaching in higher education. Bloomsbury, London, New Delhi, New York, Sydney (2015)

Atherton, C., Moore, G.: Student perspectives on the value of experiential learning in history. Australas. J. Am. Stud. **35**(2), 81–100 (2016)

Bachmann, H.: A focus on student learning. In: Bachmann, H. (ed.) Competence-Oriented Teaching and Learning in Higher Education – Essentials, pp. 12–14. Hep Verlag, Berne (2018)

Barr, R., Tagg, J.: From teaching to learning–a new paradigm for undergraduate education. Change **27**(6), 13–25 (1995)

Beard, C.: Experiential Learning Toolkit: Blending Practice with Concepts. Kogan Page, London (2010)

Beard, C., Wilson, J.: Experiential Learning: A Best Practice Handbook for Educators and Trainers. Kogan Page, London (2006)

Camp, G.: Problem-based learning: a paradigm shift or a passing fad? MEO (Med.-Ed.-Online) **1** (2), 4282 (1996)

Crawford, P., Kotval, Z., Machemer, P.: From boundaries to synergies of knowledge and expertise: using pedagogy as a driving force for change. In: Angotti, T., Doble, C., Horrigan, P. (eds.) Service-Learning in Design and Planning: Educating at the Boundaries, pp. 209–221. New Village Press, Oakland (2011)

Educating LAs: Landscape Architecture, vol. 88, no. 10, pp. 100–145 (1998)

Fink, L.D.: Creating Significant Learning Experiences: An Integrated Approach to Designing College Courses. Jossey-Bass, San Francisco (2003)

Gelmon, S.B., Holland, B.A., Driscoll, A., Spring, A., Kerrigan, S.: Assessing Service-Learning and Civic Engagement: Principles and Techniques. Campus Compact, Providence (2001)

Haridus- ja Teadusministeerium: Kõrgharidusprogramm 2016–2019. https://www.hm.ee/sites/default/files/lisa_8_korghariduse_programm_2016-2019.pdf. Accessed 05 June 2020

Huba, M., Freed, J.F.: Learner-Centered Assessment on College Campuses: Shifting the Focus from Teaching to Learning. Allyn & Bacon, Boston (2000)

Hulaikah, M., Degeng, I.N.S., Sulton Murwani, F.D.: The effect of experiential learning and adversity quotient on problem solving ability. Int. J. Instr. **13**(1), 869–884 (2020). https://doi.org/10.29333/iji.2020.13156a

Kolb, D., Yeganeh, B.: Deliberate experiential learning: Mastering the Art of Learning from Experience. ORBH working paper. Case Western Reserve University (2011). https://weatherhead.case.edu/departments/organizational-behavior/workingPapers/WP-11-02.pdf. Accessed 05 June 2020

Snyder, L.M.: The Design and Use of Experiential Instructional Technology for the Teaching of Architectural History in American Undergraduate Architecture Programs. University of California, UMI Dissertations Publishing, Los Angeles (2003)

Tucker, P.: Teaching the millennial generation. Futurist **40**(3), 7 (2006)

Tyler, K.: The tethered generation. HR Mag. **52**(5), 40–46 (2007)

Uiboleht, K.: The relationship between teaching-learning environments and undergraduate students' learning in higher education: A qualitative multi-case study, Doctoral Thesis. University of Tartu (2019)

Wadhwa, S.: Teaching and Learning Methodology in Higher Education. Sarup & Sons, New Delhi (2006)

Wagner, M., Gansemer-Topf, A.: Learning by teaching others: a qualitative study exploring the benefits of peer teaching. Landscape J. **24**(2), 198–208 (2005)

Reflection Assignment as a Tool to Support Students' Metacognitive Awareness in the Context of Computer-Supported Collaborative Learning

Aleksandra Lazareva(✉)

University of Agder, Universitetsveien 25, 4630 Kristiansand, Norway
aleksandra.lazareva@uia.no

Abstract. The present study explores the potential of a reflection assignment as a tool for supporting master's degree students' metacognitive skills in the context of computer-supported collaborative learning (CSCL). The research question (RQ) is formulated as follows: How does a regularly submitted reflection assignment support the development of students' individual metacognitive awareness in the context of CSCL? The empirical data is a text corpus (7878 words) extracted from individual students' (N = 13) reflection assignments (N = 65) submitted during one semester. Qualitative content analysis was employed to analyze the data. The results demonstrate that by the end of the course, the students significantly advanced their understanding of themselves as CSCL learners and developed a deeper understanding of collaborative learning strategies. The discussion suggests that complementing the collaboration scripting approach with a regularly occurring reflection assignment can contribute to students' awareness of the implemented learning strategies and their effectiveness in the context of CSCL. However, it is important to investigate what kind of guiding questions would be optimal for students to reflect effectively.

Keywords: Metacognition · Self-Regulated Learning (SRL) · Reflection · Computer-Supported Collaborative Learning (CSCL)

1 Introduction

Computer-supported collaborative learning (CSCL) requires a certain level of self-regulation skills from learners to effectively navigate through the learning environment, find relevant materials, engage in discussions with peers, and deliver individual and group assignments. As CSCL is becoming common nowadays in a variety of subject areas, the issue of engaging online learners is becoming even more topical. Many students do not possess a deep understanding of themselves as learners in an online setting, and do not have well-developed strategies to be a proactive online learner and collaborator. They often find CSCL contexts motivationally and cognitively demanding [1]. Common challenges in online learning include:

M. E. Auer and T. Rüütmann (Eds.): ICL 2020, AISC 1328, pp. 103–114, 2021.
https://doi.org/10.1007/978-3-030-68198-2_9

- Cross-cultural collaboration, where different participants can consider different behaviors appropriate and effective for collaboration [2, 3].
- Feeling of isolation and lack of social presence [4–6], which may negatively affect the motivation to participate.
- Insufficient exchange of information among the group participants and extra efforts paid to the coordination of actions [7, 8].
- Other problems related to time and space, such as lack of agreement around the time of online presence [9].

All these challenges put high demands on individual participants, making it necessary for them to develop an effective skill set necessary to cope with this form of learning.

A substantial part of recent CSCL research has been focused on collaboration scripts, i.e., prompts helping students have high quality group processes and internalize effective CSCL strategies [10]. Collaboration scripts have demonstrated to be effective for improving students' general collaboration and argumentation skills [11]. However, the collaboration scripting approach has received criticisms as well. One of the arguments against the collaboration scripting approach is that collaboration scripts may limit students' reflective thinking [12], which means that learners do not necessarily realize why a certain strategy is effective in a certain collaborative learning situation. Thus, the present study is focused on the development of learners' metacognitive skills in the context of CSCL.

The purpose of this study is to explore the potential of a reflection assignment as a tool for supporting the development of metacognitive skills in online students. The research question (RQ) is formulated as follows: How does a regularly submitted reflection assignment support the development of students' individual metacognitive awareness in the context of CSCL? It is expected that prompting students to reflect regularly and explicitly should lead to their deeper understanding of themselves as learners in a CSCL environment.

Importantly, it must be mentioned that CSCL research has addressed socially shared metacognitive regulation in asynchronous CSCL settings [13, 14], i.e., participants' regulation of joint cognitive processes in collaborative learning [13]. Nevertheless, research addressing the development of individual metacognitive skills in a CSCL setting is scarce.

The paper is structured as follows. Section 2 presents the theoretical framework used in the study. Section 3 describes the procedure of data collection and methodological approach. Results are presented and discussed in Sect. 4. In Sect. 5, practical implications are considered. Section 6 concludes the paper.

2 Theoretical Framework

The framing of this research study is the theory of self-regulated learning (SRL) [15]. The model developed by Paul Pintrich [15] consists of four phases – forethought/planning, monitoring, control, and reflection. Each phase, in its turn, has four areas for self-regulation – cognition, motivation, behavior, and context.

It is important to note that this model describes the *possible range* of self-regulatory activities, however, it does not necessitate them. Moreover, the phases are not necessarily ordered in a linear manner. In addition, it is possible that a learner simultaneously engages in more than one self-regulatory activity [16].

The focus of this paper is on learners' metacognition, i.e., their thinking about their own thinking. Metacognitive awareness is believed to include two major components – knowledge of cognition and regulation of cognition (see Table 1). Knowledge of cognition refers to what learners know about their own cognition and cognition in general, and can be further classified into declarative, procedural, and conditional types. Regulation of cognition, in its turn, refers to "metacognitive activities that help control one's thinking or learning" [21, p. 354]. Planning, monitoring, and evaluation are described as three essential regulatory skills.

Table 1. Components of metacognition (based on Schraw & Moshman [21]).

Component	Type	Definition
Knowledge of cognition	Declarative knowledge	Knowledge about oneself as a learner, their capabilities, skills, and factors affecting their performance
	Procedural knowledge	Knowledge on how to execute and sequence the learning steps
	Conditional knowledge	Knowledge on when and why various cognitive actions are to be applied
Regulation of cognition	Planning	Selection of strategies, setting goals, and allocation of resources
	Monitoring	One's awareness of comprehension and performance, ongoing assessment of strategies
	Evaluation	Analyzing the product and process of one's learning (e.g., the effectiveness of the chosen learning strategies)

Figure 1 summarizes the above-mentioned theoretical points and highlights the focus area of the present study.

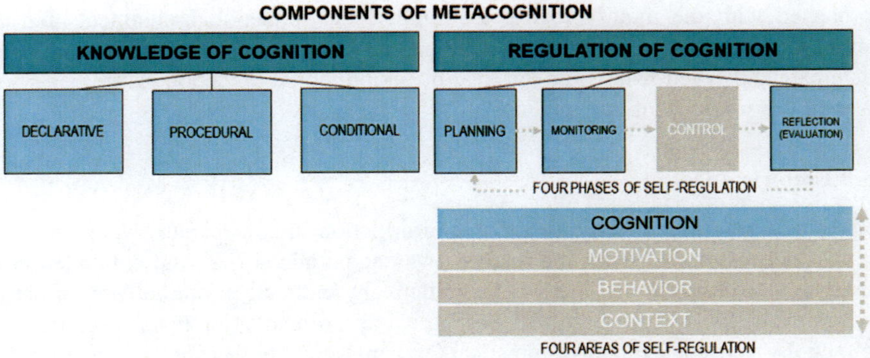

Fig. 1. Components of metacognition.

Learners' self-regulatory skills have been found to be associated with learners' self-efficacy expectations [18], i.e., expectations and judgements of their own capabilities to perform required actions to deal with a specific situation [19]. Thus, the decision to engage in SRL partly depends on one's self-efficacy expectations [17], as the learner needs to not only know about the learning strategy, but also have the confidence that (s) he is capable of accomplishing the required steps.

Learners with well-developed self-regulatory skills tend to attribute their success or failure to factors that are within their own control [20] (e.g., choice of implemented strategies, increased effort), as opposed to the factors they cannot affect (e.g., complexity of the task).

3 Research Design

3.1 Context of Data Collection

The context of the study is an online master's degree course run by a department of Information and Communication Technology (ICT). Students who took the course had background in ICT or engineering studies. The course consisted of five modules, and students worked in the assigned small groups throughout the whole course (i.e., one semester), supported by online tutors. Each module contained videos, articles, discussion tasks, and an assessed group deliverable. Each module lasted approximately three weeks and was finalized by an individual quiz and an individual reflection task. In the reflection task, students were asked to reflect on such aspects the most challenging issues in the module, their time management techniques, interactions with peers, interactions with tutors, any issues that could be improved by the teaching team, and any potential adjustments the student could work on to improve their own ways of learning. In the final module, the students were asked to revise the four earlier individual reflections, and ponder on the development of their learning strategies, as well as benefits and challenges of online collaborative learning in general. The students were also asked to do self-assessment and argument for it. Each student received individualized feedback from one of the online tutors on each submitted reflection. All the activities were carried out through the course learning management system (LMS).

After the course completion, the 18 students participating in the course were approached with the inquiry to use their individual reflection assignments for the purpose of this research. Thirteen out of 18 participants provided their consent to do so (six female and seven male), and the total of 65 individual reflection assignments was analyzed (13 assignments per module).

3.2 Empirical Data

No personal data (i.e., directly identifiable information such as names) were collected. The text corpus extracted for the analysis was anonymized (i.e., the combination of background information provided was rewritten, any indirectly identifiable information was removed). In addition, the text corpus was proofread prior to the analysis. All contractions were spelled out, grammatical (e.g., missing articles) and spelling mistakes were corrected. Naturally, this resulted in minor changes in the total amount of words. However, as the same correction procedure was applied consistently throughout the text corpus, the final numbers are still considered representative.

Thus, the empirical data of this research is a text corpus of 7878 words in total (module 1 = 1595 words, module 2 = 1101 words, module 3 = 1393 words, module 4 = 1634 words, module 5 = 2155 words).

3.3 Qualitative Content Analysis

Qualitative content analysis is "a research method for the subjective interpretation of the content of text data through the systematic classification process of coding and identifying themes or patterns" [22]. Qualitative content analysis [22–24] is employed in this study to analyze the text corpus. While being a rigorous research method, the qualitative content analysis is also a very flexible method [24]. Necessary guidelines were followed to ensure trustworthiness of the method on each of the three main stages [23]. These are outlined in the subsequent sections.

Preparation Phase. It is important to note that the data (i.e., student reflections) were not "steered" by the researcher as it can possibly happen, e.g., in the case of interviews [23]. Student reflections were an authentic learning assignment. The asynchronous nature of the assignment also gave students better time for reflection and formulation of thoughts [25], which resulted in more detailed student texts.

The choice of the unit of analysis is strongly dependent on the research questions of the study, and must be naturally occurring, identifiable, and retrievable [24]. The objective of the analysis in the present study is to capture explicit and implicit meanings and themes of students' utterances. Therefore, a logically completed expression of a thought, or a unit of meaning, was chosen as a unit of analysis. Thus, a unit of analysis could be a part of a sentence, one sentence, two or even more sentences.

Organization. A deductive approach was applied to organize the data. Each unit of analysis was coded using the categories of metacognitive awareness suggested by Schraw and Moshman [21] (see Table 1). This framework was useful in mapping out and describing the metacognitive processes ongoing in the sample, helping to provide an elaborated answer to the RQ.

The main objective of qualitative content analysis is not to provide an objective description of reality, but instead a rich description of a phenomenon that is embedded within a specific discourse. This makes the analysis and coding procedures much more interwoven in the qualitative approach, if compared to quantitative. In the quantitative approach, the quality of the analysis can be ensured by the correspondence between coders, while what is central in the qualitative approach is the conceptual consistency between the observation and conclusion [24]. In this study, one coder performed the analysis. The coding procedure consisted of three rounds:

- Initial round of coding of module 1.
- After one month – second round of coding of module 1 and comparing the results with the initial coding. The differences were closely analyzed to ensure the

consistency of the coding process. At this stage, the coder further advanced the descriptive definitions of each of the coding categories and assigned two representative examples from the data for each category. Then, the coder analyzed the rest of the text corpus.

- After one week – final revision of the text corpus as a whole and controlling for consistency of the coding.

Five components of metacognition (all three types of knowledge of cognition and two types of regulation of cognition) were manifested in students' reflection assignments through first to final module. "Monitoring" was not applicable in this analysis as it refers to an ongoing learning activity, whereas the analyzed individual reflection assignments took place at the end of each module after all activities were completed. In addition, parts of the reflection assignments were not relevant in the context of this study (e.g., students commenting on challenges related to the course registration and initial access to the LMS). Those were not coded.

In many cases, statements expressed by the students were multidimensional and could be coded under two or in few cases even three categories, which demonstrates the complexity of metacognitive processes. In such cases, the researcher coded the statement into the category that was the most prominent in the statement under question.

It must be emphasized that the metacognitive processes are difficult to observe. While the participants chose to share some of their thoughts in their written reflections, a large part of their metacognitive work remains unexpressed and unavailable for an external observer. It is thus important to keep in mind that only a small fraction of students' metacognitive processes is being analyzed in the frame of this study.

Reporting. The results are provided in an overview table. The quantitative indicators are used in the descriptive manner, while the most emphasis is on the detailed descriptions of the observed phenomena.

4 Results and Discussion

Table 2 presents an overview of the five modules, demonstrating the distribution of the five categories of metacognitive awareness in students' reflections.

Close observation of the development in each category through the first to last module revealed several qualitative changes in students' metacognitive awareness.

Table 2. Overview of the coded units of analysis in each module ("M" = module).

Category	M1	M2	M3	M4	M5	Total N of units
Declarative (N of units)	11	4	0	2	9	26 (8,7%)
Procedural (N of units)	8	10	5	4	3	30 (10%)
Conditional (N of units)	2	2	2	1	16	23 (7,7%)
Planning (N of units)	18	11	6	4	1	40 (13,3%)
Evaluation (N of units)	25	31	45	40	40	181 (60,3%)
Total N of units	64	58	58	51	69	300 (100%)

Further discussion traces the development in each of the five categories, providing specific examples for each category. The discussion is supported by direct quotes from student reflections. Each quote indicates the student number and the module where the quotation was extracted from (e.g., S07M1 stands for student's seven reflection in the first module).

4.1 Students' Knowledge About Themselves as Collaborative Learners

Through modules two to four, student occasionally and rather briefly commented on their own preferences and capabilities, as well as attitudes, as learners. However, in the first and final modules, a proportion of utterances coded into declarative knowledge category was relatively high. In their first module reflections, many students described what they knew about themselves as learners, e.g.: S06M1 "I love to have deadlines like we have had then", S09M1 "[…] it is very much out of my comfort zone in every possible way".

The reflection assignment after module 5 is especially interesting as it explicitly prompted the students to reflect on the development of their learning strategies over the course of the semester. In these reflections, the students evaluated their learning process and arrived at conclusions demonstrating that they have substantially extended their knowledge about themselves as learners. The most typical example from the present data set is asynchronous group discussions, which was a new way to learn for most of the participants. It was often evaluated as time-consuming, confusing, and/or not always relevant at the beginning of the course. In the final reflections, however, the students shared that group discussions evolved into a useful and rewarding learning tool throughout the course, e.g.: S01M5 "I am not really a person who likes to discuss, but I have learned that it can be very useful to have discussions in (preferably) small groups and get other people's views on various subjects", S12M5 "I have grown to enjoy discussions. The further we have moved into the course, the easier it is to discuss new topics branching from previous research".

Students have also extended their understanding of what online collaboration implies for a learner, e.g.: S04M5 "I was not prepared for how demanding it can be to be a part of an online learning group". One of the most important observations during the coding process was that there was a noticeable shift from students attributing their success (or failure) to external factors to discussing what they can affect themselves to improve their learning progress. At first, many of the students commented on the difficulties of orchestrating several subjects simultaneously, which led to their missing an overview of the course or even some of the deadlines (it may also be important to note that many participants were also working while taking the study) – while in modules four and five the amount of such comments reduced significantly.

4.2 Students' Procedural Knowledge on Collaborative Learning Strategies

The students had significant differences in the level of their procedural knowledge at the start of the course. Time management skills and strategies were specifically emphasized and discussed. Some students laid out a clear sequence of steps they used to tackle the learning tasks, e.g.: S08M1 "I added all the deadlines in my personal calendar with reminders on when to work on them. And when we did synchronous group assignments, we got to an agreement over social media on time and date", S13M2 "I always try to start working on discussion topics as soon as we receive them. It does not take too much time; I usually spend 1–2 h on finding information according to the topic we are given. This workflow allows me to have control over all assignments I have, not only [course name] but other classes as well". Other students had a less structured approach, but still had their learning under control, e.g.: S04M2 "I looked at the deadlines and tried to check into the discussion forum now and then to stay up to date". However, there were students who struggled finding out a systematic approach, e.g.: S07M1 "I worked with this when I had time and when someone else had been active in a discussion", S06M2 "I did the tasks when I had to".

There was not enough data to make conclusions about whether some changes in the students' time management skills happened during the course. The students did not dedicate much focus to their time management strategies in the reflections on later stages of the course. In the final reflection, only students eight and 13 explicitly commented on their time management strategies. However, these students demonstrated high quality time management skills already at the start of the course.

4.3 Students' Conditional Knowledge on Collaborative Learning Strategies

Students' thoughts on the application of the different learning strategies came through mostly in the final reflection, given one of the questions specifically asked them to reflect on group discussions and collaborative learning. Both benefits and challenges of group learning were outlined in much detail. The students highlighted specific benefits they valued in group discussions, such as acknowledgement of alternative perspectives, learning to argument for their own statements, remembering the learning content easier, and better understanding of learning material when discussing with peers who speak "the same language" (as opposed to the academic language of experts), e.g.: S03M5 "Learning from peers is a really good way to learn, as information written by experts can often be written in a complicated way and can be hard to understand. When you learn from peers, they are more or less on your level of expertise and therefore can explain the information more in a way that you are likely to understand", S01M5 "It forces you to make your own opinion and defend it". The students also demonstrated their understanding of when group discussions may not be the ideal strategy to apply, e.g., when the participants have too drastic differences in their attitudes and values. Certain challenges to cope with were also mentioned, such as time management and different working preferences and communication styles.

These utterances were coded into the conditional knowledge category, but it can be discussed that these reflections also demonstrate students' advanced procedural knowledge on collaborative discussions as a learning strategy.

4.4 Students' Planning of Their Learning in the Context of CSCL

Most of the planning utterances came through in modules one through three. Often these utterances also implied some elements of evaluation of the learning experience that had just taken place, e.g.: S01M1 "For my own improvement I could perhaps be even more proactive in the discussions", S01M1 "I should get better at planning and checking out the activity on [LMS]". What was typical for the utterances coded under the planning category was that they were lacking specificity, often remaining on a general level, e.g. S05M1 "[…] spend more time on the subject", S09M1 "[…] I should become more active in the discussion forum", S06M3 "I can start earlier and be more prepared".

However, there were some exceptions, e.g., one of the students planned for trying out a specific approach to note-taking which (s)he later tried out and evaluated in the following reflection assignment.

4.5 Students' Evaluation of Their Learning in the Context of CSCL

Most part of the utterances was coded into the evaluation category in all modules. As the course progressed, more and more units were coded as evaluative (see Table 2). Students dedicated most of the focus in their reflection assignments to how their learning progress and result was after each module. By the end of the course the evaluations also became more detailed, argumented, and critical.

In the final reflection students were also asked to assess their own performance in the course as a whole. Many students expressed uncertainty around the task of self-assessment yet providing reflective argumentation. They reviewed their own participation and commitment in the course, highlighting both the successful parts, changes in their attitudes and understanding themselves as collaborative learners (see Sect. 4.1), and the aspects that still need improvement and further work. Some of the final reflections demonstrate that students improved their self-efficacy beliefs, e.g.: S04M5 "I think we were all a little insecure about some of the group deliverables […], but since it is the first time I take this kind of course, it was a part of the learning process. […] But I think I have learned a lot and contributed with some good stuff".

5 Practical Implications

This section of the paper offers a discussion of the findings in light of earlier research in the area of collaboration scripting and socially shared metacognitive regulation in asynchronous CSCL settings.

It can be hypothesized that complementing the collaboration scripting approach with a regularly occurring reflection assignment can contribute to students' awareness of the implemented learning strategies and their effectiveness. Collaboration scripts were implemented during the online course discussed in this study. Thus, one of the scripts focused on the division of specific roles among the group participants [26, 27]. In their reflections, students explicitly reflected on the benefits of this specific approach to discussion tasks, as well as possible challenges and importance of tutor involvement on certain stages of script implementation.

Earlier research on collaboration scripts suggests that collaboration scripts must be implemented consistently over a longer period for them to have a lasting effect [10]. In line with this, it can be hypothesized that a similar approach must be implemented with reflection assignments.

The reflection assignment in this study was focused around specific questions formulated by the course instructor. It is important to explore what kind of guiding questions would be optimal for prompting effective student reflections. The collaboration scripting research demonstrates that while students need support to collaborate effectively, too much support may lead to over-scripting [28] and cognitive overload [29], causing negative effects on learning. It can thus be hypothesized that the questions in a reflection assignment should not be too closed nor too open. Evidence from the present study suggests that they should be specific enough to help students avoid repetitions and generalities in their reflections.

Another important practical concern is integrating the reflection task in the course canvas alongside with all other learning activities and assignments. For the students to approach the reflection task properly, the importance of the task should be emphasized. In the course discussed in the present study, the reflection assignments counted towards the portfolio grade which constituted 40% of the final grade in the course. The students received points for a submitted reflection automatically, and it was up to every individual student to decide how detailed or deep the reflection will be. The students in the present study demonstrated a serious approach to the task. It is important to note that the individual reflections were also a means of individualized feedback. Each reflection was commented on by one of the tutors, where the tutor followed up on the expressed challenges, made suggestions, and encouraged the learner. While appreciated by the students and being an effective tool for the tutors to follow up on each individual student's progress, it needs to be noted that providing individualized feedback on each individual reflection after each module required a large amount of time resources on the side of the tutors. It is also worth investigating how students act upon the feedback received from the tutor in their reflection assignments.

Lastly, metacognitive regulation in the context of CSCL can be best understood as a combination of individual and social processes, as individual participants' regulatory activities are interdependent and often shared in the group [13]. A major advantage of studying metacognition in asynchronous CSCL settings is that a large part of the metacognitive activities becomes visible through the discussion threads [13]. Thus, analyzing asynchronous discussion threads [13, 14] together with individual reflection notes can be an effective approach to studying metacognition in CSCL settings.

6 Conclusion

Qualitative content analysis was employed in this study to analyze student reflection assignments. The analysis revealed several qualitative changes in students' metacognitive awareness over one semester. Students enhanced their understanding of themselves as CSCL learners and developed a deeper understanding of what CSCL implies.

Moreover, the amount of external attribution expressed by the students reduced noticeably by the end of the course. It is suggested that a reflection assignment can be a complementary tool in an online CSCL course, contributing to students' awareness of the implemented learning strategies and their effectiveness, thus increasing the positive effect of collaboration scripts. In the future, carrying out a study with the experimental design would be valuable to better understand the benefits of a written, explicit reflection exercise and compare the metacognitive awareness of students in the experimental group against the metacognitive awareness of students who were not consistently challenged to explicitly formulate their reflections on their own learning.

References

1. Weinberger, A.: Principles of transactive computer-supported collaboration scripts. Nord. J. Digit. Literacy **6**, 189–202 (2011)
2. Dekker, D.M., Rutte, C.G., Van den Berg, P.T.: Cultural differences in the perception of critical interaction behaviors in global virtual teams. Int. J. Intercultural Relat. **32**(5), 441–452 (2008)
3. Popov, V., Noroozi, O., Barrett, J.B., Biemans, H.J., Teasley, S.D., Slof, B., Mulder, M.: Perceptions and experiences of, and outcomes for, university students in culturally diversified dyads in a computer-supported collaborative learning environment. Comput. Hum. Behav. **32**, 186–200 (2014)
4. Gunawardena, C.N.: Social presence theory and implications for interaction and collaborative learning in computer conferences. Int. J. Educ. Telecommun. **1**(2), 147–166 (1995)
5. Remesal, A., Colomina, R.: Social presence and online collaborative small group work: a socio-constructivist account. Comput. Educ. **60**(1), 357–367 (2013)
6. Sung, E., Mayer, R.E.: Five facets of social presence in online distance education. Comput. Hum. Behav. **28**(5), 1738–1747 (2012)
7. Pinsonneault, A., Caya, O.: Virtual teams: what we know, what we don't know. In: Kock, N. (ed.) Emerging E-Collaboration Concepts and Applications, pp. 270–289. Idea Group Publishing, Hershey (2007)
8. Powell, A., Piccoli, G., Ives, B.: Virtual teams: a review of current literature and directions for future research. Data Base Adv. Inform. Syst. **35**(1), 6–36 (2004)
9. Sarker, S., Sahay, S.: Implications of space and time for distributed work: an interpretive study of US-Norwegian systems development teams. Eur. J. Inform. Syst. **13**(1), 3–20 (2004)
10. Vogel, F., Wecker, C., Kollar, I., Fischer, F.: Socio-cognitive scaffolding with computer-supported collaboration scripts: a meta-analysis. Educ. Psychol. Rev. **29**(3), 477–511 (2016)
11. Radkowitsch, A., Vogel, F., Fischer, F.: Good for learning, bad for motivation? A meta-analysis on the effects of computer-supported collaboration scripts. Int. J. Comput.-Support. Collaborative Learn. **15**, 5–47 (2020)
12. Weinberger, A., Ertl, B., Fischer, F., Mandl, H.: Epistemic and social scripts in computer-supported collaborative learning. Instr. Sci. **33**(1), 1–30 (2005)
13. Iiskala, T., Volet, S., Lehtinen, E., Vauras, M.: Socially shared metacognitive regulation in asynchronous CSCL in science: functions, evolution and participation. Frontline Learn. Res. **3**(1), 78–111 (2015)
14. Pifarre, M., Cobos, R.: Promoting metacognitive skills through peer scaffolding in a CSCL environment. Int. J. Comput.-Support. Collaborative Learn. **5**(2), 237–253 (2010)

15. Pintrich, P.R.: The role of goal orientation in self-regulated learning. In: Boekaerts, M., Pintrich, P.R., Zeidner, M. (eds.) Handbook of Self-regulation, pp. 451–502. Academic Press, San Diego (2000)
16. Schunk, D.H.: Self-regulated learning: the educational legacy of Paul R. Pintrich. Educ. Psychol. **40**(2), 85–94 (2005)
17. Zimmerman, B.J.: Attaining self-regulation: a social cognitive perspective. In: Boekaerts, M., Pintrich, P.R., Zeidner, M. (eds.) Handbook of Self-regulation, pp. 13–39. Academic Press, San Diego (2000)
18. Schunk, D.H., Ertmer, P.A.: Self-regulation and academic learning: self-efficacy enhancing interventions. In: Boekaerts, M., Pintrich, P.R., Zeidner, M. (eds.) Handbook of Self-regulation, pp. 631–649. Academic Press, San Diego (2000)
19. Bandura, A.: Self-efficacy mechanism in human agency. Am. Psychol. **37**(2), 122–147 (1982)
20. Dweck, C.S., Leggett, E.L.: A social-cognitive approach to motivation and personality. Psychol. Rev. **95**(2), 256–273 (1988)
21. Schraw, G., Moshman, D.: Metacognitive theories. Educ. Psychol. Rev. **7**(4), 351–371 (1995)
22. Hsieh, H.F., Shannon, S.E.: Three approaches to qualitative content analysis. Qual. Health Res. **15**(9), 1277–1288 (2005)
23. Elo, S., Kääriäinen, M., Kanste, O., Pölkki, T., Utriainen, K., Kyngäs, H.: Qualitative content analysis: a focus on trustworthiness. Sage Open **4**(1), 2158244014522633 (2014)
24. White, M.D., Marsh, E.E.: Content analysis: a flexible methodology. Libr. Trends **55**(1), 22–45 (2006)
25. Serçe, F.C., Swigger, K., Alpaslan, F.N., Brazile, R., Dafoulas, G., Lopez, V.: Online collaboration: collaborative behavior patterns and factors affecting globally distributed team performance. Comput. Hum. Behav. **27**(1), 490–503 (2011)
26. Schellens, T., Van Keer, H., De Wever, B., Valcke, M.: Scripting by assigning roles: does it improve knowledge construction in asynchronous discussion groups? Int. J. Comput.-Support. Collaborative Learn. **2**(2–3), 225–246 (2007)
27. Olesova, L., Slavin, M., Lim, J.: Exploring the effect of scripted roles on cognitive presence in asynchronous online discussions. Online Learn. **20**(4), 34–53 (2016)
28. Dillenbourg, P.: Over-scripting CSCL: the risks of blending collaborative learning with instructional design. In: Kirschner, P.A. (ed.) Three Worlds of CSCL. Can We Support CSCL?, pp. 61–91. Open Universiteit Nederland, Heerlen (2002)
29. Kollar, I., Fischer, F., Slotta, J.D.: Internal and external scripts in computer-supported collaborative inquiry learning. Learn. Instr. **17**(6), 708–721 (2007)

Smart Bicycle Probe – An EPS@ISEP 2020 Project

Mélissa Boularas[1], Zuzanna Szmytke[1], Logan Smith[1], Kaan Isik[1], Juho Ruusunen[1], Benedita Malheiro[1,2](✉), Jorge Justo[1], Cristina Ribeiro[1,3], Manuel F. Silva[1,2], Paulo Ferreira[1], and Pedro Guedes[1,2]

[1] ISEP/PPorto – School of Engineering, Polytechnic of Porto, Porto, Portugal
mbm@isep.ipp.pt
[2] INESC TEC, Porto, Portugal
[3] INEB – Instituto de Engenharia Biomédica, Porto, Portugal
epsatisep@gmail.com
http://www.eps2020-wiki4.dee.isep.ipp.pt

Abstract. Air pollution kills approximately 7 million people every year and nine out of ten people are exposed to high levels of airborne pollutants. This paper describes the design of a bicycle air probe by a team of multicultural and multidisciplinary students of the European Project Semester, during the spring of 2020. This learning experience started with the analysis of the state-of-the-art, ethics, marketing and sustainability dimensions, and was followed by the design, development and simulation of a proof-of-concept solution. The result is GOairLight – a bicycle probe paired with a mobile app. The probe collects air quality, humidity and temperature data as cyclists ride, while the mobile app shares the collected data with the community, by means of a cloud database, presents relevant air quality information and suggests less polluted routes. Furthermore, it relies on a sustainable energy source – a dynamo powered by the cyclist – and automatic lighting. The latter feature improves cyclist visibility and raises the awareness towards the cyclist, contributing to increased road safety.

Keywords: Collaborative learning · European Project Semester · Sustainability · Air pollution · Community · Bicycle

1 Introduction

The European Project Semester (EPS) is a one-semester capstone programme designed by Arvid Andersen in 1995 to prepare engineering undergraduates to act and think globally. It adopts project-based learning and multicultural, multidisciplinary teamwork to foster scientific, technical and inter-personal skills. EPS is currently offered by a network of 19 European universities, including, since 2011, the Instituto Superior de Engenharia do Porto, (ISEP) of the Polytechnic of Porto.

GOairLight is a product designed by a team of five EPS@ISEP students during the spring of 2020. The team encompasses a variety of origins (Finland,

M. E. Auer and T. Rüütmann (Eds.): ICL 2020, AISC 1328, pp. 115–126, 2021.
https://doi.org/10.1007/978-3-030-68198-2_10

Belgium, Scotland, France and Poland) and fields of education (computer science, product design, mechanical engineering, electrical power engineering, and environmental sciences). EPS@ISEP provides the learning framework for the team to apply and develop the skills required for the project.

The open-ended project requirements included the design, simulation and test of a functional, ethically and sustainability-aligned bicycle probe solution. This would be compliant with applicable regulations. The cost of the materials and components required to build a prototype should not exceed 100 €.

The bicycle probe is a smart sensing bike device aimed for smart cities. The goal of GOairLight is to collect and share information about urban air quality and increase the safety of cyclists. The latter is achieved by including, in the front of the device, three bright light emitting diodes (LED). Since transportation is responsible for 31% of total energy consumption in Europe [1], the European Commission wants to promote clean, inexpensive and sustainable urban transport to reduce air pollution. The Air Quality Framework Directive 96/62/EC [3] is the leading directive towards the reduction in the atmospheric pollutants. Air pollutants that have a limit value are ozone (O_3), particulate matter (PM10 and PM2.5), and nitrogen dioxide (NO_2). The bicycle probe will collect information about the air pollution conditions in the urban space and the gathered data will be publicly disseminated and transferred to an Internet of Things (IoT) cloud so that the public can follow the air condition in real-time. In this context, GOairLight acts simultaneously as a probe and as a disseminator of urban air quality to the community, and, in the long-run, provides riders with less-polluted alternative routes.

This paper continues with the research, design, simulation, tests, discussion and conclusion regarding the bicycle probe EPS@ISEP project.

2 Background

This section analyses related products as well as the ethics, marketing, and sustainability aspects relevant for the creation of an innovative, ethically and sustainability-driven bicycle probe design.

2.1 Related Products

Related products refer to solutions already on the market and are a valuable source of information to characterise the state-of-the-art. Tech accessories for bikes can be organised into three categories:

- **Affordable devices**, like the B'TWIN 500 [9], usually display velocity, distance, and time (VDT). The most advanced, e.g., the MSW Miniac [8], provide also routes, which can be synchronised with a mobile app. The price of affordable devices reaches 65 €.
- **High-end devices** are equipped with high-quality sensors and materials. For example, Smarthalo 2 provides a compass navigation, automated lights,

anti-theft alarm among others [16]. Other products, like the Cobi app [2], are optimised for bikers. It shares weather information and can be connected with Apple Health or Google Fit Integration devices. The price of high-end options can go up to 219 €.

- **Special devices** like Speednite [17], which is a turning indicator with a tilting head, include a system with brake light and an SOS feature. Another example is the I LOCK IT GPS [13], which offers live tracking and warns the user in case of theft, thanks to a real-time alarm notification. The average price of special devices is 144 €.

Table 1 compares existing solutions. This comparison shows that there are no devices combining VDT, tracking, automated lighting and a mobile app interface. Furthermore, products with this collection of functionalities are expensive (up to 219 €). As a result, the team decided to create a solution comprising not only these, but also a set of distinctive functionalities: (*i*) air quality, temperature, and humidity monitoring; (*ii*) renewable power; and (*iii*) multi-functional packaging.

Table 1. Product comparison

Category	Product	VDT	Tracking	Automated Lighting	App
Basic devices	B'TWIN 500	✓			
	MSW Miniac	✓			
	Cannandale	✓	✓		✓
High-end options	Smarthalo 2	✓	✓	✓	✓
	Cobi	✓	✓		✓
	Beeline	✓	✓		✓
Special devices	Speednite	✓	✓	✓	✓
	I LOCK IT		✓		✓
	Garmin Varia			✓	✓

2.2 Marketing

Air pollution kills 4.2 million people every year as a result of exposure to ambient and outdoor air pollution [15]. GOairLight recommends less polluted routes as a solution to reduce biker exposure to air pollution. The goal is to create a strong value proposition because GOairLight competes with existing solutions. Furthermore, and in the long-term, the team wants to raise customer awareness towards air quality issues as well as road safety. The user must feel connected and part of the community.

GOairLight target market is represented by the middle-upper class segment, which owns smartphones and bicycles. After researching on the most likely population to use GOairLight probe, the target was the German population, where the Gross Domestic Product (GDP) is one of the highest of the world [10], and the smartphone and bike utilisation score higher in Europe [7]. GOairLight is a novel and unique product, inducing potential customers to pay for this new technology. The SWOT analysis showed that the main threat is the acquisition of both a smartphone and a bike. However, GOairLight team could partner up with municipalities to furnish municipal bikes, exploring market opportunities.

The Product, Price, Promotion and Place marketing mix (4P) helps to adapt the marketing strategy to the target market. In this respect, GOairLight will: (*i*) provide cyclists and the community with up to date urban air quality information; (*ii*) compete against related high-tech products, with prices ranging from 99 € to 219 €, and contribute to improve the health and safety of the user and community for a price of 149 €; (*iii*) be promoted mostly through social network advertising; and (*iv*) rely on a dedicated information and sales website for retail businesses and end-customers.

2.3 Ethics

The design, manufacturing and selling of a new product involves engineering, marketing, environment and liability ethics. In this regard, the main goal for the team, as future engineers, was to create a professional and trustworthy brand for the product. The marketing strategy was set to be, first and foremost, truthful to users and fair towards competitors. The brand will not adopt false advertising and will provide the public with clear and comprehensible information on the product. In terms of environmental ethics, the team adopted a power system composed of a human-operated dynamo (power generation) and rechargeable batteries (power storage). Considering liability ethics, the team must be compliant with applicable the European Union Directives:

- Measuring Instruments Directive [4] ensures that the scientific measurements (CO_2, temperature and humidity) are done accurately.
- Radio Equipment Directive [5] regulates the safety of user data sent through radio communication.
- Restriction of Certain Hazardous Substances EU Directives [6] restricts the usage of harmful substances in devices, allowing for more efficient recycling.

2.4 Sustainability

Sustainability-driven design considers the three pillars of sustainable development. Specifically, this project contributes to two of the 17 Sustainable Development Goals (SDG) defined by United Nations: SDG 11 – Sustainable cities and communities; and SDG 7 – Affordable and clean energy. To reduce the environmental impact, the team carefully chose the materials, components and designed for re-usability. The main case of the product is made of Polypropylene (PP), which is easily recyclable and resistant over time [14]. Despite common thought regarding the environmental damages caused by plastic utilisation, PP was the best compromise between efficiency, price and sustainability. The team decided, regarding re-usability, to: (*i*) create a part substitution service for customers; and (*ii*) design a reusable packaging. The part substitution service follows the principle of circular economy: the user will reuse the product by repairing and recycling instead of throwing it away (opposed to the linear economy principle). Given that, in 2017, 173 kg of packaging waste was generated per person in the EU [11], the GOairLight packaging is reusable. The shipping package, made of a bottle-recycled polyethylene terephthalate (PET) textile, can then be fitted to the middle bar of the bike and reused as a waterproof bag.

3 Proposed Solution

The creation of the GOairLight solution encompassed the following steps:

- Analysis of competing products to identify existing and missing features;
- Definition of a strong value proposition through a good, innovative concept;
- Selection of a name and logo matching the values of the product;
- Design of the control system schematics, including the selection of components and considering dimensions, power consumption, features and price;
- Creation of first drafts;
- Analysis of the product marketing, sustainability and ethics;
- Creation of communication supports (leaflet, flyer, A2 poster);
- Creation of the detailed drawings together with a 3D model of the device and packaging;
- Simulation of the control system;
- Design, development and test of the mobile application.

3.1 Concept

GOairLight probe was designed to fit the handlebar of a bike. In that position, and when the bike is moving, it collects and sends air quality data, such as $eqCO_2$ and Volatile Organic Compounds (VOC), temperature and humidity, to the GOairLight mobile app. The application not only receives and uploads the air data to an IoT cloud platform, but also suggests less polluted routes to take, preserving user well-being. The IoT cloud platform stores, for the benefit of the community, the air data collected by GOairLight devices. Moreover, the device switches on or off the 3-LED module automatically, depending on the light intensity. The goal is to improve the safety and visibility of the cyclist. The power is generated by the dynamo positioned on the side of the front wheel and is stored in the battery inside the main case. This way, the whole device depends on human energy and, thus, relies on a sustainable source of energy.

3.2 Design

The probe was designed to fit every bike. The mounting system is composed of three mounting screws (M5 25 mm) to attach the back case to the handlebar while the front case slides and locks to the back case. This system allows the user to easily assemble and disassemble GOairLight.

The front case is composed of a shade that acts like an extra protection against rain, and a light cover that protects the lighting module. The air flows through a rainproof opening, allowing the measurement of air quality. The lighting module is composed of three SSC P4 LED emitters and a reflector to provide enough light intensity for seeing and being seen. The CCS811 air quality sensor collects eqCO_2 and VOC data, and works together with the Nano 33 BLE sense board (Arduino). The battery consists of four rechargeable lithium-ion 2450 mA h units charged by the dynamo. According to Table 2, the maximum expected power consumption is less than 0.5 W h.

Table 2. Maximum expected power consumption

Component	Quantity	Voltage (V)	Current (mA)	Consumption (W h)
SSC Seoul P4 (U-bin) LED emitter	3	3.7	12–20	0.056
Arduino Nano 3 BLE sense board	1	3.3	15	0.050
CCS811 Air quality sensor	1	3.3	26	0.086
H–bridge	1	5.0	0–36	0.180
Total: 0.484 W h				

The GOairLight device weighs a total of 149 g and is made of two PP parts connected through a slide and lock mechanism. The front case, with a volume of 8 mm × 112 mm × 45 mm, holds the main device and weighs 75 g. The back case or the mounting part, with a volume of 75 mm × 112 mm × 45 mm, weighs 74 g. The slide and lock mechanism provides an elegant, user-friendly solution to attach and remove the front case and, thus, prevent theft. Figure 1 presents the exploded view (Fig. 1a) and the 3D model (Fig. 1b).

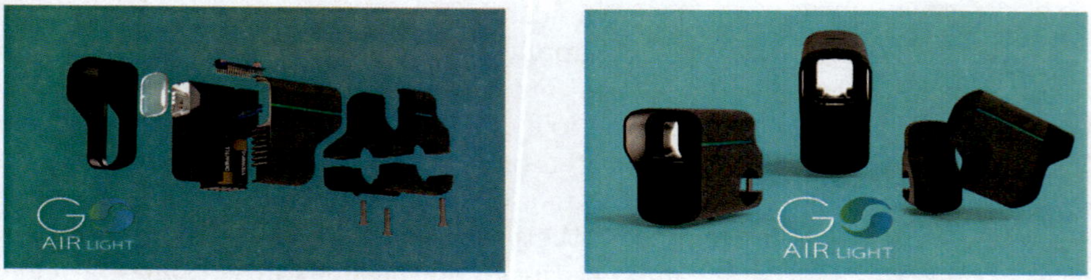

(a) Exploded view (b) 3D Model of device

Fig. 1. GOairLight design

The design of the packaging solution considered its standard protective role during transport as well as its reuse as a waterproof bag. The result was a multi-functional object that can: (*i*) hold probe and dynamo during shipping; (*ii*) store personal belongings during cycling; and (*iii*) carry, once removed, the front case. The packaging is made of recycled PET-textile, has a volume of 180 mm × 150 mm × 50 mm and weighs 90 g. After unpacking, it can be attached to the frame of the bike using three included Velcro strips. Figure 2a presents the packaging solution, including the package with the device (Fig. 1a) and the re-purposed package (Fig. 2b).

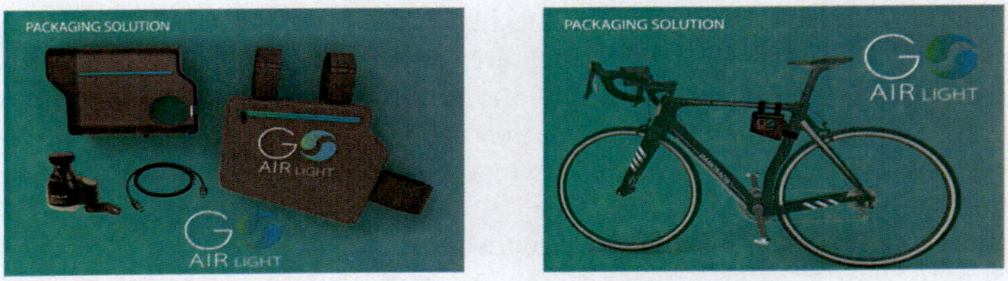

(a) Package (b) Reuse

Fig. 2. GOairLight packaging design

3.3 Application Development

The mobile application has five main purposes: (*i*) download the probe data (Bluetooth connection); (*ii*) store the probe data in its internal SQLite database; (*iii*) upload the probe data to the IoT cloud platform (online mode); (*iv*) download relevant data from the IoT cloud platform (online mode); (*v*) present the relevant data to the user. The presentation of data can take several forms: show the current probe measurement; show the data gathered by the user on a map (offline mode); and show the community data (online mode).

The application was developed in Java programming language for Android 5.1 or higher operating systems. The SQLite database was structured in two tables: the location and CO_2 level table, updated every 5 s, and the temperature and humidity table, updated every 2 min. The selected IoT cloud platform was Google Cloud Firestore. It contains a collection for each data type plus one for the users. Each data type document comprises the data, key, biker identification and timestamp. The Firebase is also responsible for handling user authentication, allowing the user to access community data via Google Analytics.

3.4 Simulation and Tests

Control System Simulation. The simulation was carried out using Tinkercad, a virtual simulation environment for electronic systems. To simulate the system, the team was forced to use the components available in the provided library. Unfortunately, it does not include any Bluetooth connectivity device.

Firstly, the Arduino Nano BLE Sense was substituted by the Arduino Uno Rev3. This led to replacing the temperature and light level sensors integrated in Arduino Nano with external ones and resigning entirely from the simulation of the humidity sensor. The CCS811 gas sensor was replaced by one with similar functionality—the Winsen gas sensor. Due to the characteristic of the software, some elements were used symbolically, like the dynamo and batteries configuration, since the batteries were registered as constantly charged. The boost converter was simulated using the Arduino itself. Figure 3a shows the simulation environment. The results of the simulation were quite promising. First, the LED reacted correctly to electric current changes and, later, to a photoresistor used to simulate light intensity changes in the environment. Next, the team tested with success the air quality sensor, visualising changes of pollution level with an additional coloured LED. Finally, the temperature sensor was tested by writing the registered values in the console and comparing with those set on the simulator. The simulation showed that the implemented Arduino code was able to read values from the gas, light and temperature sensors and actuate several LED according to the readings of the sensors.

GOairLight App. A first version of the application was developed. In the home screen (Fig. 3b) the user can view the most recent data gathered by the device. The current pollution levels are also visualised using different colours of the logo. Upon clicking the "i" button, the user obtains an explanation about the graphical representation of the pollution and the Bluetooth button. The login/logout icon is accessible in the top left corner of all screens. So far, the application is able to store and present data from the local database, which is currently populated with fake data, and upload/download data to/from the remote IoT database. IoT platforms usually use NoSQL databases to store sensor data series. This project uses Cloud Firestore that implements hierarchical structure of collections and documents. The application passed most of those tests successfully, with the exception of routing, which turned out to be imprecise due to the input data format from the maps' database.

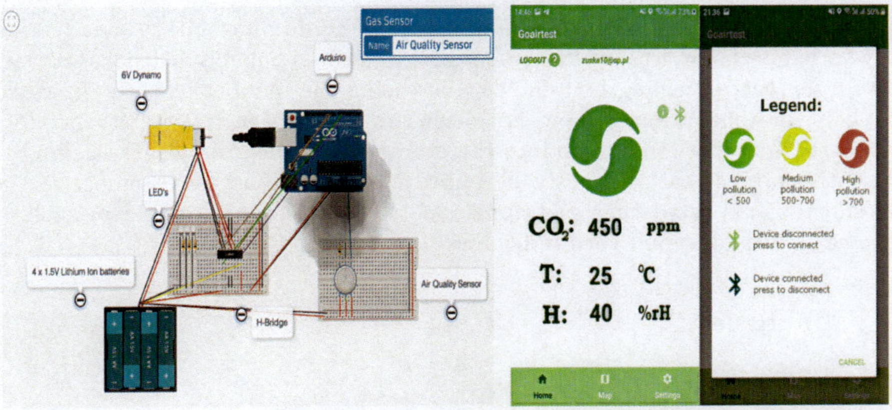

(a) Control system simulation (b) Mobile app

Fig. 3. GOairLight tests

4 Discussion

The simulation allowed a partial test of the Arduino code, namely, of the code that gathers data from the CO_2, temperature and light sensors and controls two LED according to the light and CO_2 levels. So far, the application stores and presents data from the local database, currently populated with fake data, and sends/receives data to/from the IoT database. These positive results indicate that it will be possible to expand further the concept, i.e., recommend less polluted paths from origin to destination.

The system can positively impact its users, making them more aware of the environment they live in and more conscious of their health. However, like most applications concerning health, it may lead to an unhealthy fixation with commuting only in clean air for some users. The combined results can possibly draw attention to the greater problem with pollution in cities.

By design, the data used in the application is meant to come from the probes. This means the collected data are dictated by user habits and willingness to use the system regularly. While data from frequently visited and popular places will be more accurate, some paths might never or rarely be updated. Another limitation is the dynamic nature of pollution levels, which can be influenced for short periods of time by traffic conditions, making some measurements not indicative of the general state of air in that area. Both factors could negatively affect the accuracy of the results displayed on the map. Due to the limitations of the simulator, the team was unable to test the Bluetooth connection, the humidity sensor and the dynamo operation. Finally, the team did not assemble nor test the control system, i.e., using the actual components and collecting genuine data.

Assuming a large number of users, the resulting database of measurements could possibly be studied to determine the most polluted areas in the city, with the timestamps allowing to assess sources of pollution. Since bikers usually have very varied paths, in less urban areas the results could help city authorities identify illegal waste burning households. Increasing the number of users, it would also help the application to show more accurate information and routes, therefore a great deal of effort should go into the promotion of the product. Furthermore, the entire project can obviously be improved with the actual assembly of the system. It is expected that additional safety features could be added, such as automatic backlight and turn indicators.

5 Conclusion

5.1 Project Outcomes

GOairLight proposes a crowdsourcing concept. It is based on a community approach that links citizens around the cause of urban air quality. GOairLight uploads the data collected by the probe to the cloud, where it remains accessible to the whole community (Fig. 4). To fulfil this main goal, the team designed an all-in-one sustainable device with a light module, air and light sensors, and four rechargeable batteries.

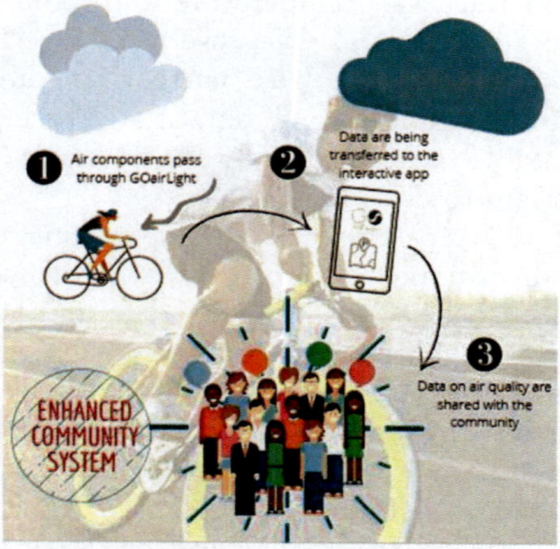

Fig. 4. GOairLight community

The proposed design would not be possible without the preliminary work on related products, marketing, sustainability, and ethics. This study led to a viable product that can be easily industrialised, including a sustainable, reusable packaging solution. Besides, communicative supports have been carried out to promote the qualities of GOairLight, like posters, leaflets, or even a self-explanatory video of the 3D model [12].

5.2 Personal Outcomes

EPS@ISEP is an international capstone programme catalysing teamwork and fostering different skills. The opinions of the five members of the team regarding this experience are summarised below:

- "My overall experience with the project was very positive. The classes accompanying the project forced me to work outside of my area of expertise, allowing me a broader look at my field of studies. I have also learned many skills connected to my field, such as knowledge on embedded systems and mobile android applications. I believe these skills will further aid me in the future both as a student and as a professional. The online learning we were forced to adopt was a challenge, especially coordinating work with distant teammates. However, I believe it was a valuable lesson on how to work with remote teams, which is not uncommon in the IT job market." – Zuzanna.
- "During the EPS, I learned how to work in a multinational team effectively and efficiently. The EPS has definitely improved my communication skills, in terms of conveying different ideas and messages across the board to the whole team. Unfortunately, the COVID-19 pandemic played a big part in our EPS semester. However, skills were still developed through the use of online teaching and video software. In the time spent completing the EPS, a part that I will look back fondly on, was the opportunity of meeting new people from all over the world and to contribute my skills in a multi-skilled group." – Logan.
- "EPS@ISEP was quite a challenge for me as the creation process from the design to the development of a proof-of-concept prototype is out of my study field. However, EPS is more than the creation of a product, it is about international teamwork. Specifically, it helped me to develop team management, searching and scientific writing skills. Thanks to this project, I now know more about the marketing field, and its important role in a company, and how to communicate efficiently through visual communication supports as well as with my team. COVID-19 situation hardened the realisation of the project and made it even more challenging, but we successfully managed it!" – Mélissa.
- "My experience with the EPS programme has been positive. I have learned how a team should work as a whole and how to improve my own doing within the group. EPS was not just focused on doing a project, but on learning other skills as well. I have learned other subjects like Portuguese, basics of coding, project management, marketing, energy and sustainability. The team was composed of students from all over the world and everyone had a different educational background. Before I used to do my work just before the deadline and now, after this teamwork experience, I realised that team partners have to work side by side to go further." – Juho.
- "I have learned a lot by working together with students from different fields and seeing their strengths and weaknesses. It ranged from self-learning (working with new tools and improving my design skills) to peer-learning (understanding how other fields are equally relevant to the design of functional

devices). Due to the outbreak of the pandemic, we were forced to switch to distance teamwork, resorting to online tools for communication. Although it was a challenge, the team was able to continue working and communicating remotely. Overall, this gave me experience and knowledge on how to work as a team without being physically together." – Kaan.

Acknowledgements. This work was partially financed by National Funds through the FCT – Fundação para a Ciência e a Tecnologia (Portuguese Foundation for Science and Technology) as part of project UID/EEA/50014/2019.

References

1. European Environement Agency. Final energy consumption by sector and fuel in Europe (2020). https://www.eea.europa.eu/data-and-maps/indicators/final-energy-consumption-by-sector-10/assessment. Accessed May 2020
2. Bosch. Smart connected - with cobi.bike (2020). https://www.bosch-ebike.com/en/products/cobibike/. Accessed May 2020
3. European Commission. Air quality - Introduction (2019). https://ec.europa.eu/environment/air/quality/index.htm. Accessed May 2020
4. European Commission. Measuring Instruments Directive 2014/32/EU (2020). https://ec.europa.eu/growth/single-market/european-standards/harmonised-standards/measuring-instruments_en. Accessed May 2020
5. European Commission. Radio Equipment Directive (RED) (2014/53/EU 2014-04-16) (2020). Available: https://ec.europa.eu/growth/sectors/electrical-engineering/red-directive_en. Accessed May 2020
6. European Commission. ROHS EU Directives (2020). https://ec.europa.eu/environment/waste/rohs_eee/index_en.htm. Accessed May 2020
7. CONEBI. European bicycle market 2017 (2017). https://issuu.com/conebi/docs/20170713_european_bicyle_industry_a. Accessed May 2020
8. World Wide Cyclery. MSW Miniac 322 GPS Bike Computer - GPS, Wireless, Black (2020). https://www.worldwidecyclery.com/products/msw-gps-322-miniac-gps-computer-black. Accessed May 2020
9. Decathlon. Van Rysel 500 wireless cyclometer - Black (2020). https://www.decathlon.co.uk/500-wireless-cyclometer-id_8382192.html. Accessed May 2020
10. Eurostat. GDP ate regional level (2016). https://ec.europa.eu/eurostat/statistics-explained/index.php/GDP_at_regional_level. Accessed May 2020
11. Eurostat. Packaging waste statistics (2020). https://ec.europa.eu/eurostat/statistics-explained/index.php/Packaging_waste_statistics. Accessed May 2020
12. Kaan Isik. 3D model video - GOairLight - Team 4 (2020). https://www.youtube.com/watch?v=D8aELmtRjLc&feature=youtu.be. Accessed May 2020
13. I LOCK IT. I Lock It GPS (2020). https://ilockit.bike/en/produkt/ilockit-gps/. Accessed May 2020
14. J.Hopewell. Plastics recycling: challenges and opportunities (2009). https://www.ncbi.nlm.nih.gov/pmc/articles/PMC2873020/. Accessed May 2020
15. World Health Organization. Air pollution (2020). https://www.who.int/health-topics/air-pollution#tab=tab_1. Accessed May 2020
16. Smarthalo. Make your bike smarter (2020). https://www.smarthalo.bike/. Accessed May 2020
17. Speednite. The first head motion controlled lighting system for bikes (2020). http://speednite.com/. Accessed May 2020

Using GitLab Interactions to Predict Student Success When Working as Part of a Team

Audrey Beatrice Ekuban(✉), Alexander Mikroyannidis, Allan Third, and John Domingue

Knowledge Media Institute, Open University, Milton Keynes MK7 6AA, UK
{audrey.ekuban, alexander.mikroyannidis, allan.third, john.domingue}@open.ac.uk

Abstract. This paper explores machine learning algorithms that can be used to predict student results in an assignment of a Software Engineering course, based on weekly cumulative average source code submissions to GitLab. GitLab is a source code version control system, commonly used in Software Engineering courses in Higher Education. The aim of this work is to create models that can be used to predict if a group of students in a team will pass or fail an assignment. In this paper, we present results from Decision Tree, Random Forest, Extra Trees, Ada Boost and Gradient Boosting machine learning models. These models were evaluated using cross-validation, with Ada Boost achieving the highest average score.

Keywords: Learning analytics · Educational Data Mining · Machine learning

1 Introduction

Two important areas of research in education are Learning Analytics (LA) and Educational Data Mining (EDM). Learning Analytics, together with Educational Data Mining, allow an institution to gain actionable insights from student data.

1.1 Overview

A regularly used definition of LA is attributed to the Society for Learning Analytics Research (SOLAR) [11]. This is stated as "Learning Analytics is the measurement, collection, analysis and reporting of data about learners and their contexts, for purposes of understanding and optimizing learning and the environments in which it occurs" [10]. In Higher Education, LA very often refers to the use of predictive models in order to optimise student success.

A definition of EDM, which predates LA by a few years, is attributed to the International Educational Data Mining Society (IEDMS) [7]. This is stated as "Educational Data Mining is an emerging discipline, concerned with developing methods for exploring the unique types of data that come from educational settings, and using those methods to better understand students, and the settings which they learn in" [1].

There is a tendency to use LA and EDM interchangeably. However, whereas there are some similarities between them, there are some differences that need to be

M. E. Auer and T. Rüütmann (Eds.): ICL 2020, AISC 1328, pp. 127–138, 2021.
https://doi.org/10.1007/978-3-030-68198-2_11

considered. One of the main differences is that EDM focuses on the creation of machine learning models to extract knowledge from education datasets, whereas LA applies Business Intelligence tools and predictive models to analyse data, in order to optimise learning [2].

In this case study, we are investigating how the results of students' team assignments are impacted by the team's interaction with GitLab. GitLab [4] is a software repository system used by Higher Education establishments. Our aim is to build Machine Learning models that predict student team results for a practical assignment. The dataset referenced in this paper was provided by The University of Manchester [9] (Data Controller) to The Open University (Data Processor). The dataset contained: (i) students' pushes to University of Manchester's software repository, GitLab and (ii) a single Tutor Marked Assignment for each student.

1.2 Research Questions

In this study, machine learning techniques will be used to address two main research questions:

RQ1: How accurately can machine learning classification algorithms identify a team assignment result when using only cumulative average pushes to a software repository in a campus-based undergraduate course at a university?

RQ2: What are the important weeks when using only cumulative average pushes to a software repository in determining team assignment failure in a campus-based undergraduate course at a university?

This case study differs from previous known work [5], where cumulative weekly averages were not considered and Area under the Receiver Operating Characteristics (AUC - ROC) [6] was not one of the metrics used to evaluate model performance.

2 Data Summary

The dataset contained data for 2 cohorts of students on a Software Engineering course that was taught at the University of Manchester. As the cohorts were independent and the data was anonymised independently, it was not possible to tell if there were any students who repeated the course. This paper assumes complete independence, i.e., none of the students had an unfair advantage.

The dataset included one team assignment per cohort. The assignment was set in the first week of the course and marked in the last week of the course. Each student was allocated to a team. Marks were initially allocated to each student based on the team assignment, i.e., all students in a team were given the same mark. If it later transpired that a student did not contribute enough to the team, then that student was given a lesser mark, whilst the marks of the other students in the team are adjusted in the positive direction. The adjusted marks create the situation where the result of student A is dependent upon the result of student B, given that student A and student B are in the same team. This paper details team analysis using the unadjusted marks, i.e. a team is treated as a single entity.

Apart from results, the only other variables were the daily number of times a team submitted (pushed) code to GitLab. The number of students in each team could be calculated. But as there were, roughly, the same number of students in a team, this value was ignored. Hence, the only independent variable was the number of weekly pushes i.e. the weekly pushes were calculated from the daily pushes.

The Data Processor was not provided with the push data of students who withheld consent. As stipulated by the Data Controller, any teams in which at least one student opted-out was not included in the analysis. This is to remove the risk of re-identification [8] of these students. Re-identification is the process by which information about an individual that exists in the public domain can be used to gather some information about that individual from an anonymized dataset.

3 Data Exploration

An initial analysis of the data showed that:

– In Year 1, out of 37 teams

- 5 teams failed (mark less than 40%)
- 5 teams passed with a mark of less than 70%
- 27 teams passed with a mark of 70% or greater

– In Year 2, out of 36 teams

- 0 teams failed (mark less than 40%)
- 7 teams passed with a mark of less than 70%
- 29 teams passed with a mark of 70% or greater

There were a low number of failures in year 1 and 0 failures in year 2. There is insufficient data to undertake any further work that would give insights into teams which attained a mark of less than 40% in year 1. Also, with only 2 years of data, it cannot be ascertained which of the years were "normal". In order to progress predictive analytics, a team with less than 70% of the marks was deemed to have failed. A team with greater than or equal to 70% was deemed to have passed. Hence overall, there were 17 teams with a fail grade and 56 teams with a pass grade (see Fig. 1).

A graph of the weekly cumulative average of the number of times that teams push to GitLab (see Fig. 2) indicate that the likelihood of a team receiving a result of 70% or more increases with the average number of pushes.

In the further work detailed below, the data for week 13 was removed as this is the week when the assignments are marked.

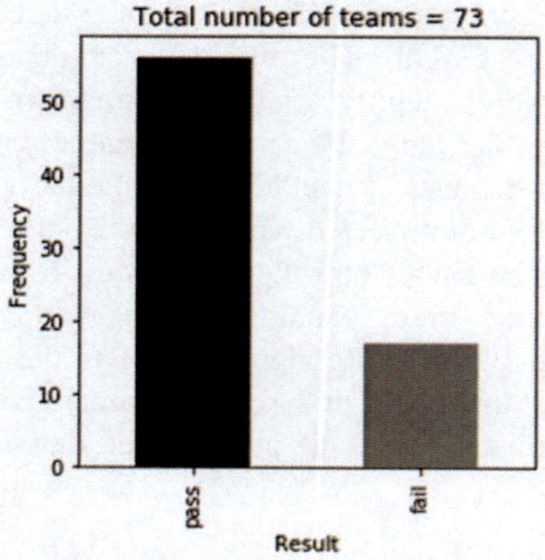

Fig. 1. Bar Plot of Team Results. Pass is attributed to teams with a result of 70% or over. Fail is attributed to teams with a result of less than 70%.

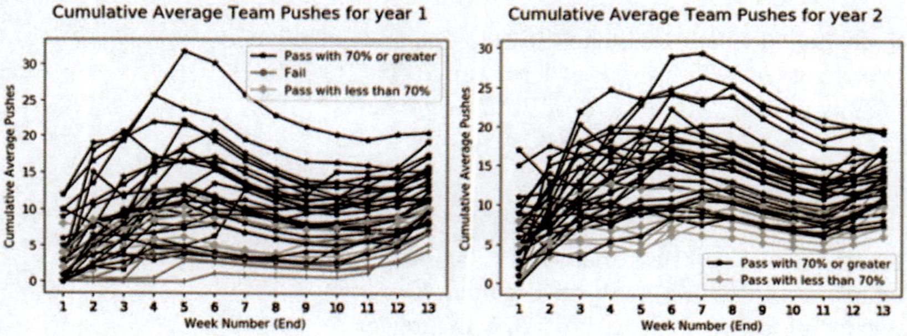

Fig. 2. Weekly cumulative average of number of times the teams submitted code to GitLab.

4 Methods

This case study uses a supervised machine learning approach. The goal was to train simple models that could be used to predict the result of a team assignment given the cumulative average number of GitLab pushes per week. The models are to be used to predict the teams that are likely to fail. The output from the classification models are 1 for fail, the positive class, and 0 for pass, the negative class.

The models were created using classifiers as implemented in the Python Scikit-learn package [3]. The following classifiers were investigated:

- Decision Tree
- Random Forest
- Extra Trees
- Ada Boost
- Gradient Boosting

More details are given in the following sections.

4.1 Model Training Considerations

In supervised machine learning, it is normal to split a dataset into training and test data. A model is created with training data, and evaluated with the test, or unseen, data. A model is said to overfit if it performs well on the training data but not on the test data i.e. the model cannot be generalized. In addition to training data (labelled data), many classifiers also make use of hyperparameters. These are parameters that are set prior to the training of the model and are used to tune a model. Model training is therefore an iterative process. However, after the first train/test cycle, the test data is no longer "unseen". Instead of splitting the data into 2, the data could be split into 3 - training, validation and test. In this case, the model is trained on the training data and evaluated using the validation data. In this case study, as the dataset is small, the whole dataset is used in training the model. Rather than manually splitting the data to create a validation set, 5-fold cross-validation is used to select the most suitable hyperparameters for each of the classifiers and to compare the performance of different machine learning models on the dataset. The dataset is divided into 5 parts (folds). In each training iteration, one fold is withheld as the validation set with training being done on the remaining 4 folds. A different fold is withheld for each cross-validation. The accuracy of the model is estimated by averaging the accuracies from all of the 5 cases of cross-validation. It is envisaged that a future year's data will be supplied in order to test the model.

4.2 Model Evaluation

The models were evaluated with a combination of accuracy score and AUROC (Area under the Receiver Operating Characteristics).

In this case study, predicting the fail class is of major importance:

- True Positive = Team fail result is correctly classified as fail
- False Positive = Team pass result is incorrectly classified as fail
- True Negative = Team pass result is correctly classified as pass
- False Negative = Team fail result is incorrectly classified as pass

Accuracy is the fraction of predictions that a classification model got correct i.e.

$$Accuracy = \Pr(correctoutcome) = \frac{Number\ of\ correct\ predictions}{Total\ number\ of\ predictions}$$

In binary classification, where there are only 2 labels:

$$Accuracy = \frac{TP + TN}{TP + FP + TN + FN}$$

where

- TP = Sum of True Positives
- TN = Sum of True Negatives
- FP = Sum of False Positives
- FN = Sum of False Negatives

A confusion matrix can be used to display these values.

A perfect model would have an accuracy score of 1 (or 100%). Accuracy can be misleading for imbalanced data sets. Consider a dataset with 95 students, who have a result of pass (positive class) and 5 students who have a result of fail (negative class). A model that classifies all results as pass will have an accuracy of 0.95 (or 95%). By relying on accuracy on its own, this model could end up being selected ahead of more useful models which have a lower accuracy.

AUC - ROC Curve

In addition to directly predicting a class, some machine learning models can also predict the probability of a data point belonging to a classification class. The probabilities can be used to determine a classification threshold to interpret the results of a classifier. By default, equal probability is given to classes in binary classifications. In reality, there is usually a trade-off between Sensitivity (missed positives) and Specificity (negatives classed as positives). Sensitivity is also known as recall or True Positive Rate.

$$\begin{aligned} \textit{True Positive Rate} &= \Pr(\textit{Predicted Fail}|\textit{Fail Result}) \\ &= \frac{TP}{TP + FN} \end{aligned}$$

$$\begin{aligned} \text{Specificity} &= \Pr(\textit{Predicted Fail}|\textit{Fail Result}) \\ &= \frac{TN}{TN + FP} \end{aligned}$$

$$\begin{aligned} \textit{False Positive Rate}(\textit{FPR}) &= 1 - \textit{Specificity} \\ &= \frac{FP}{TP + FN} \end{aligned}$$

With AUC (Area Under The Curve) ROC (Receiver Operating Characteristics) curve, the performance of a classification problem can be visualized. The ROC curve is a probability curve which considers all possible classification thresholds. AUC indicates how much a model can distinguish between classification labels. The higher the AUC, the better the model is at predicting true positives and true negatives. Values range from 0 to 1. 0.5 imply that the model performs no better than random guessing. The ROC curve is plotted with True Positive Rate (TPR) against False Positive Rate (FPR).

5 Discussion

The hyperparameters for the classifiers under consideration were selected based on the best mean AUC-ROC for the cross-validation.

Table 1 displays those mean AUC-ROC scores and standard deviation, along with the mean accuracy and standard deviation. Table 2 displays the AUC-ROC score when the predictions from the model have the mean accuracy. The sections below explore these results in more detail. Table 3 shows that altering the cut-off point of the predicted probability, in this case, 0.4, leads to an increase in Recall for both Ada Boost and Gradient Boosting. However, with this probability, accuracy increases for Gradient Boosting, but decreases for Ada Boost.

Table 1. Mean scores from 5-fold cross-validation.

Classifier	Mean accuracy	Mean recall	Mean AUC-ROC
Decision Tree	0.84 (±0.05)	0.67 (±0.42)	0.89 (±0.09)
Random Forest	0.85 (±0.08)	0.77 (±0.12)	0.91 (±0.09)
Extra Trees	0.85 (±0.03)	0.58 (±0.15)	0.92 (±0.06)
Ada Boost	0.89 (±0.09)	0.77 (±0.20)	0.93 (±0.05)
Gradient Boosting	0.79 (±0.05)	0.28 (±0.16)	0.90 (±0.06)

Table 2. Scores from cross-validated estimates.

Classifier	Accuracy	Recall	AUC-ROC
Decision Tree	0.84	0.71	0.88
Random Forest	0.85	0.76	0.88
Extra Trees	0.85	0.59	0.89
Ada Boost	0.89	0.76	0.93
Gradient Boosting	0.79	0.29	0.87

Table 3. Scores from Cross-validated estimates with probability threshold of 0.4.

Classifier	Accuracy	Recall
Decision Tree	0.82	0.71
Random Forest	0.84	0.88
Extra Trees	0.78	0.71
Ada Boost	0.51	1.00
Gradient Boosting	0.84	0.65

5.1 Decision Trees

A decision tree is usually trained by recursively splitting the data. Each split is chosen according to an information criterion which is maximised (or minimised) by one of the splitters. A decision tree is very easy to visualise.

The cross-validation mean accuracy for the Decision Tree Classifier model was 0.84 with a standard deviation of 0.05. The cross-validation AUC-ROC for the cross-validation prediction was 0.93 with a standard deviation of 0.05. The accuracy score was 0.84 with 49 of the 56 team passes and 12 of the 17 team failures being correctly predicted. The most important features were the weekly cumulative averages at week 3, 4, 9 and 11, with week 9 being the most important. The AUC-ROC for the model which gives the mean reported predicted accuracy was 0.88. (See Fig. 3).

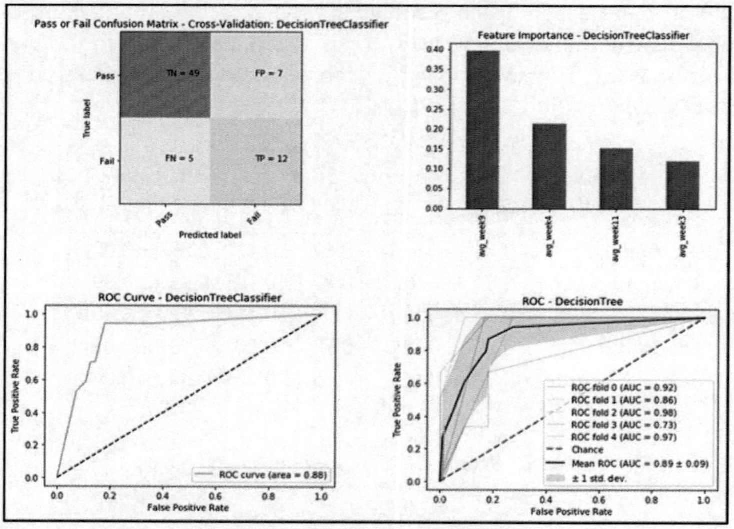

Fig. 3. Confusion matrix, important features, AUC-ROC scores for decision tree classifier.

This model performs well when considering both the individual cross-validation sets and the mean accuracy prediction. With a TPR of 0.71, there is no reason why this model should be rejected.

5.2 Random Forest

A random forest uses an ensemble approach to find the decision tree that best fits the training data. Many decision trees are created with predictions being averaged. Each decision tree is created using a random selection of features. A set of decision trees is a forest. Hence the term random forest.

In this case study, the cross-validation mean accuracy for the Random Forest Classifier model was 0.85 with a standard deviation of 0.08. The cross-validation AUC-ROC was 0.91 with a standard deviation of 0.09.

The accuracy score was 0.85 with 49 of the 56 team passes and 13 of the 17 team failures being correctly predicted. The most important features were the weekly cumulative averages at week 6, 7, 8, 9 and 11, with week 6 being the most important. The AUC-ROC for the model that gives the mean reported predicted accuracy was 0.89. (See Fig. 4).

This model performs well when considering both the individual cross-validation sets and the mean accuracy prediction. When compared to the Decision Tree Classifier, this model has the same AUC-ROC score for a higher accuracy. In addition, this model has a higher AUC-ROC mean with the same standard deviation. The Random Forest Classifier appears to be a better model than the Decision Tree model. Therefore, with a TPR of 0.76, there is no reason why this model should be rejected.

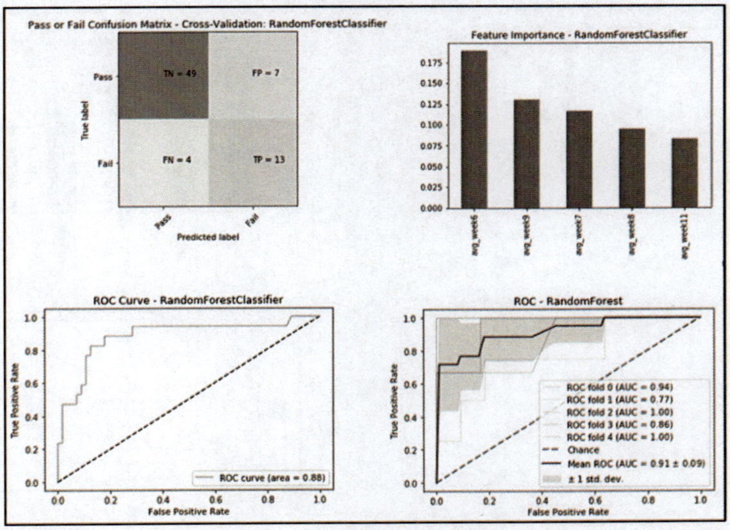

Fig. 4. Confusion matrix, important features, AUC-ROC scores for Random Forest Classifier.

5.3 Extra Trees

Extra Trees is short for extremely Randomised Trees. The main difference between random forests and extra trees is that in the random forest, a locally optimal feature/split is used, whereas in extra trees, for each feature under consideration, a random value is selected for the split.

This model gives similar cross-validated scores to the Random Forest Classifier, but for the cross-validated prediction, the Random Forest Classifier has a higher recall. This model is therefore rejected.

5.4 Ada Boost

Adaptive Boosting (Ada Boost1) is a popular boosting algorithm used for classification problems. A sequence of weak learners, e.g. small decision trees, are trained repeatedly on modified versions of the data. The predictions are then combined through a weighted majority vote to produce a final prediction.

In this case study, the cross-validation mean accuracy for the Ada Boost Classifier model was 0.89 with a standard deviation of 0.09. The cross-validation AUC-ROC was 0.93 with a standard deviation of 0.05.

The accuracy score was 0.89 with 52 of the 56 team passes and 13 of the 17 team failures being correctly predicted. The most important features were the weekly cumulative averages at week 1, 6, 3, 11 and 2, with week 1 being the most important. The AUC-ROC for the model that gives the mean reported predicted accuracy was 0.89. (See Fig. 5).

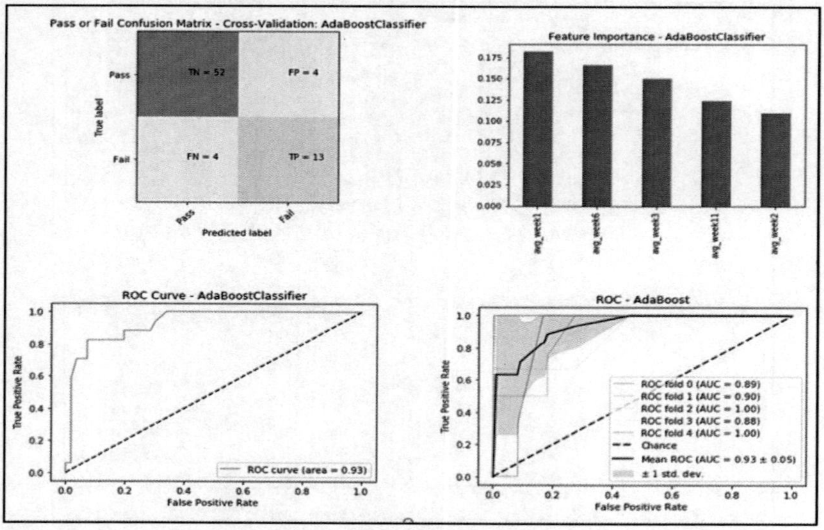

Fig. 5. Confusion matrix, important features, AUC-ROC scores for Ada Boost Classifier.

This model performs extremely well when considering both the individual cross-validation sets and the mean accuracy prediction. With a TPR of 0.76, there is no reason why this model should be rejected.

5.5 Gradient Boosting

Gradient Boosting is an Additive Model, i.e., it uses an iterative and sequential approach of adding the weak learners. Each iteration should reduce the value of a loss function.

In this case study, the cross-validation mean accuracy for the Gradient Boosting Classifier model was 0.79 with a standard deviation of 0.05. The average cross-validation AUC-ROC was 0.9 with a standard deviation of 0.06. Considering the mean accuracy and recall scores, this model does not perform well. There is therefore reason to reject this model.

6 Conclusion

This work answered the following research questions:

- How accurately can machine learning classification algorithms identify a team assignment result when using only cumulative average pushes to a software repository in a campus-based undergraduate course at a university?
- What are the important weeks when using only cumulative average pushes to a software repository in determining team assignment failure in a campus-based undergraduate course at a university?

The process started with the collection of data from University of Manchester. This data was then pre-processed to remove the teams which contained students who had opted out. The weekly cumulative averages were then calculated for each of the remaining teams. Five machine learning algorithms were chosen to explore the data with. They were Decision Trees, Random Forests, Extra Trees, Ada Boost and Gradient Boosting. The chosen evaluation metrics were accuracy, recall and AUC-ROC. As the dataset was small, a 5-fold cross-validation approach was used.

The 3 best machine learning classifiers were Decision Trees, Random Forests and Ada Boost:

- Decision Tree
  - Accuracy = 0.84
  - Recall = 0.71
  - AUC-ROC = 0.88
- Random Forest
  - Accuracy = 0.85
  - Recall = 0.76
  - AUC-ROC = 0.88
- Ada Boost
  - Accuracy = 0.89
  - Recall = 0.76
  - AUC-ROC = 0.93

There is no real consensus between the 3 selected models on the important weeks. The 2 most important features are 1) weeks 4 and 9 for the Decision Tree model, 2) weeks 6 and 9 for the Random Forests model and 3) weeks 1 and 6 for the Ada Boost model.

These models will be further evaluated using data from a future year as a test set.

Acknowledgements. This work was supported by the Institute of Coding, an initiative funded by the UK Office for Students (OfS). The Dataset for this work was provided by the University of Manchester.

References

1. Baker, R.S., Yacef, K.: The state of educational data mining in 2009: a review and future visions. J. Educ. Data Mining **1**(1), 3–17 (2009)
2. Chen, G., Rolim, V., Mello, R.F., Gašević, D.: Let's shine together! a comparative study between learning analytics and educational data mining. In: Proceedings of the Tenth International Conference on Learning Analytics and Knowledge, LAK 2020, pp. 544–553. Association for Computing Machinery, New York (2020)
3. Pedregosa, F., Varoquaux, G., Gramfort, A., Michel, V., Thirion, B., Grisel, O., Blondel, M., Prettenhofer, P., Weiss, R., Dubourg, V., Vanderplas, J., Passos, A., Cournapeau, D., Brucher, M., Perrot, M., Duchesnay, E.: Scikit-learn: Machine learning in python. J. Mach. Learn. Res. **12**, 2825–2830 (2011)
4. GitLab Homepage. https://about.gitlab.com/. Accessed 03 July 2020

5. Guerrero-Higueras, A., Matella´n-Olivera, V., Esteban-Costales, G., Fern´andezLlamas, C., Rodr´ıguez-Sedano, F., ´Angel, C.: Model for evaluating student performance through their interaction with version control systems. Learning Analytics Summer Institute (LASI), SNOLA (2018)
6. Hanley, J., McNeil, B.: A method of comparing the areas under receiver operating characteristic curves derived from the same cases. Radiology **148**(3), 839–843 (1983)
7. IEDMS Homepage. https://educationaldatamining.org/. Accessed 03 July 2020
8. Lubarsky, B.: Re-identification of "anonymized data". UCLA L. REV 1754(2010) (1701)
9. Rıos, J.C.C., Kopec-Harding, K., Eraslan, S., Page, C., Haines, R., Jay, C., Embury, S.M.: A methodology for using gitlab for software engineering learning analytics. CoRR abs/1903.06772 (2019)
10. Siemens, G.: Learning analytics: Envisioning a research discipline and a domain of practice. In: Proceedings of the 2nd International Conference on Learning Analytics and Knowledge, LAK 2012, pp. 4–8. Association for Computing Machinery, New York (2012)
11. SOLAR Homepage. https://www.solaresearch.org/. Accessed 03 July 2020

New Learning Models and Applications

Screening Executive Functions of Preschool Children via a Web Application

Nikolaos C. Zygouris[1(✉)], Kafenia Botsoglou[2], Antonios N. Dadaliaris[1], Georgios Dimitriou[1], Daniil Trontsios[1], Georgios I. Stamoulis[3], and Denis Vavougios[2]

[1] Department of Computer Science and Telecommunications, University of Thessaly, Lamia, Greece
nzygouris@uth.gr
[2] Department of Special Education, University of Thessaly, Volos, Greece
[3] Electrical and Computer Engineering Department, University of Thessaly, Volos, Greece

Abstract. The present research protocol describes the tasks of a web application that was constructed in order to assess the executive functions of preschool aged children. A total of 65 children four and five years old participated in this study. In order to screen executive functions, we used (1) a go/nogo task, (2) a visual working memory task (3) an auditory working memory task (4) a graph phoneme discrimination task (5) a phonological discrimination task and (6) a visual discrimination task. The results revealed that children at the age of four years old presented lower scores of correct answers and larger latencies in comparison to children at the age of five years. Also, females had more correct answers in the tasks of the web application in comparison to males. Additionally, the main purpose of delivering a screening procedure via internet is to make preschoolers do something useful while perceiving that they are "playing a game".

Keywords: Preschool children · Web application screener · Executive functions

1 Introduction

Infancy and early childhood are sensitive and rapid periods of brain growth that coincide with the emerge of nearly all cognitive, behavioral and social – emotional functions [1]. Throughout this period, the brain's eloquent networks are shaped and refined through the process that include myelination, dendritic arborisation and synaptogenesis and synaptic pruning [2]. These adaptive processes are modulated by neural activity and are responsive to environmental, genetic, hormonal and other influences [3]. The development and pattern of myelination follows a well described neuroanatomical arc progressing in a posterior to anterior and centre-outwards spatiotemporal pattern that corresponds to maturing cognitive functions [3]. There is a strong overlap in the emergence of a specific cognitive function and the myelination of brain regions and networks suspending that function [4]. In this line of research there

M. E. Auer and T. Rüütmann (Eds.): ICL 2020, AISC 1328, pp. 141–150, 2021.
https://doi.org/10.1007/978-3-030-68198-2_12

are several studies that present the importance of white matter and cortical myelination to cognitive development and brain plasticity [e.g. 5].

Furthermore, the first five years of life play a critical role in the development of children's executive functions, as they are adaptive in order to enable themselves to override more automatic or established thoughts and responses. Subsequently, infant changes occur both physically and cognitively. The brain of the infant grows faster than any other part of the body, and by the age of five, it weighs about 90% of the weight of the average adult's brain [6]. The two hemispheres of the brain begin to differentiate and their functions become specialized. Cerebral laterality is becoming increasingly clear during the preschool period. It is supported by the fact that the development of the brain as well as the level of myelin covering the brain neurons is related to the increased cognitive skills in preschool children. Executive functions have been strongly associated with the prefrontal and frontal cortex, which are the brain areas that present the slowest development. During the preschool period core components of executive functions develop in order to form a critical foundation that will set the stage for the development of higher cognitive processes well into later stages of life (such as adulthood) [6].

Despite the frequency with which executive functions are mentioned in neuropsychological research, they still lack a formal definition. The historical definition of "central executive" as it was presented by Lezak (1983) was described as the dimension of human behavior that deals with "how" behavior is expressed [7]. Nowadays, executive function refers to cognitive abilities associated with frontal lobe maturation, a complex construct involving a variety of higher order cognitive processes, such as inhibition, shifting attention, updating, fluency and planning. Inhibition refers to the ability to control a response or no response to a stimuli. Shifting is the ability to switch from one task to another. Updating refers to the skill to monitor and manipulate mental representations stored in working memory. Fluency is the ability to generate within a given amount of time words according to semantic categories or phonemes. Planning describes the skill to design, evaluate and select a sequence of thoughts in order to achieve a goal [8]. In general, executive functions mediate the ability to organize our thoughts in a goal – directed way and therefore are essential for success in school situations as well as everyday living [9].

Over the past 20 years, internet technology has offered opportunities to develop ongoing assessment systems which give a profile of children's learning strengths and weaknesses. The use of computers in the identification of children with learning disabilities is now well established in UK schools, with a number of different programs available [10, 11]. However, there is a lack of such screeners in Greece, with the exception of askisi [12, 13] a web delivered application screening procedure for identifying children at risk for reading disability in the Greek educational system, which assesses reading accuracy, fluency, comprehension, spelling and mathematical abilities.

The main aim of the present research protocol was to construct a battery of tests that can be delivered via internet in order to screen preschooler's executive functions. Moreover, another purpose of the study was to develop a web application screener that to the best of our knowledge does not exist. This kind of computer technology can be automatically deployed in any client that has a web browser, so it can be accessed, practically, from anywhere. It is inherent cross-platform, and in the event of any updates, there is no action that needs to be taken from the user's side. Furthermore a web application has a fast development cycle, meaning that it can be built and deployed in less time than an average desktop application, and in its purest form of development does not require proprietary software that locks it to a specific platform. Moreover, due to the huge community, emerging problems, requiring new functionalities, can be easily identified and dealt within a minimum amount of time.

The hypothesis of the present study was that Greek preschoolers four years old will present lower scores and higher time latency in comparison to participants five years old in all tasks of the web application screener. Furthermore, male participants will present different scores and altered time latency in comparison to female participants.

2 Method

2.1 Participants

A total of 65 children participated in this study. More specifically, the children recruited were 32 males four and five years old (M = 4.41 SD = 0.449) and 33 females four and five years old (M = 4.79 SD = 0.415). The two groups of the present study were formed by 26 four-year-old and 39 five-year-old preschoolers. All participants presented typical academic performance according to their teachers' ratings. Additionally, all children that participated in the study did not have a history of major medical illness, psychiatric disorder, developmental disorder or significant visual or auditory impairments, according to their medical reports of their kindergartens.

2.2 Material and Procedure

Inhibition, auditory and visual working memory, phonological discrimination, graph phoneme discrimination and visual discrimination of all participants were examined using a battery of six tasks. Before the main test procedure, children performed one training task in order to become familiar with the testing procedure. In the training task, children had to click a picture in order to use the mouse and get used to the process. During the main test, children had: (a) A go/no go task. Children had to select the target picture over a number of five pictures that were presented randomly. (b) A visual working memory task of sequences consisting of 22 sequences of numbers. The first sequence included three numbers; the second four numbers; the third five numbers; the fourth six numbers; the fifth seven numbers; the sixth eight numbers. Participants were asked to report the numbers with the use of a 0–9 num pad that was displayed. It is worth to mention that if children could not remember two series in a row or three series in general, the test was stopped (Fig. 1 upper part) (c) An auditory working memory task of sequences that was similar to the visual memory task, but the number sequences were given in auditory (Fig. 1 bottom part). (d) Graph phoneme discrimination. Children had to select out of three graphemes the phoneme that was auditory delivered (Fig. 2 upper part) (e) Phonological discrimination. Children had to listen a word and decide between two words which was the correct. It must be highlighted that the words presented auditory were commonly used and consisted of two syllables (Fig. 2 bottom part). (f) A visual discrimination task. This assessment was made up out of a series of diagrams or patterns with a part missing and children were expected to select the correct part to complete the designs from a number of options (Fig. 3).

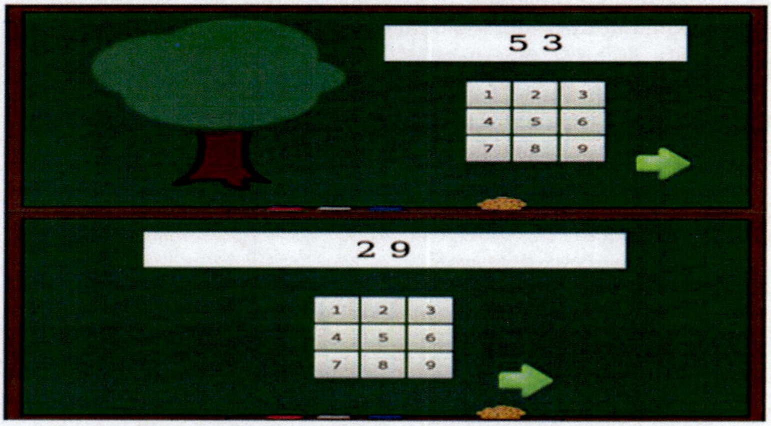

Fig. 1. Visual (upper part) and auditory (bottom part) working memory task.

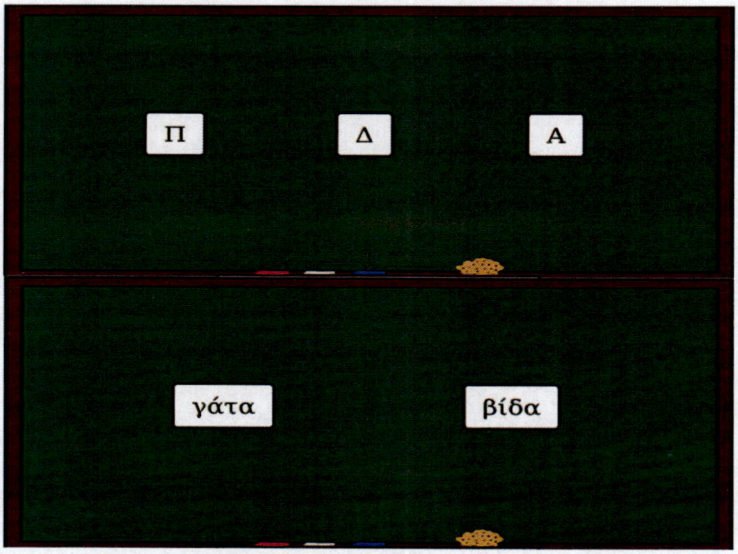

Fig. 2. Select the grapheme that fit to the phoneme (upper part) and phonological discrimination task (bottom part)

Fig. 3. Pattern with a missing part in which children have to choose the correct from the three that are presented on the right.

2.3 Implementation

In our effort to enhance the aforementioned procedure and make it accessible to a larger ensemble of students in the future, our application requires constant communication between a client and the main server that hosts it. Every time a test is finished, the user's results are saved for future inspection. Based on this, we chose to use a minimum set of basic, timeless and swift, concerning web development, technologies taking into account the reuse of the code and our ability to maintain and expand functionality to suit any future needs. More specifically we used HTML5, CSS, JavaScript and MySQL.

HTML5, which is the core technology markup language of the World Wide Web, was used for the structuring of each page and the appropriate presentation of the content. HTML5 is the immediate successor of HTML and incorporates a plethora of features that are required for the implementation of a web platform as the one we present. It supports the latest multimedia codecs, error handling so that old browsers can ignore any new HTML5 constructs, cross-document messaging for implementing communication between documents across different domains and DOM storage for storing data in the web browser, either permanently or for the duration of each session.

CSS was used for the beautification of each available web page. The graphical representation of each test is of outmost importance. The user, at any given time must be undistracted in his/her effort to complete the tests. Particular emphasis was placed in the use of a color palette, which changes depending on the sex of the user, that contain a few subtle color changes to avoid making the test a tedious ordeal.

Application functionality was implemented using the JavaScript programming language. JavaScript is a dynamic programming language and is a key component of modern browsers, which in turn have the ability of implementing client-side scripts that can interact with each user's particular choices and change the document content while it is displayed. The communication between client and server was handled by Java servlets that run over the HTTP protocol (HTTP servlets).

Moreover, a database was created using MySQL. Through this database, the examiner has the ability to check and process past results for each student, come to conclusions based on a larger scale of results and dynamically compute averages concerning every trial that is contained in the overall process.

3 Results

Descriptive statistics were performed in order to obtain mean scores and standard deviations of participants in all six tasks, as the training test was excluded from the statistical analysis. Analysis of Variance (ANOVA) was conducted to compare the scores of children according to their age group. ANOVA revealed that four-year-old children had statistically significant ($p < 0.01$) lower mean scores of correct answers in five out of six tasks compared to five-year-old participants. Table 1 presents the mean scores and standard deviations for the correct responses on the six tasks. Moreover, Table 1 presents the statistical significance of the correct answers of children according to their age.

Table 1. Mean scores and standard deviations of children's correct responses and statistical significance in all 6 tasks

	Groups				
	4 years old		5 years old		
Tasks	Mean	SD	Mean	SD	F
go/no go task	2.00	1.14	2.12	1.17	0.155
Visual working memory task	1.04	0.82	1.77	0.71	14.626**
Auditory working memory task	2.85	1.66	4.05	0.61	17.139**
Graph phoneme discrimination	4.73	1.01	5.46	0.76	11.213**
Phonological discrimination	5.96	1.15	7.33	1.33	18.564**
Visual discrimination task	4.12	0.43	5.26	0.91	35.455**

Note **$p < 0.01$

The same statistical analysis was conducted in order to compare correct answers of children according to their gender. ANOVA revealed that in all six tasks, females had more correct answers in comparison to males. However, females presented statistical significant more correct answers in three out six tasks. Table 2 presents the mean scores and standard deviations for the correct responses on the six tasks. Moreover, Table 2 presents the statistical significance of the correct answers of children according to their gender.

Table 2. Mean scores and standard deviations of children's correct responses and statistical significance in all 6 tasks

	Group				
	Male		Female		
Tasks	Mean	SD	Mean	SD	F
go/no go task	1.79	1.32	2.31	0.90	3.504
Visual working memory task	1.24	0.80	1.72	0.81	5.730*
Auditory working memory task	3.28	1.42	3.85	1.09	3.270
Graph phoneme discrimination	4.69	0.70	5.64	0.90	22.747**
Phonological discrimination	6.09	1.06	7.45	1.42	19.164**
Visual discrimination task	4.78	0.83	4.82	1.04	0.24

Note *p < 0.05**p < 0.01

The same statistical analysis was performed in order to compare four and five years old children according to the time needed in order to complete the six tasks. ANOVA revealed that only in visual discrimination task was found statistical significance as five years old children presented lower latency ($p < 0.01$) in comparison to four years old.

Next, the analysis focused on latency according to the gender. ANOVA revealed that females presented lower latency ($p < 0.01$) in order to complete graph phoneme discrimination task and visual discrimination task in comparison to males.

4 Discussion

The web application screener that was used in the present protocol revealed significant differences ($p < 0.01$) in five out of six tasks that were constructed according to the preschooler's age and in all six task ($p < 0.05$) according to their gender. Four-year-old compared to five-year-old children were found to differ substantially in five out of six executive function abilities, result which partly verifies our first research hypothesis. Furthermore, in time latency children differ statistically significantly in visual discrimination task, as five years old infants needed less time to complete the task in comparison to four years old. Those findings support the suggestion that myelination of several brain regions and networks that plays important role in attention and memory is completed at the age of five [14]. Additionally, it was found that the mean scores of correct answers between male and female preschoolers were statistically significant, as male presented lower scores of correct answers in comparison to female, which verifies our second research hypothesis. Moreover, females needed statistically significantly less time in order to complete graph phoneme discrimination task and visual discrimination task in comparison to males. It is suggested that cerebral laterality [14] grows faster in females in comparison to males. Subsequently, female preschoolers present better scores in several executive functions such as the development of language skills. It must be reminded that in auditory and visual working memory tasks, time was not measured, so it was excluded from this analysis. It must be highlighted that another important component of the present study was that all children used a web application screener, which is not used by any other study to our knowledge, suggesting that the internet can be a tool in order to screen cognitive abilities in kindergarten children.

5 Conclusion

There is a significant increase in the use of Computer-Based Neuropsychological Assessment, in clinical diagnosis practice in different specialties, such as neuroscience and cognition dyslexia screening, sports-related concussion, human computer biological signals interaction, neurologic patients. Since Computer-Based Neuropsychological Assessment offers several advantages when applied to different clinical disorders in neuroscience, it is possible to accurately control variables for measuring cognitive functions, such as latency, correct and incorrect responses, error types, and direct stimuli administration. Computer-Based Neuropsychological Assessment brings a great advantage for clinicians to dynamically manage assessment sessions, reducing the risk of errors related to sequence of tasks during an evaluation. Furthermore, it significantly simplifies quantitative analysis of the outcome results, by automatically calculating the desired data with high definition and data grouping, according to neuropsychological clinical models previously established. Administration of assessment procedures in neuropsychology in general, requires at least one measure which needs to be computed, scored or counted [for review 15]. Moreover, the main benefits of computer based screeners [16] is that they can capture and engage the interest of the user. If they are properly constructed and presented, they can help to minimize the user's frustration and loss of dignity when working on tasks once accomplished with ease. Additionally, the context of learning to use the computer can provide the user with an experience of mastery and a sense of control. The computer can measure multiple dimensions of performance (latency) at levels that are not possible for the clinician, and also provides an automated data collection and storage that can free the clinician to focus more on the treatment. Lastly, the computer is efficient at performing tasks that would otherwise require extensive setup and/or preparation time. Early screening of executive functions it is usually associated with primary – level intervention in order to address any difficulties. In more details early detection generally refers to a screening assessment that may also be carried out in the school context by a specially trained teacher in order to achieve overall better school performance and increase the sense of a positive academic experience [17].

The web application paradigm aims to engage users into an activity, which produces a common good or teaches something valuable to the "player", concealing it into a "game". The idea is that the child does something useful, which normally would not, because enjoys doing it. If the child has fun, will probably continue to "play", achieving the serious tasks of the web application [18]. It is easy to see that using the internet in order to deliver a neuropsychological screening application is extremely useful for children, because the "game" can lure players into performing the tasks with accuracy and caution. Since the web application designed and used in this research protocol is specifically intended for pre-school children, there was payed particular attention to the interface and how this can be effective in engaging children [19].

In conclusion, the web application screener for screening preschooler's executive functions that we used in this study, was found to be a feasible instrument to provide first-pass screening services and referral. However, standardization on a large-scale sample representative of the general population is necessary before widespread adoption. In addition, strict psychometric evaluation of the screening procedure is a necessary condition, before its widespread adaptation can be recommended.

References

1. Johnson, M.H.: Functional brain development in humans. Nat. Rev. Neurosci. **2**(7), 475–483 (2001)
2. Deoni, S., Dean, D., III., Joelson, S., O'Regan, J., Schneider, N.: Early nutrition influences developmental myelination and cognition in infants and young children. Neuroimage **178**, 649–659 (2018)
3. Stiles, J., Jernigan, T.L.: The basics of brain development. Neuropsychol. Rev. **20**(4), 327–348 (2010)
4. Fornari, E., Knyazeva, M.G., Meuli, R., Maeder, P.: Myelination shapes functional activity in the developing brain. Neuroimage **38**(3), 511–518 (2007)
5. Dai, X., Hadjipantelis, P., Wang, J.L., Deoni, S.C., Müller, H.G.: Longitudinal associations between white matter maturation and cognitive development across early childhood. Hum. Brain Mapp. **40**(14), 4130–4145 (2019)
6. Garon, N., Bryson, S.E., Smith, I.M.: Executive function in preschoolers: a review using an integrative framework. Psychol. Bull. **134**(1), 31 (2008)
7. Lezak, M.D.: Neuropsychological Assessment, 2nd edn. Oxford University Press, New York (1983)
8. Spencer, M., Cutting, L.E.: Relations among executive function, decoding, and reading comprehension: an investigation of sex differences. Discourse Processes, 1–18 (2020)
9. Jurado, M.B., Rosselli, M.: The elusive nature of executive functions: a review of our current understanding. Neuropsychol. Rev. **17**(3), 213–233 (2007)
10. Turner, M., Smith, P.: Dyslexia Screener. GL Assessment, London (2003)
11. Singleton, C.: Using computer-based assessment to identify learning problems. ICT and Special Educational Needs: a Tool for Inclusion, pp. 46–63 (2004)
12. Zygouris, N.C., Vlachos, F., Dadaliaris, A.N., Oikonomou, P., Stamoulis, G.I., Vavougios, D., Striftou, A.: The implementation of a web application for screening children with dyslexia. In International Conference on Interactive Collaborative Learning, pp. 415–423. Springer, Cham, September 2016
13. Zygouris, N.C., Vlachos, F., Dadaliaris, A.N., Oikonomou, P., Stamoulis, G.I., Vavougios, D., Striftou, A.: A neuropsychological approach of developmental dyscalculia and a screening test via a web application. Int. J. Eng. Pedagogy (iJEP) **7**(4), 51–65 (2017)
14. Feldman, R.S.: Development across the life span. Pearson (2016)
15. Galindo-Aldana, G., Meza-Kubo, V., Castillo-Medina, G., Ledesma-Amaya, I., Galarza-Del-Angel, J., Padilla-López, A., Morán, A.L.: Computer-based neuropsychological assessment: a validation of structured examination of executive functions and emotion. In: International Conference on Engineering Psychology and Cognitive Ergonomics, pp. 306–316. Springer, Cham, July 2018
16. Brookes, G., Ng, V., Lim, B.H., Tan, W.P., Lukito, N.: The computerised-based Lucid Rapid Dyslexia Screening for the identification of children at risk of dyslexia: a Singapore study. Educ. Child Psychol. **28**(2), 33 (2011)

17. Kokkalia, G., Drigas, A., Economou, A., Roussos, P.: Screening tools for kindergarten children. Int. J. Recent Contrib. Eng. Sci. IT (iJES) **5**(4), 76–87 (2017)
18. Zygouris, N.C., Vlachos, F., Dadaliaris, A.N., Karagos, E., Oikonomou, P., Striftou, A. Stamoulis, G.I.: New Tasks for a Dyslexia Screening Web Application. In: International Conference on Interactive Collaborative Learning, pp. 263–271. Springer, Cham, September 2018
19. Gaggi, O., Palazzi, C.E., Ciman, M., Galiazzo, G., Franceschini, S., Ruffino, M., Facoetti, A.: Serious games for early identification of developmental dyslexia. Comput. Entertainment (CIE) **15**(2), 1–24 (2017)

Project-Based Activities as Methodology for Developing Informational and Legal Competences in Future Engineers

Vladimir V. Nasonkin[1], Anna E. Serezhkina[2], Svetlana V. Barabanova[2](✉), Maria S. Suntsova[2], and Nataliya V. Nikonova[2]

[1] Russian Peoples' Friendship University, Moscow, Russia
sveba@inbox.ru
[2] Kazan National Research Technological University, Kazan, Russia

Abstract. One of the tasks to be solved by engineering education is to form the universal competences of students, which allow them to efficiently implement their knowledge and skills in their occupations afterwards [1]. Nowadays, higher education cannot be limited to classical educational methods or conventional teaching techniques. All the so-called Bologna documents emphasize that education quality assurance is a necessary condition of creating the European education zone, focus being moved from expenses to the outcomes, i.e., knowledge and competences. This, in turn, suggests stimulating the diversity of educational opportunities in terms of goals, profiles, contents, and methods. Contemporary students represent Generation Z, the beau monde society. All necessary information for them must be "moved" into the Internet. This requires some new approaches to teaching and learning. Therefore, any educational technologies must include using information technology and all its benefits today.

In Kazan National Research Technological University, Russia, the Department of Law has developed original tasks to be solved at the lessons, involving all those interested and using project-based learning methods [2, 3]. In this context, it becomes possible to develop informational and legal competences in future engineers, focusing on practice. Project-based learning allows students to get to know more about the opportunities of informational and legal support for their vocational activities. Tasks underlying the project aimed at developing informational and legal competence in future engineers are universal and can be easily modified to adapt to the major area of studies. Draft program aimed at developing informational and legal competence can be used at any engineering university, as well as within the additional professional education system.

Keywords: Informational and legal competence · Project-based learning · Engineer · Engineering activities

M. E. Auer and T. Rüütmann (Eds.): ICL 2020, AISC 1328, pp. 151–161, 2021.
https://doi.org/10.1007/978-3-030-68198-2_13

1 Context

Contemporary employers are proactively involved in discussing the matters of training future professionals and advance reasonable demands for graduates. A company would like to take on board new-generation engineers that are ready to work within the context of Industry 4.0. A future engineer is required to be prepared for performing comprehensive engineering activities, including designing and solving complex engineering problems in digitalized and multicultural environments [4]. Experience in the legal training of future engineers has proven that it is often limited to conventional teaching methods that do not allow developing the necessary competences for their successful employment and efficient vocational activities [5, 6]. The existing vocational training system at a technical university includes, to a greater extent, the general concepts of law, as well as some aspects of law in vocational activities, such as labor law, administrative law, environmental law, etc., and in general life activities of a student, such as family law, housing law, etc. Developing the digital economics imposes some obligations upon engineering education. Particularly, it is necessary to switch the training to developing the informational and legal competences of students.

This paper deals with reviewing the content of concepts, such as "project-based activities" and "informational and legal competences." We have analyzed questionnaires filled out by students. A draft program is proposed aimed at developing informational and legal competences in students majoring in petrochemistry.

2 Purpose or Goal

In the context of the global digitalization of society, the requirements are being changed regarding the structure of professional competences. One of the key elements for ensuring the competitiveness of a company taken as a whole and of an individual employee, is the active use of information technologies in their vocational activities [6, 7]. In our opinion, forming the high-level digital culture is integral to the increasingly stringent requirements for the legal literacy of a future professional and for other competences that are essential in his or her vocational activities. Understanding the above and updating the approaches to educational activities are extremely important to the contemporary universities. One of the valid methods of developing the necessary competences is project-based learning.

Topicality of implementing the project is determined by the fact that there is a growing contradiction between the social need for the high-level informational and legal competences of engineers and the incompleteness of its being developed. Percentage of companies increases, which both understand the values of project management and actively use those values in their activities. Concept of Long-Term Socio-Economic Development of the Russian Federation until 2020 notes that the modern system of education must include project-based activities in training students, based on the principles, such as openness of education to external demands, using project-based methods, competitive identification and support for leaders successfully implementing new approaches in practice, and targeting of resource-support tools and the comprehensive nature of decisions made [8].

New standards of engineering education in Russian universities include in preparing graduates for their vocational activities the tasks of the following types: Engineering task, managerial task, research task, and project task. Project-based activities in a university mean a specially organized independent activities of students, aimed at solving an important problem, the result of which is represented as an end product, technology, or services that can be used in real practical activities.

According to experts that have implemented project-based method in educational process, the following requirements shall be set for a student project [2; 3]:

- Projecting must proceed from the problem, i.e., its importance, topicality, and being in demand;
- Implementation of the full project life cycle: From concept (hypothesis) through operation (use of the knowledge acquired);
- Solution originality: Search for the uniqueness of a given project;
- Involvement in professional society: Project outcome must comply with the real requirements of professional society;
- Independency in the project implementation;
- Considering the limits of resources, such as time constraints, finances, etc.;
- Consciousness in choosing organizational solutions, i.e., individuality/teamwork, distribution of roles, and identifying obstacles and ways to overcome them; and
- Focus on educational results to be discussed by participants.

Constructing the engineering education system in the CDIO approach [9, 10] is based on the lifecycle stages of any engineering project: Conceive-Design-Implement-Operate. Training results are described in four stages:

- Technical knowledge and thinking;
- Personal and professional competences;
- Interpersonal skills, working and communicating in a team; and
- Conceiving, designing, implementing and operating the systems in company and in society.

Project-based activities can and must be implemented in a university as both as a project of an individual department and as a result delivered by the representatives of several departments, considering the needs of potential stakeholders.

3 Approach

Russian educational standards are aimed at forming universal competences, such as operational and critical thinking, development and implementation of projects, teamwork and leadership, self-organization and self-development, etc. Development of universal competencies, along with general professional and specialized professional competencies, will ensure the demand for students on the labor market. General professional competencies of experts in oil and gas upstream, midstream and downstream processes (our study involved students majoring in this area) include applications of fundamental knowledge, engineering design, cognitive management, usage of tools and equipment, research, decision making, and usage of applied knowledge.

Currently, the educational system has a small share of integrated projects in study process. This is associate with the lack of professionals capable of ensuring the development of project-based technologies and with no emphasis on real vocational activities. Applying interdisciplinarity to project implementation will allow making the students' knowledge relevant to the current manufacturing requirements and technological advances. It will also strengthen the students' motivation for their future occupation and contribute to acquiring the efficient methods of organizing project-based activities and team interactions.

According to Klaus Schwab [11], the fourth industrial revolution is notable for synergetic effect that occurs resulting from fusing different technologies, such as computer technology, IT, nanotechnology, biotechnology, etc. Therefore, blurring of distinction is forecasted among physical, digital/informational, and biological worlds.

By the projections of scientists, digitalization may cause the disappearance of 9–50% of all existing occupations within the nearest decade. Ayah Bdeir, the graduate of Media Lab at the Massachusetts Institute of Technology, states that 65% of occupations that will be demanded for in 2035 are not even existing now. For example, an expert in legal regulation of issues related to using artificial intelligence and remote-piloted vehicles, a veterinarian for robots, etc. [12].

In the context of global digitalization of society, the requirements for the composition of professional competences change significantly. One of the conditions ensuring the competitiveness of a company or a single employee is the active use of information technologies in their vocational activities. Findings of many studies show that one of the core competences is the ability to work with large amounts of information. Thus, the Target Competence Model-2025 [13] includes 3 blocks: Cognitive skills, social and behavioral skills, and digital skills. The latter ones, among other things, imply creating systems, such as programming, developing applications, designing production systems, and data processing and analysis. Analysis of sources, such as *RBC, PwC, McKinsey, Bloomberg, and The Economist*, has shown that 60% of all occupations will have been automated by 2030; therefore, 375 million people will have to learn new skills [13].

An engineering university/college student today is an engineer at a modern enterprise tomorrow, performing a complex of interrelated manufacturing, processing, designing and research tasks. And the day after tomorrow, he/she is a potential manager destined to head a manufacture in the Industry 4.0 conception. Therefore, forming a high-level digital culture and legal literacy in future professionals, as well as other key competences of their vocational activities appears to be especially important in the educational process at a university. One of the ways to develop necessary competences is the project-based method.

According to Wellingtone polls, about 20% companies use project management software products [14], while almost 80% of highly efficient projects are implemented using project management software products. In the Capterra report, the five most popular and highly demanded functions of project management software products are the following: File sharing, time tracking, e-mail integration, Gantt charts, and budget management [15].

Currently, both the informational and the legal component of the task to be solved appear to be relevant. This means knowing the regulatory and legal aspects of the matter in general and the ability to work with electronic workflows, data storage and protection using advanced digital technology and resources. Learning legal basics by the students of engineering universities must be considered as a condition of their market value as professionals.

Informational and legal competence, in our opinion, shall mean an integral characteristic of a person, including legal knowledge and the ability to apply regulatory documents to their professional activities, as well as their readiness to use advanced tools in working with information and reference legal systems and communication resources in the area of using laws.

We propose using the project-based method in the educational process of engineering universities, since this method allows acquiring knowledge within their studies and expected vocational activities. Project-based method is an independent pedagogical technology and, in our opinion, has some advantages, since it is this method that, first, allows acquiring or developing the skills of shared activities and, second, its universal nature enables forming the legal competences in students majoring in many different areas.

The program was introduced into the training process in several stages within one semester. At the preparatory stage, we studied the situation of how the information and legal competences have been developed in engineering students. For this purpose, at the first lesson, we offered our students to answer some questions regarding the legal aspects of their everyday life and in the nearest future. We also interacted with those interested to form the groups of students. Based on the findings, we selected tasks for the project. The main stage of the program consisted in implementing the project itself at practical lessons and independently, in groups. At the final stage, we repeated our poll to obtain feedback from the students and assess the efficiency of implementing this form of the project-based activities in the educational process.

4 Actual or Anticipated Outcomes

To identify the formed informational and legal competences of students in the context of their future vocational activities, we took a poll of students majoring in Oil and Gas (Table 1). 48 students participated in the poll. They were polled twice: Before and after the implementation of project-based learning.

Table 1. Results of polling the students majoring in Oil and Gas

	Project start, %	Project end, %
Legal literacy level	28	56
Lack of theoretical knowledge	52	38
Lack of practical skills	68	42
Desire to know more about their rights	76	92
Ability to defend themselves legally	20	38
Attitude to legal matters		
Intricacy of legal rules	74	56
Rapidly changing legislation	84	56
Legal disarray in the country	10	8

(continued)

Table 1. (*continued*)

	Project start, %	Project end, %
Low level of legal culture in companies	78	68
Only humanities colleges and universities should study law	56	24
Legal knowledge in the context of their future professional activities	24	36
Legal knowledge may be useful in their vocational activities	56	84
Legal issues occur in their work/studies	38	52
Ability to find and make managerial decisions in labor organizing and rating	4	36
Knowledge of regulatory documents regarding product quality, standardization, and certification	12	56
Knowledge of legal regulations in oil and gas	10	56
Knowledge of legal reference systems	44	100
"Garant"	48	64
Consultant Plus	40	70
"Kodeks"	2	2
Other legal reference systems	2	0
Information literacy	92	88
Ability to work with electronic document exchange	86	82
Ability to store and protect data using modern digital technology and resources	72	74
Knowledge of intellectual property protection	56	68
Knowledge of how to use digital data	56	74
Ability to design the team activities when solving information problems in their vocational area	48	62
Readiness to make correct decisions and efficiently solve information problems in their vocational area	88	84
Readiness to design their activities in solving specific professional problems using information systems and technologies	78	76
*Would you like to enhance your level of information and legal culture?**	*74*	
Civil law	*64*	
Labour law	*56*	
Criminal law	*40*	
Environmental law	*22*	
Constitutional law	*28*	
Administrative law	*24*	
Energy law	*24*	
Engineering law	*28*	
Copyright	*32*	
Project law	*22*	
International law	*12*	
Software law	*12*	

* polling at the beginning only

The poll has shown that, prior to start working on the project, the students had a low-level development of informational and legal competence. Their estimates of informational component were slightly higher than those of legal one. Students think they are more competent in IT matters, but not competent in legal issues. Most students had a consensus that legal disciplines must be taught in accessible and understandable terms (88%). 80% of respondents think that the Internet must make it possible to use available open information. Students expressed their wish to further use reference legal systems in their vocational activities, speaking of their universality and practicability. Future engineers must be able to use in their practice the basics of law and the regulatory documents regarding the quality, standardization, and certification of products and goods (12%), find managerial solutions in labor organizing and rating (4%), etc.

To develop in students informational and legal competences, we propose to use project-based activities. The training project developed is easily reproducible by any other interested educators, universal in terms of acquiring legal knowledge from different areas of law, it relies on interdisciplinary relationships, efficiently uses advanced information technologies, etc.

According to the results of polling the students and based on our many-year experience [16], we propose the *Draft Program for Developing Informational and Legal Competences in Future Engineers.* The program is aimed at ensuring the informational and legal support for the activities of engineers. While working on a project, students get to know about reference legal systems, i.e., online databases containing regulatory documents, judicial precedents, article-by-article comments, professional law journals, and other professional law literature.

Draft Program: Developing Informational and Legal Competences in Future Engineers

The program proposed uses a role-based (training) project pre-defining the roles of its participants and the rules of their interrelations. Prior to implementing the project, we offer to interact with the colleagues from the faculties of humanities, who teach psychological and pedagogical disciplines. Each student has already had the results of psychological testing, and the humanities teacher may provide a psychological profile of the student and identify their personality type. This, in turn, can be used in forming the project teams, distributing their roles, etc.

The draft program is built up based on the principle of sequential informational and legal skill training, using the tasks preplanned by the teacher. The process of developing the tasks for the role-based project involves the university faculty members and the managers and professionals from petrochemical companies. Since our university is one of the PJSC Gazprom flagman universities, we can always discuss the topical issues of oil & gas industry with manufacturers both at lessons and at specialized conferences. It is possible that different teams consider one problem using different branches of law.

Among the methods, we used both the advanced software products and information technologies used in project-based activities and conventional methods forming various competences, such as business simulation; case studies; brain storming; discussing problems; and exchanging experiences. We chose the Consultant Plus software product as the basic reference legal system. Advantages of this product for using it in the project: Its resources are free-of-charge in the evening and at weekends. This is convenient to our students that do not need to purchase or install the licensed software on their PCs, and therefore, they may use documents online. During academic hours, the university resources are available to students.

Goal of the Program: Understanding and using informational and legal knowledge and the formed competences of students, allowing future oil & gas professionals to optimally solve the problems on developing the projects of oil & gas facilities and manufactures.

Program tasks:

- Planning the project activities, considering inputs;
- Distributing the roles in the project team;
- Prioritizing the tasks and solutions, considering the moral aspects of activities;
- Identifying the criteria of achieving the goals;
- Forecasting the consequences;
- Finding compromise solutions;
- Considering the concerns of everyone interested in the project;
- Forecasting the project economics;
- Defining technical specifications and standards;
- Using IT;
- Developing communicative skills; and
- Presenting the project.

Form of organization – groups/teams.
Program participants: KNRTU students majoring in Oil & Gas.
Project implementation stages:

- Organizational Stage: forming teams, defining the project goal and tasks, resources, etc.;
- Informational and Processing (Core) Stage: Developing informational and legal competences through optimally solving engineering problems;
- Project Presentation Stage: Presentations of groups and individual interactive presentations; and
- Reflection Stage: Summarizing the outcomes and formulating the conclusions.

Let us consider a fragment of the role-based project (Table 2) presenting the technologies aimed at developing the informational and legal competences and at understanding their relevancy in their further vocational activities. As an example, let us take the problem of delivering and transporting petrochemicals. Students can be offered to work on real implemented projects of PJSC Gazprom, such as the Turkish Stream, the Power of Siberia, etc., information on which is publicly available.

Table 2. Fragment of the draft program: Developing Informational & Legal Competences in Future Engineers

Project tasks	Legal knowledge and skills	References	Pedagogical technology
Analysis of legal issues regarding deliveries and transportation of petrochemicals	All branches of law	https://www.gazpromexport.com/projects/transportation/	Discussion, debates on the problem, mini lecture
Search for and analysis of regulatory acts governing the deliveries and transportation of petrochemicals	Administrative law	Legal Information System “Consultant Plus”	Case studies
Studying the special regulatory approvals in oil and gas, such as rules on labor safety in oil product storage, transportation, and sales	Labor law	Rules on Labor Safety in Oil & Gas Industry/Legal Information System “Consultant Plus”	Business simulation, mini lecture
Executing contracts for deliveries and transportation of petrochemicals, adhesion contracts, etc	Administrative and international law	Legal Information System “Consultant Plus”	Business simulation, mini lecture
Types of liabilities for administrative infractions in the areas of subsoil use, environment, industrial safety, etc	Civil law	Legal Information System “Consultant Plus”	Brain storming, mini lecture
Studying the special aspects of hazardous employment	Labor law	Rules on Labor Safety in Oil & Gas Industry/Legal Information System “Consultant Plus”	Discussion, debates on the problem
Search for environmental legal requirements for transportation and storage of oil and gas	Environmental law	Rules on Labor Safety in Oil & Gas Industry/Legal Information System “Consultant Plus”	Brain storming
Considering the types of criminal and administrative punishments for infringements in storing, transporting, and selling oil products	Criminal, administrative, and environmental law	Legal Information System “Consultant Plus”	Case studies, mini lecture
Studying the issues of forming the financial strategy of the state, depending on the availability of oil on its territory	Financial and tax law	https://pravo.gov.ru/	Business simulation, mini lecture
Studying the “tax maneuver” in the oil & gas industry	Financial and tax law	https://pravo.gov.ru/	Discussion

Results of polling the students of Kazan National Research Technological University suggest obvious advantages of using the project-based method in teaching legal disciplines, as compared to classical methods. Many students decided that legal problems might arise in their future vocational activities, which they could solve independently. There is now a higher percentage of students that can find legal information using reference legal systems and want to know more about their rights. Illusion that only humanities students must study law has now become weaker.

Resulting from forming their informational and legal competences, would-be engineers must be able to use in their practical activities the basics of legal knowledge, regulatory documents on oil and gas product quality, standardization and certification, and find managerial solutions in arranging the transportation and storage of petrochemicals, etc., using reference legal systems. The project of forming informational and legal competences of future engineers includes the contents of tasks, methods, and technologies of forming such competences in students.

Along the main goal of the project, the students implementing it also acquire soft skills, i.e., abilities to plan a project considering inputs; distribute the roles in the project team; prioritize problem solutions considering the moral aspects of their activities; define the criteria of achieving the goals; forecast consequences; find compromise solutions; consider the concerns of everyone interested in the project; forecast the project economics; decide on technical specifications and standards; use information technology; develop communicative skills; and present their project. The draft program proposed helps a student enhance motivation for getting legal information, promotes integration of vocational and informational-legal knowledge, and ensures the transition from reproductive to productive method of knowledge acquisition.

5 Conclusions/Recommendations/Summary

The training project developed is easily reproducible by any other interested educators, universal in terms of acquiring legal knowledge from different areas of law, it relies on interdisciplinary relationships, efficiently uses advanced information technologies, etc. We think it would be optimal to study this course during two semesters.

Combining conventional training methods and reference legal systems in teaching legal disciplines contributes to enhancing the quality of knowledge acquisition and ensures the relationship of the knowledge acquired with the processes and phenomena of their future engineering activities. Integration of the elements of global implemented projects in the training process enhances both motivation for receiving ready knowledge from the teacher and motivation for independently searching for and analyzing information, as well as comparing their own data and conclusions to the law.

Awareness of the value of law in respect to their future occupation will be higher, if theoretical legal provisions are extended to using reference legal systems as project-based learning, where students get to know about the practical applications of knowledge acquired. The pool problem-based tasks is extended due to interactions with the representatives of manufacturing companies.

The authors are planning to further develop the projects under the Legal Support for Engineering Activities program, considering the specificity of their future occupation in the context of Industry 4.0.

References

1. Osipov, P.N.: Training competitive specialists as the priority of modern education. In: 2013 International Conference on Interactive Collaborative Learning, ICL 2013 2013, pp. 251–254 (2013)
2. Kilpatric, W.H.: The Project Method//Teachers College Record.-1918.-19 September, pp. 319–334
3. Proyektnoye obucheniye. Praktiki vnedreniya v universitetakh [Project-Based Learning. Best Practices in Universities]. In: Evstratov, E.A., et al. (eds.) M.: Open University Skolkovo, p. 152 (2018). (in Russian)
4. Tsareva, E., Bogoudinova, R., Khafisova, L., Fakhretdinova, G.: Multilingualism as a Means of Students' Technocommunicational Competence Forming at Engineering University. In: Auer, M., Hortsch, H., Sethakul, P. (eds.) ICL 2019, Advances in Intelligent Systems and Computing, vol. 1134, pp. 137–142. Springer, Cham (2020)
5. Barabanova, S.V., Nikonova, N.V., Pavlova, I.V., Shagieva, R.V., Suntsova, M.S.: Using active learning methods within the andragogical paradigm. In: Advances in Intelligent Systems and Computing, vol. 1134 AISC, pp. 566–577 (2020)
6. Valeeva, R., Ziyatdinova, J., Osipov, P., Oleynikova, O.: Development of International Academic Mobility: Success Stories. In: Advances in Intelligent Systems and Computing, vol. 1135. AISC, pp. 443–454 (2020)
7. Tselevaya model kompetentsiy 2025 [Target Competence Model-2025]. (in Russian), https://d-russia.ru/wp-content/uploads/2017/11/Skills_Outline_web_tcm26-175469.pdf
8. Concept of Long-Term Socio-Economic Development of the Russian Federation until 2020: Ordinance No. 1662 of the Government of Russia, dated November 17, 2008 [Electronic resource]. (in Russian), https://www.consultant.ru/document/cons_doc_LAW_82134/
9. Crawley, E.F., Malmqvist, J., Ostlund, S., Brodeur, D.R., Edstrom, K.: Rethinking Engineering Education: The CDIO Approach, 2nd edn. Springer International Publishing, Switzerland (2014)
10. The CDIO Syllabus v2.0. An Updated Statement of Goals for Engineering Education // CDIO Knowledge Library. Cambridge, MA; Worldwide CDIO Initiative. https://www.cdio.org/files/project/file/
11. Schwab, K.: Chetvertaya promyshlennaya revolyutsiya: monografiya [The Fourth Industrial Revolution: Monograph]. Translated from English. Moscow: Izd-vo l. "E", 208 p. ill (2017.)
12. Strange jobs that could exist by 2030. https://www.businessinsider.com/strange-jobs-2030-2016-8
13. Industriya 4.0 v 20 tsifrakh i faktakh [Industry 4.0 in 20 Numbers and Facts]. https://www.rbc.ru/trends/industry/5daef6429a7947c1bfe43006. (in Russian)
14. Sostoyaniye upravleniya proyektami [State of Project Management]. https://www.wellingtone.co.uk/wp-content/uploads/2018/05/The-State-of-Project-Management-Survey-2018-FINAL.pdf
15. Otchet ob issledovanii polzovateley upravleniya proyektami [Report on Studies of Project Management Users]. (in Russian), https://www.capterra.com/project-management-software/user-research/
16. Barabanova, S.V., Zinurova, R.I.: [New Approaches to the Formation of Legal Competence in Engineering Pedagogy]. In: Vysshee obrazovanie v Rossii = Higher Education in Russia. No. 7 (214), pp. 138–146 (2017). [in Russ., abstract in Eng.]

Teaching Project Risk Management in a BIM-Enabled Learning Environment

Emlyn Witt(✉), Theophilus Olowa, and Irene Lill

Department of Civil Engineering and Architecture,
Tallinn University of Technology, Tallinn, Estonia
emlyn.witt@taltech.ee

Abstract. Digitalization is driving changes in the architecture, engineering, construction and real estate (AEC+ RE) industry and a central feature of this digital transformation is Building Information Modelling (BIM) which refers to the integrated digital representation of all building-related information. A BIM-enabled Learning Environment (BLE) aimed at creating an experiential learning space and offering opportunities for immersive and integrated learning on the basis of real project data has been conceptualised in earlier research. This research addresses the need to design educational interventions using the BLE. On the basis of document analysis and lecturer observation, the current and proposed approaches to teaching project risk management are described and compared in terms of their activities, learning objectives, feedback and assessment strategies and contextual factors. The new approach is intended to leverage BIM in order to engage students in a learning activity that closely corresponds to industry project reality and ways of working. It also presents challenges in terms of ensuring adequate acquisition of theoretical knowledge and finding sufficient time for students to grasp the complexities of a realistic, industry project environment.

Keywords: Building Information Modelling (BIM) · BIM-enabled Learning Environment · Project risk management

1 Introduction

1.1 Background

Digitalization is driving changes in the construction industry and a central feature of this digital transformation is Building Information Modelling (BIM). BIM refers to the integrated digital representation of all building-related information from design through to demolition and it enables communication, coordination, planning, simulations, etc. to take place in a common, virtual environment.

The education of construction professionals has tended to lag industry in its deployment of BIM [1, 2] and this poses a challenge for the education of Architectural, Engineering, Construction and Real Estate (AEC+RE) students [3]. On the one hand, it must respond to the immediate industry needs by providing students with the knowledge and skills to carry out their professional tasks using BIM - this has largely been done in university AEC+RE programmes [4]. On the other hand, there are new

M. E. Auer and T. Rüütmann (Eds.): ICL 2020, AISC 1328, pp. 162–173, 2021.
https://doi.org/10.1007/978-3-030-68198-2_14

opportunities arising from BIM which have the potential to enhance AEC+RE education and benefit the industry in turn [5]. These include opportunities for improving the integration of curricula within and between construction-related disciplines using real project data and creating learning environments that more closely correspond to industry realities.

1.2 BIM-Enabled Learning Environment

Earlier research [6] established the concept of a BIM-enabled Learning Environment (BLE) for creating a common learning space that spans both higher education and industry for immersive and integrated learning on the basis of real project data, as well as greater continuity between degree studies and professional development. The BLE concept draws on Dewey's notion that education requires a social environment that allows both communication (for learning) and application (for doing what is learnt) [7]. Further development of this idea through Kolb's Experiential Learning Theory introduces the concept of "Learning Space" which elaborates transactions between learners and their environments [8, 9]. In addition, Gidden's Structuration Theory provides a framework to describe this social environment in terms of the structures which allow it to be produced and reproduced in social activity. "Structures" in this sense are the properties, rules, resources and transformational relations which allow social practices to be reproduced across time and space and which give them the form of systems [10].

As shown in Fig. 1, the BLE is embedded within the systems of education and the AEC+RE industry. Both the structures of the education system (e.g. curriculum requirements, teachers and students, learning outputs, etc.) and the structures of industry (e.g. real project data, ways of working, professional roles, etc.) are appropriated and combined in order to create the BLE. For example, BLE roles must make sense both from the industrial BIM work flow point of view - e.g. Construction Manager, BIM Coordinator, Architect - and also from the educational point of view - e.g. Learner, Instructor.

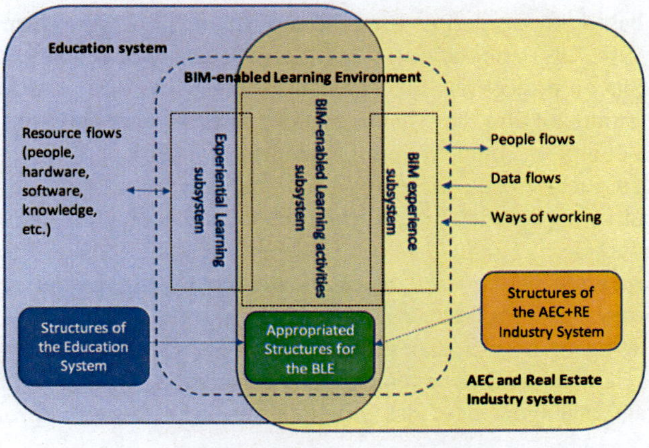

Fig. 1. The BIM-enabled Learning Environment (BLE) source: [6].

1.3 Teaching Project Risk Management

Risk in construction project management has been identified by several educational researchers as a concept that is not easily understood by students because of the difficulty in juxtaposing theory with practical reality [11, 12]. The concept of risk in construction projects originates from theoretical developments over centuries in many diverse fields including probability theory e.g. [13], economics e.g. [14], finance e.g. [15], insurance and actuarial science e.g. [16], sociology e.g. [17], etc. which has subsequently been built on by construction project management theorists. The result is a complicated concept for instructors to articulate to their students that variously relates to:

- uncertainty - having incomplete knowledge, particularly with respect to the future;
- probability - in the sense of 'rational belief' in the likelihood of a possible outcome with respect to a corpus of knowledge (evidence) which substantiates that belief;
- expectations - risk is experienced relative to expected outcomes;
- actions - in order to engage with risk, it is necessary to do something;
- time - there is a temporal dislocation between the consideration of risk and the outcome itself;
- profit/gain/loss [18].

In contrast, the practical application of risk management is relatively straightforward to understand and follows a logical process of: plan risk management, identify risks, perform qualitative and quantitative risk analyses, plan risk responses, monitor and control risks. This process, particularly the risk identification, risk analysis and risk response steps, is cyclical and repeats through the project [19].

The 'traditional' approach to teaching project risk management involves introducing the concept of risk and following this up with a (typically shallow and time-constrained) practical risk management exercise. However, the literature does report some innovations: for example, [11] developed a computer-based simulation game and applied mixed methods action research to demonstrate how both theory and practice of risk in project planning and control, using real life data, could be taught to a group of 33-students using an experiential learning approach. [12] designed and developed a similar game based on scenarios from a real-life land development. The Virtual Construction (VIRCON) simulation is another innovative development designed for educating students on project risk management. It integrates design and planning tools (with schedule simulator) for students to perform pre-contract activities including the production of a complete competitive bidding proposal [20].

BIM has also already been shown by numerous authors to be a powerful tool for enabling innovative approaches to construction education including project risk management. For example, [21] elucidate how BIM as a single repository for construction data and information can be used to eliminate the repetitive and redundant class exercises in project management courses. They argued that, with BIM, data and information relating to a particular project and context can be easily retrieved and manipulated for different class exercises and asserted that this facilitates quicker understanding and management of risk principles and applications in project execution.

1.4 Research Purpose and Paper Structure

The overall goal of this research is to fully leverage the educational opportunities of BIM in order to bring BIM-enabled construction education from concept into practice as this will offer more immersive and integrated learning experiences on the basis of real, up-to-date project data from industry. The BLE concept has been developed to achieve this but is still in the earliest stages of being applied in teaching practice. The purpose of this study is to find a solution to the question: How can project risk management be practically and advantageously taught within a BLE? Project risk management is an appropriate test learning activity because, although it is widely accepted as being important, it is often not applied systematically in the construction industry and, for good decision-making, project risk management must be based on accurate, historical project data (i.e. evidence). The greater availability and accessibility of project data with BIM raises the prospect of better project risk management provided that practitioners know how to use it effectively. Hence, experiential learning in a BLE offers opportunities to directly improve industry practices over existing project risk management teaching approaches.

In the following sections of the paper, the research methodology is outlined before two alternative pedagogical designs are presented - the case of the existing teaching arrangements and an alternative (proposed arrangement) in which the BLE concept is applied. The two cases are discussed, compared and contrasted and, finally, conclusions drawn in terms of the anticipated advantages and disadvantages of teaching project risk management in a BLE. This study is limited to the design of the new educational offering and does not extend to the implementation and evaluation of it in teaching.

2 Methodology

The study reported in this paper takes place within the broader context of a participatory action research project involving a five stage, cyclical process of Diagnosing, Action Planning, Action Taking, Evaluating and Specifying Learning [22]. The action research aims to introduce BIM-enabled education into the AEC curriculum and this study is focused on the Diagnosing and Action Planning stages, specifically the planning of an intervention with respect to teaching project risk management in a BLE.

"Diagnosing" refers to problem identification, data collection and data analysis resulting in solution proposals. In this study, the current approach to teaching project risk management within a particular project management in construction course is analysed by document analysis and lecturer observations/reflections in order to derive a baseline description which is framed in terms of its learning objectives, teaching and learning activities, feedback and assessment strategies and contextual factors.

The "Action Planning" stage relates to the detailed development of a proposed solution to the problem(s) identified during "Diagnosis". In this particular case, it refers to the design of a novel, proposed way of teaching project risk management for the same course which takes advantage of the emerging opportunities of BIM-enabled learning and, specifically, employs the BLE concept. The new learning experience is designed on the basis of current, recommended learning experience design guidance and described in the same terms of learning objectives, teaching and learning activities, feedback and assessment strategies and contextual factors so as to enable comparison with the current teaching approach.

3 Findings

3.1 Case 1: Current Approach to Teaching Project Risk Management

Description of Teaching and Learning Activities. The risk management topic is currently delivered in an approximately 90-min classroom session comprising 2 parts - a 45-min presentation in lecture format which includes some interactive illustration of points during the presentation, followed by a facilitated class exercise for which a further 45 min of class time is assigned.
The lecture covers:

- The terms and concepts of risk management;
- The basic process of risk management in projects (plan risk management, risk identification, risk analysis, risk response, monitoring and control, documentation and record keeping/learning for future projects);
- Project risk management standards;
- Tools and techniques for achieving each stage of the risk management process;
- How risk and risk management link to wider ideas in construction, science and society (such as contracts as instruments of risk allocation and transfer, Integrated Project Delivery, statistical inference, climate change and disasters, societal risk and modernity, etc.)

The facilitated class exercise is aimed at demonstrating and reinforcing key steps of the risk management process - risk identification, (qualitative) risk analysis, recommending risk response actions. It is carried out in small teams of students who assume different construction industry roles (client, designers, contractor, etc.) as follows:

- A simple construction project scenario is given;
- Students work in small teams in the format of a risk management workshop to identify risks, analyse the relative significance of those identified (using a probablility – impact matrix) and recommend actions to mitigate the most significant risks;
- Student teams report their findings;
- A general class discussion then takes place in which all teams' findings are considered and summarised and the key points of (practical) project risk management are reinforced.

Learning Objectives. The learning objectives are defined with reference to Bloom's Taxonomy as revised by [23, 24].

Table 1. Learning objectives for the current approach to teaching project risk management.

Bloom's Taxonomy (revised: [23])	Learning objectives
Remembering: Retrieving, recognizing, and recalling relevant knowledge from long-term memory	• Students are able to describe the process, tools and techniques of project risk management
Understanding: Constructing meaning from oral, written, and graphic messages through interpreting, exemplifying, classifying, summarizing, inferring, comparing, and explaining	• Students understand the concepts of risk and project risk management
Applying: Carrying out or using a procedure through executing, or implementing	• Students are able to apply the project risk management process, tools and techniques in a given (simple) project scenario
Analyzing: Breaking material into constituent parts, determining how the parts relate to one another and to an overall structure or purpose through differentiating, organizing, and attributing	• Within the given risk management process and project scenario, students are able to break up the scenario into constituent elements and analyse risks associated with each element
Evaluating: Making judgments based on criteria and standards through checking and critiquing	• Students evaluate the risks identified in terms of given (probablity and impact) criteria in order to reach a collective judgement (within their team) concerning the relative significance of each of the identified risks and what actions to take to mitigate them
Creating: Putting elements together to form a coherent or functional whole; reorganizing elements into a new pattern or structure through generating, planning, or producing	• As a whole class, a common understanding of the most significant risks and suitable responses to them is created - but this does not take place at the individual student level

Feedback and Assessment Strategies. Given the short duration of this topic within the current project management in construction course, the opportunities for feedback and formative assessment are limited within it. Direct feedback (from their peers and from the lecturer) occurs as the student teams discuss and work through the risk management process steps. Towards the end of the class exercise, when the student teams' results are collated, it is typically the case that there is notable similarity between the most significant risks identified by the different teams and their proposals for mitigating these risks. This tends to give the students some confidence in the legitimacy of their results.

Summative assessment is limited to questions in the course final exam which require students to apply the same risk management process steps to a simple project scenario and also to recall appropriate tools, techniques and process stages of risk management.

Contextual Factors. Students taking this course are typically either in their 4th year of studies in an integrated, 5-year construction-related masters degree programme or taking a 2-year masters degree programme in the case that they have completed studies to bachelor's degree level previously, in a separate programme. Their industry experience varies from none at all to highly experienced professionals. By working in teams, their experiences can be pooled to an extent.

Primarily as a consequence of time constraints, the exercise scenario is highly simplified and minimizes the use of specific, real project data. This reduces its correspondence to industry reality but makes it easier to apply the recommended project risk management process steps to it. However, this leaves a significant gap between the exercise and practical application in real industry projects.

To date, quantitative risk analysis has been briefly introduced in this course but not included in the practical exercise because it would require considerably more time both to explain and for the students to apply (even in a highly simplified scenario).

3.2 Case 2: Proposed BLE Approach to Teaching Project Risk Management

Description of Teaching and Learning Activities. As noted above, the key idea of the BLE concept is to enable immersive and integrated learning experiences on the basis of real, up-to-date project data from industry. This experiential learning takes place on the basis of a realistic industry work flow that fully utilizes BIM (see Fig. 2).

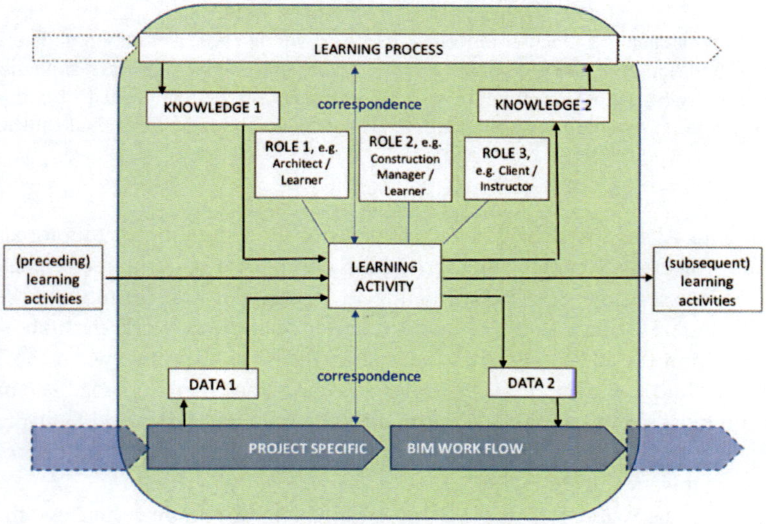

Fig. 2. The BIM-enabled Learning Environment (BLE) at the level of learning activity, adapted from [6].

BIM ensures comprehensive, organised and readily accessible project data. Much of this data is referenced directly to building objects (walls, beams, columns, windows, doors, floor slabs, pipes, etc.) which are represented in a virtual, 3D model of the building so that they can be easily viewed and understood. It therefore enables real, complicated project scenarios to be presented to and efficiently grasped by students. Using an industry BIM work flow ensures that the scenario on which the learning activity (project risk management, in this case) takes place corresponds to industrial reality and also that the data input to the learning activity (Data 1 in Fig. 2) is not contrived by the lecturer but rather exists as real project data and is drawn directly from the same sources as would be the case in industry. (It should be noted that this project data must be pre-checked and simplified to remove inconsistencies and unnecessary details which could confuse students.) Similarly, by carrying out the learning activity, the project data is further processed and the output data (Data 2 in Fig. 2) feeds directly back into the BIM work flow. The project is thus elaborated and progressed. In this way, the learning activity is intended to resemble a meaningful task in a genuine work context.

In order that students are suitably prepared and able to carry out the learning activity, they will need some pre-instruction (Knowledge 1 in Fig. 2). However, most of their learning occurs within the context of the learning activity itself (Knowledge 2 - Knowledge 1 in Fig. 2). In this way, the teaching and learning activities proceed as follows:

Initial instructions to (briefly) cover the essential pre-information necessary to commence with the experiential learning activity:

- Key steps in the process of project risk management;
- Instructions and information for participation in the learning activity.

The experiential learning activity - students work through a guided, detailed project risk management process (including both qualitative and quantitative risk analysis) on the basis of real project data within a BIM work flow. They do so in teams arranged according to typical industry roles and, in the course of the activity, they explore and discuss in detail:

- The terms and concepts of risk management;
- The process of risk management in projects (plan risk management, risk identification, risk analysis, risk response, monitoring and control, documentation and record keeping/learning for future projects);
- Tools and techniques for achieving each stage of the risk management process;
- Project risk management standards;
- How risk and risk management link to wider ideas in construction, science and society (such as contracts as instruments of risk allocation and transfer, Integrated Project Delivery, statistical inference, climate change and disasters, societal risk and modernity, etc.).

Learning Objectives

Table 2. Learning objectives for the proposed approach to teaching project risk management.

Bloom's Taxonomy (revised: [26])	Learning objectives
Remembering: (detailed description as per Table 1)	• Students are able to describe the process, tools and techniques of project risk management. With the BLE learning activity, this relates to a more realistic, detailed BIM-based process
Understanding: (detailed description as per Table 1)	• Students understand risk, project risk management concepts • As the learning activity takes place within a BIM work flow, students also acquire understanding of this work flow - which increases the learning value beyond the risk management topic
Applying: (detailed description as per Table 1)	• Students are able to apply the project risk management process, tools and techniques in a realistic project scenario based on real project data and an industrial work flow
Analyzing: (detailed description as per Table 1)	• Within the given risk management process and project scenario, students are able to break up the scenario into constituent elements and analyse risks associated with each element
Evaluating: (detailed description as per Table 1)	• Students evaluate the risks identified in order to reach a collective judgement concerning the relative significance of each of the identified risks and appropriate mitigation actions
Creating: (detailed description as per Table 1)	• Students reconsider the risk management process and the industrial work flow in order to recommend improvements

Feedback and Assessment Strategies. The intention of the BLE learning activities is that their successful completion and quality is assessed by the students themselves later in the course as the output data from each activity feeds back into the work flow and provides input data to later, subsequent activities. A poorly executed activity will lead to later process and implementation problems. The BLE thus seeks to reflect the default means of assessment of the industrial project process in the class room (see Fig. 2).

The learning activity presumes considerably more effort from students than the simplified project scenario in Case 1 does. This allows greater scope for feedback during the activity and also for formative assessment. It suggests that summative assessment should be linked to the learning activity itself as this cannot be repeated in a final exam.

Contextual Factors. Time is clearly a constraining issue for this approach. However, as the same real project data and underlying BIM work flow can be maintained throughout the course, the time for students to familiarise themselves with this learning environment may be spread across the whole course rather than impacting just a single topic.

The need for real project data presented in appropriate formats is critical to this approach and is not (yet) readily available. This requires initial investment to set up the project data and to 'clean' it of inconsistencies/errors and simplify it where necessary.

Industrial BIM work flows with respect to project risk management are by no means standardised - they are still under development and not yet widely adopted in industry. The learning activity processes are thus part of the development of BIM-based project risk management for the AEC+RE industry. On the one hand, this makes it more challenging to design the activities, on the other, the creative development of the process and work flow are currently important for industry. Although the students' industry experience varies from zero to highly experienced professionals, with respect to BIM-based project risk management, there is likely to be some novelty for all students.

4 Discussion

The two approaches to project risk management teaching described above differ fundamentally in their process. The current teaching arrangement (Case 1) focuses on explaining the concepts, terms and theoretical basis, and then engaging the students in a (simplified) scenario-based activity to demonstrate the application of theory to practice. In the proposed BLE-based approach (Case 2), the focus is on a realistic, experiential learning activity which corresponds closely to industry project reality and the theoretical consideration of the subject emerges as an extension of the practical application.

In this way, Case 2 is representative of problem-based learning (PBL) approaches which have become increasingly popular in areas of professional education [25]. The emphasis on an experiential learning context also reflects the CDIO approach which stresses engineering fundamentals set in the context of real-world systems and products [26].

The digitalization of the AEC+RE industry is radically changing the way in which buildings are designed and constructed. [3] has pointed out that BIM enables the building design process to begin with a complete model of a building which can then be adapted to suit new conditions rather than always starting from abstract concepts and ending with the building. BIM enables a similar change for risk management as it provides a database relating to the current project and also historical projects. Thus making explicit all the evidence on which project risk management should be based.

A BLE-based approach to teaching project risk management is clearly more time-consuming (at the topic level) as it requires students to navigate in a much richer project data environment than would be the case for a simple example scenario. It also favours the acquisition of wider risk concepts and ideas through collective exploration and discussion. This offers more student-centred learning experiences but also takes considerably more time to deliver than a standard lecture format. In practice, course time constraints will determine the precise balance between traditional delivery and group exploration of risk concepts on the basis of the experiential learning activity. The learning activity will, in any case, provide students with a common reference with which to frame and make sense of the concepts.

A potential danger of such a strong emphasis on correspondence to industry and industrial work flows is an unintentional 'vocationalisation' of university education where relevance to current industrial practice is overemphasised at the expense of higher level conceptualisation, professional attitudes and principles. Again, an appropriate balance is sought where education does indeed correspond to industry needs but rather the wider, future needs than current, narrower needs.

5 Conclusions

The new, BLE-based approach to teaching project risk management is intended to leverage the opportunities offered by BIM to enhance industry performance and the effectiveness of education for AEC+RE professionals. The BIM work flow provides a data-rich, real project context in which to immerse students and engage them in a learning activity which closely corresponds to industry reality and enables experiential learning. It represents the emerging, industry ways of working which students must understand so that they can contribute to developing them in their future professional careers. However, it also presents attendant challenges in terms of ensuring the acquisition of a sufficient theoretical understanding of the subject so that students are prepared to contribute at higher levels of professional activity (e.g. policy formulation, standards development, etc.) than only the implementation and application of project risk management concepts to the project context. The realistic project context requires additional time for students to become familiar with and be able to effectively learn within.

Future work will focus on the next stages of the action research cycle: action taking, evaluating and specifying learning.

Acknowledgement. This research was supported by the Integrating Education with Consumer Behaviour relevant to Energy Efficiency and Climate Change at the Universities of Russia, Sri Lanka and Bangladesh (BECK) project co-funded by the Erasmus+ Programme of the European Union. The European Commission support for the production of this publication does not constitute an endorsement of the contents which reflect the views only of the authors, and the Commission cannot be held responsible for any use which may be made of the information contained therein.

References

1. Forgues, D., Becerik-Gerber, B.: Integrated project delivery and building information modeling: Redefining the relationship between education and practice. Int. J. Des. Educ. **6** (2), 47–56 (2013)
2. Lee, N., Dossick, C.S., Foley, S.P.: Guideline for building information modeling in construction engineering and management education. J. Prof. Issues Eng. Educ. Practice **139** (4), 266–274 (2013)

3. Ambrose, M.A.: Agent provocateur – BIM in the academic design studio. Int. J. Archit. Comput. **10**(1), 53–66 (2012)
4. Gerber, D.J., Khashe, S., Smith, I.F.: Surveying the evolution of computing in architecture, engineering, and construction education. J. Comput. Civil Eng. **29**(5) (2013)
5. Olowa, T., Witt, E., Lill, I.: Conceptualising building information modelling for construction education. J. Civ. Eng. Manag. **26**(6), 551–563 (2020)
6. Witt, E., Kähkönen, K.: A BIM-Enabled Learning Environment: A Conceptual Framework. In: 10th Nordic Conference on Construction Economics and Organization. Emerald Publishing Limited, May 2019
7. Dewey, J.: Democracy and Education. Jovian Press. Kindle Edition (1916)
8. Kolb, D.A.: Experiential Learning: Experience as the Source of Learning and Development. Prentice Hall, Englewood Cliffs, NJ (1984)
9. Kolb, A.Y., Kolb, D.A.: Learning styles and learning spaces: enhancing experiential learning in higher education. Acad. Manage. Learn. Educ. **4**(2), 193–212 (2005)
10. Giddens, A. (1984) The Constitution of Society: Outline of the Theory of Structuration. Wiley. Kindle Edition.
11. Al-Jibouri, S., Mawdesley, M., Scott, D., Gribble, S.: The use of a simulation model as a game for teaching management of projects in construction. Int. J. Eng. Educ. **21**(6 PART I), 1195–1202 (2005)
12. Xia, J., Caulfield, C., Baccarini, D., Yeo, S.: Simsoft: a game for teaching project risk management. In: Proceedings of the 21st Annual Teaching Learning Forum (2012)
13. Keynes, J.M.: A Treatise on Probability. MacMillan and Co., London (1921)
14. Knight, F.H.: Risk, Uncertainty and Profit. Houghton Mifflin Company, The Riverside Press, Cambridge (1921)
15. Markowitz, H.: Portfolio selction. J. Financ. **7**(1), 77–91 (1952)
16. Denenberg, H.S., Ferrari, J.R.: New Perspectives on Risk Management: The Search for Principles. J. Risk Insur. **33**(4), 647–661 (1966)
17. Beck, U.: Risk Society: Towards a new modernity. Sage, Translated from the original German by Mark Ritter (1992)
18. Witt, E.D.Q.: Risk Transfer and Construction Project Delivery Efficiency - Implications for Public Private Partnerships. Ph.D. thesis, Tallinn University of Technology. TTU Press (2012)
19. Project Mangement Institute. Practice standard for project risk management. Project Management Institute (2009)
20. Jaafari, A., Manivong, K.K., Chaaya, M.: VIRCON: interactive system for teaching construction management. J. Constr. Eng. Manag. **127**(1), 66–75 (2001)
21. Peterson, F., Hartmann, T., Fruchter, R., Fischer, M.: Teaching construction project management with BIM support: experience and lessons learned. Autom. Constr. **20**(2), 115–125 (2011)
22. Susman, G.I.: Action research: a sociotechnical systems perspective. Beyond method: Strategies for social research **95**, 113 (1983)
23. Anderson, L.W., Krathwohl, D.R.: A taxonomy for learning, teaching, and assessing: a revision of Bloom's taxonomy of educational objectives. Longman, New York (2001)
24. Forehand, M.: Bloom's taxonomy: original and revised. In: Emerging Perspectives on Learning, Teaching, and Technology, vol. 8 (2005)
25. Fink, D.: Creating Significant Learning Experiences: An Integrated Approach to Designing College Courses. John Wiley & Sons, San Francisco, CA (2013)
26. Berggren, K.F., Brodeur, D., Crawley, E.F., Ingemarsson, I., Litant, W.T., Malmqvist, J., Östlund, S.: CDIO: an international initiative for reforming engineering education. World Trans. Eng. Technol. Educ. **2**(1), 49–52 (2003)

STACK Assessment in Mathematics Classroom: Advantages and Disadvantages

Irina Ustinova[1(✉)], Vladimir Tomilenko[2], Olga Imas[1], Evgeniia Beliauskene[1], and Olga Yanuschik[1]

[1] National Research Tomsk Polytechnic University, Lenin Avenue 30, Tomsk 634050, Russia
igu@tpu.ru

[2] Tomsk State University of Control Systems and Radioelectronics, Lenin Avenue 40, Tomsk 634050, Russia

Abstract. E-Learning is considered a new philosophy of education that is involved in all existing forms of education including full-time education. Presently, there is a variety of software to develop interactive content. This study is aimed to examine the possibility of employing STACK (the System for Teaching and Assessment using a Computer algebra Kernel) assignments in the mathematics classroom at a technical university, to compare the use of the STACK questions activity with traditional training, and to evaluate the effectiveness of this method. The experience of using STACK questions has shown that their application helps students of all forms of education in learning mathematics, and considerably facilitates the learning process of teachers, allows you to implement a point-rating (cumulative) system of assessment of knowledge, makes the learning results more visible and convenient for analysis. In addition, we note such advantages of STACK as: the STACK assignments allows you to check analytical solutions to complex types of problems and organize hints that help solve tasks, the STACK assignments saves the teacher time to check solutions and gives students individual options for tasks, tasks prepared in STACK can be used in various modes, such as training mode and monitoring one. In the course of the study, the STACK assignments were analyzed that revealed a greater number of advantages of this assessment compared to its disadvantages and demonstrated the possibility of its application in the educational process.

Keywords: Mathematics · Engineering education · STACK assignments · STACK questions · E-learning

1 Introduction

E-Learning is considered a new philosophy of education involved in all existing forms of education including full-time education [1]. During the evolution of e-learning in the world, there is already a well-established opinion that we are not ready to completely replace traditional education with e-learning, but the use of innovative technologies effectively affects both the quality of education and the economic benefits of the educational process. The introduction of new technologies in the educational process

M. E. Auer and T. Rüütmann (Eds.): ICL 2020, AISC 1328, pp. 174–182, 2021.
https://doi.org/10.1007/978-3-030-68198-2_15

has received a new name blended learning (b-learning) [2]. Modern computer technologies allow the use of animation, audios and videos, built-in learning control systems, content adaptation based on student outcomes, etc. Despite the intensive implementation of b-learning, academics are skeptical about the possibility of studying higher mathematics using the new technologies. This is mainly due to the low efficiency of acquiring subject skills. However, the appearance of a specialized software package STACK (the System for Teaching and Assessment using a Computer algebra Kernel) installed as an additional plug-in to the educational platform, MOODLE or ILIAS, can significantly advance the automation of the process of acquiring computational and subject skills of students in fundamental disciplines in General and in mathematics in particular [3].

This study examines the possibility of employing STACK assignments in the mathematics classroom at a technical university to compare the use of the STACK questions activity with traditional training and to evaluate the effectiveness of this method.

STACK evaluates the responses provided to students, which are mathematical expressions. Essentially, mathematical expressions that are responses in STACK are evaluated using computer algebra. [4]. STACK was developed by Christopher Sangwin (Loughborough University) [5]. In 2013, version 3.0 was integrated as a separate question type in Moodle. Fred Neumann and Jesus Copado (Erlangen University) developed the STACK plugin for ILIAS using crowdfunding in 2014 [6]. STACK allows implementing math test tasks in Moodle. This plugin is integrated with the Computer algebra system (CAS) “Maxima”, so you can enter character expressions as responses and evaluate them. In addition, this type of test questions offers the following features: randomize the task parameters; use random parameters throughout the task and opinion to it; automate verification of the response entered by the student; it is flexible to estimate the answer; the opportunity to assess partial results [7]. Instantaneous feedback in STACK questions takes into account all possible mistakes made by students in solving the problem [8]. All these advantages were immediately evaluated by mathematicians and used in educational activities. Thus, [9] examines the activity of students in the STACK system in preparation for the math exam and its impact on the resulting score. The authors [10] used STACK to implement two different concepts of b-learning. In both scenarios, classroom activities were partially replaced by e-learning tasks. In the first scenario, lecture hours were used to spend more time solving complex tasks in the classroom. In the second one, the number of lectures was reduced, and bonus points for the final exam could be obtained by sending additional electronic homework. Electronic homework aroused particular enthusiasm among students, but ultimately improved continuous learning throughout the semester.

One of the most important elements of mastering mathematics is timely feedback. Moreover, this element of learning could also be formalized, since a teacher, a book, a fellow student, or the Internet can provide the feedback. This problem is studied in detail in the paper [11].

2 Approach

In 2019 a workshop of mathematics teachers from two technical Universities and the classical University started functioning in Tomsk [12]. The purpose of this workshop was to explore the possibilities of STACK, in creating test tasks in mathematics. One of the priorities of the team was to develop a bank of test tasks in the Stack system, which could be used for teaching mathematics to all students, as well as for monitoring activities. A small group of teachers from Tomsk Polytechnic University conducted an experiment in the spring semester. The experiment consisted of using STACK-based tasks in the mathematics classroom. The purpose of the experiment was to analyze the possibilities, advantages and disadvantages of the STACK system in teaching mathematics.

Since 2018, students of the first two years of their course at TPU have unified training programs. This means that regardless of the specialty, all students study an equal number of subjects with an equal number of hours the first two years. This also applies to mathematics.

The experiment involved 6 groups of students of different specialties (computer science, petroleum science, ecology) learning the topic "Indefinite integral" in the second semester of the 2019/2020 academic year. More than 100 students participated in the experiment (the experimental group). In the experimental group, 6 lessons were conducted using STACK task activities and a final test was carried out. In the same specialties, control group was selected, also more than 100 students, that did not participate in the experiment, and that studied according to the traditional method. Lessons in the experimental group were conducted in the computer classroom. Students were shown the system itself, where it is located and how to work in it at the beginning of the first lesson. They were explained how to enter responses correctly.

Figure 1 shows an example of the tasks students received.

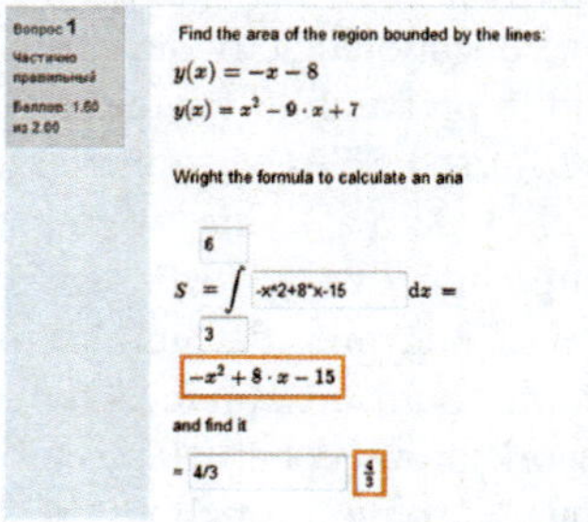

Fig. 1. The question of STACK type

As you can see from the figure, a question of STACK type is a task with prepared response fields. Students entered answers are not only numbers, but also algebraic expressions, such as the integrating function. The syntax for entering formulas is generally accepted in computer algebra. In addition, the e-course included instructions on how to enter formulas. In order for the student to control the input, the system provides a display of the entered expression (this is shown in the Fig. 1 as the highlighted rectangle). The possibility of automated algebraic comparison of mathematical expressions entered and programmed as an answer has significantly expanded the use of MOODLE test tasks with STACK type questions not only as control, but also as training elements of online courses. Figure 2 shows the system's response to the solution entered by the student.

In the first line, the student sees a general response to the decision, such as correctly, partially correctly or incorrectly, colored with the appropriate color. This is followed by lines with reviews that comment on entering responses in each field of the task. After that, you will be offered a hint and an opportunity to correct the incorrectly filled in response field. It should be noted that in this system, you can work with any gadget. There were no problems with technical equipment. Some students brought their laptops, some worked in smartphones. The majority of students preferred to work on the desktops. The experience of conducting classes has shown that on average, depending on the training of students it was possible to solve from 4 to 8 tasks per lesson. Students had to complete the unsolved class tasks at home. It turned out that those students who decided more in the classroom, decided less at home. You can use the hints to solve the task. Each task can be considered as a mini tutorial. You can find a complete solution to this problem in the tooltip, with a link to the theoretical material, or use the tooltips in stages. The assessment of the task depends on how often the student used the hints.

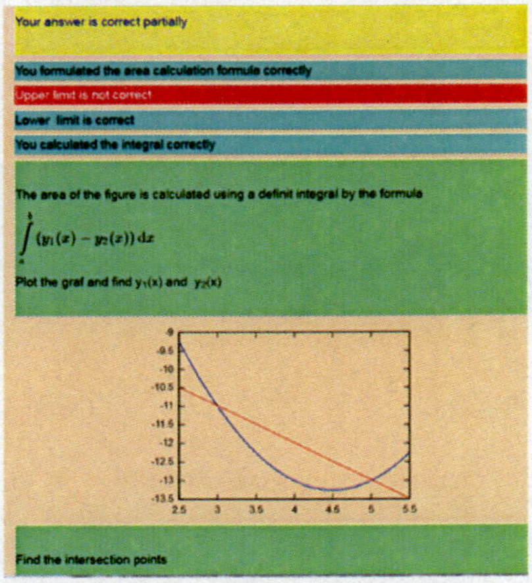

Fig. 2. Feedback at the answer in the question type STACK

3 Actual Outcomes

At the beginning of the academic year, all freshmen are independently tested in mathematics, in order to determine the level of students and more accurately configure the plan of lessons for each group. Figure 3 shows the entrance test histogram to initially compare and take into account the relative level of the experimental and control groups of computer science majors.

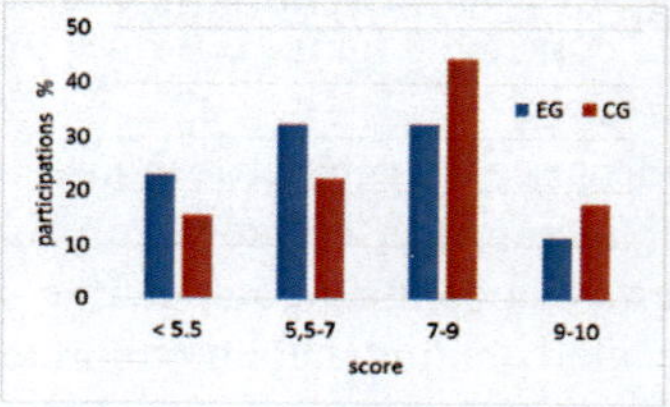

Fig. 3. Bar chart of the distribution of entrance test scores of experimental (EG) and control group (CG) of students

Figure 3 shows that the level of the control group is slightly higher than that of the experimental group, although the hypothesis about the difference in the sample mean score has not been confirmed. After conducting classes on the topic of indefinite integral in both the experimental group and the control one, tests were carried out using the traditional method, i.e. students solved the proposed tasks of the same complexity in the class for a limited time. In addition, in the middle of the semester, the center for independent testing conducted computer testing on this topic. Figure 4 shows a comparative bar chart of the distribution of the obtained estimates of the experimental and control groups.

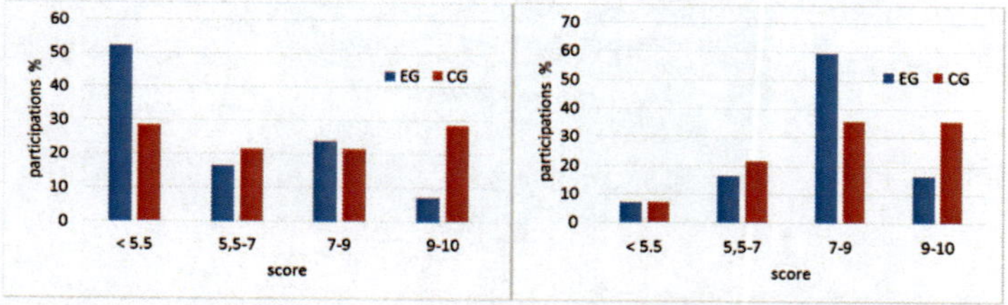

a) Test b) Control cross section

Fig. 4. The bar chart of the distribution of students' scores in the experimental and control groups

As you can see in Fig. 4a, the distribution of points in the two groups differ for the best and worst ones. At the same time, independent testing (Fig. 4b) showed similar results in the lower part of the histogram and some differences in the upper part. At the same time, the indicator of the best score for the control group significantly improved in comparison with the input test. In general, the results of both groups are quite high. The homogeneity of samples was established in accordance with the Wilcoxon criterion.

Groups of students majoring in oil and gas engineering and ecology also participated in the experiment. As can be seen from Table 1, the level of ecologists was initially significantly lower than the level of students in the fields of oil and gas engineering and computer science.

Table 1. Tests results

Focus of training	Entrance score (max 10)	Control test score (max 10)
Computer science	6.2	5.0
The oil and gas engineering	6.3	4.3
Ecology	4.4	2.5

As a result, the students have not acquired enough skills, and the test showed that almost no one has learned to integrate. In addition, the new technology assumed greater freedom of behavior of students in the classroom, greater responsibility for completing homework. Students of specialty «Ecology», not having a good enough base of school mathematics and being poorly motivated to work, they took advantage of their freedom so that they do not study independently.

We have once again seen that the new technology in teaching mathematics works well for the average student. Underachieving students require more control and more attention from the teacher. In most cases, these students have problems not so much with understanding, but with their attitude to learning. Unfortunately, new technology and questions like STACK cannot scold a student and make them pick up a pen. This is usually done by the teacher. Straight-A students usually make either the simplest computational errors, or use a non-standard solution method and get an expression for entering the answer that the teacher did not suppose when creating the questions. Unfortunately, not all equivalent expressions are recognized by computer algebra. The imperfection of the system should be compensated by the teacher in the creation of a question.

4 Conclusion

In order to study the capabilities of the STACK system in the educational process, an experiment was held, in which first-year undergraduate students of engineering specialties took part.

The experience of using STACK questions has shown that their application helps students of all forms of education in learning mathematics A survey of students was conducted, from which a conclusion was made about the convenience of STACK when performing homework, individual tasks, and performing activities provided for self-studying. It was noted that students' interest in the subject has increased. The majority of students were involved in the work. Poorly motivated students showed unsatisfactory results.

The results of control and independent testing of students who studied in accordance with traditional teaching methods and the results of testing of students who studied using tasks developed in the STACK system are analyzed. The difference in average test scores was statistically insignificant. The analysis is based on a statistical test of hypotheses about the equality of mathematical expectations against an alternative hypothesis, which is that the mathematical expectations of the average scores of test are not equal.

Based on the analysis of data and pedagogical experience of teachers, the following advantages of the STACK system in the educational process can be identified:

1) One of the important advantages of this system is the implementation of a person – oriented approach to students. It is very important that every student is involved in the work in classroom work. In the traditional approach, as a rule, several people solved problems and demonstrated their solutions at the board. The other students were silently rewriting the decisions. He did not have anyone to write off a decision. He had to decide on his own. The teacher could easily track the students ' work. Namely, how many tasks he completed and how many attempts he used, whether he used hints when solving the task. Highlighting the answers in different colors allows the teacher to see at what stage of the task the student made a mistake and explain the decision to him personally. This becomes relevant; since many students are just shy to admit that they did not understand, hoping to sort out the task at home.
2) Secondly, in the course of training with the traditional approach of conducting classroom classes, students get used to the method of conducting a lesson by one teacher, and therefore a sudden change of teacher is always perceived by students morbidly. However, when teaching methods using test tasks, the main time in the classroom students work independently, the role of the teacher is less significant and, therefore, the replacement of the teacher is not so morbidly.
3) Thirdly, the advantage of using STACK questions is that it makes it much easier for the teacher to introduce a knowledge assessment system. The STACK system allows you to implement a point-rating (cumulative) system of assessment of knowledge, makes the learning results more visible and convenient for analysis, and is open and understandable for the student. For example, after evaluating all practical tasks completed by students, the teacher sees a summary sheet with all grades and points, and each student has access to information about their progress.
4) Finally, there is an advantage to implement the teaching, such as: the STACK assignments allow you to check analytical solutions to complex types of problems; to organize hints that help solve tasks; the STACK assignments save the teacher time to check solutions and gives students individual options for tasks; the tasks prepared in STACK can be used in various modes, such as training mode and monitoring one.

There are also disadvantages of this system, such as:

1) First of all, it is poor performance of tasks by students. The ability of a student to use any mathematical software package uncontrolled in the classroom and at home. You can use it to get the task solution and enter the received solution in the task response field.
2) Secondly, non-motivated students cannot bring themselves to work either in the classroom or at home.
3) Thirdly, not all math tasks can be presented as test tasks. It is difficult to imagine as a test the tasks of theoretical plan or task to proof.
4) Fourthly, a serious disadvantage is the complexity of creating test tasks.
5) Finally, not all universities have the STACK plug-in deployed on training platforms, which hinders the exchange of experience between developers of these questions and the implementation of this training technology.

Thus, in the course of the study, the STACK assignments were analyzed that revealed a greater number of advantages of this assessment compared to its disadvantages and demonstrated the possibility of its application in the educational process. The authors' experience in creating STACK questions in various areas of mathematics can be useful for those who are engaged in the development and application of training courses. The analysis of the results of a survey of students using the STACK assignments, showed the interest of students in the assessment that improves comfort and the quality of teaching, stimulates the learning process, implements the use of active forms, the individual approach, and enhances students' motivation in acquiring new knowledge.

References

1. Vlachopoulos, D.S., Cabrera, N.: Building an inclusive definition of e-learning: an approach to the conceptual framework. Int. Rev. Res. Open Distrib. Learn. **2**, 148–149 (2012)
2. Bryan, A., Volchenkova, K.N.: Blended learning: definition, models, implications for higher education. Bull. South Ural State Univ. Ser. Educ. Educ. Sci. **8**, 24–30 (2016)
3. Sangwin, C.J.: Assessing elementary algebra with STACK. Int. J. Math. Educ. Sci. Technol. **38**, 987–1002 (2007)
4. Wester, M.: Computer algebra systems: a practical guide (1999)
5. Sangwin, C.J.: Assessing higher mathematical skills using computer algebra marking through AIM. In: Proceedings of the Engineering Mathematics and Applications Conference, Sydney, Australia, pp. 229−234 (2003)
6. Copado, J., Neumann, F.: Development of the STACK plugin for ILIAS. In: The 1st International STACK Conference in, Fürth, Germany (2018)
7. Sangwin, C.J.: Who uses STACK? A report on the use of the STACK CAA system. Technical report, Loughborough University, UK (2015)
8. Sangwin, C.J., Köcher, N.: Automation of mathematics examinations. Comput. Educ. **94**, 215–227 (2016)
9. Mäkelä, A.M., Ali-Löytty, S., Humaloja, J.P., Joutsenlahti, J., Kauhanen, J., Kaarakka, T.: STACK assignments in university mathematics education. In: Proceedings of the 44th SEFI Conference, Tampere, Finland (2016)

10. Fath, J., Hansen, P., Scheicher, Ch., Umbach, T.: E-homework with individual feedback for large lectures. Computer Aided Assessment and Feedback in Mathematics. In: Contributions to the 1st International STACK conference in Fürth, Germany (2018)
11. Grove, M., Good, C.: Approaches to feedback in the mathematical sciences: just what do students really think? Teaching Mathematics and its Applications 1 (2019)
12. Tomilenko, V.A., Lazareva, E.G., Ustinova, I.G.: Collaboration to use STACK at the universities of Tomsk. In: The 3rd international STACK Conference, Tallinn, Estonia (2020)

Developing Intellectual Competence of Future Biotechnologists in the Process of Learning a Foreign Language

Larisa A. Kosolapova[1], Margarita A. Mosina[2(✉)], Irina V. Smirnova[3], and Farida R. Khabibrakhmanova[3]

[1] Perm State University, Perm, Russia
la_kossolapova@list.ru
[2] Perm State Humanitarian Pedagogical University, Perm, Russia
margarita_67@inbox.ru
[3] Perm National Research Polytechnic University, Perm, Russia
iraivanovskaya@rambler.ru, faridhin@mail.ru

Abstract. The article discusses the methodology for the development of intellectual competence of biotechnology students in the process of learning a foreign language, and organizational and methodological learning environment that provide a comprehensive, holistic development of intellectual skills in the context of a competence-based approach to modern professional-oriented education of technical university students enrolled in the field of biotechnology. The article focuses on the analysis of the concept of intellectual competence, the components and levels of competence development, considered from linguistic and methodological points of view. As a result of the research, the role of development of cognitive processes (knowledge, understanding, application, analysis, evaluation, synthesis) in a hierarchical sequence from simple to complex as target components of educational activities in the development of the future biotechnologist's intellectual competence has been revealed.

Keywords: Intellectual potential · Intellectual competence · Biotechnology students · Organizational didactic conditions

1 Context

Due to the latest changes in the political, economic and social spheres, the importance of the professions ensuring the innovative development of our country has increased. There is a high need in specialists whose professional activities essentially include an intellectual component.

The intellectual competence of a specialist directly determines their competitiveness in professional sphere, their willingness to use their intellectual potential for the good of the future of the country.

The last decades are characterized by outstanding achievements in biotechnologies, which is actually an interdisciplinary area of knowledge based on microbiology, biochemistry, molecular biology, bioorganic chemistry, biophysics, virology, immunology, genetics, engineering sciences, and electronics (Yegorova 2003:4). In this regard,

M. E. Auer and T. Rüütmann (Eds.): ICL 2020, AISC 1328, pp. 183–193, 2021.
https://doi.org/10.1007/978-3-030-68198-2_16

there is a high demand for the specialists who are able to solve the global problems such as protection of environment, famine relief, natural resources depletion, development of new technologies and improvement of the old ones.

Training of future biotechnologists is performed at different educational levels within the system of higher professional education, including the baccalaureate level. Such a training should result in "the person himself, who has passed the educational course within the certain educational system. It is his own experience as a complex of formed intellectual, personal, behavioral qualities, knowledge and skills, which enables him to perform adequately based on his knowledge in any situation" (Zimnyaya 2006:35).

Meanwhile, in the process of mastering the discipline "Business (professional) foreign language", future biotechnologists experience certain difficulties that are associated with the development of a number of intellectual skills. This is the ability to select information in a foreign language, the ability to process and convert information from foreign sources relying on discursive operations, the ability to determine the goals and objectives, the ability to draw up a plan, evaluate and reflect on one's own foreign language activities. Thus, this research is aimed at solving this problem.

2 Purpose

The research problem, the essence of which is to identify the theoretical and methodological foundations for the development of the intellectual competence of a future biotechnologist in the process of studying a foreign language and the organizational and didactic conditions for its implementation at the university as part of professional language education, determined the purpose of the study. Therefore, the goal of the research is to scientifically prove, practically develop and experimentally test the methodology for the development of intellectual competence of future biotechnologists in the process of learning a foreign language.

3 Approach

The research has been carried out in Federal State Budgetary Educational Institution of Higher Education 'Perm National Research Polytechnic University'. The pilot study involved 48 students of the Faculty of Biotechnology, who were divided into control (36) and treatment (12) groups.

For the purpose of identifying the validity of the research there has been worked out and implemented a diagnostic and reflective complex which aimed at assessment and self-assessment of the process of the students' intellectual competence development: its motivational, cognitive, metacognitive, self-educational, research, communicative and personality components. The mathematical methods of test data analysis have been used.

4 Actual or Anticipated Outcomes

During the research, a contemporary state of the intellectual competence formation described in scientific works, has been studied. On the basis of the information obtained, definition of intellectual competence was clarified and expanded as the one that is basic and fundamental ability to master the methods of active cognitive processes on comprehension, application, analysis, summarizing, evaluation, and synthesis of information. It allows to efficiently solve the professional tasks of different complexity, using one's personal qualities based on individual experience in order to achieve the results in a certain subject area. This definition has been applied to the process of teaching biotechnologists foreign languages.

The structure of biotechnology students' intellectual competence has been defined on the basis of the works of I. A. Zimnyaya, O. N. Yarygin. It is considered as the complex of the following components: motivation-value-based, cognitive, metacognitive, self-educational, research, communicative (Yarygin 2013: 382) and personality.

The described components of intellectual competence allowed us to find out the levels of its maturity: elementary, intermediate, and advanced. Certain organizational and didactic conditions for efficient development of intellectual competence of biotechnology students, have been distinguished. The use of foreign language for developing of intellectual sphere of the future biotechnologist's personality, is considered as the main of the conditions mentioned. This statement is based on the approaches of such researchers in the sphere of neuro- and psycholinguistics as L. S. Vygotsky, A. A. Zalevskaya, A. R. Luria, L. V. Scherba, and others.

The next organizational and didactic condition follows from the definition of intellectual competence, according to which each cognitive process of intellectual competence is the target unit of educational activity.

Focusing on a specific goal mobilizes the intellectual powers of students, directs their mental activity, and gives it a certain meaning, which significantly expands the possibilities of learning, directing it to the development of intelligence and intellectual competence, in particular. Target orientation allows the teacher to determine the degree of students' progress towards the intended outcome, namely the formation of the experience of intellectual activity, and to ensure timely correction. Thus, setting and specification of goals in the teaching process while creating intellectually-provoking educational environment, and focusing on the process of mastering thinking skills, are essential parts of the educational model aimed on the intellectual competence development.

At present, the model created by the group of American psychologists and educational specialists under the supervision of Professor Benjamin Bloom, is considered to be the most developed theory of educational goals. In his classification B. Bloom divided the goals of education into three areas: cognitive (the requirements to comprehension of subject content), psychomotor (development of action-based skills and neuromuscular activity), and affective (emotion-based domain, attitude to the material a student is learning) (Bloom 1956:59). Bloom's taxonomy presents classification and categorization of educational learning where cognitive processes are arranged from simple to more complex ones.

While teaching foreign languages biotechnology students, it has been found out that focusing on a specific goal mobilizes biotechnology students' intellectual power, directs their mind activity, fills it with a certain sense, substantially expanding their ability to learn a foreign language, encouraging students to develop their intelligence and, particularly, intellectual competence.

A proper choice of the tools used for students' intellectual skills formation is considered to be the third essential condition for future biotechnologists' intellectual competence development. It is not coincidence that graphic organizers are supposed to be among such tools, as students' learning process, especially in biotechnology, involves necessity of learning huge amounts of new information. Further comprehension, processing and analysis of this information depends on the ability to compact informational content of new material for emphasizing the most general ideas and interlinks between them. In this case, graphic organizers are becoming an irreplaceable tool. Along with L. M. Levina, we consider a graphic organizer to be a visual or graphic representation of relationships between some facts, terms, concepts and ideas according to certain educational task (Levina 2010:12).

Sometimes, graphic organizers are called concept maps, cognitive organizers, conceptual diagrams, etc. According to D. S. Ausubel, graphic organizers are the visual basis of information (Ausubel 2000:41), which give teachers and students numerous and various possibilities to use them as a tool for developing new terms, arranging language and speech units, for better understanding the subject, and for more successful application of the obtained information in professional sphere in order to get desirable results and solving the tasks set. Graphic organizers, or concept maps (Novak 1990:37), can help students to sort out, simplify information, to determine relations between concepts and subjects, and teach them to manage the data fast and easily. K. Bromley and M. Modlo define graphic organizers as visual representation of knowledge (Bromley and Modlo 1999:26). As S. I. Zair-Bek and I. V. Mushtavinskaya have noted, while implementing the strategy of critical teaching, it is necessary to widely use graphic organizers (schemes, graphs, circles, tables, clusters) – the means for concepts, facts, and objects correlation (Zair-Bek and Mushtavinskaya 2011:29). The usage of graphic organizers facilitates formation of such general scientific actions as: ability to analyze, systemize, generalize, interpret, creatively process information on every stage of its learning.

Wide potential of graphic organizers (ideas generation, visualization, structuring during the process of learning and arranging of information, solving problems, making decisions (Rosch 1999:201)) determines perspective areas of their application for future biotechnologists' intellectual competence formation. Graphic organizers can be seen as one of the basic didactic means that can be efficiently combined with the technologies of project, problem and research education.

The next condition that is essential for successful intellectual competence formation, is modelling of this process.

While creating the model, we relied on the modelling theory, representing the whole process of formation as the method of scientific cognition (Fig. 1).

The proposed model of intellectual competence formation consists of the following interrelated structurally functional components:

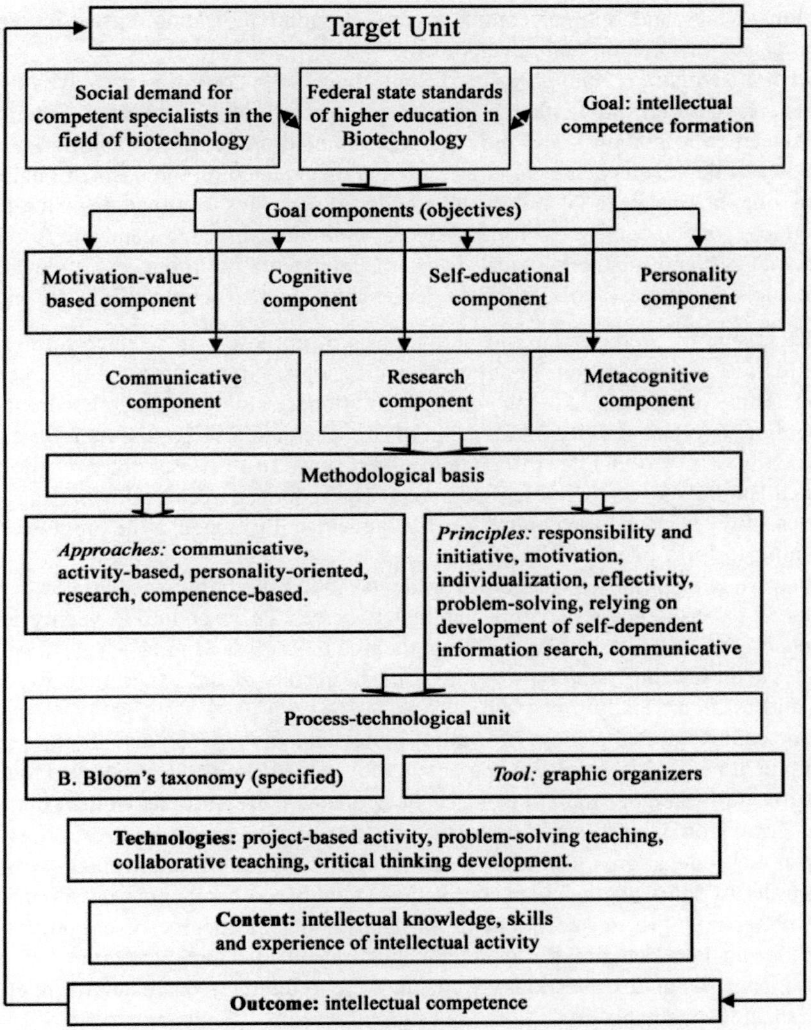

Fig. 1. Theoretical model of student's/bachelor' – future biotechnologist intellectual competence formation

- *motivational goal-setting component* carries out a stimulating, goal-setting and guiding function. It represents a set of state demand for specialists of a certain branch of industry, Federal State standard of higher education, and the main goal that is defined through a certain number of objectives that determine specific contents of future biotechnologist's intellectual competence formation process;
- *organizational technological component* ensures carrying out the performing function and is characterized by describing types, methods, methodology and means of intellectual teaching in accordance with target and content trajectory marking

rational ways and optimal control means of future biotechnologist's intellectual competence formation and development;

- *outcome evaluation component* carries out diagnostic, control and analytical functions. It describes anticipated outcome – biotechnology students' intellectual competence formation – and the means of its measurement and achievement. The quality of the results achieved is measured and evaluated on the basis of qualitative and quantitative criteria and figures under diagnostics (motivation-value-based, cognitive, metacognitive, self-educational, communicative, research, personality). Theoretical and methodological basis of intellectual training of students (approaches, principles, tools, technologies, forms) are also included into the model.

The system of exercises created on the basis of purposefully selected foreign language texts is the foundation for future biotechnologists' intellectual competence formation. This competence is being formed evolutionary step-by-step in accordance to the logic of cognitive activity. The system of exercises proposed in the training manual corresponds to this principle. It includes certain types of exercises for every level of specified Bloom's taxonomy described above. The sequence of exercises is arranged so that each of the steps is based on the knowledge and skills gained at the previous stage, facilitating to form intellectual skills.

The results of proposed method of students' intellectual competence development was tested experimentally at Perm National Research Polytechnic University among students-biotechnologists while teaching them the English language (the discipline "Business (professional) foreign language"). The results of the experiment were analyzed and mathematically processed.

For testing the efficiency of future biotechnologists' intellectual competence development method, control and diagnostic tools were developed. It allowed to define the nature of the intellectual competence components correlation as well as dynamics of their formation. Control and diagnostic tools include five scales (motivation, style of behavior, self-educational potential, creativity, critical thinking intellectual skills).

In order to define the levels of motivation-value-based component development of future biotechnologist, a questionnaire on determining the degree of cognitive motivation during intellectual skills development (created by Dombrovskaya I. S.), was used. It consists of 25 questions. Students were offered to evaluate the degree of importance of statements given in the questionnaire using a scale ranged from 0 to 4: 0 for "never", 1 for "seldom", 2 for "sometimes", 3 for "almost always", 4 for "always". The choice of this diagnostic method is based on the idea that all the statements in the questionnaire allow not only to find out the dominant type of students' motivation (cognitive or social), but also to clarify dominant motives in different types.

The assessment of mean value was performed according to the following criteria:

- 30–40: a student has a generally high level of motivation in education; he/she knows why and what for he/she studies;
- 20–30: a student has a medium level of motivation in education;
- 0–20: a student has a low level of motivation in education, he/she doesn't understand what for he/she studies (Fig. 2).

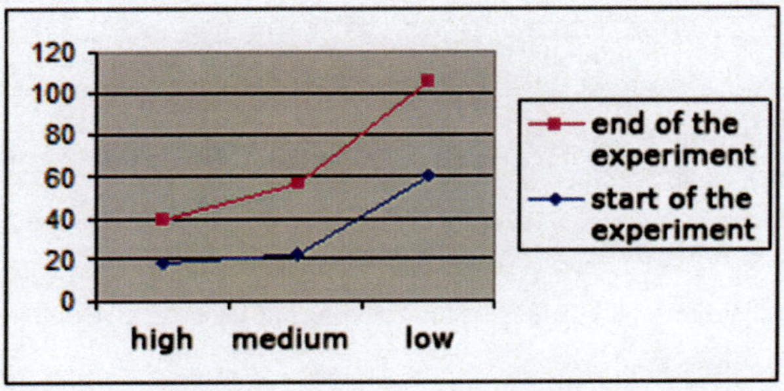

Fig. 2. The graph of dynamics of cognitive motives level change

Processing of data obtained at initial stage allowed to find out the following results: as seen on the graph, the total value of replies in the questionnaire demonstrate a low level of cognitive motivation among students. Their needs in new impressions are of a low level or entirely absent (according to L. I. Bozhovich), self-educational motivation is not well-developed. As for social motivation, the questioning revealed students' strong motives of obligatory or forced education, the absence of the motives of collaboration and social knowledge. The chosen method also helped to find out the level of students' personal attitude to the subject: only an insignificant number of respondents showed their interest in the subject; the majority of the treatment group was not interested in the subject. The answers to the last questions revealed students' fear of failure and uncertainty of half of the respondents.

Cognitive component in the intellectual competence structure is systemically important element as knowledge is one of the significant criteria for evaluation of biotechnology students' intellectual competence level. It is necessary to emphasize that usage of knowledge can be defined automatically both by its content and a creative task which is carried out by a student in a new situation every time. This component is revealed through such indicators as language knowledge (grammar, vocabulary, phonetic aspect and stylistic nuances), abilities to analyze, systemize and generalize information.

As a tool for evaluation of cognitive components, different types of tests were chosen. At initial stage of the research, 38% of students under testing showed a high level of cognitive component distinct for deep systemic, strong knowledge of the language, expressed in its different aspects (vocabulary, grammar, phonetics, stylistics). 35% of students showed a medium level of systemic knowledge; its depth and stability were demonstrated from time to time and regarding just to some aspects. 27% of respondents showed a low level of knowledge which were shown rarely and can be characterized as non-systemic, unperceived and shallow.

By the end of the experiment, a total number of students having a low level of cognitive component development, decreased (Fig. 3).

The obtained results, according to the scale of critical thinking intellectual skills formation level, indicate that students showed gradual increase in metacognitive component of intellectual competence development (Table 1).

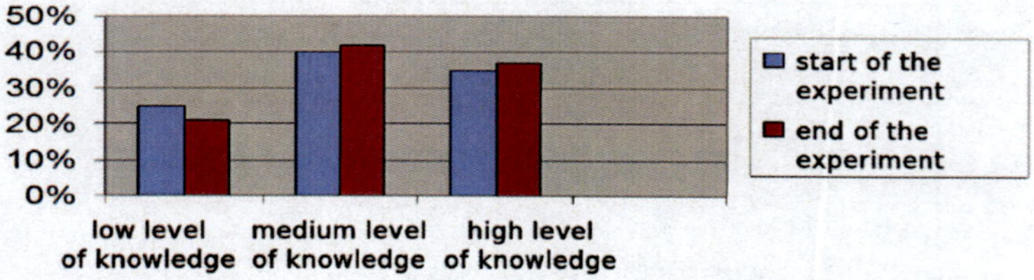

Fig. 3. The diagram of dynamics of future biotechnologist knowledge level change

Table 1. Diagnosis of the level of students' inclination towards intellectual skills

Students of treatment group	C1	C2	C3	C4	C5	C6	C7	C8	C9	C10	C11	C12
Inclination level at initial stage	56%	57%	48%	52%	53%	52%	41%	44%	53%	59%	50%	43%
Inclination level at final stage	59%	63%	50%	55%	55%	57%	44%	46%	58%	64%	53%	45%

Evaluation of intellectual competence self-educational component development was performed by means of the questionnaire created by Mensirova V. E.

For treatment and control groups, the average value (mark) was calculated; it was used for determining the dynamics of students' self-educational component development during the two stages of the experiment.

Analysis of students' intellectual competence self-educational component development at initial stage demonstrated that 36.4% of treatment group students assessed their own self-education intellectual activity skills to be rather high; 45.5% – assessed these to be of medium level; 18.1% – rather low. The questionnaire results at the final stage of the experiment demonstrated 1.5–2% increase of self-estimation level of self-education intellectual activity skills among treatment group students.

The study of research component development among biotechnology students was carried out based on the test created by V. E. Milman. It studies students' research potential, the process of task solving performed by them; it also helps to quantitatively evaluate their thinking process while solving the tasks.

The situational solution-seeking tasks are used in methodology. They are quite unusual, but contain a certain question or problem, solving of which should be interesting and non-trivial. During the search of solution, a respondent can ask a teacher any questions requiring a positive or negative answer. They have to look as a hypothesis: they should contain a certain assumption regarding some characteristic of situation or the solution itself.

The following criteria are taken into consideration when processing the results:

T – *time spent* for solving the problem (min.);

Σ – *length* of solution (total number of steps in solving the problem);

$\mathbf{\Sigma_{inf}}$ – *informative base* of solution (total number of informative, significant steps). A step is considered informative if its result discards a rather sizeable number of possible versions of solution.

Σ_{inf}/T – *informative rate* of solving the problem. Informative accumulation rate for a period of time (average per 1 min);

Σ_{inf}/Σ – *informative efficiency* of solution. Average informative accumulation per step.

According to the results of the research it has been determined that 28% of students gained a relatively high level of research (scientific) potential development by the end of the experiment. They demonstrated the ability to solve problems faster and more accurately, properly using the information obtained from the teacher. Every 6–7 informative solution steps included a constructive one, containing a real alternative solution.

41% of respondents showed a medium level. Their solutions were of a medium level and sufficiently accurate. The students rather actively used the obtained information, but showed narrowness and the lack of variety of solution areas. Their solutions were achieved with the use of a multitude of small pieces of information within a narrow activity area. They did not succeed notably, except of several last steps of solution.

31% of future biotechnologists showed a relatively low level of research potential development. Their work was slow and narrow-minded, though it was rather interesting. The students did not take into account the teacher's replies to their questions. It looked like they were more interested in communication on certain topics than actually solving the problem.

On the results obtained, the conclusion has been made that students having medium to high level of development of intellectual competence research component, are characterized by purposefulness, consciousness of educational problem solving ways and methods, self-dependence in transformation of ideas and their interrelations, and being enthusiastic about the research. They express strong interest in research, accepting their intellectual activity as a real value.

For processing the questionnaire data, four factors were used. They are closely related to creative peculiarities of a person such as: intellectual curiosity, imagination, complexity, risk proneness.

The four creativity factors were determined on the base of positive and negative replies which were evaluated at 2 points; the replies "Partially agree" were evaluated at 1 point; the replies "It's hard to answer" were evaluated at -1 point.

This scale allows to determine the person who is not creative enough, or is not confident. 12 of 50 statements in the questionnaire refer to curiosity, 12 – to imagination, 13 – to risk proneness, and 13 – to complexity factor. If all the answers match the key, then the total "raw" grade can be equal to 100, provided that there are no "Don't know" replies. If a student under the treatment gives all the "Maybe" replies, then his total "raw" grade can be equal to 50, provided that there are no "Don't know" replies.

The final quantitative result of every factor is worked out by summation of all the replies matching the key, the "Maybe" replies (+1), and subtracting all the "Don't know" replies (−1). The higher the "raw" grade of a person having positive feeling towards himself, the more creative, curious, imaginative, able to take a risk and solve a complicated problem, he is. All the personality factors mentioned above are closely related to creative abilities of a person. The result on every test factor separately, as well as total grade value, can be obtained. Separate factor grade and total grade demonstrate in the best way strong features of a student (high "raw" grade), and the weak ones (low "raw" grade). These grades can be later converted into standard marks and put into a student's personal profile.

The following results were demonstrated by students at initial and final stages of the experiment:

Creativity factors	End of the experiment	Start of the experiment
Curiosity	17.5	16.4
Imagination	16.0	15.6
Complexity	17.2	16.7
Proneness to risk	18.0	17.0
Total result	67.6	64.5

The results obtained confirm that the higher the grade of a person having positive feeling towards himself is, the more creative, curious, imaginative, able to take a risk and solve a complicated problem, he is. He is able to create extraordinary ideas, deviate from traditional thinking patterns, to solve problem situations quickly. His personality component of intellectual competence enables him to a certain entity of thinking and personal qualities, contributing to his creative manifestation. His personality potential comes up when: there is a lack of knowledge; new information is introduced into new structures and relationships; missing information is being identified; searching for new solutions and verifying them; when the results are being reported.

5 Conclusions/Recommendations

Contemporary specialists in biotechnology need to master a wide variety of professional competencies: ability to find non-standard solutions of tasks and problems, skills of collaborative work, communicative skills, creative approach to the work, high level of accuracy and focusing on details, ability to analyze statistical and technical data, skill to assess critically the results obtained, etc. All these skills comprise the matter of the term "Intellectual competence", and their formation and development should take place at every level of educational process.

The tendencies revealed allow to conclude that the proposed methodology, implemented in the process of teaching a foreign language biotechnology students, provides development of their intellectual competence in close interrelation of all its components.

Further research prospects may be associated with the application of the developed methodology for the development of intellectual competence in the context of training in the framework of master's and postgraduate educational programs.

References

Ausubel, D.S.: The Acquisition and Retention of Knowledge: A Cognitive View. Springer Science & Business Media Dordrecht, p. 261 (2000)

Bloom, B.S. (ed.): Taxonomy of educational objectives: The classification of educational goals: Handbook I, cognitive domain / B.S. Bloom, New York, Toronto, Longmans, Green (1956)

Bromley, K., Modlo, M.: Graphic Organizers (Grades K-8) Paperback. Teaching Resources, p. 120 (1999)

Levina, L.: Organizatciya samostoyatelnoi raboty studentov v usloviyach perekhoda na dvukhurovnevuyu sistemu vysshego professionalnogo obrazovaniya: metodicheskoye posobiye. – Niznii Novgorod: Nizhegorodskii gosuniversitet, p. 95 (2010)

Novak, J.D.: Concept maps and Venn diagrams: two metacognitive tools to facilitate meaningful learning. Instr. Sci. **19**(1), 29–52 (1990)

Rosch, E.: Principles of categorization. In: Margolis, E., Laurence, S. (eds.) Concepts: Core Readings, pp. 189–206. MIT Press, Cambridge, MA (1999)

Yegorova, T.: Osnovy biotechnologii: ucheb,posobie dlya vyssh. ped.ucheb, zavedenii. – M.: Izdatelskii zentr "Akademiya", p. 2008 (2003)

Yarygin, O.N.: Model intellektualnoi kompetentnosti kak obobscheniye modeli tvorcheskoi deyatelnosti // Vektor nauki TGU, №. 1(23), pp. 382–387 (2013)

Zair-Bek, S, Mushtavinskaya, I.: Razvitiye kriticheskogo myshleniya na uroke, - 2-e izd. Prosvescheniye, Moskva, p. 219 (2011)

Zimnyaya, I.: Obschaya kultura i socialno-professionalnaya kompetentnost cheloveka//Internet-zhurnal "Eidos" (2006)

Proposal for Laboratory Didactic at an Electrical Engineering Program: A Teaching-Learning Strategy for Laboratory Activities in Electrical Energy Conversion Applications

Diego Gormaz-Lobos[1](✉), Pablo Acuna[2], Claudia Galarce-Miranda[1], and Steffen Kersten[3]

[1] Faculty of Engineering, International Center of Engineering Education, Universidad de Talca, Talca, Chile
diego.gormaz@utalca.cl
[2] Faculty of Engineering, Universidad de Talca, Talca, Chile
[3] Faculty of Education, Technische Universität Dresden, Dresden, Germany

Abstract. The interaction between scientific knowledge and the repertoire of university teachers on laboratory didactic is essential for the learnings process of engineering students. Better teaching and learning methods, and resources that motivate and produce active processes of learning at the students and, at the same time, develop competencies and skills are the key components. The work at the School of Electrical Engineering of Universidad de Talca has tried to take these precepts and is specifically aimed to strengthen the work in laboratories in the area electrical energy conversion such as renewable energy integration and machine drives. The general goal of this paper is to present a proposal on laboratory didactic at an electrical engineering program at Universidad de Talca. The design of a teaching-learning strategy for laboratory activities in the area of electrical energy conversion will be discussed. Specifically, the authors of this paper will describe: (1) the scientific arguments that support the decision for designing laboratory activities based on the precepts and characteristics of the modern laboratory didactics, (2) present and describe the design of a teaching-learning strategy for laboratory activities of courses in the area of electrical energy conversion, and (3) a description of laboratory resources with expected results of the teaching-learning strategy.

Keywords: Laboratory didactics · Electrical energy conversion · Teaching-learning strategy for laboratory activities

1 Introduction

Laboratory work is in engineering programs an important part of the training of engineers. The permanent technological change and transformation of the economy and productions structure (for example through Industry 4.0 and the "smart services") create a new digital working context for them the new engineers must be adequately

M. E. Auer and T. Rüütmann (Eds.): ICL 2020, AISC 1328, pp. 194–204, 2021.
https://doi.org/10.1007/978-3-030-68198-2_17

prepared. The discussion today about the design of laboratories for engineering education in industrialized countries like Germany, is not only oriented to guided presently teaching-learning environments, but also virtual and tele operative (remote) laboratories.

At engineering laboratories, the work of engineering students is focused on putting technical knowledge on a practical plane. In general, well-planned experiments are executed to identify and understand the "connections naturally given". Kammasch (2006) describes that even technical applications are "ultimately based on the properties and laws of nature. Here, students can independently derive the natural conditions from the rules and laws themselves. Curiosity, the joy of discovery, must be used and strengthened in laboratory experiments since it guarantees the greatest and most sustainable learning success" (p. 8) [1]. Therefore, the interaction between scientific knowledge and the repertoire of university teachers on laboratory didactic is essential. Better teaching and learning methods, that motivate and produce active processes of learning at the students and, at the same time, develop competencies and skills are the key components [2]. Throughssexperiments in their university trainings, students have the chance [3]:

- to test, apply and verify knowledge,
- to establish differences between abstraction and reality,
- to have the possibility to learn that scientific work a combination of teaching and research can be,
- to see the contents in form of research knowledges and its applicability,
- to observe concrete research processes,
- to be quickly involve with equipment and methods of the working context, and
- to learn in an active learning environment, among others.

The work at the School of Electrical Engineering of Universidad de Talca has tried to take these precepts and is specifically aimed to strengthen the work in laboratory activities in the area of electrical energy conversion applications such as renewable energy integration and machine drives.

2 Laboratory Didactics in Engineering Education

2.1 Definition and Types of Experiments in Engineering Education

The word "Laboratory" comes from the Latin term "laborare", that means to work [4]. The typical definition of Laboratory is related to a working place to experiment. Experiment comes from the Latin term "experiential" that means experience, trial, experiment [5]. In his book "Learnings Methods" (Unterrichtstmethoden), Meyer (1989) defines experiment as "a scheduled controlled trial to review a problem or to clarify any unclear facts" (p. 313) [6]. Also, in general the laboratory didactics can be defined as a discipline that is responsible for the theory and practice of teaching and learning in the laboratory.

From a didactical poin of view, Eicker (1983) distinguishes between four functions of experimentations for the human learning activity [7]:

- Experimentation as cognition activity,
- Experimentation as trial and error,
- Experimentation as problem solving, and
- Experimentation as an optimization of perceptions.

For Meyer (1989) the essential features of an experiment are: (1) the scheduled execution, (2) artificial "production" or "execution" (differently of an easy observation), (3) the possibility of repetitions and (4) the possibility of the control of emerged laws or regularities.

Based on Bünning's (2006) classification of types of experiments, the authors propose a distinction of the following types of experiments for Engineering Education according to the following criteria (see Fig. 1) [8]:

1. Goals for the training of engineers
   - *Research experiments.* The main goals of this type of experiments are: to study of unknown relationships with quantitative or qualitative empirical methods, formation of hypotheses based on the experimentation, to build of objective evaluation criteria, prediction and limitation of results, among others.
   - *Teaching-learning experiments.* The main goals of this type of experiments are: to known regularities and correlations and the outcome of the experiment, to test experimental set-ups, and to introduce students at the activities at a "real" research experiment.
2. Types of problems in the scientific disciplines of the training of engineers
   - *Natural science experiments.* The main goals of this type of experiments are: to study cognitions problems, to investigate causal or conditional relationships, and to build "truth" (based on natural laws and principles) as a test criterion, among others.
   - *Technical experiments.* The main goals of this type of experiments are: to study design problems, to investigate of conditional and final relationships, and to build suitability as a test criterion.
3. The relationships to investigate
   - *Causal relationship.* The main goals of this type of experiments are the relationship between: cause and effect, and nature and phenomenon.
   - *Conditional relationship.* The main goals of this type of experiments are the relationship between: condition and conditioned, and reason and consequence.
   - *Final correlations.* The main goals of this type of experiments are the relationship between: purpose and means, and structure and function.
4. The learning goals and his orientation to personality dispositions
   - *Cognition-oriented experiments.* The main goal of this type of experiments is the acquisition of scientific and technical knowledge. This kind of experiment is strong oriented to the activation of cognitive dimension of the students.

- *Application-oriented experiments.* The main goal of this type of experiments is the development of act ability in the laboratory work. This kind of experiment require the activation of cognitive, psychomotor and emotional dimensions of the students.

5. The didactic function of experiments for the training of engineers
   - *Experiments for entry into a subject.* The main goal of this type of experiments is the factual or emotional "motivation" and "unlocking" of the students for the learning activity.
   - *Experiments to acquisition of new content.* The main goal of this type of experiments is the gaining knowledge in specific ways of cognition.
   - *Experiments to strengthen, deepen and consolidate.* The main goal of this type of experiments is made available, applicable, and transferable the developed personality dispositions.
   - *Experiments on the monitoring and evaluation of learning outcomes.* The main goal of this type of experiments is to give feedback about learning outcomes of the students and the behavior in the laboratory.

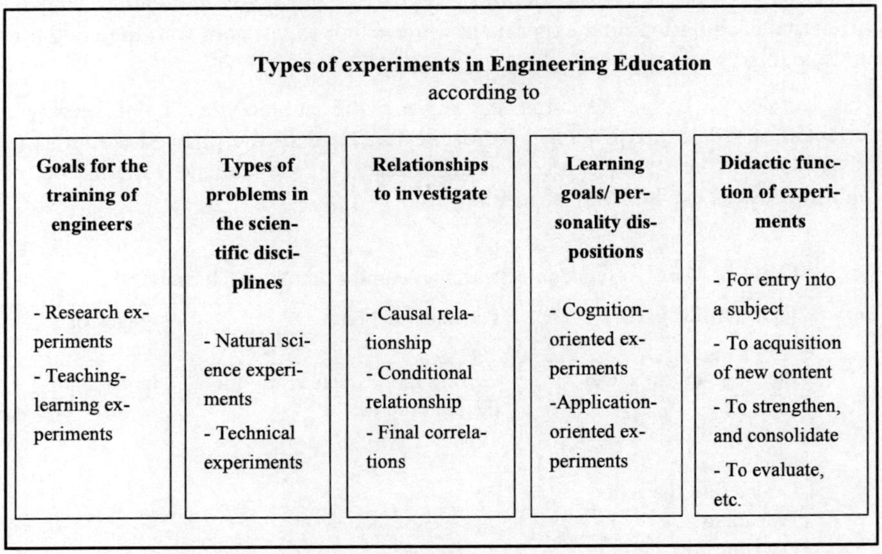

Fig. 1. Types of experiments in engineering education according different criteria.

2.2 Different Aspects About Laboratory Didactics in Engineering Education

Prof. Albert Haug was the first academic to speak scientifically about laboratory didactics (Labordidaktik) at engineering trainig programms in Germany. In his text "Labordidaktik in der Ingenieurausbildung" (laboratory didactics at the engineering training, 1980), Haug presented in a systematic and scientifically-pedagogically grounded way, the teaching in the laboratory (Laboratory didactics) [9]. The laboratory is a space for team teaching in smaller work groups, with the goals to apply and deepen knowledge and discuss "practice" [1]. The work at laboratories has a key function at the engineering training for example for (i) the development of experimental skills, (ii) the development of a scientific understanding, (iii) the learning of methods for the scientific work, (iv) the application of theory at practice, (v) the experience of the role (for the first time) as an engineering scientist, and (vi) the introduction and use of equipment and processes of the future professional life as engineer [1,9].

Kammasch, citing Haug and Bruchmüller, develops the principle of "levels" in laboratory teaching-learning activities with connection to various aspects related to the functions of the laboratory and the requirements for students [1]:

1. Level I: Reproduction. Reproductive acting of knowledge, laws, methodologies, and principles. The students received closed activities and tasks.
2. Level II: Transfer. Productive acting of knowledge and problem solving with open point of view for results. The experiments have own experimental setup with close connection to professional praxis. The students received partial open tasks.
3. Level III: Creative problem solving. Productiv-problem solving acting. Transition from the teaching-learning experiment for research experiment with high autonomy of students.

For Kersten (2016), the teaching-learning activities at laboratory in engineering can be oriented and designed trough the didactic functions of the planned activities (see Table 1) as well the stages of the learnings process of the students (orientation, performing of tasks or evaluation of results) [2].

Table 1. Didactic function of teaching-learning activities (Hortsch 2012).

Group	Didactic function	Didactic intention	Stages of learning acts
1	Starting with practical examples Getting in the mood of Motivation Orientation for action Preparation Decision for action	Getting the learners ready for the learning act	Orientation
2	Working on new contents/learning acts	Opening up the contents for the learners	Performing of tasks
3	Working on known contents/learning acts Consolidation of formerly acquired knowledge/skills	Improving the availability of learning results for the learners	Performing of tasks
4	Evaluation of results/outcomes Assessment of learner performance	Feedback on the performance level for learners and teacher	Evaluation of results/outcomes

In his text "Das Labor in der ingenieurwissenschaftlichen Ausbildung" (Laboratory in the engineering training), Tekkaya et al. (2016) presented different recommendations for the development of laboratory teaching-learning activities in engineering education. Some of them are presented in following [10]: (1) to consider the connections between theory and practice, (2) to consider aha!-effect (unexpected clash) as motivator for learning process, (3) to incorporate experiential (problem-based and research-oriented) learning, (4) students must have the possibility to work independently at research questions and hypotheses, (5) to consider the development of soft skills at students (communicational, team work, learning of English), (6) experiments must be planned with different difficulty levels, and (7) to incorporate complex problem that improve the creativity and problem-solving at a high level, among others. The authors also warned of various aspects that must be planned to avoid problems in the implementation of laboratory teaching-learning activities (see Table 2) [10]:

Table 2. Warned aspects for avoiding laboratory activity implementation problems.

Aspect	Description
Time	The amount of time students spent into laboratory activities
Resources	The availability of resources and space
Cost	The cost of technology and maintenance of laboratories is very high
Group size	The group size of students
Staff training	The increase of the number of staff for laboratory training
Teaching-learning induction	The consideration of previous activities for the activation of knowledge or induction to teaching-learning into the laboratory (that prevents rapid and efficient work in the laboratory)
Teaching-learning structure	The structure of the teaching-learning strategy at laboratory (explicit as orientation for the students)

3 Traditional Teaching-Learning Methodology

The standard teaching-learning resources used in laboratory activities in the area of electrical energy conversion includes, but not limited to:

1) use of simulation-based software,
2) use of industrial control systems,
3) use of hands-on learning training systems,
4) use of software and hardware development tools.

Each of them has its distinctive features while providing different levels of outcomes. Per example, simulation-based software such as Matlab® is used as the first approach for simulating electrical energy-conversion system models in the classroom before exploring real experiences with hardware in the laboratory. Industrial control systems like Programmable Logic Controllers (PLC) and pieces of industrial equipment are used to immerse students into an industry-related environment. Examples includes, but not limited to machines drives ranging from 1 to 5 kW (see Fig. 2), PV inverters below 1 kW when laboratory space is limited (see Fig. 3), and automated energy management systems (EMS) in the context of battery storage, smart grid. Hands-on learning training systems like Festo training systems are used to methodically build technical skills (see Fig. 2). Evaluation modules made available by manufactures for their development tools such as Texas Instruments (TI) processor platforms for digital signal processing (DSP), and system on a chip (SoC) devices, normally an ARM®-based processor with the hardware programmability of a Field-programmable gate array (FPGA), are used as a flexible and low-cost option to help students develop and test control algorithms that govern electrical energy-conversion systems (see Fig. 4). As a distinctive feature, some development kits such as the ones shown in Fig. 4 provide access to the design and construction files. The latter is considered an open-source feature, but commonly these kits are non-teaching oriented. Instead, they are available for testing and integrating products from their own product catalog such as microcontrollers, semiconductors, among others.

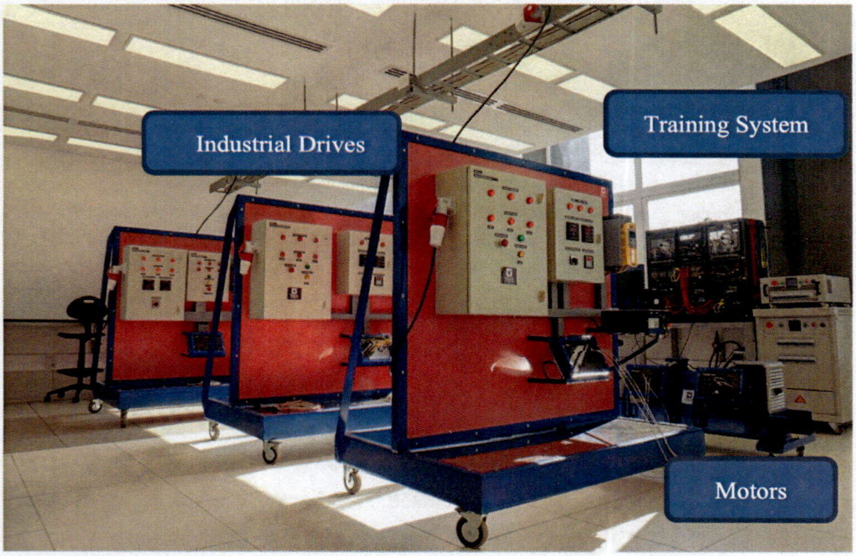

Fig. 2. Machine drives laboratory at the school of electrical engineering, Universidad de Talca.

Regardless of what or how the laboratory resources are used, the traditional teaching-learning methodology is more less the same across courses in the area of electrical energy conversion. The theory is presented in the classroom, students validates theory using simulation-based models, and finally the understanding of students is enhanced in the laboratory. There is no doubt that this “finishing touch” contributes to reach the aforementioned “Level II” due the close connection to professional praxis. Here, specific goals such as hands-on experience and practical case studying are fully accomplished, however there are still goals such as innovation, proactivity, critical and scientific thinking are not fulfilled.

4 Proposed Teaching-Learning Methodology

The proposed methodology is aimed to reach the aforementioned "Level III" taking into consideration the implementation aspects discussed in Table 2. Instead of focusing on how the availability of laboratory resources shapes the teaching-learning methodology, the authors propose that students collaborate in creating laboratory resources i.e. prototypes.

For that purpose, students are guided in the design, develop and validation process of a laboratory equipment prototype. The main features of these prototypes are set after discussing what is available in the market for teaching-oriented laboratory equipment in the area of electrical energy conversion. Each aspect described in Table 2 is taken as the design goals to be fulfilled by the prototypes. The latter is the starting point of the proposed methodology. Then, the methodology is split into a number of small tasks that follow a project pathway. Finally, the validation stage is reached, and students share their results. The whole teaching-learning methodology resembles the steps involved in R&D activities.

The results obtained for the first time the methodology is applied are shown in Fig. 5. Here, students have reached the design stage of two prototypes: a modular control platform shown on the left, and modular power electronic converter on the right. The main features obtained by the prototypes are modularity and open-source access. By adding modularity while sharing the design and constructions files as the main features, the conclusion of the students have indicated that the experienced learned and the prototypes created can be used as a powerful mitigation strategy to alleviate the negative impacts of the warned aspects described in Table 2. In this case, students were able to:

1. interact with modules more like functional parts to understand how they are designed and built,
2. work on small tasks that belongs to a particular module inside the laboratory equipment,
3. evaluate how the whole system works when one or more modules are faulty or "have a design bug",
4. propose improvements at the design level, not only changing or connecting parts,
5. avoid step-by-step methodical instructions which only help to achieve understanding-related outcomes,
6. reduce the time to production of laboratory resources while facilitating the implementation and validation of control algorithms.
7. develop a sustainable innovation vision about creating "self-evolving" prototypes

From a teaching-learning methodology point of view, the main feature is that there is no "finishing touch" as the laboratory resources will always evolve. Moreover, the starting point is also dynamic as the students have different point of view of the same problem. The student-oriented teaching-learning structure objective is reached as the students are encouraged to be protagonist in the steps of development prototypes that can be used during laboratory activities. Moreover, same prototypes can be taken by future students as the starting point for a new R&D activity.

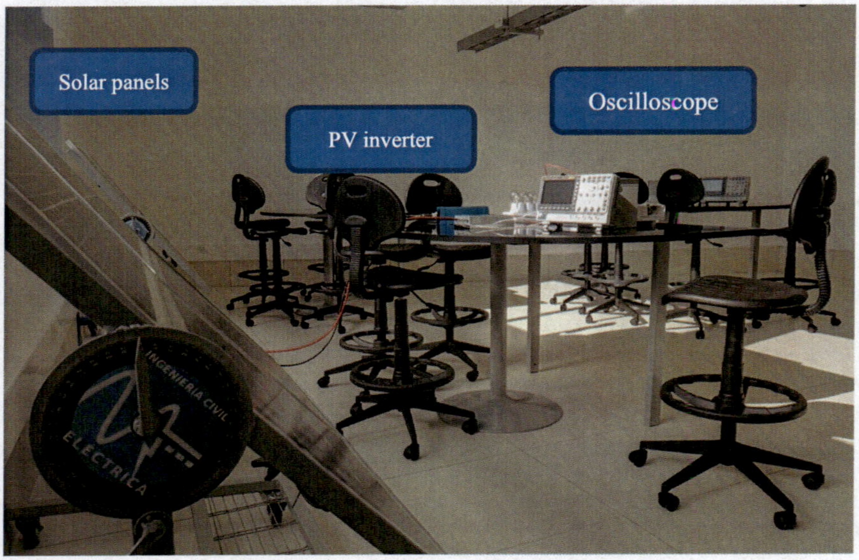

Fig. 3. Renewable Energies Laboratory at the School of Electrical Engineering, Universidad de Talca.

Fig. 4. Piccolo-based Solar Explorer Development Kit (left) and Solar Micro Inverter Development Kit (right) (source: www.ti.com).

Fig. 5. Actual prototype results (3D conceptual design) for two pieces of laboratory equipment created by undergraduate students: Modular control platform (left) and Modular Power Electronic Converter (right).

5 Conclusion

In one way or another, laboratory activities that use these industrial resources are organized and well-planned to offer students a first-hand experience with engineering equipment, taking measurements, making observations and carrying out well-defined procedures. This methodology has been forced on students for the last 30 years, which is why pretty much every energy conversion laboratory is equipped with pieces of low power industrial equipment, and in recent years with development kits. The proposed teaching-learning methodology aims to immerse the students into an innovation-related environment with the same resources they found in R&D activities.

The contribution of the proposed teaching-learning strategy for laboratory activities relies on offering a sustainable innovation vision about creating "self-evolving" resources for teaching in laboratories of electrical engineering. The proposed methodology opens opportunities for innovation by creating products and services at the university in the area of electrical energy conversion applications while offering participation of undergraduate engineering students, strengthening their autonomy and their creative problem-solving capacity through collaboration in innovation-related laboratory activities.

References

1. Kammasch, G.: Labordidaktik in der Diskussion. Das Labor und die Nutzung seiner methodischen Vielfalt im derzeitigen Umstrukturierungsprozess der Hochschulen. In: Berendt, B., Voss, H.-P., Wildt, J. (eds.): Neues Handbuch Hochschullehre. Dr. Josef Raabe Verlags-GmbH, Stuttgart (2006)
2. Kersten, S.: Das Modul Labordidaktik (Lehrmaterial). TU Dresden, Dresden (2016)
3. Gormaz-Lobos, D., Galarce-Miranda, C., Hortsch, H., Kersten, S.: The needs-oriented approach of the dresden school of engineering pedagogy and education. In: Auer, M., Hortsch, H., Sethakul, P. (eds.) The Impact of the 4th Industrial Revolution on Engineering Education. ICL 2019. Advances in Intelligent Systems and Computing, vol. 1134, pp. 589–600. Springer, Cham (2020)

4. PONS: Online Wörterbuch. https://de.pons.com/%C3%BCbersetzung/latein-deutsch/laborare. Accessed 12 Apr 2020
5. PONS: Online Wörterbuch. https://de.pons.com/%C3%BCbersetzung/latein-deutsch/experientia. Accessed 12 Apr 2020
6. Meyer, H.: Unterrichtsmethoden II: Praxisband, Frankfurt a.M. (1989)
7. Eicker, F.: Experimentierendes Lernen : Ein Beitrag zur Theorie beruflicher Bildung und des Elektrotechnikunterrichts. Didaktischer Dienst, Bad Salzdetfurth (1983)
8. Bünning, F.: Experimentierendes Lernen in der Holz- und Bautechnik. Bertelsmann, Bielefeld (2006)
9. Haug, A.: Labordidaktik in der Ingenieurausbildung. VDE-Verlag GmbH, Berlin (1980)
10. Tekkaya, A.E., Wilkesmann, U., Terkowsky, C., Pleul, C., Radtke, M., Maevus, F. (eds.): Das Labor in der ingenieurwissenschaftlichen Ausbildung. Deutsche Akademie der Technickwissenschaften, Berlin (2016)

Development of Cross-Domain Competences for Work 4.0

Galyna Tabunshchyk[1(✉)], Peter Arras[2], and Carsten Wolff[3]

[1] National University Zaporizhzhia Polytechnic, Zaporizhzhia, Ukraine
galina.tabunshchik@gmail.com
[2] KU Leuven, Campus De Nayer, Sint-Katelijne-Waver, Belgium
peter.arras@kuleuven.be
[3] FH Dortmund, Dortmund, Germany
carsten.wolff@fh-dortmund.de

Abstract. Considering the technological and the economic reality that the world is changing rapidly and is moving to a much larger degree of digitalisation than it has even been, universities also need to adapt their degree studies, in order to prepare future graduates for a digitalised, globalised and connected world. In the paper authors suggest a cross-domain core of competences which could be implemented in a variety of masters' curricula to improve employability of HEI graduates by giving them the tools and insights to work in the future economic world. The work addresses the need for (new) educational resources in order to prepare for the future workplace (work 4.0). A cross border international and multi-disciplinary approach was put forward to make a cross-border competence center in the existing consortium of EUROMPM (European Master in Project Management).

Keywords: Work 4.0 · Digitalized project management · Creativity · Master program · Cross-domain competence

1 Introduction

The way people work and live is changing due to global trends. The connected world reflects not only on technology used, but also on the way people are able to use the (social) media and the new methods of gathering and using data. On a personal level the focus of devotion to work shift more towards the combination of work and private (family) life, which makes for example the distance to the workplace an important driver when selecting a job, or the ability for working from home.

In today's economy, the ability to scale rapidly and internationally is fundamental, thereby transforming the society into part of the value-added chain. Another very important characteristic of the economy is outsourcing, which causes an increase in the numbers of temporary positions with the usage of a variety of apps and online platforms. Self-employment in the EU constitutes 16,4% of the labor market [1].

Also labor reallocation is intensifying last years. The sectors most at risk of contraction in the future world of work are those that rely heavily on routine tasks, such as low-skill manufacturing jobs, but also some craft and administrative occupations.

M. E. Auer and T. Rüütmann (Eds.): ICL 2020, AISC 1328, pp. 205–211, 2021.
https://doi.org/10.1007/978-3-030-68198-2_18

Medium risk of automation affects 5% to 70% of tasks [2]. As automation and artificial intelligence are leaping into new areas, the human-robot partnership will further transform the nature of work. Research has shown that occupations requiring strong social, interpersonal skills and non-routine analytical skills have grown dynamically since 1980 with consistent wage growth. Even in the tech industry, there are jobs that require both technical and interpersonal skills that are on the rise.

The changing world of work requires a multi-layered policy response, covering the entire life-trajectory, from cradle to retirement. Education, training and the opportunity for retraining throughout one's life needs to be the starting point. The HEI system does not fully reflect this change in their educational programs, which are still mainly oriented towards teaching academic knowledge in one domain.

2 State of Art

In 2016 the 5 major universities in Flanders (Belgium) conducted a research in the Flemish companies, with students and young professionals to search for the competences young graduates acquired, where expected to have and was perceived by the students as having/not having learned [3].

From the list of 29 main competences which are required for entering the labour market, as a young graduate the top 9 are listed below as commonly mentioned by the 3 groups of stakeholders:

- Problem-solving capacity
- Managing complex problems
- Technical knowledge
- Economical and accounting insights
- Team leadership
- Specific knowledge of sector of industry (and company specific)
- Projectized working
- Languages
- Technical-commercial skills

This list of required competences will be the core of the cross-domain competences envisaged in the suggested methodology.

By combining different concepts which individually showed their merits and results, we try to come to a workable scheme combining the best of elements to reach the competences which is necessary.

The EuroMPM-study (European master in project management at Fh Dortmund [4], is a modular based and international master which delivers strong competence buildup in projectized working, in languages and managing of complex problems. Through its' system of short mobility windows during block week in the international consortium, students acquire a lot of intercultural experiences.

The basic engineering school at TU Ilmenau [5] closes the loop and uses theory courses as supportive to the lab-making sessions. Problem solving capacity and teamwork of students is a must and technical knowledge is addressed to complete the engineering projects. This comes close and embraces the CDIO (Conceive/design/implement/operate) approach in which all is applied to products. Not only design but up to results and implementation [6].

An efficient tool for project oriented learning which can be easily introduced in a digital environment is the remote laboratory. Implementation of remotely controlled experiments into study allows teachers not only to give students the possibility to have access to the unique equipment 24/7 but allows students to self-study and (re)experiment different aspects without much time-limits. On top, the equipment is used more effective (longer time) without much personnel, as such cutting costs. Involvement of students for the development implements project based teaching approach and increase multidisciplinary knowledge. Also it is very effective tool for inclusive education.

Development of remote labs as student projects was done in a project oriented approach at NU Zaporizhzhia Polytechnic (Ukraine). Students, under the supervision of the university staff, designed and built several remote labs covering a bunch of experiments on embedded systems (ISRT) [5].

To create a cross-domain competence framework a cross border competence center was created (Fig. 1). The members of this competence center each contribute from their expertise and experience in the further refinement of the necessary competences.

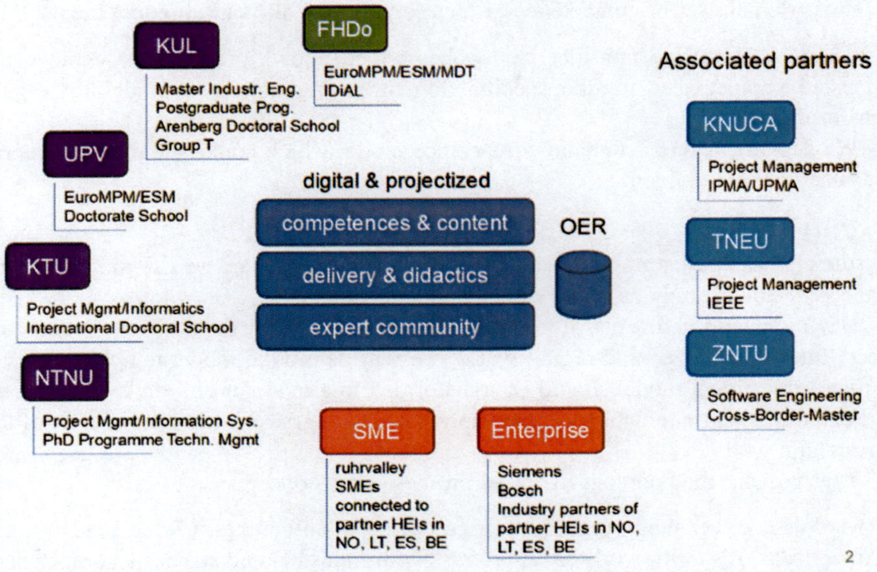

Fig. 1. Cross border competence center partnership

3 Core for Cross-Domain Competences for Work 4.0

Master programs offer the chance to move towards job-field-orientation and applied sciences. They can create real life learning situations with many interaction points between industry and academia, giving students excellent employability. They are inter- or multidisciplinary and inherently international – if HEIs from different countries are involved - fitting to today's work in interdisciplinary and international project teams.

A new co-production model for cross-domain competences with cooperation of academia, industry and professional associations in a transdisciplinary approach is needed.

The special combination of competences from engineering, project management and IT/ICT needs to be developed for students, professionals (and executives) (life-long learning). Competence delivery will be done in a projectized and digitalized way to be available anywhere and anytime (adapted to the learner). To increase employability, Work 4.0 competences need to be focused on the job-field, not on an academic domain, and need to have practical elements.

Let us consider cross domain competences on macro and micro level.

On macro level we consider.

1. Classes of competences: Social, professional, business; International, Operational, Financial, Human;
2. Way of achieving: continuous education, performance education, professional development
3. Cross-domain set of competences which is valid for all market requirements.

On micro level we consider approaches which allows graduates to achieve the requested competences for the specific domain and competence model for digital transformation.

We suggest the cross domain competence model which could integrate the micro- and macro- cross-domain.

MACRO LEVEL:
For the classes and list of competences, for engineering studies we use the list for the "Ingenieur 2020"-study as a test case.

Having established a list of competences which is required to all workers in the (near) future globalized and digital world, we will define the model to achieve this.

The educational model should be transformed to a modular and multi-disciplinary approach in which international (short) mobility is integrated as an integral part of the curriculum.

The working methodology is based on these core concepts:

– Modules deliver theoretical knowledge (with online courses, (virtual) classroom), practical skills (with workshops, projects with industry) and scientific competences (with assignments). The international, intercultural and interdisciplinary (3 x i) competencies are delivered by mixing student groups, lecturers & practitioners cross-border and cross-faculty in the workshops/projects.

- Modules involve all concepts mentioned above in a blended learning approach, not only eLearning content or online courses.
- Module development, pilot teaching, evaluation, module releases as OER are conducted in an agile process – called Scrumban [6]. It is basis of a virtual, projectized and digitalized community.

Part of the approach is the definition of Open Educational Resources (OER) which allow a broad dissemination, in universities but also for lifelong learning. Furthermore, OER allow the configuration of tailored modules for different target groups and the delivery with a blended learning approach adapted to their needs. Finally, the OER module repository, the community of practice (OpenCoP) working on the further development and the pool of lecturers who adopt the modules in their teaching and training are a very sustainable and broad output with a high impact on the digitalized and projectized lifestyle of the future.

MICRO-LEVEL.
Every module should start from the idea of closing the loop from need-design-production-feedback to ensure the real-life feeling. Other – more theoretical courses – should be added to support the closed loop approach as the opposite of having a case illustrating the theory.

The didactical approach can be used in different fields (engineering, project management).

By integrating the different concepts the curricula will better prepare the graduates for future employment in a changing and challenging world.

Therefore we suggest as a core a project semester which can be introduced after (under) graduates have a basic technical knowledge in their own field. The semester will introduce and train the competences – more by doing than by teaching.

The semester scheme can look like:

- Start with an hackathon (international) to generate ideas and to form (international) multidisciplinary teams. The competence background is delivered during mini lectures e.g. on creativity or on the business canvas-model to connect to the economic reality.
- Use the CDIO methodology [6] to do the technical side of the project, either to design and make products or processes. The aims is to reach at least a proof of concept of the generated ideas.
- Teaching on the professional skills (economic impact and choices in a design, project management, communication and teamwork, intercultural aspects of a globalised economy)
- Work with interdisciplinary and multi-national teams (engineers + business men + humanists, artists…): communication and workgroups online
- Use of short (student) mobility (in a block week concept or summer schools): at start the hackathon (generate ideas), intermediate (from concept to design) and at the end (pitching/selling the project). By preference we use industrial problems/demands as start-off for the projects)
- Short internship (2–3 weeks during the semester) to get company-work insight

An important asset in education is the digital project environment – combined in a collaboration and eLearning platform. This is important as it is part of the digitalization of education and a test case for digital transformation.

Working on the concept with the different members of the consortium is beneficial in the sense that each partner contributes in the fields of its own expertise. Furthermore, the co-creation and co-ownership by the consortium members guaranty the embedding of the materials in the curricula for the own students and as such in total strengthening the concept of international cooperation and creation of an international context for the students.

The participation of students in the short (module) mobility to other partners of the consortium can be a challenge: experiences from the past – with several summer- and winterschools and block weeks – in the EUROMPM consortium show the willingness of students to participate. The main bottle neck is getting funding or at least the means to make modules outside of the home-university affordable for all. These short mobility requires careful planning as not all academic calendars in different countries coincide.

4 Conclusion

Implementation of the successful co-production approach of lecturers, industry experts and member of professional associations the new, job-related competences for the workplace of the future (Work 4.0) will be developed. This approach will overcome the gap between academia and industry in the partner countries and equip graduates with competences to cope with the transformation of working environments, to assess where they are and where they want to go (analysis, strategy), to transform to a sustainable working environment, to consider occupational safety and health and environmental protection.

References

1. Self-employment statistics: https://ec.europa.eu/eurostat/statistics-explained/index.php/Self-employment_statistics. Accessed 6 June 2020
2. Coren, M.J.: Automation will change every job, but only 25% are on the chopping block. https://qz.com/1532006/automation-will-change-every-job-but-only-25-are-on-the-chopping-block/. Accessed 6 June 2020
3. Industrieel ingenieur 2020: Eindrapport, study by KULeuven, UAntwerpen, UHasselt, Univerity of Ghent, VUB, published internally December 2016
4. Wolff, C., et al.: Master level education in project management - The EuroMPM model. In: 2017 9th IEEE International Conference on Intelligent Data Acquisition and Advanced Computing Systems: Technology and Applications (IDAACS), Bucharest, pp. 836–842 (2017). https://doi.org/10.1109/IDAACS.2017.8095205
5. Basic Engineering school: https://www.tu-ilmenau.de/basic-school/. Accessed 6 June 2020
6. The CDIO™ INITIATIVE: https://www.cdio.org/content/cdio-standard-21. Accessed 6 June 2020

7. Arras, P., Van Merode, D., Tabunshchyk, G.: Project Oriented Teaching Approaches for E-learning Environment. In: IEEE 9th International Conference on Intelligent Data Acquisition and Advanced Computing Systems (IDAACS), pp. 317–320 (2017). https://doi.org/10.1109/IDAACS.2017.8095097
8. Reddy, A.: The Scrumban [R]Evolution: Getting the Most Out of Agile, Scrum, and Lean Kanban (Agile Software Development Series), Addison-Wesley Professional, p. 384 (2015)
9. Wolff, C., Omar, A., Shildibekov, Y.: How will we Build Competences for Managing the Digital Transformation? In: The 10th IEEE International Conference on Intelligent Data Acquisition and Advanced Computing Systems: Technology and Applications, 18–21 September 2019, Metz, France, pp. 1122–1129 (2019)

Fostering Natural and Data Science Skills of School Kids

Alexander Nussbaumer[1(✉)], Christina M. Steiner-Stanitznig[1], Silke Luttenberger[2], Sylvia M. Ebner[1], and Christian Gütl[1]

[1] Graz University of Technology, Graz, Austria
alexander.nussbaumer@tugraz.at

[2] University College of Teacher Education Styria, Graz, Austria

Abstract. While on the labour market there is a high demand for employees educated and skilled in the field of natural sciences, engineering and information technologies, a lack of interest in these fields of expertise can be identified in young people. The aim of our research is to leverage a scientifically founded and empirically validated approach and contribution towards promoting interest and motivation in dealing with data science and scientific and technical subjects. To this end a novel pedagogical approach and a related learning environment has been developed and used in a first pilot phase. This approach was implemented using research and exploration questions on weather forecasts and data from temperature measurements as an example pedagogical scenario. The pilot study was executed in real-world settings of three primary and two secondary schools. The results demonstrate high interest and motivation in the science project and related data science tool.

Keywords: Science learning · Digital literacy · Inquiry-based learning · Data Science Process

1 Introduction

While on the labour market there is a high demand for employees educated and skilled in the field of natural sciences, engineering and information technologies, a lack of interest in these fields of expertise can be identified in young people [5]. Interests of pupils in these subjects and professions is declining steadily from primary level and at the transition to the secondary level. Therefore, there is a need for innovative didactic approaches that are suitable to engage and motivate young people.

This paper describes an initiative that aims at stimulating young people's interest and motivation in dealing with natural science topics and thereby shall foster competence development in scientific and digital literacy. In contrast to other Science, Technology, Engineering, and Mathematics (STEM) education approaches, such as the use of online laboratories [6,10], our approach focuses on the integration of real world natural science experiments with data science activities.

M. E. Auer and T. Rüütmann (Eds.): ICL 2020, AISC 1328, pp. 212–223, 2021.
https://doi.org/10.1007/978-3-030-68198-2_19

While learning and teaching data science is well established at university level, there are almost no pedagogical approaches to teach and learn basic data science principles in primary and secondary schools. This paper presents a pedagogical model and an application to teach and learn basic data science skill at schools.

This approach is elaborated and implemented for different thematic areas and empirically deployed and tested in the context of primary and secondary schools. Piloting phases involving trained assistant teachers enable an iterative evaluation and refinement of the approach. The digital learning environment and didactic concept with accompanying material will be provided to schools for application in educational practice.

This paper is structured as follows: The novel pedagogical approach is described in Sect. 2. An application to a concrete pedagogical scenario with a supporting learning environment is described in Sect. 3. The evaluation of the first pilot study and respective results are reported in Sect. 4. In Sect. 5 a conclusion and outlook to future work is provided.

2 Theoretical Foundation

The key approach proposed in this paper to teach and learn data science skills is based on the combination of data science with natural science. While almost no pedagogical approaches are available to learning data science at primary and secondary school level, a lot of pedagogical research exists for learning natural sciences at this age.

2.1 Inquiry-Based Learning

A pedagogical approach often applied in the field of science teaching is Inquiry-based Learning. Anderson [1] proposes to apply scientific inquiry in the learning context. Scientific inquiry refers to the diverse ways how scientists study and explain the natural world. Applying this process to the learning context leads to an active learning with distinct learning activities. In general, Anderson reports positive effects of this way of learning and teaching on student achievements.

Four levels of inquiry are distinguished in the literature, according to the degree the teacher is involved in the learning process [2,9]. In level 0 (*verification*) the teacher provides the research questions, the data, and supports the interpretation. In level 1 (*structured*) the teacher provides the research question and data, but the student is responsible for the interpretation. In level 2 (*guided*) the teacher provides the research question and the student is responsible for data collection and interpretation of the results. In level 3 (*open*) the student is responsible for the research question, data collection and result interpretation. In a quantitative study [2] it turned out that especially level 2 approaches (*guided*) of inquiry instruction have positive effects on student learning.

A method to implement inquiry-based learning in science teaching is the use of the 5E Model [3,9]. The 5E model describes five subsequent activities (phases) that students should perform in science learning. In this way it provides a template or pattern for lesson planning, and supports the teacher in the science teaching process. In detail the activities (or phases) are:

- *Engagement.* In this phase the student's interest is stimulated. Questions are raised and connection to prior knowledge is established.
- *Exploration.* In the exploration phase students begin to investigate a problem, pose real-world questions, and develop hypotheses. Furthermore, they perform scientific and laboratory tasks.
- *Explanation.* In this phase the students develop explanations for their observations. They discuss what they have learned and internalise what they have learned.
- *Extensions.* In this phase students generalise their understanding and transfer it to real knowledge, in order to understand real-world problems.
- *Evaluation.* In this phase students compare their previous knowledge with the new knowledge. This phase increases the depth of understanding as students have to apply metacognitive skills.

2.2 Data Science Process

Data Science is an emerging field in computer science and used for many applications. Typically, it is employed to answer real-world problems by using large amount of data. These data are of different type (e.g. structured, unstructured, text). Data Science is applied in several steps, which is described as the data science process [4]. This process consists of the following steps:

- *Research goal.* In this step the problem to be solved is framed.
- *Retrieving data.* In this step the raw data are collected
- *Data preparation.* In this step the raw data are cleaned and prepared. For example, questions on the time zone, missing data, data range are dealt with.
- *Data exploration.* In this step the data is explored in order to get a first overview and insights. Initial trends can be seen and a first relation to the research goal is made.
- *Data modelling.* In this step a in-depth analysis is performed, such as machine learning or statistical algorithms.
- *Presentation and communication.* In this step the results from the analysis are documented and shared.

2.3 Data Science Learning Model

The combination of Inquiry-based Learning (i.e. the 5E Model) and the Data Science Process, as outlined above, build the foundation for a new pedagogical approch. This approach consists of five steps, whereby all activities or steps from the 5E model and the Data Science Process are mapped and integrated. An overview of this approach is depicted in Fig. 1. In detail these steps are:

- ***Engagement and research question.*** In this step the student is presented with the science problem and asked to take up a pre-defined research question.
- ***Data collection.*** In this step data is collected in the real world and entered in a computer system.
- ***Data organisation.*** In this step the collected data is reviewed and organised, which includes filtering the data according to criteria such as time range. Furthermore, students get an overview of the available data.
- ***Data exploration.*** In this step data is interactively and visually explored. Interactive data visualisations and diagrams help to understand the data and to answer the research question.
- ***Research result and presentation.*** The answers to the research questions are presented and conclusions on the overall research goal are drawn.

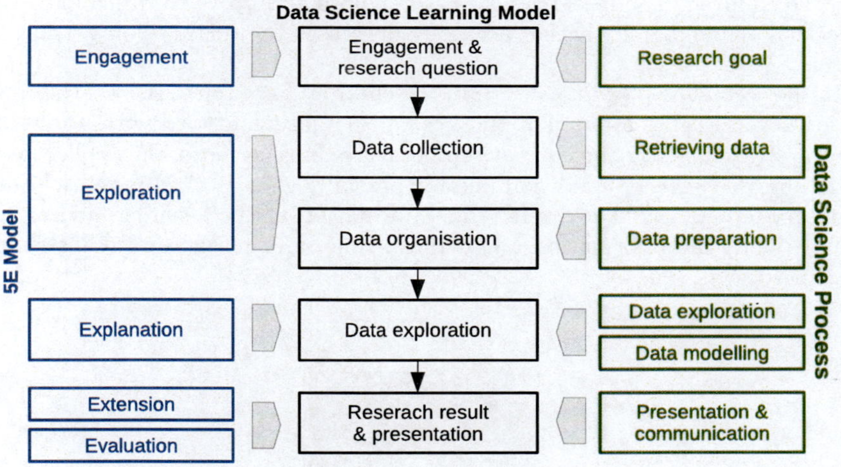

Fig. 1. This figure depicts the data science learning model with steps and activities integrated from the 5E model and the data science process.

Since previous research has shown that a guided approach to Inquiry-based instruction is most effective, meaning that a teacher should provide the initial research question, but the student should independently collect and explore data. This concept is also applied to the our Data Science Learning Model application, where we suggest a teacher providing information only for the first step (*research questions*).

3 Application and Learning Environment

This section describes how the Data Science Learning Model for learning basic data science concepts has been applied. This application consists of two parts.

First, it is explained how the science goal and science context is included in a scenario and how this is embedded in the procedure of the pedagogical approach. Second, the learning environment is presented that was used for the pupils to deal with the data.

3.1 Scenario and Procedure

The domain of meterology has been chosen for translating the data science learning model into a concrete pedagogical scenario and application. In detail, the field of weather forecast and testing its accuracy was the theme of the experiment.

In the first step (*engagement and research question*) the children are provided with a weather forecast of the upcoming week, which included the minimum and maximum temperature for each day, as well as the forecasted weather condition (e.g. sunny, cloudy, rainy). A teaching assistant discusses the research question with the children. The research goal is to explore and prove to which extent the forecast is reliable and whether accuracy decreases the further you get into the future.

In the second step (*data collection*) the children have to collect weather information several times a day. For this reason they are equipped with a micro:bit (see Fig. 2) to measure the current temperature. Furthermore, they collect information on the weather condition (sunny, partially cloudy, cloudy, rainy, snowy) by observing the sky. They enter this information in the learning environment together with the information about date, time, and measurement instrument (e.g. micro:bit).

Fig. 2. This figure shows the micro:bit together with the enviro:bit and a battery pack, as it can be used for measuring temperature.

In the third step (*data organisation*) the children get an overview of the collected data in the learning environment. The number of entered data sets (measuring points including temperature and weather condition) is shown on a time range. They can filter, remove, and update measuring points in case they contain unrealistic or false information.

In the forth (*data exploration*) step the children are provided with an interactive visualisation that displays the collected measuring points and the original weather forecast in the same diagram. The children can interactively switch between week view and day view, and they can hide or show minima, maxima, and collected measuring points. This visualisation is used to answer the research question on the forecast accuracy. Details on the visualisation are explained below.

In the fifth step (*research result and presentation*) the children present the result of their conclusion on the forecast accuracy. They are presented with specific questions which they answer in a kind of presentation. In order to create this presentation, they can download specific views of the interactive visualisation.

3.2 Learning Environment

The learning environment is the tool for the children to develop and improve their data science skills. It allows to collect, organise, and explore data, as described in the pedagogical approach (step 2–4). Thus it supports children to establish relations between the natural world, their observations, and research questions on the one hand, and collected data on the other hand.

Besides data collection and data organisation as described in Sect. 3.1, the key functionality is the data exploration tool. This tool allows to interactively explore the collected data on a time axis for one week (see Fig. 3). The measured temperatures are depicted as yellow dots that can be clicked for getting more details (time, weather condition, measurement instrument). On the same visualisation the forecasted minima and maxima are displayed as blue and red bars. This visualisation makes it easy to see if the measured temperature is within the forecast range. On top of the visualisation there are buttons to show or hide the temperature data, as well as the forecasted minima and maxima. By clicking on the weekday on the bottom, the respective weekday view is displayed (see Fig. 4). The day view visualisation shows the measured temperature on a time axis for one day and provides the same interactive features as the week view visualisation.

The learning environment provides two further main functionalities. First, it provides a user registration and management tool. Users can register either alone or as group. During the registration process, the children can select an individual icon and a name. Both an own icon and name is an instrument to stimulate motivation through a simple personalisation. Providing support for learning groups is also a method to increase motivation and learning outcome.

The second main functionality is the collection of log data. Each activity a child performs with the learning activity is recorded. Recorded activities include login, logout, enter new data set, select week view, select day view, request data set detail, show or hide forecast. These data are used for activity analysis (see Sect. 4) and for feeding the teacher tool. The teacher tool (see Fig. 5) provides an overview of the activities of all learners or learning groups. It displays how many data sets have been entered and how many activities have been performed in the learning environment. This allows the teacher to trace the activities and quickly detect if learners or learning groups are inactive and need further support.

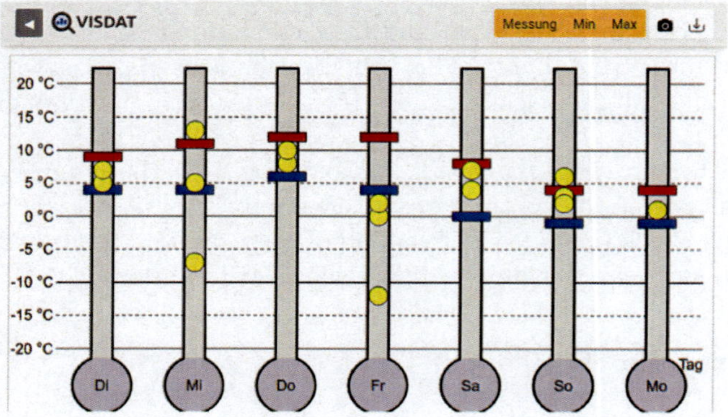

Fig. 3. This image shows the measuring points in combination with the weather forecast for one week.

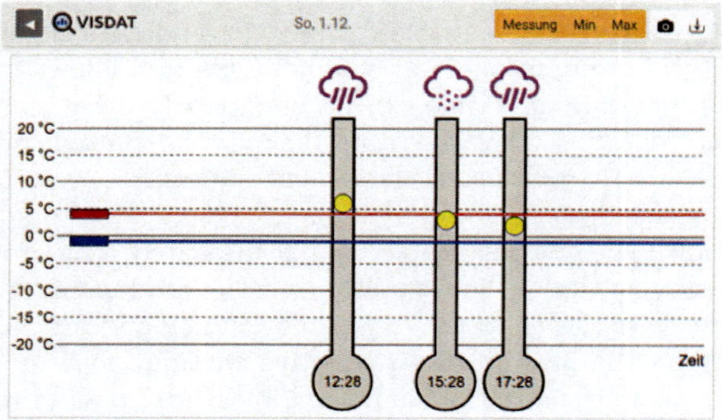

Fig. 4. This image shows the measuring points in combination with the weather forecast for one day.

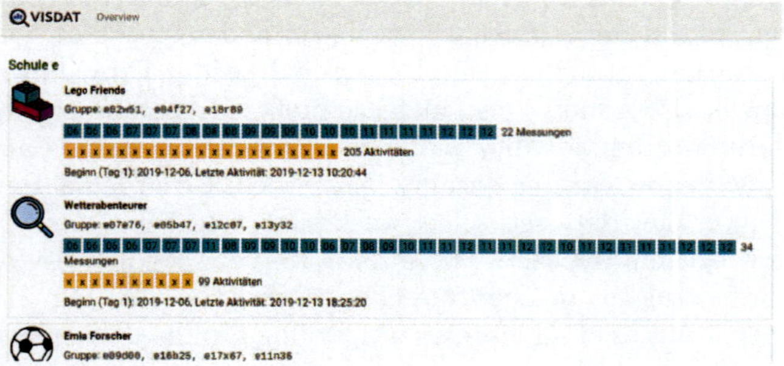

Fig. 5. This image shows the teacher tool with an overview of learning groups and their activities (number of collected data, number of performed activities).

The learning environment is implemented as Web application with a service in the back-end that stores all data. The Web application follows the responsive design paradigm and can be accessed with any Web browser on any device. It is important to note that the learning system does not collect real names or other information that can be used to identify indiviudal children. All collected information is anonymous and adheres to the GDPR.

4 Evaluation

The Data Science Learning Model (Sect. 2) and its implementation in the example pedagogical scenario and application (Sect. 3) was deployed and tested in a school context in November and December 2019.

4.1 Participants and Procedure

Children from five different schools (two primary schools and three secondary schools) participated in the evaluation. At each school two classes attended the evaluation, which resulted in ten classes. At both school types third-year grade classes were involved, which means that children at the primary schools were app. 8 years old and children at the secondary schools were around 12 years old.

The total number of children were 173, where 70 children were from primary schools and 103 children from secondary schools. All children were assigned to learning groups of three or four children per group.

The pedagogical intervention and evaluation took place during school lessons. The pilot covered 3 sessions spread over a total of 1 week and were managed by teaching assistants. At day 1 the children were provided with a weather forecast, the research goal, general weather information, and explanations how to collect data in the following week. After a few days the teaching assistants visited the classes again and helped them to enter the collected data in the learning environment in case of problems. In the final session the results were discussed and the learning groups drew conclusions on their research questions in final group presentations. In this way they followed the whole pedagogical approach. Feedback about the learning experience was collected from the school children as well as from the teaching assistants.

4.2 Interest

During the sessions, children had been asked about their situational interest on the different learning activities. The children indicated their current interest on a Likert scale from 1 (low interest) to 5 (high interest). The results are depicted in Fig. 6. As can be seen, a high interest could be identified throughout all aspects of the pedagogical scenario in primary/secondary schools (use of microbit: $M = 4.8/4.1$, use of learning environment: $M = 4.45/4.1$, data collection: $M = 4.1/3.45$). Thereby primary school children had higher interest than secondary school children and interest in the micro:bit was higher than in the learning environment and in the data collection process.

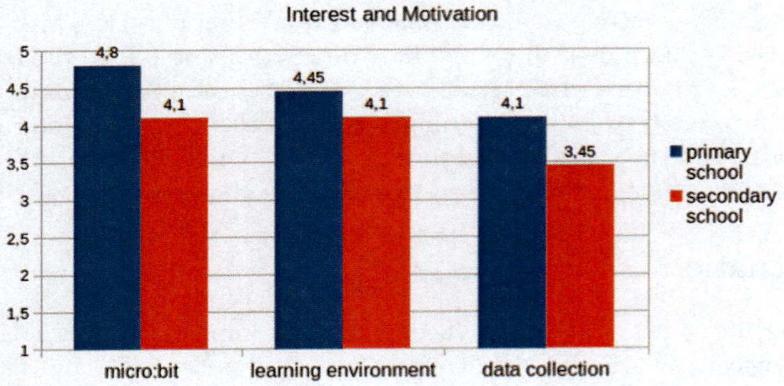

Fig. 6. This diagram shows the average interest and motivation of the children during the evaluation.

4.3 User Experience and Usability

For evaluating user experience with the learning environment in general and with respect to usability aspects, two general questions on enjoyment and an adapted and shortened version of the User Experience Questionnaire (UEQ - based on [8]) was used. In total 6 items (each item consisting of a pair of terms with opposite meanings - e.g. good - bad) of the subscales 'attractiveness', 'perscpicuity' and 'stimulation' from the version for children and adolsecents [7] were used with a 5-point Likert scale ranging from −2 to +2. The results from the UEQ are shown in Fig. 7 and represent a quite differentiated picture for the two school types. Primary school children perceived the learning environment very positively on all three aspects, while secondary school children were rather critical in their assessment, especially with respect to attractiveness. This may suggest that the visual appearance and design of the learning environment is more appropriate for younger children. The results on items gathering general assessments on enjoyment (possible score range 1–5) showed that the pupils in general enjoyed the work with the learning environment ($M = 4.58$, $SD = 1.03$ for primary school; $M = 3.69$, $SD = 1.27$ for secondary school) and that they would recommend the overall learning experience to others ($M = 4.3$, $SD = 1.09$ for primary school; $M = 3.36$, $SD = 1.25$ for secondary school), but again with better scores for primary school children.

Feedback from teaching assistants on the experience and use of the learning environment showed quite consistent assessments independent of a primary or secondary school setting. Results are therefore summarized over school types and presented in Fig. 8 (possible score range 1–5). Teaching assistants felt that learners enjoyed working with the learning einvironment the learning environment very much and that it was easy to handle for them. Difficulties in dealing with the app and resulting need for assistance were rather rare.

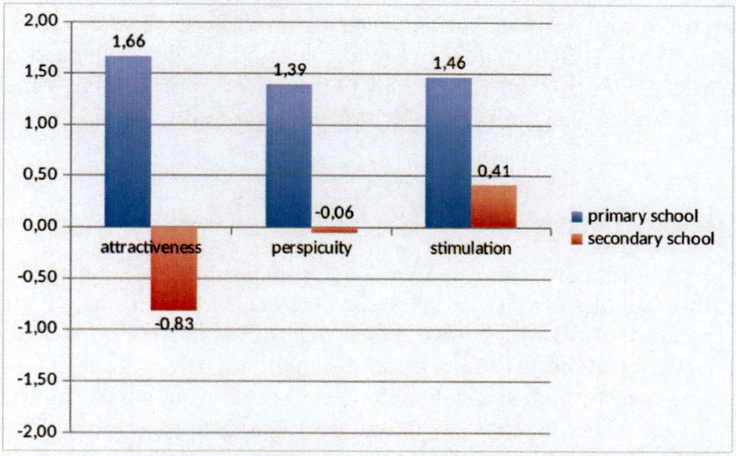

Fig. 7. Results on the subscales of the user experience questionnaire.

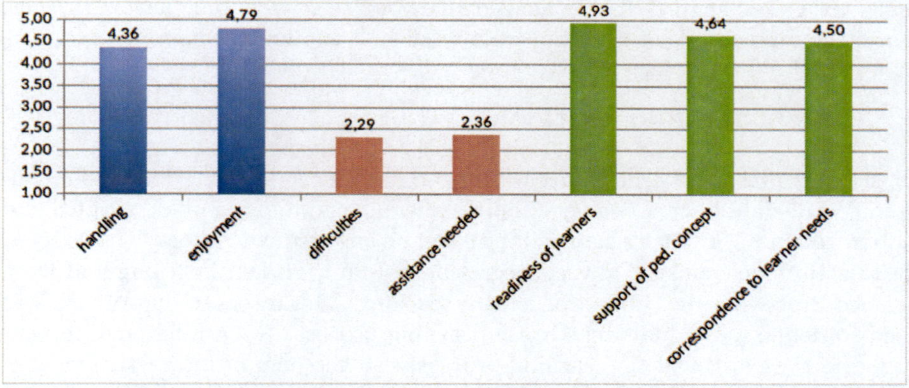

Fig. 8. Overview of results from a teaching assistant perspective.

4.4 Usefulness

From the didactic perspective the learning environment was perceived very well suitable for implementing the data science learning model and did not lack any important functionalities. The feedback obtained from teaching assistants furthermore also highlights childrens' high readiness to work with the learning environment and its good correspondence to learner needs (see Fig. 8).

Usefulness for learning had also been assessed from the school childrens' perspective. Children from both school types agreed that they have learned many new things, with primary school children benefitting more from the overall pedagogical intervention ($M = 4.4$, $SD = 0.93$ for primary school, $M = 3.54$, $SD = 1.08$ for secondary schools). Children from both school types indicated that they had understood the learning contents very well (M = 4.52, SD = 0.68 for primary school, M = 4.1, SD = 0.96 for secondary school). The evaluation outcomes thus give first evidence that the overall pedagogical approach, in general, and the learning environment, in particular, are a useful and enjoyable approach for learning.

4.5 Activity Analysis

During the pilot sessions the learning environment recorded different types of activities (number of collected weather data sets and interactions with the visualisation module). As learning took place in groups and each group used one computer (mostly tablets), the analysis was made on group level. The result of this analysis suggests high engagement. On average groups at primary schools collected 20.1 data sets (SD = 13.1) in one week and groups at secondary schools collected 16.2 data sets (SD = 7.7). The average number of all activities is 122.8 (SD = 79.1) at primary school level and 110.9 (SD = 83.2) at secondary school level. These values demonstrate active participation, whereby children at primary level were slightly more engaged than pupils at secondary level. Furthermore, the high standard deviation value indicates indiviudal differences.

5 Conclusion and Outlook

This paper presented a novel approach to data science learning for school children. While data science is a very complex field, basic steps can be identified and elaborated in a way that school children can be engaged with them. The concept for this approach consists in the integration of Inquiry-based Learning and Data Science Process in the context of science teaching. This integrated approach was used to define individual steps in the learning process. A learning environment supports these steps of the learning process with the help of interactive visualisations. A key feature is the fact that the data to be explored are collected and recorded by learners themselves, which makes them more tangible for children.

An initial deployment and evaluation in educational practice showed high interest and enjoyment of children in the learning activity and the usefulness of the overall approach. User experience of the learning environment showed that the appearance and design seem to be more approriate and therefore more positively perceived by younger learners. This also suggests the considerations of using a different and more sophisticated design and features for older learners.

While the results presented herein focus on the subjective reaction to and perceived usefulness of the learning approach, future work will incorporate also the analysis of effects on data science skills, for a more conclusive demonstration of the effectiveness of the presented approach. Such an assessment will include knowledge and understanding of diagrams, filtering and cleaning data, and understanding different views of data. Currently, the application and evaluation in the context of another pedagogical scenario on a biology topic (plant life cycle) is in peparation.

Acknowledgement. The work reported has been supported by the VISDAT project which has received funding from the Styrian Regional Government in Austria (Land Steiermark), Zukunftsfonds Steiermark, under grant agreement No. 1041.

References

1. Anderson, R.D.: Reforming science teaching: what research says about inquiry. J. Sci. Teacher Educ. **13**(1), 1–12 (2002). https://doi.org/10.1023/A:1015171124982
2. Blanchard, M., Southerland, S., Osborne, J., Sampson, V., Annetta, L., Granger, E.M.: Is inquiry possible in light of accountability?: a quantitative comparison of the relative effectiveness of guided inquiry and verification laboratory instruction. Sci. Educ. **94**(4), 577–616 (2010). https://doi.org/10.1002/sce.20390
3. Bybee, R.: The BSCS 5E instructional model and 21st century skills. BSCS, Colorado Springs, CO (2009). https://sites.nationalacademies.org/cs/groups/dbassesite/documents/webpage/dbasse_073327.pdf
4. Cielen, D., Meysman, A.D.B., Ali, M.: Introducing Data Science. Manning Publications, New York (2016). https://www.manning.com/books/introducing-data-science
5. DeWitt, J., Archer, L.: Who aspires to a science career? a comparison of survey responses from primary and secondary school students. Int. J. Sci. Educ. **37**(13), 2170–2192 (2015). https://doi.org/10.1080/09500693.2015.1071899
6. Gillet, D., de Jong, T., Sotirou, S., Salzmann, C.: Personalised learning spaces and federated online labs for stem education at school. In: 2013 IEEE Global Engineering Education Conference (EDUCON), pp. 769–773 (2013)
7. Hinderks, A., Schrepp, M., Rauschenberger, M., Olschner, S., Thomaschewski, J.: Konstruktion eines fragebogens für jugendliche personen zur messung der user experience. In: Brau, H., Lehmann, A., Petrovic, K., Schroeder, M.(eds.) Usability Professionals, pp. 78–83 (2012)
8. Laugwitz, B., Schrepp, M., Held, T.: Konstruktion eines fragebogens zur messung der user experience von softwareprodukten. In: Heinecke, A., Paul, H.(eds.) Mensch & Computer 2006—Mensch und Computer im Strukturwandel, pp. 125–134. Oldenbourg Verlag (2006)
9. Luttenberger, S., Rath, G., Paechter, M.: Forschendes Lernen. In: Fritz, U., Lauermann, K., Paechter, M., Stock, M., Weirer, W.(eds.) Methoden für kompetenzorientierten Unterricht. Budrich, Leverkusen-Opladen (in Press)
10. Maiti, A., Maxwell, A.D., Kist, A.A., Orwin, L.: Merging remote laboratories and enquiry-based learning for STEM education. Int. J. Online Biomed. Eng. (iJOE) **10**(6), 50–57 (2014). https://doi.org/10.3991/ijoe.v10i6.3997

Smart Instruction of EAP and ESP Reflecting Learner's Motivation Type

Ivana Simonova(✉), Katerina Kostolanyova, and Ludmila Faltynkova

University of Ostrava, F. Sramka 3, Ostrava, Czech Republic
iivana.simonova@gmail.com,
katerina.kostolanyova@osu.cz, faltynkova@epol.cz

Abstract. Currently, communication in a foreign language is considered one of eight key competences. Within higher education, English for Academic Purposes (EAP) and English for Special Purposes (ESP) mostly play this role. Despite the fact that smart devices are widely spread among young population and often work as a strong motivator to learning, problems may occur when the process of instruction is enhanced by smart didactic means. Logically, the question appears whether the smart instruction suits learners of all motivation types, i.e. whether they all are able to successfully acquire the learning content. Therefore, the main objective of this research is to find out whether smart instruction of EAP and ESP can be applied on learners of all motivation types. Two hypotheses were set and the quasi-experiment and ex-post-facto method were applied. Learners' motivation types were detected via the standardized Motivation Type Inventory (MTI) by Plaminek which distinguishes learners of four types: Directors, Coordinators, Explorers, Accurators. Didactic tests were exploited to monitor learners' entrance and final knowledge. The test scores were statistically processed and results considered from the view of motivation types – significant differences were not detected between motivation types. The results prove that smart instruction is entitled to be exploited in EAP and ESP as it suits learners of all motivation types.

Keywords: English for Academic Purposes · EAP · English for Specific Purposes · ESP · Smart · Motivation type

1 Introduction

Currently, communication in a foreign language is considered one of eight key competences. Within higher education, English for Academic Purposes and English for Special Purposes mostly play this role. However, the exploitation of smart devices that are widely spread among young population and often work as a strong motivator to learning may cause problems. To succeed in learning, learner's motivation type should be taken into account. Then, the question appears whether the smart instruction suits learners of all motivation types, i.e. whether they all are able to successfully acquire the learning content when smart instruction is conducted.

M. E. Auer and T. Rüütmann (Eds.): ICL 2020, AISC 1328, pp. 224–231, 2021.
https://doi.org/10.1007/978-3-030-68198-2_20

Therefore, the main objective of this paper is to introduce results of research dealing with smart instruction in English for Academic Purposes and English for Special Purposes applied on learners of various motivation types.

2 Theoretical Background

2.1 Communication in Foreign Languages

According to the European Union (EU) Recommendation on a comprehensive approach to the teaching and learning of languages [1], education systems need to be adapted to the challenges and opportunities posed by Europe's linguistic diversity. However, almost half of European countries report that their inhabitants are not able to communicate in any language other than the native one. Moreover, numerous studies show that EU Member States are not progressing fast enough towards the target that everybody should be able to speak two foreign languages from an early age [2]. Therefore, a set of recommendations was made – the first package of measures was introduced addressing key competences and digital literacy [2].

Reflecting the above mentioned, under the conditions of the Czech higher education, within improving foreign language skills, the focus was put on English for Academic Purposes, English for Specific Purposes, and the smart approach to instruction. The problem is considered from the view of learners' motivation so that learners of all motivation types were able to acquire the required learning content.

2.2 English for Academic and Specific Purposes

Numerous definitions had been set to determine what English for Academic Purposes (EAP) and English for Specific Purposes (ESP) meant and what their learning contents were. Briefly said, EAP refers to the language and associated practices that people need in order to undertake study or work in English, mostly in higher education. The objective is to learn some – mainly institutional and disciplinary – practices involved in studying or working through English as medium of instruction [3].

As summarized by Hamp-Lyons [4], the question is not what EAP is but why EAP is taught. EAP arose from the broader field of ESP, defined by its focus on teaching English specifically to facilitate learners' study or research through [5:8], [6:1]. EAP is often considered to be a branch of English Language Teaching (ELT); the tree of ELT was designed by Hutchinson and Waters [7:17]. The topmost branches of the tree show the level at which individual ESP courses occur. The branches just below them indicate that these may conveniently be divided into two main types of ESP: English for academic studies, or English for work purposes [7:10].

In this research, two branches are under the focus: EAP, the course of which deals with Academic English (AE), and ESP, where the learning content is connected to learner's future occupation (Career Development, CD).

2.3 Smart Instruction

In 1981, Doran first used the word SMART as an abbreviation meaning *Specific, Measurable, Achievable, Realistic and Time-related* in economic context [8]. These days, the term refers to latest technologies. When considering the research results of journal articles published from 2012 to 2017 in Google Scholar and Science Direct databases, Silverio-Fernández et al. propose following definition of "smartness" [9]: *"A smart device is a context-aware electronic device capable of performing autonomous computing and connecting to other devices wire or wirelessly for data exchange."* Reflecting these features, smart instruction refers mainly to new educational contexts, i.e. technology, which it reflected in methodology, including changes in the role of teachers and the shift in delivering the learning content. Most help from the technology is brought in learners' motivation and stimulation, managing lessons, assessing learners' performance, promoting interactivity and collaboration [10].

2.4 Motivation Types

Motivation is an internal process. Either we define it as a drive, or a need, motivation is a condition inside us that desires a change, either in the self or the environment [11]. It is influenced by the satisfaction of needs that are either necessary for sustaining life or essential for wellbeing and growth. Individuals differ in the strength of preference to single needs to some extent [12].

Motivation Type Inventory (MTI) [13] works as the tool for detecting learner's motivation type. The inventory consists of two parts. In Part 1, the purpose and means of motivation are monitored and respondents are differentiated on the effectiveness – usefulness scale. In Part 2, challenges and safety are under the focus, and respondents are differentiated on the stability – dynamics scale. Four motivation types are distinguished as follows: Explorers, Directors, Coordinators, Accurators [14].

3 Methodology

3.1 Process of Smart Instruction

The research was conducted in two learning contents dealing with (1) EAP – it focuses on vocabulary and grammar connected with the respondent's study, e.g. name of institution, professional terminology and explaining the content, writing essays, articles etc., and is called Academic English (AE) further on; (2) ESP – it concentrates on vocabulary and grammar connected with the respondent's future job, e.g. conducting interviews, writing C.V., letters, applications etc., and is called Career Development (CD) further on. Learners in groups of 20 as maximum attended one 90-min lesson per week and prepared for the next lesson for approximately another 90-min period of after-school work. In face-to-face lessons, attention was paid to direct conversation between teacher-student, student/s-student/s, to providing feedback on homework, answering students' questions and solving problems in learning. The process of instruction was conducted for the period of 12 weeks (one semester). Before the process of instruction started, the pretest was held monitoring the entrance level of learners' knowledge, then, at the end of semester, the final knowledge was evaluated by the posttest.

The smart instruction reflected the key features of this phenomenon defined above – autonomy, connectivity, context-awareness, as well as the exploitation of latest (smart) technologies, sources and methodologies. This approach provided opportunity to satisfy each student's needs towards acquiring the required learning content successfully.

3.2 Research Problem, Question and Objective

In this research, the problem relating to smart instruction with special attention to EAP and ESP is solved. The question appears whether using the smart approach to instruction suits learners of various motivation types. Therefore, *the main objective of the research is to discover whether the increase in learners' knowledge is statistically significant,* and thus the smart approach to EAP and ESP instruction is entitled to be exploited at higher education level.

3.3 Hypotheses

Two hypotheses were set as follows:

H1: In smart instruction of EAP statistically significant differences in test scores appear between learners of different motivation types.
H2: In smart instruction of ESP statistically significant differences in test scores appear between learners of different motivation types.

3.4 Methods and Tools

The quasi-experiment and ex-post-facto method were applied to verify/falsify the hypotheses. Two tools were exploited for data collecting: Motivation Type Inventory (MTI) and didactic tests to monitor the entrance and final level of knowledge. Data were processed by ANOVA. If statistically significant differences were detected, post hoc analysis was conducted using Fisher's LSD test.

MTI distinguishes which type of motivation prevails with the learner. Reflecting the result, four groups of learners were set (Explorers, Directors, Coordinators, Accurators). Then, increase in knowledge was calculated as the difference in pretest and posttest scores. Both in EAP and ESP the pretests and posttests included 24 tasks of open-answer, mostly the translation type. All tests were held in the first and final lessons, face-to-face, in the printed form.

3.5 Research Sample

Totally, 121 respondents enrolled at Jan Evangelista Purkyne University, Czech Republic, participated in the research. First, they administered the Plaminek's Motivation Type Inventory. However, two of them did not fill in the inventory correctly; final amount was N = 119 (43 explorers, 33 directors, 23 coordinators, and 20 accurators). Most respondents were students of Faculty of Education and Faculty of Natural Sciences. The structure of respondents was monitored from the view of *gender* (Male

7%; Female 93%; *age* (19 – 23 yrs.); *degree* (Bachelor 66%: FE 75%; NS 25%), Master (FE 100%; NS 0%) and *faculty*:

(a) Faculty of Education (FE): 82% (M 1%; F 99%) enrolled in study programmes on Teaching at pre-primary schools (39%); Teaching at primary schools (26%); Teaching at primary and SEN (Special Educational Needs) schools (13%); Czech language for media and communication (22%);
(b) Faculty of Natural Science (NS) enrolled in study programmes of Information technologies (46%), Toxicology (18%), Biology (12%), Chemistry and Biochemistry, Geography, History (12% each).

4 Research Results

The entrance didactic test both in EAP and ESP administered at the beginning of semester (pretest) and final didactic test conducted after 12 weeks of instruction at the end of semester (posttest) provided data for statistic processing in which two hypotheses were tested. Increases in learners' knowledge were calculated as difference between pretest and posttest scores in single motivation types.

First, increases in learners' knowledge between four motivation types were processed by the method of analysis of variance (ANOVA) at the significance level $\alpha = 0.05$. Results are displayed in Table 1 (part EAP). The results show that at the significance level $\alpha = 0.05$, the statistically **significant differences were not** detected between respondents of single motivation types. **Hypothesis H1** stating that *In smart instruction of EAP statistically significant differences in test scores appear between learners of different motivation types* **was rejected.** Therefore, application of further testing for differences between single motivation types via post-hoc analysis and Fisher's LSD test was not required. Differences are displayed in Fig. 1 (left).

Second, increases in learners' knowledge between four motivation types were processed by the method of analysis of variance (ANOVA) at the significance level $\alpha = 0.05$. Results are displayed in Table 1 (part ESP). The results show that at the significance level $\alpha = 0.05$, the statistically **significant differences were not** detected between respondents of single motivation types. **Hypothesis H2** stating that *In smart instruction of ESP statistically significant differences in test scores appear between learners of different motivation types* **was rejected.** Therefore, application of further testing for differences between single motivation types via post-hoc analysis and Fisher's LSD test was not required. Differences are displayed in Fig. 1 (right). The EAP learning content is marked as Academic English (AE). The ESP learning content is marked as Career Development (CD).

Table 1. Differences in increase in learners' knowledge in EAP and ESP.

	Sum of squares	df	Mean square	F	Sig
EAP: Between groups	60.225	3	20.075	.707	.550
EAP: Within groups	3265.170	115	28.393		
EAP: Total	3325.395	118			
ESP: Between groups	323.320	3	107.773	2.009	.117
ESP: Within groups	6168.378	115	53.638		
ESP: Total	6491.697	118			

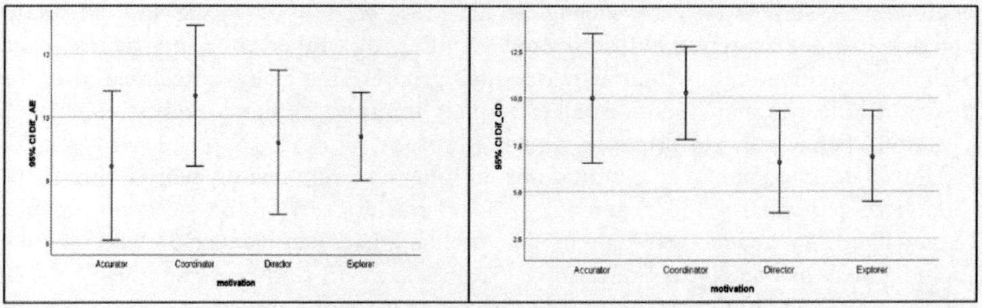

Fig. 1. Differences monitoring increase in EAP knowledge in four motivation types (left). Differences monitoring increase in ESP knowledge in four motivation types (right).

5 Discussion and Conclusion

The motivation to learning aims at learners' inner engagement into fulfilling tasks to support their further development, and also, it forms the basis for development of auto-didactic strategies. Within the instruction various individual needs of learners should be emphasized and saturated, resulting in success, not failure [2].

To sum up the research results, statistically significant differences were detected between the pretest and posttest scores in EAP and ESP. However, statistically significant difference in increase in knowledge was not detected between single motivation types. In other words, the cognitive increases all learners reached were similar; there were no learners of any single motivation type who did not meet the requirements for acquiring the learning contents in EAP and ESP. Thus the presented results *prove that the described smart approach is entitled to be applied in EAP and ESP instruction of all motivation types at higher education level.*

We expected the coordinators, who prefer stability and effectiveness, express emotional intelligence and expect the others to reflect the empathy in feedback, would show statistically higher increase in knowledge. The motivation type of coordinator is strongly required in the smart approach where the exploitation of technologies and devices might cause lack of human approach. As coordinators prefer good results, they make efforts not to fail in meeting the others' expectations; they concentrate on building positive working/learning environment.

When considering our results in the light of other works, Anthony et al.' work [15] should be mentioned. They proved the motivation of prospective teachers' trainees provides a significant influence on the process of learning, along with the delivery, performance, and evaluation. When considering the didactic means they exploit in the researched process of learning, this can be called smart, despite they describe it as a blended one. Contrary to this result, Geng et al. [16] state that learning motivation, even though motivation types are not mentioned, positively influences cognition within the process but does not have direct impact on learners' performance. Intrinsic motivation is closely connected to classroom climate and imposes strong impact on both the learners' performance and engagement. Reflecting the description of their research sample, this finding can be applied on learners of all motivation type, Wei et al. [17] conclude. As summarized by Zhang et al. [18], who applied the individualized approach in the research, they discovered significantly higher cognitive increase and strength of motivation in their experimental group. Thus they concluded that the personalized learning intervention can effectively improve learners' motivation, as well as learning behaviour, performance, and self-efficacy.

However, there are some limitations of our research results which should be mentioned, particularly those given by the characteristics of the research sample: (1) imbalance in gender (80% of female probands), (2) imbalance in participating faculties (80% of Faculty of Education), (3) imbalance in study programmes (80% of prospective teachers). On the other side, these results entitle the smart approach to be suitable for prospective teachers. This finding is very important, as the prospective teachers are those who bring the progress, in this case the smart approach to instruction. It is very positive if they have personal experience in smart approach from both the theoretical and practical point of view. In agreement with latest works, the support to teachers' motivation is a crucial task of the pre-service and in-service teacher training. Despite the fact that in foreign language instruction, the smart approach is rather wide, further researches monitoring this field are strongly required in the future.

Acknowledgements. The paper is supported by the SGS project N. SGS03/PdF/2019–2020 ICT-Enhanced Teaching English.

References

1. European Union, Council Recommendation of 22 May 2019 on a comprehensive approach to the teaching and learning of languages, 2019/C 189/03. https://eur-lex.europa.eu/legal-content/EN/TXT/PDF/?uri=CELEX:32019H0605(02)&from=EN. Accessed 18 May 2020
2. European Education Area. https://ec.europa.eu/education/education-in-the-eu/european-education-area_en. Accessed 18 May 2020
3. Gilet, A.: What is EAP?. https://www.uefap.com/bgnd/eap.htm. Accessed 18 May 2020
4. Hamp-Lyons, L.: English for academic purposes. In: Hinkel, E. (ed.) Handbook of Research on Second Language Learning and Teaching, Chapter: 6. Routledge/Taylor & Francis, pp. 89–105 (2011). https://doi.org/10.1017/CBO9780511667206.019
5. Flowerdew, J., Peacock, M.: Issues in EAP: a preliminary perspective. In: Flowerdew, P. (ed.) Research Perspectives on English for Academic Purposes, pp. 8–24. CUP, Cambridge (2001)

6. Hyland, K., Hamp-Lyons, L.: EAP: issues and directions. J. Engl. Acad. Purp. **1**, 1–12 (2002)
7. Hutchinson, T., Waters, A.: English for Specific Purposes. CUP, Cambridge (1987)
8. Doran, G.T.: There's a S.M.A.R.T. way to write manaa SMART. way to write management's goals and objectives. Gement's goals and objectives. iew. **70**(11), 35–36 (1981)
9. Silverio-Fernández, M., Renukappa, S., Suresh, S.: What is a smart device a conceptualisation within the paradigm of the IoT. Vis. Eng. **6**(3) (2018). https://doi.org/10.1186/s40327-018-0063-8
10. Innovative learning solutions. https://www.cae.net/what-is-smart-learning-and-why-does-it-interest-educational-centers/. Accessed 18 May 2020
11. Baumeister, R.F.: Toward a general theory of motivation: problems, challenges, opportunities, and the big picture. Motiv. Emot **40**(1), 1 (2016). https://doi.org/10.1007/s11031-015-9521-y
12. Reeve, J.: A grand theory of motivation: Why not? Motiv. Emot. **40**(1), 31–35 (2016)
13. Plamínek, J.: Tajemství motivace - jak zařídit, aby pro vás lidi rádi pracovali. [The secret of motivation – how to make people like working for you]. Grada, Praha (2010)
14. Škoda, J., Doulík, P., Bílek, M., Šimonová, I.: The effectiveness of inquiry based science education in relation to the learners' motivation types. J. Balt. Sci. Educ. **14**(6), 791–803 (2015)
15. Anthony, B., Kamaludin, A., Romli, A., et al.: Exploring the role of blended learning for teaching and learning effectiveness in institutions of higher learning: an empirical investigation. Educ. Inf. Technol. **24**, 3433–3466 (2019). https://doi.org/10.1007/s10639-019-09941-z
16. Geng, S., Law, K.M.Y., Niu, B.: Investigating self-directed learning and technology readiness in blending learning environment. Int. J. Educ. Technol. High. Educ. **16**(17), 22 (2019). https://doi.org/10.1186/s41239-019-0147-0
17. Wei, Y., Wang, J., Yang, H., Wang, X., Cheng, J.: An Investigation of Academic Self-Efficacy, Intrinsic Motivation and Connected Classroom Climate on College Students' Engagement in Blended Learning. In: 2019 International Symposium on Educational Technology (ISET), Hradec Kralove, Czech Republic, pp. 160–164 (2019). https://doi.org/10.1109/ISET.2019.00041
18. Zhang, J.H., Zou, L.C., Miao, J.J., Zhang, X.Y., Hwang, G.J., Zhu, Y.: An individualized intervention approach to improving university students' learning performance and interactive behaviors in a blended learning environment. Interact. Learn. Environ. **28**(2), 231–245 (2020). https://doi.org/10.1080/10494820.2019.1636078

Metalinguistic Awareness in Technical Communication

Ekaterina Tsareva(✉), Roza Bogoudinova, and Elena Volkova

Kazan National Research Technological University, Kazan, Russia
cetinas@mail.ru, rozabog@bk.ru, wolfkova@mail.ru

Abstract. The paper discusses the socio-communicative function of the engineer's foreign language training where the content should be based on materials of the global latest technological advances and understanding the essence of production culture in different countries. The structure and content of foreign languages training is revealed taking into account the potential of multilingualism, technocommunication and metalinguistic awareness. Foreign language training contributes to the formation of linguistic, communicative and metacognitive skills. International modern requirements to the students' foreign language training at an engineering university are actualized, its contradictions and features of pedagogical forms, methods and tools are identified. The content and structure of multilingualism, technocommunication and metalinguistic awareness of the students' foreign languages training at an engineering university are revealed. A methodology for the implementation of foreign language training in combination with multilingualism, technocommunication and metalinguistic awareness is devoted. The effectiveness of the proposed methodology was proved through the realization of group international study trips where students showed an understanding of technology by means of foreign languages and a metalinguistic awareness focusing on the cultural traditions of the region and local features of production. A conscious approach to deeper linguistic knowledge, cultures of different countries and technologies according to the high language and communications requirements in the field of science and technology, combining linguistic and engineering thinking in the human mind for a more complete understanding of the essence and content of engineering education is argued.

Keywords: Metalinguistic awareness · Technocommunication · Multilingualism · Foreign languages training

1 Introduction

Megatrends such as climate change, urbanization, demographic change, food and energy security on the one hand and global conditions for economic development, including globalization, digitalization, hypercompetition, informatization, industrial revolution 4.0 on the other hand are stimulating changes in all spheres of human life and are remaking strategies of high technology engineering modernization. The goal of high-tech engineering is to improve the well-being and quality of life of all people and to preserve the ecological safety of the planet. The global need for the development of

M. E. Auer and T. Rüütmann (Eds.): ICL 2020, AISC 1328, pp. 232–240, 2021.
https://doi.org/10.1007/978-3-030-68198-2_21

innovative processes, the accumulation of the newest technologies has activated international scientific and technical communications and the possibilities of technical communication. An interesting opinion was expressed by scientist Stephen Doheny-Farina from MIT University, he argued that in a large context the technical and commercial processes of converting technologies into products are communication processes [1]. One of the most relevant forms of world scientific and technical forms of interaction is the transfer of technologies. In the framework of transfer the dissemination of innovative technological solutions, intellectual activity between countries, individual industries or industries are carried out. In modern conditions of rapid transformation of socio-economic shifts this exchange gains strategic importance and shapes regional, national and international scientific and technological sustainability.

Rapid changes in the technical sphere and modern forms of scientific and economic cooperation impose corresponding requirements not only on its infrastructure but on the specialists involved in various segments of engineering. The fundamental transformation in the role of equipment and technologies in society, their invasion of the anthropological essence of man requires other key competencies from specialists than in the past. In the transition from studying at a university to the labor market in addition to technical and professional skills it is important to have personal and social skills, such as communication, organizing and managing time, knowledge of regional features of engineering, and high responsibility for making engineering decisions and social adaptation for young specialists. At the same time the socio-communicative function of the engineer's foreign language training is increasing, the content of which should be based on materials of the global latest technological discoveries and understanding the essence of production culture in different countries. This will allow students to more widely analyze current problems, evaluate the consequences of technological and social changes, predict and recognize trends and stay one step ahead.

It should be noted that on the part of students, their attitude to the study of languages has changed. It means that their understanding of the importance of communication in the modern world gives grounds for rethinking the role of foreign language training in higher education, for a deeper students' interest in learning several languages and realizing the need to adapt foreign language training to the realities of modern society [2]. The transition from foreign language training in the traditional format to a deeper understanding of the role of intercultural and technical communication in several languages in the modern context of technological change has matured.

An integral part of the scientific and technological process along with the invention and production is technical communication (technocommunication). Technocommunication is the ability of an engineering university student to professional communication strategies in order to transmit technical information in several languages, taking into account the socio-cultural characteristics of the countries producing the technical product. Technocommunication is expressed in the form of search skills, critical analysis, synthesis, distribution, adaptation, visualization information in several foreign languages, including for a wide audience of users. To successfully start-up of a product on a global and local market, an engineer requires a core analysis of customer expectations at the national and international levels and the subsequent linguistic and functional support of its quality. It deals with a simplified language describing the principles of the product in several languages, considering regional features of potential consumers [2, 3]. Despite the digitalization has allowed a large number of users to connect to the Internet, however, the majority of them do not speak English, that is why it's so important to localize (adapt) the product for international users using technocommunication and multilingualism.

In combination with knowledge of several foreign languages (multilingualism), with understanding of the principle of their functioning (metalinguistic awareness) technocommunication significantly widens the access of specialists to scientific and technological progress in the context of interstate, international scientific relations and intercultural communications. Multilingualism is considered as the acquisition of two or more foreign languages at any level necessary for intercultural, professional and technical communication, for working in multinational teams with speakers of other languages and national cultures, for understanding the national and cultural context of the countries producing the technical product and recognition the specifics of work in the engineering industries beyond boundaries. Acquisition of several foreign languages contributes to the formation of a more meaningful attitude to the language, that is to say, metalinguistic awareness. It is expressed in the ability to critically perceive someone else's culture and experience, to compare, analyze linguistic and sociocultural phenomena, independently choose strategies for studying foreign languages and linguistic reflection. Metalinguistic awareness in the process of foreign languages training allows to tap into the intricacies of linguistic and engineering thinking.

The aim of this study is to determine the content, structure, features of modern foreign language training at an engineering university, leading to the identification of the integrated potential of multilingualism, technocommunication and metalinguistic awareness.

2 Method and Materials

To achieve this goal, theoretical (analysis of psychological, pedagogical, sociological literature, normative documents and products of educational activities of students) and empirical research methods (observation, conversation) were used.

At the first stage of the study (2017), the structure and content of foreign language training was revealed considering the potential of multilingualism, technical communication and metalinguistic awareness.

Engineering education involves the formation of a professional engineering culture consisting and understanding of sociocultural and humanitarian knowledge allowed to assess the place and possible consequences of technical development in a wider sociocultural context. Therefore, we offer to consider the educational and cognitive resource of multilingualism, technological communication and metalinguistic awareness within students' foreign language training at an engineering university. By focusing on these three characteristics, foreign language training will contribute to the formation of linguistic, communicative and metacognitive skills.

The focus on multilingualism, technological communication and metalinguistic awareness in foreign language training at an engineering university was based on:

– *modern requirements* of employers to engineer's formation, to the level of their foreign languages acquisition, *recognizing* the need to learn several languages by students in order to solve professional problems;
– *contradictions* between the expansion of the exchange of products of scientific and technical activities, increasing international, intercultural, scientific and technical communications and the insufficient level of adaptation of graduates of engineering universities to professional activities in the conditions of transformation of economic relations;
– *features* of pedagogical methods, forms and means in the framework of students' foreign language training at an engineering university.

The integration of multilingualism, technocommunication and metalinguistic awareness in foreign language training at an engineering university is due to the general low level of students' ability to perceive the intercultural diversity of countries [4], to interact socioculturally in the process of caring out educational and professional tasks, the ability to show a world outlook and the ability to apply a systemic approach to solving problems, the ability to analyze scientific and technical information in several foreign languages, readiness for activity in multinational teams and the ability to self-education, self-realization and self-development [5].

At the second stage of the study (2018) the content and structure of multilingualism, technology communication and metalinguistic awareness in foreign language training of students in an engineering university were revealed.

Multilingualism includes:

- knowledge of the system of languages studied;
- knowledge of the linguistic features of different social layers, representatives of different generations, genders, social groups of various countries;
- ability to position itself as a representative of a national culture;
- the ability to mediate, interpret one culture in terms of another;
- ability to work with foreign language information and create new text in several languages;
- knowledge of the features of linguistic choice depending on the social and cultural norms of native speakers [6, 7].
- formation of tolerance [8].

Technocommunication consists of:

- knowledge of the socio-cultural foundations of society;
- the ability to produce a technical text and adapt it to a wide audience;
- strategies for transmitting technical information in foreign languages;
- skills to persuade, motivate with the help of technical information;
- the ability to take into account current trends in the development and achievements of foreign science, engineering and technology in professional activity;
- ability to collect and analyze scientific and technical information.

Metalinguistic awareness includes:

- skills of self-organization and self-education;
- linguistic reflection [9];
- regulation of personal cognitive processes, including new ways of thinking - critical, creative, non-linear and systemic [10];
- knowledge of strategies for independent study of foreign languages;
- linguistic abilities for linguistic observation and language conjecture;
- the ability to identify similarities and differences in the studied languages [11].

At the third stage (2019), a pedagogical experiment was carried out to integrate multilingualism, technocommunication and metalinguistic awareness into foreign language training of students at an engineering university, moreover theoretical and experimental results concerning the effectiveness of the proposed model of foreign language training of students at an engineering university were demonstrated.

3 Results and Discussion

The experimental work was carried out at Kazan National Research Technological University. Participants were Bachelors (second year), majoring in «Biomedical Engineering» (aged 19–20) and "Power Engineering and Electrical Engineering".

During the experiment multilingualism, technical communication and metalinguistic awareness training was realized using traditional and innovative teaching methods, comparative methods were applied to analyze the culture, communicative behavior and functional structure of different languages, differences and similarities in phonetics, vocabulary, grammar, stylistics of two foreign languages and native language were revealed [12] as one the method of intensification of professional training [13]. The use of various didactic units in the classroom (multilingual dictionaries, glossaries, contrasting assignments, authentic texts) in combination with various organizational forms of work (classroom, extracurricular and independent), methods of students' cognitive activity (explanatory, illustrative, research, method of problematic presentation of information), pedagogical technologies (training in cooperation, development of critical and systemic thinking), active and interactive teaching methods (virtual tours, brainstorming, project method, dialogue of cultures) [14], developing of tolerance focused on a deeper understanding of the importance of multilingualism, technocommunication and metalinguistic awareness. In the classes, specially developed methodological guidelines "Medical Engineering", "Multilingual Russian-English-French dictionary of medical engineering terms" and a glossary for Bachelors, majoring in "Biotechnical systems and technologies" were used.

Table 1 shows the pedagogical methods, forms, and tools used in preparing foreign students of an engineering university.

Table 1. Pedagogical methods, forms and means of developing multilingualism, technocommunication and metalinguistic awareness.

Multilingualism	Technocommunication	Metalinguistic awareness
• Multilingual dictionaries and glossaries; • parallel text; • sociodrama; • brainstorm; • virtual excursion	• Writing texts for various audiences; • schematization of information; • presentation, • explanation of the principle of operation of the device in simple language	• Contrasting tasks; • independent work; • tasks for the analysis of communicative behavior; • reflective and motivational questions; • thinking out loud

After completing the foreign language training program, students expressed a desire to independently apply for participation in the grant program "Group Study Visit to Germany" as part of the scholarship programs of the German Academic Exchange Service DAAD (2017) and take part in the internship (Germany 2019). To participate in these programs, students decided to independently study the German language, although this was not provided for by the requirements. Based on the existing linguistic experience, students individually chose the language learning strategy including the method of acquisition of two foreign languages of one Romano-Germanic group. This method is focused on communicative-cognitive and contrasting principles and implies the purposeful development of students' cognitive abilities through the new foreign language. It means the extensive support in the acquisition of the next language (L3) on similar linguistic and sociocultural phenomena in languages that are in contact in the educational process: native (L1), first (L2) and second foreign languages (L3).

During the trip in the process of the observation of the linguistic and communicative behavior of students a clear positive dynamic was revealed. At the beginning of the trip at formal meetings with the administration of universities, teachers and scientists, students adhered only to the English, for example while discussing the engineering education system in different countries or during the presentation of their university and the scientific activity of the department. Within informal conversation with foreign students, communication was carried out in several languages: in English, French and German due to a decrease in psychological linguistic barriers, lack of language control by teachers and a decline in fear of making a speech mistake. Subsequently, within the group of students, an increase in the frequency of use of phrases of everyday and technical English and German languages was noted, which is explained by their repeated listening and reproduction in an environment of everyday, business and technical communication. It is important to note that in plus with linguistic progress, a better understanding of the cultural characteristics of Germany has also formed. In the culturological aspect they became more familiar with the history, architecture of the country, production facilities, revealed differences in the behavior of Germans from Russians, recognized the characteristic features of the German mentality - punctuality, respect for the law, ecological culture.

During the trip students realized the importance of multilingualism, technocommunication and metalinguistic awareness while hosting in another country and communication with representatives of other nationalities. They understood the key role of multilingualism, technocommunication and metalinguistic awareness during all events, including a discussion seminar on the use of alternative energy sources at the University of Applied Sciences with representatives of several countries, when they tested expressions of technological innovations in several languages (L1, L2, L3). The visit of Siemens enterprise allowed them to discuss not only the latest production technologies, but also to analyze the sustainable development program, corporate ethics (Code of conduct) and social responsibility of the enterprise in terms of industrial culture, history, science and technology.

During the trip students significantly increased the level of technical communication in several foreign languages and showed accumulated knowledge and skills in situations of practical international professional communication, showed adequate communicative behavior, demonstrated their readiness for intercultural communication in the field of scientific and technical communication during all meetings, seminars, presentations and round tables at technical universities and at a German company.

Studying in an extended program of foreign language training with the inclusion of multilingualism, technocommunication and metalinguistic awareness, students have great chances and desire to learn other foreign languages in the system of additional education [15], independently or on-line [16], they actively apply for training in graduate students abroad and take part in international university programs, are part of international internships in Vietnam/China, France [17, 18], and motivate and explain to other students the possibility of participating in the university's international activities.

4 Conclusion

Multilingualism, technological communication and metalinguistic awareness in the process of foreign language training is manifested as a conscious approach to deeper linguistic knowledge, cultures of different countries and technologies, taking into account the high requirements for language and communications in the field of science and technology. Such an approach to foreign language training at an engineering university not only comes forward the language in its classical sense, but also combines linguistic and engineering thinking in the mind for a more complete understanding of the essence and content of engineering education. An engineer after such foreign language training with a systematic perception of multilingualism, technological communication and metalinguistic awareness will be able to translate the content of technical courses into several foreign languages, tapping into the essence of technological processes through linguistic communicative thinking. The appearance of technical courses in foreign languages at a university not only raises the international university attractiveness ratings, but also favors the organization of communication and interaction between the university and enterprises, business, and the state in the process of transferring knowledge, leading to increasing the academic mobility of students and teachers. In other words, within the university's mission, multilingualism, technocommunication and metalinguistic awareness should be considered as an opportunity to implementation and design of advanced education.

Because of foreign language training implies an in-depth immersion in the linguistic sphere and culture, it unites people in the framework of a common social responsibility for the preservation of the biosphere and technological safety. Specialists of all areas are connected by a single planetary thinking [19], where a special role is played by metalinguistic awareness based on knowledge of different languages and cultures [20], awareness of the social consequences of technological innovations in each country. As an illustration of the above thought, one can cite the example of the coronavirus pandemic in the current year, it clearly shows how experts in many areas have united by a common goal - the search for chemical production technologies against the threatening virus. All specialists are ready to communicate and solve one single task - to survive on our planet.

References

1. Doheny-Farina, S.: Rhetoric, Innovation, Technology: Case Studies of Technical Communication in Technology Transfer. MIT Press, Cambridge (1992)
2. Polyakova, T.Y., Karelova, D.G.: Engineering students' needs in foreign language studying in Russia. In: Teaching and Learning in a Digital World: Proceedings of the 20th International Conference on Interactive Collaboration Learning, vol. 2, pp. 481–490. Springer, Cham (2018)
3. de Jong, M.D.T., Neulen, S., Jansma, S.R.: Citizens' intentions to participate in governmental co-creation initiatives: comparing three co-creation configurations. Gov. Inf. Q. **36**(3), 490–500 (2019)
4. Volkova, E.V.: Different approaches to the problems of intercultural communicative competence: Proceedings of the 16th International conference on Interactive collaborative learning and 42-nd International IGIP Symposium on Engineering pedagogy. Book of Abstracts. Edited by Claudio da Rocha Brito, Melany M. Ciampi, pp. 456–457. Kazan, Russia (2013)
5. Geykhman, L., Kavardakova, E.: Google translate: how to use it to students advantage. Buletinul Stiintific al Universitatii de Stat Bogdan Petriceicu Hasdeu din Cahul. Seria Stiinte umanistice, **1**(7), 16–22 (2018)
6. Tsareva, E., Bogoudinova, R., Khafisova, L., Fakhretdinova, G.: Poster: multilingualism as a means of students' technocommunicational competence forming at engineering university. In: Advances in Intelligent Systems and Computing 2020, ISI, vol. 1134, pp. 137–142 (2020)
7. Franceschini, R.: Multilingualism and multicompetence: a conceptual view. Mod. Lang. J. **3**, 344–355 (2011)
8. Fakhretdinova G., Dulalaeva L., Tsareva E., Extracurricular activities in engineering college and its impact on students' tolerance formation. In: Advances in Intelligent Systems and Computing, т. 1134, с. 143–150 (ISI) (2020)
9. Blommaert, J., Rampton, B.: Language and superdiversity. Diversity. **13**(2), 2–7 (2011)
10. Kopečková, R.: Exploring metalinguistic awareness in L3 phonological acquisition: the case of young instructed learners of Spanish in Germany. Lang. Aware. **27**(1–2), 153–166 (2018)
11. Valeyeva, N.S., Kupriyanov, R., Valeyeva, E.R.: Metacognition and metacognitive skills: intellectual skills development technology. In: Metacognition and Successful Learning Strategies in Higher Education, pp. 63–84 (2017)

12. Serrano, R.: From metalinguistic instruction to metalinguistic knowledge, and from metalinguistic knowledge to performance in error correction and oral production tasks. Lang. Aware. **20**(1), 1–16 (2011)
13. Osipov, P.N.: Intensification of professional training as a pedagogical problem. In: 2012 15th International Conference on Interactive Collaborative Learning, ICL (2012)
14. Kraysman, N.V.: Dialogue of cultures and confessions at universities of the USA. In: International Conference on Interactive Collaborative Learning, ICL 2013, vol. 6644549, pp. 114–116 (2013)
15. Gazizova, A.I., Siraeva, M.N., Trofimova, G.S.: Formal and non-formal education means of mastering foreign language skills. Soc. Sci. **10**(6), 1324–1328 (2015)
16. Panteleeva, M., Sanger, P.A., Bezrukov, A.: International approaches to the development of cross-cultural education at high school. In: ASEE Annual Conference and Exposition, Conference Proceedings (2016)
17. Kraysman, N.V., Shageeva, F.T., Mullakhmetova, G.R., Pichugin, A.B.: Poster: preparation of engineering university students for academic mobility to French universities. In: Advances in Intelligent Systems and Computing, AISC 1135, pp. 713–718 (2020)
18. Shageeva, F.T., Bogoudinova, R.Z., Kraysman, N.V.: Teachers-researchers training at technological university. In: Proceedings of International Conference on Interactive Collaborative Learning 2018, ICL, vol. 1377, pp. 1699–1703 (2018)
19. Ziyatdinova, J., Bezrukov, A., Sanger, P.A., Osipov, P.: Cross cultural diversity in engineering professionals - Russia, India, America. In: ASEE 2016 International Forum, pp. 1–9 (ISI) (2016)
20. Semushina, E.Y., Ziyatdinova, J.N.: Studying English on-line as a part of a course “English for special purpose” in technological university. In: Advances in Intelligent Systems and Computing, AISC 1135, pp. 21–29 (2020)

Subtitled and Unsubtitled Movie and Listening Comprehension

Tania Trujillo, Ana Vera-de la Torre(✉), and Dorys Cumbe-Coraizaca

Facultad de Ciencias Humanas y de la Educación,
Universidad Técnica de Ambato, Ambato, Ecuador
tanniatrujillo1997@gmail.com,
{aj.vera, dm.cumbe}@uta.edu.ec

Abstract. This research project was developed with the educational purpose of investigating the effect of watching a subtitled or unsubtitled movie on EFL learners' listening comprehension. The study was developed at Universidad Técnica de Ambato with 53 students from the second semester of Pedagogía de los Idiomas Nacionales y Extranjeros major: 28 students from second semester A and 25 students from second semester B. The research project was quasi-experimental with a pre-test and post-test design. The researcher randomly assigned the second semester A to treatment and the second semester B to control. The experimental group was exposed to a movie with intralingual subtitles, while the control group was exposed to the same movie without subtitles. After watching each part of the movie, students from both groups took listening comprehension tests. To analyze the collected information the Wilcoxon test was used. Finally, the results showed that the subtitled movie had positive effects in the development of the listening comprehension.

Keywords: Subtitled movie · Unsubtitled movie · Listening skill · Listening comprehension

1 Introduction

Teaching English with movies make classes more interactive for students, increase understanding and so memory of content. A theoretical and research review presented by [1] demonstrates that video clips can be used as instructional tools because they increase understanding and provide a great cognitive and emotional impact to students; nevertheless, there are still discussions about whether or not to use subtitles in order to develop English skills, specifically listening skill. Some researchers are against the use of subtitles because they considered them as distracting elements that slow down the development of listening comprehension; for instance, [2] report that subtitles create a negative effect, that is, subtitles improve the visual aspects but reduce the aural perception. On the other hand, other researchers, such as [3] supports the positive effects of subtitles by arguing that learners are provided with text-support of authentic language, that way, learners can establish associations for retention and use of language. To investigate the effect of watching a subtitled or unsubtitled movie on EFL learners' listening comprehension, a quasi-experimental research was applied: one group was

M. E. Auer and T. Rüütmann (Eds.): ICL 2020, AISC 1328, pp. 241–252, 2021.
https://doi.org/10.1007/978-3-030-68198-2_22

assigned to treatment (movie with intralingual subtitles) and the other group was assigned to control (movie without subtitles). In order to obtain measurable results, both a listening pre-test and post-test were given; in addition, three listening comprehension tests were designed and given after watching 35 min part of the movie. The study lasted 5 sessions: (1) pre-test, (2) first listening comprehension test, (3) second listening comprehension test, (4) third listening comprehension test, and (5) post-test. That is, the pre-test determined the students' level of listening skills prior to the exposure to the movie, the listening comprehension tests measured the progress during the use of intralingual subtitles, and the post-test determined whether the students from both the experimental and control group had improved their listening comprehension skills. That way, it was established a strong and visible link between the pre-test, the use of subtitles, and improvement in the post-test. All the information gathered was analyzed, thus it contributed to come to conclusions.

2 State of the Art

[4] in their research entitled "Film Subtitles and Listening Comprehension Ability of Intermediate EFL Learners" aimed to determine whether there were any differences between the levels of comprehension between the group exposed to an intralingual subtitled film and the group that was not exposed to it. Through the results, the researchers concluded that the students who were exposed to videos with English subtitles got higher understanding of L2. There are two main factors from Saed, Yazdani & Askary's study that are associated with the present one: the population they worked with; that is, intermediate EFL learners; and the skill they chose to measure learners' improvement. Therefore, it can be said that using movies in English with English subtitles leads towards raising listening comprehension skills in this level. It also allows to carry out further research with a population at a lower level in the langauge, as in this research that worked with pre-intermediate learners.

[5] in their research "Implicit Instruction, Subtitles, Vocabulary and Listening Comprehension" aimed to investigate the impact of English subtitles and Bahasa Indonesian subtitles on content comprehension of movies through the students' listening comprehension. The results showed that English subtitles had better impact than Bahasa Indonesian subtitles. Even though, the present research does not address subtitles in the L1, it is important to compare the results from previous studies between the use of English subtitles, L1 subtitles and none subtitles; that way, it can be understood why this mode (L1 subtitles) was not selected in order to carry out the research. To provide learners with comprehensible language input, which [6] explained in their theory of second language acquisition, does not mean to present authentic English videos in the L1, but at the appropriate or just a little beyond learners' level in the L2. Interlingual subtitles can allow content comprehension, which can be labeled as a process regardlees language learning because of visual input, but create listening interference, because learners focus on reading subtitles; on the other hand, intralingual subtitles assist listening learning by indicading written form of spoken speech.

[7] in the research "The impact of keyword and full video captioning on listening comprehension" aimed to investigate the effect of full and keyword captioning on

listening comprehension. It was concluded that full video captioning helped to enhance listening comprehension. Besides the outperformance of the full captions group over the keyword captions and without captions groups, the choice of selecting full captions instead of keywords captions for this research is supported by the level in which it was applied; that is, elementary. The learners for this research are pre-intermediate level, but it is important to analyze how even lower levels succeeded at listening tests by using subtitles. Additionally, keywords captions lack practicality while full captioned video materials are easier to produce or get from online platforms.

[8] in their research "The Effects of Captions on EFL Learners' Comprehension of English-Language Television Programs" aimed to investigate the comprehension of Japanese university students when watching a television program with and without captions. The results of the study indicated that the captions led to increase comprehension of authentic television; however, it also pointed out that participants who watched the episodes without captions were also able to improve their comprehension. What the researches concluded shows that it is important to select the correct source of comprehensible input according to EFL learners' level; and,as it has been explained in the previous paragraphs, this source can also be a little beyond learners' level to make it more challenging and aid learners reach the next level of language learning process. That is why, in order to carry out the present research the selected source, in this case a movie, was neither too easy nor too difficult for pre-intermediate EFL learners. The experimental group was expected to take more advantages of the intralingual subtitles when scenes were more challenging to understand.

[9] in his study "Gist watching can only take you so far': attitudes, strategies and changes in behavior in watching films with captions" aimed to focus on changes in learners' viewing behavior when watching with captions. It was concluded that there were changes in behavior, attitudes, characteristics and strategies over time, especially in the participants who maximized successfully language learning gains. It is important to understand learners' perceptions, behaviors, attitudes, and strategies towards the use of subtitles, due to it allows to predict whether or not subtitles will work. Some people will argue that subtitles are distracting while others will find them useful. In addition, the present research does not intent to state subtitles as a technique that should be used forever for learning English, but as long-term technique which will help learners develop strategies to maximize their use over a period of time, as well as confidence to start watching videos without subtitles.

3 Research Design

3.1 Participants

The participants of this research included 53 students: 28 students from second semester "A" and 25 students from second semester "B" from Pedagogía de los Idiomas Nacionales y Extranjeros major at Universidad Técnica de Ambato.

3.2 Materials

Pre-test and Post-test
For the pre-test, a listening part was taken from The Preliminary English Test (B1 level, Cambridge English Level Certificate in ESOL International-Entry 3) to determine the students' level of listening skills prior to the exposure to the movie. Likewise, for the post-test, another listening part was taken from The Preliminary English Test (B1 level, Cambridge English Level Certificate in ESOL International-Entry 3) to determine whether the students from both the experimental and control group had improved their listening comprehension skills.

Movie Selection
The fantasy comedy movie "Bruce Almighty" (2003), directed by Tom Shadyac, was selected for this research. The running time of this movie was approximately 101 min, which were divided into 3 parts of 35 min each one; each part was presented to the students in one hour of class. The movie was selected based on both basic criteria and linguistic-instructional criteria adapted from [10].

Basic criteria. First, the movie rated PG-13 of maturity level[1], so it was appropriate for the students' age. Second, the movie had good picture quality, clear and sequential presentation. Third, the movie had access to intralingual subtitles, where writing speech was the same as the audio. Fourth, the movie addressed an interesting content.

Linguistic-instructional criteria. First, the movie presented realistic styles of speech: a balanced mixture of various speech styles ranging from colloquial, semi-formal, and formal that pre-intermediate learners can understand. Second, the speech was authentic (redundancy, reduced forms, stress, rhythm, intonation, American accent, etc.); however, the pronunciation was intelligible. Third, the movie presented proper speed according to pre-intermediate learners' level. Fourth, the movie presented proper scope of grammar according to EFL learners' level; reasonable or acceptable amount of grammar beyond EFL's level. Fifth, the movie presented proper scope of vocabulary and idioms according to EFL learners' level; reasonable or acceptable amount of slang, jargon, cultural references. Sixth, the movie presented considerable percentage of speech/ dialogue.

Listening Comprehension Tests
In order to measure the progress during the use of intralingual subtitles and analyze it between the experimental group (A) and the control group (B), listening comprehension tests were designed and given after each 35 min part of the movie. There were two parts: in the first part, the students had to read 10 statements and decide if each sentence was true or false; in the second part, the students had to read 10 multiple-choice questions and circle the correct answer. The questions were written in English and focused on assessing responsive, selective and extensive listening performance [11]; thus, the students had to listen for main ideas, for details and for making inferences.

---

1 https://www.filmratings.com/Search?filmTitle=Bruce+Almighty&x=18&y=5

Procedure and Data Collection

Step 1: Both groups; that is, the students from second semester of Carrera de Pedagogía de los Idiomas Nacionales y Extranjeros "A" and "B", were ensured to be equal in their proficiency level on English, according to the Common European Framework, by interviewing two teachers and analyzing the book they were using for English which was Top Notch A2 level by Pearson Education.

Step 2: One whole group was randomly assigned to treatment (students from second semester "A") and the other whole group was randomly assigned to control (students from second semester "B").

Step 3: Both groups took the listening pre-test.

Step 4: One movie was selected and divided into three parts of 35 min each one.

Step 5: Lesson plans were designed in order to present each part of the movie for both groups: first, the students were presented the overall information about the movie (plot and characters); second, the students were presented the key vocabulary through a matching activity; third, the students were given the listening comprehension test to read and familiarize with the questions; fourth, the students were told to answer the questions while watching the movie along with other instructions (they kept the vocabulary sheet while watching the movie and were allowed to consult their dictionaries); fifth, they were given 5 extra minutes to select or check their answers. It is important to mention that the students were not allowed to talk to their classmates to compare answers in the listening comprehension test. The only difference in the lesson plan was that the experimental group (A) was presented each part of the movie with English subtitles and the control group (B) watched them without subtitles. This process was followed in 3 sessions.

Step 6: Both groups took the listening post-test. The students who did not take the pre-test, one of the comprehension tests or the post-test were not taken into account to carry out the data analysis.

The research project was quasi-experimental with a pre-test and post-test design. It was quasi-experimental because, according to [12], it lacks one of the two characteristics of a true experimental research: randomization. In this case, quasi-experimental research involves using intact groups in an experiment, rather than assigning individuals at random to research conditions. One whole group was assigned to treatment (students from second semester "A") and the other whole group was randomly assigned to control (students from second semester "B").The experimental group was exposed to a movie with subtitles, while the control group was exposed to a movie without subtitles. The information was collected in order to make a comparison and an analysis through the Wilcoxon signed-rank test. As [13] explains, the Wilcoxon signed-rank test determines if one population is shifted with respect to another before and after an intervention by calculating the differences between their ranks.

4 Results and Discussion

4.1 Pre-test

Table 1. Pre-test

	Experimental	Control
Pre-test	10.64	11.04

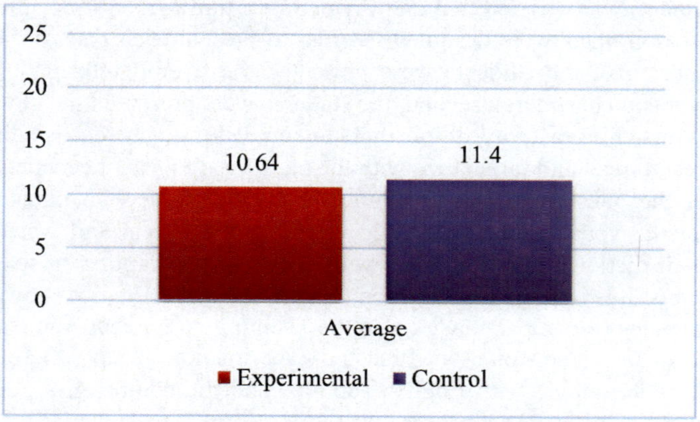

Fig. 1. Pre-test

Figure 1 represents the average score over 25 that the students from both groups got in the pre-test. In the pre-test, the experimental group' average was 10.64 points over 25, while the control group got 11.04 points over 25. By analyzing the results, it was evident that the students from both groups had a low level of listening skills (Table 1).

The difference in average between them is the only 0.4 points. The group selected as the control group got a relatively higher score. Therefore, as both groups got almost the same low score, either could had been selected as the experimental group in order to measure progress afterwards.

4.2 Listening Comprehension Tests

Table 2. Listening comprehension tests

	Experimental	Control
1st listening test	14.46	13.84
2nd listening test	14.68	12.92
3rd listening test	15.39	14.44

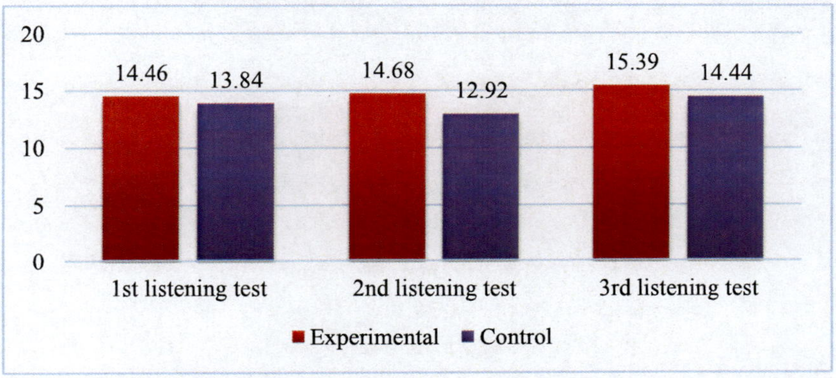

Fig. 2. Listening comprehension tests

Figure 2 represents the average score over 20 that the students from both groups obtained in each of the three listening comprehension tests. In the first listening comprehension test the experimental group scored 14.46 points out of 20, while the control group scored 13.84 points out of 20. The experimental group outperformed the first test by a difference of 0.62 points over the control group. In the second listening comprehension test the experimental group scored 14.68 points, while the control group scored 12.92 points. Again, the experimental group outperformed the second test by a difference of 1.76 points over the control group. In the third listening comprehension test the experimental group scored 15.39 points, while the control group scored 14.44 points. The experimental group outperformed the third test by a difference of 0.95 points over the control group (Table 2).

In the first listening comprehension test the experimental group scored 14.46 points, in the second test it scored 14.68 points, and in the third test it scored 15.39 points. From the first to the second test, the experimental group improved by 1.07%, and from the second to the third test the experimental group improved by 3.57%. In the first listening comprehension test the control group scored 13.84 points, in the second test it scored 12.92 points, and in the third test it scored 14.44 points. From the first to the second test, the control group decreased their performance by 4.60%; however, from the second to the third test the group improved by 7.60%.

4.3 Post-test

Table 3. Post-test

	Experimental	Control
Post-test	15.04	10.92

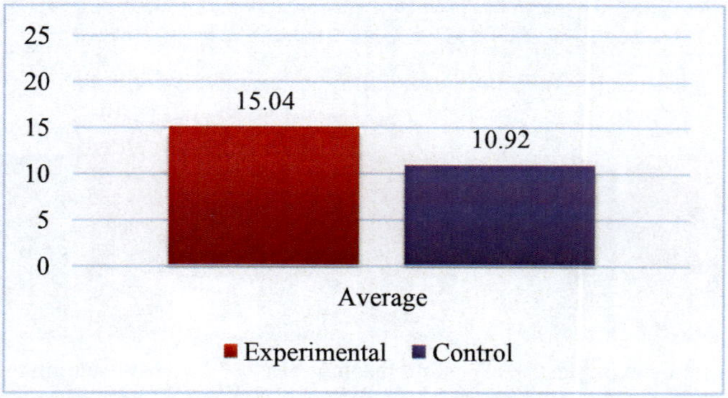

Fig. 3. Post-test

Figure 3 represents the average score over 25 that the students from both groups got in the post-test. In the post-test, the experimental group' average was 15.04 points over 25, while the control group got 10.92 points over 25. The students from the experimental group performed better than the control group in the post-test. The difference in average between both groups is 4.12 points (Table 3).

4.4 Tests

Table 4. Tests

	Experimental	Control	Over
Pre-test	10.64	11.04	25
1st listening test	14.46	13.84	20
2nd listening test	14.68	12.92	20
3rd listening test	15.39	14.44	20
Post-test	15.04	10.92	25

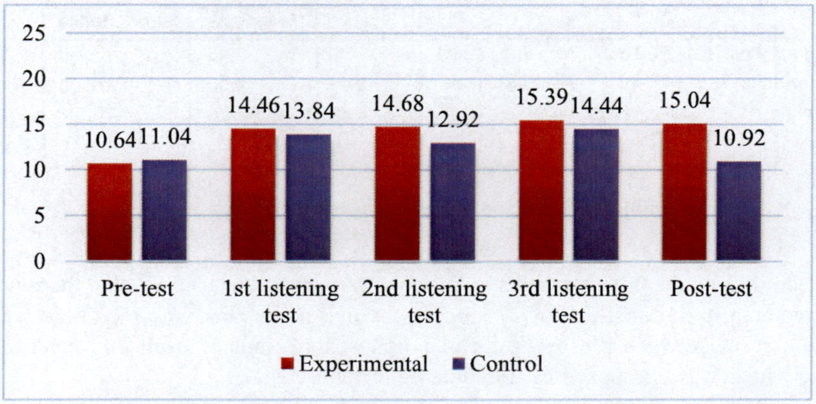

Fig. 4. Tests

Figure 4 represents the average scores that the students from both groups got in the pre-test, in the three listening comprehension tests and in the post-test. Comparing the results within each group, the experimental group scored 10.64 points in the pre-test, while in the post test it obtained 15.04 points; thus, the experimental group improved by 17.57%. On the other hand, the control group got 11.04 points in the pre-test, while in the post-test it obtained 10.92 points; thus, the control group declined their performance by 0.48%. Taking into account the pre-test and post-test results and the listening comprehension test results, it can be seen that both the experimental and the control group had low average scores in the pre-test; however, in the first listening comprehension test, both groups increased their performances (Table 4).

The experimental group improved progressively from the first listening comprehension test until the third test, while in the post-test it can also be shown a significant improvement from the pre-test; it is important to mention that the circumstances of how they performed the pre-test and post-test and the listening comprehension test were different: it means that while in the pre-test and post-test the students only had the aural input, in the listening comprehension tests they also received visual input, having in mind that the experimental group received additionally written input.

On the other hand, the control group did not show consistency: as prove of that, they got their lowest score in the second listening comprehension test. Also, they performed worse in the post-test than in the pre-test.

4.5 Wilcoxon with SPSS: Experimental Group

The results obtained through the pre-test and the post-test were analyzed by using the Wilcoxon test (Fig. 5).

		N	Mean Rank	Sum of Ranks
PostTest - PreTest	Negative Ranks	0[a]	,00	,00
	Positive Ranks	27[b]	14,00	378,00
	Ties	1[c]		
	Total	28		

Fig. 5. Wilcoxon signed ranks test (Experimental)

According to the positive ranks of Wilcoxon, it can be determined that there was an improvement in the post-test of 27 students, which mean rank was 14. There was not any alteration between the pre-test and post-test in 1 student from the experimental group, which was composed of 28 students in total (Fig. 6).

Test Statistics[b]

	PostTest - PreTest
Z	-4,564[a]
Asymp. Sig. (2-tailed)	,000

Fig. 6. Wilcoxon test statistics (Experimental)

According to the analysis of Wilcoxon, it was obtained a bilateral significance of −4.564, considering a p-value of 0.000, which is less than 0.05; therefore, the null hypothesis is rejected, and the alternative hypothesis is accepted: English subtitles did affect to the development of the listening comprehension in students from the experimental group.

4.6 Wilcoxon with SPSS: Control Group

		N	Mean Rank	Sum of Ranks
PostTest - PreTest	Negative Ranks	9[a]	9,00	81,00
	Positive Ranks	8[b]	9,00	72,00
	Ties	8[c]		
	Total	25		

Fig. 7. Wilcoxon signed ranks test (Control)

According to the negative ranks of Wilcoxon, there was a decrease in the performance in the post-test of 9 students, which mean rank was 9. According to the positive ranks of Wilcoxon, it can be determined that there was an improvement in the post-test of 8 students, which mean rank was 9. Finally, there was not any alteration between the pre-test and post-test in 8 students from the control group, which was composed of 25 students in total (Fig. 7).

Test Statistics[b]

	PostTest - PreTest
Z	-,225[a]
Asymp. Sig. (2-tailed)	,822

Fig. 8. Wilcoxon test statistics (Control)

According to the analysis of Wilcoxon, it was obtained a bilateral significance of −,225, considering a p-value of ,822, which is greater than 0.05; therefore, it is not statistically significant and indicates strong evidence for the null hypothesis. The null hypothesis is accepted and the alternative hypothesis is rejected. There were no differences in the development of the listening comprehension from the control group before (pre-test) and after (post-test) watching the movie without English subtitles (Fig. 8).

5 Conclusions

Watching the subtitled movie, that is, with intralingual subtitles (English), had positive effects on listening comprehension by indicading written form of spoken speech, it means, the students from the experimental group received more comprehensible input that they used in order to perform better at tests. On the other hand, watching the unsubtitled movie did not help develop listening comprehension to the control group, as prove of that, they did not show any improvement. Therefore, this research suggests that teachers use L2 subtitles with pre-intermediate students in order to help develop their listening comprehension skills, that way, they will acquire confidence progressively to start watching videos without subtitles.

The experimental group got 10.64 points over 25 in the pre-test, while in the post test it obtained 15.04 points over 25; thus, the experimental group improved by 17.57%. On the other hand, the control group got 11.04 points over 25 in the pre-test, while in the post-test it obtained 10.92 points over 25; thus, the control group declined their performance by 0.48%. Therefore, there was a significance improvement in the experimental group which could get a higher score in the post-test without relying on the subtitles or visual input, in comparison with the control group.

Through the listening comprehension tests, it was established a strong and visible link between the pre-test, the use of subtitles, and improvement in the post-test. The experimental group performed better in the three listening comprehension tests in comparison with the control group. While the experimental group improved their listening skills progressively until achieve 15.39 points over 20, the control group did not show consistency nor progressive improvement, as prove of that, in the second test they got the lowest score. It is important to mention, however, that both groups got better scores in the listening comprehension tests than in the post-test: this happened because in both cases they received visual and aural input in the listening comprehension tests, and the experimental group received additionally written input; while in the post-test they only received aural input. Hence, it is important to select the appropriate visual-aural input for students by following a criteria; movies must be appropriate and interesting for students' age and level. In addition, these movies must have access to intralingual subtitles, that is, the writing speech is the same as the audio, so it might not confuse students.

References

1. Berk, R.A.: Multimedia Teaching with Video Clips. Ambato (2014)
2. Bryant, R., Constante, J., Manna, M., Serechia, T., Starks, T.: Visual Reaction to Subtitling in Television and Media (2004)
3. Vanderplank, R.: Captioned Media in Foreign Language Learning and Teaching. Palgrave Macmillan, London (2016)
4. Saed, A., Yazdani, A., Askary, M.: Film subtitles and listening comprehension ability of intermediate EFL learners. Int. J. Appl. Linguist. Transl. **3**(2), 29 (2016)
5. Kusumawati, E., Hasan, H.: Implicit instruction, subtitles, vocabulary and listening comprehension. Alphabet **1**(2), 129–135 (2018)
6. Krashen, S., Terrell, T.: The Natural Approach (1983)
7. Bensalem, E.: The impact of keyword and full video captioning on listening comprehension. J. Teach. Engl. Specif. Acad. Purp. **3**(4), 453–463 (2016)
8. Rodgers, M.P.H., Webb, S.: The effects of captions on EFL learners' comprehension of english-language television programs. CALICO J. **1**(34), 20–38 (2017)
9. Vanderplank, R.: Gist watching can only take you so far': attitudes, strategies and changes in behaviour in watching films with captions. Lang. Learn. J. **47**(4), 407–423 (2019)
10. Kwon, W.: Selecting Movies for English Language Learning (2014)
11. Brown,D.: Teaching by principles. In: An Interactive Approach to Language Pedagogy, p. 103. Pearson Education Limited, New York (2007)
12. Campbell, D.T., Stanley, J.C.: Experimental and Quasi-Experiment AI designs for research. Houghton Mifflin Company All, Boston (1963)
13. McDonald, J.: Handbook of Biological Statistics. Sparky house publishing, Baltimore (2014)

Technology for L1 Kichwa/L2 Spanish Speakers' Vowel Sound Production of English as Their L3

Soledad Chango, Mayorie Chimbo, Wilma Suárez, and Ana Vera-de la Torre(✉)

Facultad de Ciencias Humanas y de la Educación, Universidad Técnica de Ambato, Ambato, Ecuador
soledad.chango@gmail.com, {elsamchimboc,wilmaesuarezm,aj.vera}@uta.edu.ec

Abstract. The current research shows the results of a study developed in an indigenous area of highlands Ecuador. The investigation tried to determine the influence of Kichwa, as a mother tongue (L1) and Spanish as a second language (L2) over the pronunciation of vowel sounds in English as a third language (L3). Twenty-six students participated in this study. The participants were homogeneous in terms of age, (17–18 years old); level of instruction in English, (elementary level according to CEFR); but hey were heterogeneous in terms of gender (11 males and 15 females). Participants showed difficulty in the pronunciation of English vowel sounds, since their L3 is linguistically different from their L1 and L2, in terms of pronunciation. The limited number of vowel sounds, both in the students' L1 and L2 impedes accurate production of vowel sounds in their L3. Furthermore, the lack of formal instruction in the English phonological aspects has created confusion in the students when pronouncing some vowel sounds. To measure the students' pronunciation, a reading paragraph was designed for them to read. A survey was also conducted to find out the reasons why students have pronunciation problems. The findings indicate that both, L1 and L2, influence the pronunciation of the L3. Similarly, these results have proved that the English vowel sounds, which are distant from the mode and point of articulation of the students previously acquired languages are the ones which cause more difficulty when they are pronounced. Furthermore, survey results show that there is not enough pronunciation instruction in the participants' English as a Foreign Language (EFL) classroom. The implementation of technological materials with an oral-based approach through formal instruction is required to overcome the student's pronunciation difficulties.

Keywords: English pronunciation · Vowel sounds · Kichwa · Spanish · Technology

M. E. Auer and T. Rüütmann (Eds.): ICL 2020, AISC 1328, pp. 253–262, 2021.
https://doi.org/10.1007/978-3-030-68198-2_23

1 Introduction

Communication is a complex process which involves more than a simple exchange of messages by a speaker and an interlocutor; the application of phonological, lexical and grammatical features is fundamental to prevent the message to be misunderstood (Deterding and Mohamad 2016). In the development of good pragmatically-appropriate speaking skills in a language, good pronunciation is also needed [1].

Pronunciation is one of the most important aspects in communication in foreign language learning since it influences learners' communicative competence and performance [2]; however, there is not much information regarding L3 pronunciation [3], even in English pronunciation of indigenous people whose native language is Kichwa. For this reason, the main aim of this investigation was to establish the influence of Kichwa, as a mother tongue (L1) and Spanish as a second language (L2) over the pronunciation of vowel sounds in English as a third language (L3). The research questions that guided this study were: Do L1 and L2 interfere the L3 acquisition in terms of pronunciation? What vowel sounds are the most difficult for Kichwa/Spanish speakers to produce when they are learning English as a third language? How can technology be used to improve students' English pronunciation?

In an EFL environment, communication provides enjoyment to students, since they use the L2 language spontaneously [4]. In order to communicate effectively, it is necessary to master some skills; one of them is to be able to pronounce sounds accurately so listeners can understand the message in an appropriate way. Thus, pronunciation plays an important intelligibility role to listen and understand others, and correctly respond to them. [5] argues that speech must be intelligible, so speakers should be able to convey an intended message. Consequently, the sounds of the target language should be articulated and modulated effectively to let the listener perceive them accurately.

2 State of the Art

2.1 Approaches in Pronunciation Improvement

Even though in the past pronunciation and comprehension were neglected, today they are essential aspects in the development of oral skills, so non-native speakers must be trained in this field to become competent speakers in the target language [6]. In Peru, is structured and implemented an application with Augmented Reality (AR) for the learning of Quechua in preschool children, which will give children the opportunity to relate to their environment, their own history and their ancestors in a fun way. They can grow with a different awareness of the others, in addition it provides linguistic stimuli for learning other languages [7]. In Iowa, United States was built a set of web-based tools to facilitate the L2 pronunciation learning, these tools give the students the opportunity of record themselves and instantaneously compare their voiceprint to that of a native speaker, thanks to that they can practice and practice again to improve their suprasegmental features [8].

In the EFL context, pronunciation has been studied only a short time before the beginning of the twentieth century. Celce-Murcia, Brington, Goodwin and Griner (2010) as cited in [9] describe two approaches to the teaching of pronunciation. The first one is the Intuitive-Imitative Approach which is based on the learner's ability to listen to and imitate the rhythms and sounds of the target language. This approach is based on good models and the use of different types of audio recorded material. The second approach is the Analytic-Linguistic Approach, which is based on the use of tools such as a phonetic alphabet, articulatory descriptions, charts of vocal apparatus, contrastive information, and other aids to supplement listening, imitation, and production (Celce-Murcia, Brington, Goodwin and Griner, 2010 as cited in [9]. This approach supports teaching pronunciation in an explicit way to complement the Intuitive-Imitative Approach. So, technology places an important role to develop pronunciation since learners may use podcasts and other online audio files as examples to imitate the production of sounds which are not found in the EFL learners' native languages. A speech recognition engine has been used on the mobile phone to identify spoken words in order to correct the pronunciation mistakes [10]. Most students prefer to use technological devices to learn English [11].

According to Kolabinova, Zaripova, Zaglyadkina, and Yusupova (2015) when the learner uses his L1 or other previously learned languages in the acquisition of his L2 or other additional languages, this process is called language interference. Weinreich (1953) as cited in [12] states that interference was initially used to describe the negative influence of L1 on L2 production. Nevertheless, more recent research has shown that L1 influence on L2 can also be positive, so the term "transfer" started to replace "interference". Transfer or interference from L1 to L2 can be either positive or negative [13]. It depends on the learners' age, immersion and language proficiency. L2 acquisition is, then, expressed as a dynamic process which depends on continuous L2 usage [13]. Results of a study done by [14] indicate that bilinguals utilize both languages during third language acquisition. Consequently, some psychological and sociolinguistic factors like the age, gender, proficiency level, sequential or simultaneous acquisition play an important role in the learning of a new language [15]. Taking into consideration that learners have already acquired their first and second language in a natural way, from the time they were born, a positive role may be applied with the use of teaching methodologies, based on the learners' prior knowledge [16].

2.2 Sound Production Among Languages

[17] make words, sentences, and phrases to be understood by others. The problem emerges when learners are not able to pronounce the target language effectively; then, communication stops, and learners feel frustrated to fulfill this goal. In this context, the problem Kichwa/Spanish speakers face when they are learning English as a foreign language is not only related to the syntactic and semantic aspects of the language, but also the prosodic features. It is essential to study the phonological problems students face when they are learning a third language; and this is because learners must deal with the phonological differences of their mother tongue, their second, and third languages. One of the areas in which the influence of the mother tongue is most noticeable and long-lasting is pronunciation when compared with grammar and vocabulary [18].

The mother tongue is a factor that influences the pronunciation of the L2 [19]. Thus, the pronunciation of the target language is facilitated if its pronunciation features are like the L1. Otherwise, when the pronunciation features are not similar in the L1, then there is interference with the acquisition of the L2 Approach (Celce-Murcia, Brington, Goodwin and Griner, 2010 as cited in [9]. Regarding L3 pronunciation, some experiences told by multilingual speakers prove that their prior linguistic knowledge is useful when acquiring a third language since their previous language experience allows them to create a sound inventory [20].

According to Gass (1979) as cited in [21] L2 learners have problems to distinguish L2 sound categories when their mother tongue has a scarce number of sounds (Nowacka, 2010; Flege, 1991 cited in [5]. This is the case of the students who participated in this study since the two languages they use, Kichwa and Spanish, have fewer vowel sounds than English. See table below.

Table 1. Phonological-phonetic comparative vowel production

Language	Vowel	Phonetic script	Sample word	English Translation (for kichwa and spanish words)
Kichwa	A	/a/	atil	Hen
	I	/i:/	inty	Sun
	U	/u/	tupu	Pin
Spanish	A	/a/	alma	Soul
	E	/e/	elefante	Elephant
	I	/i/	iguana	Iguana
	O	/o/	operación	Operation
	U	/u/	último	Last
English		/I/	Fish	
		/i:/	Tree	
		/ɑ:/	Car	
		/æ/	Cat	
		/ɒ/	Clock	
		/ɔ:/	Horse	
		/ʊ/	Full	
		/ʊ:/	Boot	
		/ə/	Computer	
		/3:/	Bird	
		/e/	Egg	
		/ʌ/	Up	

2.3 Technology and Language Learning

In the twentieth century, technology has become an essential tool for communication and education [22] and it can be used in the English language learning process. In fact, ICT provides a lot of benefits in the development of learners' linguistic and communicative competence in language learning [23]. Intelligent systems can be used in the field of education, specifically in foreign language teaching [10]. Indeed, this author states that mobile devices can be a good alternative for education. In the field of language learning, students may find mobile technology applications useful and motivating for language learning. Additionally, [24] states the benefit of using both, Computer-Assisted Language Learning (CALL) and Task-Based Language Teaching (TBLT) when learning an L2 since students get involved in the use of technology. In fact, technology helps learners to strengthen their self-expression, and interaction, using the target language [22].

[25] suggest pairing digital technology and language teaching in activities in which students are producing the target language in an oral way. According to [26] as cited in [27], Computer-Assisted Pronunciation Teaching (CAPT) can be very useful to teach pronunciation. In the same way, Pennington (1996) as cited in [27] establishes that CAPT can be better to teach pronunciation than a teacher or a phonetician since a computer can analyze users' pronunciation and provide feedback in a faster and more accurate way than a person could do. The author also mentions that another advantage of using technology to teach pronunciation is that it can give users different types of input so they can become more aware of their pronunciation to improve it.

3 Methodology

3.1 Participants

The participants of this study were 26 indigenous high school students from a bilingual Kichwa-Spanish community in a rural area, located in the central part of Ecuador. Their ages range from 16–18 years old. 15 of them are women and 11 are men. All of them speak Kichwa as their native language and Spanish as their second language; and they are learning English as their L3. They have an A1 level of English according to the Common European Framework of Reference. Participants belong to the third year of a secondary public school where learning English is mandatory. They were selected randomly.

3.2 Data Collection

This investigation used quantitative analysis in order to answer the research questions about L1 Kichwa and L2 Spanish transfer in L3 English learning. The selection and design of the instruments for this study were based on the pronunciation diagnostic guide by [28]. The instruments used for this investigation were a short paragraph and a checklist of the 12 English vowel sounds. The paragraph was related to a topic students were familiarized with, to avoid problems related to lexical or structural knowledge; and it included target words for all the English vowel sounds. Students were asked to read the paragraph silently to themselves first. Then, they were requested to read it aloud while they were recorded.

The technological device used to record students was a smart cell phone. The audio files were analyzed by some language experts who used a checklist to identify the vowel sounds which negatively affect intelligibility.

In order to have a better understanding of the reasons why students' pronunciation was not good, a survey was conducted to gather information about the strategies used by the English teachers to improve students' pronunciation, as well as the use of any technological devices employed for this purpose. Some of the strategies used by teachers included controlled guided oral practice, such as memorization, choral drills, chants and repetition [17].

4 Results

According to the articulatory system, there are three dimensions of articulation: voicing, place of articulation and manner of articulation. Thus, Kichwa has three vowel sound positions, Spanish five, and English twelve. As shown in Table 1, there are seven English vowel sounds that do not exist in Kichwa and Spanish. Consequently, the differences in the number of vowel sound positions makes it difficult for Kichwa native speakers, who speak Spanish as a second language, to acquire the English phonological system (Table 3).

Table 2. English vowel sounds not present in Kichwa and Spanish

Vowel sound	English words	Meaning in spanish
I	Fish	pez
Ǣ	Cat	gato
ɒ	Clock	reloj
ʊ	Full	lleno
ə	Computer	computador
3:	Bird	pájaro
Λ	Up	arriba

Table 3. The most complex English vowel sounds to pronounce in Kichwa and Spanish

Sound	Number of students who pronounce the sounds correctly	Percentage	Number of students who pronounce the sounds incorrectly	Percentage
/I/ Near- close front unrounded vowel	7	27%	19	73%
/æ̃/ near open front unrounded vowel. The highest point of the tongue is positioned in front in the mouth without creating a constriction	5	19%	21	81%
/ə/ mid – central vowel	10	38%	16	62%
/ʌ/ open-mid back unrounded vowel	8	31%	18	69%

From the twelve existing English vowel sound positions, four are the most complex for students to pronounce. These sounds are /I/, /æ/, /ə/ and /ʌ/. The manner of articulation of these vowel sounds occurs in different positions as shown in Table 2. Additionally, from these four vowel sounds, the one that causes the most pronunciation difficulties is /æ/ with 81% of students who pronounced it incorrectly. Students do not form a tight circle with their lips to correctly produce this sound. Likewise, the wrong pronunciation of this sound may be because it combines two sounds in one, and students can only pronounce these sounds individually as in their native and second languages.

Regarding the vowel sound /I/, 73% of the participants of this study mispronounced it using the /i/ sound present in Kichwa and Spanish. Likewise, another phonological problem, with 62% of wrong pronunciation, is the reduced vowel sound called schwa /ə/ which is the most common vowel sound in spoken English. Schwa is a quick, relaxed and neutral vowel sound whose purpose is to allow unstressed syllables to be said more quickly, so the main beats of spoken words are easier to place on the stressed syllables. Schwa does not have an exact and standard pronunciation, so this is a difficult sound even for monolinguals and bilinguals because it is present in words which contain the letters A, E, I, O, and U.

Finally, the vowel /ʌ/ which is an open-mid back unrounded and neutral sound is also difficult to produce. 69% of the students pronounced it in a wrong way. This could be because of the difficulty to produce this sound since the tongue should be flat and resting, the lips should be relaxed (not spread or rounded) and the jaw should be relaxed too. As long as the strategies used by the teachers to improve students' pronunciation were not enough, it is recommended to use technological applications, depending on the students' needs.

5 Discussion

The main aim of this investigation was to find out how the L1 Kichwa and the L2 Spanish influence the pronunciation of L3 English vowel sounds. Although there are not any previous studies of bilingual Kichwa/Spanish speakers who are learning English as a third language, the results of this research can be compared with what is stated by [20] who mentions that second language learners have an advantage over first language learners since the L2 learners have a broader linguistic knowledge, metalinguistic awareness, and better use of learning strategies in the process of third language acquisition.

Furthermore, [20] points out that there are some studies during the 60s and 70s which show that bilingual speakers have some advantages over monolingual speakers in phonetic discrimination skills and auditory discrimination. However, this study shows that Kichwa speakers have difficulty pronouncing some vowel sounds because of external factors such as teacher's methodology or lack of technological resources used for pronunciation practice.

Spanish and Kichwa have just one high front vowel /i/ and Spanish and Kichwa speakers often use this vowel for both the /ɪ/ vowel in WILL and the /iː/ vowel in WEEKS. One ‘i’ in English is normally the lower /ɪ/ vowel. Furthermore, Spanish and Kichwa speakers often make the vowels in HUT /hʌt/, HAT /hæt/ and HEART /hɑːt/ into the Spanish /a/ – they should be made in different positions in English. Consequently, students should be trained in the correct pronunciation of these vowel sounds to avoid intelligibility problems.

Statistical analysis of the data shows that from the 12 existing English vowel sounds, there are four that are more complex for students to pronounce. These sounds are /I/, /æ/, /ə/ and /ʌ/. The findings show that students had more difficulty in the pronunciation of these simple vowel sounds which are in a distant articulatory point of the vowel sounds that the participants of this study master in their L1 and L2. The students did not have any problem with the five English sounds which are phonologically and phonetically similar to the Spanish vowel sounds (/i/ /e/ /a:/ /u/ /o/). It means there is positive influence of the students’ L1 and L2 over the pronunciation of these five English vowel sounds. This is because the L2 and L3 five vowel sounds have the same manner and place of articulation. Consequently, the English corpus used for the investigation shows that the following vowels in students’ L1, Kichwa: /a/ which is central, long and open; /i:/ front and close; and /u:/ back and close; have the same pronunciation in their L2, Spanish; and L3, English.

On the other hand, the problem participants presented in the English corpus pronunciation is in the production of the short vowels in English like: /æ/ front and open; / I/ half close and front; and /ɒ/ back and open, and /ʌ/ open mid back since they do not exist in their L1 and L2. It means that there is negative interference of the L1 and L2 in the pronunciation of the L3, especially on the sounds which do not exist in the students previously learned languages. So, in order to improve the students’ pronunciation, both, the Intuitive-Imitative Approach and the Analytic-Linguistic Approach should be used in the EFL classroom. Technological mobile devices and applications could enhance pronunciation when applying the Analytic-Linguistic Approach.

References

1. Yates, L.: Learning how to speak: Pronunciation, pragmatics and practicalities in the classroom and beyond. Lang. Teach. **50**(20), 227–246 (2017)
2. Akyol, T.: A study on identifying pronunciation learning strategies of Turkish EFL learners. Procedia - Social Behav. Sci. **70**, 14561–21462 (2013)
3. Tremblay, M.: L2 influence on L3 or acquisition of native pronunciation: The case of voice onset time in the L3 Japanese of L1 English-L2 French bilinguals, Ottawa (1993)
4. Illés, É., Akcan, S.: Bringing real-life language use into EFL classrooms. ELT J. **71**(1), 3–12 (2017)
5. Lipińska, D.: Production of L3 vowels: Is it possible to separate them from L1 and L2 sounds? Res. Lang. **13**(1), 77–92 (2015)
6. Marzá, N.E.: Pronunciation and comprehension of oral English in the English as a foreign language class: Key aspects, students' perceptions and proposals. J. Lang. Teach. Res. **5**(2), 262–273 (2014)
7. Zapata-Paulini, J., Soto-Cordova, M., Lapa-Asto, U.: A mobile application with augmented reality for the learning of the quechua language in pre-school children. In: 2019 IEEE 39th Central America and Panama Convention (2019)
8. Garcia, C., Kolat, M., Morgan, T.: Self-correction of second - language pronunciation via online, real- time, visual feedback. In: Levis, J. (ed.) Proceedings of the 9 Pkronuncitation in Second Language Learning and Teaching Conference, pp. 54–65 (2018)
9. Çimenli, B.: On pronunciation teaching and semiotics. Soc. Behav. Sci. **199**, 634–340 (2015)
10. Cavus, N.: Development of an intellegent mobile application for teaching English pronunciation. Procedia Comput. Sci. **102**, 365–369 (2016)
11. Sriudomkij, P., Sopirak, S.: Factors influencing the development in English pronunciation skills training in primary students of schools in the primary education services area office in Bangkok. Procedia Soc. Behav. Sci. **103**, 396–399 (2013)
12. Kassaian, Z., Esmae'li, S.: The effect of bilinguality on L3 breadth of vocabulary knowledge and word reading skill. Theory Pract. Lang. Stud. **1**(8), 966–974 (2011)
13. Chen, Y., Li, X.: Effects of L1 transfer on English writing of Chinese EFL students. Cross-Cult. Commun. **12**(4), 1–4 (2016)
14. Bartolotti, J., Marian, V.: Bilinguals' existing languages benefit vocabulary learning in a third language. Lang. Learn. **1**(67), 110–140 (2017)
15. Tavakol, M., Jabbari, A.: Syntactic transfer in the initial stages of adult third language and fourth language acquisition. Indonesian J. Appl. Linguist. **5**(2), 186–198 (2016)
16. Kolabinova, T., Zaripova, A., Zaglyadkina, T., Yusupova, L.: English as an additional language for Russian university students learning German as the first foreign language. Procedia Soc. Behav. Sci. **199**, 334–339 (2015)
17. Yu, X.: Domestication of Audiolingualism in ELT in China : from the perspective of cultural accommodation. Stud. Lit. Lang. **3**(13), 22–27 (2016)
18. Keshavarz, M.H., Abubakar, M.K.: An investigation into pronunciation problems of Hausa-speaking learners of English. Int. Online J. Educ. Teach. (IOJET) **4**(1), 61–72 (2017)
19. Gilakjani, A.P.: What factors influence the English pronunciation of EFL learners? Mod. J. Lang. Teach. Methods **6**(2), 135–327 (2016)
20. Cenoz, J.: The influence of bilingualism on third language acquisition: focus on multilingualism. Lang. Teach. **46**(01), 71–86 (2011)
21. Bian, F.: The influence of chinese stress on English pronunciation teaching and learning. Engl. Lang. Teach. **6**(11), 199–211 (2013)

22. Chang, S., Shak, P., Gregory, N.: Students' perception on the use of instructional pronunciation video in improving pronunciation. In: International Conference on Education and E-Learning, (EeL), pp. 170–175 (2016)
23. Bilyalova, A.: ICT in teaching a foreign language in high school. Procedia Soc. Behav. Sci. **237**, 175–181 (2017)
24. Blake, R.: Technology and the four skills. Lang. Learn. Technol. **2**(20), 129–142 (2016)
25. Pop, M.M., Giurgiulescu, M., Măgheruşan, V.: The perception of micro-expressions while learning English and Italian through computer techniques. Procedia Soc. Behav. Sci. **237**, 816–822 (2017)
26. Freundlich, J.: Cross-linguistic influence in third-language acquisition: Learning mandarin chinese (L3) through the medium of English (L2). In: Foreign Languages in Education, Science and Culture, Krakow, p. 186 (2016)
27. Gomes, C., Levis, J., Cabral, V.: The design of an instrument to evaluate sotware for EFL / ESL pronunciation teaching. Ilha Do Desterro **69**, 141–160 (2016)
28. Gerhiser, A., Wrenn, D.: Second language Pronunciation assessment handout packet (2007)
29. Anchorena, D.J.: Gramática Quechua. En Idioma del Imperio de los Incas (1874). https://books.google.com.ec/books?hl=es&lr=&id=JXUSAAAAYAAJ&oi=fnd&pg=PA1&dq=la+pronunciaci%C3%B3n+del+idioma+quichua&ots=1fnusuBkDA&sig=c3QDtOhWid1XJxAXtDkeAYLiGv4#v=onepage&q=la%20pronunciaci%C3%B3n%20del%20idioma%20quichua&f=false
30. Wang, T., Hashimoto, T.: Learning third language improves phonological perception of second language. Assoc. Nat. Lang. Process. **22**, 1–4 (2016)

Professional Development of Educators of General and Professional Education Systems in University Complex: Convergent Educational Environment

Mansur Galikhanov, Ljubov' Ovsienko, Dinara Kulikova, Irina Zimina, and Alina Guzhova(✉)

Kazan National Research Technological University, Kazan 420015, Russian Federation
mgalikhanov@yandex.ru

Abstract. Convergent approach means development of input sort testing system for educators to estimate lacking competencies and pedagogical deficiency in order to design individual educational path for every learner depending on the requirements. Testing system gives recommended models of individual educational programs. 7 professional development programs are suggested for educators. Each program consists of core course modules and a wide range of elective course modules. Graduation thesis of an educator is a project on the program topic that comes up with a solution of his own educational institution challenge or problem.

Keywords: Convergent educational environment · Continuing education · Educator

1 Introduction

In the context of globalization, educational policy of any country should contribute to economic growth acceleration, technological modernization and social sustainability. New education development projects should correspond to global trend of digitalization that cannot be stopped or ignored. For example, one of the main tasks of the Russian Federation National Project "Education" is enhancing professional skill level in the format of continuing education of teaching staff in the systems of general, additional and professional education [1].

Today educational system faces following challenges:

- overcoming learning failure;
- talent growth and support;
- formation of students' universal skills and positive social attitudes (communication, cooperation, creativity, critical thinking), self-organizing and entrepreneurial competencies) [2];
- developing new set of knowledge and skills of students required to use all abilities of modern civilization (digital, legal and financial).

M. E. Auer and T. Rüütmann (Eds.): ICL 2020, AISC 1328, pp. 263–271, 2021.
https://doi.org/10.1007/978-3-030-68198-2_24

To meet this challenges new approaches and requirements for educational staff should be applied.

Modern technology development reduces periods of production technology replacement. Now it takes about 5–7 years, while it took 15–20 years quarter of the century ago [3]. Nowadays educational system goes behind new economic challenges.

Digital economics is penetrating in all life spheres, it offers new challenges for educational system and prompts to consider reality of digital society:

- new technologies and digital environment;
- new requirements that economy sets for staff;
- "digital generation" as a new type of students [4–6].

1.1 Professional Development of Educators in Russia

There is a range of projects in Russia that are aimed at increasing educators' level of excellence, e.g.:

- project of the Agency for Strategic Initiatives "Digital School" is focused on creating efficient network methodological community of educators to increase prestige and competitiveness of young educators;
- project of the federal education network "School League "Rosnano" develops and implements additional educational programs for educators and children in the sphere of innovative entrepreneurship, promotes development of students' project and research activities, holds competitions for teachers [7];
- project "Kurchatov's Center of Continuous Interdisciplinary Education" is oriented to the development of convergent education principles in interdisciplinary educational environment for class and out-of-class activities of pupils based on cooperation with National Research Center "Kurchatov Institute" [8];
- project "Engineering class in a Moscow school" combines efforts of teachers from Moscow schools that have engineering classes, resources of all network organizations of Education and Science Department of the Moscow City, centers of education technological support and the most experienced university educators. The objective of the project is to enable development of science education in engineering, to motivate children to choose engineering professions, etc. [9].

Universities of the Russian Federation are more readily taking on the task of improving pedagogical skill of school teachers, since they are considered as agents of influence on human capital. Interaction with the teachers provides expected quality of school education.

1.2 Kazan National Research Technological University Experience

Kazan National Research Technological University (KNRTU) has successive system of general, additional and professional education that comprises active professional community of mentors, wide list of public and competitive contests, practice of business entity participation in educators' professional growth development.

To the benefit of general education institutions Institute of Lifelong Education and Institute of Additional Professional Education of KNRTU organize collaborative activity according to signed agreements and conventions with Russian Fund “Talent and Success”, Tatarstan Institute of Education Development, Independent Nonprofit Organization Kazan Open University of Talents 2.0, Independent Nonprofit Organization School League, Public Joint Stock Company Gazprom, Association of secondary-school students’ project education support (founding parties technopark Idea and technopolis Himgrad).

“University profession-oriented classes” project has been run for 12 years. It is aimed at increasing quality of teaching of natural science subjects. KNRTU signed cooperation agreements with 320 schools of the Republic of Tatarstan. Under the terms of the project lessons of high-school students on the most intricate topics in chemistry, physics and mathematics are given by teachers together with academicians from KNRTU. Elective courses (47 in total) and resource books were developed to form suprasubject competencies of students and introduce them to new scientific achievements and basics of professional activity.

Annually KNRTU organizes republic conferences and seminars for teachers on support of gifted pupils and methodology of teaching natural science subjects. KNRTU supported by the Republic Chemical Society after D.I. Mendeleev founded in 2007 Republic Scientific Methodological Union of Chemistry Teachers and Educators. In collaboration with the Union KNRTU held 12 scientific conferences for teachers. In 2017 Conference “Growing” System of Children Giftedness: Methodology and Practice” was held, in 2018 – “Development of Humanitarian Technologies of Mentoring for the Benefit of National Technological Initiative”. The title of the conference in 2019 was “Digital Didactics of Professional Education: challenges and tasks of the Transition Period”. Now preparations are in progress for the conference “Three Pillars” of Natural Science Literacy of Pupils: Knowledge, Competencies and Research” (October 2020).

Best Teaching Practice Bank was formed by annually published proceedings and scientific methodological guidelines. KNRTU for 6 years has been holding Competition of the best scientific and methodological results of teachers and academicians “All Facets of Chemistry”. The partner of the competition is OAO “Tatneftekhiminvest-holding”. Its specialists participate in defining competitive assignments and arbitration.

In 2016–2017 KNRTU hosted network testing site of the Federal Education Development Institute “Integration of Engineering Education in School Environment The best 17 schools in Kazan participated in the project. The topics of the research and development were following: new model of learner-centered approach in education; development of interdisciplinary skills of learners; creating model of developing leadership skills, cognition, self-education skills, research activity, efficient communication and team-work; forming special educational environment for professional orientation; development of professional probing methods. This testing site created community of teachers able to form new design and mechanisms of educational environment organization.

Purpose of the Paper
The system of additional professional education of an educator is the basis for determining future needs in new competencies: determining the trends of changes in education content, educational technologies, education quality indicators and its evaluation methods. The development of an educator's professional skills, competence, individual growth, effectiveness of professional experience should be open, flexible and advancing system.

The purpose of this versatile work is to create and test new model of partnership with schools and direct involvement in educator training that organizes open modern system of advanced training for teaching staff of general and professional education. The role of educator changes – he/she becomes an organizer of learning, project and research activities and educational trainings; consultant and mentor; researcher; project manager; and "navigator" in educational and digital environment. Teacher drives qualitative transformations in school and professional education system.

2 Convergent Approach for Professional Development of Educators

Effectiveness of educator's work will increase due to application of new technologies and designed digital resources enabling to personalize educational path in the best interests of every child and, thus, improve the quality of education.

Professional development of educators of general and professional education systems in university complex implies convergent approach. It means development of input sort testing system for teaching staff to estimate lacking competencies and pedagogical deficiency in order to design individual educational path for every educator depending on the requirements. Testing system gives recommended models of individual educational programs.

Besides, at stay-at-home period and temporary switch to distant learning we got unique experience to assess our suggestions on opportunities and challenges of digital education; we realized how difficult it is to redefine teaching practice and change key role of an educator. Information technologies in cooperation with teaching ones generate new type "digitally-born" technologies.

We surveyed teachers of secondary vocational education from KNRTU scientific-educational cluster, that comprises 38 professional educational institutions in the Republic of Tatarstan and Russia. There were 1612 respondents who work as secondary vocational education teachers, average age was 43.5 years, 24.6% were men, 75.4% were women.

We conducted the survey to reveal so called teaching deficiency for better training system development. About 40% of teachers had lack of IT and e-learning process organization knowledge (see Fig. 1). However, this survey was meant for solving current problems there and then, immediate help for teachers.

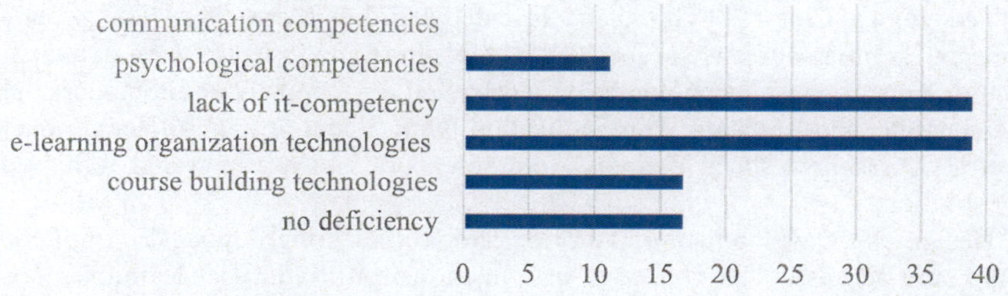

Fig. 1. Teaching deficiency of educators revealed by implementation of online programs

On the other hand, the lack of it- competencies, in our opinion, resulted in great popularity of such communication tools as WhatsApp and ZOOM in educational process, since they are user-friendly (see Fig. 2).

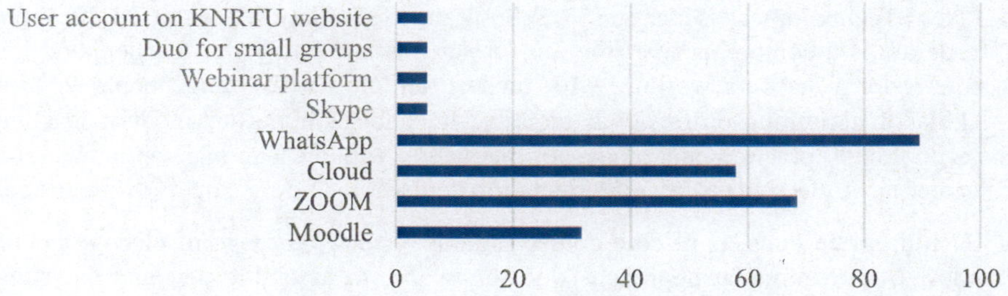

Fig. 2. The most convenient it-resources for e-learning

2.1 Professional Development Programs

There are 7 professional development programs suggested for educators:

1) "Convergent Educational Environment: Philosophy and Didactics". This program is intended for innovative educators. It teaches methodology for designing new educational space as interdisciplinary, metaprofessional, erasing barriers scientific and technological knowledge; forms competencies of pedagogical guiding project, learning and research activities of students and pupils.
2) "Basic Technologies of Learning Process Digital Transformation". The program is targeted at wide range of educators. It provides methods and tools to develop competencies of new e-literacy and design educational resources and knowledge assessment using digital didactics approaches. It explains how to form "digital footprint" of learning process.
3) "Pedagogical Design: Principles, Tasks and Models" program teaches the approaches of pedagogical design based on analysis of students' educational demands, competencies and expected results of learning; organizing learning process with open architecture of teaching environment.

4) “Pedagogical Culture of Gifted and Talented Children Mentoring”. This program is meant for broad audience of teachers and lecturers and instructs them in innovative culture distribution, technologies of accelerated development environment creation for pupils and students; communication with youth in non-directive manner; attraction of new social partners - entrepreneurs, businessmen and development institutions.
5) “Design and Development of Tutoring and Volunteering Among Young People”. Program addresses the challenge of transition from traditional learning model to creating educational environment for the development of a successful self-sufficient individual. Program members will be provided with tools for creating environment for development and cooperation, personalization and advancing “success skills”.
6) “New Methods of Learners’ Teaching and Training of the 21st Century Skills”. This program enables educator to develop and update project and research activity in an educational institution, maximize its potential applying for socializing, teaches methods of project activity, gamification, developing hackathons, competitions, etc.
7) “New Technological Education in School and Secondary Vocational Education” program is a comprehensive solution for science and handicraft teachers and lecturers, form teachers working with modern children and young people to form skills of conscious professional identity. It enables to master modern teaching technologies, practices and models of interaction with teenagers (career navigation, career days, elective profession orientation courses, including careers of the future).

Each program consists of core course modules and wide range of elective course modules. Due to modular approach educator is able to assemble personalized study program. MOODLE platform was used for on-line study of each module that can be accessed from anywhere in the world. Core courses of the programs include modules on convergence in education, digital literacy, digital generation psychology, “The four C’s skills” term, etc. The core component of the program is within 25% of the total hours. Educator assembles his own learning content from a large number of elective modules. Thus, phenomenon of “Education upon request” is formed, when study product is constructed by an individual.

Face-to-face trainings of educators are held as project sessions, team-works, and discussion clubs with invited experts and leading scientific schools’ representatives. Graduation thesis of an educator is a project on the program topic that comes up with a solution of his own educational institution challenge or problem.

Competencies Formed by the Programs

Following universal competencies will be formed during training programs:

- cognition (search and systematization of information, implementation of the best solutions);
- openness, initiative, entrepreneurial attitude (offer own initiatives, actions, search and attraction of required resources, monetization and sale of services to achieve goals);
- project management up to outcome (advancing ideas from “idea design” to “project waste disposal” with a result guarantee, “I-project” management);

– team-work and interaction efficiency (interaction and collaboration with people in order to achieve personal and common results);
– leadership (anticipation, generation and pushing new ideas, new results, ability to create something first of a kind).

Professional and personal interaction of young and experienced educators in a university complex is based on principle of creating comprehensive-activity educational environment, forming integrated professional educational space. Sharing experience between educators motivates to enhance and upgrades professional skills in all professional activities, promotes opportunities to exchange experience at both regional and international scale.

Educators master new tools, novel approaches and technologies for organizing teaching process based on convergence, digital, psychological and emotional literacy, creativity, self-study ability, customization of educational path in in the best interests of every child, etc.

Implementation Plan

Implementation of this work includes following actions:

1) *Upgrading infrastructure of the university to execute the project:*
  - Equipping a classroom for developing and implementation of online additional professional education programs for educators of general and professional education;
  - setting up co-working space, design of learning space;
  - creating regular methodical consultation center for educators;

2) *Creating training programs for teaching staff of general, professional and additional education:*

– program testing;
– educational path modeling;
– developing content of educational program (theoretical and experiential parts, program structure, evaluation tools, studying content and technologies, learning materials);
– organizing project sessions, discussion clubs for professional community, work with the experts;
– turning program into e-learning form.

3) *Testing and replication of the developed training programs:*

– developing training programs for educators of general and professional education system in university complex;
– working out guideline for educators on modeling technology of personalized educational path of a learner;
– organizing and work on efficient practices bank;
– consulting of educators;

– marketing activities to promote training programs;
– publishing scientific papers on project implementation and obtained results in top-rating journals.

One of the key ideas of the work is post-project tracking and support of educators. On one hand it has scientific relevance allowing to estimate different effects of the project implementation. On the other hand, it has real-life impact since suggested ideas can be applied to improve work of a specific educational institution.

3 Conclusion

Thus, this paper describes model of teachers' training that builds up open modern system of professional development for general and professional education teaching staff. Professional growth of an educator in university complex requires convergent approach. It enables to design individual educational path for each teacher depending on his or her requests.

Nowadays the most crucial task of education development policy is to apply created potential for the benefit of the country development – to contribute to acceleration of economic growth, technology modernization and social sustainability.

Human capital in XXI century is the key factor of economy and society development. It is an enabler for consolidation of the Russian Federation position in the context of global competition. Education is now considered as the sphere of investments, it determines scale and quality of economic growth, contributes to the wellbeing of the country and individual. Besides being the key factor of economic growth education is the growing sector of economy. Scale of the educational services market increases as well.

The project we suggest changes the requirements for professional role of an educator– he/she becomes an organizer of learning, project and research activities and educational trainings; consultant and mentor; researcher; project manager; and "navigator" in educational and digital environment. This project with the system of post-project tracking and support will motivate innovative changes in educational institutions. Graduation thesis of an educator is an individual project implemented in his/her educational institution to promote transformation.

References

1. Address of the Russian Federation President on 01.03.2018. https://www.kremlin.ru/acts/bank/42902, Accessed 30 May 2020
2. Valeeva, E., Ziyatdinova, J., Galeeva, F.: Development of soft skills by doctoral students. In: Advances in Intelligent Systems and Computing, AICS, vol. 1135, pp. 159–168 (2020)
3. The report of Center for Strategic Research and Higher School of Economics "12 Solutions for New Education". https://www.csr.ru/ru/publications/ekspertnyj-doklad-12-reshenij-dlya-novogo-obrazovaniya/, Accessed 30 May 2020
4. Yamamoto, J., Ananou, S.: Humanity in the digital age: cognitive, social, emotional, and ethical implications. Contemp. Educ. Technol. **6**(1), 1–18 (2015)

5. Esenina, E., Blinov, V., Sergeev, I.: Digital Didactics: New field in the Education's Philosophy. TVET@Asia, vol. 12 (2019). https://www.tvet-online.asia/issue12/blinov_etal_issue12
6. Blinov, V., Herkner, V.: Analytische Ubersicht der Zusammenarbeit Russlands und Deutschlands im Bereich der Berufsbildung: in zwei Teilen. Teil 1. Historischer Blick auf die Entwicklung der Berufsbildung in Deutschland und Russland. Urait, Moscow (2019)
7. School League "Rosnano". https://www.schoolnano.ru/, Accessed 30 May 2020
8. Kurchatov's Center of Continuous Interdisciplinary Education. https://profil.mos.ru/kur.html#/, Accessed 30 May 2020
9. Engineering Class in a Moscow School. https://profil.mos.ru/inj.html#/, Accessed 30 May 2020

QRing Specialties' Mobile Course Elements to Support VET Learners' Cognitive Level

Charilaos Tsihouridis[1](✉), Marianthi Batsila[2], and Anastasios Tsihouridis[3]

[1] University of Thessaly, Volos, Greece
hatsihour@uth.gr
[2] Directorate of Secondary Education, Ministry of Education, Athens, Greece
marbatsila@gmail.com
[3] Democritus University of Thrace, Komotini, Greece
tsabtsih@gmail.com

Abstract. This study aims to investigate the effectiveness of QR (Quick Response) codes in teaching and learning. More specifically, the study explores the extent to which, the use of QR codes, embedded in the learners' class material, can improve their linguistic performance in English and enhance their interest and participation in their specialty lessons. For the purposes of the study, a number of 132 secondary vocational randomly selected school learners from four schools and eight classes, aged between 17–21 years, and four teachers took part in the research process. The study was implemented in five stages and had duration of three months. The research employed a mixed method using a quantitative and qualitative approach. Learners' linguistic level before and after the intervention was measured with the help of the balanced state KPG test for foreign languages, while four focus groups, one per class, were also implemented. According to the results, it can be argued that the integration of the QR codes in the teaching material and process enhances both receptive English language skills but mainly reading. What is more, it seems that VET (Vocational and Technical Education) students found the use of QR codes interesting, innovative and helpful for their studies.

Keywords: ICT · QR codes · ESP · Mobile learning

1 Introduction

The student population of today can be called a "tech" generation as their digital skills are at an unusual or remarkably high level. Nowadays, students are very familiar with all kinds of digital tools and applications which interest them to a high extent. Technology is used in the classroom to motivate students and turn lessons into an interesting or game-based like process, enhancing the effectiveness of students' skills and their participation in the lesson [1–5]. Information and Communication Technology (ICT) is undoubtedly an important tool in the hands of today's teachers and an infinite source of input. As learners are well familiarized with ICT, it comes as no surprise that they use it with impressive easiness and that they immediately become fully adapted to the

M. E. Auer and T. Rüütmann (Eds.): ICL 2020, AISC 1328, pp. 272–283, 2021.
https://doi.org/10.1007/978-3-030-68198-2_25

operational requirements of all kinds of digital devices. Due to this high interest and dexterity, ICT new application emergence never takes us by surprise as new ones appear every day opening new horizons in learning, providing us with a variety of data, filling knowledge gaps, eliminating distances in education and bringing students together through social media and contemporary communication tools and platforms (skype, facebook, twitter, e-mail, Internet, etc.). Thus, in order for teachers to train learners into being literate and global citizens of the 21st century, ICT is suggested to be integrated in school/academic curricula, among which in the teaching of the English language as well. In recent years, many educators and researchers have investigated the effectiveness of a variety of ICT devices and applications [6]. Especially for the teaching of English for specific purposes (E.S.P.), ICT has been found to be an important tool for educational reasons. E.S.P. refers to language programs designed for a specific reason and needs of English language learners [7]. E.S.P. courses are part of Vocational and Technical Education (VET) schools curricula, because English is considered an important qualification for professional purposes. For this purpose, VET learners need to be trained accordingly in order to prepare for their future English language requirements in the labor market. Therefore, VET teachers need to display updated specialty skills and knowledge, be inventive, innovative and properly trained in order to provide their learners with such material, methods and techniques that will help them successfully meet the needs of their future professions [8]. Mobile teaching has lately been another teaching method that exploits mobile devices and applications that aim at mobile learning (m-learning). Mobile learning is defined as "Any activity that allows individuals to be more productive when consuming, interacting with or creating information, mediated through a small digital portable device that transports a person to a regular base, has reliable connectivity and fits in a pocket or wallet"[9]. These devices have increased significantly, and are very popular especially among young students. This is due to their portability and easiness of use, the accessibility of applications and their ability to download material or information without much difficulty. In addition, many of the mobile apps are free to use/download from the web and can be used for a variety of purposes like personal, social, academic or professional. Mobile learning can be achieved anywhere and anytime. What is needed is simply internet connection and a mobile device. Such devices are mobile and smart phones, netbooks, tablets and i-pads, multimedia devices (podcasting and video), game consoles or e-book readers. Research shows that mobile learning has been effective in teaching, [10] developing students' language and communication skills [11], dealing with certain educational problems [12], being able to introduce innovations in pedagogy, offering personalized learning [13] and differentiated activities [14] and supporting teaching practices through inquiry-based experimental activities[15].

Among ICT applications we discern QR (Quick Response) codes. QR Codes are a form of barcodes, which have a smart, easy and efficient way to connect via a mobile, offline environment with links and information stored online. They have become very popular due to their large storage capacity and fast readability. Nowadays we find QR codes in magazines, various publications, store cards, buildings, and more. It has been shown that the ability of QR Codes to connect people with Multimedia content is useful for many reasons [16]. In essence, there is no limit to the size of information that can be shared with a QR Code. It can be a link, a Video or an entire eBook. In education, QR codes can be embedded in teachers' material and used to support learning. They can be a useful tool that makes the lesson material more interactive and attracts students' attention. The codes, which look like a square image resembling a bar-code, have the ability to transfer connected digital information when scanned through ICT applications. Based on research, by using QR codes in the classroom the separation between the classroom and the "real world" is blurred, and students become more accountable for their own learning by applying their language skills to a wide variety of English stimuli using their personal devices [16]. Research shows that the use of QR codes in education facilitates lessons preparation and classroom management [17], supports learning in different contexts, enhances both independent and collaborative learning and can motivate and engage learners in the learning process [18]. Additionally, QR codes encourage students persevere with language problems, thus, lead to good learning outcomes [19]. Furthermore, QR codes motivate learners to complete homework in a variety of non-traditional settings and increase their communication with their classroom teacher in the learning process [20]. Finally, research findings report the great potential of QR codes in education [21], and their overall beneficial results [22].

2 Rationale of the Research

Today's classes display a number of varied characteristics like multiculturalism and plurilingualism. Such classes are usually mixed ability due to the diversity of the student population, their different cultural and cognitive background, age, needs, or skills. In the area of the research, over the past years, the number of students with different background gradually increases, due to different groups of people (i.e., immigrants, refugees) that come to the country from different areas. These diversified groups are a usual phenomenon especially for Vocational Schools (both with morning or evening classes) for a number of reasons: one, many students, (teens and adults) choose to come to VET schools because they also work in several businesses and firms, during the hours that do not attend lessons. These learners work either to make a living, supplement their income or to train and practice for their working skills and gain experience for their future careers.

However, and according to teachers, it is a frequent phenomenon to see them miss classes because they are tired, have an unexpected working duty to attend to, or due to lack of interest; two, immigrants or refugees over eighteen years old that come to the country, usually attend VET schools, due to the fact that, such schools are considered by them more accessible and easier to cope with the lessons' cognitive input. This is because VET curricula are not very cognitively demanding and mainly skills orientated as opposed to those of General high schools curricula that are more academically orientated and all aim at college/university entry. For all the above reasons, and the diverse features of the classes, it seems urgent for teachers to employ differentiated instruction - the process of "tailoring instruction to meet individual needs" making it "a successful approach to instruction", [23] - in order to address all students' needs. Driven by the aforementioned points, and by the fact that there is minimum research on VET Secondary education on the issue (the use of QR codes), the authors of this paper decided to investigate the extent to which the use of QR codes, embedded in VET learners' class material can improve their linguistic performance in English and enhance their interest and participation in their specialty courses.

3 Research Methodology

3.1 Research Purpose and Research Questions

The purpose of the research is to investigate the extent to which the use of QR codes in the learning process can improve VET learners' cognitive level and their participation in the teaching practices. In particular, the research aimed to look into the extent to which QR codes, impeded in the class material, can enhance learners' English language performance in reading and listening skills, their interest and active participation in their specialty courses. The research was conducted following a mixed method approach (qualitative and quantitative) and the research questions are as follows: 1.To what extent can the integration of QR codes in the class material support learning? 2. To what extent can the use of QR codes enhance learners' reading and listening skills? 3. To what extent can the application of QR codes in teaching enhance learners' participation and motivation for learning?

3.2 Research Participants

A number of four randomly selected suburban Vocational Secondary Schools participated in the research. Two classes per school with 132 learners altogether took part in the research process along with four teachers (a teacher per school-each teacher taught both classes in each school). The students' age ranged between 17–21 years. The average level of the students is B1 according to the CEFR in relation to the English language and they are multilingual in relation to nationalities and mother tongue. Apart from the native students (Greek) there are also refugees from Syria, as well as immigrants from other countries (i.e. Romania, Morroco, Georgia). Many of them (especially the refugees) wish to travel to northern Europe to work or live. Many students also work for a living (mainly adult students) and wish to improve their English in order to cope with their professional needs. Students attend specialties of car mechanics, engineers, refrigerants, electricians. Due to the fact that many of them work, they often miss classes. This, according to their teachers, results into having gaps in knowledge. What is more, especially the non-native learners are shy or afraid to ask for clarifications or explanations in the fear of being rejected, or misunderstood by others. Participants in this research are also the four teachers of the target students. They are English language teachers who teach ESP courses in the particular VET schools. Their experience in ESP teaching ranges between 12 to 24 years of teaching in public and private education.

3.3 Research Method and Research Tools

The research was conducted following a mixed method approach (qualitative and quantitative). In order to provide answers for the second research question, learners' reading and listening skills competence was measured with the use of the National Foreign Language Exam System (KPG test), which comprises four modules (reading comprehension and language awareness, writing and written mediation, listening comprehension and speaking and oral mediation). KPG is a stabilized test, used to assess learners' skills competence in foreign languages (Table 1). The maximum possible score at B1 level candidates can gain is 100. The pass mark for the KPG, as set by law, is 60. Candidates must gain at least 30% of the maximum possible marks in Modules 1–3; there is no minimum mark required in Module 4, although the marks candidates receive are included in their total score.

Table 1. KPG Exam specifications.

Module	Type of items/tasks	Number of items/tasks	Weighting	Mark Max	Mark Min	Time (minutes)	Text size (words)
1	Choice	40	0.5	20	8	65	700-1000
	Completion	10	0.5	5			
2	Semi guided written production	2-3	-	30	9	45	160-200
3	Choice	15	1	15	8	15-20	Not defined
	Completion	5	2	10			
4	Semi guided oral production	3	-	20	-	15-20	Not defined
Minimum total grade required for B1 level certification					60		
Highest possible grade for B1 level certification				100			

For the purposes of research questions one and three and in order to shed more light into the research issues, four focus groups were conducted with 10 randomly selected students derived from all specialties per school/focus group (total 40 students from all specialties), upon the end of the interventions as well as a focus group with the four teachers of the target schools.

3.4 Research Process

The research comprised five stages and it had duration of three months with two teaching hours per week. All phases started and finished the same date for all four schools. The four specialties (engineers, mechanics, refrigerants, electricians) exist in all four schools and the students are taught the same amount of hours in all specialties. Additionally, all students in all four schools are taught the same material/state textbook per specialty.

1st phase:

The first phase took place before the intervention, where all schools, teachers and learners had been explained the process and the purpose of the research and consented to their voluntary and anonymous participation. The participants were also explained that they could withdraw anytime they wished and that the data received would serve only for the purposes of this research.

2nd phase:

During the second phase, all students were given a pre KPG test to detect their level of reading and listening skills. They were also explained what QR codes are, how to install a free code reader on their smart devices and how to read/scan the codes appropriately. They were also informed that for the next three months they would work using QR codes which would be embedded in the worksheets uploaded for them on e-class where they could download them from and use them alone or in the class.

3rd phase:

The third phase was the actual intervention, where the teachers of each of the four different specialties (mechanics, engineers, refrigerants, and electricians) in all four schools and the eight classes created the same worksheets for their lessons (based on the textbooks material and the instructions of the researchers) with embedded QR codes using the tool "code generator" [24]. The codes were used for a variety of rich basic or supplementary material like videos, vocabulary, texts, pictures, definitions, recorded class presentations, syllabi content and more. These worksheets were uploaded on the national school e-class platform for the students to have access in or out of the class. Students could download the material, they were able to scan the codes and have immediate access to their tasks. The majority of the students used this material also at their work, where they were able to find valuable information for their working needs on the spot, by just scanning the relevant code which corresponded to useful data (equipment design, detailed pictures of tools and appliances, explanations of texts or vocabulary/terms in English and mother tongue, equipment appropriate use guidelines, advice on operational skills, problem solving tasks, case studies and how to handle them, instructions on using devices and tools, tutorials, videos, online dictionaries links for each specialty and so on).

4th phase:

Upon the end of the intervention, students took a similar post National KGP test to detect any differentiation in their level. What is more, four focus groups – one group per class with ten randomly selected learners per class/group-were also held to provide answers mainly for the first and third research questions but also for clarification of all research issues. A focus group with the four teachers also took place right after those of the students'. Students' focus groups had a duration that ranged between forty four to fifty three minutes, whereas the teachers' focus group had duration sixty seven minutes. All focus groups discussions were held in the school libraries upon the end of the school day so as not to cause problems with the lessons and disturb the participants or the schools. All participants gave their voluntary consent and were explained the process, that they could withdraw if they chose to and that all discussions would be anonymous.

5th phase.

In this last phase students took a follow up KPG similar test three weeks later to detect any differentiation in their level compared to the results of the post test.

4 Results

For the analysis of the KPG tests the SPSS statistical package was used. For the analysis of the interviews the "content analysis" method was employed. After the transcription, the repeated listening and reading of the focus groups discussions content, the most significant parts that linked to the research questions were isolated and recorded and the data analysis units/key words were determined. The key words were "mobile devices", "interesting", "motivating", "difficult", "easy", "helpful", "needs", "level", "competence", "reading skills", "listening skills" that were included in the participants' answers, in relation to the aims of the research and research questions. Due to the large body of information, all data were classified in groups, based on which the content analysis took place.

KPG Test Results

Due to the bulk of the KPG tests data received, and the limited space given in this paper, the full analysis of the tests is not presented here but it will soon be presented analytically, in a subsequent paper, thus, adding to the rest of the research findings. Therefore, it was decided that this paper will include only the qualitative research results. However, based on the results received from the National tests, it can be argued that there was a difference between learners' pre and post-tests performance. This showed that the integration of the QR codes in the teaching material and process enhanced both receptive English language skills but mainly reading where there seemed to be statistically significant results.

Students' Focus Groups Results

The results of the focus groups discussions indicated that the use of QR codes had a positive effect in learning. The students argued that they found the use of QR codes interesting, innovative and helpful. Specifically, and according to their answers, the use of QR codes was something they enjoyed and got easily familiar with. As they explained, before the actual intervention, they were hesitant and rather indifferent thinking that this would be "just another tool" but they gradually became aware of its usefulness and realized that they could use it to navigate for a lot of information, which, as they argued, had they had to do it alone, might had taken them much longer, with questionable results, as they might have had to search in "the wrong direction". They admitted that they preferred this sense of "security" for "where to look for information", and having "easily accessible data" that "take very little space" when they needed it for educational or working purposes. Especially for the students that worked, the use of QR codes "saved them valuable time", which they did not waste looking for information on their own in order to "remedy work problems", when those appeared, thus, allowing them to offer "more time to communication with clients".

Other students replied that the use of QR codes gave them the ability to "follow the lesson" they had missed, "fill in gaps" and "understand the lesson better", which, had it been otherwise, they would have had to "search elsewhere" like "call friends, search the internet, but not the teacher", as they admitted. Others, especially younger students, found it fascinating to have "the whole lesson and workplace world together" just by scanning a code with their mobile phones. What is more, when asked, students replied that they had a "good load of reading practice" and "did a lot of listening comprehension" which they enjoyed much. However, what really fascinated them, and mainly the students that worked, was the fact that the use of QR codes in the worksheets allowed them to "address problems of work on the spot" without having to waste time "searching for information for ages". They admitted that the material was so rich that "it had it all", meaning a big selection and variety of tasks to practice or exploit for class or work purposes. In addition to that, the students claimed that QR codes make the activities and the learning process appealing due to the color, sound, and image impeded in the tasks, making lessons more exciting and motivational for them.

It is noteworthy that they shared a sense of "relief" in relation to their class obligations. When asked for clarifications, they explained that they enjoyed this sense of "mobile student-teacher-work-home and school interaction" and felt they were supported a lot this way. Additionally, and according to the learners, QR codes gave them "a sense of being technologically upgraded, a sense of success and personal achievement". In particular, the working students argued that they liked this "different way" to have lessons and think that the "mobile kind of learning" is very effective. They also suggested that "it would be nice to use the tasks this way more times throughout the year. This way, we won't have to worry if we miss the lesson" and commented that "It is a strange feeling of confidence. It is like having your own personal portable teacher at work with you!". Nevertheless, some problems were discussed by the participants. One of them was lack of data in their phones or lack of internet connection, mainly at work, which resulted into failing to read the codes. Another issue they discussed was the fact that some mobile phones had a small screen that impeded them to read the material clearly.

Teachers' Focus Group Results

According to the teachers' focus group comments, QR codes are easy to use and share the prepared material with their learners. As they argued, QR codes facilitate communication between teacher and student and among learners themselves and promote student engagement in the teaching and learning process. QR codes offer teachers the opportunity to "manage the load of material" they want to give to the learners as it does not take so much space as tasks in printed form. It also saves them from photocopying endless pages of worksheets with many activities that might not be able to fit in just a few pages. In relation to that, and based on the teachers' answers, this is a great advantage, when schools face financial difficulties and it is not easy to buy paper. They also considered the use of QR codes an advantage because too often, photocopiers either malfunction or paper is not always enough for every teacher to use. Instead, just by scanning the code, they have access to a lot of useful supplementary material that "would take a lot of space or money to print outside school otherwise". What is more, QR codes make the activities more interesting due to the possibility to add a variety of features like images, videos, recorded items, links to internet games and interactive activities, thus, "augmenting the value of the aim of the tasks" (i.e. this not just being cognitive but also "pedagogical/creative/recreational"). Another benefit is that QR codes allow teachers to control in a way their learners' web inappropriate or irrelevant internet surfing because they offer students web pages that they themselves have selected and "have approved in terms of cognitive and pedagogical aspect".

What is more, teachers are able to provide internet addresses that are relevant to students' tasks, "lessening the burden learners feel of having to find the right address amid so many", thus, saving them valuable time, especially for the working students. Thus, the teachers serve both as providers or facilitators, but at the same time offer their learners scaffolding techniques and lead them to self-independent learning. Additionally, and according to the teachers, students do not have to type long web links with possible mistyping problems due to symbols or repetitive letters and numbers in a web link. According to their answers, this benefit was particularly helpful for the students with dyslexia as "they were not afraid of misspelling or writing in anagrams the link to find information". Instead, by just reading the codes, all the information was offered to them and this gave them a lot of reassurance and self-esteem.

Overall, the use of QR codes is considered a lot of fun as within seconds "the whole world is right in front of the students' eyes". Teachers also admitted it is very easy to make a QR code and generate them. What they liked a lot was that QR codes are free to use and that they "don't have to pay for any subscription or signing up" to create them. Additionally, they argued that they found the process really exciting and innovative as it helped them "change their normal everyday classroom routine". Furthermore, as they argued, it made their life and teaching easier and interesting. In addition to the above, the easiness of use and very quick access of information helped teachers and students "avoid boredom for having to wait long", meaning the need to think of what you want to use and then wait for a long time to download it which now was already done, appropriately and quickly. Students were able to know in seconds what they needed to do, so they had more time to organize and plan their work. What is more, they found the possibility of QR codes, to allow them have immediate feedback during the lesson, an important advantage in the learning process.

Teachers also said that another benefit of QR codes is that it does not require background knowledge either by learners or by them. This implies that everyone, even those that are not digitally skilled, are able to use them inside or outside classes with "no operational difficulty". What is more, due to the portability features they display, QR codes can be used with any portable device anywhere, anytime. These characteristics give teachers and learners "a sense of freedom and flexibility" as QR codes can be used not only within the walls of a classroom but also at work, in private space/home, on the road/commuting and so on. This in turn, gives "a sense of authenticity of tasks" as learning may happen in real life contexts, and gave examples like "when QR codes were used to explore a place, a situation or a problem at work, like the working students have done". Additionally, teachers have the ability to enrich their material, especially boring and insufficient content that is included in non-updated textbooks.

What is more, problems like lack of computer laboratories can be addressed effectively because mobile phones or tablets for instance, can substitute labs when these do not exist or because due to their limited number they are rarely available to all teachers in school when they want to use them. Overall, teachers admitted that the experience was very positive, and they were very satisfied to see that learners were anxious to repeat this type of tasks. They noticed a positive change in their reaction in class and admitted that students even gave them ideas for how the worksheets could be formed. Some students even informed them of other tools to create codes and some other students suggested that the students themselves should be assigned to create their own codes, as part of a task. This zeal and enthusiasm was a very encouraging sign for the teachers, who, as they explained, it was clear to them that "we, as teachers cannot be still or indifferent but need to go by students' preferences and era changes". They also stressed the satisfaction of the working learners with "this mode of mobile classwork availability and fast information". According to their answers, what surprised was the interaction between them and these students, which was "unexpectedly better than before", explaining that the working students sent them feedback regularly, as opposed to the past when tasks were incomplete.

However, teachers also expressed their concern for a number of issues. For instance, one problem they referred to was the prohibition of mobile phone use by students at school. As they explained, after the research, they would they have to take the principal's permission in order to use them for class purposes. Another difficulty they mentioned was internet connection problems they had to encounter in class sometimes. Additionally, in the beginning, they had to explain to the learners that phones should not be removed quickly from the codes as it needed some seconds for the reader to decode them. Another difficulty encountered-though for few students (three students)-was the lack of smart phones, which impeded them from reading the codes and doing the activities assigned. Finally, some codes did not appear correctly in some mobile devices and therefore, there were problems identifying the material. However, as they stated, "the whole process was a challenge for them and their learners" and "confessed" that they had already tried to persuade colleagues that teach other courses in their schools to use QR codes in their material as well, even offering to "do some kind of training" to them. They added however, that there is always a small number of teachers that "never want to change and get out of their comfort zone" but this did not discourage them from trying.

5 Discussion

In this research we have tried to look into the extent to which the use of QR codes, impeded in VET learners' material, can enhance their English language performance in reading and listening skills and their interest and active participation in their specialty courses. A number of 132 VET students from specialties such as mechanics, refrigerants, electricians or mechanics, as well as their four teachers participated in the research. All students took a pre state EFL test to determine their actual level before the intervention as well a post-test upon the end of the study. Due to the large amount of data received and space limit, it was decided not to present the full tests results in this paper. However, briefly discussed here, the findings revealed that there was a statistical difference in students' performance in relation to reading skills, although there seemed to be an improvement in listening skills as well. The students' focus groups results revealed that students found this method of EFL task delivery an interesting, innovative and helpful process for their academic and professional needs as many of them worked in parallel to their studies. According to them the use of QR codes enhanced their linguistic skills and their interest in their specialties courses. The students found the codes very easy to navigate and were able to keep a record of the lessons they were unable to attend for various reasons. Especially the working students found both the mobile mode of teaching and the integration of the QR codes in their worksheets very helpful for their work and admitted they exploited the information to find problems that would come up while at work. Overall, the students stressed the portability, easiness of use and interactivity of the QR codes which they found exciting and expressed their wish to continue their lessons through such tasks and method in the future.

Teachers' focus group discussion results shed valuable light into the research issues, sharing their own opinion on this teaching mode and tool and on learners' performance and attitude. The teachers expressed not only their excitement and interest in using this application and teaching method more often but also their intention to act as "ambassadors" of QR codes integration in the lesson, in an effort to motivate their colleagues towards their use. They particularly stressed the fact that QR codes gave them a sense of flexibility and originality of ideas making their lessons more appealing to learners. They also emphasized the positive impact this QR codes had for the working students who were able to exploit them through mobile learning which was very supportive as a method.

The findings of this research are in agreement with other findings, based on which students and teachers found the use of QR codes helpful, interesting, innovative, interactive and effective for the learning process [16, 19, 20, 22]. What makes this research different however, is that it addresses VET Secondary education, an area which is rather under-investigated on the topic. We are hoping that the results of this paper could possibly give an insight into the impact of QR codes on VET secondary education and how they could best be exploited for the benefit of VET learners. We are also hoping that the results may constitute a motive for further research in VET education in relation to Quick Response systems and ICT in general and their impact in student learning.

References

1. Askari, A.J.: The effect of ICT-based teaching method on medical students' ESP learning. J. Med. Educ. **4**, 81–83 (2004)
2. Korhonen, A., Malmi, L., Myllyselka, P., Scheinin, P.: Does it make a difference if students exercise on the web or in the classroom? In: Proceedings of the 7th Annual SIGCSE/SIGCUE Conference on Innovation and Technology in Computer Science Education, ITiCSE 2002, Aarhus, Denmark (2002)
3. Mumtaz, S.: Children's enjoyment and perception of computer use in the home and the school. Comput. Educ. **36**, 347–362 (2001)
4. Leone, S.: The use of new technologies in advanced Italian classes. In: Proceedings of the Emerging Technologies Conference, University of Wollongong, pp. 18–21 (2008)
5. Padurean, A., Margan, M.: Foreign language teaching via ICT. Revista de Informatica Sociala **7**(12), 97–101 (2009)
6. Crystal, D.: Language and the INTERNET. Cambridge University Press, Cambridge (2001)
7. Dudley Evans, T., St John, M.: Developments in ESP. A Multi-Disciplinary Approach. Cambridge University Press, Cambridge (2001)
8. Batsila, M.: Linking Vocational Education to the World of Employment: the Case of Business English in Greece. EdD thesis The Open University (2017)

9. Baker, A., Dede, C., Evans, J.: The 8 essentials for mobile learning; success in education. In: QLIALCOMM, Wireless Reach, pp. 1–39 (2015)
10. Koutroumani, O.: Mobile computational devices in pre-school age, Postgraduate diploma, Faculty of sciences, Department of mathematics, Patras (2014)
11. Amanatidis, N.: Mobile learning, learning through mobile devices, "Digital and Internet application in education". In: Proceedings of the 2nd Panhellenic Educational Conference of Imathia, Veria, Nousa, 23–15 April 2010 (2010)
12. Hashemi, M., Azizinezhad, M., Najafi, V., Nesari, A.J.: What is Mobile Learning? Challenges and Capabilities. Procedia Soc. Behav. Sci. **30**, 2477–2481 (2011)
13. Darmi, R., Albion, P.R.: Enhancing oral communication skills using mobilie phones among undergraduate English language learners in Malaysia, In: Murphy, A., Farley, H., Dyson, L. E., Jones, H. (eds.) Mobile Learning in Higher Education in the Asia-Pacific Region, vol. 40, pp. 297–313. Springer, Singapore (2017)
14. West, M.D.: Mobile learning: transforming education, engaging students, and improving outcomes, Center for Technology Innovation at Brookings (2013)
15. Oller, R.: The future of mobile learning, EDUCAUSE, Center for Applied Research (2012)
16. Thorne, T.: Augmenting classroom practices with QR codes. TESOL J. **7**(3), 746–754 (2016)
17. Cruse, D.T.H., Brereton, P.: Integrating QR codes into ELT materials. In: Clements, P., Krause, A., Bennett, P. (eds.) Language Teaching in a Global Age: Shaping the Classroom, Shaping the World. JALT, Tokyo (2018)
18. Rikala, J., Kankaanranta, M.: The use of quick response codes in the classroom. In: Specht, M., Sharples, M., Multisilta, J. (eds.) Proceedings of the 11th Conference on Mobile and Contextual Learning, pp. 148–155 (2012)
19. Rikala, J., Kankaanranta, M.: Blending classroom teaching and learning with QR codes. In: Sánches, I.A., Isaías, P. (eds.) Proceedings of the 10th International Conference Mobile Learning, pp. 141–148 (2014)
20. McCabe, M., Tedesco, S.: Using QR codes and mobile devices to foster a learning environment for mathematics education. Int. J. Technol. Inclusive Educ. **1**(1), 37–43 (2012)
21. Law, C.-Y., So, S.: QR codes in education. J. Educ. Technol. Dev. Exch. **3**(1), 85–100 (2010)
22. Kossey, J., Berger, A., Brown, V.: Connecting to educational resources online with QR codes. FDLA J. **2**(1), 11–21 (2015)
23. Tomlinson, C.A.: The Differentiated Classroom: Responding to the Needs of all Learners. Association for Supervision and Curriculum Development, Alexandria (2000a)
24. Code generator: https://gr.qr-code-generator.com/

Digital Transformation of Interdisciplinary Engineering Education

Brit-Maren Block(✉), Benedikt Haus, Anton Steenken, and Torge von Geyso

Institute of Product and Process Innovation, Leuphana University Lüneburg, Lüneburg, Germany
{block,haus}@leuphana.de, {anton.steenken,torge.geyso}@stud.leuphana.de

Abstract. Global transformation processes and sustainability issues will continue to yield a rapid increase in problems at the boundary between technical and non-technical disciplines in higher education. Furthermore, new fields of work emerge due to the digital transformation. Graduates need to be prepared to identify and describe problems and to develop appropriate solutions in teams in order to contribute to change processes related to the future in a smart world. Engineering sciences have to take up the challenge to provide suitable educational programs for a broader target group, i.e. non-technical students, especially in light of the current shortage of qualified specialists. This paper contributes twofold to that discourse; (1) by a novel theory-based teaching and learning concept for an engineering course for bachelor students of non-engineering disciplines (e.g. sustainability sciences) and associated empirical findings of implementation, and (2) by innovative project-based laboratory experiments that encourage interdisciplinary approaches. As a specific contribution to the innovative practice of engineering education, part (1) outlines the student-centered lecture scheme "Electrical and Automation Engineering" (four semester hours per week). The framework-based development, the objectives and the didactic design of the bachelor course as well as the engineering key topics in the context of smart technologies and sustainability are presented. Part (2) details novel practices in the area of engineering education by presenting two specially designed lab experimentation platforms. Starting from the theory framework, the paper contributes to a theoretical understanding and educational practice of engineering courses designed for a specific group of students at the crossroads of engineering and other disciplines.

Keywords: Theory-based student-centered course design · Engineering for non-engineering students · Transformation processes in engineering education

1 Introduction

In today's world, which is characterized by global transformation processes and sustainability issues, engineering sciences continue to be an important shaper

M. E. Auer and T. Rüütmann (Eds.): ICL 2020, AISC 1328, pp. 284–296, 2021.
https://doi.org/10.1007/978-3-030-68198-2_26

of these change processes. In addition to digitalization as a major influencing factor, further changes lead to challenges and modified framework conditions that the engineering sciences have to face [1–5], including:

- changing requirements for the professional field, the subject content and the competence development of students,
- increase in complex global problems, dealing with constant change,
- changes in the institutional framework and in the roles of universities and teaching staff,
- dealing with diversity, heterogeneous target groups, opening up new target groups to secure young talent and developing interdisciplinary solutions,
- need for new teaching and learning arrangements (including digital, individualized, collaborative).

Furthermore, new fields of work emerge due to the digital transformation. Those novel fields of work can be characterized by high degrees of complexity and responsibility as well as the need for a broad knowledge on topics where disciplines intersect. Graduates need to be prepared to identify and describe problems and to develop appropriate solutions in teams in order to contribute to change processes related to the future in a smart world, cf. [1,3,6].

Engineering sciences have to take up these challenges to provide suitable educational programmes for a broader target group (e.g. non-technical students); especially in light of the current shortage of qualified specialists. In this context there is a constant demand for research on theory-based development of targeted didactic concepts for this specific target group and associated empirical findings of implementation.

This paper contributes to that research discourse on transformation processes in engineering education. In Sect. 2, the model of Educational Reconstruction as a theoretical framework for a student-centered conception and implementation of engineering lectures for non-technical students is presented. Based on the framework, Sect. 3 introduces the theory-based lecture scheme "Electrical and Automation Engineering" as a contribution to the translation of educational research to practice. The objectives and didactic design of the bachelor course are shown as well as the curricular implementation. Section 4 presents two innovative project-based laboratory experimentation platforms that focus on interdisciplinary teaching in the engineering sciences in the context of digitalization. Finally, the paper outlines the experience of the first implementation of the course in wintersemester 2019/20 and further discusses future perspectives within the discipline of engineering education and research.

2 Theory-Based Framework

The research objectives in this paper are standing for both theoretical understanding and educational practice. With a focus on the engineering sciences, it deals particularly with the theory-based course-design for students who do not

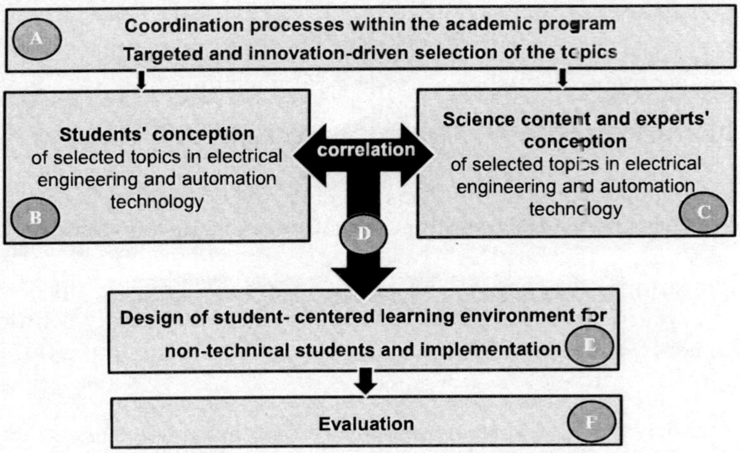

Fig. 1. Model of Educational Reconstruction with focus on engineering and automation technology and non-technical students, based on [7–9]

come from a technical field. Furthermore, the presented framework for student-centered course design in engineering is adjusted to constant change (e.g. changes in occupational fields and subject contents, changes in institutional frameworks, increasing complexity and globalization), and to heterogeneous groups of students. The framework is based on Educational Reconstruction, see [10]. The model of Educational Reconstruction is a research framework that is used in scientific education, as in [7,8]. It is still common in engineering sciences to base the teaching on technical matters. Other aspects like the teaching objective or the perspective of the learners are often considered secondary. To counter this imbalance in the didactic work and to take the students seriously as an active starting point for the construction of knowledge, the model of Educational Reconstruction could be a major help, see [9,10], especially for that specific target group. In the model of Educational Reconstruction, scientific concepts and the perspective of the learners are related to each other. A conclusion about the design of student-centered learning environment is drawn from the comparison. This is particularly important for the didactic construction of engineering courses for non-technical students. In this case, it is a challenge that the majority of students are new to the field and not familiar with it. Therefore, it can be concluded that teaching contents may not simply be dictated in a scientific manner but have to be "created" in a pedagogically useful manner through the conception of the learners themselves, as in [11]. By constructing teaching contents in this way, there are three elements that interact as teaching methodology triplet, [10, p. 4]: the students' perspectives, the clarification of experts' conceptions, and the didactical structuring. As result, a theoretical guided course concept is derived. Figure 1 shows the generic model adapted for the field of engineering sciences and non-technical learners. To implement the model, the research steps A to F must be completed. For further details on these steps, see [12]. The model in all its sequences provides research data that is used for the consistent implementation of study-centered education for non-technical students in engineering education. The first step is a coordinated and innovation-driven selection of content that fits the challenges of digitalization and Industry 4.0 (step **A** in Fig. 1). Then, the experiences and the students' conceptions about electrotechnical concepts, particularly in context of Industry 4.0, have to be collected (step **B** in Fig. 1) as well as the industrial innovations and scientific key concepts have been identified and analyzed from industry applications and the literature, e.g. [13–18], (step **C** in Fig. 1). In step **D**, the correlation between learners' conceptions and scientific perspectives lead to the theory-based development of the design of student-centered learning environment (step **E** in Fig. 1). This design phase of an engineering course for non-technical students is presented in Sect. 3. Further influences are theoretical frameworks like constructivist learning theories, e.g. [19,20], and gender theories focusing on STEM education, e.g. [21–23]. As a final evaluation (step F in Fig. 1), in the first implementation phase we gathered student feedback on the course. The feedback from the students will be incorporated into the process of improvement of the course design. In the next development stages, an evaluation in mixed-method design is planned to get empirical evidence about the effectiveness of the course. Based on the phases of the model, the theory-based development of the teaching and learning concept "Electrical and automation engineering" for non-engineering bachelor students is illustrated. This bachelor course (4 semester hours per week) is one of six courses of the minor (i.e. secondary subject of the study program) "Engineering Fundamentals", among others Mechanical Engineering, Information and Communication Technologies. The minor "Engineering Fundamentals" provides an overview of the most important technologies and technology-oriented processes in the manufacturing industry. This minor can be studied in combination with different majors, e.g. Economics, Environmental Sciences and Digital Media. In times of continuous mechanization and digitalization, a basic understanding of technology is becoming increasingly important in order to actively shape transformation processes at the interface of society, economy, technology and the environment. As described in the introduction, interested non-engineering students have to be empowered to deal with the most important technical disciplines in context of industrial systems, too. The general objective of the new course is to impart a basic understanding of technology in a context relevant for engineering and professional practice. Furthermore, it aims at the following targets:

- imparting knowledge and skills of selected technical basics in electrical engineering, metrology and sensor technologies (including optics) as well as control and drive systems,
- developing close links between basic technical knowledge and possible applications in technical innovations, use of selected examples of systems (e.g. e-mobility, smart sensors, VR/AR), and

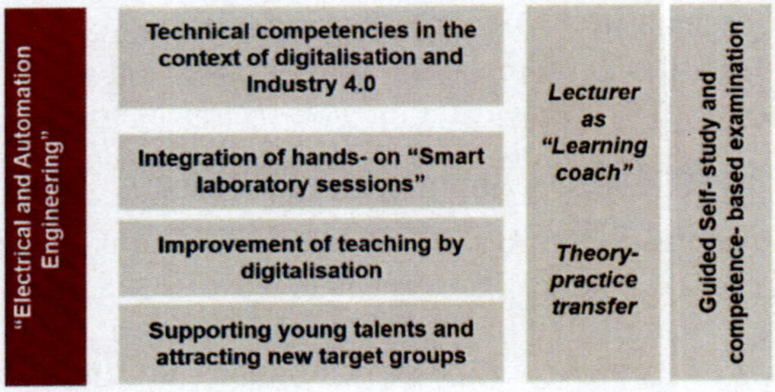

Fig. 2. The theory-based concept of the bachelor course "Electrical and Automation Engineering" for non-engineering students.

- improving the transfer between theory and practice by implementing hands-on sessions ("smart" lab sessions) as well as use of selected digital approaches of teaching.

The strategy for pursuing these objectives is set out in the next section.

3 Novel Theory-Based Teaching and Learning Concept

In correspondence to the model of Educational Reconstruction, the development and implementation of the learning environment is presented below. As mentioned before, the student-centered course design originates in constructivist learning theories, e.g. [19,20] and gender-sensitive teaching concepts, as in [21–23]. Both theoretical approaches emphasize three essential aspects of the process of learning and the acquisition of competences as well as the conditions promotive to this process: the importance of active engagement and problem orientation, the importance of motivation and success and the importance of scientific relevance and transfer. These theoretical principles have been incorporated into the different parts of the concept (Fig. 2). The framework-based course design process mainly focuses on two targets. First, the didactic layout of the course should optimally foster learning processes in the highly diverse group of students and make the course attractive to students. Second, the teaching and learning concept actively addresses innovative technologies. Based on a solid fundamental knowledge of engineering, course graduates should be prepared to identify and describe problems and to develop appropriate solutions in teams in order to actively contribute to necessary change processes. These goals and the limited time resources require a focus on paradigmatic innovative technologies and technical content, shown in Table 1. Students get basic knowledge of selected systems, models, and parameters in the range of automation technology (e.g. electrical engineering and electronics, control engineering, and actuator technology) in the context of digitalization and Industry 4.0. They are proficient in methods for calculating simple electrical circuits, acquire practical skills in the analysis of selected automation systems and in measuring relevant

Table 1. Topics of the course "Electrical and Automation Engineering"

Sequence of topics	Technical contents	Practical integration of technological innovations and industrial trends in the context of Digitalization
1.	Electrical engineering basics (DC and AC technology)	Renewable energy, solar cell
2.	Measurement and sensors technology	Smart Sensors, VR and AR, Auto Identification (including RFID)
3	Control and acuator technology	E-Mobility

process variables. Students acquire professional and methodical expertise that enables them to successfully develop suitable solutions for complex, and at least to some extend technical, problems.

The highly diverse group of students requires an innovative teaching and learning concept to integrate different disciplinary cultures and to specifically support those "non-traditional" students in acquiring competences. The course takes up on innovative trends, interaction of technical components in complex and interlinked systems as well as on a strong focus on transfer between theory and practice. For this purpose, the concept integrates "smart laboratory sessions" as digital approaches to strengthen practical relevance in class. The experiments developed specifically for this purpose are presented in Sect. 4. The process "know-comprehend-apply", as well as a strong focus on practice, enables students to transfer individual topics into the context of complex issues analytically and systematically. This is supported by reinforcing interdisciplinary and systematic competences of students, who are encouraged to utilize their individual academic background (e.g. business administration or environmental sciences) to work on problems and case studies. Working on interdisciplinary problems, and getting to know new technical fields independently, prepares students for their future professional life in industry with interdisciplinary and diverse teams. In summary, the concept incorporates the structural elements as seen in Fig. 2. As depicted there and in Table 1, innovation-based key topics in the context of digitalization, Industry 4.0 and smart laboratory sessions are the centerpiece of the teaching and learning concept. They are expected to allow immediate transfer between theory and practice as well as quick access to understanding complex systems. There are four modular and smart (IT-based, intelligent) laboratories, which are closely related to the concept of Industry 4.0 [18]. The smart laboratories are useable for both practice sequences during lecture sessions as well as practical experiments and projects of students. In addition, concepts of digitalization are integrated to improve the teaching and learning arrangement. For example, modelling and simulation via Matlab and Simulink software by The MathWorks, Inc. pick up mathematical system models

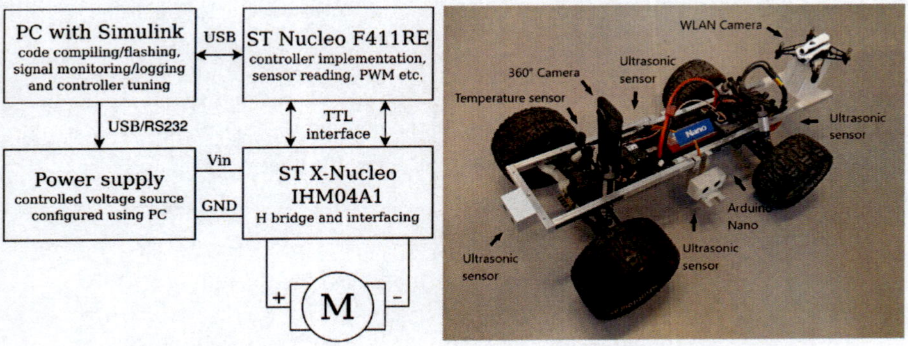

Fig. 3. Components of the experimental setup (without encoder interfacing, left) and AVC vehicle with components (right)

from the smart laboratories and provide an authentic computer simulation and digitally supported failure analysis, while also allowing for graphical software development for embedded systems that interact with the real systems. These approaches are explained in detail in the following section.

4 Project-Based Laboratory Experiment

Instead of defining a standardized, fixed laboratory experiment, as it has been customary in engineering education for the past decades, flexible experimentation platforms can be used for project-based experiments that are both adapted to the specific students' background and also iteratively developed further by the same students. Two of such platforms, especially designed for the later topic of the presented course (see Table 1), will be introduced in the sequel and one of them explained in detail.

4.1 DC Motor Control

Regarding the third topic indicated in Table 1, "Control and acuator technology" and "E-Mobility", an experimentation platform from the broad field of electrical powertrain technology was developed. The common denominator of these disciplines obviously is the electric motor. Its simplest possible form (in terms of usage) is the brushed DC motor, one of which is controlled using an interesting new hardware platform. It consists exclusively of low-cost hardware which is, nonetheless, capable of hardware-in-the-loop operation in real time. A Nucleo F411RE microcontroller development board by STMicroelectronics forms the hardware base of the system. It is, in form and function, comparable and compatible to the popular Arduino boards, yet more powerful, even more affordable, and rather targeted at industrial R&D facilities. It is extended using an STM X-Nucleo IHM04A1 shield for DC motor control, see Fig. 3. At the software side, the H bridge that constitutes the core of the IHM04A1 is PWM-controlled using a program created with Matlab Simulink. Apart from simulation, Simulink is able to translate the intuitive, graphical, signal-flow-based programs to C code, which is then flashed into the fast ARM microcontroller of the Nucleo board, and runs in real time, all with a single click and continuously communicating with Simulink (for online monitoring/tuning and signal logging). Using Simulink for the programming facilitates a smooth integration of lab experiments which are based of this platform into the lessons. For example, the differential equations describing the DC motor dynamics can be derived from physical foundations like Faraday's law, Newton's laws of motion and the Lorentz force within the lessons, and then implemented and demonstrated in simulation using Simulink, after which the real motor is controlled with the same program, comparing the results in real time. Thanks to the high versatility and low cost of the described innovative setup, it contributes towards a sustainable transformation of agile engineering education. More details on the DC motor experimentation platform and on its adaptation to changing requirements are given in [24].

4.2 AdvancedVisionCar (AVC)

The other experimental framework, called "AdvancedVisionCar" (AVC), was created with the goal to familiarize students with the use of virtual reality (VR), augmented reality (AR) and mixed reality (MR), fitting in the second topic indicated in Table 1. With this, not only the handling of the different "Realities" can be trained, but also the students have the opportunity to bring to bear their previous knowledge in the field of sensors. The AVC can be seen in Fig. 3, based on a naked aluminum frame for easy expansion with new peripherals. It is an RF remote-controlled vehicle with several sensors and camera systems. This vehicle is supposed to bridge the gap between Industry 4.0, the different "Realities" and a sensible application in the industrial/corporate context. This can be achieved by using the prototype vehicle to practice or simulate preventive and acute maintenance work in hard-to-reach places, e.g. of harsh industrial environments. The driver has the option of controlling the AVC via radio-transmitted video outside of their sight. They receive this video stream from the WiFi camera at the rear of the vehicle. It currently is transferred to a monitor, which also displays the sensor data. In addition, they have the option of recording a 360° video which can then be examined for evaluation purposes after the journey. The sensors on board the vehicle also offer the possibility of detecting temperatures. This significantly improves error analysis. The recorded video can be examined retrospectively with the help of VR glasses and thus give the user the feeling of having been the passenger themself, which improves the chance of error detection. Microsoft HoloLens is also used to enable the combination with mixed reality. This is used to set waypoints on the course that should be checked regularly. In addition, the HoloLens offers the option of binding error reports to parts of the examination object so that they are available digitally and on site. This means that a user wearing a HoloLens can access these reports as soon as they approach the object

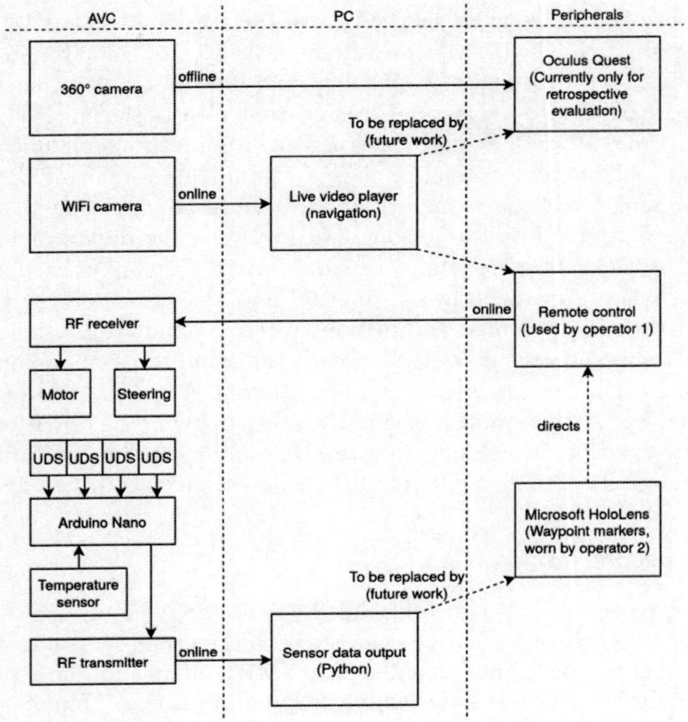

Fig. 4. Schematics, components, and dependencies of the VR/AR experiment

under investigation. The complete architecture of the experimental framework can be seen in Fig. 4.

The first preliminary implementation of the AVC platform in the teaching context was realized with the help of a trail that the vehicle has to follow. In Mixed Reality, relevant points were also marked out on the route at which measurements were to be carried out. While the driver navigates this course, they must pass obstacles. At the points where there is a marking in the MR, they stop for a measurement. Variables to be measured include temperature, and on some markings also labels that must be recognized by using the video of the 360° camera. The AVC is equipped with four ultrasonic distance sensors (UDS) and one temperature sensor. The sensors used are four HC-SR04 (ultrasonic sensors) and one TMP36 (temperature sensor). The control and monitoring of the sensors was carried out with the help of an Arduino Nano, which is operated on a 9 V block and was programmed with the standard Arduino IDE. Regarding optics, the AVC has a WiFi camera and a 360° camera. The WiFi camera is currently used for live image transmission to a computer, which the driver uses to orient themself, while the 360° camera records a video of the entire environment. The distance sensors are specially designed to be able to navigate the vehicle in

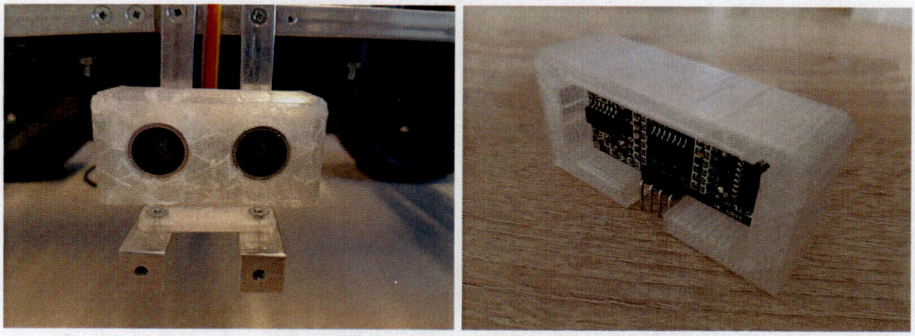

Fig. 5. UDS in the printed case

remote-controlled mode without crash. The temperature sensor is located above the vehicle for maintenance and analysis purposes. All sensor data is transmitted live to the computer where it is displayed next to the live image. The 360° video is loaded onto an Oculus Quest, which are VR glasses that allow the user to intuitively rotate and look around in the video. The MR data is transferred to a Microsoft HoloLens which another user can wear. Using the data and waypoints in the HoloLens, the other user helps the driver navigate. The waypoints are currently only displayed in the HoloLens.

In the future, it is planned to transmit the live image of the WiFi camera and/or the 360° camera to VR or MR glasses, so that the driver can see the entire all-round view while driving, ideally also including the waypoint markings. In addition, thanks to the ultrasonic distance sensors mounted on the vehicle (see Fig. 5), there is the possibility of linking a neural network to the vehicle control which automatically approaches the marked points and enables measurements. So, the operator could concentrate solely on observing the environment of the AVC, e.g. to identify faults or perform troubleshooting. Another possible improvement is the integration of a capacitive and/or inductive sensos for material detection and material specific measurements. Furthermore, the sensor data could be transferred directly into virtual reality so that the operator can see live what the sensor is detecting, like a head-up display in the cockpit of a plane. In addition, this data is to be transferred to the HoloLens in mixed reality, so that a second operator can carry out analyzes in parallel from the outside.

5 Implementation and First Experiences

The course "Electrical and Automation Engineering" was implemented for the first time in the winter term 2019/2020. 13 bachelor students (4 female, 9 male) of the major study programmes business administration, economics, environmental sciences and business informatics participated. For the following round of the Minor Program "Engineering Sciences Fundamentals" 35 students have registered. The aforementioned variety of new fields of work and professional profiles requires high degrees of knowledge and skills from the students with regard to both their minor and their respective majors. Furthermore, the students need the particular skill to systematically form networks of the interdisciplinary knowledge. Imparting the methodological knowledge on working interdisciplinary is an essential part of the course “Electrical and Automation Engineering”, and of the minor “Engineering Fundamentals”. This was achieved by the inclusion of different discipline-specific perspectives of students' majors to develop solutions for scientific questions or case studies. Students and lecturers willingness and motivation to “bring to life” the interdisciplinary discourse, and their readiness to deal with interdisciplinary problems as well as new scientific fields independently, leads to an profound preparation of students for their future career in interdisciplinary and diverse teams. The lab experiments, including the presented one about AR/VR/MR applications, also proved popular among the students. The small group sessions were highly interactive and agile, due to many interposed questions by students with an advanced technical background, about side aspects and other applications. For others, working with the example systems turned out as a first close contact with technology and a journey with a steep learning curve. Being able to react to such diverse educational preconditions, especially incorporating the skills and perspectives of the non-technical students in a constructive way instead of considering them as the “weakest link”, requires a high degree of flexibility both from the lab equipment and from the teaching concept. According to the students' positive feedback on the new course, these goals were achieved. In fact, as justification, they referred in particular to the possibility of combining the engineering fundamentals with their non-technical main subjects.

6 Conclusions and Future Work

The objective of the paper is to represent an interdisciplinary engineering course concept for bachelor students who do not come from a technical field. A suitable framework for the student-centered course design in engineering education is introduced. The didactic concept actively addresses the diverse backgrounds of the students in this course. This professional diversity carries an important potential to take up an interdisciplinary point of view in class sessions in order to prepare students for globalized scientific and working environments. To facilitate a quick grasp of basic technological aspects, laboratory experiment systems were developed. They cover AR/VR technology (i.e. the experiment presented in this contribution) as well as other fields, like electrical drives, sensors, intuitive programming, control engineering and other related fields. After successful implementation of the combined concept, activities to further establish and extend the concept to other courses (e.g. within the first semesters of the masters degree programme) are planned. Evaluation results will be used to improve the concept continuously and in a systematic manner. First experiences from the first implementation could be presented at the ICL 2020 conference.

As part of the research project “Innovation Plus”, the development of the smart technology lab-sessions is funded by the Ministry for Science and Culture of the German state of Lower Saxony in 2019 and 2020.

References

1. Harteis, C.: The impact digitalization in the workplace: an educational view. Springer, Berlin Heidelberg, New York, NY (2018)
2. Rauhut, I.: VDI-Studie Ingenieurausbildung für die Digitale Transformation [Engineering education for digital transformation]. https://www.vdi.de/ueber-uns/presse/publikationen/details/vdi-studie-ingenieurausbildung-fuer-die-digitale-transformation March 2019
3. Auer, M.E., Kim, K.S. (eds.): Engineering Education for a Smart Society: World Engineering Education Forum & Global Engineering Deans Council 2016, Advances in Intelligent Systems and Computing, vol. 627. Springer International Publishing, Cham (2018)
4. Bauer, W.: Weiterbildung und Kompetenzentwicklung für die Industrie 4.0 [continuing education and competence development for Industry 4.0]. In: Vogel-Heuser, B., Bauernhansl, T., ten Hompel, M. (eds.) Handbuch Industrie 4.0 Bd.1: Produktion [Handbook Industry 4.0, vol. 1: Production]. Springer, Berlin (2017)
5. Barth, M.: Implementing Sustainability in Higher Education: Learning in an age of Transformation. Routledge studies in sustainable development, Routledge, London, New York (2015)
6. Uskov, V.L., Howlett, R.J., Jain, L.C.: Smart Education and e-Learning 2017. Springer International Publishing Imprint, Springer, Cham (2018)
7. Niebert, K., Gropengiesser, H.: The model of educational reconstruction: a framework for the design of theory-based content specific interventions. the example of climate change. In: Plomp, T., Nieveen, N. (eds.) Educational design research – Part B: Illustrative cases, pp. 511–531. SLO, Enschede, the Netherlands (2013)
8. Duit, R., Gropengiesser, H., Kattmann, U., Komorek, M., Parchmann, I.: The model of educational reconstruction - a framework for improving teaching and learning science. In: Jorde, D., Dillon, J. (eds.) Science Education Research and Practice in Europe, pp. 13–37. SensePublishers, Rotterdam (2012)
9. Block, B.M.: Educational reconstruction as model for the theory-based design of student-centered learning environments in electrical engineering courses. In: 2016 IEEE Global Engineering Education Conference (EDUCON), pp. 105–113 (April 2016)
10. Kattmann, U., Duit, R., Gropengiesser, H., Komorek, M.: Das Modell der Didaktischen Rekonstruktion - ein Rahmen für naturwissenschaftsdidaktische Forschung [The model of educational reconstruction - a framework for educational research and development within natural sciences]. Zeitschrift für Didaktik der Naturwissenschaften **3**, 3–18 (1997)
11. Krapp, A., Weidenmann, B.: Pädagogische Psychologie [Educational psychology]. Beltz, Weinheim (2006)
12. Block, B.M., Haus, B., Steenken, A., von Geyso, T.: Interdisciplinary engineering education in the context of digitalization and global transformation processes. In: Proceedings of the 48th SEFI Annual Conference (2020), to appear
13. Weissgerber, W.: Elektrotechnik für Ingenieure [Electrical engineering for engineers]. Springer Fachmedien Wiesbaden, Wiesbaden (2015)

14. Albach, M.: Elektrotechnik [Electrical Engineering]. Pearson, München (2014)
15. Lunze, J.: Regelungstechnik [Control Engineering]. Springer, Berlin Heidelberg, Berlin, Heidelberg (2010)
16. Hagl, R.: Elektrische Antriebstechnik [Electrical Powertrain Technology]. Hanser-Verlag, München (2015)
17. Kruse, D., Frerich, S., Petermann, M., Ortelt, T.R., Tekkaya, A.E.: Remote labs in ELLI: lab experience for every student with two different approaches. In: 2016 IEEE Global Engineering Education Conference (EDUCON), pp. 469–475. IEEE, Abu Dhabi, United Arab Emirates April 2016
18. Bauernhansl, T., ten Hompel, M., Vogel-Heuser, B. (eds.): Industrie 4.0 in Produktion, Automatisierung und Logistik [Industry 4.0 in production, automation and logistics]. Springer Fachmedien Wiesbaden, Wiesbaden (2014)
19. Deci, E., Ryan, R.: Theory of self-determination and motivation and its importance for pedagogy. Pedagogy **39**(2), 223–239 (1993)
20. Mandl, H., Reinmann-Rothmeier, G.: Die konstruktivistische Auffassung vom Lehren und Lernen [The constructivist approach to teaching and learning]. In: Schneider, W., Knopf, M. (eds.) Entwicklung, Lehren und Lernen. Zum Gedenken an Franz Emanuel Weinert [Development, Teaching and Learning. In memory of Franz Emanuel Weinert], pp. 366–403. Hogrefe, Göttingen (2003)
21. Mills, J., Ayre, M., Gill, J.: Gender Inclusive Engineering Education. Routledge research in education, Routledge, New York (2010)
22. Gill, J., Sharp, R., Mills, J., Franzway, S.: I still wanna be an engineer! women, education and the engineering profession. Euro. J. Eng. Educ. **33**(4), 391–402 (2008)
23. Wächter, C.: Interdisciplinary teaching and learning for diverse and sustainable engineering education. In: Beraud, A., Godfroy, A.S., Michel, J. (eds.) GIEE 2011: Gender and Interdisciplinary Education for Engineers, pp. 47–63. SensePublishers, Rotterdam (2012)
24. Block, B.M., Haus, B.: New ways in engineering education for a sustainable and smart future. In: Proceedings of the 50th Annual Frontiers in Education (FIE) Conference (2020), to appear

Poster: Technological University Faculty Attitudes Towards Online Education

Gulnara F. Khasanova(✉) and Mansur Galikhanov

Kazan National Research Technological University, Kazan, Russia
gkhasanova@mail.ru, mgalikhanov@yandex.ru

Abstract. The paper discusses the implementation of a research project funded by the Russian Science and Higher Education Ministry during which 20 blended faculty continuing education programs were designed and delivered at IDPO KNITU in November – December 2019. Two surveys of faculty members were conducted within 15 months, the first in October 2018 and the second in March 2020. The paper focuses on the impact that participation of the faculty in the IDPO KNITU project has had on instructors' assessment of factors that impede and motivate participation in online learning as well as their assumptions about the conditions for effective online delivery and its best practices.

Keywords: Faculty training · Professional development programs · Online teaching · Blended learning

1 Introduction

Online education is rapidly growing in the modern world. Its main advantages including accessibility and flexibility both in terms of time limit and location are even more valuable to adult learners participating in professional development programs. Meanwhile, the success of online teaching and blended learning will depend not only on latest tools and innovative technologies but mainly on how instructors are prepared for performing changing roles and responding to new challenges [1–3].

This paper examines changes in university faculty attitudes to online teaching, its advantages and disadvantages that can affect their participation in online and blended learning. The research was intended to register a shift in faculty attitudes towards online education over the period of a dozen years, taking as a starting point the methodology and data from the 2007 research held by Panda and Mishra at Indira Gandhi National Open University [4].

Two surveys of instructors participating in further education programs of the Faculty Development Institute at the Kazan National Research Technological University (IDPO KNITU) were conducted in October 2018 [5] and March 2020. The data obtained were compared to the 2007 research results. The data from 2018 and 2020 surveys were also compared so that to reveal possible changes in faculty attitude towards online education following participation in a project during which 20 blended further education programs were delivered at IDPO KNITU in November – December 2019. The project was funded by the Russian Science and Higher Education Ministry.

M. E. Auer and T. Rüütmann (Eds.): ICL 2020, AISC 1328, pp. 297–302, 2021.
https://doi.org/10.1007/978-3-030-68198-2_27

The contingent of 1101 listeners included university faculty members and in-company trainees from Russia's innovative enterprises.

Programs were delivered in blended format, and at least a half of each program was implemented online. Blended format included various options: some programs started with online modules and continued face-to-face; others began with face-to-face classes and finished online; there were programs where learners studied on-campus and online modules in parallel.

Although IDPO KNITU delivered previously all these programs numerous times [6], it was for the first time that at least some of their elements were presented online. Thus, such an experience required revision of programs curricula, content and teaching materials. Digital educational resources including video lectures and tests were developed.

2 Methodology

The data sources in the study included faculty survey, end-of-course debriefing reports on the all delivered courses, and personal observations from the authors during course delivery and participation in the project. We conducted surveys of IDPO faculty based on the questionnaire by Panda and Mishra [4], which was translated into Russian with minor changes of replacing "E-learning" with "online learning".

The questionnaire included personal data entries and a list of 22 statements evaluating online teaching: (1) Online learning will never replace other forms of teaching and learning. (2) Online learning makes me uncomfortable because I do not understand it. (3) Online learning is a de-humanizing process of learning. (4) Online learning can solve a lot of our educational problems. (5) I feel intimidated by online learning. (6) Online learning will bring new opportunities for organizing teaching and learning. (7) Online teaching is difficult to handle and therefore frustrating to use. (8) There are unlimited possibilities of online learning that have not yet been thought about. (9) Online learning saves time and effort of both teachers and students. (10) Online learning increases access to education and training. (11) Online learning will increase my efficiency in teaching. (12) Online learning enables collaborative learning. (13) Online learning can engage learners more than other forms of learning. (14) Online learning increases quality of teaching and learning because it integrates all forms of media: print, audio, video, animation. (15) Online learning increases the flexibility of teaching and learning. (16) Online learning improves communication between students and teachers. (17) Online learning enhances the pedagogic value of a course. (18) I get a sinking feeling when I think of trying to use online learning for my courses. (19) Online learning is not effective for student learning. (20) Online learning experiences cannot be equated with face-to-face teaching or even distance education. (21) It is essential that online learning material be of high quality. (22) Universities should adopt more and more of online learning.

We also used parts of the questionnaire concerning motivators and inhibitors of faculty participation in online teaching, but the analysis of those data is not included in this paper.

So that to obtain compatible data, we used the five-point Likert-type measuring scale and replicated the same measuring procedures. In the scale, «1» meant «strongly disagree» while «5» - «strongly agree». Seven negatively worded items were also reverse scored. Mean scores were determined using the following scale: ≥ 4.5 – "positive," 3.5–4.49 – "moderately positive," 2.5–3.49 – "neither positive nor negative," 1.5–2.49 – "moderately negative," ≤ 1.49 – "negative."

In the first survey, paper copies of the questionnaire were distributed to 21 full-time faculty members of IDPO, and 21 responses were collected (response rate 100%, 71% female and 29% male). The second survey was conducted online. Messages containing a link to a survey form were sent to 75 faculty members, both full-time and part-time, who took part in the project. Fifty-three answers were uploaded to the form (response rate 71%, 70% female and 30% male).

Over 80% of respondents in the first survey have not participated in online teaching before, while in the second one, the rate of participation increased to over 80%. Although all the respondents took part in delivering blended programs, part of them were involved just in face-to-face modules.

3 Results of Surveys on Faculty Attitudes Towards Online Education

The data were analyzed for descriptive statistics, and Fisher angular transformation coefficient tests were conducted to evaluate differences in opinions between the 2018 and 2020 respondents. The data from the 2007 study, on the one hand, and from the 2018 and 2020 studies, on the other, were compared based on average values.

The range of scores showed that the integral attitude of IDPO faculty (3.25) was at the "neither positive nor negative" score, being lower than in the 2007 study (3.81) that was at "moderately positive" score. Between two surveys, IDPO faculty attitude became even more negative – 3.17 (2020) as compared to 3.33 (2018). Almost all the scores are lower than in the 2007 study.

The number of items at "neither positive nor negative" score decreased from 14 to 9, while no item in both surveys was at "negative" or "positive" score.

While in 2018, a single item was at a "moderately negative" score ("Online learning experiences cannot be equated with face-to-face teaching or even distance education"), in 2020, there were four ("Online learning saves time and effort of both teachers and students", "Online learning increases quality of teaching and learning because it integrates all forms of media: print, audio, video, animation", "Online learning improves communication between students and teachers") (see Fig. 1).

Seven items in 2018 were at a "moderately positive" score ("Online learning will never replace other forms of teaching and learning", "Online learning makes me uncomfortable because I do not understand it", "I feel intimidated by online learning", "Online learning will bring new opportunities for organizing teaching and learning", "There are unlimited possibilities of online learning that have not yet been thought about", "Online learning increases access to education and training", "It is essential that online learning material be of high quality"). In 2020, the list increased by two ("Online teaching is difficult to handle and therefore frustrating to use," and "I get a

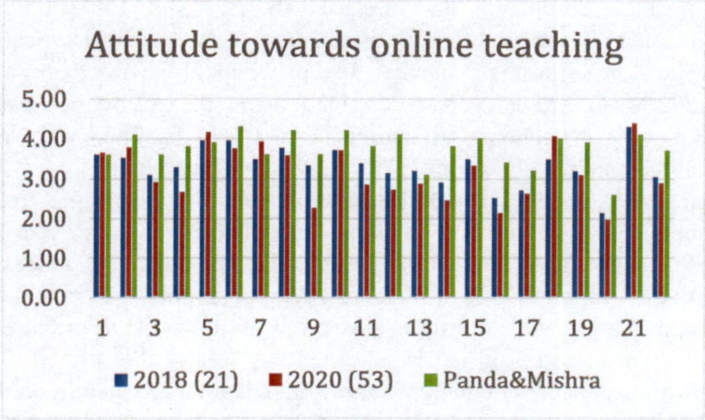

Fig. 1. Attitude towards online teaching

sinking feeling when I think of trying to use online learning for my courses"). Item 21 was at the highest score in both polls.

An analysis was made of the significance of the differences in the ratings of 22 statements about online learning based on the Fisher angular transformation coefficient. The number of those who agreed and fully agreed with the statement "Online learning can solve a lot of our educational problems" reduced ($\varphi^*_{emp} = 2.094$, $p < 0.05$), and the number of those who disagreed and strongly disagreed with it grew even more significantly ($\varphi^*_{emp} = 2.8$, $p < 0.01$). The number of those who agreed and fully agreed with the statement "Online teaching is difficult to handle and therefore frustrating to use" increased ($\varphi^*_{emp} = 2.063$, $p < 0.05$). The biggest differences appeared in the attitude towards the statement "Online learning saves time and effort for both teachers and students." The number of those who agreed and fully agreed with it significantly decreased ($\varphi^*_{emp} = 4.363$, $p < 0.01$). The approval degree of the statement "I get a sinking feeling when I think of trying to use online teaching for my courses" grew ($\varphi^*_{emp} = 2.063$, $p < 0.05$).

The share of those who disagreed and strongly disagreed with the statement "Online learning will increase my efficiency in teaching" decreased ($\varphi^*_{emp} = 2.42$, $p < 0.01$). The share of those who disagreed and strongly disagreed with the statement "Online learning enables collaborative learning" reduced as well ($\varphi^*_{emp} = 2.067$, $p < 0.05$). The proportion of those who did not agree and totally did not agree with the statement "Online learning increases quality of teaching and learning because it integrates all forms of media: print, audio, video, animation" has grown significantly ($\varphi^*_{emp} = 2.978$, $p < 0.01$). The disagreement degree with the statement "Online learning increases the flexibility of teaching and learning" increased ($\varphi^*_{emp} = 2.083$, $p < 0.05$).

4 Discussion

An important conclusion of the research is that the IDPO faculty attitude towards online teaching appeared to be more negative compared to the 2007 study. It is evident that such a comparison has certain restrictions, connected to the differences in the “E-learning” and “online-learning” terms, differences of specific socio-economic and sociocultural conditions, but still, we believe that it is of interest.

The negative dynamics in the attitude towards online teaching between the two surveys conducted with an interval of fifteen months is also noteworthy. This dynamics can be explained by conditions in which the project has been implemented.

Specifically, the fact that most faculty members did not have any experience in online teaching created a stressful situation, which was aggravated by the large number of implemented programs and tight deadlines. Faculty had to adapt quickly the content to online learning, including recording video lectures, preparing presentations, textual learning resources, interactive materials and links. In fact, it was a true learning by doing experience for the faculty.

The instructors described changes in their pedagogical practice. They had to revise and re-design the content of their programs, getting rid of unnecessary elements and thus making the content more precise and informative. More detailed and specific guidelines for learners were developed. The introduction of new assessment tools, such as online testing, made content more focused on the training results.

The stressful nature of the first experience of participating in online learning affected the assessments of respondents in the 2020 survey. At the same time, modular program design, a blended teaching format, and technical and managerial support allowed faculty members to cope with the ICT barrier and provided sufficient flexibility for the successful implementation of programs. Intense communication between colleagues, the exchange of innovative solutions and the unification of efforts in solving new problems were also listed as an important factor that led to a greater cohesion and mutual support.

Evaluation by faculty of their online instruction experience presented in the study can provide insight for higher education institutions that increase online segment of their continuing education programs and train faculty to deliver online instruction. Research findings on instructors' evaluations of best online teaching practices can help higher education institutions develop strategic models aimed at promoting faculty readiness for quality online instruction.

Acknowledgment.

Co-funded by the
Erasmus+ Programme
of the European Union

The publication is co-funded by the Erasmus+ Programme of the European Union.

The European Commission support for the production of this publication does not constitute an endorsement of the contents which reflects the views only of the authors, and the Commission cannot be held responsible for any use which may be made of the information contained therein.

References

1. Coppola, N.W., Hiltz, S.R., Rotter, N.G.: Becoming a virtual professor: Pedagogical roles and asynchronous learning networks. J. Manag. Inf. Syst. **18**(4), 169–189 (2002)
2. Bawane, J., Spector, J.M.: November) Prioritization of online instructor roles: Implications for competency-based teacher education programs. Distance Educ. **30**, 383 (2009)
3. Neely, P.W., Tucker, J.P.: Unbundling faculty roles in online distance education programs. Int. Rev. Res. Open Distance Learn. **11**, 20–32 (2010)
4. Panda, S., Mishra, S.: December) E-Learning in a mega open university: faculty attitude, barriers and motivators. Educ. Media Int. **44**(4), 323–338 (2007)
5. Khasanova, G.F., Galikhanov, M.F.: Poster: Development of faculty competences in online teaching. In: Auer, M., Hortsch, H., Sethakul, P. (eds) The Impact of the 4th Industrial Revolution on Engineering Education, ICL 2019. Advances in Intelligent Systems and Computing, vol. 1134, pp. 376–381. Springer, Cham (2020)
6. Galikhanov, M.F., Guzhova, A.A.: Complex approach for preparation and implementation of continuous professional education programs in technological university. In: 2013 International Conference on Interactive Collaborative Learning, ICL 2013, art. no. 6644535, pp. 54–55 (2013)

Emergency Remote Learning During COVID-19: Socio-educational Impacts on Portuguese Students

Luciana Oliveira[1](✉), Anabela Mesquita[2], Arminda Sequeira[1], and Adriana Oliveira[1]

[1] CEOS.PP, ISCAP Polytechnic of Porto, Rua Jaime Lopes Amorim s/n, Matosinhos, Portugal
lgo@eu.ipp.pt

[2] Polytechnic of Porto and Algoritmi RC, Rua Jaime Lopes Amorim s/n, Matosinhos, Portugal

Abstract. The pandemic caused by COVID-19 is not just a global crisis, it is 'the first' global crisis, and as the mandatory confinement shifted all education to Emergency Remote Instruction/Teaching/Learning, Higher Education Institutions were faced with the heavy task of balancing the immediate massive technological-pedagogical request by teachers and providing students with the socio-educational support that they need. This paper analyses the socio-educational impacts of the current confinement period on student's lives and how they are responding to implemented ERL solutions, specifically in a stage of abundant pressing changes in which critical challenges are mostly felt. A close-ended questionnaire was built, comprising six dimensions of issues that may impact ERL: educational and organizational issues, technological and working conditions, social issues, family-related issues, psychological issues, and financial issues. Results were collected right after the first month of ERL and reveal that the most severe problems reside on the pedagogical and psychological domains.

Keywords: Emergency Remote Learning · COVID-19 · Socio-educational analysis

1 Introduction

In the current context of global confinement due to the COVID-19 pandemic crisis, higher education institutions have shifted from face-to-face or blended instruction to fully remote instruction, to ensure the bare minimum provision of education to its students, adhering to what has been recently coined as Emergency Remote Teaching (ERT) and Learning (ERL). According to Hodges et al. [1], ERT can be defined as a "temporary shift of instructional delivery to an alternate delivery mode due to crisis circumstances". This approach "involves the use of fully remote teaching solutions for instruction or education that would, otherwise, be delivered face-to-face" and "return [ing] to that format once the crisis or emergency" is over. As such the aim of ERT consists of providing temporary access to instruction and instructional support, in

M. E. Auer and T. Rüütmann (Eds.): ICL 2020, AISC 1328, pp. 303–314, 2021.
https://doi.org/10.1007/978-3-030-68198-2_28

synchronous and asynchronous formats, in reliable ways during a particular moment and is essentially a shift of delivery modes, methods and media which are progressively adjusted to the available settings, resources and limitations of organizations, teachers and students. It is distinct from most elaborated forms of technology-mediated instruction, such as e-learning, b-learning and m-learning, which consist of well-planned online learning experiences supported by robust educational ecosystems. Global awareness of this distinction is critical for organizations, teachers, students, and society so that an assessment of current practices does not collide with the principles of the various long-term established models of technology-mediated education. Having said that, what students are experiencing all over the world, during the Spring of 2020, is not online learning but remote learning, defined as a "quick, ad-hoc, low-fidelity mitigation strategy" [2].

The generalized absence of contingency plans in the transition from the face-to-face instruction to the ERL and ERT has pushed organizations and teachers to focus on easy-to-implement operational strategies that facilitate the transmission of instruction, much in the fashion of guaranteeing the continuity of the provision of a service, but with a considerable, and yet somewhat inevitable, level of unawareness and inattention towards the receiver - the student. Undoubtedly, all educational measures implemented so far are aimed at resuming instruction and supporting students. However, there has not been much space or opportunity to investigate how the current crisis is impacting students lives, particularly in aspects that may pose additional challenges or impediments to ERL. Therefore, this paper aims to analyse the socio-educational impacts of the current confinement period on students lives and how they are responding to implemented ERL solutions, specifically in a stage of abundant pressing changes in which critical challenges are most felt.

2 Background

The confinement period imposed by the pandemic crisis caused by COVID-19 forced HEI to close all in-person activities and to ensure the provision of education through any technological means. The urgent and unplanned transition to fully remote instruction posed real challenges to all actors involved in the process, namely teachers, students, and staff, requiring an immediate and unprecedented digitalization of education. The impacts of the transition to fully remote instruction cannot, however, be evaluated only at the educational level, particularly for students, because it consists of the broader social shift with multilevel repercussions on educational/pedagogical, technological and working conditions, social, family-related, psychological and financial issues.

2.1 Educational/Pedagogical Issues

The pandemic situation had a profound impact on education and pedagogy. The loss of face-to-face contact and direct interactions with both peers and teachers may potentially stunt students' development as students are being taught topics related to traditional practical-based learning materials or models [3]. That may be the case of Anatomy learning without cadavers which is generally seen as less favorable [4]. There are applications and simulators that replace these practical classes but there is also a learning curve to use them, besides the costs associated.

Another challenge is related to the interaction. Knowledge is socially created, and this means that it hatches from interactions. In face-to-face instruction, the teacher needs to consider and promote interaction between students, students and the content and students and the teacher. It is the presence of all these types of interaction that increases the learning outcomes. This means that there is a need for careful planning for online learning which should include, not just identifying the content to cover, but also tending to how teachers can support different types of interactions that are important to the learning process. This approach recognizes learning as both a social and a cognitive process, not merely a matter of information transmission [1]. As an attempt to reduce the distance between learners, teachers are using chat rooms or real-time tutorials, but these solutions always have some kind of limitation.

In the transition from face-to-face teaching to online delivery has a serious impact on assessments and evaluation. Although technology has been used earlier to support teaching and learning, the assessment aspect is often under-developed [5]. Applying assessments online on those courses designed for face-to-face learning is a challenging task. In cases where the assessment was done, mostly, with a written examination and oral viva (for instance, disciplines in medicine), the move to online examinations requires the preparedness of teachers and students. If scores are lower, citizens might think that these professionals are not as good as those who graduated in the past. Moreover, students, as well as faculty members, are uncertain about the procedure for administering assignments, projects, and other continuous assessments [6, 7]. Teachers must rethink assessment and develop monitoring and antiplagiarism strategies [8].

Another challenge relies on the readiness of teachers and required technological support. The quick transition to ERL did not allow teachers, students, and schools to prepare in advance nor to immediately provide the required massive technological and pedagogical support. This means that teachers and students were often left alone in solving the problems. As a result, keeping quality is a struggle and not all problems are solved, which has been reported as additional stress for the teachers [1]. Moreover, issues regarding the technological proficiency of teachers [9, 10] and universities not having enough infrastructure or resources to facilitate online teaching with immediate effect [11], have also been raising.

Another aspect resides on lectures consisting of an instructional component of an overall ecosystem designed to support learners with formal, informal, and social resources. Efficient online education requires an investment in an ecosystem of learners' support, which takes time to identify and build. Of course, it is possible to use a simple online content delivery which can be quick and inexpensive. But it can be confusing and not robust. This means it is important to build a whole ecosystem integrating all sorts of instruction. And for that, planning a design process is the key [1]. Thus, issues of online design must be taken into consideration as well. In a period of crisis and crisis response, there is an increased risk of diminishing the quality of the courses delivered. If there was already a stigma concerning the online instruction, with people thinking it has lower quality, this prejudice may be exacerbated with this quick move online, without the proper preparation of all the key actors involved. Usually, a full-course development project can take months when done properly. The need to "just get it online" is in direct contradiction to the time and effort normally dedicated to developing quality education.

2.2 Technological and Working Conditions

In the current context, technology is central for teachers and students but it may be less accessible for students, as some of them do not have computers, and others may have to share the one computer they have, as parents and siblings are also working remotely during the crisis. Furthermore, the lack of internet access is exacerbating the digital divide [1].

Pandemic instigates a digital revolution in academia and higher education [12]. Social distancing, months-long quarantine, and economic shutdown will broader the understanding of the majority of people working in academia and higher education. It will propel the transition to the fully functional and operational online instruction but also the integration and management of online admissions, online examination, assessment and vivas, as well as the understanding that online academic jobs are as effective and as meaningful as those performed "in real life". Digitalization in higher education allows streaming lectures online or enables professors and students to interact in the virtual environments but not everyone is ready for this and even young people, who usually spend a lot of time in online activities, confess that they would have preferred being lectured in traditional classrooms and universities' facilities.

On the plus side, due to the crisis, innovations in academia and higher education that would have normally taken several years to introduce due to the contradictory administrative regulations, are now introduced quickly in a matter of days. This is a clear example of the Schumpeterian 'creative destruction' in making and will forever change the status quo in academia and higher education.

2.3 Social Issues

The pandemic situation has brought a sudden disruption in the everyday life of schools, colleges, and universities, influencing more than 1.7 billion students worldwide. Closing schools introduced several challenges to students and their families and one of the most important concerns is related to how the government and education system handled the pandemic crisis with the study-from-home approach. Surveys suggest that students experience severe limitations on particular subjects that benefit from physical interaction with the materials, and tend to lose the "pacing mechanism" of scheduled lectures, thus having a higher chance of dropping out than those in traditional settings [13, 14].

Another concern is related to the impact of the pandemic situation on this years' university graduates [10]. These future graduates are experiencing major interruptions in teaching and assessment in the final part of their studies. This can imply, for some of them, to graduate later than they expected and also entering the job market later. These young graduates will have to face the severe challenges of the global recession caused by the COVID-19 crisis, including career challenges and not so well-paid jobs, besides unemployment.

2.4 Family-Related Issues

Issues concerning the family pertain to two different scopes: on one side, it relates do structure and stability of family income and, on the other hand, it relates with the conditions of the physical space that the family has to share, which may be too small once, predictably, all members may be in confinement and having to share a space that may prove to be too small, which raises questions of privacy but also questions of equipment (desktops, laptops or iPads) and internet access sharing. If the family is experiencing income shrinking, the continuation of studies may be threatened. The stability of family income was also a significant factor in students' experiencing anxiety during the COVID-19 crisis, which could be explained by increased psychological and economic pressure [15]. Moreover, issues of privacy and poor physical overall conditions may have an impact on the ability to attend online classes and develop remote activities.

2.5 Psychological Issues

One of the most evident psychological problems concerns anxiety [10, 16], followed by fear, worry, depression and traumatic stress, the last two more visible among people working in the healthcare system [10]. These psychological issues have their origin in the feeling of uncertainty about what is going to happen, namely, on the studies [17], future employment [18], and psychological health of students [19, 20].

This anxiety and stress may have been caused by the gradually increasing distance between people resulting from the quarantine and affects all students and in particular, those staying far from home as they are not only worried about their health, safety, and education but they also have concerns for the wellbeing of their families. Moreover, it is known that anxiety disorders are more likely to occur and worsen in the absence of interpersonal communication [21, 22], which is the case. The significant shortage of masks and disinfectants, the overwhelming and sensational news headlines, and erroneous news reports also contributed to this effect [23].

Studies also show that the anxiety regarding the epidemic was associated with the place of residence of the students, source of parental income, whether living with parents and whether a relative or an acquaintance was infected with COVID-19 [16]. Living with parents is a favourable factor against feeling anxious. Moreover, social support not only reduces the psychological pressure during the epidemics but also changes the attitude regarding social support and help-seeking methods. This result suggests that effective and robust social support is necessary during public health emergencies [24].

In some countries, for some individuals, the psychological impact of the pandemic situation was suggested to have been greater than the physical health danger posed by the diseases themselves [25] affecting the mental health of college students.

2.6 Financial Issues

In the context of an epidemic crisis coupled with an economic crisis, students will experience more difficulties, even if classes take place at distance as they will need to invest in good quality equipment and internet access. Moreover, families are not prepared to perceive distance education as having the same quality as face-to-face instruction and so they will not be prepared to pay the same amount of money for the tuition fees. In fact, a reduction or refund of the tuitions has already been discussed on the media. According to The Guardian, the UK National Union of Students (NUS) believes that students should get a year's refund of tuition due to Covid-19 crisis, which is something that HEI, in general, are not willing to do [26]. In Portugal, some HEI are providing extended deadlines for the quarterly payments. An NUS study also states that 85% of working students will need additional financial support because they have lost their jobs as a result of the pandemic and subsequent lockdown [26].

In addition, the economic crisis and the sanitary measures to combat the epidemic will produce a significant personal income shrinkage [27]. This might result on an increase of the number of working students, as it is often college students who report the highest levels of financial strain, defined by their perceptions of economic stress or lack of financial support from their families, that feel most compelled to work during their undergraduate studies [28, 29]. On the other hand, as distance education does not require the student to move to the city/region where the university is located, this means that those who previously could not afford or did not want to relocate can now apply for admission in a university in another region [27]. Finally, it is known that students' financial situation has a high impact on a student's dropout decision [30].

3 Methodology

The adopted methodology consists of an exploratory survey-based research. Exploratory research is characterized by employing a single data collection method to obtain an initial view of the issues being analysed. Provided the scarceness of literature and systematization on the socio-educational impacts of the current confinement period on student's lives and on how they are responding to implemented ERL solutions, the authors propose an evaluation instrument, close-ended questionnaire, comprised of 67 items organized in six dimensions of issues that may impact ERL, as follows: (A) Educational and organizational issues – 33 items, (B) Technological and working conditions – 5 items, (C) Social issues – 11 items, (D) Family-related issues – 4 items, (E) Psychological issues – 6 items, (F) Financial issues – 8 items. Each item on the questionnaire was presented in a labelled 4-point Likert scale. Items on the scale were anchored at 1 = never, 2 = rarely, 3 = frequently, and 4 = always. Higher scores represent the higher frequency of the item.

According to the literature review, these are the current critical dimensions affected by the confinement period in students' lives and that there is a high degree of correlation and interchanges among these. The research strategy is, therefore, based on tackling a multidimensional analysis of the students contexts, by focusing on a broader picture, rather than on an isolated specific dimension. At this stage of research, a multidimensional analysis is considered more beneficial, even in detriment of some depth, which may be later explored, as very little is known about how students are coping.

The survey was disseminated among students enrolled in HEI in Northern Portugal, through social media channels, namely in institutional public pages and open groups. Survey data were downloaded and transferred into IBM SPSS Statistics 26.0 statistical analysis software package for analysis.

4 Results

Overall, 197 questionnaires were excluded because they were not fully answered. At the end, 360 valid answers were obtained. The scores of the counter items were reversed prior to conducting the analysis.

4.1 Demographics and Transition to ERL

The great majority of the participants are female (74.72%), aged 18 to 22 years (83.06%), living in districts located in Northern Portugal. Non-working students make up 78.33% of the sample, though the percentage of working students (20%) is relevant.

Regarding financial aid, 36% of the students are currently benefiting from a scholarship and approximately 76% of these believe they will still need it in the future. The percentage of students is not receiving financial aid is substantial (60%), however, 25% of these believe they will need it in the future. Around half of the respondents ($\approx$ 42%) has changed residence during the confinement period, which indicates that a significant amount of students who were displaced have returned to their family (or other) homes, having abandoned student lodging.

A portion of 78% of the students indicates having unlimited access to the Internet connection, while the remaining 20% have conditioned access or (1%) none at all. However, approximately 95% ($n = 341$) of the students indicate that they are currently involved in some form of online learning during the confinement period, which can be read as a rapid large scale response from schools to ERL. Among these students, ($n = 341$) only about 8% state having had contact with any form on online learning before the current context ERL imposed by COVID-19. Also, worth to notice that, more specifically, *a priori*, 19% of the students attending online learning have connection restrictions, which may add to the functional difficulties that may build on top of any other learning or personal challenges. Finally, the most used devices are laptops (95%) and smartphones (93%), followed by tablets (31%) and desktops (20%).

On average, the respondents for the subsample of students attending online learning ($n = 341$), were enrolled in 7 regular courses in the second semester of the school year 2019–2020, with most answers varying between 5.53 and 8.47 ($\sigma = 1{,}47$). Regarding the courses in which students had no access to ERL, these vary between 0 (*Min*) and 12 (*Max*), with a low average of 1,70 and with most responses comprised between 0 and ≈ 2 ($\sigma = 1.95$). This means that students were provided with access to ERL in about 76% of their courses within the first month of the confinement imposed the COVID-19 pandemic.

Regarding the delivery methods for ERL, students were provided with simplified descriptions of the modes synchronous ("Courses based essentially in videoconference classes"), asynchronous ("Courses based essentially in autonomous readings, watching videos, doing exercises and other autonomous prescribed work") and mixed ("A combination of the two above situations"). Results show that, on average, the synchronous mode is the most frequent ($\bar{x} \approx 3$), followed by mixed ($\bar{x} \approx 3$), and asynchronous ($\bar{x} \approx 2$). For most students, the majority of courses delivered in synchronous mode vary between 2 and 5, the courses delivered in mixed mode vary between 1 and 4 and, despite the average value, the courses delivered in asynchronous mode vary between 1 and 5, moving this mode up to second favourite (*Mo* = 2).

Finally, the most used systems in ERL are, in decreasing order, Colibri Zoom, Moodle and Microsoft Teams. The other systems referred by students are, Skype (2,93%), BlackBoard (1.17%), Google Hangouts (1.17%), Cisco Webex (0.59%) and Email (0.29%).

4.2 ERL Impacts on Students' Lives

Cronbach's alpha, a measure of intercorrelation among measurement items used as an indicator of internal consistency was calculated for the non-demographic Likert-scale questions. The items worded negatively were reverse coded and results show Cronbach's $\alpha = 0.89$, indicating acceptably high internal consistency ($\geq$ 0,70 is considered the minimum acceptable for use in research [31]). For all items, the maximum value is 4 ("Always").

Educational Context. The first 12 items of the survey (A1-A10) focus on assessing information and work proposed by teachers and on the teacher-student relationship. On average, students report that teachers have developed strategies that permitted to (A1) resume the work that was held in the classroom ($\bar{x} = 3.28$), with no reports of this not being possible at all (*Min* = 2), that (A10) teachers are using adequate software and resources for ERL ($\bar{x} = 3.11$), and that teachers have frequently provided (A2) clear information on how ERL is implemented ($\bar{x} = 2.34$), and that they (A7) are available to answer inquiries ($\bar{x} = 3.44$), though rarely (A12) the diversity of course materials ($\bar{x} = 2.36$) and the (A11) amount of materials has changed ($\bar{x} = 3.15$). With less consensus, but on average, students also believe that (A3) the classwork proposed by teachers is adequate ($\bar{x} = 2.79$) and that teachers have the necessary competencies to manage ERT ($\bar{x} = 2.91$).

Concerning the teacher-student relationship, students report that teachers have, on average, frequently (A6) created alternative mechanisms to keep in touch with students (such as chat rooms on social media) ($\bar{x} = 2.86$), that (A5) this contact is aligned with their needs ($\bar{x} = 2.80$). The most critical aspects in these subcategories, concerning the teacher-student relationship, rely on how frequently (A9) they feel close to their teachers ($\bar{x} = 2.70$ and $\sigma = 0.78$), though no one answered "never", and (A8) students feel that teachers worry about their well-being ($\bar{x} = 3.13$ and $\sigma = 0.73$).

Regarding assessment, on average, students indicate that frequently (A13) they have participated in some form of online assessment ($\bar{x} = 2.93$) and that (A14) the assessment tools are adequate ($\bar{x} = 2.66$). The great majority indicate that (A15) they were not taken the possibility to pursue continuous assessment ($\bar{x} = 1.48$) and they (A16) did not request their assessment mode to be changed ($\bar{x} = 1.32$). The critical aspect regarding assessment relies on the fact that, frequently, teachers (A17) altered the assessment components in such a way that it is not most beneficial to students ($\bar{x}$ = 2.48).

Regarding the institutional support, students believe that the HEI has frequently (A20) provided clear information on the implementation of ERL ($\bar{x}$ = 2.95) and that the (A21) HIE is compromised with their academic success ($\bar{x}$ = 2.91), which lead them to (A21) increase the usage of institutional digital channels, such as their school email inbox ($\bar{x}$ = 3.04).

Among the remaining items, as the most positive, students report that, frequently, on average, (A19) it is easy to work with their classmates ($\bar{x}$ = 2.60), (A27) the online classes are interactive allowing to actively participate ($\bar{x}$ = 2.96), and they have (A28) all the competencies necessary for participating in ERL ($\bar{x}$ = 3.27). Most of the students are rarely (A29) preoccupied with their privacy in online learning ($\bar{x}$ = 2.08).

The most critical aspects regarding the educational dimension rely on students feeling that ERL is (A25) more exhausting that traditional classes ($\bar{x}$ = 2.79), that they (A24) do not learn as much as they did in traditional classes ($\bar{x}$ = 2.21), and that (A26) the workload is heavier ($\bar{x}$ = 2.08). As a consequence, students feel (A23) less motivated for ERL ($\bar{x}$ = 2.42), less (A30) optimistic regarding their academic success ($\bar{x}$ = 2.35), and (A33) less likely to be interested in enrolling in any form of online learning in the future ($\bar{x}$ = 2.23). Despite these, students have frequently been (A31) trying to maintain their work/study routine ($\bar{x}$ = 2.29) and believe that (A32) the bare minimum quality of their education is preserved ($\bar{x}$ = 2.74).

Working Conditions. Most of the students report (B34) having all the necessary equipment ($\bar{x}$ = 3.63) and (B37) a proper workspace for participating in online classes ($\bar{x}$ = 3.19). Also, they rarely (B35) need to share their equipment with others ($\bar{x}$ = 1,60) and, although as not unanimously, they (B36) do not need to share their workspace ($\bar{x}$ = 2.04) nor (B38) are disturbed by other people living in the house ($\bar{x}$ = 2.18).

Psychological and Physical Wellbeing. This is the dimension in which the overall evaluation is significantly lower and troubling. Most students report (C44) increased sleeping disorders ($\bar{x}$ = 2.87, with *Mo* = 4), (C40) lower motivation ($\bar{x}$ = 3.13), (A41) increased anxiety ($\bar{x}$ = 3.05), (C39) increased exhaustion ($\bar{x}$ = 3.04), (C43) feeling more nervous ($\bar{x}$ = 2.95), and (C42) feeling sadder than usual ($\bar{x}$ = 2.84). Adding to this, most students have not had (C46) more time for hobbies ($\bar{x}$ = 2.66; σ = 0.87) though some have been practicing exercise ($\bar{x}$ = 2.69). Students have not been receiving any kind of psychological support from their schools, such as support emails or online psychology appointments ($\bar{x}$ = 1.42). On a positive note, students do not feel that (C49) when their schools closed they lost access to a safe space ($\bar{x}$ = 2.11) and (C48) thinking about returning to regular classes rarely makes them anxious ($\bar{x}$ = 2.13).

Social Interactions. It is almost unanimous that students always (D50) miss their school friends/colleagues ($\bar{x}$ = 3.45; *Mo* = 4.00) Most of them frequently (D52) contact with friends/colleagues ($\bar{x}$ = 3.33), as some believe that (D51) their well-being depends on it ($\bar{x}$ = 2.81). Students feel that the suspension of traditional classes has rarely (D52) caused them any relationship difficulties with their school friends/colleagues ($\bar{x}$ = 1.98).

Family Context. For most students, their (E54) family context is frequently beneficial for their participation in ERL ($\bar{x}$ = 3.02 and A59 with $\bar{x}$ = 1.38). Answers indicate that (E55) their families respect their personal space and work/study hours ($\bar{x}$ = 3.40; *Mo* = 4), they (E56) feel valued in their family circle ($\bar{x}$ = 3,29), and they feel that they (E58) can talk about their problems and concerns ($\bar{x}$ = 3.06). The COVID-19 pandemic has rarely (E57) required students to become informal caregivers of family members ($\bar{x}$ = 1.70; *Mo* = 1).

Financial Situation. On average, students (F60) feel frequently worried about their financial situation ($\bar{x}$ = 2.81; Mo = 3). On a positive note, students (F61) did not have to buy equipment/devices specifically for ERL ($\bar{x}$ = 1.28) and there are not such impediments to ERL ($\bar{x}$ = 1.24). Even though students indicate that during the confinement period they were able to (F64) save money in transportation ($\bar{x}$ = 3.70), they were rarely able to save money in (F65) food ($\bar{x}$ = 2.73) and (F63) school supplies ($\bar{x}$ = 2.04) and were never able to save money in (F66) rent ($\bar{x}$ = 1.42).

5 Conclusion

Taking into consideration the unforeseen and deeply impactful character of the situation caused by the COVID-19 pandemic, results show an overall better scenario than expected, as there are dimensions in which the negative impacts are fewer (not so notable). In the social dimension, despite missing face-to-face interaction, interactions appear to keep flowing at distance, which is a characteristic of Gen Y and Z. The family context also appears to be beneficial to student's wellbeing and their ERL activities, as most students have the necessary conditions for ERL. Moreover, the financial situation does not seem problematic, although, after returning home students kept paying for accommodation near schools. In the educational context, there are two main aspects. The provided structures, resources, and tools for ERL appear sufficient. The administrative side of ERL seems to be working well, with clear indications, instructions, and guidelines, provided by teachers and schools.

The negative impacts are mostly felt at two levels: the management of the classwork and the social-educational interactions (educational), and the psychological well-being. Although provided with all the necessary educational resources students lack the social interaction with teachers and colleagues and refer to low personal involvement from teachers. Also, ERL is notably more exhausting and not as fulfilling, as students report on a much heavier workload and on perceiving that they learn less. Consequently, students feel less optimistic regarding their academic success and less interested in pursuing online learning in the future, despite recognizing that the bare minimum quality of education is preserved. The most problematic dimension is the psychological one. A significant number of students revealed sleeping disorders, lower motivation, increased anxiety, incremented nervousness, sadness and, maybe due to the workload, no time for hobbies. Students state that schools have not granted access to any kind of psychological support or online appointments.

These results characterize the main difficulties of the first month of transition to ERL and, considering the main risks identified, are relevant for HEI and educators in designing organizational support and pedagogical interactions.

Acknowledgments. This work is financed by Portuguese national funds through FCT - Fundação para a Ciência e Tecnologia, under the project UIDB/05422/2020.

References

1. Hodges, C., et al.: The difference between emergency remote teaching and online learning. EDUCAUSE Review (2020). https://er.educause.edu/articles/2020/3/the-difference-between-emergency-remote-teachingand-online-learning
2. Craig, R.: What Students Are Doing Is Remote Learning, Not Online Learning. There's a Difference. - EdSurge News. 2020 2020-04-02. https://www.edsurge.com/news/2020-04-02-what-students-are-doing-is-remote-learning-not-online-learning-there-s-a-difference
3. Franchi, T., The Impact of the Covid-19 Pandemic on Current Anatomy Education and Future Careers: A Student's Perspective. Anatomical Sciences Education, 2020.
4. McLachlan, J.C., et al.: Teaching anatomy without cadavers. Med. Educ. **38**(4), 418–424 (2004)
5. Timmis, S., et al.: Rethinking assessment in a digital age: opportunities, challenges and risks. Br. Edu. Res. J. **42**(3), 454–476 (2016)
6. Kearns, L.R.: Student assessment in online learning: challenges and effective practices. J. Online Learn. Teach. **8**(3), 198 (2012)
7. Raaheim, A., et al.: Digital assessment–how does it challenge local practices and national law? a Norwegian case study. Eur. J. High. Educ. **9**(2), 219–231 (2019)
8. Watson, G.R., Sottile, J.: Cheating in the digital age: Do students cheat more in online courses? (2010)
9. Lee, K.: Coronavirus: universities are shifting classes online – but it's not as easy as it sounds (2020) https://theconversation.com/coronavirus-universities-are-shifting-classes-online-but-its-not-as-easy-as-it-sounds-133030
10. Sahu, P., Closure of Universities Due to Coronavirus Disease 2019 (COVID-19): Impact on Education and Mental Health of Students and Academic Staff. 2020, Cureus.
11. Dill, E., et al., As Coronavirus Spreads, the Decision to Move Classes Online Is the First Step. What Comes Next? 2020.
12. Strielkowski, W.: COVID-19 pandemic and the digital revolution in academia and higher education (2020)
13. Fedynich, L.V., Teaching beyond the classroom walls: The pros and cons of cyber learning. Journal of Instructional Pedagogies, 2013. 13.
14. Morrison, G.R., et al.: Designing effective instruction. John Wiley & Sons (2019)
15. Woodgate, R.L., et al.: The experience of the self in Canadian youth living with anxiety: a qualitative study. PLoS ONE **15**(1), e0228193–e0228193 (2020)
16. Cao, W., et al.: The psychological impact of the COVID-19 epidemic on college students in China. Psychiatry Res. **287**, 112934 (2020)
17. Cornine, A.: Reducing Nursing Student Anxiety in the Clinical Setting: An Integrative Review. Nursing education perspectives (2020)
18. Wang, C., et al.: A novel coronavirus outbreak of global health concern. Lancet **395**(10223), 470–473 (2020)

19. Al-Rabiaah, A., et al.: Middle East Respiratory Syndrome-Corona Virus (MERS-CoV) associated stress among medical students at a university teaching hospital in Saudi Arabia. J. Infect. Public Health **13**(5), 687–691 (2020)
20. Kafka, A.: Shock, Fear, and Fatalism: As Coronavirus Prompts Colleges to Close, Students Grapple With Uncertainty. 2020 2020–03–12. https://www.chronicle.com/article/Shock-FearFatalism-As/248240/
21. Kmietowicz, Z.: Rules on isolation rooms for suspected covid-19 cases in GP surgeries to be relaxed. British Medical Journal Publishing Group (2020)
22. Xiao, C.: A novel approach of consultation on 2019 novel coronavirus (COVID-19)-related psychological and mental problems: structured letter therapy. Psychiatry Invest. **17**(2), 175 (2020)
23. Ayittey, F.K., et al.: Economic impacts of Wuhan 2019-nCoV on China and the world. J. Med. Virol. **92**, 473–475 (2020)
24. Yin-xia, B., et al.: correlation between psychological changes of the community cro wd and the social support in grave public health event. Nei Moivgol Med. J., 4 (2005)
25. McBride, O., et al., Monitoring the psychological impact of the COVID-19 pandemic in the general population: an overview of the context, design and conduct of the COVID-19 Psychological Research Consortium (C19PRC) Study (2020)
26. Weale, S., Students 'should get a year's refund due to Covid-19 crisis' (2020)
27. Sinelnikov-Murylev, S.G.: Prospects for the higher eeducation system's development in the pandemic. monitoring of Russia's economic outlook. Moscow. IEP **2020**(6), 15–19 (2020)
28. Adams, D.R., Meyers, S.A., Beidas, R.S.: The relationship between financial strain, perceived stress, psychological symptoms, and academic and social integration in undergraduate students. J. Am. Coll. Health **64**(5), 362–370 (2016)
29. Peltz, J.S., et al.: The role of financial strain in college students' work hours, sleep, and mental health. J. Am. College Health **2020**, 1–8 (2020)
30. Gupta, S.K., et al.: Lean Six Sigma for reducing student dropouts in higher education – an exploratory study. Total Qual. Manage. Bus. Excellence **31**(1–2), 178–193 (2020)
31. Cronbach, L.J.: Coefficient alpha and the internal structure of tests. Psychometrika **16**(3), 297–334 (1951)

Cybertraining – A Development by Using a Holonic Control Structure

Dorin Isoc(✉), Amalia-Hajnal Isoc, and Teodora Surubaru

GRAUR (Group for Reform and University Alternative),
400626 Cluj-Napoca, Romania
Dorin.Isoc@graur.org

Abstract. School reform is process that develops from within, but under the pressure of external influences. Externally, social demands and the increasing pressure of technologies and research that find a field of application in education are acting. From within can act the desire for modernization, the accumulated competence. The article combines the two forms of pressure in the form of solutions already implemented in the CyberTrainer™service (www.cybertrainer.online), a professional service to assist learning processes. It is justified, for the beginning, the need to integrate information technology in the school concept. Subsequently, the set of requirements of the specification is built and the details of developing and implementation of the solution are provided. The chosen path is to treat the school as a second-order cyber system and to build a holonic control system. In this system, students and instructor are holons. The main features provided by the control system are the student's autonomy and the high degree of cooperation during the learning process. The effects of applying the solution indicate a sketch of the change of school idea: the teacher becomes an instructor, the student becomes more responsible and involved in his work, and the school activity more and more efficient and adaptable.

Keywords: Cybertraining · Second order cybernetics · Holonic system · Training group · Educational technology

1 Introduction

The COVID 19 pandemic has once again, clearly, shown that the use of e-learning is far from a real alternative of school. The main goal of e-learning services consists, in fact, in ensuring the preserving of the traditional role of the teacher through technical solutions that use computer applications and electronic communication resources. The pandemic also showed that the technical solutions covered by the use of teleconferencing computer applications mean both a depletion of resources and a decrease in the emphasis for student-centered learning.

The pandemic acted as an examination that confirmed what some researchers have reported earlier, more or less explicitly: in reality, at this time, learning is

M. E. Auer and T. Rüütmann (Eds.): ICL 2020, AISC 1328, pp. 315–326, 2021.
https://doi.org/10.1007/978-3-030-68198-2_29

not done by integrating the technique but by simply joining the technique to a traditional and outdated school.

In the described situation, there are, entirely, appreciations from the works of some researchers [6].

> ... technology is providing tools that provide radical new opportunities for education, but simply adding technology to the existing mix is not enough. We need to use technology to develop better pedagogies, and most importantly, to redesign educational organization at all levels, from the course to the national system, to allow potential benefits to be realized ...

The consequences are immediate: the performance of assisted learning methods, whatever they may be, is comparable to the personal performance of the teacher.

The following are excluded from the assisted learning applications: a) the real and personalized involvement of all the students of the study group; b) the way of generating the authentic problems, specifically for the students of the study group; c) an objective, systematic, personalized assessment of the students.

A synthesis of Murray [10] captures at least three essential aspects of current education: a) the static character of knowledge, b) the dynamic character of training; c) the importance of communication and training through communication.

> ... learning is constructed in communication in the relationships we build and the connections we make with our environment, which includes other living systems. In this view, knowledge is not formed by the senses taking information in, but as a whole body changing in dynamic reciprocal interaction in an environment. In fact, learning may be more accurately described as engendering knowing rather than some kind of static, stored knowledge. In this cybernetic view of the world it appears that learning happens to us as we communicate in an environment. It enables us to go on living ...

Thorough research give a good synthesis, in a salutary way, as requirements for the transition from conventional to student-oriented education [12]. There are some special ideas to remember, including the need to reconsider the balance of power, the need to place the responsibility for learning on the student and the need for an objective assessment. The shift of emphasis to this paper is to emphasize that there are strategies in parallel with point research [2,7].

Before approaching the reported research, we note a new suggestive passage from [6] that once again summarizes the artisanal way of doing school:

> ... this structure constrains the educational options available to learners, who have to be fitted into its structure. They typically need to choose from lists of available courses, and have to be assessed for their suitability to study these. This usually results in so-called ability groupings, and

> typically, teaching proceeds as if all students in a cohort were identical, separated only by final examination results. But people do not easily fit these categories. They have different personal histories, aptitudes, interests, desires and preferred ways of learning. Fortunately, the efforts of teachers and the willingness of learners to forgo their uniquely individual requirements allow the system to continue ...

Here are all the evidences that the school can be described better as *lack of school* and purely bureaucratic arguments are put forward for what an entire system is unable and unwilling to do. And again, the system one thanks the individual who does what the institution cannot: improvise.

This paper proposes the framework of an alternative of a new school. It is a framework already implemented and verified through the CyberTrainer service. The essential element of the framework is the action of framing and integrating a dedicated IT service in the school structure. Two elements are highlighted with priority: a) the fact that the purpose of the school, i.e. the learning of each student simultaneously with the learning of the group of students, must be functionally ensured; b) the fact that the school, as a social system, can be approached only through the means of second order cybernetics. The modeling of the school is done by introducing the holons associated with the students. Holons are the ones who ensure the learning activity at the individual level and at the level of the training group.

The holonic approach requires the construction of the virtual instructor. The virtual instructor is specified and his attributions in the matter of the organization of the learning activity and in the matter of ensuring the relations of autonomy and cooperation of the holons are retained. The organization of the learning activity has as its backbone the binomial built-problem - a solution of a built problem in which the student is directly involved and to which are added activities of analysis and direct assessment of problems and solutions. In this activity, the instructor is involved in the level of obligations of the students. Ensuring the relations of autonomy and cooperation of the holons is done by adjusting the position of the teacher who is functionally divided into virtual student, head instructor and authority. Subsequently, a complete reticular structure is built that includes the students of the training group together with the virtual student. The virtual instructor ensures the publication of all student contributions, their anonymization as well as the operations required by the analysis and evaluation process. The training session or lesson ends by communicating the results determined automatically regarding the activity of each student.

In the first chapter are given reference points from which it results that the current education contains a series of irreconcilable functional contradictions in the conditions in which the school avoids to integrate, in a real way, the information technologies. In the second chapter, a connection is made between the functioning of the school and the need to attach a goal in order to find cybernetic control solutions of the social institution. The third chapter details the group of students as an essential operative element that becomes the object of a holonic structure. In the fourth chapter we proceed to the specification of the holonic control structure and the first implementation details are given. A fifth chapter examines the effects of introducing the new vision and building the virtual instructor for school reform. The chapter of conclusions emphasizes the importance of introducing the concepts of second-order cybernetics and the holonic mechanisms for the redesign of the education system.

2 Functional Integration of Information Technology in School

A correct treatment of the conventional school in the face of the technical revolution requires a paradigm shifting. For this it is necessary to resume and update an image of the conventional school. As principle, it consists, generically, of a teacher, a group of students, an institution, a material base. If the teacher, the students and the institution are regarded as a social ensemble, together with its components, then the material basis is always an auxiliary reality without initiative.

The emergence of computer applications dedicated to school use has created a new segment of the material base. We will conventionally call this new segment "school technique" by its similarity, for example, to medical technique.

The evolution of the school technique followed the general evolution of the technique and became more and more, from a useful annex, a component part of the school.

2.1 A Motivation of Cybernetic Treatment of the School

In the evolution of technology, it is not the first time that machines have come to be assimilated, at least in part, as human partners.

By having a goal or direction, the new situation of the school, as institution and technology, makes it even more eligible for treatment through cybernetics-specific means [3].

The existence of a purpose, of a direction raises the problem of keeping them. The action of keeping a certain purpose in the face of the action of disturbances corresponds to a controlling or monitoring activity [13]. Wiener introduces the concept of system and its description through mathematical variables. In this way, the reduction of the effect of disturbances on the purpose of the system is done by introducing the feedback as a negative reaction. The feedback can be summarized as the action of reducing the effect of the disturbance when it causes a departure from the intended purpose. The feedback involves both the mechanism for determining the effect of the disturbance and the mechanism for correcting the evolution of the observed system. The technical mechanism solving these tasks is called controller.

A brief analysis shows that we are not in the specific situation of first-order cybernetics, specific to the observed systems, even if the presence of control to achieve the goal is also a necessity here. If we were in such a situation, the control problem could be solved by algorithmic design and by using mathematical models associated with the observed systems. These models are usually dynamic models described by the methods of differential equations. The main quality sought is that of dynamic stability, i.e. keeping the purpose of the system in time in the face of all disturbances that may occur.

In this situation, however, we are in a social system, made up generically of a man and a machine or exclusively of people. In this ensemble, the man is a part but also an agent who knows his role as a part and who thinks about the status, respectively about his model in the system. The new system corresponds to the situation that can be treated only through second-order cybernetics in which man, agent and observer try to build a model of another cybernetic system.

In this new context, the individual's initiative, his interests and feelings, his ability to control himself no longer allow the construction of a simple and comfortable mathematical model.

It is much more convenient to work with entities for which it is only noted that they have the ability to have a certain behavior and that they can act based on a set of rules.

2.2 A Holonic Model of School

The current school can be functionally described, in the new context, as a set of holons, an informational structure that represents a more adequate description of the social reality [8], including at the level of this institution. We check here Koestler's definition [4] according to which by holon is meant any entity that is socially or biologically stable and displays a behavior governed by rules.

In this way, the holon becomes the constitutive functional element of all activities in which control can manifest itself and which refers to behaviors in which the human being is involved. Each holon is seen both as an individual and as part of a whole of which he is a part. It should be noted that each holon belongs simultaneously to two different levels so that, in one level, he is a member, and for a lower level he has the position of chief.

In a synthesis of Mella [9] it is stated that by generalization of the Koestler definitions:

> ... the term *holon* indicates any object or concept observable on three levels: (1) as an autonomous and independent unit that acts according to its own behavioral *canons*; (2) as a superordinate unit, possessing emerging properties, with respect to the component parts that it transcends; and (3) as a subordinate unit in that it is part of a vaster whole that conditions it ...

In this way, the holon becomes the basic element in the materialization of second-order cybernetics.

In the case of the school, both the education system and the students and the instructor are found as holons. Together with their mode of association, holons make up what is called a holarchic structure or holarchy. In the holarchy, holons are considered along with all their control attributes. In this way, control becomes an attribute of a social reality. Holarchy is, by the specificity of its elements, that is, holons, a structure devoid of conventional hierarchy, with ascending control paths and descending control paths. Such a control structure will also be called heterarchy to distinguish it from the well known path of hierarchy.

The reason for this analysis is to find a way to integrate the machine and the social environment. We must not forget that along with people and the social system, like holons, there is the school technique.

Compliance with the objectives of the school as a second-order cybernetics system can be ensured only through a control system. Due to the special nature of the description made by means of holons, any action that refers to the organization of holons in order to ensure a purpose becomes an alternative of holonic control.

3 Training Group as a Holonic System

As a holonic system, the training group must be open, modular and standardized. The structure of the group is one open by the common nature of the participants in the form of training. In this situation, by the nature of individual interest and common knowledge, students prove to be interchangeable. The modular nature is intrinsically ensured by the identity of the people involved. Finally, standardization consists mainly in providing an interface, through which interchangeable holons must be harmonized so that they can be integrated through communication possibilities in the considered framework [1].

For the general case of a school based on learning groups, each student can be regarded as a holon.

Holons all have the same ability to ensure the success of training activity. This capacity is a form of manifestation of interest in acquiring competence, as a group, but also as an individual, regardless of the form of its manifestation. For a good understanding it will be admitted that, mainly, the competence will be found by ensuring the good understanding of a given lesson or a good acquisition of the required skills.

The integration of the holon activity is done through an integration and control holon and this is the instructor. Its role is regarded through the quality and performance requirements imposed on training and not through hierarchical attributes.

As a holon, each student represents an autonomous entity that is set to cooperate with other similar entities to achieve the overall goal. Each holon has its own goals that sometimes, conflict with the goals of other related holons. These conflicts must be negotiated. The effectiveness of negotiation will have a significant effect on the efficiency of the whole system [1].

A closer analysis of the conflicts that occur in the activity of holon students highlights that they are specific to any human group activity. Moreover, here the conflicts are oriented towards those aspects that lead to competition, thus to the increase of the efficiency of the group's activity. Beneficial conflicts between members of a holarchy occur, either due to natural causes or are stimulated by the activity of the holon-instructor.

A natural conflict occurs when there are differences in individual traits such as pride, desire to win, desire for superiority in action.

A stimulated conflict can be the correct classification of objectives in such a way that their approach leads to different awards. It is enough for a problem to solve to be declared a priority and the desire to win makes the students of the group enter a direct competition.

Another situation of conflict stimulated by the activity of the holon-instructor is the scattering of the value of the awards in a sufficiently wide range, simultaneously with the limitation of the number of prizes. In this case, it is found that holon-students end up building real strategies to approach the activity towards alternatives with high chances of success.

On the contrary, a moderation of conflict states is obtained when risky situations are defined. The simple elimination of the name, the anonymization, which should be associated with the results of the work of the holons and their exposure to the interested competent vote of the members of the holarchy leads to an increase of the degree of responsibility. At the same time, there is an increase in the objectivity and quality of training activities.

4 School – Structure of Holonic Control System

In the first phase, the technical conditions for performing the holonic control will be analyzed. In this control structure, stability and accuracy are no longer priorities. Instead, it is necessary that the performance of the control system be achieved by ensuring the autonomy of the holons, i.e. their ability to act autonomously in the face of unpredictable circumstances and the ability to cooperate with other holons [5].

Obviously, the holonic control structure aims at associating holons in holarchies that contradict traditional hierarchies and appears to be without defined ascending or descending control pathways.

Through the interventions imposed by the design of the holonic control, the observance of the characteristics of the holonic systems derived from the observance of the behavior of the systems and human societies will be followed. A human holon simultaneously exhibits two basic, rather opposite, behaviors: a self-affirmative tendency and an integrative tendency. The self-affirmative tendency is the dynamic expression of the holon as a whole, while the integrative tendency is the expression of its part condition [11].

The achievement of the holonic control system supposes the building of a virtual instructor, technical component realized as computer service. Its role is that, during a lesson, to bring together students-holons with the instructor-holon in such a way that the performances of autonomy and cooperation are achieved in conditions of efficiency of the training process.

The specification of the virtual instructor is based on the requirements:

(a) A structure of the holon set is created without any conventional hierarchy. For this, we opt for a heterarchy based on the regular reticular structure, i.e. a structure that includes sides between each two participating holons.
(b) Any detail by which he can intervene through actions of excessive authority is removed from the *functionality* of the instructor. Functionally, this is achieved by separating the possible actions of the holon instructor into three categories: a) actions specific to the interaction with students and included in a character or actor with the role of virtual student; b) specific actions of monitor and coach character included in a character or actor with the role of head instructor; c) actions specific to the investiture of a representative of the authority for decisions regarding the confirmation of the impact analyzes of the students, actions associated to a character or actor with the role of authority.
(c) Learning activity is limited to student actions and interactions between students completed through contributions i.e. built problems, solutions to built problems and analysis reports on peer contributions.
(d) The entire control of the quality of the contributions is transferred to the students' vote, which manifests itself in front of some public contributions, problems or solutions of problems, without revealing the names of the authors.
(e) A unitary assessment module of the students' results is built based on the actually submitted activity and on the quality of this activity appreciated in the group.

Through holonic control, developed and implemented inside of CyberTraner service, an unnecessary hierarchy was removed from the school structure and in this way all control decisions are made at the local level. In the new situation, without the possibility of a global optimization of the school's functioning, an increased flexibility of the learning system is reached. It is also noted that, in the new framework, all school technology systems are integrated with the reason of the school as a social system.

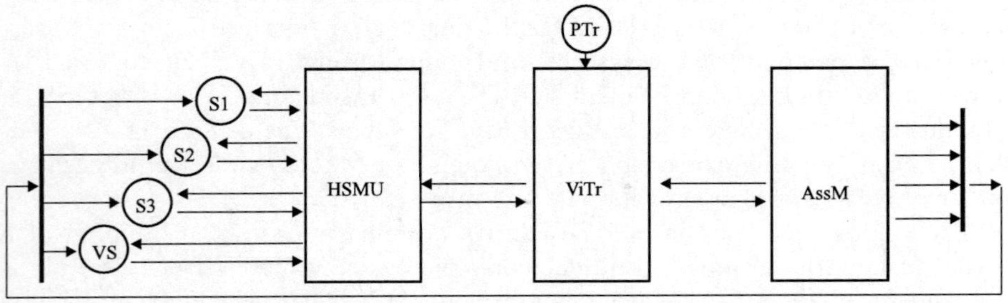

Fig. 1. A sketch of the holonic control of the learning activity.

In Fig. 1, a group of three students, $S1$, $S2$, and $S3$, respectively, form a holonic structure under the action of an $HSMU$ coordination unit. The holonic control structure is under the influence of the virtual instructor $ViTr$ and the results of the activity from the $AssM$ assessment unit are evaluated and communicated later to each student.

A peculiarity of the proposed solution is the exploitation of the heterarchy characteristic of the holarchic structure [3]. For this, another solution was exploited that meets Weimer's recommendation [12] on changing the balance of power in student-centered systems. Specifically, the person of the instructor was decomposed into three characters with different roles that intervene in the learning process in different places. The three hypostases of the instructor are: as a virtual student (VS), as a head instructor ($H.Tr$) and as a representative of the authority or instructor-decision maker ($A.Tr$). In this way, Fig. 1 is completed by including the virtual student in the holistic learning structure. Specific to the holonic structure is that it is by way of construction a heterarchy, i.e. in the learning process there is no pre-established hierarchy.

The schematic diagram in Fig. 1 is the basis for the operation of the CyberTrainer service implemented and installed at www.cybertrainer.online.

Figure 2 a) describes the idea of separating the three actors to take over the functional responsibilities of the conventional teacher. Along with the separation, there is a resettlement of these roles/actors in the reticular network associated with the holonic control of the learning system.

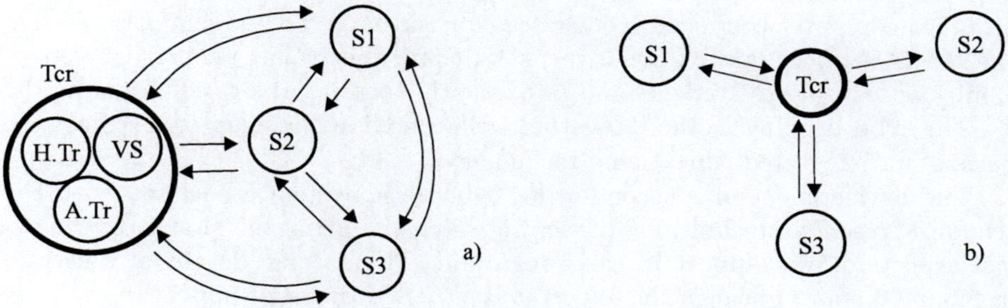

Fig. 2. Reconsideration of the instructor role by separating the teacher's attributions: a) separating the instructor's roles in the Cybertrainer service and repositioning him in the group of students; b) the vision of the conventional school where functionally, the teacher is located in the center of the activity.

It is assumed that the training group is composed of students $S1$, $S2$, respectively $S3$ who cooperate with a trainer Tr. Following the functional separation, the teacher Tcr becomes a group of actors: VS - virtual student; $H.Tr$ - the main instructor and $A.Tr$ the representative of the authority. Unlike the case of the conventional school as in Fig. 2 b) following the change, the teacher Tcr functionally changes his central place in the relationship with the students. Students $S1$, $S2$, $S3$ and the virtual student VS form a complete reticular structure, where a predefined hierarchy is lacking.

5 Outcomes and Interpretations

By implementing the holonic control system, the new school, the cybernetic school is composed of students, virtual students, instructors and a virtual instructor service. The virtual instructor is the technical component meant to implement the rules that ensure holonic control for students considered holons in a specific holarchy. Holistic control is built to allow through uniform and systematic rules the autonomous action of students and their permanent cooperation.

The same virtual instructor service is the one that manages the duration of the training session or lesson. This duration is divided into four active stages designed to build problems, solve built problems and analyze and assess the quality of problems, respectively of solutions. All student contributions are public and available to all participants in the session. The authors of the contributions are permanently anonymous to the participants in the session. The analysis of the contributions is finalized by the quantitative vote of the students.

The instructor can intervene, in the training activity, only from a position equal to that of a student, as a virtual student. Thus he can indefinitely introduce problems, contradictory or correct solutions, without his name being known among the other participants to learning session. Technically, in this way the effect of heterarchy specific to the holarchic structure is obtained.

Cybertraining is compatible with all other forms of training and learning currently practiced. Cybertraining and the virtual instructor service recognize all other computer products and services for learning. The advantages of using the virtual service appear when it turns out that cybertraining shifts the responsibility of learning to students and to explicitly covering their individual needs.

The cybertraining method, together with the virtual instructor, becomes the basis of an integrated educational technology.

The development of a second-order cybernetic system associated with the school represents a technical approach in which the pedagogical and psychological aspects do not appear in the foreground. The specific details of a certain discipline do not appear in the foreground either. It does not appear in the foreground, nor does the working language matter. In this way, one can speak of the fact that the cybernetic school is an "abstract machine" intended for the "learning industry".

But what appears permanently in the cybernetic school is first of all the individual, it is the student and then the training group. The personal interest of each student through the way of reporting to the subject to be learned and the cross-interaction in the training group are the driving elements of the cybernetic school.

The instructor's effort is reduced to the achievement of the best selection of knowledge that can be associated and that can guide the acquisition of an imposed theme. This selection of knowledge, called support of the training session, is only a necessary reason to initiate the "search movement" of the target of learning, understanding, learning skills, but also innovation, design.

Although no form of appeal to the instructor is excluded, such an approach is no longer considered the school's endurance piece, just as teaching is no longer infallible, which is considered a completely obsolete and unnecessary form of didactical activity. One highlights the individual activity, driven by their own needs, influenced and peer supervised.

6 Concluding Remarks

Cybertraining, together with the virtual instructor, determines a new educational technology. The operation of the ensemble respects the principles of second-order cybernetics in which the parties involved have, at the same time, the role of observed part and observer part. Modeling students as holons and the training group as a holarchy allows the construction of the appropriate controller that can address individuals and the learning process as a goal pursued and ensured. The purpose of learning is thus pursued flexibly and specifically for each individual and the group as a whole.

The cybernetic school, i.e. the school that operates through the extended application of the educational technology of cybernetic training, possesses a significant synergistic resource resulting from the distributed character of the holonic control. To this is added the active and always fresh action between the initiative of one's own interest and the pressure of the integrated interest of the group.

References

1. Datta, V., Hodgson, R., Polikow, I., Saxena, R.: Computer controlled system for other publications modularizing the information technology structure of a holonic manufacturing scheduling: architecture, cooperation business enterprise into self-contained modules. US Patent 7,281,235 (2007). Filed: 9 January 2002, Issued: 9 October 2007
2. Derntl, M., Motschnig-Pitrik, R.: The role of structure, patterns, and people in blended learning. Internet High. Educ. **8**(2), 111–130 (2005)
3. Heylighen, F., Joslyn, C.: Cybernetics and second-order cybernetics. In: Encyclopedia of Physical Science & Technology, vol. 4, pp. 155–170 (2001)
4. Koestler, A.: Beyond atomism and holism–the concept of the holon. Perspect. Biol. Med. **13**(2), 131–154 (1970)
5. Leitão, P., Colombo, A., Restivo, F.: A formal specification approach for holonic control systems: the ADACOR case. IJMTM **8**(1/2/3), 37–57 (2006)
6. Liber, O.: Cybernetics, eLearning and the education system. Int. J. Learn. Technol. **1**(1), 127–140 (2005)
7. Magno, C.: Assessing students critical thinking and approaches to learning. Int. J. Educ. Psychol. Assess. **12**(2), 19–32 (2006)
8. Martinez, A.: Complexity and the universe of education. Technical report (2008)
9. Mella, P.: The Holonic Revolution: Holons, Holarchies and Holonic Networks: The Ghost in the Production Machine. Pavia University Press (2009)
10. Murray, J.: Cybernetic circularity in teaching and learning. Int. J. Teach. Learn. High. Educ. **18**(3), 215–221 (2006)

11. Van Brussel, H.: Holonic manufacturing systems the vision matching the problem. In: Proceedings of the 1st European Conference on Holonic Manufacturing Systems, vol. 1. IFW, Hannover (1994)
12. Weimer, M.: Learner-Centered Teaching: Five Key Changes to Practice. Wiley, Hoboken (2002)
13. Wiener, N.: Cybernetics or Control and Communication in the Animal and the Machine. MIT Press (1948)

E-inclusion Prediction Modelling in Blended Learning Courses

Ieva Vitolina(✉) and Atis Kapenieks

Riga Technical University, 1 Kalku Street, Riga 1658, Latvia
ieva.vitolina@rtu.lv

Abstract. This study aims to address the e-inclusion problem related to digital skills improvement and meaningful use. In professional work, more and more jobs require the use of digital skills. Combining e-learning and face-to-face training is a convenient and affordable way to learn new digital skills. However, the problem is the low number of e-learning graduates and the even lower number of those who use the newly acquired skills for professional or personal purposes. Machine learning approach is used to predict student achievement and other events. However, no comprehensive study has been conducted, analyzing how to improve digital skills training by ensuring that newly acquired skills are meaningfully used in professional life.

This paper explores how to predict students' learning impact outcomes - the use of newly acquired digital skills. We are using classification algorithms and machine learning approaches. We have compared five different classification algorithms and selected the ones with the best performance: LMT and lazy. LWL. We concluded that at least 81.60% of learners identified as *not e-included* are correctly predicted with lazy.LWL.

Keywords: E-inclusion · Digital skills · Machine learning

1 Introduction

This study aims to address the e-inclusion problem related to digital skills improvement and meaningful use. Improvement of digital competences is one of the priorities in the European Union Digital strategy [1]. According to the study of the McKinsey Global Institute (2018), demand for technological skills will increase by 55% by 2030 [2]. Failure to respond to this challenge could lead to digital exclusion and widening of the digital divide in society. E-inclusion aims at reducing gaps in information and communication technology (ICT) usage and promoting the use of ICT to provide as many individuals as possible access to the benefits of ICT.

Combining e-learning and face-to-face training is a convenient and affordable way to learn new digital skills. However, the problem is the low number of e-learning graduates and the even lower number of those who use the newly acquired skills for professional or personal purposes. According to a CEB Global (2014) study, for the average organization, 45% of learning investments are scrap learning - learning that is delivered but not applied back on the job [3].

M. E. Auer and T. Rüütmann (Eds.): ICL 2020, AISC 1328, pp. 327–337, 2021.
https://doi.org/10.1007/978-3-030-68198-2_30

There is a number research on student dropout reduction in on-line courses [4–6]. Machine learning approach is used to predict student achievement and other events [7]. Predictive models for blended courses are still limited [8]. Conijn et al. (2017) analyzed 17 blended courses and concluded that the portability of the prediction models across courses is low [9].

Research on the application of newly acquired skills focuses more on the process of cooperation with the employee's workplace than on the employee's training process [10]. No comprehensive study has been conducted, analyzing how to improve digital skills training by ensuring that newly acquired skills are meaningfully used in professional life. There is a need to promote individuals' e-inclusion and reduce scrap learning which has a negative impact on corporate finances. There is no special technology for e-inclusion prediction.

This paper explores how to predict students' learning impact outcomes - the usage of newly acquired digital skills. We are using classification algorithms and machine learning approaches. The study uses data on teachers training performance in three blended learning courses on the application of technology in school education. The courses were on (1) Mobile Technologies; (2) Video Technology and Design; (3) Robotics. Each learner could choose one of the courses. This study contributes to research of meaningful ICT use in a blended learning environment. The goal of the study is to create a machine learning based model for learning impact and e-inclusion prediction for blended courses.

The paper is organized as follows. Section 2 includes a description of the methodology used for prediction. Section 3 introduces results and discusses prediction algorithms. Section 4 contains the conclusions part.

2 Methodology

We did our research based on the following steps: Data acquisition; Data preprocessing; Attribute selection based on knowledge management theory and previous research; Application of the selected algorithms to create a prediction model; Model evaluation and analysis of results.

2.1 Data Acquisition

Data were obtained from our programme "Modern Education of Interests" on the e-learning platform Moodle. Our programme has three versions/courses: Mobile Technologies, Robotics and Video Technology and Design. Additionally, data were obtained from questionnaires filled in by teachers throughout the course of six to seven months after completing the course. Learners were asked about usage of the skills acquired in the courses for their professional needs.

2.2 Data Pre-processing

We cleaned the data by finding missing values and constructed new attributes by calculating the average values.

We created two types of data sets for e-inclusion prediction. The first data set contained all students. We prepared data sets for each course with the following number of learners: Video Technology and Design course - 82 learners' records; Mobile Technologies course - 58 learners' records; Robotics course - 84 learners' records. This data set includes learners with no availability of ICT technology (digitally technologically excluded). We assume that the learners are digitally technologically excluded if they do not have the technology to use their newly acquired digital skills. For example, if a learner does not have a video camera to use the newly acquired skills in a Video Technology and Design course after completing this course then this learner has a technological obstacle. Therefore, we created another data set type that included only students who had access to technical equipment to use the newly acquired skills. We used both of these data sets for prediction for each algorithm within each course (Video Technology and Design, Mobile Technologies, Robotics). The least available equipment was for the use of robotics skills, only 46.43% of the students had access to robots (Arduino or LEGO MINDSTORMS). The equipment needed for video technology was the most available - 92.68% of the students. 87.93% of the students had access to mobile technology equipment.

2.3 Attribute Selection

Selection of the relevant attributes that affect the prediction result was done based on the Nissen (2005) knowledge management theory and previous research of factors predicting the degree of e-inclusion [11–13]. We used seven numeric attributes for the prediction of e-inclusion: learner's interest in learning, learner's general digital skill level, learner's ability to learn, learner's evaluation of e-learning materials, instructor's willingness to share knowledge, learner's assessment of e-learning environment, learner's predicted use of newly learned skills. Moreover, the selected attributes correspond to factors causing scrap learning mentioned by Pike (2018) [10].

Also, we have an attribute *observed usage of newly learned skills*. This attribute has 2 values: *e-included* and *not e-included*. We defined that the value is *e-included* if we observed that the learner uses newly learned skills. The value is *not e-included*, if we observed that the learner doesn't use newly acquired skills. We labeled each learner's record in the data set as *e-included* or not *e-included*.

2.4 Application of the Selected Algorithms to Create a Prediction Model

We trained 5 classifiers: NaiveBayes, SimpleLogistic, lazy.LWL, OneR, and LMT using the Waikato Environment for Knowledge Analysis (WEKA) platform [14]. We chose these 5 classifiers since they are techniques which use different approaches to solve a classification problem. According to the WEKA platform, we categorized these techniques in 5 categories (Table 1): Bayes, Functions, Lazy-learning Algorithms, Tree-based Learning Algorithms, Rules-based Learning algorithms.

Table 1. Algorithms used to predict e-inclusion.

Category	Algorithm	Description
Bayes	Naive-Bayes	Class for a Naive Bayes classifier using estimator classes. Numeric estimator precision values are chosen based on analysis of the training data [15]
Functions	Simple-Logistic	Classifier for building linear logistic regression models [16]
Lazy-learning	lazy.LWL	Locally weighted learning uses an instance-based algorithm to assign instance weights which are then used by a specified Weighted Instances Handler [17, 18]. LWL can do classification (e.g. using naive Bayes) or regression (e.g. using linear regression)
Tree-based Learning	LMT	Classifier for building 'logistic model trees', which are classification trees with logistic regression functions at the leaves [16]
Rules-based Learning	OneR	One Rule is a simple method based on a 1-level decision tree described in 1993 by Rob Holte [19]

2.5 Evaluation Metrics

There is no consensus on which classifier performance metric is better to use [20]. In a binary classification the performance of the classifiers is assessed using the standard measures of Recall (1), Precision (2), Accuracy (3), Balanced Accuracy (4) where:

$$\text{Precision} = \frac{\text{tp}}{\text{tp} + \text{fp}} \tag{1}$$

$$\text{Recall} = \frac{\text{tp}}{\text{tp} + \text{fn}} \tag{2}$$

$$\text{Accuracy} = \frac{\text{tp} + \text{tn}}{\text{tp} + \text{tn} + \text{fp} + \text{fn}} \tag{3}$$

$$\text{Balanced Accuracy} = \left(\frac{\text{tp}}{\text{tp} + \text{fn}} + \frac{\text{tn}}{\text{tn} + \text{fp}}\right)/2 \tag{4}$$

tp is true positive (*not e-included*), *tp* is the amount of positive (*not e-included*) data that is predicted right. *tn* is true negative (*e-included*), *tn* is the amount of negative (*e-included*) data that is predicted right. As our goal is to identify students who are at risk of not being e-included, we consider *not e-included* as the positive class and *e-included* as the negative class. *fp* (false positive) is the amount of positive data that is predicted wrong. *fn* (false negative) is the amount of negative data that is predicted wrong.

Precision indicates what percentage of the predictions on the positive class *(not e-included)* is correct. Recall shows the proportion of the positive class (*not e-included*) instances that are correctly classified. Precision and recall in this study are calculated with respect to the positive class, which is the rare class (*not e-included*). Accuracy is defined as the ratio of correctly classified instances by the total instances. Balanced accuracy is especially useful when the classes are imbalanced [21]. Balanced accuracy is defined as the average of the ratio of correctly classified instances of each class individually.

2.6 Evaluation Strategy

For evaluation of the e-inclusion prediction model we used 10-fold cross-validation that is a widely accepted model evaluation approach in machine learning [22]. The main principle in cross-validation is model training with some instances of the data set and model testing with some other instances after model has been trained.

2.7 Data Balancing

The classification algorithm becomes less efficient if the distribution of data in the classes is imbalanced. The distribution of *e-included* and *not e-included* learners was imbalanced for all courses (Table 2).

Table 2. The distribution of *e-included* and *not e-included* learners.

Class/course	All students			Students with access to technology		
	Video	Mobile	Robotics	Video	Mobile	Robotics
e-included	71.95%	65.52%	33.33%	77.63%	74.51%	71.79%
not e-included	28.05%	34.48%	66.67%	22.37%	25.49%	28.21%

Synthetic Minority Over-Sampling Technique (SMOTE) is specifically designed and widely used for learning from imbalanced datasets to generate synthetic minority class examples [23]. In our research we used the WEKA platform and SMOTE to achieve oversampling of the minority class. As a result, we obtained 6 balanced data sets for three courses (Video Technology and Design, Mobile Technologies and Robotics) for all students and for students with access to technology.

3 Results and Discussion

In this section we present the results on e-inclusion prediction by 5 classification algorithms. For each algorithm we used an original and balanced data set to improve evaluation metrics as precision and recall. We also present the prediction results for two different data sets: (1) the data set for all students; (2) the data set of students who have access to technology. The results of e-inclusion prediction are presented for each of the 3 courses: Video Technology and Design Course, Mobile Technologies and Robotics.

3.1 Results in the Video Technology and Design Course

Figure 1 summarizes the evaluation of the 5 models trained to predict student e-inclusion after completing the Video Technology and Design Course. The LMT algorithm has the highest score 0.802 for average of all metrics in case of balanced data of students with access to technologies. The accuracy is 0.797, precision is 0.778, recall is 0.831. LMT has the highest accuracy value compared to other algorithms. Precision indicates that 77.80% of the *not e-included* learners predictions on the positive class *(not e-included)* are correct. Recall shows that 83.10% of the *not e-included* learners are correctly classified. According to the predicted accuracy 79.70% instances of the total number of instances are correctly classified.

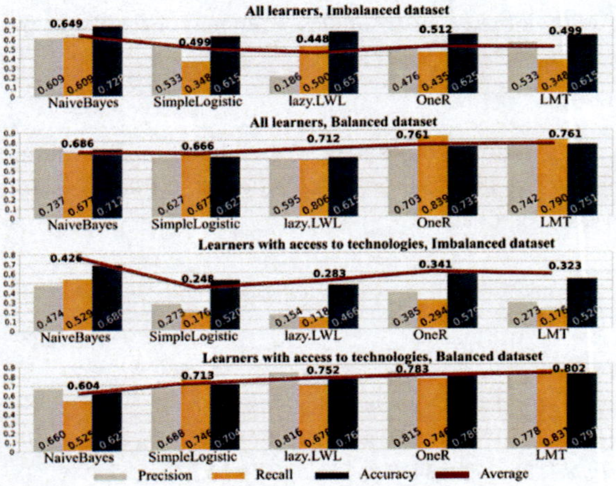

Fig. 1. Video technology and design course. Results: metrics of 5 classification algorithms.

The second highest rating is 78.30% for the OneR algorithm for balanced data of students with access to technologies. But the third highest rating 76.10% is for the LMT and OneR algorithms for balanced data of all students. The OneR algorithm has the highest recall rate 0.839. The lazy.LWL algorithm has the highest precision rate 0.816 for balanced data set in case of learners that have access to technologies.

Imbalanced data sets have the lowest scores, for example, with SimpleLogistic the score is only 24.82% for average of metrics in case of imbalanced data for students with access to technologies. Algorithms perform better in case of balanced data sets for students with access to technologies.

The lazy.LWL algorithm has the highest rate of improvement after balancing the data sets. The average of precision, recall and accuracy increases by 0.305. The NaiveBayes algorithm is an exception. After data balancing and exclusion of students without access to technology, this algorithm has lower performance results, the average of precision, recall and accuracy decreases by 0.045.

3.2 Results in Mobile Technologies Course

Figure 2 presents accuracy, precision and recall measures of 5 algorithms in the Mobile Technologies Course.

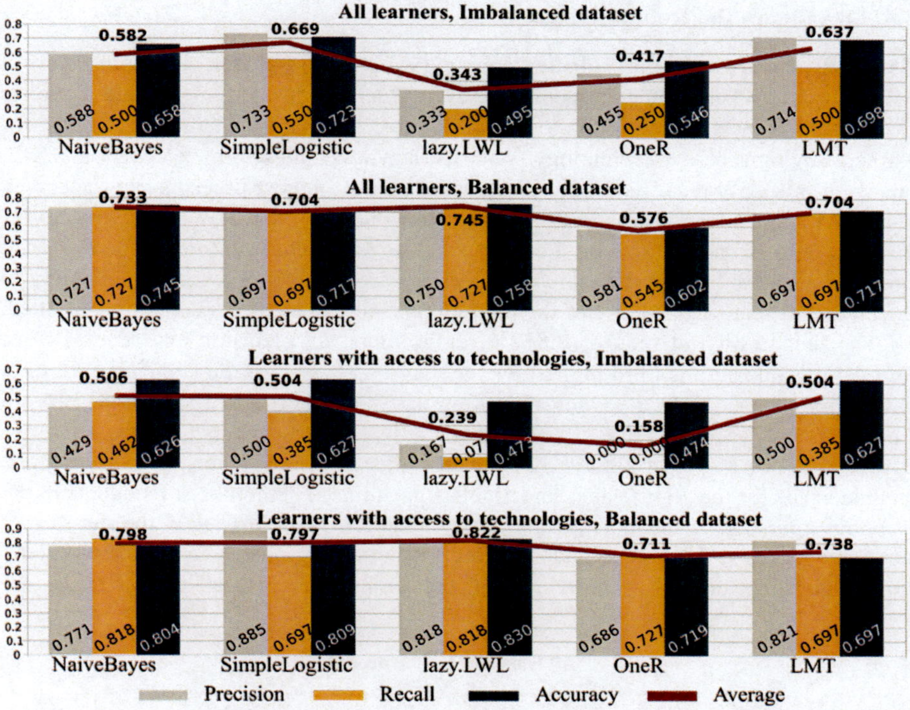

Fig. 2. Mobile technologies course. Results: metrics of 5 classification algorithms.

lazy.LWL has the highest average score of 0.822 for Mobile Technologies Course, accuracy is 0.830, precision is 0.818, recall is 0.818. Precision indicates that 81.80% of the *not e-included* learners predictions on the positive class *(not e-included)* is correct. Recall shows that 81.80% of the *not e-included* learners are correctly classified. According to predicted accuracy, 83.00% of the total number of instances are correctly classified.

The second highest score is 79.80% (average of recall, precision and accuracy) for NaiveBayes. The highest metrics are in case of the balanced data set for learners with access to technologies. SimpleLogistic has the highest precision 0.885. The highest recall is 0.818 for NaiveBayes and lazy.LWL . The highest accuracy is 0.830 for lazy. LWL.

As in the Video Technology and Design Course, the model performance improved after balancing the data sets and excluding students who did not have access to technologies. The lazy.LWL algorithm has the highest improvement of accuracy, precision and recall (0.480). lazy.LWL also has the highest rate of improvement in the Video course. However, lazy.LWL has only the third highest average metrics score in the Video Technology and Design Course.

3.3 Results in the Robotics Course

As in other courses, in the Robotics course the highest score for algorithms is for a balanced data set for students having access to technology (Fig. 3). In the Robotics Course, the highest average score of 0.843 is for lazy.LWL in a balanced data set for learners having access to technology. With the lazy.LWL algorithm, accuracy is 0.843, precision is 0.857, recall is 0.828. lazy.LWL has the highest precision and accuracy values compared to other algorithms. Precision indicates that 85.70% of the *not e-included* learners predictions on the positive class *(not e-included)* are correct. Recall shows that 82.80% of the *not e-included* learners are correctly classified. According to the predicted accuracy 84.30% of the total number of instances are correctly classified.

The second highest score is 0.812 (average of recall, precision and accuracy) for SimpleLogistic. But the third highest average score is 0.797 for the LMT algorithm. In case of imbalanced data, the OneR algorithm has the highest recall value 0.875 for all learners.

The scores of several algorithms decrease after data set balancing. For example, they decrease for the NaiveBayes and OneR algorithms. The number of students in the Robotics course who did not use the newly acquired skills exceeded the number of students who used the new skills. Therefore, it is understandable that after balancing some algorithms decreased the recall and precision values. Recall and precision are

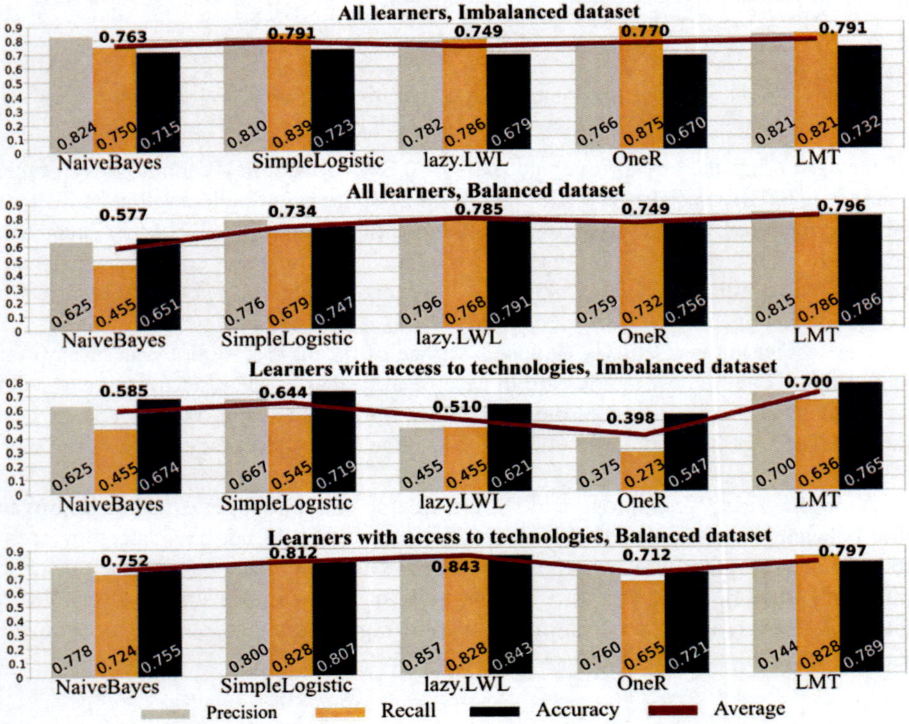

Fig. 3. Robotics course. Results: metrics of 5 classification algorithms.

computed for the *not-included* class, but in the Robotic course this class contains the largest number of instances. However, as in previous courses, the lazy.LWL algorithm has the largest average metrics increase (0.094) in performance after balancing.

Figure 4 presents the average performance evaluation for all courses for the balanced data set. The lazy.LWL algorithm indicates the highest value, 80.56% for learners having access to technology. The lowest score is for the NaiveBayes algorithm - 66.53%.

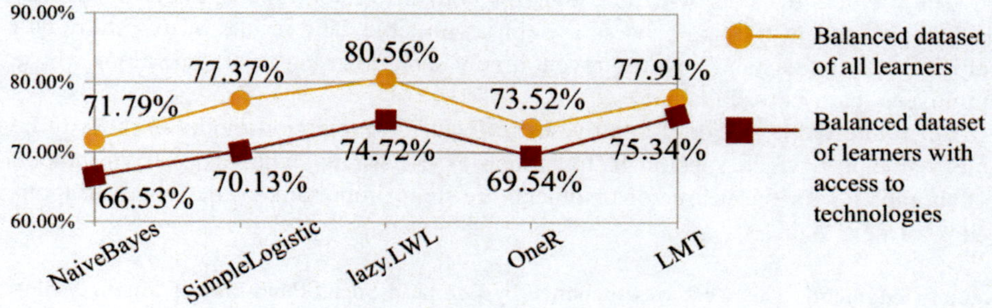

Fig. 4. The average of performance evaluations for all courses. Comparison of algorithms used a data set balanced with the SMOTE method.

4 Conclusions

In this paper, we have presented an e-inclusion prediction model based on machine learning methods. We have compared five different classification algorithms and selected the ones with the best average performance: LMT - 77.91% and lazy.LWL - 80.56%. LMT is a classifier for building logistic model trees, which are classification trees with logistic regression functions at the leaves. lazy.LWL is locally weighted learning and a very accurate function approximation method where it is easy to add new training points.

We concluded that the lazy.LWL algorithm has the highest average performance score for all three courses in the balanced data set that we analysed. Thus, we concluded that the lazy.LWL algorithm can be applied to various digital skills development courses. It is possible to predict *not e-included* learners with the average performance score ranging from 75.20% to 84.30% in case of balanced data sets of learners that have access to technologies. To achieve higher performance, the lazy.LWL model is limited by the fact that people need to have access to the technologies they will learn about. lazy.LWL performance metrics differ for each course. For all courses, lazy.LWL algorithm recall ranges from 0.678 to 0.828, precision ranges from 0.816 to 0.857, accuracy ranges from 0.763 to 0.843. The highest scores are for the precision values. At least 81.60% of learners identified as *not e-included* are predicted correctly with lazy.LWL.

We concluded that data set balancing before classification procedure improves performance results, especially for the lazy.LWL algorithm.

Classification models predict which class the learner will belong to - users or non-users of the newly acquired skills. One part of the non-users are learners who do not have the appropriate technology to use the newly acquired skills. Access to technology is an essential precondition for predicting whether a learner will use the newly acquired skills. Before starting the training, it is important to record the learner's opportunities for the future availability of technology. The availability of technology could be an issue that learners could address together with the organization that sends the learner to courses. This could be the way how the organization could reduce scrap learning.

The predictive model will be useful for both instructors and learners to identify risks in the student training process as soon as possible. The results of this study will help us better understand how to prevent timely scrap learning in organizations whose employees learn new skills.

For future work we could improve the e-inclusion prediction model in order to be able to identify which specific factors, such as e-learning materials, e-environment, collaboration with the instructor or others, are more important to the practical application of new skills.

Acknowledgment. This work was supported by European Social Fund Project "Strengthening of Academic Staff of Riga Technical University in Strategic Specialization Areas" No. 8.2.2.0/18/A/017.

References

1. European Commision: The European Digital Strategy (2020). https://ec.europa.eu/digital-single-market/en/content/european-digital-strategy. Accessed 24 Apr 2020
2. McKinsey Global Institute: Skills Shift: Automation and the future of the workforce (2018). https://www.mckinsey.com. Accessed 24 Apr 2020
3. CEB Global: Confronting Scrap Learning (2014)
4. Ortigosa, A., Carro, R.M., Bravo-Agapito, J., Lizcano, D., Alcolea, J.J., Blanco, Ó.: From lab to production: lessons learnt and real-life challenges of an early student-dropout prevention system. IEEE Trans. Learn. Technol. **12**(2), 264–277 (2019)
5. Limsathitwong, K., Tiwatthanont, K., Yatsungnoen, T.: Dropout prediction system to reduce discontinue study rate of information technology students. In: 2018 5th International Conference on Business and Industrial Research (ICBIR), Bangkok, pp. 110–114 (2018)
6. Kashyap, A., Nayak, A.: Different machine learning models to predict dropouts in MOOCs. In: 2018 International Conference on Advances in Computing, Communications and Informatics (ICACCI), Bangalore, pp. 80–85 (2018)
7. Tamada, M.M., Netto, J.F.D.M., de Lima, D.P.R.: Predicting and reducing dropout in virtual learning using machine learning techniques: a systematic review. In: 2019 IEEE Frontiers in Education Conference (FIE), Covington, KY, USA, pp. 1–9 (2019)
8. Gitinabard, N., Xu, Y., Heckman, S., Barnes, T., Lynch, C.F.: How widely can prediction models be generalized? Performance prediction in blended courses. IEEE Trans. Learn. Technol. **12**(2), 184–197 (2019)
9. Conijn, R., Snijders, C., Kleingeld, A., Matzat, U.: Predicting student performance from LMS data: a comparison of 17 blended courses using moodle LMS. IEEE Trans. Learn. Technol. **10**(1), 17–29 (2017)
10. Pike, B.: Want to eliminate scrap learning? Training **55**(4), 61 (2018)

11. Nissen, M.E. (ed.): Harnessing Knowledge Dynamics: Principled Organizational Knowing & Learning. IGI Global, Hershey (2005)
12. Vitolina, I., Kapenieks, A.: E-inclusion measurement by e-learning course delivery. Procedia Comput. Sci. **26**, 101–112 (2013)
13. Vitolina, I., Kapenieks, A.: User analysis for e-inclusion in a blended learning course delivery context. In: Proceeding of the International Scientific Conference, 23th–24th May 2014, vol. 2 (2014)
14. Hall, M., Frank, E., Holmes, G., Pfahringer, B., Reutemann, P., Witten, I.H.: The WEKA data mining software: an update. ACM SIGKDD Explor. Newslett. **11**(1), 10–16 (2009)
15. John, G.H., Langley, P.: Estimating continuous distributions in Bayesian classifiers. In: Eleventh Conference on Uncertainty in Artificial Intelligence, San Mateo, pp. 338–345 (1995)
16. Landwehr, N., Hall, M., Frank, E.: Logistic model trees. Mach. Learn. **95**(1–2), 161–205 (2005)
17. Atkeson, C., Moore, A., Schaal, S.: Locally weighted learning. AI Rev. **11**, 11–73 (1996)
18. Englert, P.: Locally weighted learning. In: Seminar Class on Autonomous Learning Systems (2012)
19. Holte, R.C.: Very simple classification rules perform well on most commonly used datasets. Mach. Learn. **11**, 63–91 (1993)
20. Seliya, N., Khoshgoftaar, T.M., Hulse, J.V.: A study on the relationships of classifier performance metrics. In: 21st IEEE International Conference on Tools with Artificial Intelligence, Newark, NJ, pp. 59–66 (2009)
21. Brodersen, K.H., Ong, C.S., Stephan, K.E., Buhmann, J.M.: The balanced accuracy and its posterior distribution. In: 2010 20th International Conference on Pattern Recognition, pp. 3121–3124. IEEE, August 2010
22. Yadav, S., Shukla, S.: Analysis of k-fold cross-validation over hold-out validation on colossal datasets for quality classification. In: 2016 IEEE 6th International Conference on Advanced Computing (IACC), Bhimavaram, pp. 78–83 (2016)
23. Jishan, S.T., Rashu, R.I., Haque, N., Rahman, R.M.: Improving accuracy of students' final grade prediction model using optimal equal width binning and synthetic minority over-sampling technique. Decis. Anal. **2**(1), 1–25 (2015)

Project Based Learning

Poster: Project-Based Learning in the Mathematics Course for First Year Students at a Technical University

Svetlana Rozhkova[1,3], Irina Ustinova[1(✉)], and Olga Yanuschik[1,2]

[1] National Research Tomsk Polytechnic University, Lenin Avenue 30, Tomsk 634050, Russia
igu@tpu.ru

[2] Tomsk State University of Control Systems and Radioelectronics, Lenin Avenue 40, Tomsk 634050, Russia

[3] National Research Tomsk State University, Lenin Avenue 36, Tomsk 634050, Russia

Abstract. Project activity is a form of educational and cognitive activity of students, which consists in motivational achievement of a consciously set goal. Project-based learning is increasingly being used in the various academic disciplines around the world. However, the question of when to start using this method effectively remains open. Therefore, the purpose of this article is to study the possibility of using project-based learning in the study of mathematics in the first semester in the first year at a technical University. To study the effectiveness of this method, we conducted a statistical test of the hypothesis of equality of average scores in the intermediate testing, scored by students who studied using the project method and the traditional one. The result of a survey of students about their knowledge of the project method before using it in a mathematics course at the University is presented. It is established that the use of the project-based learning can increase the interest of students in the learning process. The project-based learning has increased the average score of students at the end of the semester by 16.19% compared to the control group.

Keywords: Mathematics · Engineering education · The project-based learning

1 Introduction

Project-based learning is increasingly being used in the various academic disciplines around the world. Project activity is a form of educational and cognitive activity of students, which consists in motivational achievement of a consciously the goal [1–3].

The traditional approach to teaching is that the teacher transmits scientific knowledge to students, and students perceive or not this knowledge. In other words, the role of the teacher is active, he must have a large baggage of systematic knowledge and be able to clearly and interestingly present them, and the role of students is passive. Using project-based learning assumes that students choose the project topic themselves and self-developed it [4]. At the same time, a distinctive feature of this educational

M. E. Auer and T. Rüütmann (Eds.): ICL 2020, AISC 1328, pp. 341–348, 2021.
https://doi.org/10.1007/978-3-030-68198-2_31

technology is that at the end of work on the project, the mandatory stage is the protection of the project [5, 6]. When using the project method, the role of students in learning changes, as they work on their own on the project, they research, make decisions, reflect and all this allows them to expand knowledge. When working on a project, it is assumed that students use the knowledge and skills that they received earlier and simultaneously acquire new knowledge [7].

Working on a project is a form of activity that is aimed at studying a certain educational topic that is part of a standard course or several courses. There are the following types of projects: research projects, creative projects, game projects, information projects, and application projects [8]. Regardless of the type of project, there are mandatory stages of work on the project such as: problem statement, work on the project, getting the final product and presentation of the project.

There are undoubted advantages of project-based learning such as:

- Firstly, it is based on the principle of individual learning.
- Secondly, it implements an activity-based approach to training.
- Thirdly, it is based on the principles of problem-based learning.
- Fourthly, it contributes to the development of internal motivation to learn [9, 10].

"Mathematics is the Queen of the sciences," these are the words of the great mathematician Gauss [11]. It is the organization of students' project activities in the study of mathematics that is the main factor that helps to increase interest in mathematical science, make it exciting, entertaining and useful. Students should realize that mathematics is not scary, mathematics is interesting. How to revitalize the learning process and create an atmosphere of joyful elation that accompanies search and creativity? How do I make learning activities joyful, exciting, and interesting? How to awaken the students' desire for knowledge?

Many math teachers are concerned about the answers to these questions. We have a way. This is the project activity of students.

2 The Project Activities in TPU

Now mathematics combined with computer science is becoming interdisciplinary tool that allows you to perform two main functions. The first function is training the future specialist in the ability set the goal correctly, as well as to determine the conditions and restrictions for achieving it. The second function is analytic, so the ability to model possible situations and get optimal solutions. The criterion for success of a future specialist today is not so much performance in the study of academic disciplines as the acquisition of personal and professional experience in the learning process, the development of the desire and ability to independently extract and use new knowledge. This effect is almost unattainable with a traditional approach to education. The most promising of all technologies is project-based learning. It allows you to teach students to search for information correctly, develops critical thinking, cognitive interest, independence, and so on.

Not random people embark on Tomsk Polytechnic University. Most of them are initially aimed at getting an engineering education. "This is difficult and just too abstract. It will not be useful either in the study of our special disciplines, or in life, - students of the first year of training often think about the discipline "Mathematics". After a couple of years, they usually change their minds when it turns out that the ability to set tasks, solve them and analyze the process, all that mathematics teaches is necessary in any field. Our goal at the University is to broaden the horizons of students, teach them to think outside the box and find solutions to other problems using mathematics.

To date, the Department of mathematics and computer science of the National Research Tomsk Polytechnic University has accumulated extensive experience in the development of various educational technologies in the discipline "Mathematics" for various engineering specialties [12, 13].

Since 2019, a study is being conducted, the purpose of which was to identify the impact of project activities for mastering of the course of mathematics [9]. The method helps students to successfully master mathematical knowledge and, most importantly, teaches them to apply this knowledge in a variety of situations. Last year, during the semester, teachers "tested" the method of projects in the course of studying the topic "Multiple integrals". We noticed that students who drafted on projects became more involved in practical classes. They began to perceive mathematics as a universal tool that allows them to solve practice-oriented problems. Students who implemented the project, when performing control tasks on the topic "Contour integrals and multiple integrals" showed higher results. Having received the approval of the school management, it was decided to extend the method of projects further to other sections of the discipline "Mathematics". This academic year, first-year students were also engaged in project activities.

2.1 Project Activity in the First Course

For the experiment, first-year students (4 groups) were selected for the specialty: "electric power and electrical engineering". In the first semester, they study the discipline "Mathematics 1". The Mathematics 1 module is part of the basic part of the mathematical and natural science cycle of the curriculum. This module of the discipline is necessary for the development of other disciplines of the mathematical and natural science cycle and the disciplines of the professional cycle of the curriculum. The development of the discipline (module) is aimed at the formation of students' following competencies: to possess mathematical apparatus for describing, analyzing, theoretical and experimental research and modeling of physical and chemical systems, phenomena and processes, use in training and professional activities. Project activities were offered to students after studying the topics:

- Linear algebra.
- Vector algebra.
- Analytical geometry.
- Introduction to mathematical analysis.
- Differential calculus of functions of one variable.

Work on projects was carried out in groups of 3 to 4 people. Students were offered a choice of two types of projects: application and informational. Student's independent study consisted of several stages. Firstly, it was necessary to establish the correspondence of the content of the problem and the sections of mathematics studied in the current semester. Secondly, it was necessary to create a description of the problem condition in terms of the corresponding section. Thirdly, students should learn the methods of solution themselves. Here are examples of some tasks.

2.1.1 Derivative in Electrical Engineering

The task is formulated here. To collect the garland, there is a battery with electromotive force of E = 120 V and an internal resistance of r (Om) and a large number of light bulbs designed for a voltage of U_0 (volts) and a power of W_0 (watts). The task is to calculate the maximum number of bulbs that will burn at full intensity when connecting in a garland.

Let us consider the solution to this task. We calculate the maximum utility of the battery power for the resistance R in the circuit according to the Joule-Lenz law $W = I^2R$, where I is the current in the circuit (see Fig. 1) which can be found from the Ohm's law of the complete circuit $I = E/(r + R)$. Thus, the power released in the external circuit depends on the resistance R: $W = E^2R/(R + r)^2$. Draw an approximate graph of the function $W(R)$ (see Fig. 2). Let us find the derivative of the function $W(R) = E^2R/(R + r)^2$ in order to find the extremum. This is $W'(R) = E^2\frac{r-R}{(R+r)^3}$. We get from this equation $R_{opt} = r$ and $W_{max} = \frac{E^2}{4r}$.

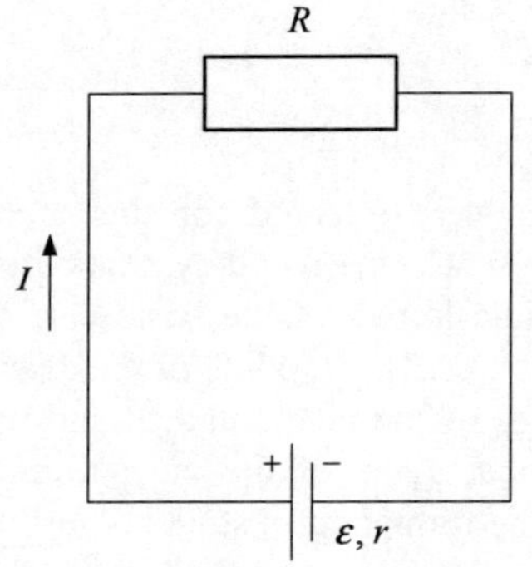

Fig. 1. This is the electrical circuit for the task at hand.

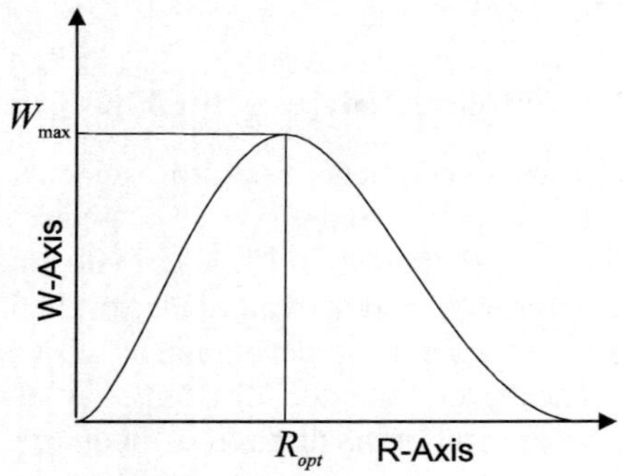

Fig. 2. The graph of the function $W(R)$.

2.1.2 Application of Linear Algebra in Electric Power and Electrical Engineering

An electrical circuit is given (see Fig. 3). You need to determine the currents in the branches, using Kirchhoff's laws. Parameters of electric circuit elements are as follows: R_1 = 45 Ω, R_2 = 15 Ω, R_3 = 45 Ω, R_4 = 75 Ω, E_1 = 60 V, E_2 = 450 V.

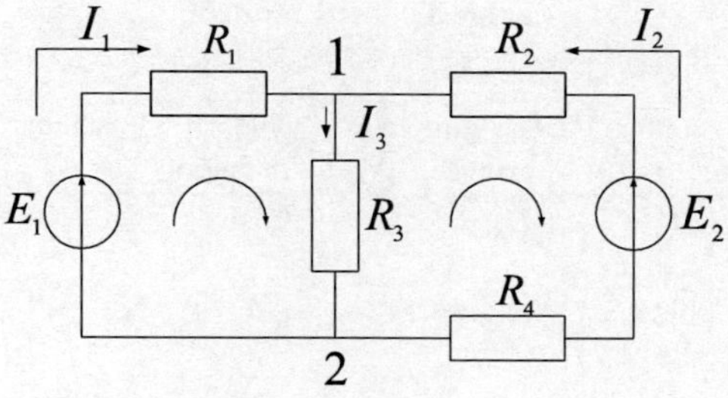

Fig. 3. Diagram of an electric circuit

Let us consider the solution to this task. Choose the positive directions of the unknown currents and mark them on the diagram. We make up the equations using the first Kirchhoff law for node 1. After selecting the direction of contour traversal, we write down the equations according to the second Kirchhoff law. As a result we get a system of three linear equations with respect to unknown I_1, I_2, I_3:

$$\begin{cases} I_1 + I_2 - I_3 = 0, \\ I_1 R_1 + I_3 R_3 = E_1, \\ -I_2(R_2 + R_4) - I_3 R_3 = -E_2. \end{cases}$$

This system can be solved by one of the three methods studied in the course linear algebra and analytical geometry, namely: by the Kramer method, the matrix method or the Gauss method [14]. The system has the only solution that students have found using the Kramer method: I_1 = 1.2 A, I_2 = 3.73 A, I_3 = 2.53 A.

3 Some Results of Work on the Project Method

At the end of the first semester, students defended their projects during the credit week, where they reported on the results of their activities. All the reports were accompanied by presentations, discussions and suggestions on how to improve their work. After the defense of the projects, a survey was conducted. A total of 66 people were interviewed. 19 respondents used the project method while still in school. This is 28.8% of all survey participants. Of those who previously studied using the project method, only 26.3% worked on a math project, which are 5 people. If we consider all respondents, 5 people are only 7.6%.

4 Statistical Analysis of the Results of a Pedagogical Experiment

We compared the achievements of students in four groups involved in project work with the achievements of students who studied in a similar program earlier (2018), but without using the project method. The results are shown in Table 1.

Table 1. Tests results.

Initial testing				
	The sample mean	The sample variance	The sample standard deviation	The number of students
Project-based learning	$\bar{x} = 8.91$	$D_x = 5.18$	2.28	$n = 96$
Traditional training	$\bar{y} = 9.025$	$D_y = 6.74$	2.6	$m = 40$
Intermediate testing				
Project-based learning	$\bar{x} = 8.65$	$D_x = 12.19$	3.49	$n = 93$
Traditional training	$\bar{y} = 7.25$	$D_y = 5.87$	2.42	$m = 38$

Let's test the hypothesis that the average scores of groups trained using the project method and traditional one differed significantly during the initial testing. To do this at the significance level of 0.05 we will check the null hypothesis H_0: M[X] = M[Y] about the equality of mathematical expectations of test scores with the competing hypothesis $H1$: M[X] $\neq$ M[Y] [15], where X is the entire assembly of test scores of students who studied using the project method, and Y – the entire assembly of test scores of students who studied using the traditional method. The sample sizes are equal accordingly: $n = 96$, $m = 40$. Observed value of the criterion is $Z = \frac{\bar{x}-\bar{y}}{\sqrt{\frac{D_X}{n}+\frac{D_Y}{m}}} = -0.245$. From the table of values of the Laplace function we find the auxiliary point z by the formula $\Phi(z) = \frac{1-\alpha}{2} = 0.475$. Then $z = 1.96$. Since $|Z| < z$, there is no reason to reject the null hypothesis, that is, the sample means differ insignificantly. Similar calculations are performed for the data of the intermediate testing. In this case, $Z = 2.59$, and the auxiliary value of z is the same. Since $|Z| > z$, we reject the null hypothesis, that is, the average scores of students who studied in different technologies differ significantly. The average score of students who studied using the project method is significantly higher than the average score of students who studied by traditional one. Let's see how the experiment affected the average score of students who studied using the project method. In other words, let's compare the sample mean of the initial test and the intermediate one of students who studied using the project method. In this case, $Z = 0.61$, and the auxiliary value of z is the same. Since $|Z| < z$, there is no reason to reject the null hypothesis, that is, the average scores of students before and after the pedagogical experiment were approximately the same. As for the control group, $Z = 3.17$, and the auxiliary value of z is the same. Since $|Z| > z$, we reject the null hypothesis, that is, the average score of students who studied traditionally decreased significantly. Thus, the application of the project method allowed improving students' achievements by 16.19% compared to the control group.

Another interesting question is whether the type of project affects student achievement. To answer this question, we will compare the sample mean of scores of testing depending on the type of project being run. Two groups of four groups were engaged in application projects, and the remaining two were engaged in information ones. The results of testing, depending on the type of project, are shown in Table 2.

Table 2. Test results depending on the project type

Initial testing				
	The sample mean	The sample variance	The sample standard deviation	The number of students
Application project	$\bar{x} = 9.15$	$D_x = 4.64$	2.15	$n = 48$
Information project	$\bar{y} = 8.67$	$D_y = 5.72$	2.39	$m = 48$
Intermediate testing				
Application project	$\bar{x} = 8.62$	$D_x = 12.76$	3.57	$n = 47$
Information project	$\bar{y} = 8.7$	$D_y = 11.86$	3.44	$m = 46$

Let's test the hypothesis that students who worked on an application project have a significantly different average score for the initial test and the intermediate test. Let's find $Z = 0.87$ and the auxiliary value of z is the same. Since $|Z| < z$, therefore, there is no reason to reject the null hypothesis, that is, the average scores of students engaged in an application project differ slightly. We get a similar conclusion about the scores of students who were engaged in an information project. The initial test also demonstrated that the groups of students who worked on an application project and on an information one are equal in strength, since $Z = 1.02$, means that $|Z| < z$, there is no reason to reject the hypothesis of equality of the average scores of the initial test. Having calculated the value of $Z = -0.11$ for the intermediate test, we also have come to believe that the average scores of students differ slightly.

Finally, we conclude that the use of the project method (regardless of the type of project) has had a positive impact on students' achievements. This conclusion is consistent with [16, 17].

5 Conclusion

No one claims that project work will help solve all problems, but it is an effective remedy for monotony, boredom, it contributes to the development of the student, self-awareness as a member of the group. The project is also a real opportunity to use the knowledge gained in other subjects in the process of studying mathematics.

References

1. Vasconcelos, T.: Using the project approach in a teacher education practicum. ECRP **9**(2), n2 (2007)
2. Stauffacher, M., Walter, A.I., Lang, D.J., Wiek, A., Scholz, R.W.: Learning to research environmental problems from a functional socio-cultural constructivism perspective: the transdisciplinary case study approach. Int. J. Sustain. High. Educ. **7**(3), 252–275 (2006)
3. Pellegrino, J.W., Hilton, M.L.: Education for Life and Work: Developing Transferable Knowledge and Skills in the 21st Century. National Academies Press, Washington (2012)
4. Otake, M., Fukano, R., Sako, S., Sugi, M., Kotani, K., Hayashi, J., Noguchi, H., Yoneda, R., Taura, K., Otsu, N., Sato, T.: Autonomous collaborative environment for project-based learning. Rob. Auton. Syst. **57**(2), 134–138 (2009)
5. Spronken-Smith, R., Kingham, S.: Strengthening teaching and research links: the case of a pollution exposure inquiry project. J. Geogr. High. Educ. **33**(2), 241–253 (2009)
6. Larmer, J., Mergendoller, J.R.: Gold standard PBL: Essential project design elements. Buck Institute for Education (2015). www.bie.org
7. Hanney, R., Savin-Baden, M.: The problem of projects: understanding the theoretical underpinnings of project-led PBL. Education **11**(1), 7–19 (2013)
8. Harmer, N.: Project-based learning. Plymouth University (2014)
9. Rozhkova, S.V., Ustinova, I.G., Yanuschik, O.V., Korytov, I.V.: The use of the project-based learning in the study of the course of mathematical analysis. In: Proceedings of the 22nd International Conference on Interactive Collaborative Learning (ICL2019), vol. 1, pp. 871–879 (2020)
10. Ford, A., Kluge, D.: Positive and negative outcomes in creative project-based learning: two EFL projects. J. Nanzan Acad. Soc. **98**, 113–154 (2015)
11. Biihler, W.K.: Gauss. A Biographical Study. Springer, Heidelberg (1981)
12. Pokholkov, Y.P., Rozhkova, S.V., Tolkacheva, K.K.: Practice-oriented educational technologies for training engineers. In: International Conference on Interactive Collaborative Learning (ICL 2013), pp. 619–620 (2013).
13. Rozhkova, S.V., Rozhkova, V.I., Chervach, M.Y.: Introducing smart technologies for teaching and learning of fundamental disciplines. Smart Innov. Syst. Technol. **59**, 507–514 (2016)
14. Konev, V.V.: Linear Algebra, Vector Algebra and Analytical Geometry. TPU Press, Tomsk (2009)
15. Larsen, R.J., Marx, M.L.: An Introduction to Mathematical Statistics and its Applications. Prentice Hall, Boston (2012)
16. Boaler, J.: Mathematics for the moment, or the millennium? Educ. Week **17**, 30–34 (1999)
17. Bell, S.: Project-based learning for the 21st century: skills for the future. Clearing House **83**, 39–43 (2010)

Module Proposal in Mechanical Engineering Based on Project Based Learning

Diego Gormaz-Lobos[1(✉)], Gonzalo Pincheira-Orellana[2], and Claudia Galarce-Miranda[1]

[1] Faculty of Engineering, International Center of Engineering Education, Universidad de Talca, Talca, Chile
diego.gormaz@utalca.cl

[2] Faculty of Engineering, Department of Industrial Technologies, Universidad de Talca, Talca, Chile

Abstract. Despite many changes in the productive sector of Chile and the Region of Maule, engineering programs and learning strategies continue to be very traditional. The most contemporary teaching strategies are still mainly centered on the teacher, and the development of competencies are frequently remote from the reality of Chilean society and the demands of the students of the 21st century. Compared to the typical teaching methodologies in engineering, Project-Based-Learning (PjBL) is believed to offer a number of advantages. It not only permits students to develop technical and methodological knowledge, gain confidence, and become more independent learners, but it also helps them to develop professional skills, e.g. leadership, team working, planning competencies, communication, among others. Donnelly et al. (2005) define PjBL as a learning method based on the principle problem-solving as a starting point for the acquisition and integration of knowledge and skills development. Considering the need to innovate in the way of teaching and at the same time, strengthening the development of diverse competencies of the engineering students, the authors of this work want to expose in this paper the following aspects: (1) the scientific arguments that support the decision to develop a module of the mechanical engineering training program at the University of Talca, (2) present the module and the PjBL proposal that supports him; and (3) the results of an evaluation survey (satisfaction) by the students who visited the module.

Keywords: Project based learning in mechanical engineering · PjBL at University of Talca · Project based courses

1 Introduction

Compared to the typical teaching methodologies in engineering, Project-Based-Learning (PjBL) is believed to offer several advantages. Thus, PjBl not only permits students to develop technical and methodological knowledge, gain confidence, and become more independent learners, but it also helps them to develop some relevant professional skills for their future professional performance, e.g. leadership, team working, planning competencies, decision-making abilities, time management, communication, among others. Donnelly et al. (2005) define PjBL as a learning method

M. E. Auer and T. Rüütmann (Eds.): ICL 2020, AISC 1328, pp. 349–361, 2021.
https://doi.org/10.1007/978-3-030-68198-2_32

based on the principle problem-solving as a starting point for the acquisition and integration of knowledge and skills development. The fundamental characteristics of PjBL are e.g.: (a) learning as a student-centered process, (b) learning occurs in small groups, (c) teachers are facilitators or guides in this process, (d) problem solving is the focus and approach for learning organization and motivation, and (e) projects are a vehicle for developing resolution skills from problems [1]. Based on these characteristics and the PjBL strategy, a group of academics developed a module for students of mechanical engineering at the University of Talca (UTalca, Chile) with the purpose to improve personal and professional competencies at the students and prepare them for the new demands of the Chilean productions structures [2].

2 PjBL in Engineering Education

2.1 PjBL Definition and Characteristics

In general, Project-Based Learning (PjBL) can be understood as a teaching-learning strategy that organizes learning around projects. Behind this definition is a concept of learning that is important to note: learning as an "active process of investigation and creation based on the learners' interest, curiosity, and experience and should result in expanded insights, knowledge and skills" [3]. A central aspect of the PjBL approach is the idea, that learning is more effective when students put theory into practice activities: in other words, the conception of "learning by doing" from Dewey and Kilpatrick learning theory [4].

In Engineering Education there is a general consensus that presented PjBL as a teaching-learning strategy organized on project work, for the development of competences and skills of engineering students. The fundamental starting point of the PjBL process is the problem (which the students wish and should solve) [5]. Kolmos et al. (2007) consider the following key elements for PjBL [6]:

1. Ill-structured and complex questions based on real-world scenarios.
2. Student-centered active learning occurs.
3. Learning occurs in small groups, (teamwork) considering and reviewing solutions to open-ended problems.
4. The teacher becomes a facilitator.
5. Self-assessment increases efficacy.

Kolmos et al. (2009) summarized the main learning principles of PjBL in three approaches (see Table 1) [7]:

1. A *cognitive learning approach* means that learning is organized around "real-world" problems and will be carried out in projects. A problem provides the starting point and the motivation for the learning process, situating learning in a context and basing learning on the learner's experience. PjBL means that students have to work with a unique task involving complex and situated problem analysis and problem-solving strategies.

2. A *contents approach* especially concerns interdisciplinary learning, which not only stresses but also spans traditional subject-related boundaries and methods. It is an exemplary practice in the sense that the learning outcome provides a good example of the overall objectives. Furthermore, it supports the relationship between theory and practice by demonstrating the fact that the learning process involves an analytical approach using theory in the analysis of problems and problem-solving methods.
3. A *social approach* (team-based learning). Team learning considers the learning process as a social act in which learning takes place through dialogue and communication. Furthermore, the students learn not only from each other, but they also learn to share knowledge and organize the process of collaborative learning.

Table 1. Approaches on PjBL in engineering education.

Approach	Main characteristics
1. Cognitive learning approach	- Problem-oriented - Project work - Based on learner's experience - Context related problems or "real-world" problems
2. Contents approach	- Interdisciplinary learning - Exemplary practice - Support relation "theory-practices"
3. Social approach	- Team-based learning - Participant directed learning

The benefits PjBL for student learning are presented in different scientific papers and empirical works related to Engineering Education. Some benefits that PjBL offers are [8–10]:

- experience of authentic engineering problems and professional practices,
- experience of problem-solving and the design process,
- experience and development of a team and collaborative work,
- self-motivation and student ownership of the problem, solution, and learning,
- development of self-regulation, agency, commitment, time management, and another management competences,
- development of reflective skills,
- presentation of the multi-disciplinary and systemic nature of engineering problems,
- development of written, oral and other communication skills, among others.

However, different authors warn that there are also risks by the implementation of PjBL: (i) PjBL needs effective learning environment, (ii) students must be trained to work in teams and for the collaborative work, (iii) before and during the project, students needs to be prepare for conflicts management between members of their team, (iv) for making group decisions, (v) for sharing out tasks, and (vi) for the necessary organizational preparations [8]. In addition, Palmer et al. (2011) summarized 'needs to improvement' (reported by students) for the work at PjBL, that considered: (i) Need for an introduction to teamwork, and preparation for that; (ii) Need for instruction on engineering/design report writing; (iii) Strategies to deal with group members who did not pull their weight, and (iv) High time demands of project work [11].

2.2 PJBL in Chilean Context

The Chilean Ministry of Education (MINEDUC, 2019) defines PjBL as a proposal of teaching-learning for all educational levels, organized around a problem or need that can be solved by applying different perspectives and areas of knowledge (multidisciplinary). To find the solution, students need to mobilize knowledge, skills, and attitudes throughout the process, until a solution in a result or product is expressed [12].

Since Chile is a member of the OECD, the educational policy is strongly oriented for the goals of this organization (for instance see OECD Future of Education and Skills 2030 project, among others) [13, 14]. MINEDUC (2019) established a list of competencies for the 21 century (see Fig. 1) with close recommendations to the planning and implementation of PjBL.

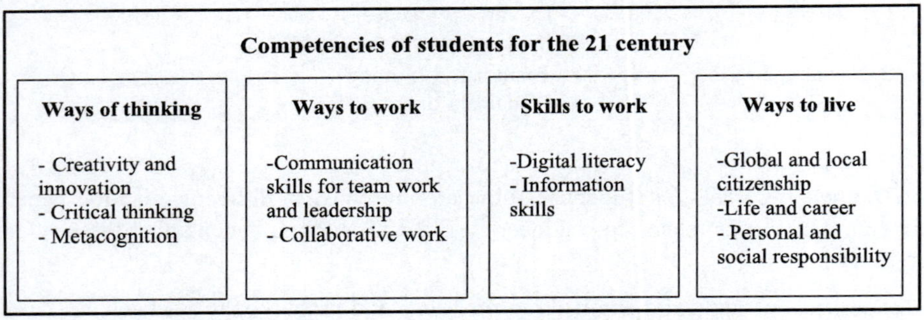

Fig. 1. Competencies of students for the 21 century, MINEDUC 2019.

MINEDUC recommends the use and application of PjBL at all educational levels because allows students [12]:

1. to enhance these skills and attitudes in the joint to search for a problem solution,
2. to develop flexibility, creativity and innovations competencies to find answers to the problem,
3. to reflect with others and consider different perspectives,
4. to generate teamwork and collaborative work, with commitment and responsibility
5. to develop communication skills and critical and creative thinking, among others.

Despite the fact that in Engineering Education in Chilean context there are no official engineering courses or careers that organize their engineering programs under PjBL (as is the case and the showed international experience at the University of Aalborg in Denmark and the National Technical University in Trondheim in Norway), there are several courses in Chilean engineering schools that take some elements of PjBL as a learning strategy inside the study courses. For instance, traditional project-based modules implemented at the University of Talca (UTalca) do not include PjBL as a strategy, these courses assess the development of an engineering project or product, but they are not especially focused on the whole learning process of the engineering students at the project work.

3 A PjBL Proposal on Mechanical Engineering

3.1 Design of PjBL Module at University of Talca

This module is part of the curriculum of Mechanical Engineering of UTalca, which is taken by the students in the fifth and ninth semester of the career. During the first four semesters in the undergraduate program, students receive a basic science training (including materials science, 3D design, manufacturing process, etc.) and soft skills (teamwork, oral and written communication, self-learning, etc.), and then, in the next four semesters, fundamental courses of mechanical engineering as a solids mechanics, thermodynamics or numerical simulations must be taken by students. The current PjBL module has been designed including different formal aspects as: (i) Objectives and knowledge; (ii) Types of problems; (iii) Progression and size; (iv) Academic staff and facilitation, and (v) Assessment and evaluation.

3.2 Objectives and Knowledge

The module is organized in weekly teacher-student meetings and divided into a working group of 3 students, which must present their progress weekly. The main aim of the module is the development of a basic engineering project, where the individual and group performance of students can be evaluated. For this, according to the specific objectives of the module, students should be able to: organize and systematize information of the project, show communication skills, developed interdisciplinary solutions considering technical, social and environmental requires, design a 3D model and engineering drawings, define technical specification and costs, and prototyping, among others.

3.3 Types of Problems

Generally, the problems to be solved in this PjBL module come from real problems in industries or a current context that can be approached from engineering to provide a solution, which is presented by the tutor of the module. Several times, it is presented through a problematic and not with the direct problem: in this way students must and need to define the specific problem. During different presentations of the groups, tutor and external expert or academics from specific areas, evaluate possible solutions and pre-design presented.

3.4 Progression and Size

PjBL Module must be calendarized in 18 weeks according to one single semester. Five different phases in the PjBL structure are defined to achieve goals and aims. The first phase considers the introduction and formalization of the problem. In the second phase, technical specifications, and costs are defined, followed by the third phase where 3D design, engineering drawing, and performance video are prepared. Finally, in the fourth phase prototype is manufactured and in the final phase, a final evaluation, a validation of performance must be carried out. The following table summarizes this information (Table 2).

Table 2. Progression and size

Weeks	Phases	Result
1–3	Phase 1: Introduction and formalization	Report and oral presentation
4–6	Phase 2: Technical specifications and costs	Report and oral presentation
7–9	Phase 3: 3D design and engineering drawing	Report and oral presentation
10–17	Phase 4: Prototyping	Prototype
18	Phase 5: Final evaluation	Report and oral presentation

3.5 Academic Staff and Laboratories

The academic staff is formed by two tutors and one assistant student. The role of academics in this module is to model the PjBL methodology and give positive feedback to students, in particular, one staff member has a pedagogical training in PjBL and, the second one has a mechanical engineering formation. The assistant student has a role in helping groups in technical and technological applications or developments. This staff structure allows: designing PjBL module, modeling by example, and developing of an engineering solution. Students have available the mechanical workshop and the prototyping laboratory to manufacture a working prototype. The Prototyping Laboratory is a new place in Mechanical Engineering, designed and created for students to work in an environment that stimulates innovation and creativity. In it, they have 3D printers, test machines, composite manufacture and relax and entertainment zones, these are considered fundamental in a design process.

3.6 Assessment and Evaluation

Students' performance is continuously assessed and evaluated with individual and group work. During the semester they have five evaluations according to the phases described above. Phases have different worth: phase I 10%; phase II 20%; phase III 20%; phase IV 40% and phase V 10%. Independently, each phase is separated by 50% for the report and 50% for oral presentation. Finally, oral presentation is divided in 70% individual and 30% group performance. In addition, students are weekly assessed using: progress presentations, group discussions, and particular activities that enable goals to be achieved in the PjBL process. Both control systems have immediate feedback from academic staff. This is fundamental in PjBL, because it promotes that students improve and develop effective communicational and engineering skills. To achieve this the academic staff designed different rubrics for reports and oral presentation, including around to 40 concepts, such as: (i) Problem, aims and action plan are connected and well defined, (ii) Information is presented in a clear and organized way, (iii) Theoretical framework and technical specification are properly developed, (iv) Design and solution are adequate to the proposed objectives, (v) Vocabulary, times and body expressions are used correctly, (vi) Students follow formats and basic rules for scientific work, among others.

4 Students' Perceptions of PjBL Initiative in Mechanical Engineering

4.1 Methodology

The survey was designed using a mixed model of qualitative and quantitative methods, using a concurrent triangulation strategy. The aim is to use two different survey methods to confirm, supplement, or validate the research results [15]. The main goal of the design was to integrate the perceptions of the students from close questions with the answers of the open questions of the instrument (questionnaire). Based on the work of different authors [4, 7, 8, 10, 11] about the benefits of PjBL for engineering students, a questionnaire about the perceptions of students at the PjBL course in mechanical engineering was developed.

The main goal of the instrument was to identify the perceptions of students regarding their experience in a mechanical engineering course designed under a PjBL strategy. In general, the instrument and indicators seek to obtain information about: (i) the characteristics of students (years old, career semester, etc.), (ii) experience with PjBL, (iii) developed personal and interpersonal skills in PjBL course, (iv) developed communication skills, (v) organizational skills and, (vi) the identification of strengths and weaknesses at the experience at the PjBL course (see Table 3).

Table 3. Instruments Categories and Indicators about Student's perceptions of PjBL.

Categories	Indicators* (simplified version for this publication)
I. Use and application of knowledge	I.1. Application of theoretical knowledge I.2. Learning of new engineering knowledge I.3. Learning of new skills required by engineers I.4. Assessment of learning achieved
II. Organizational capacity	II.1. Development of capacity for individual organization II.2. Developing organizational skills for teamwork II.3. Knowledge to plan and to develop projects
III. Communicative skills	III.1. Communication skills to present written scientific work III. 2. Dialogical communicative processes for learning
IV. Personal skills	IV.1. Analytical and critical thinking skills of engineering topics IV.2. Synthesis capabilities of specific engineering topics IV.3. Personal responsibility in the learning process IV.4. Times management for the learning process
V. Interpersonal skills	V.1. Conflict resolution skills V.2. Ability to analyze the social scope of the project to be developed
VI. Solutions solving skills	VI.1. Problem solving skills focusing on the expected solution VI.2. Analysis of social and environmental scope of the project
VII. Scientific work	VII.1. Organization of scientific written and oral work VII.2. Structuring of the written scientific work
VIII. Use of technological and digital tools	VIII.1. Use of teaching resources and information and communication technologies

4.2 Population and Procedure

The sample of the study was composed of 20 students of the Mechanical Engineering school at the University of Talca, Chile (Table 4 presents a characterization of the sample). The questionnaire was implemented in online und anonym form. The first part collected general information of the participants (age, gender, semester, city, etc.). The second and third parts correspond to information collection of the closed questions. The statistical analysis applied was exploratory-descriptive to raise problems. The fourth part consists of open questions. These questions are analyzed through a textual content analysis by codifying the discourse of each student, based on the item generating conceptual categories. The instrument was individually applied, considering the ethical aspects according to the Chilean social sciences research criteria.

4.3 Characterization of the Sample

Table 4. Characterization of the sample.

Categories	Sample
Number of respondents	20
Age	
Between 18 and 23 years old	9
Between 24 and 29 years old	11
Gender	
Female	3
Male	17
Course	
Disciplinary project	9
Interdisciplinary Project	11
Previous experience at PjBL	
Yes	10
No	10

4.4 Results of Student's Survey

Closed Questions. This part presents the results about the perceptions of students regarding their experience in a mechanical engineering course designed under a PjBL strategy. It was asked, "*Do you consider that your participation in the course facilitated…?*" For this section, 22 aspects were considered based on the indicators of Table 3.

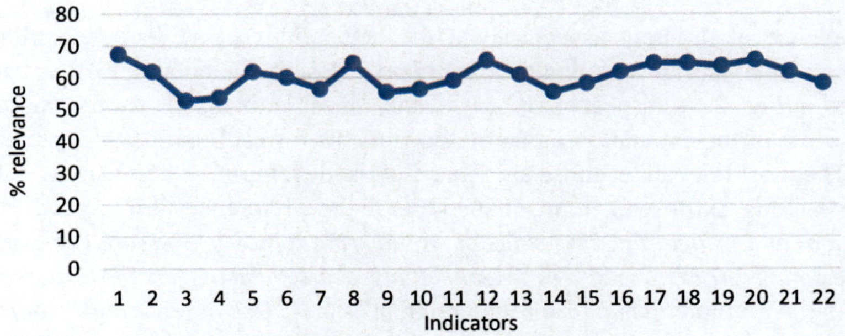

Fig. 2. Relevance of different aspects about perceptions of students at PjBL (general overview).

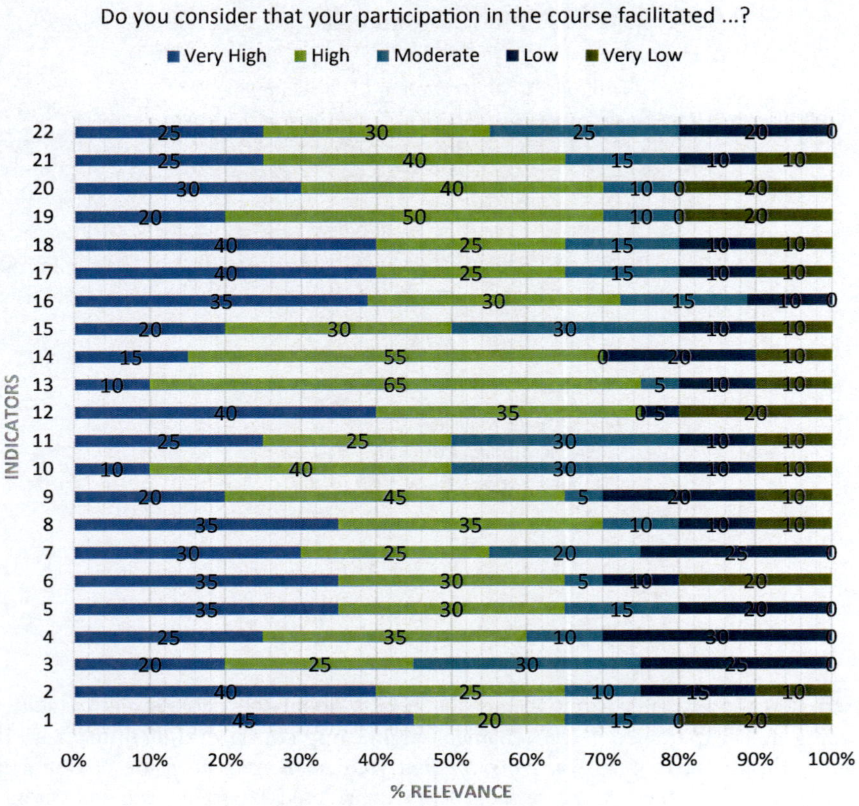

Fig. 3. Relevance of different aspects about perceptions of students at PjBL by valuations criteria.

The relevance of different aspects about the perceptions of students regarding their experience in a mechanical engineering course designed under a PjBL strategy is presented in Fig. 2. All aspects have a relevance of more than 50% for the participants. An overview of the evaluation made by the students for each indicator is presented in Fig. 3. The most relevant of those are related to the development of personal skills, and solution solving skills with more than 62% of the preferences for aspects such as: "*Analytical and critical thinking skills of engineering topics*", "*Personal responsibility in the learning process*" and "*Problem-solving skills focusing on the expected solution*". Then with more than 57% of the preferences are "*Use of teaching resources and information and communication technologies*", "*Analysis of social and environmental scope of the project*", "*Communication skills to present written scientific work*" and "*Learning of new skills required by engineers*".

The aspects considered less relevant (less than 55% of the preferences) were: "Knowledge to plan and to develop projects", "Skills for conflict resolution" and "Synthesis capabilities of specific engineering topics".

For the authors is relevant to present, that students with or without previous experience at PjBL had clear differences in relating the valuation of each indicator. Students with previous experience at PjBL gave a high valuation (average of 15%) in each indicator of the questionnaire (see Fig. 4).

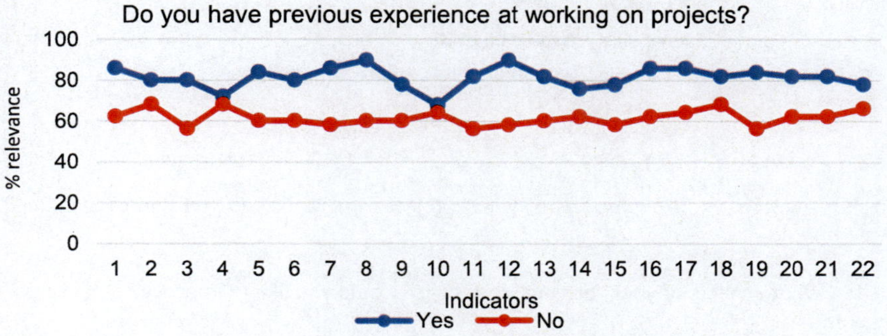

Fig. 4. Relevance of different aspects about perceptions at PjBL, differentiate by previous experience.

Open Questions. This part presents the results of the open questions about the perceptions of students regarding their experience in a mechanical engineering course designed under a PjBL strategy. It was asked four questions: (1) "What aspects of your "project work skills" do you find outstanding?", (2)"What aspects of your "project work skills" would you like to improve?", (3) "Do you consider project work useful? (Give your arguments) and, (4) "What specific aspects/thematic areas do you consider were the most outstanding in the module?". The answers for questions 1 and 2 are presented in Figs. 5 and 6 respectively. In question 3, all the participants consider useful the work at projects for different reasons (for example the development of organizational and communication skills, learning of new engineering knowledge, etc.). For question 4 refers to the most outstanding aspects of the participation in the course, the students mention:

The feedback and assessment of teachers and other students for the work at the project,

The development of oral and written communication skills (presentations, informs, project plane, etc.),

The development of organizational skills (time management, distributions of tasks, responsibilities, resources determination, etc.),

Use of different types of knowledge: from engineering sciences to other disciplines, among others.

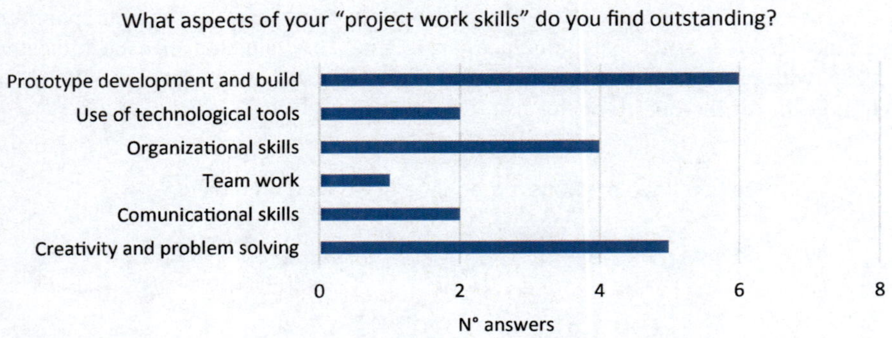

Fig. 5. Answers to "What aspects of your "project work skills" do you find outstanding?"

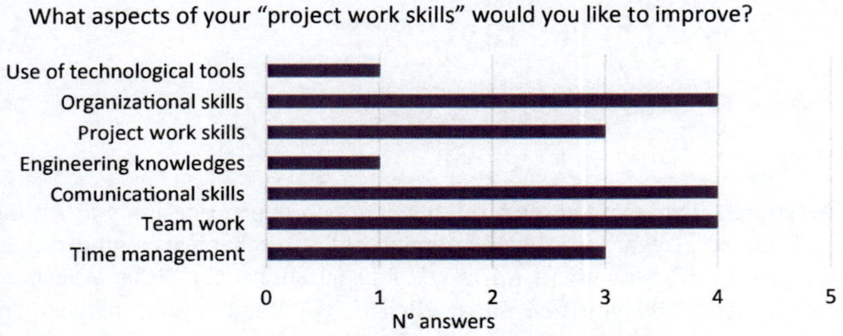

Fig. 6. Answers to "What aspects of your "project work skills" would you like to improve?"

5 Conclusion

One of the strengths of PjBL as an educational strategy is the central role of students at the teaching-learning process. Sometimes students take the solutions of their problems through a project much further than their teacher expects and manage to achieve much more in a very short time than the teacher ever anticipated. PjBL is a methodology that is developed in a collaborative way that confronts students with situations that lead them to make proposals to confront certain engineering problems. The authors have observed along the semesters (not only through the evaluation of the students but also through their work and commitment), that the PjBL has many effects in the development of personal, organizational, technical, and social competences of the students. For this reason, they are committed to continue to improve this module in Mechanical Engineering.

References

1. Donnelly, R., Fitzmaurice, M.: Collaborative project-based learning and problem-based learning in higher education. In: O'Neill, G., Moore, S., Mcmullin, B. (eds.) Emerging Issues in the Practice of University Learning and Teaching. AISHE/HEA, pp. 87–98 (2005)
2. Hortsch, H., Gormaz-Lobos, D., Galarce-Miranda, C., Kersten, S.: Needs-oriented engineering pedagogy - research projects in Chilean Universities. In: Auer, M., Tsiatsos, T. (eds.) The Challenges of the Digital Transformation in Education. ICL 2018. Advances in Intelligent Systems and Computing, vol. 917, pp. 741–753. Springer, Cham (2019)
3. Kolmos, A.: Reflections on project work and problem based learning. Eur. J. Eng. Educ. **21**(2), 141–148 (1996)
4. De Graaff, E., Kolmos, A.: Management of Change: Implementation of Problem-Based and Project-Based Learning in Engineering. Sense Publishers, Rotterdam (2007)
5. Boud, D.: Problem-based learning in perspective. In: Boud, D. (ed.) Problem-Based Learning in Education for the Professions. Higher Education Research and Development Society of Australasia, Sydney (1985)
6. Kolmos, A. Problem based learning, tree – teaching and research in engineering in Europe (2007). In: Gavin, K: Case study of a project-based learning course in civil engineering design. Eur. J. Eng. Educ. **36**(6), 547–558 (2011)
7. Kolmos, A., De Graaff, E., Du, X.: Diversity of PBL-PBL Learning Principles and Models: Research on PBL Practice in Engineering Education, pp. 9–21. Sense Publishers, Rotterdam (2009)
8. Frank, M., Lavy, I., Elata, D.: Implementing the project-based learning approach in an academic engineering course. Int. J. Technol. Des. Educ. **13**(3), 273–288 (2003)
9. Mills, J.E., Treagust, D.F.: Engineering education – is problem based or project based learning the answer? Aust. J. Eng. Educ. **3**, 2–16 (2003)
10. Helle, L., Tynjälä, P., Olkinuora, E.: Project-based learning in post-secondary education – theory, practice and rubber sling shots. High. Educ. **51**(2), 287–314 (2006)
11. Palmer, S., Hall, W.: An evaluation of a project-based learning initiative in engineering education. Eur. J. Eng. Educ. **36**(4), 357–365 (2011)
12. MINEDUC: Metodología de aprendizaje basado en proyectos. https://curriculumnacional.mineduc.cl/614/articles-140166_recurso_pdf.pdf, Accessed 16 Feb 2020
13. OECD: Knowledge for 2030. https://www.oecd.org/education/2030-project/teaching-and-learning/learning/knowledge/Knowledge_for_2030_concept_note.pdf, Accessed 12 Mar 2020
14. OECD: Future of Education and Skills 2030. https://www.oecd.org/education/2030-project/about/E2030%20Introduction_FINAL.pdf, Accessed 12 Mar 2020
15. Creswell, J.W.: Research Design: Qualitative, Quantitative, and Mixed Method Approaches. Sage Publications University of Nebraska, Lincoln (2003)

Introducing Real-World Projects into a Chemical Technology Curricula

D. Sultanova[1(✉)], P. A. Sanger[2], and A. Maliashova[1]

[1] Kazan National Research Technological University, Kazan, Tatarstan, Russian Federation
econsultan@mail.ru
[2] Purdue Polytechnic Institute, Purdue University, West Lafayette, Indiana 47907, USA

Abstract. This article describes the experience of introducing real-world projects into the curriculum of the Department of Innovation in Chemical Technology of Kazan National Research Technological University. The authors examined the challenges of developing competencies of practical project management in undergraduate students, and applied new teaching methods to develop and implement projects in practice. This technique allows the graduating students to build their competencies and satisfy the requirements of their future employers. Therefore, the task was set: to form teams to work on real projects for chemical enterprises with a full immersion in the design process.

Students were assigned to project teams according to their area of interest. Students were given surveys on their personality and style of decision making to learn about themselves. Based the student's interest and on these surveys, a self-portrait of each participant was developed and used to help organize the team, define roles and establish corresponding project tasks. During this implementation, students developed projects offered by Egida, one of Russia's leaders in polyurethane foam production for mattresses. Having mattresses fit into the 2030 goal of a circular economy is a looming challenge for this industry. Egida divided this complex problem into several key areas and challenged the teams to develop innovative solutions. The solutions developed by the students in the process of their project helped the enterprise highlight new ways of development for itself. Students were given a survey to get feedback and help improve the approach.

Keywords: Project-based learning · Project management · Competency building

1 Introduction

It is clear that creativity, team working, leadership, problem solving, inter-disciplinary integration, and project management have become essential skills if engineering and technology students are to remain in high-demand and be globally competitive. [1, 2] These competencies go far beyond the technical knowledge on which many programs concentrate [3, 4]. Industry constantly complains that new graduates require several years of practical training before they can be effective employees. Curricula that only

M. E. Auer and T. Rüütmann (Eds.): ICL 2020, AISC 1328, pp. 362–370, 2021.
https://doi.org/10.1007/978-3-030-68198-2_33

gives theorical knowledge does not prepare the student for the real world. [5] Unfortunately, many educational programs in Russian universities are devoid of practical, applied based instruction. [6, 7] Project based learning (PBL) offers part of a solution to this pedagogical problem. PBL consists of complex tasks and challenging questions or problems that stimulate the students' problem solving, decision making, investigative skills, and reflection. Research suggests that the PBL learning experience tends to have a stronger long-term positive influence on the students building their useful competencies. While PBL has been around for more than 50 years, PBL is not used in many Russian universities. Furthermore, in our innovative approach, the projects are real coming from industry and are not created by the instructors. Real-world research questions and industrial problems are great candidates for PBL projects. But PBL alone is not sufficient. Learning and applying the tools of project management tools must be part of the pedagogical content.

2 Innovation and Project Management

2.1 Innovation in Chemical Technology

The training of specialists in the Innovation area in the field of Innovation Management dates back to 2007, when the first enrollment of students under this program was carried out. A program of study was developed with a focus on innovative chemical technologies. In 2012, it was decided to allocate a new department with a program in Innovation in Chemical Technology (ICT) to ensure interdisciplinary training strengthened with organizational and managerial disciplines [8]. The development of the program also made a gradual transition to include project management training. The program has retained its traditional interdisciplinary nature. Education begins with the study of the basics of chemistry, physics, materials science, engineering graphics and moving into disciplines in industrial technology and innovation, the basics of polymer chemistry, mechanics and technology. Disciplines related to innovation management are mainly introduced in the 2nd year. Starting from the third-year students study the disciplines of project management: the economy of an enterprise, strategic management in innovative organizations, and analysis of the financial and economic activities of the enterprise. The need to increase the competitiveness of graduates determines the importance of maintaining the main development trends: interdisciplinary training, project training, focus on technological entrepreneurship. The visit to KNRTU via a U. S. Fulbright scholar offered the program an opportunity to strengthen the practice-oriented training for the students and to enhance the interaction with the local business community

2.2 Project Management/Innovation Spine

Project management and the tools that support it are more important today than ever before. Global multi-cultural project teams are the norm and project management is essential to efficient team operation. The concept of a multi-year curricular spine has been used in many engineering programs to teach and reinforce particular thematic elements of their programs such as design, product development and project management [9–11]. The concept of a comprehensive spine of training in project management integrates very well with the established ICT curriculum at KNRTU. As part of a Fulbright grant, one of the authors proposed to create and teach a cross-disciplinary, four-course spine in project management and innovation (PMI) offered at the undergraduate level. This PMI spine would have several threads: project management tools such as work breakdown structures and Gantt charts, teaming skills and interpersonal skills, communications in all aspects, and innovation/market defining skills.

During these classes, tools from http://itpmetrics.com are used to enhance the students team experience as well as to provide a window into understanding the team dynamics [12]. The first two of these survey tools, a personality survey and a conflict management survey, allowed the students to see their own individual traits that were being brought into the team. The spine courses break down as follows:

For first year students: basic skills in teaming, brainstorming and group decision making with a hands-on exercise.

For second year students- Personality and conflict management surveys, project work organization, timeline generation, project economics and finances and a competitive team challenge.

For third year students: work breakdown structure, Gantt chart and task relationships, quantitative decision matrix, failure/risk management and the requirement to plan a real project that will change people's lives.

For fourth year students: reinforcement of the project management processes and systems with teams completing a real project for a customer. The academic schedule and content are based on an established curriculum with presentations and three formal reviews modeled after the senior capstone program described in [13].

During the project, several additions to the PMI spine were made. The first change for the 3rd year students added to their scope of work the task of securing funding for the project. It was intended that the second semester would include a funding effort to enable carrying out the project. As will be discussed later, the 4th year class was supported by a company primarily with materials but the 3th year students needed funding to allow them to execute their project. The second change to the PMI spine was to include explicit training on how to make a professional presentation. Most of the students were used to *Powerpoint* slides with all their text on the slides, and reading from the slides with little interaction with the audience. This approach was changed to slides with only key information. The teams had several rehearsals for their final presentations to develop communicative skill.

During the Fulbright grant, this PMI spine was offered to all the students of ICT program as well as invitations were extended to a multi-disciplinary group of students from all the disciplines of the university. However, the full execution was interrupted by the COVID19 pandemic in the early part of the second semester. This paper reports on the experience with the 3rd and 4th year students through the first semester.

3 Applied Projects for 3rd and 4th Year Students

3.1 Projects Addressing Community and Societal Problems for 3rd Year Students

Thirteen students from the third year ICT program were exposed to the third year PMI spine approach. The students in the 3rd year class were asked to create projects on topics that address problems in their community. This class was gender balanced with 7 women and six men in the class. As is usual in Kazan, the class was mixed culturally with five Russian, five Tatar, one Uzbekistan and two Chuvash. After brainstorming with their classmates, two topics emerged. The first topic focused on the need for students to have places to congregate and collaborate with each other on the university campus. If project-based learning and team projects were going to be part of the curriculum, then a place to meet and work together on campus was going to be important. The second topic focused on educating children on the need for recycling. The lack of adherence to recycling of household waste is a big problem in Russia and this team felt that educating school age students would help to support the nascent recycling effort.

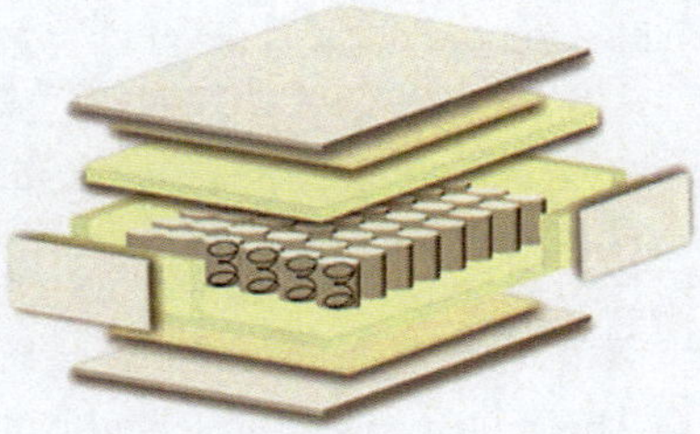

Fig. 1. 95% of mattresses in Russia are constructed with an inner core of springs surrounded by polyurethane foam.

The tasks for the 3rd year teams were to:

1. Research and understand the problem.
2. Develop the requirements for an effective solution.
3. Identify possible solutions and select the best way based on a decision matrix
4. Develop a plan for executing the project.

Both teams made three presentations according to the phase in which they were: first, a review of the requirements, followed by a conceptual design review and then the final proposal review. For the collaboration/communication space, the students identified a suitable location, designed the space with the multi-media capability and areas for team meeting and decided to focus on plastic recycling as the theme for the décor to support getting funding from the plastics manufacturers in the region. The second team developed a series of competitions and activities for middle school children and a curriculum with hands-on modules for education and found a school interested in being the first school to engage in the program.

3.2 Projects Addressing the Recycling of Mattresses Containing Polyurethane Foam

Projects for the 4th year students were provided by Egida, a major supplier of polyurethane (PU) foam for mattresses and who operates a production plant in the suburbs of Kazan. The challenge for this company is to achieve the goals of a circular economy by 2030. Generally, in the world, disposal of mattresses is a major problem. In Russia, the primary mattress design (95%) is a core of springs surrounded by PU foam primarily glued together and enclosed in a cloth cover as shown in Fig. 1. Mattresses are usually not recycled, little of the foam is reused and lots of mattresses each year are put into landfills worldwide. This problem is fertile ground for innovation and Egida challenged the class to develop approaches to solve the problem.

The following problem areas were identified:

1. Financially viable process for mattress collection
2. A process for recycling PU foam,
3. New products/uses for recovered PU foam from mattresses, and
4. New more recyclable approaches to mattress fabrication.

The students were presented with the problems and then asked to choose the areas that they were interested in pursuing. Then students were divided in six teams based on their interests. At the same time, the students completed the itpmetrics' personality and conflict management surveys to help in organizing their teams. The 4th year class consisted of twenty-seven students completing their degree. This class was not gender balanced with eight men and nineteen women. Ten students were Russian and seventeen Tatar. All the teams were led by women. The teams followed a path similar to the 3rd year team: research, requirements, conceptual design and final design. However, the goal of the 4th year projects was to execute the project in the second semester. Due to COVID19, this second phase did not happen and the projects ended with the final design review.

4 Student Feedback

Following the final project presentations, the students were asked to complete a 19-question survey on their experience, rating their experience from 5 to 1, strongly agree to strongly disagree. The results are shown in Table 1.

Table 1. Results of a feedback survey with all students

Question	4th year class				3rd Year class	
	Female (n = 18)		Male (n = 9)		All (n = 8)	
	Average	Std. Dev	Average	Std. Dev	Average	Std. Dev
1. Our team works well together	4.29	0.77	4.89	0.33	4.43	0.53
2. I learned about my style in a team	3.82	0.88	4.00	1.00	4.00	1.00
3. This project gave me a chance to be creative	3.59	0.71	3.78	0.97	4.14	0.90
4. I enjoyed the team I worked with	3.92	0.79	4.56	0.73	4.29	0.95
5. I wanted a different project	2.24	0.83	2.33	1.00	2.71	1.25
6. I improved my project planning skills	4.05	0.66	3.67	1.12	4.29	0.76
7. It was difficult to understand English	2.94	1.14	2.78	1.09	3.14	0.90

(*continued*)

Table 1. *(continued)*

Question	4th year class				3rd Year class	
	Female (n = 18)		Male (n = 9)		All (n = 8)	
	Average	Std. Dev	Average	Std. Dev	Average	Std. Dev
8. The professor did not explain things well	1.33	0.49	1.89	1.05	1.71	0.95
9. I learned how to make a professional presentation	3.35	1.00	3.33	1.22	3.71	0.49
10. I was afraid of having to build something	2.29	0.85	2.22	1.09	2.14	0.69
11. I was able to make a good plan for the project	3.65	0.70	3.67	0.71	3.57	0.53
12. My team was well organized	3.88	0.99	4.44	0.53	3.86	0.90
13. My team had a strong leader	3.94	0.90	4.11	0.78	3.71	0.95
14. Everyone did their share on the project	4.29	0.92	4.67	0.50	4.43	0.98
15. I was happy with my team	4.00	1.12	4.67	0.50	4.57	0.53
16. I was happy with what my team accomplished	4.47	1.07	4.00	1.22	4.43	0.53
17. I want to do another team project	3.29	0.92	3.78	0.83	3.86	0.38
18. I felt that my opinions were heard and respected by the team	4.35	0.79	4.44	0.53	4.43	0.53
19. I felt that my project was not important	1.88	0.86	2.22	1.20	1.71	0.49

Table 2. Student self evaluation of their conflict management styles during their project

Conflict management styles	3rd year		4th year	
	Average	Std dev	Average	Std dev
Dominate	13.00	8.00	19.33	7.60
Integrate	34.17	7.50	24.74	6.86
Compromise	25.83	7.44	28.44	7.28
Avoid	8.83	4.50	10.00	6.07
Accommodate	18.17		17.48	6.13

Take note that some of the questions are written in the negative tense. It should be noted that over 90% of these students have not worked on a team project even though many have worked in collaborative groups on assignments. The results were quite encouraging. The student had high rating for many of the important team qualities with scores above 4.

1 Team worked well together
4 Enjoyed the team
14 Shared in the work
15 Happy with the team
16 Happy with accomplishment
18 Opinions were heard

A good correlation ($R^2 = .53$) was found between having a strong leader and being organized. This correlation was strongest with the men. ($R^2 = .71$). No correlation was found between having a good plan and everyone doing their fair share which was surprising. All the lectures were given in English, often with the assistance of an interpreter, and one concern was how much this affected the class. While there is a large variation in comfortable level among the students with the males having the most trouble, no correlation was found between English language ability and the ability to understand the concepts that lecturer was explaining. 86% of the students rated strongly disagree with question 8. Another concern of the authors was the project topics. Over 50% of the students would have liked a different project with the strongest feelings in the 3rd year class. Nevertheless, there seemed to be a strong feeling that their project was important (note the statement is a negative one). While this is an area for improvement, one of the goals of the PMI spine is to acquaint the student with real world situations. [14] For many students, this is first time they are asked to tackle real world messy problems where no one knows the correct answer and this situation typically makes them uncomfortable.

According to the research done by Google on successful teams [15], they have five traits: dependability, clear goals and roles, personal significance, importance of the project and psychological safety. Our data supports some of these traits in these teams. A good correlation was found between question 19 (the perceived importance of the project) and question 16 (their feeling of accomplishment in the project) suggesting a good element of impact and project importance. Additionally, being happy with the team (question 15) and team members doing their share (question 14) had a good correlation and suggests overall team member dependability. The high responses ($\sim$4.4) and tight distributions ($\sim$.5) to question 18 suggests a reasonable amount of psychological security. In addition to the 19 questions, the students were asked to evaluate the way that they managed themselves during conflicts in the team. Using 100% to account for all their time, allocate points to amount of time spent in each of the itpmetrics.com' conflict management styles. The results are shown in Table 2. The data indicates that there is a large difference between the classes regarding Dominating and Integrating styles with the 3rd year class tending toward Integrating and the 4th year class tending toward Dominating. No correlation was found between the higher presence of the dominating and compromising styles in the 4th year class and the feeling toward their leadership (question 13).

5 Conclusions

The program was a successful launch of a new approach to the Innovation in Chemical Technology curriculum. The students demonstrated very good teaming skills and implemented program management tools with good competency. The survey data suggest that the traits needed for successful projects were present and indicate a good amount of psychological safety in their teams. From the authors observation, this was true.

However, the results of the projects were disappointing especially regarding innovation. The students seemed reliant on looking for and adopting the ideas of faculty instead of having confidence in their own ideas. Creating a sense of security in their own ideas should start earlier in the program and be reinforced throughout the curriculum.

However, there are several areas where adjustments to the program are indicated. It was clear that project topics and the expectations from the project need to more closely match the skills and interests of the students. Student commitment to the project topics are essential for long term success. The experience in this initiative indicates that the 4th year students do not have room for a full project. The project interferes with the thesis that must be completed in the second semester. The program could be accelerated and have the 3rd year students complete a full two semester project instead of doing it in the 4th year. Finally, sources of funding to complete the project need to addressed earlier in the program.

Acknowledgements. The authors wish to thank Timofei Kalinin from Egida for his willingness to support this challenging new initiative with his valuable time, expertise and guidance. In addition, each of teams were supported by faculty mentors from the ICT department. Thank you all because, without mentorship, the students could not be successful.

References

1. Ziyatdinova, J., Sanger, P., Sultanova, D., Burylina, G.: Approaches to Entrepreneurship and Leadership Development at an Engineering University (2016). https://doi.org/10.18260/p.26271
2. Lefterova, O., Giliazova, D., Valeeva, E., Ziyatdinova, J.: Poster: computer-aided translation course for students majoring in engineering. In: Advances in Intelligent Systems and Computing, vol. 1135 AISC, pp. 154–158 (2020)
3. Bezrukov, A., Sultanova, D.: Development of a "smart materials" master's degree module for chemical engineering students. In: Advances in Intelligent Systems and Computing, vol. 1135 AISC, pp. 169–180 (2020)
4. Bezrukov, A., Sultanova, D.: Application of microfluidic tools for training chemical engineers. In: Advances in Intelligent Systems and Computing, vol. 1135, pp. 496–504. AISC (2020)
5. Valeeva, E., Ziyatdinova, J., Galeeva, F.: Development of soft skills by doctoral students. In: Advances in Intelligent Systems and Computing, vol. 1135, pp. 159–168. AISC (2020)
6. Fakhretdinova, G., Dulalaeva, L., Tsareva, E.: Extracurricular activities in engineering college and its impact on students' tolerance formation. In: Advances in Intelligent Systems and Computing, vol. 1134, pp. 143–150. AISC(2020)
7. Valeeva, R., Ziyatdinova, J., Osipov, P., Oleynikova, O., Kamynina, N.: Assessing intercultural competence of engineering students and scholars for promoting academic mobility. Adv. Intell. Syst. Comput. **917**, 815–825 (2019)
8. Sultanova, D., Maliashova, A., Bezrukov, A.: Consistent development of the training program "innovation management". In: Advances in Intelligent Systems and Computing, vol. 1135, pp. 234–243. AISC (2020)
9. Shepard, K., Gallois, B.: The Design Spine: Revision of the engineering curriculum to include a design experience each semester. In: 1999 ASEE Annual Conference, Charlotte, North Carolina (1999). https://peer.asee.org/collections/4
10. Lulay, K., Dillon, H., Doughty, T.A., Munro, D.S., Vijlee, S.Z.: Implementation of a Design Spine for a Mechanical Engineering Curriculum, University of Portland, Pilot Scholars, Engineering Faculty Publications and Presentations (2015). https://pilotscholars.up.edu
11. Frank, B., Strong, D., Sellens, R., Clapham, L.: Progress with the professional spine: a four-year engineering design and practice sequence. Australas. J. Eng. Educ. **19**(1), 63–74 (2015)
12. Sanger, P.: Integrating project management, product design with industry sponsored projects provides stimulating senior capstone experiences. Int. J. Eng. Pedagogy **1**(2), 13 (2011)
13. O'Neill, T.A., McLaron, M.J.W, Hoffart, G.C., Woodley, H.J., Allen, N.J.: The structure and function of team conflict profiles, J. Manage. https://doi.org/10.1177/0149206315581662'

14. Osipov, P., Ziyatdinova, J., Girfanova, E.: Factors and barriers in training financial management professionals. Adv. Intell. Syst. Comput. **916**, 167–175 (2020)
15. Schneider, M.: Google Spent 2 Years Studying 180 Teams. The Most Successful Ones Shared These 5 Traits, (2017). (https://www.inc.com/michael-schneider/google-thought-they-knew-how-to-create-the-pe

Poster: Practice of PBL Education in Collaboration with Printing Company

Akiyuki Minamide(✉), Kazuya Takemata, and Satoshi Fujishima

International College of Technology, Kanazawa, Japan
minamide@neptune.kanazawa-it.ac.jp, {takemata, fujishima}@naptune.kanazawa-it.ac.jp

Abstract. Implementation of PBL (Project Based Learning) is important for engineering education. However, PBL programs have the problem that it is difficult for teachers to manage class and to set the standard for student evaluation. Furthermore, students' problem-solving conclusions may be weak and may not have educational effects. To solve these problems, project themes were limited to one set by the company. Furthermore, in order to improve the level of problem solving for students, the project aims to design products as an outcome of problem solving. After the project, about 60% to 90% of the students felt that the 12 abilities defined by the Ministry of Economy, Trade and Industry in the "Fundamental Competencies for Working Persons" improved their abilities. This paper describes the development project of paper toys and paper crafts using special stickers that can be freely attached and removed as many times as possible, by 4th grade students of Kanazawa International College of Technology, and its educational effects.

Keywords: Engineering design · Active learning · Project based learning

1 Introduction

In recent years, as a typical example of the introduction of project-based learning (PBL) and active learning in Japanese higher education institutions, social implementation education [1, 2] aimed at solving concrete social problems has been attempted. In social implementation education, collaboration with organizations, companies, and communities outside the school is essential. Even if a student discovers a problem in a wide range of areas of society, it is difficult to find an external organization to work with in the process of finding a solution. Therefore, in our study, we tried to implement it in a limited field as the initial stage of introducing social implementation education.

Students' PBL activities may be a shallow survey and the thinking of a solution may end up being simple. In other words, the project is often completed by the students' self-satisfaction. To prevent this problem, it is effective to let students create their own ideas as prototypes. In general, introducing prototyping in short-term projects is difficult due to the problems of student skills, required tools, and time constraints. Therefore, in this study, we obtained the cooperation of a printing company that manufactures paper products that make prototype production easy.

M. E. Auer and T. Rüütmann (Eds.): ICL 2020, AISC 1328, pp. 371–378, 2021.
https://doi.org/10.1007/978-3-030-68198-2_34

Since 2014, the Department of Electrical and Electronic Engineering of International College of Technology, Kanazawa has been conducting social implementation education in collaboration with local printing companies. [3] This paper describes the projects and educational effects that were implemented in FY2018 in collaboration with local printing companies.

2 Outline of Educational Practice

This education is being given as one of the themes of "Design and Drawing" for the 4th year students of the Department of Electrical and Electronic Engineering. The course included one 100-min lesson each week for 30 weeks. Since the target students were unfamiliar with the project activities of the team, we lectured for the first 20 weeks to give students the skills necessary for team activities.

In the first 15 weeks, students were taught to use a freehand sketch technique called communication drawing [4, 5] to smooth out activities and to organize and summarize ideas; in the next 5 weeks, students were taught techniques for analyzing engineering design processes and problems, generating ideas, and organizing ideas. Project activities in collaboration with printing companies took place in the last 10 weeks.

2.1 Engineering Design Process

The project was basically executed in the following engineering design process [6]: (1) discovering the problems, (2) clarifying the problems, (3) creating ideas, (4) evaluating/selecting ideas, and (5) realizing the idea Before engaging in actual project activities, we taught students how to proceed with team projects, how to analyze problems, and how to devise and organize ideas.

2.2 Project Activities

The local printing company, Wellco Holdings Co., provided students with a theme. Therefore, the first step, identifying the problems in the engineering design process, was done by the company, and students started with the second step, clarifying the problems. The theme provided by the company was "Proposed toys with stickers that children can stick or peel off freely in casual restaurants." This special sticker, called a magic sticker (MS) [7], has the following features: It is printable on both sides of the sticker, is printable on release paper, and has partial paste on the back of the sticker.

Many families with children visit casual restaurants in Japan. Gifts for children tend to be simple toys, and children play with them while they wait for their food to arrive. The company challenged students to propose a toy using MS that can be enjoyed by children younger than primary school age.

3 Student Project

The student project was carried out using the following procedures.

(1) Discovering the problems (Providing the company's theme to students)

At the beginning of the project, engineers from Wellco Holdings Co. came and provided the company's selected theme to students. They explained to students the features of the MS, problems, examples of its present use, and so on.

(2) Clarifying the problems
After listening to the explanations, students were divided into seven teams to begin team activities. Each team consisted of three or four members. Students extracted features of the MS and clarified the problems of this project.

(3) Creating ideas
Students created many ideas using the brainstorming method.

(4) Evaluating and selecting ideas
Figure 1 shows an idea evaluation sheet written by students. In this process, each student team initially established five criteria for evaluating ideas, a number of the ideas devised were scored in accordance with the criteria, and the team selected their best ideas. In the example of Fig. 1, Ease of realization, Interesting, Novelty, Easy to use, and Clarity of target children are listed as five evaluation items, and four ideas are evaluated. As a result, the origami with quizzes got the highest score and became the team's best idea.

(5) Implementing and improving ideas
A prototype was produced at the end of the project activity, but in order to refine its ideas, the team was organize its ideas into B1-sized posters and present them to printing company engineers and other students. In this presentation, the students were able to get advice from others outside the team, and the ideas were improved. Figure 2 shows a photograph presented using a poster created by a student and presented to a engineer of the company for advice.

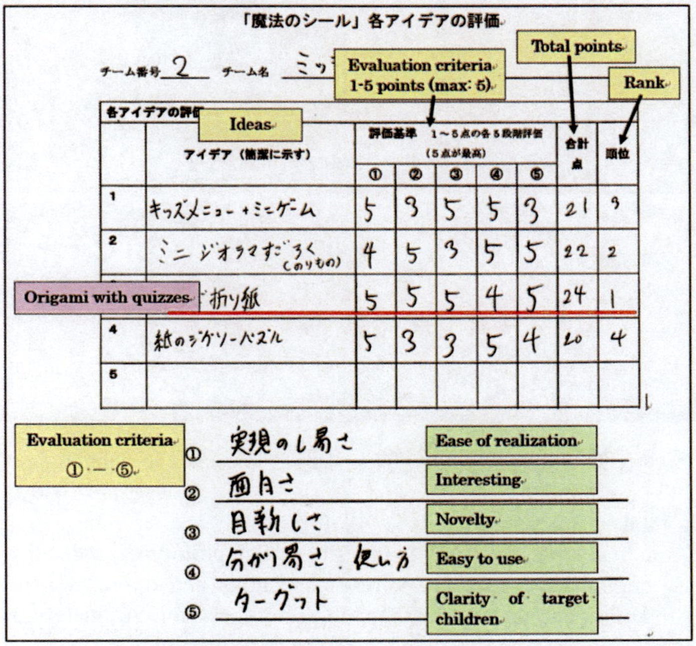

Fig. 1. Idea evaluation sheet.

Fig. 2. Posters produced by students.

(6) Producing a prototype
To confirm their ideas, students made prototypes using simple materials such as printing paper, glue, tape, and colored pencils. Prototyping using paper has the advantage that you can create prototypes with only simple tools such as cutters and scissors, and you are not restricted by time, place, or student skills. These strengths are very important because the project aims to get students to think deeply. Figure 3 shows prototypes created by students.

Fig. 3. Prototype created by students. (Origami with quizzes)

(7) Making final presentations
Each team made presentations using slides and prototypes created by printing company engineers and fielded various questions and opinions from the engineers. Students gave presentations using prototypes, not just materials created in PowerPoint. Figure 4 shows pictures of the final presentation.

(a) (b)

Fig. 4. Final presentation by students. (a): State during presentation and (b): Engineers checking the prototype.

4 Survey of Educational Effects

In order to verify the educational effect of this project, we conducted a questionnaire survey for students. As a questionnaire item, "Fundamental Competencies for Working Persons" defined by the Ministry of Economy, Trade and Industry (METI) was used. METI defines this basic abilities as follows. The basic abilities of working persons can be categorized into skills related to thinking, behavior and teamwork. Thinking skills are divided into skills that include problem detection, planning, and innovation. Action skills are divided into the ability to take initiatives, perform actions, and influence others to do certain things. Teamwork skills are divided into the ability to communicate with others, listen to others, respond flexibly to changing situations, understand situations, adhere to the rules, and control stress.

Table 1. Ability items investigated.

No.	Ability
A1	Independence
A2	Approach power
A3	Executive power
A4	Problem recognition
A5	Planning
A6	Creativity
A7	Transmission power
A8	Harking power
A9	Flexibility
A10	Situation recognition
A11	Discipline
A12	Stress control ability

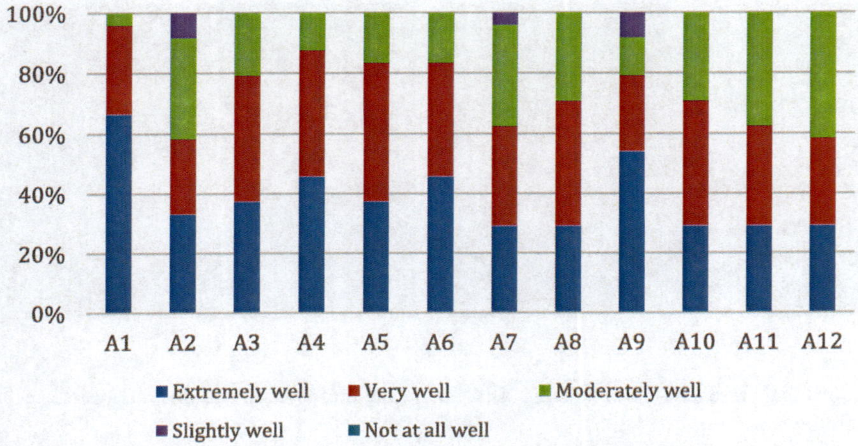

Fig. 5. Questionnaire survey results regarding changes in 12 abilities defined by "Fundamental Competencies for Working Persons". (n = 24 students)

A subjective evaluation was made of how much the students themselves felt that their various abilities (Table 1) improved through this project. The survey target was 24 students who had taken the classes. Figure 5 shows the results of the survey.

In all abilities, about 60% to 90% of students felt their abilities improved. In particular, many students felt that their Independence (A1), Problem recognition (A4), Planning (A5), and Creativity (A6) abilities were relatively improved. On the other hand, there were relatively few students who felt improvement in Approach power (A2), Transmission power (A7), Discipline (A11), and Stress control ability (A12).

Furthermore, we investigated to see what factors contributed to the improvement of students' abilities. We asked the student the following question. "Do you think that each of the following factors contributed to your ability improvement?" The results are shown in Fig. 6.

Q1 Project theme was given by the company.

Q2 Present results to corporate engineers.

Q3 Get advice on your ideas from engineers.

Q4 Expressing your ideas in the form of prototypes.

In each factor, 81% to 96% of students answered positively. In particular, it became clear that getting advice from the engineers of the company and shaping their ideas as prototypes contributed to capacity building. Furthermore, the engineers' participation in the students' projects significantly raised the quality level of students' projects. Since students were able to come up with ideas that engineers had not previously thought of, we determined that social implementation education is effective for engineering education.

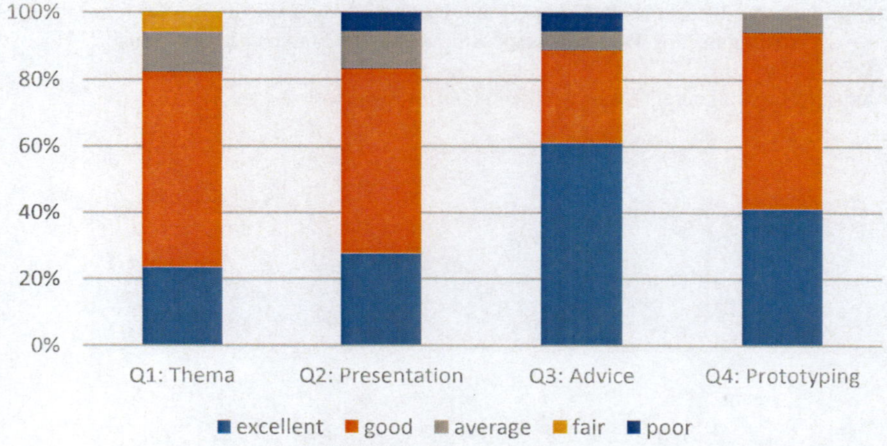

Fig. 6. Results of the survey

5 Summary

In this paper, we described social implementation education practiced in cooperation with a local printing company. Some of the ideas proposed by students were of high quality that led to commercialization. From the results of the questionnaire, about 60% to 90% of the students that the 12 abilities defined by the Ministry of Economy, Trade and Industry in the "Fundamental Competencies for Working Persons" improved. Therefore, it can be said that PBL in collaboration with local printing companies is effective in improving students' ability.

Acknowledgment. We would like to thank the engineers at Wellco Holdings Co. for helping our students in their project work. This work was partially supported by a Grant-in-Aid for Scientific Research from the Ministry of Education, Culture, Sports, Science, and Technology.

References

1. Yagihita, H., Fujio, M.: Incorporating a social implementation program into a manufacturing education program in Japan: case study in collaboration with a medical facility. Procedia Manufact. **10**, 1054–1065 (2017)
2. http://www.innovative-kosen.jp/Innovative-Japan-Project-by-KOSEN/
3. Minamide, A., Takemata, K.: Practice of social implementation project in engineering education. In: Proceedings of 10th International Conference on Education and New Learning Technologies, pp. 8639–8643 (2018)
4. Nakamura, S., Matsuishi, M.: Education of drawing courses and students' achievements (how to develop and make the best use of freehand sketch skills). In: The 3rd International Conference on Design Engineering and Science, Pilsen, pp. 43–48 (2014)

5. Nakamura, S.: Idea Drawing: How to Draw. Tanaka & Shobundo Graphic Art Co., Ltd (2011)
6. Cross, N.: Engineering Design Methods–Strategies for Product Design. WILEY, Hoboken (2008)
7. https://www.well-corp.jp/factory/all/seihin26/ [in Japanese]

Poster: Project Based Learning Using Digital Storytelling: Educational Program for Students Before Learning Full-Scale PBL Practice

Kazuya Takemata(✉) and Akiyuki Minamide

International College of Technology, Kanazawa, Ishikawa 921-8601, Japan
takemata@neptune.kanazawa-it.ac.jp

Abstract. This paper describes on the education program that we have developed for colleges of technology in Japan. The college of technology in Japan is Japan's unique educational system that aims at high-school and college-age students. The five-year consistent education system specializing in engineering has garnered attention under the name of KOSEN from Asian nations. Thus, this paper assumes the second-grade students at our college of technology to be students who have not been introduced to PBL yet. The group activity-based PBL (Project Based Learning) process requires the ability to express one's own ideas to others (communication skill), the ability to abstract problems and develop one's thoughts (computational thinking skill), and the ability to move ahead with tasks in a planned manner (project management skill). Therefore, targeting these students, this study examines whether digital storytelling practice is feasible as an education program that broadens the above mentioned three skills. It is required in digital storytelling to condense one's thoughts and present them as digital contents. We think that the affinity is strong between digital storytelling and PBL, as the processes that learners experience through digital storytelling activities are similar to learners' experience in the processes of the PBL program that is based on our engineering design process. According to a questionnaire survey conducted after the class, our education program received positive feedback from 90% of the students who took the course.

Keywords: Digital Storytelling · Engineering design · Computational thinking

1 Introduction

In the classroom setting where the PBL (Project Based Learning) is employed, educators can take a variety of approaches to creative education using PBL according to the grade of learners by setting a subject to cover in class as a "familiar issue," an "issue that the local community is facing," an "issue to tackle in collaboration with companies," or the like. The PBL exercise attaches weight to solving issues through teamwork, which contributes to enhancement of students' generic skills required after they start working for companies. As PBL is performed by teams composed of students from various departments in our college, it is expected that it will improve students'

M. E. Auer and T. Rüütmann (Eds.): ICL 2020, AISC 1328, pp. 379–385, 2021.
https://doi.org/10.1007/978-3-030-68198-2_35

capabilities to think about matters from a multifaceted perspective. PBL is one of the most effective educational methods available in higher education, and a number of studies have been conducted regarding more effective, PBL-based classroom practice [1–3]. Thus, this research explores introductory education programs that will enable learners to carry out PBL exercise in a smooth and more satisfactory manner in higher education.

Our practical experience of education has shown that the PBL process based on group activities requires the ability to express one's own ideas to others (communication skill), the ability to abstract problems and develop one's thoughts (computational thinking skill) [4–6], and the ability to move ahead with tasks in a planned manner (skill to see the big picture of matters). If educators can successfully provide students with an education program that nurtures these skills before they are actually introduced to PBL exercise, we believe that the students by themselves will develop more challenging PBL practice afterward. In addition, these three skills will have favorable impacts on students' learning through "the effective process by which ideas that could lead to problem solving are put forward," which is referred to as design thinking that is utilized in PBL exercise.

This paper reports on the education program that we have developed for colleges of technology in Japan. The college of technology in Japan is Japan's unique educational system that aims at high-school and college-age students. The minimum period required for graduation at the college of technology in Japan is five years, meaning it is a higher educational institution established in a manner to combine the Japanese three-year high school system and two-year junior college system. Students at a college of technology in Japan will join companies after graduation, or go on to other universities as seniors. The five-year consistent education system specializing in engineering has garnered attention under the name of KOSEN from Asian nations as an educational model of developing human resources that are linked directly to the industrial fields [7].

Thus, we gave our attention to digital storytelling as a teaching material that develops these three skills before students work on PBL exercise [8, 9]. We think that the affinity is strong between digital storytelling and PBL, as the processes that learners experience through digital storytelling activities are similar to learners' experience in the processes of the PBL program that is based on our engineering design process. This paper assumes the sophomores at our college of technology to be students who have not been introduced to PBL yet. These learners will be required to engage in more advanced PBL processes in higher grades (their fourth and fifth years) at our college and after they get transferred to university. Therefore, targeting these students, this study examines whether digital storytelling practice is feasible as an education program that broadens the above mentioned three skills.

2 Digital Storytelling Exercise

Digital storytelling is an activity of producing and presenting "digital stories that represent people's thoughts and feelings," which was started in the 1990s in the United States [10]. The length of each digital storytelling work is about two to four minutes and a number of universities in Europe have adopted digital storytelling in their respective curricula. Digital storytelling has been employed in Japan, too, as an activity of expressing and presenting ideas of learners, ranging from junior high school students to college students.

We offered "Digital Storytelling Exercise" in "Creative Design II" that sophomores of the department of global information at our college took in academic year 2018. A total of 30 participating students were divided into a first group of 15 (for the spring and summer quarters) and a second group of 15 (for the fall and winter quarters). When one group was attending the digital storytelling class, the other group was engaged in another course (computer graphic design). For both groups, the period of the course was 15 weeks, with about two hours allotted to each week (two 50-min lectures per week).

2.1 How to Make an Animation

In the first six lessons of 15-week course (from the first week to the sixth week), the students learn animation theory by creating two assignments as follows:

Scanimation
In the first three lessons of our 15-week course (from the first week to the third week), the students produced animations based on the principle of the scanimation techniques. The students took on challenges for "entertaining the audience" with their works. Figure 1 is the work of one of the students. The students and teachers together watched all the animations created by each student at the end of the third week. The teachers appraised all the works submitted and made comments on each of them.

Fig. 1. Student's works, "A girl dancing in waving dress", produced in a Scanimation production class.

Flip-Book Animation
In the next three lessons of our 15-week course (from the fourth week to the sixth week), the students created flip-book animations. A flip-book is a work where an array of slightly different pictures is drawn on each page of sticky notes in a way to make the pictures seem to be moving due to the residual effect brought about by turning the pages quickly. The aim of this assignment is to offer the students an opportunity of

giving it their all to "communicate some messages." In the same way as the previous assignment, the teachers appraised all the works submitted and made comments on each of them. Figure 2 is the work of one of the students.

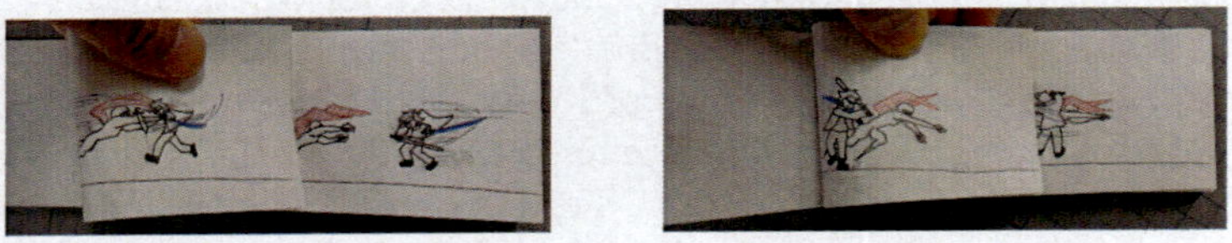

Fig. 2. Student's works produced in a flip-book animation production class.

2.2 Digital Storytelling Exercise with Stop Motion Animation

The last nine lessons of the 15-week course (from the seventh week to the fifteenth week) were dedicated to digital storytelling. Digital storytelling here means an activity of producing two to three-minute-long stop motion animation with a well-constructed plot according to the theme that the teacher gave. The production processes are to 1) sort out problems with the theme and collect materials, 2) establish a story (create storyboards), 3) photograph static images for creating stop motion animation, and 4) edit the pictures for producing a final short video.

Stop motion animation is produced by gradually moving static objects and take several pictures for each film frame, editing the pictures as a short video using film editing software, such as Windows Media Player, and adding music and sound effects as needed. The students were instructed to take about 4 pictures per one second of their videos to produce an about two-minute-long animation work. This means that they had to finish photographing approximately 500 images in total within a predetermined time frame. The students efficiently took pictures of still images and completed their works based on their respective storyboards that they prepared in advance. Figure 3 shows the work of one of the students.

Fig. 3. Student's works produced in a stop motion animation production class.

The students presented their works at an interim reporting session held around the tenth week, and received advice from the teachers. After the session, some students were inspired by other students' animations and thus rethought and made improvements in their own works. In the fifteenth week, both the students and the teachers watched all the students' works in the same manner as mentioned above in the sections of scanimation and flip-book animation. Based on five criteria, which are 1) meticulousness of picture drawing, 2) movement of objects and living things, 3) camera angle changes, 4) the plot, and 5) whether the characteristics of stop motion was fully exhibited, the teachers appraised all the works submitted and made comments on each of them.

3 Questionnaire Survey After the Class

After the class, a questionnaire survey was carried out, targeting 26 students who had submitted the above the work. 26 out of 30 students gave valid answers (Fig. 4). The following questions were asked, in order to grasp the situation of students after this class.

- Q1: After this exercise, do you think that you became able to summarize and convey your idea to others?
- Q2: After this exercise, do you think that you became able to reconsider your design while developing it and redesign it (to embody your idea)?
- Q3: After this exercise, do you think that you became able to plan and engage in something within a limited period of time?

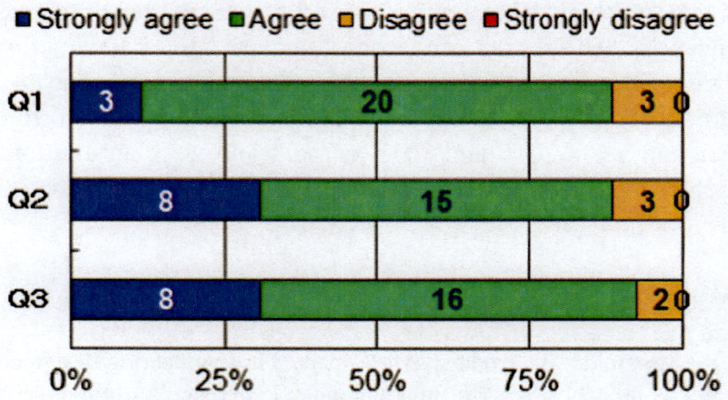

Fig. 4. Questionnaire survey conducted targeting students.

Q1 is a question regarding "to convey ideas to others," and about 88% of students gave positive answers. Q2 are regarding "the abilities to develop ideas and embody them." For the question, about 88% of students gave positive answers. Q3 is a question about project management. About 92% of students answered these questions positively, "I certainly became able to do so" and "I became able to do so." These results are reasonable as an introductory course.

The results of the questionnaire survey carried out after the class have revealed that about 90% of the students gave positive feedback in response to all questions. The results have revealed that about 30% of the students gave strong agreements in response to some questions (Q2 and Q3), such as "Have you learned how to work on assignments in a planned manner?" and "Have you learned how to develop your thoughts?". However, only about 10% of students showed a strong consensus on the question (Q1), such as "Have you learned how to communicate your thoughts?" It is considered that this is attributable to these students' high expectations about the quality of their own animation works, which seem accordingly make them assess that their own "communication skill (of delivering their messages) was not satisfactory."

4 Conclusions

The group activity-based PBL (Project Based Learning) process requires skills such as communication skill, computational thinking skill and project management skill. In this study, we gave our attention to digital storytelling as a teaching material that develops these three skills before students work on full-scale PBL exercise. In most cases PBL for students in programming design courses provides students with the materials to create computer programming. Hence in our research, we adopted digital storytelling instead of computer programming to assist learners in various ways.

Before working on digital storytelling using stop motion animation techniques, the students engaged in two animation creation assignments as preliminary steps to learn about how animations are produced. The results of the questionnaire survey carried out after the class have revealed that about 90% of the students gave positive feedback in response to questions. The results of the questionnaire survey have allowed us to get the sense that this education program is feasible as a preliminary education program to run before introducing students to PBL. We have already conducted this educational process in academic years 2016 and 2017 as well, and we plan to submit a comprehensive research report on three-year implementation of the program.

References

1. Takemata, K., Minamide, A., Kodaka, A., Yamada, H.: Engineering design education based on the CDIO approach. In: 19th International Conference on Engineering Education, pp. 759–766 (2015)
2. Bonwell, C.C., Eison, J.A.: Active learning; creating excitement in the classroom. Jossey-Bas (1991)
3. Hsu, T.C., Chang, S.C., Hung, Y.T.: How to learn and how to teach computational thinking; suggestions based on a review of the literature. Comput. Educ. **126**, 296–310 (2018)
4. Wing, J.M.: Computational thinking. Commun. ACM **49**(3), 33–35 (2006)

5. Miller, L.D., Soh, L.K., Chiriacescu, V., Ingraham, E., Shell, D.F., Ramsay, S., Hazley, M.P.: Improving learning of computational thinking using creative thinking exercises in CS-1 computer science courses. In: 2013 IEEE Frontiers in Education Conference (FIE), pp. 1426–1432 (2013)
6. Kelleher, C., Pausch, R.: Using storytelling to motivate programming. Commun. ACM **50** (7), 58–64 (2007)
7. Isami, H.: Education Reforms in the National Institute of Technology (KOSEN). Tokkatsu; the Japanese educational model of holistic education, World Scientific Publishing, Singapore, pp. 283–298 (2019)
8. Kordaki, M., Kakavas, P.: Digital storytelling as an effective framework for the development of computational thinking skills. In: EDULEARN17 Proceedings, pp. 6325–6335 (2017)
9. Tsuchiya, Y.: The global diffusion of digital storytelling; a cross-frontier and transformative media practice. Studies in Media and Society, Nagoya University, no. 5, pp. 77−84 (2015)
10. Robin, B., Pierson, M.: A multilevel approach to using digital storytelling in the classroom. In: SITE 2005 (Society for Information Technology & Teacher Education International Conference), pp. 708−716 (2005)

Good Practice of Using Project Education in Technical Subject Teaching

Ivana Tureková and Alena Hašková(✉)

Faculty of Education, Constantine the Philosopher University in Nitra, Nitra, Slovakia
{iturekova, ahaskova}@ukf.sk

Abstract. In the subject structure, the subject of Risk Management belongs to the main topics of the essential knowledge in the field of study Safety Sciences. The survey carried out in the expert public showed that the weak preparation of students to solve practical problems was one of their most significant lacks and disadvantages. The objective to improve the preparedness of students for real practice was the main challenge for innovations in teaching this subject. The implementation of the activating method of project education helped to achieve better knowledge and improve skills and competences of students. Outsourcing of real tasks, application of the specific technique in analysing and assessing of risks, teamwork and self-evaluation of achieved results, comparing of outputs, persuasive argumentation represent the results of project education, and they were valid also in this subject. A questionnaire survey, which respondents were the students, verified this project education. After implementing this innovative form of teaching, students compared their knowledge at the beginning and the end of the teaching process. The obtained results confirmed that the method of project education increased students' interest in this subject. The subject of Risk Management has become more attractive for students, and their preparation for the lessons has improved, as well. The students also confirmed that the method of project teaching had increased their knowledge in the given subject. At the same time, the teamwork has also brought along a more objective mutual evaluation of students between themselves.

Keywords: Innovativeness of teaching · Project education · Risk management

1 Introduction

Graduates of the study programme Occupational Safety and Health have the subject of Risk Management included in the structure of their subjects, as well. Content of this subject is focused on looking for risks, their assessment and evaluation in real practice and specific situations. Time allocation of the subject is two exercise-style lessons per week Traditionally teaching of the subject has been carried out in a classroom and has been based on frontal solving of the assignments given to students.

A survey findings showed that the expert public sees the lack of these graduates in their weak preparedness to solve practical tasks [1]. The primary challenge of how to improve the preparedness of the graduates for their future occupation was the implementation of project teaching into their subjects, as a method used also in managerial

M. E. Auer and T. Rüütmann (Eds.): ICL 2020, AISC 1328, pp. 386–396, 2021.
https://doi.org/10.1007/978-3-030-68198-2_36

projects [2]. It is an efficient way of teaching where we can apply some progressive didactic methods such as problem teaching, cooperative teaching and discussion [3–5]. Although it is difficult to acquire the explicit understanding of how to apply formal methods and tools in unique situations we deal with in most managerial projects [6], it was the subject of Risk Management that seemed to be suitable for applying this approach.

The primary objective of project teaching is to involve students actively in the learning process of acquiring knowledge. The characteristic feature of this process is its openness. The teacher brings about problematic scenarios and questions which force students to think about what they are learning. The project scenarios have only the framework character, and they are completed during the problem-solving itself. The realisation of the project depends on students, their creativity, phantasy, critical thinking, inner motivation, as well as their interests and needs [7].

The educational and formative objectives of project teaching primarily focus on developing students' abilities and habits:

- to work independently and creatively,
- to plan and finish their work,
- to be responsible for their work and to overcome obstacles,
- to work with information (e.g. books, encyclopaedias, internet),
- to defend their work, to present it in public, and to express themselves correctly,
- to use arguments,
- to cooperate, communicate and accept other opinions,
- to evaluate their work and work of their colleagues [8–10].

2 Background of the Stated Problem

The subject of Risk Management aims at students' acquisition of risk assessment methods. Students learn to choose the most suitable method for a specific project (activity), to implement this method in practice and teamwork, to interpret results of risk acceptability, and to propose relevant measures [11–13].

The risk management phase begins with revealing procedures of risk factors. In most cases, we use the expert evaluation method [14]. The quality of performed work in this phase depends on the information provided by the procedure, the level of experts' qualification, and the risk culture level of respondents. As a result of processing the identified risk factors by different methods (What if, HAZOP, Delphi method), the management risks are identified according to the ranks assigned to them [15]. The analysis includes not only quantitative risk assessment, but also low-rank risk analysis, which involves developing management measures in place based on experience from previous years [16–18].

Figure 1 depicts the gradual steps in risk assessment that represents the subsystem of risk management [19].

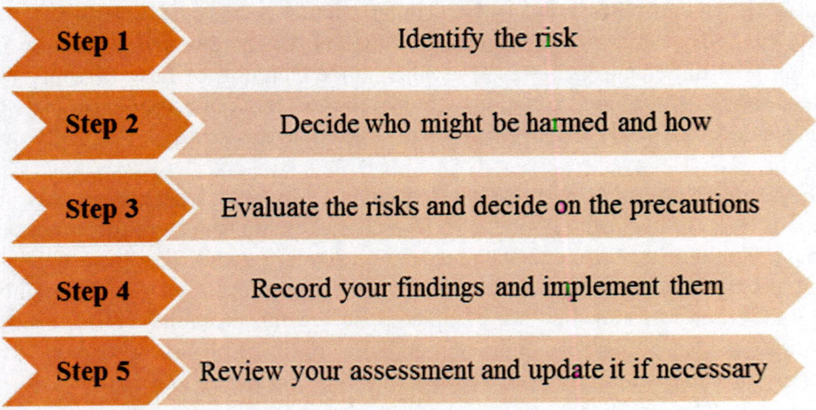

Fig. 1. Risk assessment steps.

Risk assessment is defined as "a process that begins with risk identification and analysis, through which we establish the probable severity of harm or damage. Subsequently, we estimate probability of the incident or exposure occurrence, and conclude it with the risk statement" [20].

3 Teaching Risk Management Based on the Use of Project Education

As it is above-mentioned as a weakness of the graduates of the study programme Occupational Safety and Health, among others also in relation to the subject of Risk Management, was identified the graduates' weak preparedness to solve practical tasks. In relation of the subject Risk management we decided to eliminate this weakness through implementation of the activating method of project education into this subject teaching.

Implementation of the project education into the subject was based on the exact definition of the problem which the students were expected to solve within their projects, main outputs of which were posters showing implementation of the particular method of risk assessment which was the matter of the subject teaching during the given lesson. The whole implementation (in relation to each lesson and the concerned method taught during the lesson) consisted of four phases which in a graphical form are presented in Fig. 2.

According to the scheme presented in Fig. 3 teaching the subject of Risk Management based on the use of project education was carried out in hereinafter described way.

Lectures had a presentation-problem character, and they took place in an adequately long time before the practical lessons. The lecturer primarily aimed to provide students with enough information about the specific topic and sources for further study. The lecturer explained the principles of particular method of risk assessment. Students had

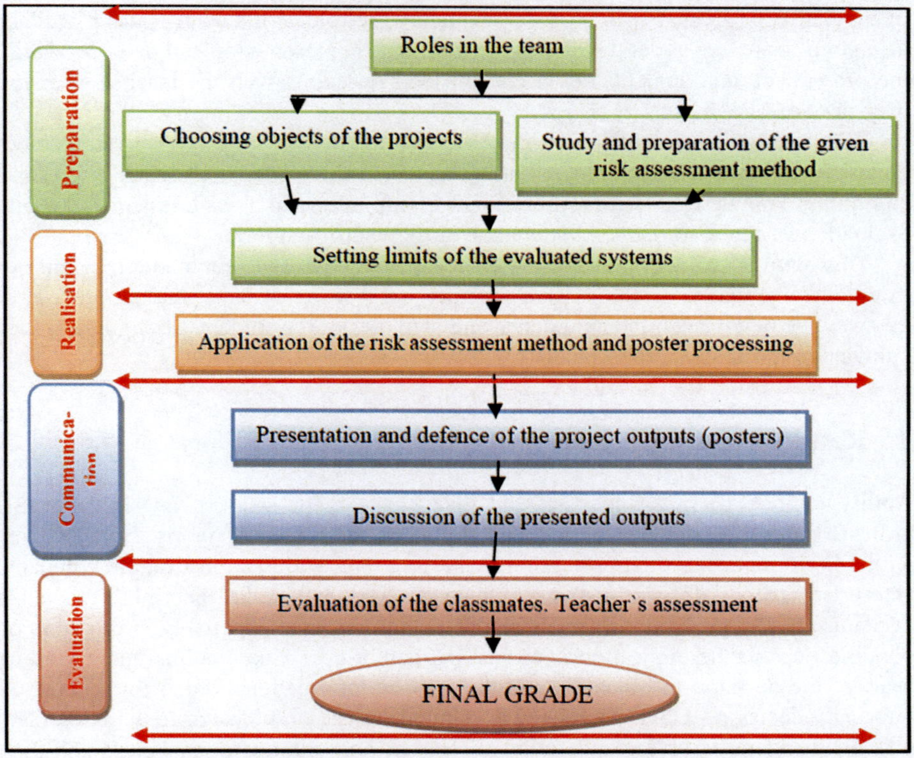

Fig. 2. Scheme of the implementation of the project education into the subject Risk Management teaching.

to know both positives and negatives of the method, which the teacher presented (taught). Moreover, the students needed to understand how to implement the presented method in their project which they were expected to prepare. They had to be aware of their time complexity, and they were to know the ways to ensure sustainability of the project.

The practical lessons always began with an excursion to the workplace which was the object of the problem tasks solving. Students with their teacher visited the workplace where a music pavilion was under the construction. At the build ground there were working 20 workers. Students observed the activity of all of them. In this environment there were discussed and deeper explained the exact definitions of the project assignments and task to process a poster showing implementation of the concerned method of risk assessment (explained within the previous lecture). The projects were processed in teams of 3–4 students with a chosen team leader and set time for project processing. In dependence on the type of the given assignment there might be also other roles given to the particular team members, as e.g. project manager, construction manager, foreman, safety engineer, etc. The main principle of a student nomination into a particular role was rotation of the given position (role) among the students (members of the particular teams) within the term (mainly the role of the team leader). Students also got information about the date of the final project presentation and defence. During the project solving, students could consult their questions with the teacher whenever they needed.

Efficiency of this innovative method of teaching the Risk Management subject was done by means of a questionnaire survey in which the students evaluated their input and output knowledge. The questionnaire was administrated at each lesson and a final evaluation of the recorded data was done at the end of the term.

This innovative way of teaching was focused not on the subject matter transmission from an active teacher to the passive students, but it was a dynamic mutual process. In this way the newly acquired experience and activities of the students strengthened their knowledge in the concerned field.

4 Results

Ability to look for possible risks and to evaluate them is a crucial skill of OSH professionals [21] and creates a precondition of their success in risk management within performance of their profession. Assessing risks within an institution, company, enterprise enables decision-makers to manage risks and make the right reasonable decisions. Safety professionals must be able to communicate effectively information on possible risks to the highest decision-making authorities. This requires understanding heart of the decision to be made, and the specific information needed for making an informed decision. OSH professionals should select and design risk assessment methods to identify, assess and communicate not only the operational risks and their controls but also the resulting trade consequences and subsequent impacts.

The questionnaire, by means of which efficiency of teaching the subject Risk Management based on the project education was carried out, consisted of closed items to which the students responded using the assessment scale identical with the evaluating criteria of tertiary education (A for excellent, B for very good, C for good, D for average, E for satisfactory and FX for insufficient). Students evaluated their input and output knowledge (i.e. their knowledge at the beginning and at the end of each of the practical lessons) and as well they rated their presentations of each of their projects (i.e. they continuously evaluated all projects they prepared to each of the taught risk assessment method). The questionnaire consisted of three parts.

a) students' knowledge at the beginning of the lesson (how students evaluated their knowledge of the given issue);
b) evaluation of the poster presentation (quality of presentation, oral expression, use of didactic aids, explicitness and correctness of the presented problem solution);
c) subjective evaluation of acquired knowledge.

Moreover, the students could express in the questionnaire (which was always anonymous) also their opinions, attitudes and criticism to the carried out lessons and taught issue.

As a representative lesson hereinafter we present results recorded after the seventh week of the term when the problem task was application of the "What if" risk assessment method.

Purpose of the given questionnaire was to find out:

a) evaluation given by students to their classmates presenting their projects,
b) how students perceive improvement of their knowledge based on the lectures, presentation and discussion at which they participated.

Sixteen students filled in the questionnaire and four students presented their project. Table 1 shows overview of the results of the respondents (students) according to the project phases. Graphs in Fig. 3 present results of the students' self-evaluation.

Table 1. Overview of the questionnaire results.

Evaluation sheet							
Presenting team (names):							
Classification (grade)	A	B	C	D	E	FX	
Success rate (%)	91–100	81–90	71–80	61–70	51–60	Less than 50	
Unit	(%)	(%)	(%)	(%)	(%)	(%)	Guidelines to the questionnaire survey
Knowledge at the beginning of the lesson	0	43.7	25.0	18.7	6.3	6.3	What grade would I give myself considering my preparation, participation in the lecture, study of other materials, experience?
Quality of the team presentation	12.5	12.5	25.0	25.0	12.5	12.5	Presentation without aids, reading the text
	6.3	25.0	18.7	18.7	25.0	6.3	Articulation and communication skills
	25.0	25.0	25.0	18,7	6.3	0	Method explanation
	18.7	25.0	18.7	18.7	12.6	6.3	Suitability of examples
	25.0	18.7	12.6	18.7	18.7	6.3	Clarity
	31.3	18.7	12.6	18.7	18.7	0	Visualisation
	18.7	18.7	18.7	12.6	25.0	6.3	Suitability
Average evaluation	19.7	20.5	18.7	18.7	17.0	5.4	
Knowledge at the end of the lesson	6.2	50.0	25.0	12.6	6.2	0	What grade would I give myself? Did I comprehend the method with the use of examples? Am I able to work independently using this method?

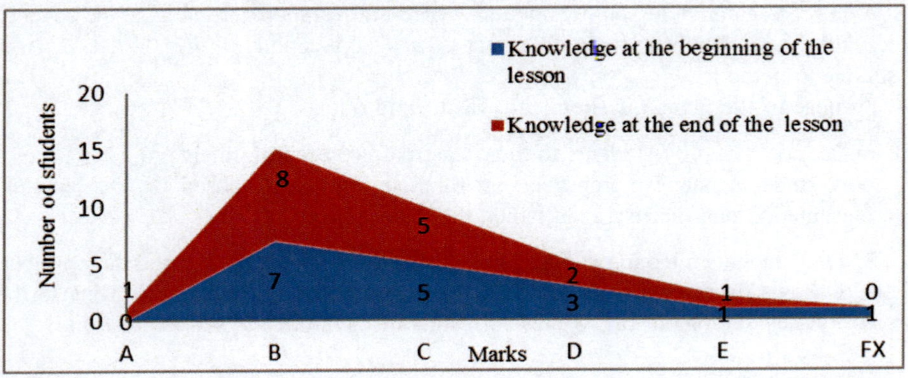

Fig. 3. Subjective evaluation of the knowledge acquisition related to the taught topic of WHAT IF method.

5 Discussion

Purpose of the implementation of project teaching into the subject of Risk Management teaching was:

- to increase interest of students in the subject by giving them attractive tasks,
- to improve preparation of students for lessons,
- to help students to recognise problems and to learn principles of solving them,
- to increase competitiveness and objective self-evaluation among students,
- to improve IT skills of students by means of making posters,
- to familiarize students with underlying pedagogical principles of presenting,
- to train students for team working and brainstorming,
- to develop at students their personal responsibility for solving the given task,
- to evoke constructive discussions on the presented topic,
- to increase quality of bachelor theses and
- to improve employability of the OSH graduates in the labour market.

From the students' outputs, it is evident that in the project teaching none of the students evaluated their knowledge as insufficient (grade FX), although one student stated such evaluation at the beginning. One-half of the students rated their knowledge above average and a quarter of the students achieved average results. Two students (12.6%) evaluated themselves satisfactorily and one student achieved excellent results (grade A). One student thought that his knowledge met the minimal criteria, although he did not have minimal knowledge at the beginning of the lesson (we did not examine his reasons, but we think that he had not prepared for the lesson).

During the presentation, students prepared questions about the presented method and subsequently, they asked their questions to the presenting team. Then they also evaluated whether the obtained answer fulfilled their expectations. We have to evaluate positively also the atmosphere during the presentation, students' interest in the subject, their motivation and competitiveness.

Fig. 4. Examples of the posters created by students within the project teaching of the subject Risk Management.

The team leader played a significant role in the whole process. In the preparation phase, s/he managed the whole team, assigned tasks to the particular team members and checked how the tasks were fulfilled. S/he was responsible for the whole work, including the final version of the project, its presentation and defence. The team leader also proposed grades to the particular members of his/her team according to their participation and involvement in the work. This task corresponds with real practice in production companies. The team leader has to have managerial skills, experience and knowledge about the assessed process. S/he has to be familiarized with the used method of the risk assessment and to be skilled in its use. The leader guides a professional team, which represents relevant knowledge and appropriate experience with regard to the assessed process. In practice, according to the project character, experts like a designer, process and engineer, safety engineer, operational and maintenance representative are used to be included in the expert teams to conduct risk assessment.

The teacher acted as an impartial observer and he entered into the discussion only if it was necessary to guide or lead the discussion from the teacher's view.

Figure 4 shows examples of the student teams project outputs – created posters related to the use of risk point method, FMEA, HAZOP and WHAT IF method.

6 Conclusion

Based on the findings from the survey and experiences from the one term, we can conclude that students improved their IT skills, communication skills, ability to discuss, and objectivity of their self-evaluation [22]. Moreover, project education influenced in a positive way also quality of students' bachelor theses where students could apply suitable methods of risk assessment in solving specific tasks. In discussions with students, we could notice their acquisition of pedagogical skills which are indispensable in their preparation for their future profession of safety technicians in practice [23]. In this professionally-oriented occupation, the first contact of safety technicians with employees is the initial training. The primary objective of this training is to inform and explain the results of the risk assessment and management.

The obtained results brought along several significant conclusions for the way forward in the OSH study programme: it is necessary to apply new teaching methods into the teaching process. At the same time, we have to focus on solving non-fictitious, real problems from practice to provide students possibilities to obtain adequate experience. This approach will be possible only if students participate in professional internships for several weeks so that they could understand the complexity, content and tasks of their future occupation.

The social recognition of OSH experts will be higher if people working in these positions are better prepared and more competent when taking up their duties. This assumption requires a high-quality education at universities. We have to primarily focus on those subjects which represent the core of knowledge in this study field. At the same time, we have to incessantly improve the educational system through implementation of new progressive methods and keeping up to date with the development of new technologies.

Acknowledgement. This work has been supported by the Cultural and Educational Grant Agency of the Ministry of Education, Science, Research and Sport of the Slovak Republic under the project No. KEGA 021UKF-4/2018.

References

1. Tureková, I., Bánesz, G.: Innovation of the occupational health and safety study programme. In: International Conference on Interactive Collaborative Learning, pp. 742–747. IEEE (2015)
2. Lošonczi, P., Nečas, P., Kováčová, L., Vacková, M.: Implementácia projektového vyučovania v bezpečnostnom vzdelávaní v predmete technické bezpečnostné prostriedky. Acta Scientifica Academiae Ostroviensis, Sectio A, Nauki Humanistyczne, Społeczne i Techniczne **7**(1), 492–500 (2016)
3. Duchovičová, J., Fenyvesiová, L.: Applying of strategies of critical and creative thinking by teachers according to the teaching subject and degree of education. Ad Alta: J. Interdisc. Res. 9(1) (2019)
4. Herreid, C.F.: Case study teaching. New Dir. Teach. Learn. **2011**(128), 31–40 (2011)
5. Allen, D.E., Donham, R.S., Bernhardt, S.A.: Problem-based learning. New Dir. Teach. Learn. **2011**(128), 21–29 (2011)
6. Peterson, F., Hartmann, T., Fruchter, R., Fischer, M.: Teaching construction project management with BIM support: experience and lessons learned. Autom. Constr. **20**(2), 115–125 (2011)
7. Orbánová, D., Velichová, Ľ.: Možnosti využívania projektového vyučovania v pedagogickej praxi stredných škôl na Slovensku. Media4u Magazine, 14(1) (2017)
8. Petrášková, E.: Projektové vyučovanie. Metodicko-pedagogické centrum, Prešov (2007)
9. Ojiako, U., Ashleigh, M., Chipulu, M., Maguire, S.: Learning and teaching challenges in project management. Int. J. Project Manage. **29**(3), 268–278 (2011)
10. Serafin, C.: Information science in technical education process in Czech Republic. Int. J. Eng. Pedagog. (iJEP) **9**(5), 89–102 (2019)
11. Anttonen, H., Pääkkönen, R.: Risk assessment in Finland: theory and practice. Saf. Health Work **1**(1), 1–10 (2010)
12. Popov, G., Lyon, B.K., Hollcroft, B.: Risk assessment: a practical guide to assessing operational risks. John Wiley & Sons, Hoboken (2016)
13. Badri, A., Gbodossou, A., Nadeau, S.: Occupational health and safety risks: towards the integration into project management. Saf. Sci. **50**(2), 190–198 (2012)
14. Yacov, Y., Haimes, Y.: Methods of risk management at work during process university education. John Wiley & Sons, Hoboken (2005)
15. Coneva, I.: Vzdelávanie študentov v oblasti PO na katedre PO FŠI ŽU v Žiline. In: Fórum mladých odborníkov požiarnej ochrany, pp. 7–14. TU, Zvolen (2011)
16. Pacaiova, H., Nagyova, A., Oravec, M.: Risk-based thinking methodology and its influence on OHS process. In: International Conference on Applied Human Factors and Ergonomics, pp. 267–276. Springer, Cham (2019)
17. Gaduš, J., Hašková, A.: Elimination of safety risks in biogas facilities. J. Technol. Inf. Educ. **4**(2), 71–76 (2012)
18. Jaďuďová, J., Kanianska, R., Marková, I.: Eco-innovation as a part of the green economy in the Slovak Republic. In: SGEM 2017, pp. 571–576. STEF9, Sofia (2017)

19. Kuempel, E.D., Geraci, C.L., Schulte, P.A.: Risk assessment and risk management of nanomaterials in the workplace: translating research to practice. Ann. Occup. Hyg. **56**(5), 491–505 (2012)
20. ANSI/ASSE: Risk management principles and guidelines (ANSI/ASSE Z690.2–2011). Des Plaines, IL: ASSE (2011)
21. Lundgren, R.E., McMakin, A.H.: Risk Communication: A Handbook for Communicating Environmental, Safety, and Health Risks. John Wiley & Sons, Hoboken (2018)
22. Brečka, P.: Súčasný vývoj audiovizuálnych technológií v školstve. Didaktika **1**, 27–29 (2013)
23. Hrmo, R., et al.: The research of the engineering pedagogy. In: 2015 International Conference on Interactive Collaborative Learning (ICL 2015), pp. 503–507 (2015)

Poster: Lesson Effects of PBL Education Promoted the Use of AI

Yunosuke Kawazu[1(✉)], Kazuya Takemata[2], Akiyuki Minamide[2], and Hirofumi Yamada[1]

[1] Kanazawa Institute of Technology, Ogigaoka 7-2, Nonoichi, Ishikawa 921-8501, Japan
kawazu@neptune.kanazawa-it.ac.jp

[2] International College of Technology, Hisayasu 2-270, Kanazawa, Ishikawa 921-8601, Japan

Abstract. Now there is a widespread demand for human resources who can apply Artificial Intelligence (AI) to problem solving. Kanazawa Institute of Technology has a major goal of making it possible for students to naturally use AI to solve problems. Therefore, in 2019, the university opened "AI Basics" for first-year students. The authors tried to see how AI was applied to the problem solution of Project Based Learning (PBL) education in two subjects, Project Design I (PD I) and "AI Basics", which the author was in charge of. The purpose of this study is to understand the current situation of AI utilization in the first year and to find out some reflections to subjects in 2020. In practice, many PD I students had not yet taken the AI Basics, so the author gave supplementary lectures on AI and encouraged the use of AI for solutions. Also, in "AI Basics", in the group activity of the last lesson, the author asked students to use AI just learned as a problem solution. The correspondence between each solution and the AI utilization categories were summarized in tables for understanding the current situation. As a result, it was confirmed that AI was widely used for problem solving in both subjects. In particular, the students learned "AI Basics" could propose an appropriate solution in a short time. This means lesson effects of "AI Basics" exist in problem solutions.

Keywords: Project based learning · Artificial Intelligence · First-year education

1 Introduction

AI (Artificial Intelligence) is evolving day by day, and is now positioned as a general-purpose technology [1] that has the potential to transform all operations and create new businesses. The Japanese government formulated "AI Strategy 2019" [2] in 2019 to get international competitiveness in this new era. One of the major goals of this is to ask all college students, regardless of arts or sciences, to receive AI elementary education, and to set up a specialized course for adults. In response to this, various efforts have already begun in the public and private sectors. For example, the Ministry of Economy, Trade and Industry of Japan has launched a problem-solving human resource development program, called "AI Quest", which has begun to develop human resources who can

M. E. Auer and T. Rüütmann (Eds.): ICL 2020, AISC 1328, pp. 397–404, 2021.
https://doi.org/10.1007/978-3-030-68198-2_37

implement AI in society in order to strengthen not only education as mere knowledge but also industrial competitiveness. This program modeled after the French programming school "42" [3], has the goal of cultivating 2000 new programmers a year by learning from each other. Behind these movements of the Japanese government, there is a sense of crisis that there is a critical shortage of AI human resources due to the rapid development of advanced technologies such as big data and robots.

Kanazawa Institute of Technology has begun preparations for AI education along these government-led movements, and established a new course called "AI Basics" in 2019 as an elective course for first-year students and working adults, which will be a compulsory subject for all first graders in 2020. In addition to basic courses, advanced specialized courses have already been offered, and an organizational education system has been established. A major goal of the university is to bring up engineers who can use AI to solve social problems. Therefore, in PBL education, which is the main educational center of the university [4], a member of the CDIO Initiative, it is expected that all students will be able to naturally connect AI utilization as a solution to their problems.

The authors have been actively engaged in the development of PBL education in various ways at the university [5, 6]. In 2019, the author got the opportunity to take charge of Project Design I (PD I) at the same time as taking charge of the "AI Basics" course for the first time. This means that the author was in charge of basic education related to AI, and gained a position to see how much AI penetrates PD education for first-year students. Taking advantage of this position, the author investigated how AI was applied to the team solution of PD education for two different courses. The purpose of this study is to clarify the current situation for AI utilization in the first year and to find out the items to be reflected in 2020.

2 Implementation Method

2.1 General

The two different courses are PD I and "AI Basics" that the author took charge of in the second semester of 2019. The former is a traditional course of the university, which was a compulsory subject of 2 credits for first graders. The latter is the first course to be offered, and it was positioned as an elective course for first-year students in 2019. In addition, this course is a one-credit course consisting of seven lectures, and two classes, autumn and winter, were opened during the semester. From the above, all students learning "AI Basics" also took PD I at the same time, but these two courses are independent and have no curriculum relevance. Here, the author decided to set the group discussion place in the last lesson of "AI Basics" as the place of association with PD I. In other words, it was decided to confirm the educational results of "AI Basics" by setting the theme of the group discussion as "What kind of ideas would you apply the learned AI to PD I solutions".

2.2 PD I

PD I for the 1st grade is supposed to deal with everyday problems (PD II, which is given in the 2nd grade, deals with social issues). The class in charge consisted of 32 students of mixed departments, and was organized into 5 teams of 6 to 7 so as to be evenly distributed. The number of students taking "AI Basics" in this class was limited to two, and most of the students lacked basic knowledge about AI. In order to deal with this situation, the following auxiliary efforts on AI were made.

(1) Introduced the introduction of "AI Basics" and examples of its use in the class.
(2) The main theme of PD was set as a comprehensive one, "To live a fulfilling campus life", to avoid to reach a simple problem solution.
(3) The author gave appropriate advice when considering the solution.
(4) After the interim report, it was conducted to have hearings on the solution by a teacher in the Information and Computer Science who specializes in AI (see Fig. 1), and refined the content so that it could withstand evaluation from a professional standpoint.
(5) In a mixed-disciplinary team, the class schedules of members are different, so it is difficult for the team to conduct extracurricular activities. To avoid this, we fixed the weekly gathering time for extracurricular meetings, and prepared a meeting place where activities were well prepared.

Fig. 1. Hearing by a teacher specialized in AI

The rest of the course management procedure was the same as that of the general PD I, including two poster sessions, and the final oral presentation. As for the AI utilization solution, since many of the students did not receive a formal lecture on AI, we emphasized the idea of AI utilization rather than the technical rigor and possibility of implementation. For this reason, even if it is not feasible at present, it is widely accepted as having potential in the future. In order to evaluate the solution, based on the final report, it was clarified which category of the AI application the solution falls into and summarized in the table.

In the final lesson of the class, an anonymous questionnaire survey was conducted to investigate students' awareness of PD activities in general and AI utilization in particular.

2.3 AI Basics

The author was in charge of fall classes (30 students) and winter classes (28 students) of the Faculty of Engineering, which was a mixed department. The lessons are learned through practical training on basic functions and practical examples of AI, and specifically, AI ethics, image recognition, and natural language processing. Since it is the basis, it does not pursue the technical depth, and the goal of education is to experience that AI is not a special technology but a tool that anyone can use normally.

At the final stage of the group discussion, the author had them tackle a group activity for 30 min under the theme "What kind of ideas would you apply the learned AI to PD I solutions", and then let them make a presentation in 3 min. Since PD I is a compulsory course that all students take, and it makes them think about a solution again based on their learning experience in "AI Basics" for the same problem, and aimed at confirming the effectiveness of the lesson. At the time of the activity, a worksheet of A3 size was distributed to each group, and hand-drawn materials with drawings and pictures were made, which was shown on the screen with a video camera for presentation (see Fig. 2). This work was excluded from the grades to encourage free thinking.

Fig. 2. The presentation of the group activity by a team

3 Results and Discussions

3.1 PD I

Table 1 shows the PJ themes, solutions, and AI utilization categories for each team. Here, the seven categories of AI utilization are image recognition, voice recognition, natural language processing, chatbots, future prediction, personality analysis, and image/voice synthesis. It can be seen from this table that all teams have utilized AI in some way for solutions. For example, the 1st team set the PJ theme to improve their English conversation ability. Its problems were that (1) motivation did not continue (2) the native speaker was not nearby (3) they were nervous when speaking. At this time, the team made a solution to these problems by creating a conversation partner of their choice on the screen and combining it with a mechanism such as Cortana or Siri that enables English conversation. In this case, it was determined that the utilization categories were image recognition, voice recognition, natural language processing, chatbot, and image/voice synthesis. As shown in 2.1, we do not attach importance to technical rigor, so if we allow ambiguity in the evaluation of utilization categories, these results will lead to the intention that AI will be used to solve PD activity issues. It means that the purpose was almost achieved.

Table 1. Relationship between PJ theme, solution and AI utilization categories (PD I)

	PJ Theme	Solution	AI utilization categories						
			a	b	c	d	e	f	g
1	To improve English conversation ability during university days	Man to machine English conversation system, like "Siri", with a created partner on the screen	●	●	●	●			●
2	To relieve anxieties about the future, especially getting occupations after graduation	System that Measures and advises on the difference between current abilities and those required by your desired occupation.					●	●	
3	To use 120% of university facilities, equipment, events, etc.	System that inform you how to use the university optimally for needs.		●			●		
4	To self-manage during university days	System that keeps track of daily schedules and room conditions, and gives optimal advice for living an ideal life.	●	●	●		●	●	
5	Housework is difficult if you live alone	System that makes cooking, laundry, cleaning, etc. enjoyable according to personal taste.	●	●			●	●	

a: Image Recognition b: Voice Recognition c: Natural Language Processing.
d: Chatbots e: Future Prediction f: Personality Analysis g: Image /Voice Synthesis.

Figure 3 shows the answer result of the anonymous questionnaire, "How did you feel that you were instructed to use AI as a solution? Positive evaluation (Good, and Generally Good) was a little less than half 46%, negative evaluation (Generally bad and Bad) was 15%, and neutral evaluation (Neither) was 39%. With regard to free description, the positive evaluations were mainly those who expressed satisfaction with the use of new tools called AI, such as "AI felt close to me"and "Thinking from a new perspective". On the other hand, there were voices complaining that "the idea was biased" in the negative evaluation, and "there was a feeling of force" and "the field of view was narrowed" in the neutral evaluation. It was confirmed that some students perceived that they had impaired their free thinking about the method of demanding AI utilization from the beginning. This would be an improvement point for the next year.

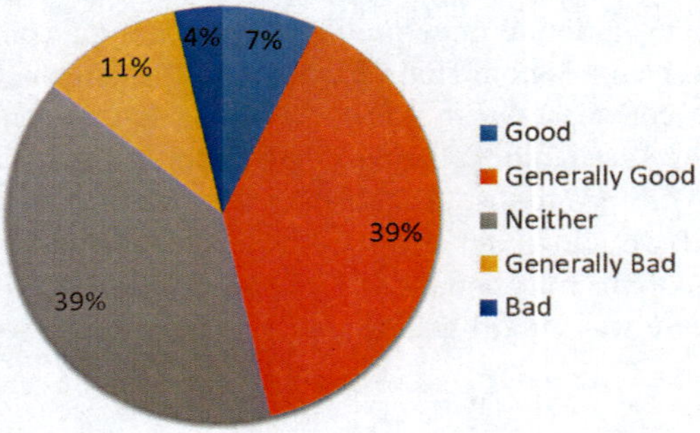

Fig. 3. The answer result of the anonymous questionnaire "How did you feel that you were instructed to use AI as a solution"

3.2 AI Basics

Table 2 summarizes the results of the group activities for "What kind of ideas would you apply the learned AI to PD I solutions" for the winter class of "AI Basics". The PJ theme of each group was the PD I theme of the group representative, and the solution was derived as a result of learning "AI Basics". The AI utilization categories for the solutions are the same as in Table 1. From this table, it was found that all groups devised new solutions using AI that had just been learned, apart from the solutions devised in the actual PD I class. The group presentation was made by showing a hand-drawn A3 worksheet on the screen, but each group tried to give an easy-to-understand explanation by skillfully using the illustrations. Considering that this group activity was carried out in a short time of 30 min without any prior instruction, the author felt that many groups have reached AI utilization ideas that exceeds expectations. "AI Basics" lessons were mainly about basic contents, and there was almost no time to introduce examples of using AI, but this result revealed that the lesson effects of "AI Basics" also exist in applied situations. Similar results were obtained for the fall class.

To some groups, the author tried to confirm the difference between the solution devised in the PD I class and the solution utilizing AI. The solutions of groups 3 and 4 in Table 2 were to authenticate individuals with student IDs when lending umbrellas and towels to students, but here they are replacing it with a solution that incorporates image recognition technology called face recognition. Furthermore, when asked the question "Which idea do you choose now?", they replied that they would choose the latter. This suggests that the use of AI is likely to naturally emerge as a candidate when considering PD solutions by incorporating "AI Basics".

Table 2. Relationship between PJ theme, solution and AI utilization categories (AI Basics)

	PJ Theme	New solution with AI	AI utilization categories						
			a	b	c	d	e	f	g
1	To extend healthy life expectancies by improving diets	Determining deficient nutrients from food images and giving advice	●			●			
2	To improve English conversation ability during university days	Man to machine English conversation system, like "Siri", with a created partner on the screen	●	●	●	●			
3	In order not to lose the umbrella	Prevent mistakes in receipt by depositing your umbrella with face recognition	●						
4	Automatic towel rental device that can prevent forgetting to return	Easy to borrow and return with automatic towel lending device with face recognition	●						
5	There are many errors in answering problems in math textbooks	Image recognition of textbook problems, guidance of answers, and correction of mistakes in textbook answers	●		●			●	
6	Effective use of one-week activity history that must be submitted	Analyze past student portfolios and advise on current student behavior improvements	●		●		●	●	
7	Effective use of empty classes in my class	Propose the most effective usage of time based on personality diagnosis			●	●	●	●	
8	Pillows for comfortable sleep	Pillows that read physical condition and advise on optimal sleep and physical condition management		●	●	●	●		
9	Prevention of contact accidents at intersections	Navigation that collates past accident data at intersections and calls attention in advance	●	●	●		●		●

a: Image Recognition b: Voice Recognition c: Natural Language Processing.
d: Chatbots e: Future Prediction f: Personality Analysis g: Image/Voice Synthesis.

4 Conclusions

In the first year of "AI Basics", we investigated the extent to which AI was used to solve problems in PBL education in the PD I and "AI Basics" classes taught by the author.

In the PD I class, most of the students did not take "AI Basics" classes, so the author gave supplementary lectures on AI and encouraged the use of AI for solutions. As a result, (1) Each team devised a solution that uses AI, although there is a range to some extent. (2) Hearings by teachers specializing in AI were effective in ensuring technical support. (3) A little less than half of the students were positive about encouraging the use of AI as a solution, but there are also negative students who hindered their free thinking, so it is necessary to devise a method that does not impose a solution.

In the "AI Basic" class, we asked them to consider using AI that had just been learned in the PD I solution at the group discussion. As a result, many groups devised a solution that exceeded expectations, considering there was no prior instruction and the work time was 30 min. It was suggested that the basic content of "AI Basics" classes is fundamental, the lesson effects of "AI Basics" also exist in applied situations.

Since "AI Basics" will be compulsory for all first-year students in 2020, it would be payed attention what kind of change will occur in the solution of PD I as a result.

References

1. Peter Buxmann. https://e3zine.com/artificial-intelligence-technology/, Accessed 25 May 2020
2. Prime Minister's Office of Japan. https://www.kantei.go.jp/jp/singi/ai_senryaku/pdf/aistratagy2019en.pdf, Accessed 25 May 2020
3. Homepage. https://www.42.us.org/, Accessed 25 May 2020
4. KIT president. https://www.kanazawa-it.ac.jp/ekit/about/strategy/index.html, Accessed 25 May 2020
5. Takemata, M., Minamide, A., Matsuishi, M.: Engineering design education through the CDIO approach. J. JSEE **60**(2), 15–21 (2012)
6. Minamide, A., Takemata, K.: Design of PBL educational program in collaboration with printing company. In: The Impact of the 4th Industrial Revolution on Engineering Education, pp. 749–754. Springer, Heidelberg (2019)

Developing Cross-Cultural Communicative Competence of University Students in the Globalized World

Elena Volkova(✉), Elena Yurievna Semushina, and Ekaterina Tsareva

Kazan National Research Technological University, Kazan, Russia
wolfkova@mail.ru

Abstract. Today globalization has created the conditions in educational space of the Russian Federation, which stimulate the direct dialogue of students with persons of the same age from other countries. So communication develops a cross-cultural communicative competence of students for decision of humanitarian tasks in professional activity. The aim of our research was to reveal factors of this environment, which influence the elements of a structure of cross-cultural communicative competence, promote to its development or degradation. Globalization in education creates the basis for developing linguistic competences of technical universities students through the international exchange programs, which are used actively by Russian universities for organization of educational process and development of cross-cultural communication among students, through the Internet, which destroys borders and barriers in communication between students. This aims the decision of specific targets:

1) to work out the structure of cross-cultural communicative competence of technical universities students;
2) to reveal factors having an influence on the development of cross-cultural communicative competence of students;
3) to create and test the pedagogical conditions in which cross-cultural communicative competence is formed and developed;
4) to work out and implement the methods of cross-cultural communicative competence development.

Keywords: Cross-cultural communicative competence · International exchange programs · Teaching and learning process · Intercultural communication · Mobility · Grants · Globalization

1 Introduction

The theme of formation of intercultural communicative competence of technical university students is currently central and very important because resettlement of people made our societies multicultural and multinational. The professional growth of young people, their safety and welfare of community, as a whole, depends on whether each of us will be tolerant, sociable. Today our society needs the specialists, qualified in further

M. E. Auer and T. Rüütmann (Eds.): ICL 2020, AISC 1328, pp. 405–416, 2021.
https://doi.org/10.1007/978-3-030-68198-2_38

independent life activity, who are competent for solving practically important and professional problems [8]. Undoubtedly, a person who knows languages and has a high level of education is unlikely to want to enter into conflicts. Most likely, this person will try to resolve all existing disagreements peacefully. That is why it is so important to know the mechanisms of the formation of the ICC and implement them in higher education institutions.This investigation was carried out with participation of students of the Kazan national research technological university (KNITU) and American pupils, which studied Russian language under the program of NSLI-y on the base of health improvement camp of KNITU "Green pine wood". At the beginning of the investigation we made the hypothesis:

1. Russian students have the middle or low level of formedness of cross–cultural communicative competence.
2. The main components of cross–cultural communicative competence must be included in the teaching and learning process in order that students can realize their potential in all parts of the development of cross–cultural communicative competence.

Concerning intercultural communicative competence development in the system of additional education of foreign countries, we identified problems of interculturalists training, (specialists, possessing the intercultural communicative competence) and organization of translation practice, which is considered as one of the methods of dipping in foreign environment [7]. The development of the definite level of cross–cultural communicative competence can be realized only in the terms of language immersion when interacting with native speakers. Today globalization has created precisely these conditions in the educational space of the Russian Federation which teachers use actively for developing professional competences by means of foreign languages [6].

2 Methods and Approaches

The methodological basis of our study consisted of two main scientific directions, communicative and competency building approaches. In the framework of the competency building approach, we determined the general principles of education to achieve the goals of our study, namely: the selection of the education content, the organization of the educational process both in classroom form and in the form of organizing face-to face contact at the university camp during the period of linguistic immersion and evaluating of educational results. Using a communicative approach gave the opportunity to track the results of educational activities, namely, how, who and when uses the language and for what purposes and functions. As part of the communicative approach, we used interactive immersion technologies to obtain information about language changes depending on a particular communicative situation among its participants. We formed and developed the ability to create, read and understand texts of various types and character (for example, stories, interviews, dialogs, reports), the ability to maintain a conversation even with a limited lexical and grammatical base.

In our research we used testing as a form of measuring of student knowledge in communication based on using pedagogical tests. It includes the preparation of tests, the actual conducting of testing and the subsequent processing of the results, which gives an assessment of the learners' skills [2]. These methods are widely used both in distance learning and in ordinary practice [3]. Qualitative and quantitative testing was carried out at the beginning of the session and final testing was put into practice at the end of the session, and the results were analyzed. During all course of education, teachers and the psychologist gave English and special classes for students for getting the results in changes of orientations on future plans, ideals, solution to conflict and others. American teenagers studied these questions in Russian and Russian students learned with Americans in one group. In our research we compared the results in groups where students passed the language practice together with participants of the international program with the results in groups of students who didn't take part in this program.

3 Progress of Work

The first stage of the experiments was carried out on the basis of the summer linguistic camp on the bank of the river Volga. A summer language practice was organized in the camp for students of the Kazan national research technological university who were taking the program of additional education "A translator in the sphere of professional communication" and for American teenagers - pupils studying Russian.

The second series of experiments took place during the training year with participation of Russian students after finishing summer language practice, who were divided by faculties during the educational process. In this case, a differentiated analysis was carried out about identification and influence of the correlation dependence of components for the formation and development of ICC of students.

In accordance with the logic and culture of the realized experiments we have identified two concepts: intercultural communication and intercultural competence. In our point of view, intercultural competence is a set of abilities which are necessary for effective interaction with people who differ from each other in their culture and linguistic affiliation. Foreign researchers distinguish four basic levels in the questions connected with formation of the intercultural communicative competence:

Level I: "An educated traveler" - refers to participants in short-term exchange programs (1–2 months)

Level II: "A temporary resident" - refers to participants who are involved in extended cultural immersion (for example, longer work experience for a period of time, including service programs (3–9 months)).

Level III: "A professional" - corresponds to people working in intercultural environment, for example, a staff in international institutions or organizations such as FEIL and MOs.

Level IV: "An intercultural/A multicultural specialist" - these are trainers and educators involved in the training or consulting of students from different countries [9].

In our experiment, we were talking about the formation of the base level №I. The main indicators of the ICC level will be sociocultural and linguistic skills. Thus, during the survey, experts asked questions regarding the presence or absence of selected basic features of intercultural communicative competence. When testing, questions related to educational aspects, which are integral components for the formation of the students ICC, were discussed. In order to determine the dynamics of intercultural competence development, students were tested in two stages - at the beginning of summer language practice and at the end. The participants of the experiment were 65 Russian students and 13 were foreign students who also attended communicative classes in the English language, only they did their tasks in Russian.

Test number 1 Mark the following knowledge, skills and abilities on a ten-point scale (from 1-I do not speak well, up to 10-I know perfectly):

- I know individual foreign words
- I am able to use them in word combinations and sentences
- I am able to maintain a conversation
- I can understand the partner in conversation
- I know traditions and patterns of Russian and American culture
- When communicating, I take into account the level of education, upbringing, social status of the partner

What are the most effective ways to learn a foreign language for you personally? Mark on a ten-point scale (from 1 - not effective, to 10 - the most effective)

- Performing lexical and grammar exercises
- Communicative games
- Songs
- Role-plays
- Communication with native speakers
- Internet chat

In what spheres do you plan to realize your foreign language skills?

- In business
- In production
- In the sphere of education
- In the service sector
- Tourism
- Write your own version

During the first testing and processing of the first results indicators of the formation and development of sociocultural and linguistic aspects of the intercultural communicative competence by means of a correlation analysis, the following dependencies were identified having an impact on the course of this process:

- those students who like communicative games in English classes think about using their language knowledge in the service sector in the future;
- those students who know and learn new words and are able to use them in phrases and sentences, as a rule, are able to maintain a conversation and emphasize communication with a native speaker as the main factor influencing the formation of communicative abilities;

- those students who know the traditions and patterns of Russian and American culture distinguish communication with native speakers as one of the main ways of learning a foreign language;
- those students who know how to maintain a conversation, support Internet communication actively;
- students who understand the partner in communication, know the traditions and patterns of Russian and American culture, want to use their knowledge in the sphere of education;
- those students who communicate with native speakers consider Internet communication as one of the most effective ways of communication;
- students who know the traditions and characteristics of a foreign language culture, as a rule, must take into account the level of their partner when communicating;

This distribution indicates a certain interest in learning of a foreign language for most students and effective ways to master it.

After finishing the summer translation practice and multiple occupations with American students, and sharing classes, and participating in cultural events, testing No. 2 were offered to Russian students in order to identify other dependencies and mechanisms that affect the development process of the intercultural communicative competence, as well as to approve the existing ones. During the summer classes in English, the psychologist simultaneously held a training of effective communications, the purpose of which was to remove the psychological barrier in communication and successfully implement intercultural communication.

Test number 2 How much did the English classes help you to improve your language skills (mark each statement from 1 point to 10 in degrees of effectiveness):

- To learn new foreign words
- To use them in phrases and sentences
- To be able to maintain a conversation
- To understand the partner in communication
- To learn the traditions and characteristics of Russian and American culture
- When communicating, take into account the level of education, upbringing, social status of your partner

What methods of foreign language learning were the most effective for you personally? (mark the degree of effectiveness from 1 point to 10 opposite each statement):

- Performing lexical and grammar exercises
- Communicative games
- Songs
- Roleplays
- Communication with native speakers
- Internet - communication

Have you expanded your beliefs about the areas of application of a foreign language after communicating with native speakers? If so, in what spheres do you plan to realize your language skills?

- In business
- In production
- In the sphere of education
- In the service sector
- Tourism
- Write your own version

How did communication with native speakers help you to improve your language skills (mark each statement from 1 point to 10 in terms of effectiveness):

- To learn new foreign words
- To use them in phrases and sentences
- To be able to maintain a conversation
- To remove the barrier in speaking
- To learn the traditions and patterns of Russian and American culture
- When communicating to take into account the level of education, upbringing, social status of the partner in conversation

How has the training of effective communications helped you to improve your communication skills (mark each statement from 1 point to 10 in terms of effectiveness):

- to learn new foreign words
- to use them in phrases and sentences
- to be able to maintain a conversation
- to remove the barrier in speaking
- to learn the traditions and characteristics of Russian and American culture
- to take into account the level of education, upbringing, social status of the partner in conversation when communicating

An analysis of the second test revealed new dependencies, increased motivation and activation of students in terms of learning a foreign language.

- those students who believed that new foreign words help in improving their language skills, are keen to use them in communication after communicating with foreigners;
- those students who use new words in phrases and sentences aim to maintain a conversation;
- those students for whom the study of new words was one of the main ways of learning a foreign language, after joint classes in English with the Americans changed their priorities in the ways of learning towards communication with native speakers;
- students who considered lexico-grammatical exercises as one of the effective ways in language learning, were keen to use new words in phrases;

After joint classes, students increased their awareness in learning English. Each stage of class work, which was perceived as a boring and optional element previously (for example, performing lexical and grammatical exercises), began to be perceived by a fairly large group of students (about 60%) as a necessity which should be reflected in practice.

- those students who understood the partner in communication, began to consider the performance of lexical and grammatical exercises as one of the effective techniques for the successful implementation of intercultural communication;
- those students, who understanding the partner in communication, expanded their language concept and gained inner confidence, which caused a desire for communicating with foreigners to apply their knowledge not only in the service sector, as before, but also in the tourism, education and business sectors;

The second test confirmed already established dependencies, such as:

- understanding of the partner in communication, respectively, entailed the ability to maintain a conversation;
- the perception of speech by ear led to the removal of the barrier in speaking;
- those students who possessed knowledge of the traditions and patterns of Russian and American culture, considered communication with native speakers as one of the main ways to achieve intercultural communication and learning of foreign language;
- knowing traditions, respectively, leads to the ability to maintain a conversation and to intensify Internet communication;

After given classes to American students, Russian students expanded their perception of a foreign language and gained confidence in their abilities. So, those students who knew the traditions and patterns of Russian and American culture before communicating with foreigners, thought about application of their knowledge only in the service sector. After joint training, they intended to use their knowledge in education, tourism and business.

Those students who considered communicative and role-playing games in English classes as one of the effective ways to achieve intercultural communication, when communicating took into account the level of education, upbringing and social status of the partner in communication;

- Students who considered communicating with native speakers as the most effective way to achieve intercultural communication took into account the level of education, upbringing and social status of the partner in communication;
- All students who took into account the level of the partner in communication, planned to apply their knowledge in the business sphere;
- Those students who sang in a foreign language recognized that it was easier for them to maintain a conversation and to remove the linguistic barrier in speaking;
- Those students who had removed the barrier in speaking planned to realize their linguistic skills in the workplace;
- Singing students used new foreign words in phrases and sentences more often;
- Communication with native speakers increased motivation to learning new words and phrases and intensified their use in sentences;

Thus, based on the correlation analysis of input and control testing, the studying of the obtained dependencies during the experiment, we confirm our suggested hypotheses.

The level of the intercultural communicative competence is weakly expressed and absent in most students. We can give findings based on the analysis of correlation dependencies. About 30% of the surveyed students associate the effective studying and using the language in speech with Internet communication, knowledge of traditions and peculiarities of linguistic cultures. They do not see further continuation, consolidation and application of the results when performing traditional lexical and grammatical exercises, learning new words and their using. Low self-esteem of second language skills creates a vision of its use only in the service sector, that is, at the consumer grade. In fact, the main body of students does not go abroad and the language environment, in fact, is inaccessible for them, so our students perceive the traditions and features of a foreign language culture only from the point of view of the value system, norms and traditions of their native culture. Accepting traditions and patterns of a foreign language culture only from the students position leads to a certain hidden restriction in communication, as far as students begin to take into account the social status of the partner, and this is the reason for locking intercultural communication in a certain extent. These facts explain the lack of skills of fluent speech, tolerance, empathy in most students, which creates a certain barrier in communication. However, after a joint staying in the framework of the summer language practice, all students received certain knowledge about another culture, about the interaction between representatives of different cultures in practice and were able to compare cultural differences and to discover some similarities. Of course it is impossible to form the intercultural communicative competence in 2 weeks, but we can outline the positive dynamics and ways of the formation and content of the intercultural communicative competence on the basis of the existing theoretical and experimental data.

Thus, for the successful formation and development of the intercultural communicative competence, it is necessary to create pedagogical conditions, which consist of organizing direct communication with representatives of a foreign language culture. This condition can be created and implemented with the participation of students in international programs, in the organization of foreign internships, in the organization of student clubs with the mandatory participation of representatives of a foreign language culture.

To determine the professional orientation and development levels of the intercultural communicative competence, we put forward a hypothesis about the dependence of the formation and development of the intercultural communicative competence on the basic level of students' language training. For checking up of this statement, the next stage in the study was carried out.

Students were divided into three large groups in accordance with faculties. The first group consisted of students of the Polymer Faculty and the Faculty of Social and Humanitarian Technologies, the Faculty Management of Economics and Law. That is, the part of young people who took the program of additional education "A translator in the sphere of professional communication" for the second year of study or studied in a group at the Intermediate level (Middle). The second group consisted of students of food engineering and food technology faculties, who took one year training course in the above-mentioned program of additional education. And finally, the third group of students consisted of students from mechanical, petroleum faculties and information technologies who were in the camp according to a volunteer program, attended classes in English, talked with foreign students, but did not take the professional retraining program "Translator in the sphere of professional communication".

The results of the differential correlation analysis have already helped to identify and to form into groups the identified correlation dependencies, taking into account the level of language proficiency.

So, for example, the following dependencies were revealed in the first group of students who studied under the program of additional education at the Intermediate level and in the group of the second year students of training in the same program.

- learning new words and their using in phrases and sentences leads to a more detailed understanding of the partner in communication and cultural knowledge of the traditions of a foreign language and thus expands communication with native speakers;
- Internet communication helps to maintain a conversation;
- those students who understand the partner in communication and know how to carry on a conversation in English, assume the application of their knowledge in the sphere of education and production;
- communication with native speakers helps to learn the traditions and patterns of Russian and American culture;

Students of the second group who took the first year of the program "Translator in the sphere of professional communication" are first-year students of the food technology and engineering faculty, the level is initial and Pre – Intermediate. The following correlation dependencies were revealed for this category of students:

- understanding the partner in communication helps to remove the barrier in speaking;
- using new words in speech leads to a deeper understanding of the partner in communication;
- the ability to maintain a conversation helps to learn the traditions and cultural patterns of a foreign language;
- knowing new words allows you to maintain a conversation and in such a way to learn more about the traditions and patterns of other cultures;
- knowing of the cultural patterns of a foreign language leads to the fact that students are bound to take into account the level of the interlocutor. That is, the possession of communicative and linguistic competencies leads to a certain socialization in communication or to expansion, or to limitation, or maybe, on the contrary, to more detailed communication taking into account professional orientation;

Students in this group perceived that performing lexical and grammatical exercises helps in learning of a foreign language and emphasized the importance of this work type, which, in turn, is one of activities type that leads to the removal of the language barrier.

Most of the students in this subgroup determined the importance of communicative games in the foreign language classes and their role in the ability to maintain a conversation, emphasized the importance of songs in English in learning new words. It is necessary to note that 30 min of daily classes in English were devoted to learning of songs in the original language for Russian-speaking students and in Russian for American school children.

It is notable that the students of this subgroup are future production workers, therefore, a correlation dependence associated with production has been identified during processing questionaries of these students.

Those students who take into account the level of the interlocutor, learn the traditions of a foreign language culture, aim to break down the barrier in speaking connect the use of their knowledge in the language with production in the future.

And, finally, the weakest third group of students who did not take the program "Translator in the sphere of professional communication", having passed the test, showed the following results:

- learning new words and using them in sentences helps in communicating with native speakers, in understanding the interlocutor and studying the traditions of a foreign language culture;
- a low level of foreign language proficiency, studying of a foreign language according to the work program and the general curriculum of higher professional education during 2 h classes a week creates students' conviction that their knowledge can be applied only in the service sector. This conclusion can be made on the basis that most students consider doing lexical and grammar exercises in the classroom as one of the typical stages of working with a textbook and associate the application of gain knowledge in the process of standard training only in the service sector. And limited communication with native speakers leads to the fact that Internet communication is perceived by trainees as one of the main ways in the formation of skills to maintain a conversation and to understand the interlocutor.

4 Conclusion

The culture study of other country lets involve students in "the dialogue of cultures", acquaint them with universal values, enrich base knowledge with the help of language means. The study of culture gives the great opportunities for providing with cognitive motivation and adaptation of students in other language culture. Talented professionals are becoming more mobile and many authors observe a significant increase in academic and labor mobility [5]. To become more global, it is important to study the most successful experiences [1].

General results of carried correlation analysis showed that participation of students in the international programs is the one of the most important pedagogic conditions for development of cross–cultural communicative competence of students. Learning in the terms of international educational programs, which are as a part of curriculum, gives the opportunity to students "to find legs" in intercultural communication, to raise motivation in studying of foreign languages and foreign culture, to get to a whole new level and in such a manner to extend boundaries of interpersonal space. One of the main tasks of a teacher is the necessity of using the interactive technology to a social cultural component of education in the sphere of foreign language. The results of correlation analysis proved the existence of well-defined structure of cross–cultural communicative competence and interdependence of its components such as strategic, language-specific and sociocultural in teaching and learning process. All these conclusions prove the correctness of our developed hypothesis at the beginning of the research.

At the same time, it is necessary to attract the elements of the students native culture for comparison of world perception features by the representatives of different language cultures.

Unfortunately, we haven't enough time for full value realization of all components during lectures. In this case, programs of additional education in foreign language are one of the effective means of intercultural communication. These programs realize not only practical, developing and educational tasks, but and promote better understanding and communication in foreign language. The influence of globalization process, current economic and political situation set the task of training students in multicultural environment, as well as developing their ability to communicate with people of different religions and ethnicities to ensure peaceful coexistence [4, 10].

References

1. Valeeva, R., Ziyatdinova, J., Osipov, P., Oleynikova, O., Kamynina, N.: Assessing intercultural competence of engineering students and scholars for promoting academic mobility. Adv. Intell. Syst. Comput. **917**, 815–825 (2019)
2. Semushina, E.Y., Ziyatdinova, J.N.: On-line testing of engineering students as a form of assessment when studying English in distant form. Adv. Intell. Syst. Comput. **716**, 475–480 (2018)
3. Semushina, E.Y., Ziyatdinova, J.N.: Studying english on-line as a part of a course "english for special purpose" in technological university. In: Auer, M., Hortsch, H., Sethakul, P. (eds.) The Impact of the 4th Industrial Revolution on Engineering Education. ICL 2019, Advances in Intelligent Systems and Computing, vol. 1135, Springer, Cham, pp. 21–29 (2020)
4. Fakhretdinova, G., Dulalaeva, L., Tsareva, E.: Extracurricular activities in engineering college and its impact on students' tolerance formation. Adv. Intell. Syst. Comput. **1134**, 143–150 (2020)
5. Valeeva, R., Ziyatdinova, J., Osipov, P., Oleynikova, O.: Development of international academic mobility: success stories. In: Auer, M., Hortsch, H., Sethakul, P. (eds.) The Impact of the 4th Industrial Revolution on Engineering Education. ICL 2019, Advances in Intelligent Systems and Computing, vol. 1135, pp. 443–454 (2020)
6. Valeeva, E.E., Kraysman N.V.: The impact of globalization on changing roles of university professors. In: Proceedings of 2014 International Conference on Interactive Collaborative Learning, ICL 2014, 7017901, pp. 934–935 (2014)
7. Volkova, E.V.: Different approaches to the problems of intercultural communicative competence. In: da Rocha Brito, C., Ciampi, M.M. (eds.) Proceedings of the 16th International conference on Interactive collaborative learning and 42-nd International IGIP Symposium on Engineering pedagogy. Book of Abstracts 2013, pp. 456–457. Kazan, Russia (2013)

8. Volkova, E., Zinurova, R., Tuzikov, A.: Formation experience of sociocultural competence in the system of additional foreign education. In: Lee, G., Schaefer, G. (eds.) Social sciences, 4th International conference on Social Sciences and society (ICSSS) 2015, pp. 144–147, Paris, France (2015) http://apps.webofknowledge.com/full_record.do?product=WOS&search_mode=GeneralSearch&qid=1&SID=2DD7begH5NWdz2TjcsY&page=1&doc=1
9. Alvino, E.F.: About Intercultural communicative competence. Vermont, Brattleboro (1995)
10. Astafieva, A., Tsareva, E., Khafizova, L.: Development of postgraduate students' foreign language communicative competence. In: 14th International Technology, Education and Development Conference, Proceedings, INTED, pp. 4043–4049 (2020)

Experience in an Application of Project-Based Learning to Teaching of Mechanical Engineering and Energy Technology Processes Control

Aleksei Hõbesaar(✉), Tatjana Baraškova, and Veroonika Shirokova

School of Engineering, Virumaa College, Tallinn University of Technology, Järveküla tee 75, Ida-Virumaa, 30322 Kohtla-Järve, Estonia
aleksei.hobesaar@taltech.ee

Abstract. Project-based learning (PBL) is a modern learning methods, in which the various projects are successfully applied in different courses in the educational programs of different disciplines. The engineering programs, such as Mechanical Engineering and Energy Technology Processes Control requires not only the theoretical knowledge, but also the teamwork skills to solve the real engineering tasks. Learning methods are needed to be incorporated in student-centered team-based learning pedagogy such as project-based, case-based, inquiry-based and problem-based scenarios. The used approach in this education program is project-based learning. In this paper the advantages and problems of using the PBL method for teaching the special disciplines or courses was discussed based on got experience and students feedback.

Keywords: Project-based learning · Mechanical engineering and energy technology · Demonstration models

1 Introduction

Interactive lectures have become a part of our education, being an effective teaching tool. Unfortunately, only a small part of the information from the lecture can be remembered. For better memorization and assimilation of the theoretical material of the lecture, this material should be used in a practical way. For this reason, it was decided to use the project learning method, which would help to more effectively anchoring the learning material during the course of the project, and could also motivate students to actively attend lectures and ask the lecturer questions [1, 2].

In this article, the author confirms that the innovative PBL environment can be applied under certain conditions in the teaching of engineering disciplines. For example, PBL can be a real school task (development and construction of demonstration stands). Students develop a task and are guided in the process of working with teachers.

In the 21st century, the best success requires the right knowledge, skills, tools, and support. The same can be said about the principle of project-based learning (PBL). There are gold standard of PBL which relies on "4C" rule (communication, critical

M. E. Auer and T. Rüütmann (Eds.): ICL 2020, AISC 1328, pp. 417–428, 2021.
https://doi.org/10.1007/978-3-030-68198-2_39

thinking and problem solving, collaboration, creativity and innovation) [3]. This rule helps students solve real problems by developing a specific model. In [4], it is stated that PBL consists of six stages. At the first stage, the teacher gives students the necessary knowledge and skills to successfully complete the project. The teacher can then indicate which part of rule 4C should be used first. In the second stage, the teacher explains the task itself and the goal, which should be reached at the end of the project. At the third stage, the teacher engages students in the learning process, which takes place not only in the classroom, but also outside of it. At the next stage, students have the opportunity to think and analyze the quality of their work and how to overcome possible problems. At the fifth stage, students are given time to self-analyze and critically evaluate their work. After all, at the last stage, the final model is presented.

This article describes the authors' experience in applying the PBL methodology in technical disciplines at the College of the University of Technology in Estonia.

The authors rely on their experience, on the results obtained, on the assessment of students' performance before and after applying this method, and on the feedback of students to assert that PBL can be effectively used in teaching technical as well as theoretical subjects.

In the literature, you can find a lot of articles about the use of the PBL method in Electrical Engineering. For example, [5] describes the application of this method in the course Power Electronic and Drives; [6–9] cites the positive experience of this method in electrical power system, elementary circuit analysis, analog electronic and electrical power system engineering.

In these examples, PBL was applied in specific individual courses within the program. However, in this article, the PBL method is used not only in teaching individual disciplines, but also in the framework of interdisciplinary interaction.

2 Using the PBL Method in the Study of Engineering Disciplines

2.1 Principles of PBL Methodology

Recently, there is a huge need for specialists in technical fields who are able to solve complex professional tasks. The lack of a sufficient number of qualified specialists in enterprises forces educational institutions to change their programs and courses structures.

Students of Mechanical Engineering and Energy Technology Processes Control speciality in Virumaa College of Tallinn University of Technology live and work in industrial regions where different factories are located: chemical, energy and mechanical plants. At the end of this educational program, students must have the necessary knowledge and skills. This program is also constantly updated. The latest modification includes 30% of subjects with implemented PBL methodology. The PBL method has been used in technical disciplines such as Theoretical Mechanics, Machine Elements, Electric Circuits, Renewable Sources of Energy, Synthesis of Mechatronics Systems.

The PBL methodology was based on problem-based learning, which included different types of projects (projects with a specific task, specialized projects, problem projects). This training ended with a working model, documentation, and technical drawings.

Our PBL methodology can be represented as a cyclic chain (see Fig. 1). The Chain consists of three stages: I-description of the problem, study of technical terminology, setting goals and tasks; II-self-guided learning; III-solution of the problem, making summaries and conclusions and presentation of results [10].

Before the PBL methodology, the simulation method was used. This type of training allows you to simulate the real process, thereby facilitating project training. Since we are talking about specific stands, the Matlab r2020 software environment was used for the simulation.

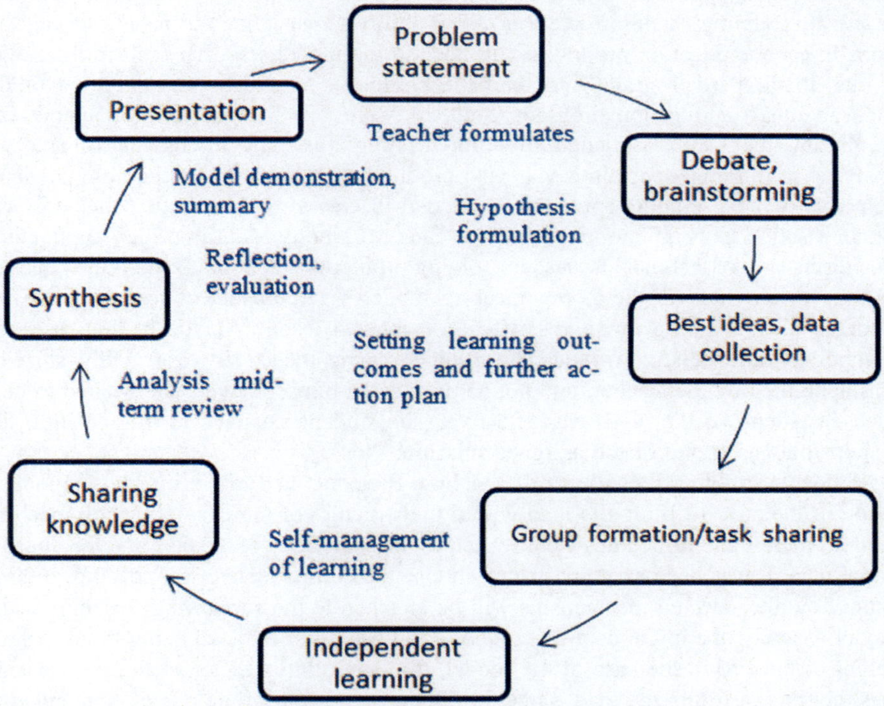

Fig. 1. Evaluation of the learning process, problem solving, and group work [10]

2.2 Applying of PBL Method in Technical Courses

Two stands were designed by two large teams at the same time. Each team was made up of students who studied in different programs and took different courses at the College. Students studying the subject of Theoretical Mechanics have made technological contributions to the design of the model a Chain reaction machine. The theory of integration of equations of non-holonomic mechanics was of interest. In other words, non - holonomic relationships are imperfect relationships. The theory itself is not as fully developed as in the case of ideal connections. First, the equations of non-holonomic mechanics are more complex, even than the Lagrange equation. The problem of rolling the ball on the surface with sliding was solved. The results of the theoretical solution had to be clarified when designing the above-mentioned model, selecting the material for the guides of the movement of the steel ball, with a diameter of 25 mm. To reduce the vibration when the ball slides, it is best to use wooden knotless corners of the size (25 × 25 × 3000) mm. The practice of experimental verification of the correctness of solving theoretical problems is more effective, since it gives a more objective picture of students' knowledge. In addition, the emotional aspect plays an important role when the problem is solved after a long experiment.

To calculate the geometry of the mechanism, students studying the subject of Synthesis of Mechatronics Systems used not only special computer programs, but also practical modeling (hand-on learning). The ability to visualize the results of calculations in each application program is limited, so the model was first built in the Solidworks 2019–2020 program, and then the complete Assembly was exported to the MatLab r2020 application program. Students were very interested in analyzing and comparing the results obtained in different programs. The interaction of students studying in different specialties was also productive. Information technology students helped install the graphic applications needed to create the kinematic scheme of the Chain reaction machine model in the Simscape Multibody MatLab r 2020 application. The discussion of technical issues and solving problems was conducted on the pages of various forums and in the Zoom meeting mode, if circumstances required it. Most students liked working in groups with a designated leader. Maybe in this case, the psychology of leadership worked. The attitude towards the leader was friendly, since he was appointed by the teacher, and not he nominated himself. Everyone wanted to help such a student, so the work was effective. The student assigned to this position felt responsible and proud of the assigned mission.

Students studying the subject of Machine Elements had to help in the implementation of the task of lifting the load within the structure of the Chain reaction machine model. Interest and difficulties were related to the specificity of the task and the unusual initial data. It was necessary not just to choose the components of the lift, but to think about how the selected mechanisms will be attached to the supporting structure of the model. Usually, the initial data was given as the weight of the load being lifted and the height of the lift. In the case of the model, the determining factor of the ideal design was the time of lifting the load, since the Chain reaction machine model is intended to demonstrate the duration of different types of object movement.

Students studying the subject of Electric Circuits had to ensure the continuity of the object's movement. Knowledge of feedback types and the theory of control and automation helped in solving this problem. Limit switches were used to control the position of the object, and a programmable controller AL2-24MR-D ALPHA Mitsubishi Electric was used to control the movement.

When the Chain reaction machine model was close to completion, there was a lack of a spectacular conclusion about the readiness of the stand. A competition was announced, according to which the movement of the object should end with a light phenomenon. The 4 × 4 × 4 led cube was most suitable for this task. To make a led cube, you need basic knowledge in electronics. LEDs were selected for the required brightness and color. Students performed installation work by studying the basics of installation, repair and maintenance of electrical installations within the subject of Study practice.

The "Green Energy" stand is also designed to test solutions to theoretical problems.

At the moment, there is a problem of predicting wind speed in atmospheric physics. There is the Navier-Stokes equation, which gives an approximate solution in large-scale processes, but in a specific place it is impossible to use this equation. Therefore, modeling using the "Green Energy" stand will allow you to collect practical data that will enrich the theory of atmospheric physics.

3 Practical Implementation

The built stand "Chain reaction machine" became a model that was used to test the solution of the problem of non-holonomic mechanics. Theoretically, it is difficult to solve such problems, since a non-holonomic system cannot be characterized by a single function of its state and time. The equations of non-holonomic mechanics generally do not have an invariant measure. The problem is related to the equation of motion of the ball on the surface. In contrast to the traditional approach in mechanics, which uses a coordinate system that is rigidly connected to the body, it is more convenient to write the equations of motion in a fixed coordinate system when studying the motion of a homogeneous ball.

The joint movement of two bodies is considered, namely:

Body A-conditionally, rectangle (parallelepiped) with sides of 350 × 500 mm. The body makes oscillatory movements (swings) relative to the center of rotation, located at a distance of 1500 mm.

Body B-conditionally circle (ball). Performs translational movements (up and down) relative to the body A. the length of the relative path is 400 mm.

Weight: Body A-20 kg, body B-2 kg (see Fig. 2).

The initial data is as follows:

Body A is moved to the leftmost position (at an angle of 30° from the vertical axis). Body B is in the upper position.

Let go of the body A. At the same time the body B begins to move down. When the body A is in the vertical position, the body B will shift to the lowest position (time spent ~ 0,5 s).

During the next quarter of the cycle, the body A is moved to the extreme right position (at an angle of 30° from the vertical axis). During this time, the body B will move to the topmost position.

The movement in the opposite direction is similar. Then the next cycle is repeated. The trajectory of relative movement of the body B is parallel to the line connecting the body A to the axis of rotation.

It is necessary:

To evaluate the change in amplitude. If it decreases, how much? When will it reach the values of 10° and 5°?

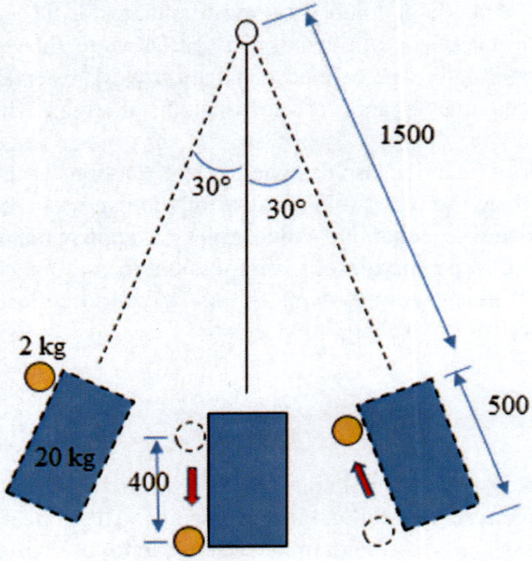

Fig. 2. Non-holonomic mechanics problem

If we assume that the mass of the body A is concentrated in the center of mass of the body, mass m = 20 kg, then we will consider the movement of the material point. A material point, forced to remain on a circle of constant radius $\vec{r}$, makes a movement in the field of parallel forces similar to the movement of a body on a stretched thread, when the size of the body is small compared to the size of the thread. The speed is tangent to the trajectory, and the trajectory is a semicircle.

The equation of change in the kinetic moment

$$\frac{d}{dt}(m\vec{r} \times \vec{v}) = \vec{r} \times \vec{F} \tag{1}$$

$$\omega^2 sin\varphi + \ddot{\varphi} = 0 \tag{2}$$

has a non-zero projection only on the perpendicular to the plane of the circle. The radius vector $\vec{r}$ and velocity $\vec{v}$ are mutually perpendicular, and $v = \mathrm{r} \cdot \varphi$. After the transformations the equation of change of the kinetic moment for the body A will take the form (see Fig. 3).

Now consider the motion of the body B in a non-inertial reference system (see Fig. 4) associated with the body A. It is necessary to apply the force of inertia to the body B, which is directed in the opposite direction to the acceleration of the body A. When the body A is at the top point, it begins an accelerated movement to the equilibrium position, so the acceleration is directed perpendicular to the speed vector.

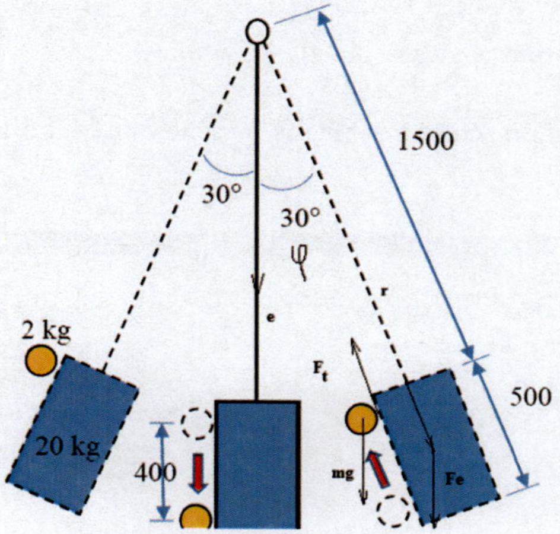

Fig. 3. Ball movement in to surface

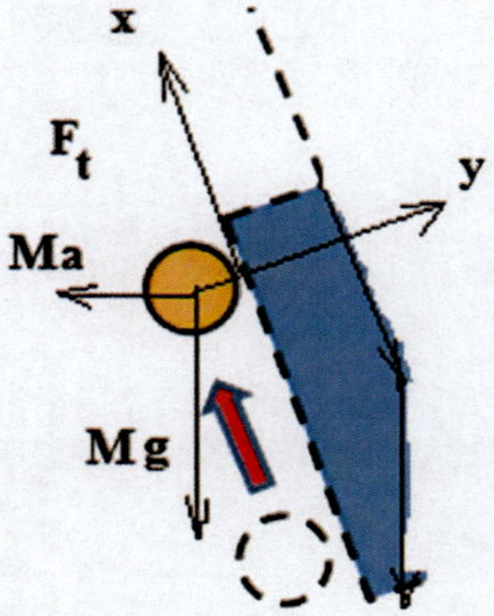

Fig. 4. Ball movement in the non-inertial system

Let's now consider the movement of a body with mass M = 2 kg. The inertia force Ma is determined by the acceleration a from Eq. (1). The movement of the body B should be considered in two directions, along the x and y axes. We define the projections of all forces along each direction. In Fig. 3, the inertia force is directed in the opposite direction. We also need to add the reaction of the support N, which is directed back to the y axis

a_x this is the acceleration of the body B along the x axis

a_y (this acceleration of the body B along the y axis) is zero.

The friction force Ft = μN, where the coefficient of friction μ depends on the quality of the contact surfaces of the bodies A and B.

From the lower Eq. 3, the force N is found, the friction force is determined, and is substituted into the upper equation. The acceleration a_x is found, on which all amplitudes depend.

$$r\omega^2 sin\varphi(\mu cos\varphi - sin\varphi) - g(\mu sin\varphi + cos\varphi) = \frac{d^2x}{dt^2} \quad (3)$$

Simulation of chain reaction machine is presented in Fig. 5 and Fig. 6.

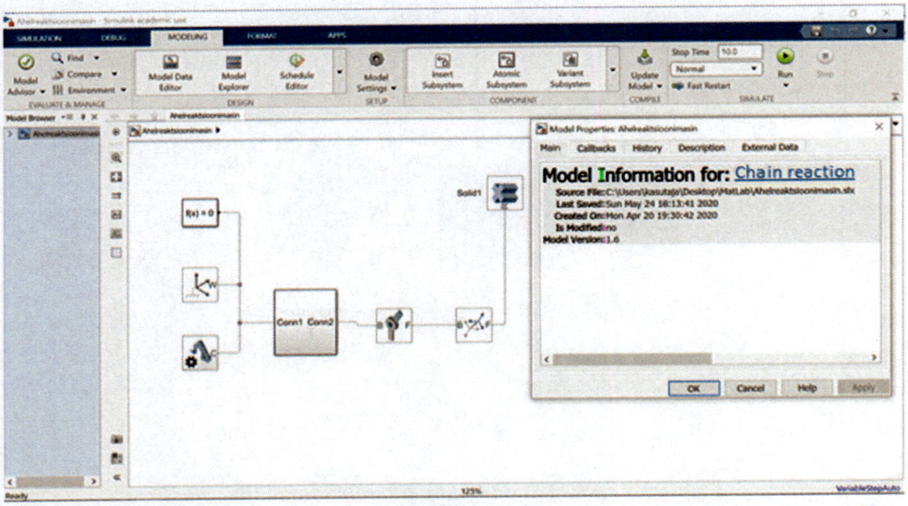

Fig. 5. Realization in simscape multibody MatLab r 2020 of the chain reaction machine model

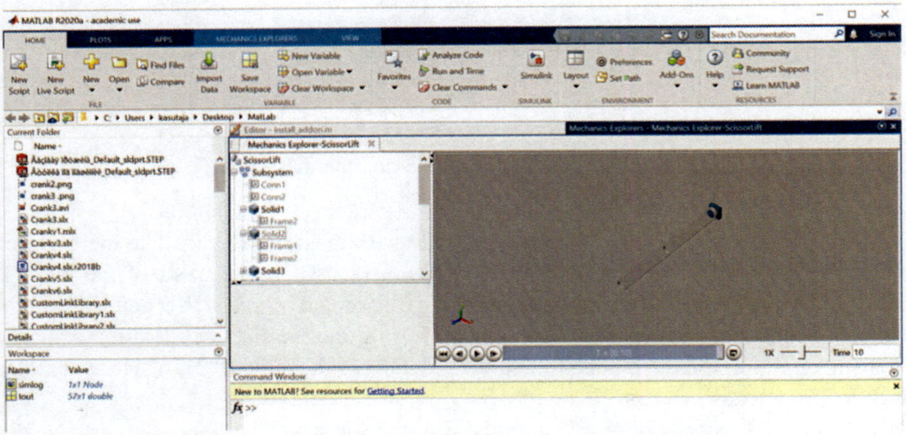

Fig. 6. Simulation in simscape multibody MatLab r2020 of the chain reaction machine model

4 Results

As part of this project, students' academic performance before and after the implementation of the PBL method was studied. Since this method was included in the second-year subjects, 24 students participated in the analysis, of which 5 students did not participate in the project, since their program did not include subjects that fit the PBL method. The tasks were distributed among the participating students: 5 students participated in the construction of the final model, the remaining 19 used the simulation method in the Matlab program for modeling and calculating individual elements. To analyze the results, we selected a comparison of students' average score, based on assessment criteria (see Table 1), before applying the PBL method and after.

Table 1. Assessment criteria

Grade	%
5	91–100 (Excellent)
4	81–90 (Very good)
3	71–80 (Good)
2	61–70 (Satisfactory)
1	51–60 (Sufficient)

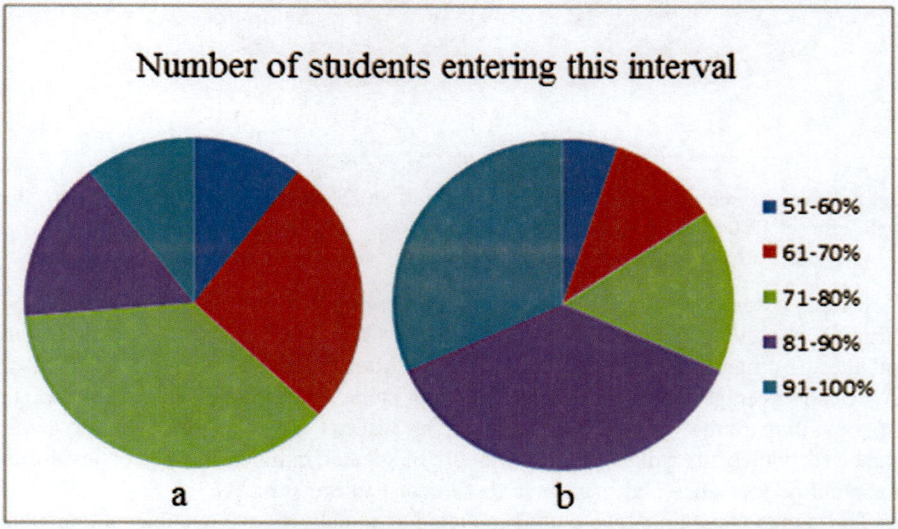

Fig. 7. Average score of participating students: a) before application of PBL; b) after application of PBL

Comparing the obtained results of assessments before and after implementation, we can say that the average academic performance of participating students after the introduction of the PBL method showed a significant improvement in grades (see Fig. 7). The number of students who have an average score of "4" or higher has increased by 2 times. While working on the project, it was noticed that the students were actively engaged in the project outside of school hours, which shows their interest. The motivation to study the lecture material and the desire to attend consultations has also increased in order to better understand the material and successfully complete the assigned part of the project.

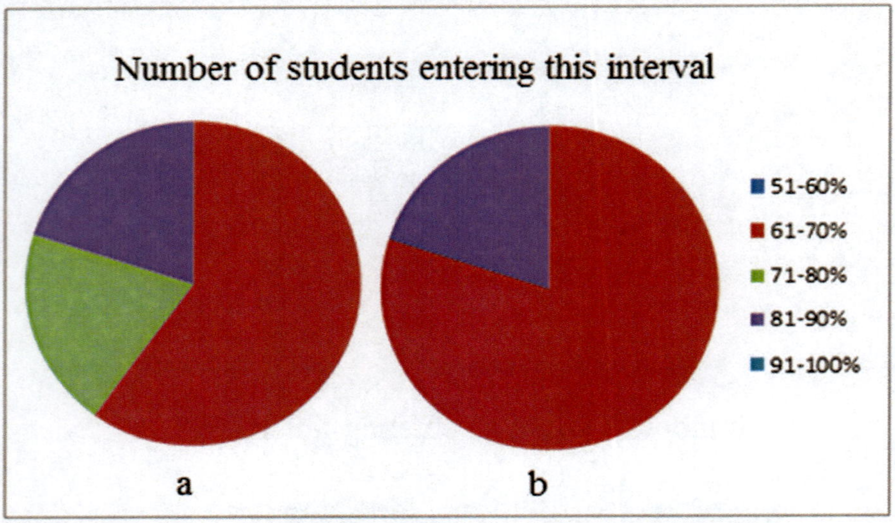

Fig. 8. Average score of non-participating students: a) before application of PBL; b) after application of PBL

Based on the two graphs presented, it can be concluded that the majority of non-participating students had an average score of 61–70% (see Fig. 8). You can also see that these estimates are not higher than 70%. After the implementation of the PBL project, the average grades of non-participating students significantly deteriorated and the prevailing range of 61–70% increased, as well as the number of higher grades decreased, which may indicate that students in greater numbers have worsened their academic performance and may have lost interest in the subjects.

The stands (see Fig. 9) were used in the educational process on subjects to explain the basic knowledge and laws of the profession, as well as to demonstrate the application of the theory in the modern world.

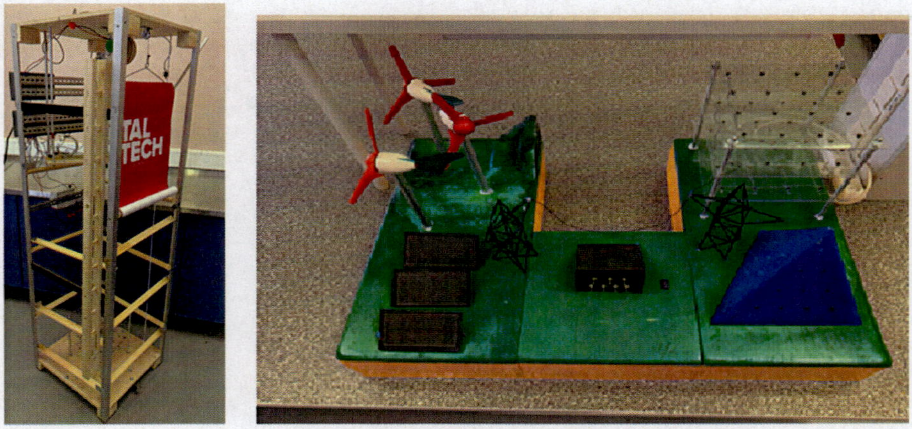

(a) (b)

Fig. 9. Mobile stands: a) "Chain Reaction"; b) "Green Energy"

The stands are mobile and can be used to promote the curriculum in schools, open University doors and educational exhibitions.

5 Conclusion

Since an increasing number of universities and colleges have recently applied PBL project-based learning, our curriculum has also taken into account the experience of using the latest educational technologies. This article shows that optimal integration of traditional teaching methods with practical verification of calculation results is of great benefit to both students and teachers. We have chosen to create two stands, "Green Energy" stand and "Chain reaction machine", to implement PBL project training. This experience will be used to accelerate the implementation of innovations in different training programs for college students.

Acknowledgments. This work was supported by Good Lecturer Development Program (Teaching development grant for teaching personnel to develop their teaching skills and learning activities), which is coordinated and issued by the Office of Academic Affairs.

References

1. Oliver, R.: Developing e-learning environments that support knowledge construction in higher education. In: Presented at the 2nd International We-B Conference, pp. 407–416. Perth, Western Australia (2001)
2. Bani-Hani, E., Shalabi, A., Alkhatib, F., et al.: Factors affecting the team formation and work in project based learning (PBL) for multidisciplinary engineering subjects. J. Proj.-Based Learn. High. Educ. **6**(2), 136–143 (2018)

3. Miller, A.: Designing PBL project to increase student literacy (2014). http://www.literacyworldwide.org/docs/default-source/member-benefits/e-ssentials/ila-e-ssentials-8060.pdf
4. Treadwell, S.M.: Making the case for project-based learning (PBL) in physical education. J. Phys. Educ. Recreation Dance **89**(1), 5–6 (2018)
5. Hong Chu, R., Dah-Chuan Lu, D., Sathiakumar, S.: Project-basedlab teaching for power electronics and drives. IEEE Trans. Educ. **51**(1), 108–113 (2008)
6. Valdez, M.M.T., Agreira, C.I.F., Ferreira, C.M.: Electrical power system security analysis using problem-based learning. In: Proceedings 43rd UPEC, pp. 1–4 (2008)
7. Costa, L.R.J., Honkala, M., Lehtovuori, A.: Applying the problem based learning approach to teach elementary circuit analysis. IEEE Trans. Educ. **50**(1), 41–48 (2007)
8. Mantri, A., Dutt, S., Jupta, J.P., Chitkara, M.: Design and evaluation of a PBL-based course in analog electronics. IEEE Trans. Educ. **51**(4), 432–438 (2008)
9. Hosseinzadeh, N., Hesamzadeh, M.R.: Application of project-based learning (PBL) to the teaching of electrical power systems engineering. IEEE Trans. Educ. **55**(4), 501–495 (2012)
10. Delaney, Y., Farrell, A., Hack, C.J., Lawlor, B., McLoone, S.C., Meehan, A., Phillips, D.T., Richardson, I.: An Introduction to Enquiry/Problem-based Learning, Maynooth: Facilitate and the All Ireland Society for Higher Education (AISHE). T&L (2015)

Game Based Education

Polytech WebQuest as an Organization Form of Students Project Activities

Pavel Kozlovskii[1], Anastasia Tabolina[2](✉), Olga Kunina[2], Veronika Fokina[3], Inna Yudina[4], and Pavel Sataev[5]

[1] Advanced Manufacturing Technologies Center (National Technology Initiative), Scientific Laboratory of Strategic Development of Engineering Markets, Peter the Great St. Petersburg Polytechnic University, St. Petersburg, Russia
pavelspb@mail.ru

[2] Institute of Humanities, Department of Engineering Education and Psychology, Peter the Great St. Petersburg Polytechnic University, St. Petersburg, Russia
asyatabolina@gmail.com, okun96@mail.ru

[3] Institute of Humanities, Higher School of Mediacommunications and PR, Peter the Great St. Petersburg Polytechnic University, St. Petersburg, Russia
fokina_vv@spbstu.ru

[4] Leading Specialist, Research Laboratory "Systems of Data Streaming Processing", Peter the Great St. Petersburg Polytechnic University (SPbPU), St. Petersburg, Russia
inna.yudina@spbpu.com

[5] Scientific Laboratory of Strategic Development of Engineering Markets, Peter the Great St. Petersburg Polytechnic University, St. Petersburg, Russia
sataevpa@gmail.com

Abstract. Project activities are an important component of the innovative development trend. Skillful project team building influences the efficiency of all work, so it is critical to understand roles of team members. This understanding will help put together an efficient group and improve the quality of decision-making on group management. Roles can be defined through a number of tests, but actual practical activities represent a more reliable source of information. A WebQuest is an umbrella term for various online and offline events in this format based on game mechanics. A number of solutions within the WebQuest allow defining students' roles in the team in the course of practical activities with a high level of engagement in the game. This approach has been studied in this article. The article presents assessments of roles with the help of the following indicators: the player's speed of response as compared with other team members; differences between answers within the group; the order of answers given, etc. Additionally, the project allows testing during the game, and that makes it possible to find out the participants' (players') opinions about the current distribution of roles in the team. Aside from studying the results of the event and collecting data on the aforementioned hypotheses, it is planned to compare the obtained results with other generally accepted team role tests, and the actual results of student team project activities in the framework of a special university course – Introduction to Project Management. Therefore, the main

M. E. Auer and T. Rüütmann (Eds.): ICL 2020, AISC 1328, pp. 431–438, 2021.
https://doi.org/10.1007/978-3-030-68198-2_40

focus of the study is to check the hypothesis that roles of participants in university project teams can be selected through student games in a specially arranged short-term playing environment at the university.

Keywords: Cloud Quest · WebQuest · Students · Project activities

1 Context

A. V. Tabolina notes that a WebQuest is a joint learning, educational, creative and playing activity with the purpose to reach a common result in solving a certain program. A WebQuest is a complex problem-based task that requires using information resources of the Internet global network to solve it, as well as obligatory application of role play elements.

According to M. V. Olennikova, a WebQuest is an effective didactic means for students of technical universities used to form communicative competencies in the process of their professionally oriented learning.

According to V. N. Kruglikov, the interactive learning model with the use of online resources in general and WebQuests in particular are based on fundamental principles of the learner-centered approach.

In the modern context, an active educational process cannot be effective without implementation of feedback technologies. First and foremost, the educational process should be clear to all parties involved in it and must be aimed at acquiring practical knowledge and skills in a convenient form [1].

According to P. S. Kozlovsky, Cloud Quest, as an example of active and constructive joint efforts with the use of online resources, serve as a driver for students of technical universities that motivates them to get involved in project activities.

According to V.V. Fokina, WebQuest is primarily a tool for teaching and developing applied skills in the field of information and analytical work in an interdisciplinary environment, which is of especial importance for technical students [2]. Cloud Quest can be used in the scope of on-boarding into the corporate policy of the University for creation of special projects for undergraduate applicants and first-year students in order to build University-related knowledge, and also as an additional tool to Citizen Science [3].

According to I. Yudina, Cloud Quest applied to project activities is, primarily, a technology for involving participants into a project through combining knowledge with entertainment in an accessible environment. This combination enables integration of diverse approaches, technologies and training methods, implies working in groups and promotes interaction between participants of the process [4].

2 Purpose

The purpose of the project is to review Polytech WebQuest and Cloud Quest as an Organization Forms of Students Project Activities. The result of the study is checking the hypothesis that roles of participants in university project teams can be selected through student games in a specially arranged short-term playing environment at the university.

Objectives:

1. Compare Cloud Quest and WebQuest approaches.
2. Define the need for such tools and consider their applicability.
3. Identify structural components of a WebQuest.
4. Describe WebQuest implementation stages in SPbPU.
5. Define main constituents of Cloud Quest under implementation in SPbPU.
6. Study participants' engagement in SPbPU 2019 Cloud Quest.
7. Draw conclusions.

3 Approach

3.1 WebQuest

WebQuest is a training tool focused on using web interface and quest mechanics. Cloud Quest is an umbrella term for various online and offline events included in this format and based on game mechanics. A variety of Cloud Quest solutions help identify roles of students in the team during their practical activities which require a high level of involvement in the game. A WebQuest is an inquiry-oriented lesson format in which most or all the information that learners work with comes from the web. The model was developed by Bernie Dodge. The basic structural components of a WebQuest are as follows. According to B. Dodge, WebQuest is a research assignment which requires students to use information obtained mainly from Internet sources [4]. (Table 1).

Table 1. Structural components of SPbPU WebQuest

WebQuest component name	Component content
Introduction	Includes key information concerning the project objectives and motivates students to perform the assignment. Introduction clearly describes the roles of the participants, the work plan, and an overview of WebQuest content. Besides, it poses the question to be answered by the WebQuest participants
Task	This is the key part of a WebQuest, where the final result of the research activities of the participants is clearly defined. The task must be clear, feasible, and relevant to the interests of students
References	List of information sources, links, or a database of the educational environment
Process	Description of the work implementation procedure and the algorithm of actions
Evaluation	Description of WebQuest evaluation criteria which depend on the training task established for the participants
Conclusion	The result of the research task implementation

WebQuest is actively used in disciplines where students need to be introduced to the subject and involved into the process of gaining knowledge. Being initially a project intended for school and extra-curriculum school education, this technology has firmly established itself in the university environment.

1. The model described below is one of the effective and widely recognized applications of the technology used in teaching foreign languages for the technical specialties of Peter the Great St. Petersburg Polytechnic University [6]. First-year technical students are being involved in the process of learning English based on dedicated assignments to search information relevant to their academic interests. In a preliminary questionnaire survey, the teacher asks the students about their preferred options for scientific research and prepares a list of the English-language content based on the obtained answers. The student gets an opportunity to search for information in a new language environment and becomes involved into the language practice in the most efficient way. The obtained information is the more valuable, the more successful is the completion of language tasks in the previous stage.
2. Another environment of WebQuest technology application in SPbPU is the dedicated zone for youth science and engineering known as FabLab Polytech [7]. The platform is committed to promoting the importance of initiative and free scientific research among student. One of the forms of involving new participants in projects is interactive assignments and tasks that, while being solved, incorporate the participant into project teams.
3. There is a number of disciplines that, in the students' opinion, lack the appeal, to the extent that the relevance of these subjects is often doubted. They include, inter alia, "Basics of life safety" and first aid courses. In the meantime, these subjects are of significant importance in the current context of the international health care crisis, an increase in the number of emergencies and industrial disasters. The interactive format, even if partially introduced into teaching these disciplines, could vastly change the attitude of students to the subjects and improve the resultant knowledge.

3.2 Cloud Quest

Cloud Quest is under development in SPbPU since 2016. The Cloud Quest concept combines online and offline assignment performance formats. In 2019, the Knowledge Day project for first-year students of the university has made a strong emphasis on the diagnostic constituent. The Quest is a team competition across 80 game stations. It involves 2,500–3,000 participants in teams consisting of 10 to 15 persons. Assignments on each station present a challenge in quick wits, execution speed and teamwork. Within 2.5 h, the first-year students complete assignments at stations (stations and assignments are both real and virtual), answer quiz questions and participate in an Instagram competition. Interaction between participants and with the Quest organizers occurs through a web interface via smartphones and tablets. Receipt of assignments and evaluation of results occur "in the cloud", therefore this is a "Cloud" quest. Based on the results of the game, participants are awarded in various nominations (with recognition of the best university teams).

The goal of the quest for the players is to gain as many game points as possible. In the project, the participants use web interface containing the main functionality, i.e. questions and a tool for entering answers. In addition, players must physically visit certain places to which the questions are directly related. In the scope of the reviewed event – the Knowledge Day, the Cloud Quest focuses on the entertaining nature of the mechanics. The educational questions and business game elements may be incorporated into the assignments. The analytic blocks are built into the mechanics and do not distract the participants from the basic process. The shift in the focus provides high engagement of the participants.

The Cloud Quest consists of three activity units:

1. Stations – physical places in the campus which the players must visit to complete the intended assignment. Depending on the format of the event, there may be a need for station hosts. On the Knowledge Day, they were students studying in various departments of the university. The stations are an offline activity.
2. Quiz questions – questions prepared in advance and fully presented in the web interface, so that one does not need to search for the answers offline.
3. Instagram competition – an entertaining activity where participants are asked to make theme-based photos and hashtag them as required. The results are subsequently evaluated either by the participants themselves or by members of the jury.

The Knowledge Day Cloud Quest participants are divided into the following groups:

1. Players: first-year students of the University, participants in the event.
2. Teams: study groups of first-year students.
3. Captains: students who are members of the Onboarders public institution. Their role is to accompany the team members and help them directly in their onboarding.
4. Station hosts: they prepare their stations; interact with the players at the stations and if necessary, assign game points to the team.
5. Organizers: they are responsible for the project methodology, subject matters, development of the web interface, involvement of station hosts, etc.

On the 2019 Knowledge Day, the organizers received a positive feedback from the players, captains and onboarders. The question was as follows: "Are you interested in joining the organizing group of the project to make the next quest better?" This question reflects involvement of the participants (Table 2). The table below shows that 313 players were ready to participate in the development of the project. It shows a high rate of interest in the event, which was achieved due to the special nature of the Cloud Quest project, namely due to its focus on the game. Many participants associated their favorable impression of the educational process with the opportunities to participate in team games, projects, various types of professional or project activities [8]. Moreover, modern universities are interested in shaping a consistent favorable impression in the students while they adapt to the learning process and while they become involved in project activities as the basis for future professional competencies, building the corporate culture and improving the value of the university brand.

Table 2. Involvement of the 2019 cloud quest participants

Survey participant's answer	Player	Captain	Host
Yes	313	40	38
No	475	105	55
Total respondents	788	145	93

3.3 Differences Between Cloud Quest and WebQuest

The key differences between Cloud Quest and the WebQuest format are discussed further. The discussion is based on the approaches adopted in WebQuest and highlights the role of these elements in the Cloud Quest.

1. Introduction. In fact, the introduction is instructions received by the players. In Cloud Quest, these are the rules and algorithms of actions for all categories of participants listed above.
2. Task. For Cloud Quest, the assignment includes three key activities: stations, quiz, and Instagram. For WebQuest, all assignments have different focuses and are associated with diverse areas of knowledge.
3. List of information sources. One of the key differences is that players in the Cloud Quest have to search for information in a variety of sources or rely solely on their knowledge and skills. On the other hand, in the Cloud Quest, the assignment is dedicated not to a specific subject, but to a whole set.
4. Description of the work implementation procedure. It is provided in the algorithms and rules.
5. Description of the criteria for evaluating a WebQuest. For a Cloud Quest, this is a system of internal evaluation of assignments in each of the three activities, considering their complexity and correctness of completion, as well as a system for standardization of scores gained in the three kinds of activities.
6. Conclusion. A significant difference of the Cloud Quest is that, unlike in the WebQuest, it is impossible to immediately announce its result and evaluate the demonstrated knowledge of the academic material. In the Cloud Quest, only winners of the event according to their scores can be announced immediately. Next begins the analysis of the collected data, which goes in parallel with collecting feedback and preparing new events.

3.4 Difficulties in the Interpretation of the Results

Further, there is a problem of interpretation of the results of the answers at the stations and in the quiz questions. The Knowledge Day Cloud Quest is a team game for first-year students. Teams of 10–15 persons, who also belong to the same academic group, compete between each other. This event is held in the form of a game and ensures high involvement of the participants in introducing many people (4–6 thousand participants) to the university, to the capabilities of the university departments in a short time (1.5–3 h), as well as in teambuilding between the new students. In addition, the event provides initial data on the students that contributes to preparation of individual educational paths and forming the project teams.

To enable the individual assessment, each participant needs to be granted access to the web interface, i.e. to the answers' entering functionality. This option is likely to cause internal confrontations within the team where the need for a group decision may conflict with an individual choice of a player who may disagree with the opinion of the team. Accordingly, one of the parameters that can be tracked after the game is the differences in the answers to the same question within one team.

The next indicator is the answer input speed. Here the feature of interest is the sequence of entering answers to one question by different players of the same team throughout the game. The hypothesis is that the frequency of answering among the first players will correlate to the degree of the player's leadership.

Other indicators will depend on the type of the selected assignment. The Cloud Quest involves development of multi-level tasks where the participants will be offered a choice between one of the several sub-tasks. Analysis of the players' choices in favor of a particular type of assignment, as well as the degree of success in its completion, can provide a basis for conclusions about proneness to a certain cognitive type.

The interaction between the Cloud Quest and WebQuest formats can be built as follows. Cloud Quest will be the first source of building the player's profile with the focus on the entertainment component to increase the engagement, as well as on the diversity of assignments to identify the types of the behavior and provide the first team building advice. WebQuest is an additional tool for subsequent checks where assignments will be formed depending on the results of Cloud Quest. Further, teams can be formed regardless of the academic group but based on the advice derived from the analytics; the Cloud Quest project for such teams can verify the previous hypotheses.

4 Conclusions

Firstly, the article compares the Cloud Quest and WebQuest approaches implemented in SPbPU, justifies relevance of such tools and considers the opportunities for their application. It also provides an analysis of the concept of WebQuest, which has the following common features with the Cloud Quest: use of the quest concept; online mechanics of answers to the prepared questions; relation to the educational process. The article compares the concepts and concludes that WebQuest can be used in combination with Cloud Quest.

Secondly, the structural components of a WebQuest are defined, the stages of implementing the WebQuest in SPbPU are described, the main units of the Cloud Quest implemented at SPbPU are defined, and engagement of participants in the scope of the 2019 Cloud Quest is reviewed.

Thirdly, the article shows that the application of Cloud Quest and WebQuest tools helps collect and analyze data that can further be used to shape an individual approach to each student's education. First of all, it means development of individual educational paths and providing team building advice.

Fourthly, the obtained data will help universities and the market develop a new effective solution for assessment and subsequent building of project teams. The solution will ensure high engagement of participants, convenience for the target audience and the possibility of integration into the educational and extra-curriculum processes. Further studies may include identification of the best balance between game and non-game features of Cloud Quest depending on a specific task, as well as further application of this format for new market segments.

Fifthly, modern research in the field of consumer behavior indicates the tendency to form trends for impressions as the basis of social activity and life experience [9]. WebQuest and Cloud Quest technologies can be largely perceived as elements of a new trend in the field of "shaping impressions" as they help actively demonstrate one's success via social media and serve as the basis for building relations with the educational environment through direct participation.

References

1. Kruglikov, V.N., Kasyanik, P.M.: The role of active learning in the concept of global engineering education. Scientific and technical statements of the St. Petersburg State Polytechnic University. Hum. Soc. Sci. **3**(227), pp. 159–168 (2015)
2. Trostinskaia, I., Popov, D., Fokina, V.: Communicative competence of engineers as a requirement to the future professions. In: Chernyavskaya, V., Kuße, H. (eds.) Proceedings of the 18th International Conference PCSF 2018. EpSBS, vol. 51, pp. 1672–1678. Future Academy, London (2018)
3. Ruiz-Mallén, I., Riboli-Sasco, L., Ribrault, C., Heras, M., Laguna, D., Perié, L.: Citizen science: toward transformative learning. Sci. Commun. **38**(4), 523–534 (2016)
4. Tabolina, A., Kozlovskii, P., Popov, D., Yudina, I., Snegirev, N., Tretyakov, D.: Socio-psychological program for the selection of students in the adapters public institute. In: Anikina, Z. (eds) Integrating Engineering Education and Humanities for Global Intercultural Perspectives. IEEHGIP 2020. Lecture Notes in Networks and Systems, vol. 131 (2020)
5. Dodge, B.: Some Thoughts About WebQuests. http://www.webquest.org/sdsu/about_webquests.html. Accessed 20 May 2020
6. Rubtsova, A.V., Almazova, N.I.: Productive model of foreign languages learning: realities and prospects. In: International Conference Communicative Strategies of Information Society (CSIS 2018). Advances in Social Science, Education and Humanities Research, vol. 289, pp. 319–324 (2018). https://doi.org/10.2991/csis- 18.2019.65
7. Olennikova, M.V., Tabolina, A.V.: Psycho-pedagogical support of students project activities in multi-functional production laboratories (fab lab) on the basis of technical university. In: Advances in Intelligent Systems and Computing, vol. 917, pp. 732–740 (2019)
8. Obukhova, I., Popov, D., Tanova, A., Veronika, F.: The evaluation of university's impact on «human resource potential» of alumni. In: E3S Web of Conferences, vol. 164 (2020)
9. Just do it: the experience economy and how we turned our backs on 'stuff'. https://www.theguardian.com/business/2017/may/13/just-do-it-the-experience-economy-and-how-we-turned-our-backs-on-stuff. Accessed 20 May 2020

sCool: Impact on Human-Computer Interface Improvements on Learner Experience in a Game-Based Learning Platform

Martin Sackl[1(✉)], Alexander Steinmaurer[1], Christopher Cheong[2], France Cheong[2], Justin Filippou[3], and Christian Gütl[1]

[1] Graz University of Technology, Graz, Austria
martin.sackl@student.tugraz.at
[2] RMIT University, Melbourne, Australia
[3] University of Melbourne, Melbourne, Australia

Abstract. As many concepts in programming are difficult for novices to learn, it can be frustrating for them to stay motivated. For this purpose, alternative approaches like game-based learning can help increase their motivation. sCool is a serious game that supports learning computational skills through conceptual aspects and practicing programming, but is also used in revision activities as it provides an engaging way for students to review their understanding of programming. In a revised version of sCool, we focus on obtaining an improved human-computer interface (HCI) by adapting several interface aspects. Thus, our research goal is to determine the effects of the improved interface on novice student programmers' revision of learning programming, but also how experienced student programmers perceive the interface. In a preliminary study on the revised version of sCool, 11 novice and 9 experienced student programmers played the game and completed a survey providing data to answer the defined research questions. We are able to show that the improved interface has a positive impact on novices' revision of learning programming as well as on the game-play of sCool.

Keywords: Serious game · Mobile learning · Human-computer-interaction · Coding

1 Introduction

In modern society, the value of programming and computational thinking is more important than ever, as computational methods let us solve problems and build systems accordingly, and this is not only in the area of computer science. Computational thinking enables people to recognize problems and solve them in efficient ways. It involves algorithmic thinking, evaluation, decomposition, abstraction and generalisation [3]. To be able to understand computational

M. E. Auer and T. Rüütmann (Eds.): ICL 2020, AISC 1328, pp. 439–451, 2021.
https://doi.org/10.1007/978-3-030-68198-2_41

thinking, there are several approaches which improve engagement in learning programming. This study deals with the perception of the improved human-computer interfaces used in a game-based learning approach, which has two purposes - to be fun and to be educational [1].

For this study, we initiated, developed and researched the serious game sCool [2], a mobile learning tool for novice student programmers to help them with the basics in the world of coding and computational thinking. Previous research [9] on sCool showed that students were motivated while playing the game, but often got distracted by interface issues. The goal for this study is to improve the human-computer interface in such a way, that the learning process is not hindered by interaction problems. In this paper, we argue the importance of a concise and well-designed human-computer interface (HCI). We investigated how the user interface affects novice student programmers' revision of learning programming as well as how experienced student programmers experience the improved interface. Students should be able to write code in a basic way and thus be able to concentrate on specific tasks, as well as get more helpful feedback and guidance from the system.

The remainder of the paper is organized as follows. In Sect. 2, we outline the existing tools and approaches in the area of game-based learning and which HCI aspects are important in this area. We then introduce sCool in detail and specify how it is used in contrast to other learning tools. We also discuss improvements that were made in order to obtain a concise and well functioning interface. The next section covers a detailed evaluation of the game's new version, in which we tested the game and specifically its interface on novice and experienced student programmers. This includes a description of the setup of the experiment, the evaluation instruments, procedures and participant details, as well as a discussion of the results. We then summarize our findings and discuss future work with sCool.

2 Related Work

Mobile games have become increasingly popular over the last few years and are even an influential part of the learning behaviour of students. Such games are not only useful instruments for learning specific content, but they can also leave an impression on students [4]. In general, students are more motivated while playing games than when using other learning approaches and thus they can be totally immersed in the game, because of game's complex graphics and sounds as well as instant responses to player actions [5].

Shabalina et al. [6] introduced a role playing game, developed for learning programming languages, which is mainly used in computer science classes at technical universities. Users can be further classified into beginner students and students who want to improve their skills. The authors proposed three basic concepts, on which the game is based on: 1) learners should get the course content through their interpretation of the game world, 2) they must see results of their programming in a game context and 3) the results should influence the game. This approach is believed to improve students' skills in programming and increase their motivation. Another learning tool is the game *iPlayCode*, proposed by Zhang et al. (2014) [10], which was developed for non-programmers to learn programming in C++ in a cycle-learning mode. Great value is placed on the game's simple interface, which they state is a core component to improve the user experience. This tool includes a reward system that awards or deducts points, depending on the player's answers. Besides functional testing and performance testing, the game's usability was tested, divided into learnability, efficiency, memorability, errors and satisfaction. Results showed that users were able to manage the system's interface easily at the first use.

The serious game *sCool* [2] is a learning tool, which was developed as a research collaboration between Graz University of Technology (Austria) and Westminster University (UK). This educational game is designed for novice programmers to learn computational thinking and programming in Python in a playful and exploring way. Based on an engaging story, it offers different game types to learn concepts as well as practical coding exercises. Additionally, educators are able to adapt the learning content using a web interface. With concept-learning tasks, players can explore maps and learn new programming concepts, which can then be applied in a practical mode. Users write actual Python code in order to develop a program that leads a robot to the destination on a squared chessboard playground. Therefore, basic commands are given in form of buttons (Move, Print, If, Var and For), but students are free to edit code blocks as well.

Previous research projects with sCool identified issues regarding its interface and controls and thus the game did not provide the best learning experience. The practical mode of the game had the most issues. To write code, a virtual in-game keyboard was initially implemented, which had some serious issues such as: overlapping other content, confusing layout and prohibiting the use of a native keyboard. Further issues included players not being able to write the first lines of code on their own, but rather had to edit a code block's content. The drag and drop functionality also caused confusion as players often did not know how to use these elements properly. Players also identified issues with the game menu (missing descriptions of navigation and ambiguous named buttons) as well as tutorials and hints, which many users often skipped. As the design of hints did not stand out from the editor, players tended to gloss over them [9].

In more abstract observations, while using games in educational areas, the interaction between students and the system, the human-computer interaction (HCI), is an important aspect. Therefore, simple and well designed interfaces are essential for a high learning rate. To achieve this, HCI evaluations and usability testings are a significant part of contemporary applications [8].

According to Looi et al. (2010) [7], improvements on serious games platforms in order to obtain best results from students and high success rates can be achieved by HCI evaluations. Such evaluations can be used to overcome low engagement of students or repetitive nature of tasks, and to accomplish a successful educational game, which motivates students over a longer time period. In their work they stated that players' engagement is based on how close the interaction between the players and the game appears. To evaluate this, they used the video-diary method, which investigates the gamers' perspectives. Thereafter, questions about the student's level of boredom during the game, whether they could sustain longer time periods and whether they prefer non-learning associated activities were evaluated.

3 sCool - Interface Improvements

Based on the issues revealed from previous sCool research projects [9], there was a significant need to adapt the game's interface so that users on any device are able to write code in a simple way and do not get distracted by other reported interface problems. These issues concerned **i)** code editor, **ii)** controls/keyboard, and **iii)** menu/instructions/hints.

3.1 Requirements

Code Editor. There is a strong need to improve the code editor. For easier use of code blocks, users should be able to write the first line of code on their own, rather than using the drag and drop functionality. In general, they should be able to do this anytime. To show players that elements like code blocks can be dragged and dropped, a visible indicator (like an icon) should be displayed next to these elements.

Controls and Keyboard. To achieve a well-functioning multi-platform game, the initial virtual keyboard should be replaced by the device's native keyboard, since it showed some earlier mentioned problems. Especially in the Windows version, players were frustrated as they had to click the keyboard's buttons to write code.

Menu, Instructions and Hints. Once a player pressed the "Menu" button during a robot mission, s/he would be taken back to the menu right away, without confirming the action and all the written code would be gone. Thus, the data loss should be avoided, since it is an important aspect of system feedback. Another feature in that respect is the design of hints and descriptions during the game; they should be better emphasised so that they stand out more from the editor and background. Tutorials are shown when a player enters the practical mode and can also be displayed anytime. Some hints however appear only once and are never shown to the player again. The goal therefore is to ensure that players read and understand these messages, as they tend to gloss over them.

3.2 Implementation of Interface Optimizations

One of the most significant issues identified is regarding the **controls and the virtual keyboard**. A simple solution therefore is to remove the virtual keyboard completely and allow the use of native keyboards for all devices. The most effective way therefore is to use input field elements instead of ordinary text components as they automatically display the device's native keyboard when they are in focus (see Fig. 1).

In order to enable a better coding experience, several components of the **code editor** were improved. Since there was no feedback from the system when a player clicked predefined drag and drop buttons, an alternative method was implemented which marks the selected button and enables an "Add" button. Drag and drop functionalities inside the code editor left out an important aspect of usability design, which is a visible indicator that a user can interact with an element. Users may not know that code lines are draggable and for this purpose, such an indicator in the form of an eight dotted icon was placed inside a draggable element. Since users are only able to edit existing code blocks, an empty code line including a placeholder was added to the editor. This line is placed at the end of the code at any time, so that the player can write additional lines of code one after another by pressing the "Enter" button.

To improve **hints and tutorials**, a simple approach to encourage players to pay more attention to in-game tutorials was implemented. Some examples are the descriptions of the different tab views (task description, code editor and output panel), which are important. These descriptions are only shown once to a player. To get the player focused on the tutorial and give the feeling that the game is paused momentarily, we added a darkened background canvas and an animation of the tutorial pop-up. Further, all of the code a player had written could be lost as a result of a single errant click, which is a grave interaction issue. This was resolved by adding a confirmation screen after clicking the "Menu" button. This functionality was added only in cases of data loss or major changes.

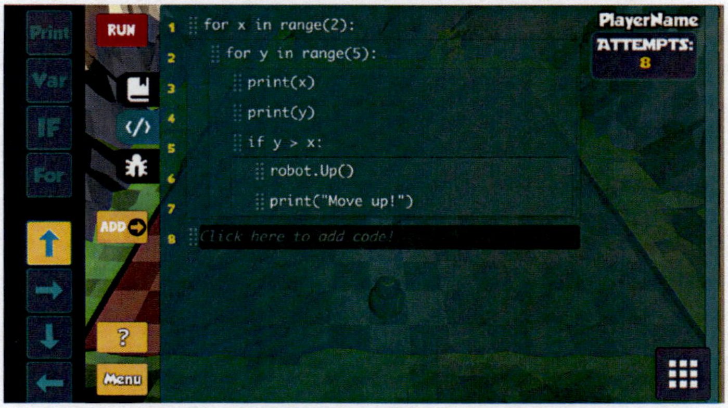

Fig. 1. Improved user interface of the code editor.

4 Evaluation

Based on the improved user interface (see previous section), an evaluation was performed to answer the following research questions:

RQ1: What are the effects of the improved human-computer interface on novices revising learning programming?
RQ2: How do improvements to the in-game code editor and communications (such as hints and tutorials) affect player's game experience?

4.1 Setting and Instruments

Participation in this study was voluntary and was conducted

1. in an in-class activity in an Australian business information systems course for novice student programmers and
2. with students from different programs in computer science at a technical university in Austria addressing those with programming skills.

The instruments used involved the **revised version of sCool**, an **online questionnaire** and the **usage data** (collected on a server). The novice group experienced the game in a supervised setting while the experienced group used it in an unsupervised setting.

Novice Student Programmers. The study on novices was initially planned in classrooms, but due to the COVID-19 global pandemic restrictions, these classes were held online. There were 5 online classes in a Python introduction course at an Australian university, with around 25 students per class. The first half was a regular class held by the tutor. In the second half, the tutor left the classroom (due to the ethics approval) and an external supervisor joined to conduct the experiment. Due to occurring connection problems, we pre-recorded an introduction video showing the use of sCool. The students then had 30–40 min to play the game and were asked to answer an online questionnaire afterwards.

Experienced Student Programmers. We also had to conduct the study for experienced student programmers online. Computer science students from a technical university in Austria were asked to participate. As the experiment was not conducted in regular classes, we sent the participants an information document about the study, the introduction video and game download links. The participants were able to play the game anytime they preferred, so no supervisor was present. They were told to take 40–50 min for this experiment.

Online Questionnaire. Regarding usability testing and analysing, we included the Computer Emotion Scale [12] (CES) and the Game Experience Questionnaire [11] (GEQ) in our survey, which was hosted on the Qualtrics survey platform.

The GEQ holds 12 adjectives associated with the four basic emotions: happiness, sadness, anxiety and anger and allows the participants to mark how often they felt an emotion while playing the game by choosing from a range from "1 - none of the time to 4 - all of the time". The CES includes 19 statements, which describe the engagement of a user during the game. These can be associated with immersion, presence, flow and absorption; answers can be made using a Likert scale from "1 - strongly agree to 5 - strongly disagree". The mean and standard deviation from the CES and GEQ results were calculated using the open source programming language R.

We also asked the participants to choose from a Likert scale on how much they agree on eight statements about usability and interface, which are shown in Table 1. Besides two open-ended questions about HCI improvements, the participants were given seven game-related statements (involving the game's language, the difficulty of levels and the encouragement of sCool), which they are required to rate as well on a Likert scale ranging from 1 to 5.

Table 1. Statements about usability and interface.

ID	Statement
S1	The game's control was easy to use
S2	The code editor was easy to use
S3	Writing code was easy
S4	I liked to us the drag and drop function for coding
S5	I liked to write code on my own
S6	Error messages were clear and understandable
S7	The tutorials helped me to understand what to do
S8	Hints helped me to complete levels

Usage Data. While playing the game, the system stores the player's usage data (relevant information about player, player's performance and log data) securely on a Microsoft Azure SQL database server. To observe how well students performed in the in-game course, the database provides information about all players' attempts, unlocked skills and solved tasks. Analysis and visualization of the usage data, as well as the overall results from the online questionnaires, was done with Microsoft Excel.

4.2 Procedure

The experiments for novices and experienced student programmers were structured similarly, with the main difference that a supervisor was only present at the novices experiment. Both groups were introduced to sCool and the research project and given time to play the game. The in-game course, which was created explicitly for this experiment over the web platform included three parts:

1. Fundamentals of Python and Simple Data Types
2. Complex Data Types
3. Constructs and Logic

Each part of the game included three concept-learning tasks and three practical exercises and students had to solve all tasks in order to unlock the next skill. This covered the Python introduction course's content of the weeks before. After students revised their programming knowledge in the concept-learning section, they had to apply it in the practical part.

In the first part of the *robot missions* (practical part), students had to solve basic tasks related to fundamental programming concepts (commands and sequences) in Python which introduced them to the programming environment. The second part covered tasks related to data types and variables. Players learned this concepts using the so-called robot storage which serves as an internal memory where external data from the web application can be provided into the scope of the game. Also, participants were faced with arithmetic tasks. In the last part, students used loops and conditional statements in order to complete the missions. Table 2 shows an overview of the tasks given for the robot missions.

Table 2. Overview of practical tasks with Rob the robot.

Skill	Task
Introduction	1. Use basic commands to move Rob
	2. Move Rob and avoid obstacles
	3. Print a phrase to the console
Data types	1. Access the robot storage, add a value and print it
	2. Print the robot storage and (optionally) collect coins
	3. Use the robot storage and arithmetic operators for a calculation and print it
Constructs and logic	1. Use a loop to print the list in the robot storage
	2. Use conditional statements to print right content of the robot storage
	3. Compute the Fibonacci Sequence and print the index of the list in the robot storage, where the lists differ

4.3 Participants

As part of the study, 49 undergraduate students used sCool. Of these students, 39 gave their consent for sCool to store their in-game data to be used in the research and 11 completed the questionnaire. We also invited 20 experienced student programmers to the study, of which 9 participated and all of them completed the questionnaire. In total, 20 students fully answered the questionnaire.

Table 3 shows the details of the participants divided into the two groups: novice and experienced students. According to this table, the main difference between the two groups are the programming experience and age of students.

Table 3. Overview of participant demographics.

Criteria	Possible responses	Novice student programmers	Experienced student programmers
Gender	Male	8	8
	Female	3	1
Age	18–24	10	0
	25–34	1	8
	35–44	0	1
Played an educational game before	Yes	0	1
	No	11	8
Programming Experience	Non existent	2	0
	Novice	6	1
	Intermediate	3	1
	Advanced	0	5
	Expert	0	2

4.4 Results and Discussion

To be able to answer RQ1, we analyzed the collected usage data of the novice student programmers. According to this data, of the 39 novice participants, 34 (87.18%) were able to solve all concept-learning tasks and 26 (66.67%) all practical tasks of the first skill and thus, unlocked the second skill. In the second part, 18 (46.15%) passed all of the concept-learning and 6 (15.38%) the practical tasks. This means that 6 (15.38%) novice student programmers were able to unlock the last skill, in which 2 (5.13%) solved all concept-learning and 1 (2.56%) all practical tasks. In contrast to previous experiments with the former sCool interface, in which no student reached the third part, this shows a slightly improved performance. This means, in respect to RQ1, that the improved interface indeed had a positive effect on the revision of novices as it improved their performance on a small scale.

On the other side, all 9 experienced programmers were able to reach the third part and thus solved 100% of the concept-learning and practical tasks of the first two skills. Regarding the third skill, 6 (66.67%) passed all concept-learning and 5 (55.56%) passed all practical tasks.

In general, both novice and experienced student programmers showed similar results according to the Game Engagement Questionnaire and Computer Emotion Scale. Results are shown in Figs. 2 and 3. Concerning the GEQ, both groups showed a high level of immersion (M = 3.45, SD = 0.82 for novices and M = 3.78, SD = 0.83 for experienced) and presence (M = 3.09, SD = 0.88 and M = 3.22, SD = 1.12). Since immersion can be described as a state of consciousness where the player is totally engaged in the game environment and presence the regular state of consciousness (Brockmyer et al. (2009) [11]), the novice players were aware of the virtual environment, but also engaged into it. Their level of flow (M = 2.56, SD = 1.14 and M = 2.69, SD = 1.10) indicates the balance degree between challenge and skill, which leads to a good learning experience [13].

The CES results show that most of the participants had positive feelings while playing the game. This can be observed from happiness levels (M = 2.55, SD = 0.91 for novices and M = 2.69, SD = 0.68 for experienced students). Levels of sadness, anxiety and anger were almost equal for both groups, however, experienced student programmers showed a slightly higher level of anger than novices (M = 1.40, SD = 0.50 and M = 1.58, SD = 0.64). Compared to evaluations from previous versions of sCool, the level of happiness increased while the level of anger decreased, which points to a positive effect regarding RQ2.

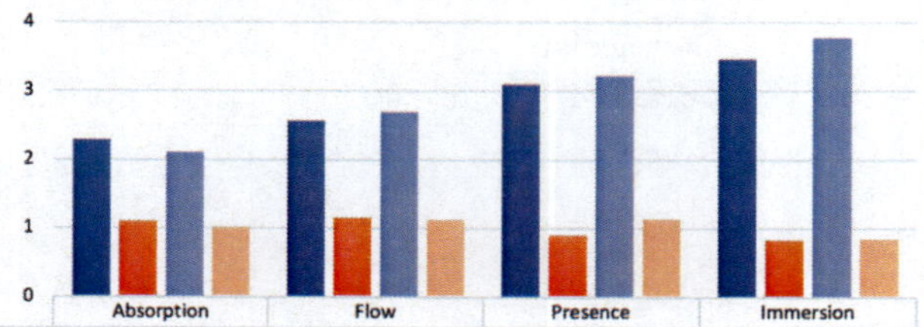

	Absorption	Flow	Presence	Immersion
Mean (novices)	2,29	2,56	3,09	3,45
SD (novices)	1,10	1,14	0,88	0,82
Mean (experienced)	2,11	2,69	3,22	3,78
SD (experienced)	1,01	1,10	1,12	0,83

Fig. 2. Results of the game engagement questionnaire.

According to answers on sCool's interface specific statements, the novice student programmers liked the possibility to write code on their own (S5) more than using the drag and drop functionality (S4). Their opinion on the usage of the code editor for writing code (S2 and S3) was positive. Thus, we can observe that the improvements had a reasonable impact on novices revision of programming (RQ1). Rather negative answers were collected about error messages and hints in the game (S6 and S8), which means that regarding to RQ2, communications between the system and the player did not improve for novices. Experienced student programmers also tend to like writing code (S5) more than using buttons (S4) and rated the simplicity of writing code (S3) very positively. However, they also stated overall positive feelings concerning the code editor (S2). Like the

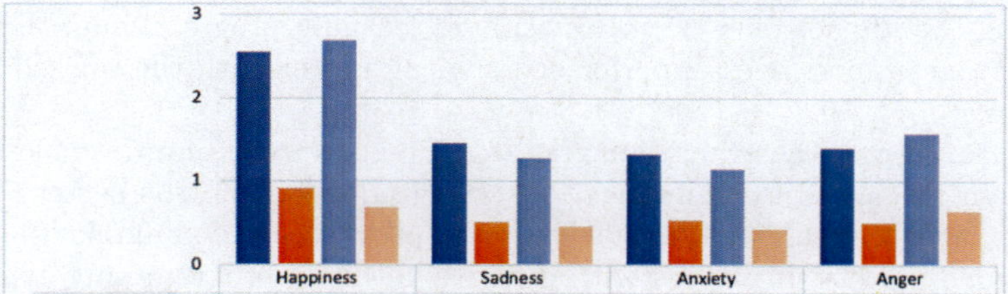

	Happiness	Sadness	Anxiety	Anger
Mean (novices)	2,55	1,45	1,33	1,40
SD (novices)	0,91	0,51	0,53	0,50
Mean (experienced)	2,69	1,28	1,14	1,58
SD (experienced)	0,68	0,46	0,43	0,64

Fig. 3. Results of the computer emotion scale.

other group, they disliked error messages and hints (S6 and S8), even more than novices. We can therefore state, that the improvements had a bigger impact on novices (RQ1) than they had on experienced students (RQ2).

Regarding open-ended questions, participants mostly had concerns and improvement suggestions for specific tasks and the "planet" exploration mode (*"I couldn't move from the starting point of the map to the discs and enemies"*, *"Change controls (A is for shooting AND going left)"*). Answers categorized as "Interface" mostly concerned indentations and error messages (*"Errors are a little hard to understand sometimes"*, *"Make hints easier to understand"*, *"Make adding indentations for desktop users more intuitive"*). Also several suggestions regarding tutorials and hints were given. Overall, only a few answers were provided regarding the improvements implemented in the interface.

5 Conclusion

In the revised version of sCool, our goal was to implement an improved human-computer interface by adapting several issues identified with the interface. In two similarly-structured experiments, but with different levels of participants' programming experience, our aim was to show how these improvements affect the game-play and whether they improved novices' revision of learning programming.

Since participation and completion of the survey was completely voluntary, the evaluation was limited to 20 participants. The results of the evaluation showed that players did not have as many problems with the interface as before, but also that the improvements had a bigger impact on novices than experienced student programmers. Compared to *iPlayCode* [10] many of the novice but almost all of the experienced participants managed to interact easily with the interface. Observations of the GEQ and CES showed an increased level of happiness and a lower level of anger compared to previous research studies. On the other hand, we were not able to identify improvements on communications between system and player in the form of error messages and hints, but rather (like the approach from Moreno-Ger et al. [8]) respective communication issues.

Regarding the impact on novices' revision of learning programming, the novice student programmers overall performed slightly better with the revised version of sCool.

For future work, the evaluation results indicated further improvements of the interface in the area of communications and feedback from the system such as error messages and hints. Our goal is to improve these communications such that a broad variety of players with different levels of programming experience are provided a more self-explanatory game-play experience and receive more understandable messages from the system.

Acknowledgment. We would like to gratefully acknowledge the support of Graz University of Technology for funding and RMIT University's School of Business IT and Logistics for hosting Martin Sackl. We would also like to thank participants for volunteering in our research project.

The research with students from the Australian university was conducted with ethics approval from the RMIT Business College Human Ethics Advisory Network under register number 22675.

References

1. Bellotti, F., Kapralos, B., Lee, K., Moreno-Ger, P., Berta, R.: Assessment in and of serious games: an overview. Adv. Hum.-Comput. Interact. (2013). https://doi.org/10.1155/2013/136864
2. Kojic, A., Kojic, M., Pirker, J., Gütl, C., Mentzelopoulos, M., Economou, D.: sCool – a mobile flexible learning environment. In: Online Proceedings from 4th Immersive Learning Research Network Conference, Missoula, Montana (2018)
3. Wing, J.M.: Computational thinking. Commun. ACM **49**(3), 33–35 (2006)
4. Gros, B.: Digital games in education: the design of games-based learning environments. J. Res. Technol. Educ. **40**(1), 23–38 (2007). https://doi.org/10.1080/15391523.2007.10782494
5. Beck, J.C., Wade, M.: The Kids are Alright: How the Gamer Generation is Changing the Workplace. Harvard Business Review Press (2006). 9781422166499
6. Shabalina O., Vorobkalov P., Kataev A., Tarasenko A.: Educational Games for Learning Programming Languages. Institute of Information Theories and Applications FOI ITHEA, 1313-0455 (2008). hdl.handle.net/10525/1136
7. Looi, Q.E., See, S.L.: Effectively engaging students in educational games by deploying HCI evaluation methods. In: Proceedings of the World Congress on Engineering and Computer Science, vol. 1 (2010)
8. Moreno-Ger, P., Torrente, J., Hsieh, Y.G., Lester, W.T.: Usability testing for serious games: making informed design decisions with user data. Adv. Hum.-Comput. Interact. (2012). https://doi.org/10.1155/2012/369637
9. Steinmaurer, A.: Revising a game-based learning platform for computational skills in education. Master's thesis, Graz University of Technology, Graz, December 2019. https://doi.org/10.13140/RG.2.2.35247.76967
10. Zhang, J., Lu, J.: Using mobile serious games for learning programming. In: The proceedings of The Fourth International Conference on Advanced Communications and Computation (INFOCOMP 2014), Paris, France, pp. 20–24 (2014)

11. Brockmyer, J.H., Fox, C.M., Curtiss, K.A., McBroom, E., Burkhart, K.M., Pidruzny, J.N.: The development of the game engagement questionnaire: a measure of engagement in video game-playing. J. Exp. Soc. Psychol. **45**(4), 624–634 (2009). https://doi.org/10.1016/j.jesp.2009.02.016
12. Kay, R.H., Loverock, S.: Assessing emotions related to learning new software: the computer emotion scale. Comput. Hum. Behav. **24**(4), 1605–1623 (2008). https://doi.org/10.1016/j.chb.2007.06.002
13. Csikszentmihalyi, M., Abuhamdeh, S., Nakamura, J.: Flow. In: Flow and the Foundations of Positive Psychology: The Collected Works of Mihaly Csikszentmihalyi, pp. 227–238. Springer, Dordrecht (2014). https://doi.org/10.1007/978-94-017-9088-8_15. ISBN 978-94-017-9088-8

"LimStorm" – A Didactic Card Game for Collaborative Math Learning for Gen Z Students

Szilvia Szilágyi and Attila Körei(✉)

University of Miskolc, Miskolc, Hungary
{matszisz,matka}@uni-miskolc.hu

Abstract. All engineering and informatics BSc programs include a basic course in Mathematical Analysis or Calculus. Thus, all students enrolled in the first year of these majors encounter the concepts of sequences, convergence and limit during the first semester. Understanding and processing this topic is usually a serious problem for most students and it is difficult to keep their attention and interest in the long term. In order to motivate our students we created a mathematical game that is suitable to introduce the concept of sequence limit in an entertaining way and thoroughly practice it, presenting various examples. The card game LimStorm appears to meet the needs of Z-generation students and effectively supports the teaching and learning processes.

Keywords: Didactic game · Gen Z students · Gamification · Small-group education

1 Introduction

Today, the vast majority of students in higher education belong to Generation Z. They grew up in the 21st century, an era in which the pace of technical and digital development has been faster than ever before. Teachers at all levels of education must have been confronted with the fact that there is a need for different communication and different teaching methods for Gen Z students than those used for the previous age groups. Generation Z people are also often referred to as digital natives, while their teachers are generally considered digital immigrants. Bridging the gap between them is not easy, but it is not impossible, as Generation Z students are not less motivated than their predecessors and are not passive at all, contrary to popular belief. They like to take on roles and responsibilities and be active in whatever they do. They are happy to take the initiative, which was not a characteristic of the previous generation at all [1]. However, their monotony tolerance is poor, so the frontal teaching method should be gradually pushed out of classrooms and given way to small-group, student-centred learning. A teacher explaining at a blackboard is not able to maintain the attention of today's students for a long time anyway, and even catching their attention is not an easy

M. E. Auer and T. Rüütmann (Eds.): ICL 2020, AISC 1328, pp. 452–463, 2021.
https://doi.org/10.1007/978-3-030-68198-2_42

task. The avarage attention span of Generation Z students is 8 s which filter helps them quickly choose the content they are interested in [2]. Generation Z students have an opinion on education (and about everything in general), and it is not hidden, so teachers have the opportunity to find out what students expect from them and the education system [3].

Teachers need to allow room for the autonomy and creativity of the Gen Z students. It is necessary to articulate exactly what the task to be completed is and the motivation behind the task, but students like to find a way to a solution on their own. If they achieve some partial success they expect a reward and positive confirmation. Gen Z students grew up with computer games where they had to become effective problem solvers if they wanted to go through different levels of difficulty in the game. Following the example of video games, learning programs can be created in which the goals to be achieved are built on each other and in which the experience of success in solving the subtasks encourages the student to move on and complete the next level. This is one example of how to make education playful; we need to find opportunities for gamification and game-based learning at all levels of education [4].

All engineering and informatics bachelor programs at university include a mathematics course among the basic courses. The concepts of sequences, convergence and limit are to be learned by all students enrolled in the first year of technical or IT education. Maths courses are usually difficult to complete, as significant amounts of learning material are processed week by week. In order to keep students' attention and interest in the long term, gamification tools provide powerful support because one of the keys to motivating Gen Z students is playing.

The idea of using card or board games for educational purposes is not new. Many well-known and popular games can be integrated into different segments of education almost without change and there are specific didactic games designed mainly for pupils in elementary and secondary grammar school [5]. Unfortunately, mathematics teachers in higher education rarely use gamification tools due to lack of time or because special maths games are not available for them. For this reason, we consider it is important to create specific didactic games that can be integrated into the topics of Mathematical Analysis and are useful in expanding and strengthening students' knowledge.

2 Didactic Games and Gamification in Higher Education

According to the definition created by Deterding et al. in 2011, "Gamification is the use of game design elements in non-game contexts" [6]. The purpose of gamification is to make the work and task to be done more exciting and fun. Especially in education we want to make students more active and motivated to learn. Gamification does not mean that students learn with digital games, but rather that teachers incorporate the working principles of games into their previous methods of teaching. In contrast to the traditional system, game-based teaching methods support students' acquisition of knowledge, motivating them to continue working even in the event of failure.

Gamification can appear in education on two levels. Structural gamification seeks to make the education system itself play-like, for example, by giving points to students for solved tasks and written essays, and they can move from level to level by collecting points. Content gamification tries to make the learning material more colourful by smuggling mysterious elements, stories, challenges into homework and tests [7]. Educational technical tools that implement the gamification of a well-defined subject unit essentially support content gamification.

In recent years a lot of high-quality products have appeared on the market in connection with gamification and game-based learning, specifically with the aim of helping to develop certain skills, knowledge or competencies. There has also been a significant increase in the number of games designed specifically for educational use [8,9]. The games that we build into the teaching-learning process are called *didactic games.* They differ from spontaneous games in that they are assigned to students and serve the well outlined educational purpose of the teacher. Didactic games can be grouped in several ways. We can form groups according to the purpose of the game, the number of players, the type of reaction to the game, the number of usage, and the need for equipment. Specific games, which are developed for a certain subject, obviously can be divided into several categories; their general feature is that they have a special need for equipment, otherwise they cannot be played. Here we distinguish two further subgroups: games that require an ICT device and games that require traditional devices (board games, card games, and so on). A common feature of didactic games is that they have well-defined rules, and they require coordination, control and evaluation. When using didactic games, it is worth keeping in mind that they require thorough preparation, so before the first use it is worth applying a test phase and then some fine-tuning can be needed based on the experiences. Learning cannot be preceded by the game, so the knowledge transfer step cannot be left out and it is not advisable to alternate the learning and playing phase in a well-defined educational unit. It is important that the game be experiential and varied, because even the most exciting game becomes boring if we play with it all the time, so we can boldly vary the tools and techniques used in the lessons. Continuous feedback and regularity is also essential, which can be provided by playing the selected didactic game regularly.

The application of didactic games is not new, but with the spread of gamification, this method of teaching has gained new strength. In recent years, a number of articles have been published in which the authors present specific games and techniques developed on their own, so it can be said that the educational palette is coloured in almost all areas of higher education, keeping in mind the student-centred approach [10–12]. In our experience, Z-generation students are happy to participate in the testing and piloting of gamified educational content. In addition to ICT-based gamification solutions, we also consider it important to develop and create traditional device-based didactic games, because beside the predominant digital technologies today's young people demand traditional didactic tools too, sometimes redesigning them.

When we talk about the learning process, we are essentially planning the learning experience of the students and usually focus on two areas: classroom events and out-of-class learning. When applying didactic games, we obviously have to adapt to these two segments. In the case of didactic games with special equipment needs, it is an advantage to be able to provide students with a tool that they can incorporate into the learning process, either during home study or during small-group lessons.

In this article, we present the didactic card game called LimStorm, which we designed for small groups of students to learn and practice the notion of limits of real sequences. Figure 1 shows the printed cards of the game.

Fig. 1. The complete LimStorm deck

In the following chapters, the basic concepts and results related to sequences are briefly described, because they provide the theoretical background of our game.

3 Briefly About Limits

The concepts of sequences and limits are fundamental to Mathematical Analysis.

A sequence of numbers is a function whose domain is the set of positive integers [13]. This means that to each natural number $n \in \mathbb{N}$ a real number a_n is uniquely corresponded:

$$n \longmapsto a_n, \quad a_n \in \mathbb{R}.$$

Notation: $\{a_n\}_{n=1}^{\infty}$, briefly $\{a_n\}$, or $a_1, a_2, ..., a_n, ...$ $a_i \in \mathbb{R}, i \in \mathbb{N}$. The number $a_n \in \mathbb{R}$ is called the nth term (or the general term) of the sequence. A real sequence is considered to be given, if the nth term of the sequence can be uniquely determined for all $n \in \mathbb{N}$, i.e. we know the formula of the function describing this term. It is conventional to use the letter n for the independent variable in the defining rule, so we used this notation on the LimStorm cards and throughout this paper.

Real sequences can be distinguished by their convergence or divergence. We say that the sequence $\{a_n\}_{n=1}^{\infty}$ converges to the real number A if to every positive number ε there corresponds an index $n_0(\varepsilon)$ such that

$$|a_n - A| < \varepsilon \quad \text{for all } n \geq n_0(\varepsilon).$$

The number A is called the *limit* of the sequence. Notation:

$$\lim_{n\to\infty} a_n = A \quad \text{or} \quad a_n \to A.$$

A sequence a_n is called *convergent* if there exists a limit to which it converges. If no such limit exists we say that a_n diverges, or a_n is *divergent.* The convergence of a_n means that if N is sufficiently large then all terms after the Nth lie between $A - \varepsilon$ and $A + \varepsilon$, i.e. only finitely many terms are located outside the ε neighbourhood of A. It can be verified that if a sequence converges, its limit is unique.

If we make basic operations by convergent sequences, the limit of the obtained sequence can be easily calculated using the following rules. Suppose that the sequences $\{a_n\}_{n=1}^{\infty}$ és $\{b_n\}_{n=1}^{\infty}$ converge and

$$\lim_{n\to\infty} a_n = A \text{ and } \lim_{n\to\infty} b_n = B,$$

then the sequences $\{c \cdot a_n\}_{n=1}^{\infty}$, $c \in \mathbb{R}$, $\{a_n + b_n\}_{n=1}^{\infty}$, $\{a_n - b_n\}_{n=1}^{\infty}$, $\{a_n \cdot b_n\}_{n=1}^{\infty}$ and $\left\{\dfrac{a_n}{b_n}\right\}_{n=1}^{\infty}$, $(b_n \neq 0,\ B \neq 0)$ are also convergent and

$$\lim_{n\to\infty} c \cdot a_n = c \cdot A \qquad c \in \mathbb{R};$$

$$\lim_{n\to\infty} (a_n \pm b_n) = \lim_{n\to\infty} a_n \pm \lim_{n\to\infty} b_n = A \pm B;$$

$$\lim_{n\to\infty} (a_n \cdot b_n) = \lim_{n\to\infty} a_n \cdot \lim_{n\to\infty} b_n = A \cdot B;$$

$$\lim_{n\to\infty} \frac{a_n}{b_n} = \frac{\lim\limits_{n\to\infty} a_n}{\lim\limits_{n\to\infty} b_n} = \frac{A}{B}, \qquad B \neq 0.$$

There is a natural way of extending the definition of convergence considering the symbols $+\infty$ and $-\infty$ as limits in this extended sense. If a sequence diverges to plus (minus) infinity it means that its terms outgrow every positive (negative) number. More precisely we say that the sequence $\{a_n\}_{n=1}^{\infty}$ *diverges to* $+\infty$ if for every finite number $K > 0$ there is a positive integer n_0 such that $n \geq n_0$ implies $a_n > K$. Similarly the sequence $\{a_n\}_{n=1}^{\infty}$ *diverges* to $-\infty$ if for every finite number $K > 0$ there is a positive integer n_0 such that $n \geq n_0$ implies $a_n < K$. Notations: $\lim\limits_{n\to\infty} a_n = +\infty$ and $\lim\limits_{n\to\infty} a_n = -\infty$.

4 Commonly Occurring Sequences and Limits

The following limits arise so frequently that they must be familiar to all first-year students majoring in engineering or computer science.

- The limit of the constant sequence (its terms are equal):

$$\lim_{n\to\infty} c = c,$$

  where $c \in \mathbb{R}$.
- If $k > 0$ the sequence $\{a_n\}_{n=1}^{\infty} = \{n^k\}_{n=1}^{\infty}$ diverges to infinity

$$\lim_{n\to\infty} n^k = +\infty.$$

- The sequence $\{a_n\}_{n=1}^{\infty} = \left\{\frac{1}{n}\right\}_{n=1}^{\infty}$ converges to 0:

$$\lim_{n\to\infty} \frac{1}{n} = 0.$$

- The sequence $\{a_n\}_{n=0}^{\infty}=\{q^n\}_{n=0}^{\infty}$, where $q \in \mathbb{R}$ is called *geometric sequence.* Its limit:

$$\lim_{n\to\infty} q^n = \begin{cases} 0, & \text{if } |q| < 1 \\ 1, & \text{if } q = 1 \\ \text{does not exist,} & \text{if } |q| > 1 \text{ or } q = -1 \end{cases}$$

- The limit of the sequence $\{a_n\}_{n=1}^{\infty} = \{\sqrt[n]{a}\}_{n=1}^{\infty}$ is:

$$\lim_{n\to\infty} \sqrt[n]{a} = 1, \text{where } a \text{ is a positive real number.}$$

- The limit of the sequence $\{a_n\}_{n=1}^{\infty} = \{\sqrt[n]{n}\}_{n=1}^{\infty}$ is:

$$\lim_{n\to\infty} \sqrt[n]{n} = 1.$$

- The sequence $\{a_n\}_{n=0}^{\infty} = \left\{\left(1+\frac{1}{n}\right)^n\right\}_{n=1}^{\infty}$ is convergent and

$$\lim_{n\to\infty} \left(1+\frac{1}{n}\right)^n = e,$$

  where e is Euler's number. This is an irrational real number and

$$e \approx 2.718281828.$$

- If $\alpha \in \mathbb{R}$ then

$$\lim_{n\to\infty} \left(1+\frac{\alpha}{n}\right)^n = e^{\alpha}.$$

Although the list can be continued we decided to use only these eight basic limits to create the cards of the game **LimStorm**.

5 The Structure of the LimStorm Deck and the Rules of the Game

The idea of the didactic game LimStorm was inspired by the popular card game Solo. We found the original game suitable for further development, so we created a JAVA application that allows us to produce our special cards in a ready to print form. In some details we left the concept of the base game unchanged, for example, the cards are made in four colours as in Solo; however, a significant difference is that there are limits on the cards instead of numbers. Similarly to the original game, the LimStorm deck contains action cards, some of them have the same function as the corresponding Solo card (miss a turn, take 2 and so on), but new types of action cards were also designed. We also had to change some rules to better meet the didactic goals.

The complete **LimStorm** deck consists of 110 cards, of which 80 coloured base cards contain limits of real number sequences. These base cards have yellow, green, blue or red backgrounds. All cards have a purple back that shows a fractal. The same fractal also gives the background of the fronts of the base cards in the four colours listed earlier. All base cards were made in duplicate, so we have 4×10 different cards containing limits of sequences described in a white base ellipse tilted 45° (see Fig. 2). With a small routine the results of the tasks written on the cards can be easily calculated using frequently occurring limits and the limit laws. We tried to make varied tasks which are not too difficult to solve. It is not easy to find the right balance because tasks that are too easy can make the game boring while too complicated tasks are frustrating. Obviously the difficulty level of the tasks on the cards is not uniform. With easier-to-calculate limits, we can

Fig. 2. Some base cards of the LimStorm deck

quickly give a sense of success, while with more difficult-to-calculate tasks that require the synthesis of more theoretical results, we aimed to maintain interest.

For example the yellow series contains the following limits:

1. $\lim\limits_{n\to\infty} \dfrac{5-3^n}{n^3} = -\infty$
2. $\lim\limits_{n\to\infty} \left(\dfrac{\cos(n\pi)}{n^2} - 2\sqrt[n]{n}\right) = -2$
3. $\lim\limits_{n\to\infty} \dfrac{(-1)^n - n}{n} = -1$
4. $\lim\limits_{n\to\infty} \dfrac{(-1)^n + 2n}{n^2} = 0$
5. $\lim\limits_{n\to\infty} \dfrac{\cos n}{n} = 0$
6. $\lim\limits_{n\to\infty} \left(\dfrac{n}{n+1}\right)^n = \dfrac{1}{e}$
7. $\lim\limits_{n\to\infty} \sqrt{\dfrac{n+e}{n}} = 1$
8. $\lim\limits_{n\to\infty} (1+\sqrt[n]{\pi}) = 2$
9. $\lim\limits_{n\to\infty} \dfrac{e}{\sqrt[n]{10n}} = e$
10. $\lim\limits_{n\to\infty} \sqrt{n} = +\infty$

Here, after the equal sign, the calculated limit is also indicated, which of course is not presented on the cards (Fig. 3). Of the 10 sequences, two are divergent and the others are convergent. Two null sequences (whose limit is zero) are placed in each series making the game more complex and also emphasising the importance of this type of sequence. Not only Euler's number but also its reciprocal is included in the limits written on the cards, so there are also irrational numbers between the limits to be calculated. For all four series, the results as limits are the same numbers as shown in the case of the yellow series, but as limits for different sequences, so that the limits for 40 different real number series are displayed on the cards. The momentum and variety of the game are provided by the action cards. There are 30 rainbow-coloured action cards included in the set.

Fig. 3. Limits of the yellow series

The game is designed specifically for groups of 4–10 people, but it can be enjoyable for two players too. After shuffling the complete deck the dealer deals seven cards to each player. The other cards form the draw deck, which is placed in the centre of the table. The top card is turned over and put in a place visible to everyone not too far from the draw deck. This card will be the first card in the discard pile. The starting player is selected by the group. He throws first, then the player to his left follows, and so on. The object of the **LimStorm** game is to get rid of the cards possessed by the player as soon as possible. Each turn, a player can put a card matching in colour or limit of the card at the top of the discard pile. The players can discard at most one card in each round, except for one option. A player can only discard more than one card at a time if he has more than one card of exactly the same colour and limit. This can mean putting two, three or four cards on the discard pile at a time. The probability of occurrence of the last two cases is very low, because there are only null sequences duplicated in each colour series, i.e. the limit of 0 for each colour can be counted as a limit on two cards. When a player places a card on the discard pile, he must always say out loud the limit displayed on the discarded card. This is the learning phase of the game, as the player has to calculate the limit on the card even if he puts down his card in case of colour matching or answering to a rainbow-coloured action card. Saying the limit out loud means that visual information is reinforced with auditory information to fix the units to be learned.

It is also allowed to use *jump-in* during the game. This can happen if someone has a card in his hand that is the same colour and value as the last discarded card. This card can be put down by the player even if it is not his turn. After a jump-in, the player next to the player who took the opportunity to jump-in will be next. Due to the fact that there are two zero limits in a given colour series, the jump-in can be validated even if the player has a null sequence on his card of the same colour that is different from the null sequence on the top card of the discard pile.

The player who cannot put down any card must pick one up from the draw deck, and the game can be continued by the next player.

LimStorm is diversified by the *action cards*, which can be used to gain an advantage or can lead to a setback. Action cards can be played when the player deems it necessary and the player can only put down one of this type of card at a time. This means that simultaneous playing of multiple cards is not allowed even in case of two completely identical action cards. The deck contains six types of rainbow-coloured action cards (see Fig. 4):

- *Miss a turn* If you play this card the next player misses his turn, he cannot throw and pick up a card.
- *Swap cards* You can choose one of the players and swap all of your cards for his.
- *Take 2* With this card you force the next player to draw two cards from the draw deck. The card *Take 2* can be covered by another *Take 2*, in this case the next player must draw four cards from the draw deck if he does not have this type of card.

- When throwing the *Take 3* card the next player must take three cards, which can be avoided if he possesses another *Take 3* card and use it. A *Take 2* card can be covered by a *Take 3* card, but it is not allowed in the reverse order.
- When you throw a *Present* card you can give an arbitrary card from your deck to one of the players.
- The action card *Four-leaf clover* has several functions. The player who places the card may request a circular exchange of cards or reverse the direction of the game. In addition, it can be used for a colour request.

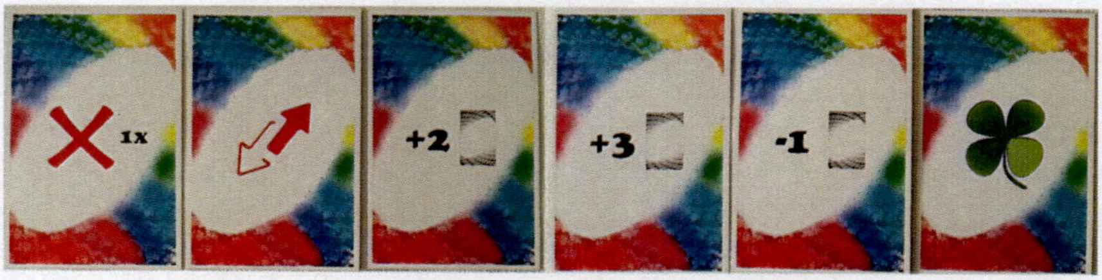

Fig. 4. The six action cards of the LimStorm deck

The action cards are rainbow coloured, so after playing them, once the instruction on them is validated, the next player in turn can continue playing by throwing a card of any colour.

As soon as a player has only two cards left in his hand and can play one of them, he is obliged to say out the word *Storm*, warning the other players that he has only one card left. If he forgets to say it and someone notices it, he has to pick up two cards as a penalty. If the player who said *Storm* is in turn again and he can place his last card, he is the winner of the game. The other players can then stop, or they can continue playing until only one of them has a card left. The team or the instructor has to decide how long the game will last.

6 Possibilities of Integrating the Didactic Game LimStorm into Education

The game **LimStorm** has been developed specifically for the small-group form of education, so it is no coincidence that the best results can be achieved in this case. Adapted to the training structure, we consider the game can be used in practical classes or consultation classes. In the lectures the theoretical foundations of the topic of sequences are presented and after that, the tasks related to the topic will be solved during the practical classes. We recommend writing a short test after the classic problem solving. By solving the 10-question test, it is possible to revive the necessary theoretical background material and the frequently occurred limits. The test can be solved individually, or a group solution is also possible. In the first case it is important not to miss the checking phase, because we have to give the correct answers to all questions in order to prepare for the game properly. The rules of the game **LimStorm** can be explained in some minutes, so the game can start after forming groups of 4–10 with students of mixed levels. Groups can also be self-organizing. If the base game is not known to the students, it may be worth playing an introductory game with *Solo*. This is not absolutely necessary, as in our experience there are very few students who are unfamiliar with some version of the base game. We recommend playing at least three games of **LimStorm**, and a game takes about 10–20 minutes depending on the quickness and knowledge of the players. The games make it possible to record and deepen the processed learning material and provide an excellent opportunity to practice. In the case of students who are not so good at maths, it is advisable for the instructor to get involved in the game, because in this way the incorrectly stated limits can be corrected immediately. When forming heterogeneous groups, this is not necessary because the group provides control and feedback. We can schedule a quize at the end of the lesson or at the beginning of the next lesson.

Students can play with the **LimStorm** deck not only in practical classes; it is advisable to provide students with the opportunity to play while preparing for exams. To test the game, we ordered three decks from the printer, so we can currently play with a group of students of up to 30 people. We are also planning to order additional decks so that we will be able to support learning at home.

The game can be further developed. Extra cards are planned to print in two new colours, making the game expandable with an additional 20 sequences. Easy-to calculate limits are placed in the brown series, while more difficult ones are placed in the turquoise series. With additional cards, you can create a way to further differentiate, as well as make the game more varied. Our other idea is to gamify the calculation of the limit of real functions with the help of **LimStorm**. These ideas are yet to be implemented.

7 Conclusion

The application of innovative learning and teaching techniques in different areas of higher education aims to increase the level of motivation of students and their successful participation in the learning process, as nowadays it has become obvious that Z-generation students learn in different ways and use different tools than previous generations. Therefore, their effective teaching requires careful consideration of the appropriate educational strategy and tools that can be used to support the chosen form of work. It is a fact that the best results can be achieved in the case of individual training, but it is also a fact that higher education is currently unable to implement individual training due to the large number of students. However, in the case of Z-generation students, efforts should be made to develop individual abilities, so frontal education should be combined with or replaced by small-group work. The centre of small-group education need not be only the teacher; groups can also be organized around a well-chosen tool to support education. In this case, several smaller groups can work effectively under the supervision of a single instructor. The educational support tool can also be a specific didactic game. LimStorm proved to be an excellent tool in connecting elements of the teaching-learning process, since it can be reused several times during the acquisition, practice, monitoring and evaluation phases. The card game does not require lengthy preparations, it has easy-to-learn rules and teaches and entertains with fast, dynamic and exciting action. In our experience the LimStorm game can not only serve to transfer and practise content it can also effectively develop logical thinking, memory, problem-solving, concentration, counting skill, efficient scheduling and teamwork ability.

References

1. Mohr, K.A.J., Mohr, E.S.: Understanding Generation Z students to promote a contemporary learning Environment. J. Empowering Teach. Excellence **1**(1), 9 (2017). https://doi.org/10.15142/T3M05T
2. Meet Generation Z: Forget everything you learned about Millenials, Sparks & Honey www.slideshare.net/sparksandhoney/generation-z-final-june-17 (2014)
3. Selingo, J.J.: The New Generation of Students: how colleges can recruit, teach and serve Gen Z. Chronicle of Higher Education (2018)
4. Faiella and F., Ricciardi, M.: Gamification and learning: a review of issues and research, J. E-Learning Knowl. Soc. **11**(3), 13-21 (2015). https://doi.org/10.20368/1971-8829/1072
5. Vankús, P.: Didactic Games in Mathematics, Faculty of Mathematics, Physics and Informatics. Comenius University Bratislava (2013). https://doi.org/10.13140/2.1.3138.9120
6. Deterding, S., Dixon, D., Khaled, R., Nacke, L.: From game design elements to gamefulness: Defining "gamification". In: Proceedings of the 15th International Academic MindTrek Conference. ACM Press, New York, USA, pp. 9–15 (2011)
7. Kapp, K.M., Blair, L., Mesh, R.: The Gamification of Learning and Instruction Fieldbook: Ideas into Practice. Wiley, San Francisco (2014)
8. Kalmpourtzis, G.: Educational Game Design Fundamentals: a Journey to Creating Intrinsically Motivating Learning Experiences. CRC Press, Boca Raton (2018)
9. Pellas, N., Fotaris, P., Kazanidis, I., et al.: Augmenting the learning experience in primary and secondary school education: a systematic review of recent trends in augmented reality game-based learning. Virtual Reality **23**, 329–346 (2019). https://doi.org/10.1007/s10055-018-0347-2
10. Damsa, A., Fromann, R.: Gamification and Gameful approaches in education, business, and IT. Informatika **18**(1), 28–33 (2016)
11. Lee, J.J., Hammer, J.: Gamification in education: what, how, why bother? Acad. Exch. Q. **15**(2), 1–5 (2011)
12. Rigóczki, C., Damsa, A., Györgyi-Ambró, K.: Gamification on the edge of educational sciences and pedagogical methodologies. J. Appl. Tech. Edu. Sci. **7**(4), 79–88 (2017). https://doi.org/10.24368/jates.v7i4.12
13. Thomas Jr., G.B., Finney, R.L.: Calculus and Analytical Geometry, 9th edn. Addison-Wesley, Reading (1996)

Effects of Competitive Coding Games on Novice Programmers

Konrad Fischer, Sarah Vaupel, Niels Heller, Sebastian Mader(✉), and François Bry

Institute of Informatics, LMU Munich, Munich, Germany
konrad.fischer@campus.lmu.de

Abstract. Coding platforms provide challenges for users to solve in order to improve their programming skills, utilizing elements of both *Technology Enhanced Learning* and *gamification*. Inspired by previous approaches, this article presents a coding platform focused on competition, and its evaluation in the context of tertiary programming education. The system provides programming challenges for two users to compete against each other in real time. We seek to answer the question whether this notion of one-on-one competition can increase motivation to solve programming tasks. Two user studies were conducted in order to evaluate the effects of the system's competitive component on users' self-assessment, enjoyment, and motivation.

The results of these studies suggest that this kind of competitive environment in programming leads to learners assessing themselves more distinctively, while their enjoyment decreases. Furthermore, learners' motivation drops in competition with others, although it remains on a high level. Learners who already achieved a relatively high level of programming knowledge are generally more inclined to enjoy competitive game components in programming.

We conclude that competition in programming education does not necessarily have solely positive effects on learners' experiences. The inclusion of competition in tertiary computer science education might prove to be useful, but further research needs to be conducted to identify the best way of integrating it into educational concepts.

Keywords: Technology Enhanced Learning · Gamification · Competition · Programming education

1 Introduction

This article touches on the subject of coding platforms, which utilize elements of *Technology Enhanced Learning* (TEL) and *gamification* to enhance the learning experience of aspiring programmers. We contribute to this topic by introducing a coding platform with focus on real-time competition, and its evaluation in the context of tertiary programming education.

M. E. Auer and T. Rüütmann (Eds.): ICL 2020, AISC 1328, pp. 464–475, 2021.
https://doi.org/10.1007/978-3-030-68198-2_43

An example of an online coding platform is *Codewars* (https://www.codewars.com/), where players solve small programming exercises and test their solution against a set of unit tests. However, on *Codewars* players do not interact with each other while solving challenges. Such interaction might be desirable though, as it might further motivate users to solve the challenges. Real-time competition could be a particularly effective form of interaction, as it gives the players immediate feedback on their performance compared to others. However, previous work indicates that competition may have both positive and negative effects on intrinsic interest [12].

In this research paper, we seek to answer the question whether real-time competition is able to increase motivation to solve programming challenges. We implemented the coding platform *SuperDevBros*, which provides programming challenges that can be solved by two users simultaneously, where the user who first solves the challenge "wins". Two user studies were conducted using our platform, where the participants were presented with both singleplayer and multiplayer versions of small programming challenges. The evaluation of the studies showed that the introduction of a competitive component did not have a clearly positive effect on the users' self-assessment, enjoyment, or motivation. This leads to the conclusion that competition in programming education does not necessarily have solely positive effects on learners' experiences.

In Sect. 2, we first give an overview over research articles and platforms that motivated our approach. A description of our platform is provided in Sect. 3. In Sect. 4, we describe the user studies we conducted and present the results and their interpretation. To conclude, we summarize our findings and give an overview of features we envision as future work in Sect. 5.

2 Related Work

This section discusses related work. Previous research on motivation in programming education is described, followed by an overview of research on gamification in education, and a report on similar coding platforms.

Previous Research on Motivation in Programming Education. Jenkins investigated in their master thesis [16] the motivation of students during their first year programming courses. Among a first (larger) group of students, the most frequently reported attitude to their studies was their (*intrinsic*) "own satisfaction" [p. 97], followed by their (*extrinsic*) will to "get a good job" [p. 97]. Notably, the number of students who only wanted to pass the module rose significantly over time. Among a second (smaller) group of both experienced and novice programmers, all interviewees looked back on the course with a constant theme of achievement. Among the novices in the group however, the majority reported a largely negative overall experience. All members of the second group passed the course, but there was a recurring theme among the novice subgroup of "initial enthusiasm being gradually eroded until the course [was] viewed in a wholly negative light" [p. 163].

Law et al. conducted a study in [17] on factors influencing students' motivation to learn programming. The authors' system *PASS* uses test cases to examine students' solutions to programming tasks. Among the factors of motivation the authors investigated, "'individual attitude and expectation', 'clear direction', and 'reward and recognition'" [p. 226] were affecting students' motivation the most. They also discovered that "'individual attitude and expectation', 'challenging goals', and 'social pressure and competition'" [p. 226] were significantly improving efficacy, i.e. "what a person believes he or she can do in a particular learning task" [p. 220]. Another insight gained was that the teacher evaluating solutions in real-time and presenting good programs made the participants feel a small amount of "peer pressure and competition" [p. 227]. The authors also considered extending their system with a real-time component showing students the advance of their peers towards a solution, with which the authors hope to add "an element of 'peer pressure and competition'" [p. 227].

Previous Research on Gamification in Education. In [1], Butt conducted a study on game-based learning with the coding platform *CondinGame* (https://www.codingame.com/) among undergraduate software engineering students, which were asked to work on 3–5 challenges. The results of the study showed that most of the subjects were interested in game-based learning and in getting more chances to use it. However, fewer students believed that game-based learning could replace conventional learning techniques.

In [10], Dicheva et al. systematically reviewed existing work on gamification in education in 34 papers. Among the most widely used gamification design principles were visual status, social engagement (e.g. competitions), freedom of choice (e.g. among types of challenges), freedom to fail (e.g. allowing students to re-submit assignments), and rapid feedback. The most popular game mechanics used were points, badges, and leaderboards. The majority of reported results of the experiments were positive, such as increased engagement of students, attendance, participation, and material downloads, positive effects on the quantity of contributions (without a reduction in quality), increased passing ratios, and participation in voluntary activities. Dicheva et al. concluded their study with the observation that true empirical research on the effectiveness of game elements in learning environments was still scarce and most previously conducted empirical studies had not included a proper evaluation.

Similar Approaches to Coding Platforms. This section introduces similar coding platforms some of which acted as inspiration for the system presented in this article.

On *Codewars*, players use an editor in their browser to craft a solution to a challenge. These challenges are created by other players and then reviewed by the community [6]. *Codewars* validates the player's solution to a challenge using tests that are executed on their server. After the player has completed a challenge, they will be shown the solutions other players have come up with and can then vote on these [4,7]. Each challenge also has a dedicated web page where users can discuss its content and seek help [4].

CodinGame offers (like *Codewars*) an interface players can use from any browser and uses a set of tests to validate the players' solutions [9, 11]. Players can also discuss individual challenges on the *CodinGame* website, view other players' solutions and vote on them [8]. One unique feature of *CodinGame* is that there is a simple visual representation of a game in the interface. The challenge represents a part of the logic of that game and each time the players test their solution, the game is executed with the logic built by them [1, 11]. Furthermore, *CodinGame* includes a multiplayer mode called "Clash of Code" where the players face off against multiple opponents. One type of these matches requires the player to be the first to submit code that passes a set of tests [2], which is the same principle our platform uses.

Coderbyte (https://coderbyte.com/) and *CodeAbbey* (https://www.codeabbey.com/) also provide challenges for the users to solve in their browser [3, 15]. *Exercism* (https://exercism.io/) too provides such challenges but no environment to solve them in [13].

3 Implementation

On our platform, users can not only play challenges by themselves, but also directly compete against other users in the form of a one-on-one multiplayer mode that adds a notion of competition to the solving of programming tasks. In this section, we present this platform and its main features.

The Lobby Page: Creating and Joining a Game. From the Lobby Page, users can create and join game sessions. They can select the desired programming language, difficulty level, and optionally a particular challenge. One can also choose whether to play the challenge in singleplayer or in multiplayer mode.

The Game Page: Playing Games. After a game has started, the player will be presented with the Game Page, which for multiplayer games will contain the two players' code editors beside each other. If the current game is a singleplayer game, there will only be one code editor.

The players' code editors are conventional code editors with syntax highlighting included. The opponent's code editor shows an obfuscated version of the opponent's code with each character replaced by special characters. This editor is synchronized so that it always displays an approximation of the opponent's current code length, similar to the considered real-time component of *PASS* as envisioned by Law et al. [17, p. 227].

When players submit their code it is sent to the server to be evaluated. The server will execute a set of unit tests on the player's code. If some of the tests failed, the respective player will be shown the output of the tests and will therefore receive a hint on which parts of their program should be improved. Users can submit their code for testing as often as they wish without any negative consequences.

Upon the regular end of a game (a game that ended as a result of one player's code passing all tests), both the winner and the loser are shown statistics on the match. A user who won a particular challenge gains access to all solutions that have been submitted for this challenge.

4 Evaluation

To evaluate the effects of the previously described one-on-one programming challenges on the users' self-assessment, enjoyment, and motivation, two user studies were conducted among undergraduate students. This section presents the key aspects of our evaluation of these two user studies. The complete version of this article features a broader and more in-depth evaluation [14]. A description of the scope and the execution of the studies is provided, followed by a statistical analysis and an interpretation of the results.

Scope: Computer Science Students in Their First or Second Year. For the evaluation of the system, two groups of Computer Science students were surveyed: a (first) group of 7 students taking part in a practical course on game development utilizing *JavaScript*, and a (second) group of 16 students attending a lecture on programming and modelling utilizing the functional programming language *Haskell*. Both groups represent the main target group of the system: programmers learning a new language or new techniques in a programming language they already know. Improving motivation and providing immediate feedback, as our systems aims to do, is particularly important in the first stages of the learning process.

Method. The conducted user studies were designed to collect two types of data on the participants: subjective feedback from questionnaires, and objective data on the participants' interaction with the system. These two types of data were collected in turn and repeatedly in order to introduce the participants to the system's features step by step, and to evaluate how the participants' attitudes changed while using the platform. Both user studies were similarly structured, see Table 1.

Table 1. Structure of the conducted user studies.

Step	Name	Purpose
1	Pre-questionnaire	Demographical information, general attitude
2	Singleplayer phase	Introduction to the system in singleplayer mode
3	Singleplayer questionnaire	Feedback on the singleplayer mode
4	Multiplayer phase	Introduction to the competitive mode
5	Multiplayer questionnaire	Feedback on the competitive mode
6	Freeplay phase	Freedom of choice between modes
7	Post-auestionnaire	General feedback the platform, general attitude

Results. In the following section, a statistical analysis of the results is provided. After touching on possible limitations to validity and generalizability, we give insights into the (subjective) questionnaire answers, and a comparison of those answers to (objective) data that was collected during the user studies (Table 2 gives an overview over all games played).

Table 2. Numbers of games played per study and phase, including the ratio of multiplayer games (MP) for the freeplay phases.

Phase	First study	Second study	Total
Singleplayer phase	8	30	38
Multiplayer phase	7	34	41
Freeplay phase	20 (20.00% MP)	57 (21.05% MP)	77 (20.78% MP)
Total	35 (31.43% MP)	121 (38.02% MP)	156 (36.54% MP)

Limitations to Validity and Generalizability. This section contains results that utilize the ratio of multiplayer games won per participant, which is computed from the participants' recorded amount of multiplayer games won/lost. Because there are edge-cases where the loser of a game could not be determined, the true ratios could be smaller than the ratios stated in this section.

Additionally, caution needs to be exercised when generalizing the results: Possible limitations to the generalizability include the potentially biased selection of study participants (as students of computer science), the participation in the study being on a voluntary basis (though reward was given in the form of a lottery), and the limited time range of the study (two hours).

Comparison of Answers to Trend Questions. A number of so-called *trend questions* was asked twice in different phases of the study to gain insights about the impact of the platform on the participants' assessments throughout the study.

Figure 1 illustrates the change of participants' responses to selected trend questions. The responses of all 24 participants from both studies are included.

As can be seen in Fig. 1a, the numbers of participants assessing their skills as both better than 75% and 0% of the group increased. Furthermore, before being confronted with our platform, 10 participants rated their programming skills on the lower half of the spectrum, while 14 participants rated themselves on the upper half. This distribution was inverted after using the platform. When comparing the participants' answers to this question from the pre- and post-questionnaires, a shift of opinions towards the center of the scale can be observed, as can be seen in Fig. 1b. Figure 1c shows that there was some decrease in motivation among the participants after being confronted with the multiplayer mode. Decreased levels of calm and confidence after being confronted with the multiplayer feature can be observed in Fig. 1d.

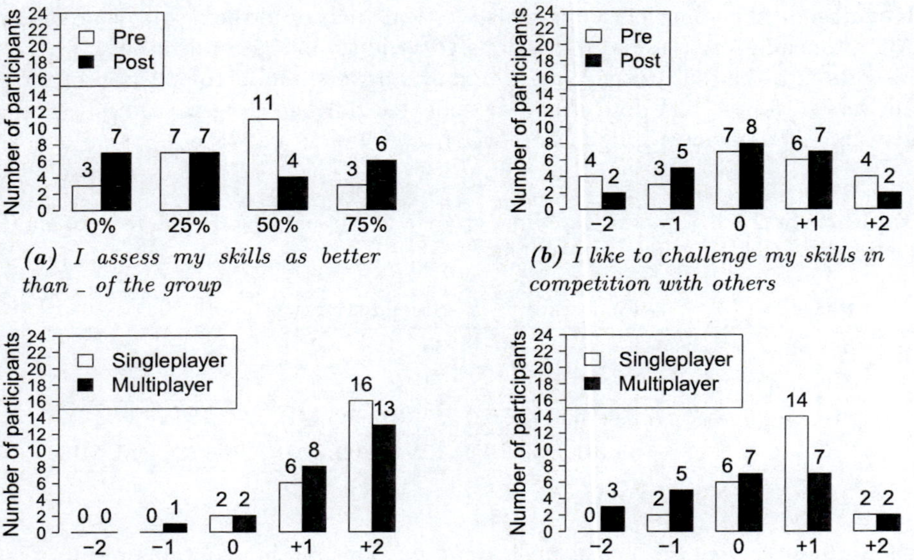

(a) I assess my skills as better than _ of the group

(b) I like to challenge my skills in competition with others

(c) I was motivated to solve the programming tasks

(d) I felt calm and confident while programming

Fig. 1. Distributions of the participants' answers to selected trend questions. Possible answers range from "strongly disagree" (= −2) to "strongly agree" (= +2).

The average answer for each question and phase can be seen in Table 3. The most prominent change (−0.67) represents how calm and confident the participants felt after being confronted with the multiplayer feature compared to before. The participants' motivation to solve the programming tasks decreased by 0.20 after the addition of the competitive element, which is the second most prominent change. The average assessment of participants' skills decreased by 3%. Apart from this, the averages varied only slightly between the phases of the study.

Comparison of Questionnaire Answers and Performance. Figure 2 shows the participants' performance against others over their reported desire for competition and motivation to solve the programming tasks.

Figure 2a shows the participants' readiness to challenge their skills in competition with others in relation to their ratio of multiplayer games won. One noteworthy aspect is that 13 participants of varying desire for competition did not win any multiplayer games, while 11 participants (also of varying subjective desire for competition) won at least one multiplayer game. Not to win any multiplayer game in this context does not necessarily mean that all of the started games were lost, but that all of the started games were either lost or aborted. Another observation is the distribution of the quadrant sizes (neutral answers (= 0) are included both in the left and in the right quadrant). On one hand, there are as many participants in the bottom left quadrant as in the bottom

Table 3. Comparison of the response mean values before and after using the platform (Pre vs. Post) and before and after confronting the participants with the multiplayer feature (Singleplayer (SP) vs. Multiplayer (MP)).

Question	Pre	Post	Change
I assess my skills as better than _ of the group	39%	36%	−3%
I like to challenge my skills in competition with others	+0.12	+0.08	−0.04
Question	SP	MP	Change
I was motivated to solve the programming tasks	+1.58	+1.38	−0.20
I felt calm and confident while programming	+0.67	0.00	−0.67

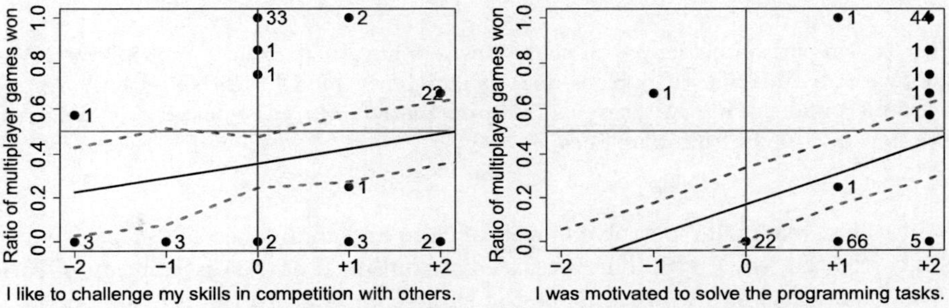

***(a)** Ratio of games won over desire for competition*

***(b)** Ratio of games won over motivation solve the programming tasks*

Fig. 2. Comparison of the participants' answers and their ratio of multiplayer games won, with a regression (thick black line) using a linear model and the respective 89% confidence interval margins (red punctuated lines). The data can be separated into four quadrants (thin blue lines).

right quadrant, i.e. among the participants who won less than 50% of their multiplayer games, and their desire to compete with others spreads across the entire spectrum. On the other hand, there is only one participant in the top left quadrant, but four participants in the top right quadrant. This means that, among the participants who won more than 50% of their multiplayer games, there are more participants who like to compete with others than participants who do not.

In Fig. 2b, the participants' motivation to solve the programming tasks in relation to their ratio of multiplayer games won is shown. This plot can again be divided into four quadrants (analogous to Fig. 2a). The top left quadrant has only one member while the top right quadrant has nine. This indicates that among the participants who won more than 50% of their multiplayer games, there are nine times more participants rating their motivation as high than who rate it as low. The bottom left quadrant is empty (excluding neutral answers), while the bottom-right quadrant has 12 members.

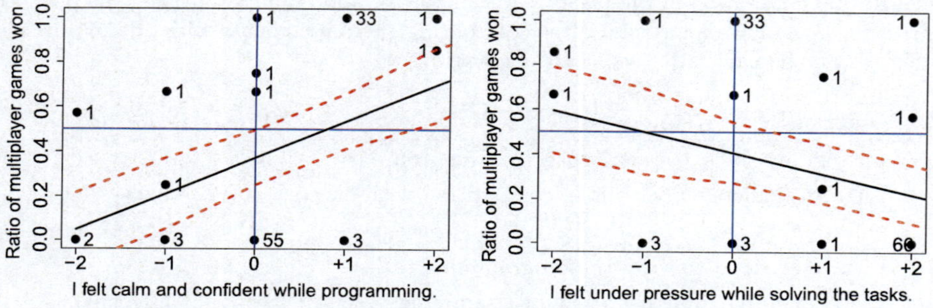

(a) Ratio of games won over perceived calm and confidence

(b) Ratio of games won over perceived pressure

Fig. 3. Comparison of the participants' answers and their ratio of multiplayer games won, with a regression (thick black line) using a linear model together with the respective 89% confidence interval margins (red punctuated lines). The data can be separated into four quadrants (thin blue lines).

Next, the participants' perceived levels of calm and confidence as well as pressure is investigated, see Fig. 3. The vertical distribution (i.e. the distribution of ratio of multiplayer games won) remains the same compared to the previous plots.

As can be seen in Fig. 3a, a positive correlation between perceived calm and confidence and ratio of multiplayer games won can be observed. A majority of participants who won less than 50% of their multiplayer games either reported neutral or negative levels of perceived calm and confidence.

In Fig. 3b, among all participants who did not win any multiplayer games (13 participants, about 54% of all participants), approximately 46% (six participants) stated that they felt highly pressurized (+2) while solving the tasks.

Participants' Enjoyment of Different Modes. Apart from perceived values such as the participants' self-assessment, motivation, and pressure, the effects of the competitive environment on their enjoyment are also investigated. The participants were asked two questions addressing this issue in the post-questionnaire, the answers to which are shown in Fig. 4. The (blue) line separates the plot into two regions: above the line are participants who had more fun solving the tasks in competition with others, while data points below the line stand for participants who enjoyed solving them alone more. Data points on the line represent equal enjoyment of both modes. 10 participants stated that they had more fun solving the tasks alone, 10 participants reported equal levels of enjoyment, and four participants reportedly enjoyed solving the programming tasks in competition with others more.

Interpretation. In this section, we interpret the results gathered in the previous section regarding competition's effects on self-assessment, fun, and motivation.

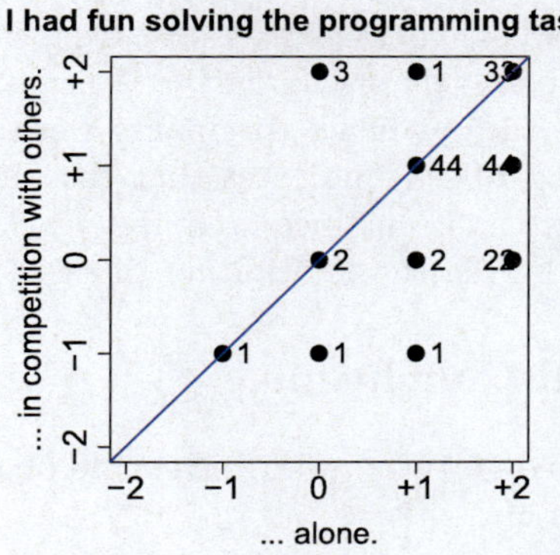

Fig. 4. The participants' enjoyment gained from solving the programming tasks in competition with others (multiplayer) over solving them alone (singleplayer).

Effect on Self-Assessment. After the experiment, more students reported either very high or very low self-assessments, while fewer students reported medium self-assessments (see Fig. 1a). One possible explanation of this polarisation effect is that the average participant did not get a chance to evaluate their skills before being confronted with our system. Without a point of reference, they might have been inclined to give a neutral answer. However, after competing with others, they might have been driven to more distinctive answers depending on their (subjective) impression of how well they performed. This can be seen as a positive effect, as participants might know their true skill levels better after using the platform. Knowing one's true skill level is important feedback on where one can still improve and where one has already achieved some level of understanding. On the other hand, the slight decrease of the participants' average perceived skill level (see Table 3) is a cause for discussion. This decrease might be caused by the high number of participants who won a minority of their multiplayer games compared to the low number of participants who won a majority of the multiplayer games they played.

Effect on Fun. Regarding Fig. 1c, the participants' opinion whether they like to compete with others seemed to shift towards the center of the scale after using our platform. In Fig. 2a, a slight positive effect of desire for competition on the participants' performance (i.e. their ratio of multiplayer games won) can be observed. Additionally, Fig. 4 shows that a majority of the participants enjoyed solving the tasks alone more than while competing with others. A different observation that matches the previously described one is the high number of participants who reported high levels of pressure under competitive conditions (see Fig. 3b). One could argue that the participants experienced this pressure because they had trouble understanding their assignment, but this conflicts with our observation from Table 3 that they perceived the tasks as clear and comprehensible.

Effect on Motivation. Instead of experiencing a positive form of competitive pressure, participants seem to have felt pressurized in a negative way. However, an initial average response of +1.58 when participants were asked for their motivation (see Fig. 1c, see also Table 3) indicates that the motivation to solve the tasks was high to begin with, and remained on a high level even after decreasing by 0.20 to an average of +1.38 in the multiplayer phase.

5 Future Work and Conclusion

Finally, this section touches on future work which could be done on the platform, and a conclusion of this article.

Future Work. There are a number of features that could be added to our system. One of these is a rating system for challenges and their solutions, allowing the users to give feedback on the quality of a challenge or solution. A more advanced rating system, similar to the one offered by *Codewars* [5], could allow users to rate specific aspects of a challenge (e.g. educational value or difficulty) or a solution (e.g. brevity or exemplariness). Additionally, the possibility to comment on solutions in order to discuss them and e.g. suggest improvements could be implemented.

Furthermore, more game elements could be added to the system. One such game element is experience points, which users gather by solving challenges. Experience points could then allow the users to rise in ranks, providing an additional extrinsic motivator for them to solve more challenges. Another element that could be added are badges, which are awarded for fulfilling certain requirements (e.g. solving a number of challenges using a specific language). These could also be displayed on the users' profile page. The effects of the suggested improvements would have to be evaluated in further user studies.

Conclusion. The goal of this research paper was to study the effects of competitive environments on novice programmers. To achieve this goal, a coding platform was implemented, which differs from existing platforms in adding real-time competition to solving programming tasks. This platform was then evaluated in two user studies among undergraduate students of computer science in the first years of their study program. The results suggest that the users' self-assessment shifted towards extremes when being confronted with a competitive environment, while the users' enjoyment seemed to follow an opposite trend. Furthermore, motivation dropped in competition with others, although it remained on a relatively high level.

We conclude from our findings that competition in programming education does not necessarily have solely positive effects on learners' experiences, as our results confirm its potential for both positive and negative effects on intrinsic interest as mentioned in previous work [12]. The inclusion of competition in learning environments might prove to be useful, but it is not a silver bullet to increase the learners' motivation. Further research needs to be conducted to identify the best way of integrating it into educational concepts.

References

1. Butt, P.: Students' perceptions of game-based Learning using CodinGame. In: International Conference on ICT in Education, ICICTE; Conference date: 07-07-2016 Through 09-07-2016, p. 151 (2016)
2. [CG]Thibaud Clash of Code: Challenge your Friends to Short Coding Battles (2020). https://www.codingame.com/blog/clash-of-code-time-has-come-for-clash/. Accessed 10 Feb 2020
3. Coderbyte Coderbyte (2020). https://coderbyte.com/. Accessed 17 Feb 2020
4. Codewars About Codewars (2020). https://github.com/Codewars/codewars.com/wiki/About-Codewars. Accessed 10 Feb 2020
5. Codewars Codewars (2020). https://www.codewars.com/. Accessed 10 Feb 2020
6. Codewars Kata (2020). https://github.com/Codewars/codewars.com/wiki/Kata. Accessed 10 Feb 2020
7. Codewars Kata Solutions & Voting (2020). https://github.com/Codewars/codewars.com/wiki/Kata-Solutions-&-Voting Accessed 10 Feb 2020
8. CodinGame CodinGame (2020)(2020). https://www.codingame.com/ Accessed 10 Feb 2020
9. CodinGame FAQ (2020). https://www.codingame.com/faq. Accessed 10 Feb 2020
10. Dicheva, D., Dichev, C., Agre, G., Angelova, G.: Gamification in education: a systematic mapping study. J. Educ. Technol. Soc. **18**(3), 75–88 (2015)
11. Dillet, R.: With CodinGame, Learning To Code Becomes A Game (2020). https://techcrunch.com/2015/11/11/with-codingame-learning-to-code-becomes-a-game/ Accessed 10 Feb 2020
12. Epstein, J.A., Harackiewicz, J.M.: Winning is not enough: the effects of competition and achievement orientation on intrinsic interest. Pers. Soc. Psychol. Bull. **18**, 128–138 (1992)
13. Exercism Exercism (2020). https://exercism.io/. Accessed 17 Feb 2020
14. Fischer, K., Vaupel, S.: Effects of Competitive Coding Games on Novice Programmers (2020). https://www.en.pms.ifi.lmu.de/publications/projektarbeiten/Konrad.Fischer_Sarah.Vaupel/Konrad.Fischer_Sarah.Vaupel.pdf. Accessed 24 Jun 2020
15. Gorkovenko, R.: CodeAbbey (2020). https://www.codeabbey.com/. Accessed 17 Feb 2020
16. Jenkins, T.: The motivation of students of programming. In: Proceedings of the 6th Annual Conference on Innovation and Technology in Computer Science Education, Association for Computing Machinery, ITiCSE 2001, New York, NY, USA, pp. 53–56 (2001)
17. Law, K.M., Lee, V.C., Yu, Y.: Learning motivation in e-learning facilitated computer programming courses. Comput. Educ. **55**(1), 218–228 (2010)

Developing Blitz Games and Using Them in Teaching Engineering Disciplines

Irina V. Pavlova[1], Irina A. Strelnikova[2], Nataliya V. Nikonova[1], Maria S. Suntsova[1], and Svetlana V. Barabanova[1](✉)

[1] Kazan National Research Technological University, Kazan, Russian Federation
sveba@inbox.ru

[2] National Research University Higher School of Economics, Moscow, Russian Federation

Abstract. Modern system of vocational education is focused on enhancing its quality and increasing the competitiveness and investment attractiveness of secondary vocational education. Russian National Project "Education" poses a challenge of modernizing this level of education, including by implementing adaptive, practice-focused and flexible educational programs.

Russian educators recognize the importance of expanding the range of active learning methods used in studying engineering subjects. Implementing them into teaching/learning process allows students not to be passive narratees; moreover, it activates their thinking and develops independence in making their decisions. This helps increase their motivation and interest in their studies and ensures more efficiently forming their professional and general cultural competences.

One of simple but efficient learning methods are blitz games that are interactive by their nature and presuppose free exchange of opinions regarding the ways and techniques of solving certain problem situations among those involved in educational activities. Blitz game makes learning exciting and interesting for both the teacher and the students.

Keywords: Blitz game · Professional and personal skills · Active learning methods

1 Blitz Games as a Means in Interactive Learning

1.1 Increasing Interest in and Motivation for Learning Using Blitz Games

One of simple but efficient teaching methods is that of blitz games. It is interactive and suggests a free exchange of ideas regarding how to solve some or other problematic situations the parties of educational activities may have. Blitz game makes the teaching/learning process exciting and intriguing for both those who teach and those who study. Play-based methods are widely used in extracurricular activities arranged for college students [1] and actively implemented in developing communication skills in both college and university students [2]. Play-based manufacturing design method

M. E. Auer and T. Rüütmann (Eds.): ICL 2020, AISC 1328, pp. 476–486, 2021.
https://doi.org/10.1007/978-3-030-68198-2_44

activates studying academic subjects and enhances learning performance due to developing various skills necessary for the expected production engineering activities of students. Later, this will allow them to solve complex methodical, engineering, designing, and organizational problems in a more efficient manner [3–5].

The main goal of a game is to enhance the likelihood of successfully achieving the expected results in their further vocational activities due to pre-training the perception mechanisms, decision making, and forming/adjusting their action plans. Obviously, success in solving such problems depends on how illustratively and comprehensively one gaming model or another reflects the real conditions of the student's expected vocational activities. Developing a blitz game includes the same stages as that of any other educational technology. Traditionally, there are nine of such stages.

1.2 Stages in Developing Blitz Games

As said above, blitz games are usually developed in the same stages as any other educational technology. "Educational technology," in this case, shall mean a systematic approach to designing, implementing, assessing, correcting, and subsequently reproducing the teaching/learning process [6].

Normally, development of a blitz game starts well ahead of using it at a lesson. In literature, nine stages of game technology development are described. Let us consider them regarding engineering disciplines taught within the secondary vocational education (SVE) system:

Stage 1. Analysis of the expected vocational activities of students. At this stage, we gather information on the types of labor activities and operations and state problems to be solved based on the knowledge acquired. Then the game content is selected: Actions and operations forming the basis of the expected vocational activities.

Stage 2. Deciding on syllabus. This stage is based on making the game-based training content as specific as possible, considering conventional requirements, such as availability, visibility, continuity, etc.

Stage 3. Evaluation of time and loads on students when using this method of organizing their learning. Strict timing is an integral part of a blitz game, since loss of time may remit the competitiveness among the students involved.

Stage 4. Selecting the most favorable forms of arranging teaching/learning processes to successfully implement the method. The outcome represents a description of such forms selected.

Stage 5. Preparing materials, cases, and texts to motivate the participants while implementing the method. Original motivation texts and situations should be developed at this stage and included in the previously defined syllabus.

Stage 6. Developing a system of training exercises aimed at acquiring disciplines, with predefined quality indicators. In the context of blitz games, the outcome of this stage will be a game script containing the goal hierarchy, training information, game introduction, and forms and conditions of playing the game.

Stage 7. Developing materials and texts to perform the objective quality control of how the content is acquired. The simplest way to assess the game results is to sum up the points scored while performing individual game activities. Final total is achieved by ranking the participants and identifying the group of leaders, as well as discussing on the results.

Stage 8. Developing the contents and structures of extramural training activities. This stage is used for students to independently prepare for the game to play.

Stage 9. Trying and testing the training game, including checking whether the teaching/learning process is complete, evaluating the results obtained, and correcting afterwards, where necessary.

1.3 Blitz Games for Engineering College

Study Goals.

Introducing game-based procedures in training is necessary to arrange the process in such a manner that students can feel their own success, intellectual worthiness, and engagement in the activities shared with other students and in thought exchange and material perception techniques, which forms a productive cognition.

This study is aimed at justifying the need for and identifying the efficiency of using game-based procedures for the students of an engineering college, as well as at developing and testing some blitz-games adapted to the students of this college.

In developing blitz games to be used in teaching/studying engineering disciplines within the secondary vocational education system, the authors followed four basic stages.

Stage 1, preparatory: Gathering and analyzing information regarding the expected vocational activities of students. Curriculum and steering documents of engineering disciplines are studied, and the blitz game plan is drawn up.

Stage 2, game developing: Creating the plot and general description of the game and preparing the material support – the more interesting the materials are, the more exciting the process of playing will be. In-depth development of the instruction sheet and hand-outs allows us not to spend or minimize time spent on additional explanations during playing.

Stage 3, game deployment: Problem statement, discussing game conditions, rules, and regulations; distribution of roles; forming the groups; and consulting.

Stage 4, final: Analyzing the outcomes and, if necessary, correcting the game script. At this stage, self-reflections should be performed, i.e., it is important for the students to realize what new is that they have got to know, such as terms or operations; what they have learnt, such as skills or competences; and why it is useful for their expected vocational activities.

To successfully implement blitz games into educational process, some factors have to be taken into consideration, such as training timing and blitz game goals, i.e., whether the game is used as introduction or as an illustration for a specific matter, or as a summary of a certain topic [7].

We developed some blitz games for the 3- and 4-year university college students majoring in the area titled Installation and Technical Operation of Industrial Equipment and being a part of the Machine Engineering integrated group of training areas; full-time mode of study; qualification: Mechanical technician.

Area of professional practice for the graduates includes arranging and performing activities in installing, testing, operating, maintaining, and repairing industrial equipment, as well as organizing the activities of relevant functional units. Objects of their vocational activities are industrial equipment; materials; tools; process rigs; workflows in repairing, manufacturing, restoring, and assembling central mechanisms; and design and process-control documents.

Research Methods

In our study, we have used the following methods of research:

1. Theoretical methods, such as analysis of foreign and national scientific literature on our research topic, studying and analysis of best practices in education, analysis of students' performance, classification, and comparative analysis;
2. Experimental methods, such as questionnaires, interviews, and observations; and
3. Interpretation of experimental findings, such as scaling.

2 Actual Outcomes

2.1 Introducing Blitz Games in Teaching Engineering Disciplines Within the SVE System

The authors have developed some blitz games that we are planning to use in the training process. In this paper, we are going to consider in more details two games introduced in the training process, on which games we have performed our studies. More games will be researched in our further publications.

"Coordinator". Our first blitz game was developed for the professional module discipline titled Arranging and Performing Activities in Operating Industrial Equipment. This professional module (PM.02) program is intended to last 3 years of studies with totally 564 academic hours, including 420 h of the maximum academic load, of which 280 h are intramural, 140 h are extramural, and 144 h are practical and on-the-job training.

We extracted from the professional module steering document the topic titled Maintenance, which is studied in the third year of studies. It includes:

- Basic elements of equipment maintenance;
- Acceptance and run of equipment;
- Equipment maintenance types and intervals;
- Equipment maintenance contents;
- Special aspects of equipment maintenance; and
- Forms of organizing the equipment maintenance at an enterprise.

Based on the topic contents, we developed and successfully performed blitz game titled "Coordinator" at a lesson with the third-year students of group 461.

Purpose: Studying the special aspects of equipment maintenance and developing professional and communicative skills, as well as individual and teamwork skills.

Conducting recommendations: First stage of the game is independent work. Each participant fulfills independently the defined game conditions. At most 10 min are allocated for this stage.

Second stage of the game is group work. A leader should be elected from among the group members, which makes the summary table in discussions with the group. Then their table is compared to the correct answers.

Third stage represents comparing the individual results with the correct answers.

Fourth stage represents summarizing the results.

Duration of the blitz game: 30 min.

Game situation: Imagine you are the CEO of company N. You should distribute the duties among maintenance staff, such as mechanical repairman, electrician, and operator, to perform basic activities related to the NC equipment maintenance. Put a plus at the operation that must be performed by the relevant staff member. More than one plus can be in one line.

Abbreviations:

MO – maintenance operations.

MO&R – maintenance operations and repairs (Table 1).

Table 1. Schedule of maintenance and repair operations to be performed by the staff

Operation	Machine component	Maintenance staff		
		Mechanical repairman	Electrician	Operator
Time-scheduled MO				
Routine inspection	Mechanical and electrical			
Shift inspection	Mechanical			
Shift cleanliness control	Entire machine			
Shift lubrication	Mechanical			
Refilling and changing the lubrication (after 40 h)	Entire machine			
Machine adjustment	Mechanical and electrical			
Machine geometry and processing accuracy check				
Corrective MO&R				
Replacement of suddenly failed parts	Mechanical and electrical			
Restoring of sudden failures in the adjustment of devices and joints	Mechanical and electrical			

Correct answers:

Mechanical repairman: Routine inspection; shift inspection; machine adjustment; machine geometry and processing accuracy check; replacement of suddenly failed parts; restoring of sudden failures in the adjustment of devices and joints.

Electrician: Routine inspection; machine adjustment; replacement of suddenly failed parts; restoring of sudden failures in the adjustment of devices and joints.

Operator: Routine inspection; shift inspection; shift cleanliness control; shift lubrication; refilling and changing the lubrication (after 40 h).

This blitz game is recommended to be used in studying the special aspects of equipment maintenance, as well as its types and intervals. It can also be used as an introduction or as an illustration for a specific problem.

The second blitz game was developed for the other professional module (PM.01): Arranging and Performing the Operations of Mounting and Repairing Industrial Equipment. Totally, 1,120 academic hours are allocated to the acquisition of this program, including 904 h of the maximum academic load, of which 607 h are intramural, 297 h are extramural, and 216 h are practical and on-the-job training.

To develop a blitz game, we extracted from the steering document the topic titled Preparations for Equipment Repairs. It includes:

- Preparing the equipment for being repaired;
- Procedures of transferring the equipment to/from repairs; repairing drawings;
- Tools and machinery used in repairing the equipment; and
- Mechanization of repairing operations.

Blitz game titled Repairing a Machine was developed for the third-year students. We performed the game with the students of group 461 in studying the topic of Procedures of Transferring the Equipment to/from Repairs.

Blitz Game: Repairing a Machine

Purpose: Studying the preparatory work for repairing the equipment and developing professional and communicative skills, as well as teamwork skills.

Recommendations on performing the game:

Stage 1. Students get into 2 groups, each of which performs the tasks given. The outcome is represented as a list of operations to be performed to repair the machine. Average time allocated for this stage is 10 min.

Stage 2. Each team presents its list. Number of answers matching with the correct ones for each item in the list scores one.

Stage 3. Counting the scores and summarizing. Group having the highest score is the winner.

Blitz game lasts 20 min.

Game situation: When repairing a machine, you should follow a certain sequence of operations to ensure the clearest and most efficient repairs. Decide on the correct sequence of such operations by filling out the right column in the Table 2.

Table 2. Sequence of operations in repairing a machine.

Operations	Your sequence
Identifying the defects in the mechanism	1
Repairing the parts	2
Disassembling the mechanism into assembly units and parts; washing them	3
Checking and adjusting of the material assembled	4
Identifying the sequence of disassembling the mechanism	5
Assembling mechanisms with selecting and adjusting the parts	6
Identifying the nature and extent of the parts wear and their defects	7

Correct answer: 1;5;3;7;2;6;4.

This blitz game is intended for the fourth-year students.

The blitz games above can be used in teaching students majoring in the integrated group of areas (15.00.00), since those majoring in machine building have to deal with machine parts, units, and assemblies, as well as with mechanisms used in general machine engineering, power machine engineering, transport engineering, special-purpose machine engineering, and other technical areas. Therefore, the blitz games developed will also be useful in studying special subjects in such areas of studies. Blitz games can also be widely used in teaching adjacent engineering disciplines for studying the similar topics: Machine Maintenance; or Equipment Repairs: General Assemblies and Parts.

2.2 Analysis of Survey Findings

Questionnaires are the most common type of polling, in which the surveyor and the respondent communicate indirectly via the questionnaire texts [3].

Students of Kazan Technological College participated in our survey, namely the 25 students of group 461. They are all young men, 23 of them are 21 years old and 2 of them are 22 years old. 80% of respondents work, 74% of them work at enterprises focusing on the areas the students are majoring in, and 90% of respondents are planning to work in their major areas, which follows from personal conversations. Each questionnaire form was considered individually, and 100% of respondents provided their answers to all the questions.

In our study, we used two questionnaires. Full-time students were offered to fill out the form before the blitz games were introduced into the studying process to estimate their attitude to using active game-based learning methods. The first questionnaire included 5 questions. The second questionnaires included 7 questions and was given to students after the blitz games had been used in teaching engineering disciplines in their college.

Our first questionnaire allowed us to identify the students' general attitude towards game-based educational methods, particularly to blitz games. Students were asked to answer the following questions:

1. Are game-based learning methods of interest to you?
2. Do you know what blitz games are?
3. Have you ever attended lessons where blitz games were used?
4. Would you like to try and study using blitz games?
5. Do you think game-based methods will be efficient in studying these subjects?

When answering, the respondents had to specify the degree of their agreement/disagreement to the statement specified, using the Likert-type scale. Likert-type scale is a rating scale named after its developer, Rensis Likert. Such scale is used in questionnaires aimed at measuring somebody's attitude towards a problem. When working with the scale, the respondent shall estimate the degree of his/her agreement/disagreement to each statement, using the following answers: "Strongly agree," "Agree,", "Neither agree nor disagree," "Disagree," or "Strongly disagree."

Our findings are shown in Fig. 1.

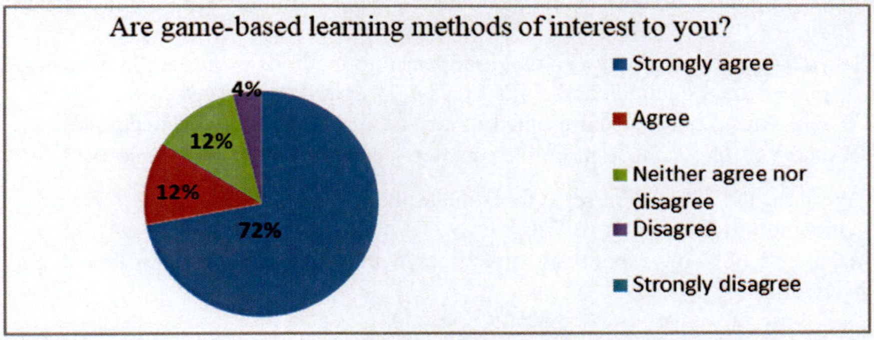

Fig. 1. Students' attitude towards game-based learning methods.

47% of the respondents answered question 2 in the affirmative, 29% neither agreed nor disagreed, which proves that this learning method is poorly known by students. Answering question 4, 72% of respondents wrote they would like to study using this method (Fig. 2).

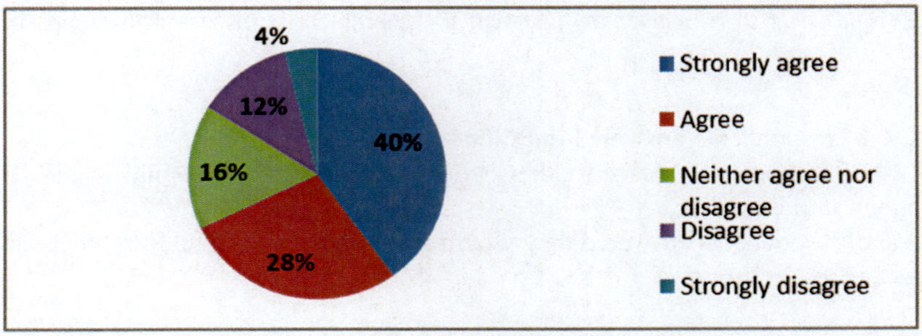

Fig. 2. Do you think game-based methods will be efficient in studying these subjects?

Thus, analyzing the findings, we can see the emerging interest in game-based learning methods, especially in blitz games. Although blitz games were used extremely rarely in the educational process, students would like to try and study using this game-based method. About 70% of responding students think that blitz games will be efficient in studying the following disciplines: Arranging & Performing the Works on the Industrial Equipment Operation and Arranging & Performing Installation and Repairs of Industrial Equipment.

The second questionnaire consisted of seven questions to see the students' attitude towards the blitz games already performed and towards the further use thereof in the college educational process. Students had to answer the following questions:

1. Did you like the Coordinator blitz game?
2. Did you like the Repairing a Machine blitz game?
3. Is it easier to remember the materials, if the problem is discussed in groups and blitz games?
4. In your opinion, do blitz games promote the increased motivation for and interest in studying?
5. In your opinion, do active game-based learning methods promote the development of professional qualities and skills to a larger extent than classical ones?
6. Would you like to see more blitz games at the lessons in these disciplines?
7. Would you like to have more blitz games when studying other subjects?

We used the Likert-like scale to evaluate the results.

Our findings are shown in Fig. 3.

80% and 68% of respondents answered in the affirmative to questions 1 and 2, respectively.

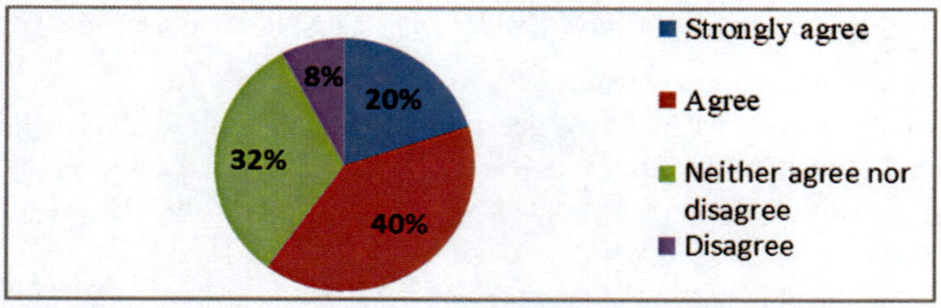

Fig. 3. Is it easier to remember the material, if the problem is discussed in groups in blitz games?

72% of respondents answered question 4 in the affirmative.

80% of respondents answered they would like to have more blitz games in the disciplines studied.

76% of respondents answered they would like to have such games in other subjects (Fig. 4).

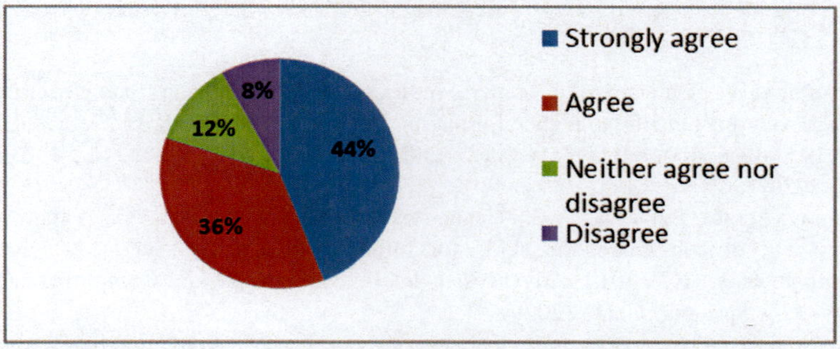

Fig. 4. In your opinion, do game-based methods promote the development of professional qualities and skills more than the classical methods do?

3 Conclusions

The work performed has shown that using game-based procedures activates the thought of students. The blitz games developed enhance the activity, motivation, and emotional perception of the training process by the engineering college students. Using game-based procedures also allows developing the cognitive and creative activities of students, enhancing their performance, and developing/assessing their professional competences, since these blitz games form the necessary set of soft and vocational skills and competences in accordance with the national standards for secondary vocational education [8, 9].

We consider the introduction of blitz games in the educational process to result in the following: Better learning experience and the intensification/activation of educational process. Solving the problems stated have developed the students' motivation for learning, and they have managed to appraise the importance of practice-focused techniques in their professional training.

Experimentally, based on questionnaire method, we defined the general characteristics of the game-based procedures functionality, i.e., their relevance, performance, and interest awaken in the students. We also identified how introducing the blitz games developed influenced upon the development of the students' personal qualities.

Testing results have shown the 25% increase in numbers of students who have got good and excellent grades when using the blitz-game module. Blitz games awoke interest in over 70% of students. Our conclusions allow suggesting that it would be reasonable to include blitz games in teaching/learning.

Thus, using game-based procedures allows increasing the students' motivation, self-confidence, interest in the training, and performance. The blitz games developed can be efficiently used for both the other majoring areas inside the college and in other engineering colleges in Kazan and the Republic of Tatarstan.

References

1. Fakhretdinova, G., Dulalaeva, L., Tsareva, E.: Extracurricular activities in engineering college and its impact on students' tolerance formation. In: Auer, M., Hortsch, H., Sethakul, P. (eds.) ICL 2019, Advances in Intelligent Systems and Computing, vol. 1134, pp. 143–150. Springer, Cham (2020)
2. Khusainova, G.R., Astafeva, A.E., Gazizulina, L.R., Fakhretdinova, G., Yakimova, J.Y.: Development of communication skills for future engineers. In: Auer, M., Hortsch, H., Sethakul, P. (eds.) ICL 2019, Advances in Intelligent Systems and Computing, vol. 1134, pp. 229–124. Springer, Cham (2020)
3. Usenkova, E.Y.: Aktivnyye metody obucheniya – za i protiv [Active Learning Methods: Pros and Contras]. In: Vestnik MGUKI (Herald of Moscow State University of Culture), vol. 16, no. 9–10, pp. 84-85 (2015). (in Russian)
4. Yushko, S.V., Galikhanov, M.F., Barabanova, S.V., Kaybiyaynen, A.A., Suntsova, M.S.: International network conference as an efficient way to integrate universities and businesses in the context of digital economy. In: Advances in Intelligent Systems and Computing, vol. 1134, pp. 663–673. AISC (2020)
5. Osipov, P., Ziyatdinova, J., Girfanova, E.: Factors and barriers in training financial management professionals. In: Advances in Intelligent Systems and Computing, vol. 916, pp. 167–175 (2020)
6. Barabanova, S.V., Nikonova, N.V., Pavlova, I.V., Shagieva, R.V., Suntsova, M.S.: Using active learning methods within the andragogical paradigm. In: Advances in Intelligent Systems and Computing, vol. 1134, pp. 566–577. AISC (2020)
7. Chechet, V.V.: Aktivnyye metody obucheniya v pedagogicheskom obrazovanii [Active Learning Methods in Pedagogical Education]. In: Chechet, V.V., Zakharova, S.N. (eds.) Minsk: BGU, p. 127 (2015). (in Russian)
8. Polyanskaya, E.K.: Aktivnyye metody obucheniya kak sposob povysheniya effektivnosti obrazovatelnogo protsessa [Active Learning Methods as a Way to Enhance the Efficiency of Educational Process]. In: Orenburg: OGPU, pp. 16–27 (2015). (in Russian)
9. Kolokolnikova, Z.U.: Tekhnologiya aktivnykh metodov obucheniya v professionalnom obrazovanii: ucheb.posobiye [Technology of Active Learning Methods in Vocational Education: Textbook]. In: Kolokolnikova, Z.U., Mitrosenko, S.V., Petrova, T.I. (eds) Krasnoyarsk: Siberian Federal university, p. 176 (2007). (in Russian)

Improving the Teaching-Learning Process in Engineering Through a Game-Based Web Support System: Edutrivias

Renzo Angles[1], Luis Silvestre[1], Gonzalo Pincheira-Orellana[2], Claudia Galarce-Miranda[3(✉)], and Diego Gormaz-Lobos[3]

[1] Faculty of Engineering, Department of Computer Science, Universidad de Talca, Talca, Chile

[2] Faculty of Engineering, Department of Industrial Technologies, Universidad de Talca, Talca, Chile

[3] Faculty of Engineering, International Center of Engineering Education, Universidad de Talca, Talca, Chile

claudia.galarce@utalca.cl

Abstract. In this article, the authors propose the use of trivia games as a tool to support the teaching-learning process in engineering, and present results of the use of trivia games in the Chilean university context. Specifically, we describe a web system called Edutrivias (https://edutrivias.cl) which allows the interaction between teacher and engineering students through game-based learning. Edutrivias allows teachers to create trivia games that can be played by students, as many times as they want, within determined goals and a period of time defined by the teacher. Every time a student plays a trivia, he/she receives information regarding his/her level of progress. At the same time, the teacher can verify the participation of the students, as well as the level of the group and individual advancement. From a pedagogical point of view, teachers deliver knowledge through the trivia games, while the students "play" trivia games to acquire, increase and apply their knowledge through an intermittent reinforcement learning program. The main goals of this article are: (1) to present the foundations of the use of game-based learning in the competence development of engineering students, (2) to describe the main components and functionalities of Edutrivias, and (3) to present the results of an exploratory case study of Edutrivias with information of Chilean students of an engineering school at the University of Talca.

Keywords: Game-based learning in engineering · Gamification in engineering careers

1 Introduction

Nowadays, one of the factors that has most affected the teaching-learning process is the massive use of technological elements such as smartphones, social networks and online courses, among others. The popularity of these technologies has generated that traditional methods of teaching fail to attract the interest of the students, losing their

M. E. Auer and T. Rüütmann (Eds.): ICL 2020, AISC 1328, pp. 487–498, 2021.
https://doi.org/10.1007/978-3-030-68198-2_45

applicability and effectiveness [1]. Many universities are no strangers to this phenomenon and try to implement strategies to correct learning problems in courses associated with the basic sciences related to Mathematics, Physics, and Chemistry [2]. These problems are usually associated with a low level of understanding, knowledge, retention, reflection, and application of theory in real cases [3, 4]. In response to the above problems, many academics and researchers are developing teaching-learning methodologies, trying to take advantage of information technologies and virtual environments [5, 6]. An important example of that is the use of online education systems (e-learning) which allow to improve the traditional and formal educational systems. An e-learning system grants flexibility, permanence and synchrony [7]. Other relevant resources are the educational technologies and tools for formal and non-formal learning processes like educational applications in the form of simulators or educational games. The use of games as a learning tool has been extensively studied in the literature, and it has been confirmed that education games are an effective and attractive way to improve the learning process in the students [8].

2 Game-Based Learning in Engineering Education

2.1 Game-Based Learning

A general definition presents Game-based learning as a type of game play with defined learning outcomes [9]. In cognition psychology, the role of play as a key factor for cognitive development, don't have discussion. For instance, Piaget describes the role of play for the cognition development of children (from concrete to abstract stages); and for Vygotsky, play is a "leading factor" in children's development and responsible to create a zone of proximal development for the child [10].

Plass et al. presented a number of arguments for the use of game-based learning [10]:

- Motivation. A game motivates learners to stay engaged over long periods through a series of features that are of a motivational nature.
- Player Engagement. Games allow for a wide range of ways to engage learners (cognitive, affective, behavioral and sociocultural engagement). The goal of all these types of engagement, however, is to foster cognitive engagement of the learner with the learning mechanic.
- Adaptivity. Adaptivity is the capability of the game to engage each learner in a way that reflects his or her specific situation. This can be related to the learners' current level of knowledge, to cognitive abilities, to the learners' emotions, or to a range of other variables.
- Graceful Failure. The lowered consequences of failure in games encourage risk taking, trying new things, and exploration. They also provide opportunities for self-regulated learning during play, where the player executes strategies of goal setting, monitoring of goal achievement, and assessment of the effectiveness of the strategies used to achieve the intended goal.

In addition, Prensky describes twelve arguments for the use of games in the learning process. All this argument are in close relationship with the characteristics and goals of the (digital) game-based learning (see Table 1) [11]:

Table 1. Arguments for the use of games-based learning.

Arguments for games	Effects for learning process
1. Games are a form of fun	Enjoyment and pleasure
2. Games are a form of play	Intense and passionate involvement
3. Games have rules	Structure
4. Games have goals	Motivation
5. Games are interactive	"Doing" or activity
6. Games are adaptive	Flexibility
7. Games have outcomes and feedback	Learning's evidences
8. Games have won states	Gratification
9. Games have conflict/competition/challenge/opposition	Resilience
10. Games have problem solving	Creativity
11. Games have interaction	Social interaction
12. Games have representation and story	Emotional engagement

In a simple description (see Fig. 1), the educational games have: (i) an input phase with instructional content and determined game characteristics; (ii) a process phase in form of a game cycle (the players interacted with tasks, the different characteristics and the mechanic of the game, and receive systems feedback); and (iii) the output phase with the results of the interaction [12].

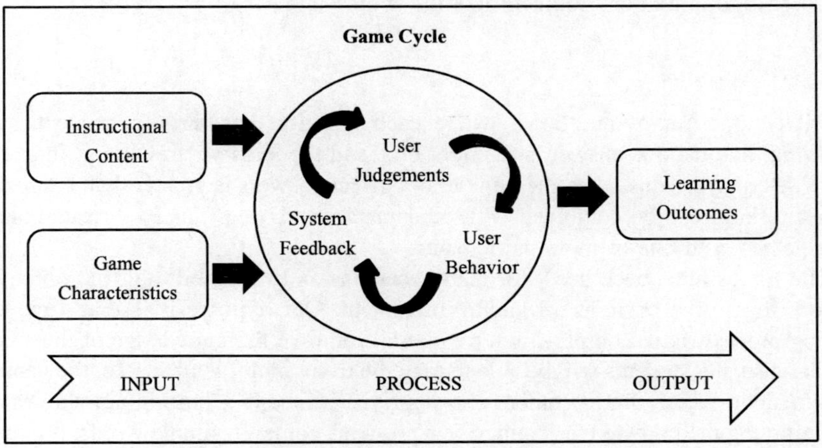

Fig. 1. Input-process-output game model [12]

2.2 Game-Based Learning in Engineering

The use of game-based-learning in many engineering fields has been investigated by different authors. The specialized literature on Engineering Education presents experiences at mechanical engineering [13], software engineering [14], and electrical engineering [15] among others. Specifically, for civil engineering, Hartmann et al. [16] presents research's results about the motivational effects of games at engineering's training process. With the results at the use of seven software tools for civil engineering (GasSolution/T-Xchange, RiskSwitch/T-Xchange, RAMSes, HighwayStakes/TX change, among others), the authors concluded that different motivational forms can co-exist when engineering students play the educational games: self-determined motivational forms (intrinsic motivation and identified regulation) and non-self-determined motivational forms (external regulation and motivation) [16].

Although engineering education in many countries is strongly oriented to a traditional lecture-based form (probably for the specific contents related to the knowledge areas and the development of technical skills), the recent literature shows different proposals with the use of education games [16]. Among the effects and contributions of game-based learning to engineering education are: (i) the cognitive growth and digital literacy; (ii) the social-emotional growth; (iii) the soft skills development; (iv) enhanced decision making and problem-solving skills, as well as critical thinking; (v) improvement of the collaboration with others; (vi) generation of a positively competitive environment; (vii) the build of a progressive learning through experience; and (viii) facilitation of feedback driven and student-centered learning [17–19].

3 Edutrivias

In this section we describe Edutrivias, a software application that follows the game-based learning paradigm. Edutrivias uses trivia games to support the teaching-learning process. We present the trivia game as a training tool for learning, and its implementation in a digital environment like the Web.

3.1 Trivia Games

A "trivia" is a set of questions, where each question has three possible answers, satisfying that just one answer is the right one, and the rest two are wrong. In a "trivia game", the player wins when the number of correct answers is greater than a threshold. Such threshold can be combined with other parameters (e.g. time) to create complex trivia games with one or more participants.

The trivias have been used, for many years, as an evaluation tool. In contrast, we propose to use the trivia as a training instrument. Our hypothesis is that a repetitive training process, based on playing a trivia, will improve the knowledge of the student. In this sense, the students will be able to play the trivia multiple times, i.e. the game has many matches. Each time a student plays a trivia, he/she is acquiring new information or reinforcing the knowledge acquired in a previous game. In simple words, the student will learn by repetition.

The training process described above enables an important feature: instant feedback. Each time the student finishes a match, he/she can receive specific (for the match) and general (for the game) information about the training process. This information is also valuable for the teacher to be aware about the activity of the student, and his/her level of training. Moreover, the teacher can analyze the performance of the student, and provide additional feedback. A teaching-learning process based on trivias could be tedious and costly to implement in a real environment (classroom), because it implies to create the same material multiple times (i.e. printed trivias). Moreover, the revision and feedback activities imply a lot of time and effort for the teacher. In contrast, a trivia game is feasible to implement as a software application working in a virtual environment like the Web.

3.2 The Teaching-Learning Process in Edutrivias

Edutrivias (https://edutrivias.cl) is a Web application designed to support the teaching-learning process by using the notion of trivia as the central resource. In general terms, Edutrivias considers the participation of teachers and students, whose interaction consists in a group of students playing a trivia created by a teacher. A complete description of the teaching-learning process is presented next.

Assume that the teacher and the students are registered in the system. The process begins when the teacher creates a "Group", and associates each student with the group (once a time). Additionally, the teacher can generate a group-code and send it to the students, so each student is able to relate itself with the group by using such code.

After the creation of the group, the teacher creates the trivia, that is, a set of questions. Each question is composed of a statement (or question declaration), a picture (which is an optional resource), and three possible answers, such that one answer should be correct, and the other two should be wrong. In order to facilitate the creation of the trivia, the teacher is able to "copy" an existing trivia and take it as the starting point.

Once the trivia is created, the teacher must link the trivia with the group of students. This connection is called a "game". A game has a "start date-time" and a "finish date-time", parameters that define a period of time during which the students are able to play the trivia as many times as they want. Based on this, the status of a game could be "created", "open" or "closed". Once the trivia is "open" for the students, they are able to play matches. A "match" refers to a (random) subset M of the complete set of questions that compose the trivia. Hence, the students must play multiple matches to review all the questions of the trivia. The size of M must be defined by the teacher during the creation of the game. Additionally, the teacher could define a "timer" that defines a waiting time to get the answer of a query.

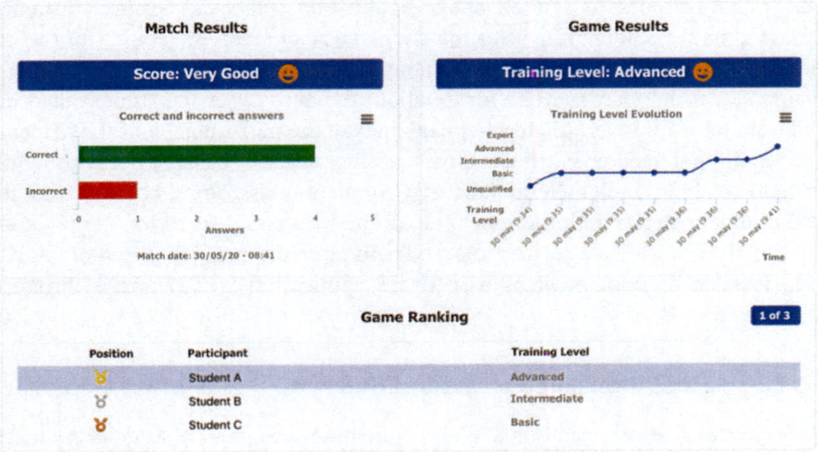

Fig. 2. Report shown to the student after completing a trivia match in Edutrivias.

Each time a student plays and finishes a match, the system presents a report like the one presented Fig. 2. The box in the top-left side shows the score for the match, the number of correct answers, and the number of wrong answers. Internally, the system maintains a cumulative score for each question of the trivia. The score of the match is given by the number of correct answers. The box in the top-right side shows the training level of the student in the game, which is given by the cumulative score of all the questions in the trivia. There are five training levels: unqualified (No califica), basic (Básico), intermediate (intermedio), advanced (avanzado) and expert (experto). The box in the bottom shows the ranking of the student in the game. The ranking considers the training level, the number of matches, and the time of participation of the student.

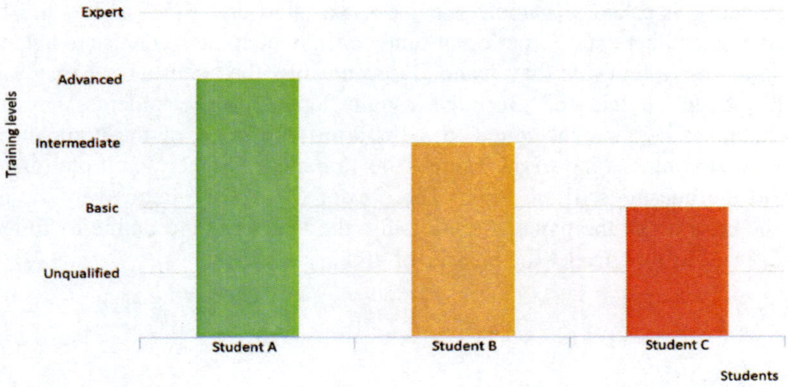

Fig. 3. Report that shows (to the teacher) the training level obtained by the students in a group.

Anytime, during the life of a game, the professor can monitor the activity and evolution of the students, either individually or as a group. For example, Fig. 3 shows the report of the training level obtained by each student in a group. Additionally, there is a report that shows the percentage of correct and wrong answers for the whole group of students. This report provides an insight of the concepts that require less or more training.

4 Exploratory Case Study

Several theories on case study as a research instrument have appeared in the literature [20]. Yin defines a case study as an empirical method of analysis of "a contemporary phenomenon within its real life context" [21] taking into account multiple information sources (e.g., qualitative and quantitative data combination). Yin distinguishes four types of case study designs: single case, multiple case, embedded, and holistic. We conducted an exploratory single-case study where we evaluated the improvement of the teaching-learning process using Edutrivias.

The research questions addressed in this research are the following:

RQ1: What is the interest of the students to use Edutrivias?
RQ2: What is the impact on student scores using Edutrivias?

4.1 Subjects Selection

The exploratory case study was carried out during the academic year 2019 with undergraduate students at University of Talca. Specifically, Edutrivias was used in the Database Design module. This module is taken by the students in the fifth semester of the career of Computer Science Engineering at University of Talca. The module is evaluated with two theoretical/practical tests according to teaching units. Edutrivias was used to support traditional techniques in the educational learning process. In this sense, engineering students applied a game-based learning paradigm through competition quizzes (better ranking) for training the database design concepts.

4.2 Evaluation Characteristics

The goal of this exploratory case study is evaluating the interest and score impact of students using Edutrivias. The use of Edutrivias can be analyzed from different points of view. First, the interest and use that generate on students, through i) number of students using Edutrivias and ii) time spent in using the software. Second, the effect on academic score that generates on students, through: iii) comparison of evaluations vs training level obtained in Edutrivias, and iv) detection of critical areas.

In this sense, we consider the following evaluation criteria: (1) "interest" is measured as the number students and games using Edutrivias, (2) the "score impact" is measured as related to score test and training level in Edutrivias.

4.3 Data Collection

In this exploratory case study, we conduct an experimentation in software engineering as the method for data collection [22]. In the database design course, theoretical evaluations are designed to measure students' knowledge from different formal techniques, for example: tests, exams, mind maps, quizzes and others. The tests and exams may consider practical questions, multiple-choice questions, true/false questions, and others. We design two trivia that is associated with each unit of the module, thus, Unit 1 Trivia (U1T) has 56 questions and Unit 2 Trivia (U2T) has 42 questions. We use Edutrivias for building a trivia game that were shared with the engineering students. Each student had the opportunity to play several games for a period defined by the teacher and their use was voluntary. In this sense, students can practice or study for the formal evaluation test. The formal evaluation test considers a theorical and practical section. Finally, we decide to use some questions/sentences of Edutrivias in the formal theorical section for comparing results.

Table 2 and Table 3 show the collected data that considers 44 students (7 U1T and 37 U2T). Both tables consider student ID (anonymous), number of games, training level (obtained in Edutrivias) and score test (evaluation). Table 2 shows a score test (column 4) with 10 questions where every student has a number of good answers. On the other hand, Table 3 shows a score test with 5 questions (column 4).

Table 2. Data collection U1T

Student ID	Numbers of games	Training level - Edutrivias	Score test - evaluation (%)
13	21	Advanced	6/10 (60%)
17	27	Advanced	9/10 (90%)
24	12	Basic	4/10 (40%)
31	72	Advanced	10/10 (100%)
33	34	Advanced	8/10 (80%)
36	8	Basic	5/10 (50%)
44	27	Advanced	9/10 (90%)

Table 3. Data collection U2T

Student ID	Numbers of games	Training level - Edutrivias	Score test - Evaluation (%)
1	16	Expert	5/5 (100%)
3	37	Advanced	5/5 (100%)
4	18	Advanced	5/5 (100%)
5	44	Expert	5/5 (100%)
6	105	Expert	5/5 (100%)
7	5	Basic	1/5 (20%)
8	50	Expert	5/5 (100%)
9	6	Basic	2/5 (40%)

(*continued*)

Table 3. (*continued*)

Student ID	Numbers of games	Training level - Edutrivias	Score test - Evaluation (%)
10	45	Advanced	4/5 (80%)
11	6	Basic	3/5 (60%)
12	26	Advanced	5/5 (100%)
13	27	Expert	5/5 (100%)
14	49	Expert	5/5 (100%)
15	19	Expert	5/5 (100%)
16	76	Expert	5/5 (100%)
17	28	Expert	5/5 (100%)
18	49	Advanced	3/5 (60%)
19	14	Expert	4/5 (80%)
20	19	Expert	4/5 (80%)
21	20	Expert	5/5 (100%)
22	23	Expert	4/5 (80%)
24	30	Expert	5/5 (100%)
26	11	Intermediated	4/5 (80%)
27	19	Advanced	4/5 (80%)
28	36	Advanced	5/5 (100%)
29	62	Expert	5/5 (100%)
30	24	Expert	4/5 (80%)
31	43	Expert	4/5 (80%)
32	41	Advanced	4/5(80%)
33	21	Advanced	5/5 (100%)
37	26	Expert	3/5 (60%)
38	9	Basic	3/5 (60%)
40	15	Advanced	5/5 (100%)
41	14	Intermediated	2/5 (40%)
42	13	Intermediated	4/5 (80%)
43	47	Expert	5/5 (100%)
44	26	Expert	5/5 (100%)

4.4 Data Analysis

To answer each of our research questions, we analyze the number of students and matches reported in Edutrivias. Moreover, we analyze the relation between score test and training level.

Number of Students and Matches Using Edutrivias. In U1T, 7 of 44 students used Edutrivias (15.9% of students), generating 201 matches or interactions. In U2T, 37 of 44 students used Edutrivias (84.1% of students), generating 1119 matches or interactions. Our information shows an increment in the number of students. Probably, students increase the use of Edutrivias for various reasons: complexity of Unit 2, academic motivation, interest in Edutrivias, and others. However, we can see that in general the students with best training levels (expert and advanced) in U1T have better performance (Table 2); the positive results increase the interest of other students in Edutrivias. Moreover, we observed that students recommended the use of Edutrivias.

Score Test and Training Level Relation. In U1T, 5 students have intermediate training level, and 2 students have basic training level. In general, the intermediate level has good performance (80%–100% good answers) in the evaluation test. In U2T, 20 students have expert level, 10 students have advanced level, 3 students have intermediate level, and 4 students have basic level. In general, the expert, advanced and intermediate level have good performance (80%–100% good answers) in the evaluation test.

Table 4. Relation between score tests and training level (obtained in Edutrivias)

Training level	Score test vs training level				
Expert	14 students with excellent answers (100%)	5 students with good answers (80%)	1 student with neutral answers (60%)	0 students with fair answers (40%)	0 students with bad answers (40%)
Advanced	6 students with excellent answers (100%)	3 students with good answers (80%)	1 student with neutral answers (60%)	0 students with fair answers (40%)	0 students with bad answers (40%)
Intermediate	0 students with excellent answers (100%)	2 students with good answers (80%)	0 students with neutral answers (60%)	1 student with fair answers (40%)	0 students with bad answers (40%)
Basic	0 students with excellent answers (100%)	0 students with good answers (80%)	2 students with neutral answers (60%)	1 student with fair answers (40%)	1 students with bad answers (40%)

Table 4 shows three behaviors in the relation between score test and training level: "Consistent behavior", the students with expert, advanced and intermediate levels have excellent/good/neutral performance (4 students of U2T, 28 students of U1T); "Positive behavior", the students with intermediate and basic level have good excellent/good performance (4 students of U2T), "Negative behavior", the students with expert and advanced levels have bad or fair performance (1 student of U2T, 2 students of U1T). Moreover, 2 students of U1T and 3 students of U2T with basic and intermediate levels have fair or bad performance (consistent behavior).

Finally, these results indicate that students increase the interest in Edutrivias in terms of the number of students and games (RQ1). The students have a consistent impact on their test score in terms of relation between their score test and training level (RQ2).

5 Conclusions and Future Work

The use of trivia games provides various benefits that allow the teacher invest more time in active class activities (e.g., problem-based activities) and generate a better discussion (e.g., debates, conversations). On the other hand, students can take advantage of using trivias to applicate and evaluate knowledge in an entertaining way. In addition, the constant practice of a trivia allows students to achieve a better understanding of concepts and provide students with supplementary material or resources. Constant feedback is another characteristic of Edutrivias, allowing teachers and students to know the achievement levels in a trivia game. Hence, students can be aware of their progress, and are constantly motivated to meet learning objectives. At the same time, teachers could monitor students' progress, and deliver additional and personalized feedback.

Although the study's results and the students' comments show positive effects of the use of Edutrivias at the learning process of engineering students, the researchers are agreed about the need of a bigger number of experimentations and a more detailed analysis of the results.

References

1. Fleer, M.: Technologically constructed childhoods: moving beyond a reproductive to a productive and critical view of curriculum development. Australas. J. Early Child. **36**(1), 16–24 (2011)
2. Mendoza, D., Flores, E.: Relación entre el rendimiento académico de los estudiantes universitarios en el área de matemática y la praxis docente mediadora. INNOVA Res. J. **3**(6), 21–31 (2018)
3. Jenaro-Rio, C., Castano-Calle, R., Martin-Pastor, M., Flores-Robaina, N.: Rendimiento académico en educación superior y su asociación con la participación activa en la plataforma Moodle. Estudios sobre educación, vol. 34 (2018)
4. García, A.E.: Estilos de aprendizaje y rendimiento académico. Revista Boletín Redipe **7**(7), 18–228 (2018)
5. Hsin, C.-T., Li, M.-C., Tsai, C.-C.: The influence of young children's use of technology on their learning: a review. J. Educ. Technol. Soc. **17**(4), 85–99 (2014)
6. Hortsch H., Gormaz-Lobos D., Galarce-Miranda C., Kersten S.: Needs-oriented engineering pedagogy - research projects in Chilean Universities. In: Auer, M., Tsiatsos, T. (eds.) The Challenges of the Digital Transformation in Education, ICL 2018. Advances in Intelligent Systems and Computing, vol. 917, pp. 741–753. Springer, Cham (2019)
7. Mayer, R.: The Cambridge Handbook of Multimedia Learning. Cambridge University Press (2005)

8. Kebritchi, M.: Factors affecting teachers' adoption of educational computer games: a case study. Br. J. Edu. Technol. **41**(2), 256–270 (2010)
9. Shaffer, D.W., Halverson, R., Squire, K.R., Gee, J.P.: Video games and the future of learning (WCER Working Paper No. 2005-4). University of Wisconsin–Madison, Wisconsin Center for Education Research, Madison (2005)
10. Plass, J., Homer, B., Kinzer, C.: Foundations of game-based learning. Educ. Psychol. **50**(4), 258–283 (2015)
11. Prensky, M.: From Digital Game-Based Learning, McGraw-Hill (2001)
12. Garris, R., Ahlers, R., Driskell, J.E.: Games, motivation, and learning: a research and practice model. Simul. Gaming **33**(4), 441–467 (2002)
13. Chang, Y., Aziz, Z., Zhang, M., Esche, S.K.: Evaluation of a video game adaptation for mechanical engineering educational laboratories. In: 2016 IEEE Frontiers in Education Conference (FIE), pp. 1–6 (2016)
14. Valle, P.H.D., Toda, A.M., Barbosa, E., Maldonado, J.C.: Educational games: a contribution to software testing education. In: 2017 IEEE Frontiers in Education Conference (FIE), pp. 1–8 (2017)
15. Callaghan, M., Savin-Baden, M., McShane, N., Eguiluz, A.: mapping learning and game mechanics for serious games analysis in engineering education. IEEE Trans. Emerg. Top. Comput. **5**(1), 77–83 (2015)
16. Hartmann, A., Gommer, L.: To play or not to play: on the motivational effects of games in engineering education. Eur. J. Eng. Educ. (2019)
17. Vaz de Carvalho, C.: Is game-based learning suitable for engineering education? In: Proceedings of the 2012 IEEE Global Engineering Education Conference (EDUCON) (2012)
18. Darling, J., Drew, B., Joiner, R., Iacovides, I., Gavin, C.: Game-based learning in engineering education. In: International Conference on Innovation, Good Practice and Research in Engineering Education, Loughborough University, England (2008)
19. Anastasiadis, T., Lampropoulos, G., Siakas, K.: Digital game-based learning and serious games in education. Int. J. Adv. Sci. Res. Eng. **4**(12) (2018)
20. Eisenhardt, K.: Building Theories from Case Study Research. The Academy of Management Review, vol. 14, no. 4 (1989)
21. Yin, R.: Case Study Research. Design and Methods, pp. 13–83. Sage, Newbury Park (1989)
22. Wohlin, C., Runeson, P., Höst, M., Ohlsson, M.C., Egnell, B., Wesslen, A.: Experimentation in Software Engineering. An Introduction. Kluwer Academic Publishers, Boston (2000)

About the Effectiveness of Different Game Design Elements for an Introductory Programming Course

Andreas Schwarzmann(✉), Dieter Landes, and Yvonne Sedelmaier

Faculty of Electrical Engineering and Informatics, Coburg University of Applied Sciences and Arts, 96450 Coburg, Germany
{andreas.schwarzmann, dieter.landes, yvonne.sedelmaier}@hs-coburg.de

Abstract. Learning a programming language is a fundamental part of nearly every technical study program. Nevertheless, even students of computer science face challenges before they master sufficient programming skills. Therefore, learning to program computers is tough especially for non-computer science students. A highly promising approach towards alleviating this problem is the integration of game design elements in a non-game context in order to improve user engagement. Gamification comes along with a variety of game elements with different and unclear motivational impact. Consequently, this paper aims at identifying those game elements which tend to be suitable for an introductory programming course. To do so, the potential relationship between game features and intrinsic motivation will be examined first. Subsequently, a collection of game features will be analyzed to meet predefined selection criteria. As a result, the work revealed a set of competence related game features that are likely to fit better to a course for programming than other types of features.

Keywords: Introductory programming · Gamification · Game design elements · Game features · Self-determination theory · Psychological need satisfaction · Intrinsic motivation

1 Introduction

Common academic search engines easily return more than one million results for queries that look for scientific references to problems that students encounter while learning programming, mostly irrespective of the particular language chosen. A closer look at these search results reveals that these difficulties are already around for several decades.

Almost forty years ago, programming has been declared as the new Latin of the curriculum. Similar to Latin, programming was initially considered as a tool to develop minds and improve problem-solving skills. Unfortunately, reports from teachers as well as several empirical studies revealed significant problems that students encounter during introductory programming [8].

M. E. Auer and T. Rüütmann (Eds.): ICL 2020, AISC 1328, pp. 499–509, 2021.
https://doi.org/10.1007/978-3-030-68198-2_46

Some twenty years later, it is still commonly accepted that learning a programming language is very difficult. Furthermore, the Computer Science Education (CSE) community confirms this problem and largely agrees that programming is one aspect that contributes to higher drop-out rates [2]. In order to examine this issue more closely, a study [2] aimed at finding the average failure rate of introductory programming courses around the world. According to this study, the average failure rate is as high as 33%. This figure may only be a vague indicator because only 63 institutions worldwide provided data for this study, while the US alone had 413 universities at that time. In addition, high failure rates were only reported off the record, possibly since universities were embarrassed by their high failure rates [2].

Twelve years later, the same authors decided to repeat the evaluation in order to get an updated and more precise failure rate based on a larger data set. After collecting data from 161 universities this time, the result showed a slightly lower average failure rate of 28%. Yet, this figure needs to be dealt with carefully as there are several threats to validity in the study: for one thing, most participants of the study were already aware of the problem and concerned about the improvement of learning and teaching. Nevertheless, the study concluded that a failure rate of 28% is not alarmingly high especially in relation to poor rates of other subjects [1].

In any case, high failure rates are still a problem due to the enormous demand for computer science specialists these days. Personal experiences indicate that computer science students receive many job offers on platforms like XING or LinkedIn even before they finish their university degree.

A very promising approach towards lowering failure rates in introductory programming is gamification, i.e. the use of game design elements in non-gaming contexts [3]. This concept mainly aims at improving user engagement through the undoubted motivational power of games [7]. Despite the scarcity of empirical studies with respect to the effectiveness of gamification, many authors agree that the proper use of gamification can improve learning [4]. Due to the mainly positive reputation of gamification, the concept will be used to enhance an introductory programming course.

Gamification offers many game design elements with different and unclear motivational impact. Consequently, this paper aims at answering the following question: Which game design elements tend to be suitable for an introductory programming course? To do so, the potential relationship between game features and intrinsic motivation will be examined first. Subsequently, a collection of game features will be analyzed to meet predefined selection criteria.

In the remainder of the paper, Sect. 2 provides introductory information on the concept of gamification. Section 3 uses a concept from motivation research for figuring out how game design elements may address intrinsic motivation. Section 4 presents personal suggestions towards a selection of game design elements for an introductory programming course. Section 4.3 discusses the results of the previous section and derives a tentative set of appropriate game features. The last section summarizes the paper.

2 The Concept of Gamification

This section provides some basics of gamification, in particular a widely used definition, which consists of different terms that will be explained in detail. In addition, several game design elements are reviewed in this section as well.

2.1 Definition of Gamification

The first documented use of the term "gamification" dates back to the year 2008, but the term was unpopular till the second half of 2010. At that time, many terms appeared, which addressed similar or identical concepts. In order to distinguish the existing academic work within the realm of games and learning better, gamification was defined as "the use of game design elements in non-game contexts" [3]. The definition builds upon the four terms (1) "game", (2) "elements", (3) "design", and (4) "non-game contexts".

1. The term **"game"** can be described by four characteristics that all games have in common: a goal that players want to achieve; rules that determine how players can achieve the goal; a feedback system that returns information about the current progress towards achieving the goal; and voluntary participation, which implies that every player deliberately agrees with the goal, the rules, and the feedback [5].
2. The term **"elements"** helps to differentiate gamification from serious games, which are fully developed games serving a distinct non-entertainment purpose. In contrast to this, gamification refers to the explicit use of specific elements of games embedded in real-world contexts [3]. Several of these elements will be introduced subsequently.
3. The term **"design"** helps to distinguish game design elements from game-based technologies. Game-based technologies refer to the underlying technical principles of games such as the used game engine, whereas gamification refers exclusively to a specific design process [3].
4. The term **"non-game contexts"** specifies that the application domain of gamification is not restricted to a certain area. The only context that is excluded by definition is the use of game design elements within game contexts itself [3].

In summary, gamification can be described by characteristics shared by all games: a goal, rules, a feedback system, and voluntary participation. Gamification can be applied universally except within game contexts itself and uses certain game design elements extracted from games without deploying a fully developed game. It refers to a specific game design process and not to underlying game-based technologies.

2.2 Game Design Elements

Game design elements are the fundamental building blocks of gamification. They hold specific characteristics of games that can be used to gamify applications [9]. In the realm of gamification many authors, e.g. [7, 9, 11], presented more or less comprehensive lists of game design elements. These lists differ considerably, although they exhibit some overlap. This is due to the fact that the elements on the list are the authors' subjective choice rather than a comprehensive and objectively evaluated collection of all reasonable elements.

This paper will not present yet another list of elements. Similar to [7], only those game design elements will be considered that are most often discussed in literature.

1. **Points** can be found in a multitude of games and gamified applications. They are earned by completing specific tasks within the game. Moreover, points indicate the players' progress in a numeric way. Points come in different flavors, e.g. experience points, redeemable points, or reputation points, according to distinct purposes. A major benefit of points is that they return an immediate and continuous feedback to the players [7].
2. **Badges** are graphically visualized achievements that can be earned or collected by choice. They do not only confirm a successfully finished task, but also symbolize the effort to unlock the badge. They can be earned by reaching a certain amount of points or by performing specific activities within the game. Achieving a badge can be a fixed goal if the players knows what they need to do in order to unlock it. A badge that was hard to achieve may also serve as a virtual status symbol for the players [7].

Just like points, badges return a feedback to the players. Furthermore, this game element may influence the players' behaviors to select certain routes or challenges. Since badges indicate the membership of a group of those who hold a specific badge, they can exert social pressure on the players [7].

3. **Leaderboards** list players with regard to their relative success towards defined success criteria such as points. The resulting overview helps to quickly identify the best and worst performing players within a certain discipline. It contrasts a player's own performance with the performances of others. Since leaderboards create competition and, consequently, social pressure, they may promote the players' engagement [7].

Nevertheless, leaderboards are known as double-edged swords. If players are close to a better position, leaderboards can increase motivation. Yet, they may be huge demotivators for players that are far behind all other competitors [7].

4. **Performance graphs** can usually be found in simulation games. They hold information about the personal performance over a fixed period of time. In contrast to leaderboards, they do not relate a player's own performance to those of others. Players focus on improvements by visually seeing their individual performance over time. According to motivation theory, performance graphs promote mastery orientation, which has a positive effect towards learning [7].
5. **Storytelling** does not refer to any performance. Storytelling is a mechanism to embed a narrative context into a gamified application in order to give activities and characters a new meaning. Storytelling can alter contexts to be more inspiring and exciting. Narrative contexts may simulate real-world contexts as well [7].
6. **Avatars** graphically represent the players within the gamified environment. The players themselves usually choose or create their favorite avatar. Avatars can be simple pictures or more complex graphics like three-dimensional animations. Their

main purpose is to identify the players and to set them apart from other characters within the game. With avatars players can create or adopt identities and become part of an in-game community [7].

7. **Levels** are defined steps of the players' progress within the game [9]. In roleplay games for example, the next higher level can usually be achieved by earning a certain amount of experience points. The needed amount of experience points for the next level typically depends on the player's individual level.

3 Addressing Motivation with Gamification Features

This section introduces a concept from motivation theory. Within this field of research, various perspectives are defined to analyze the relationship between motivational mechanisms and different gamification features.

3.1 Intrinsic Needs with the Perspective of Self-determination

Game design elements are likely to address certain motivational mechanisms. In motivation theory, these mechanisms are defined within different motivational perspectives, which focus on distinct aspects. These perspectives are the behaviorist learning perspective, the trait perspective, the cognitive perspective, the perspective of interest, the perspective of emotion, and the perspective of self-determination. Each perspective relates to a different approach towards the motivational mechanisms that can be addressed by specific game design elements. The motivational mechanisms can be used to examine the motivational tendency of certain game design elements and should be considered carefully while creating gamification environments [6].

The primary focus in gamification research lies on the perspective of self-determination which postulates that the satisfaction of three intrinsic needs can induce intrinsic motivation. This perspective is particularly popular since intrinsic motivation is generally regarded as the most efficient factor to affect human behavior positively. According to the self-determination theory, autotelic behaviors can be promoted when the gamified environment supports a feeling of autonomy, competence, and relatedness. Therefore, users are likely to be intrinsically motivated to engage within the gamified domain if they experience the fulfillment of such intrinsic needs by interacting with certain gamification features [10].

3.2 From Self-determination Theory to Game Design Elements

Scientific literature in motivation research distinguishes the needs for autonomy, competence, and relatedness [7]. Likewise, the literature related to games and gamification often relies on a distinction between immersion, achievement, and social dimensions [10].

A closer look shows a striking similarity between those concepts. This observation leads to the common assumption that gamification has the ability to address specific intrinsic needs through certain gamification features. This approach helps to conceptualize the potential relationship between intrinsic need satisfaction and gamification. Furthermore, it also enables the distinction between immersion, achievement, and social-related game design elements. Figure 1 summarizes the relationship between the intrinsic needs of self-determination theory, characteristics of gamification features, and finally their associated game design elements.

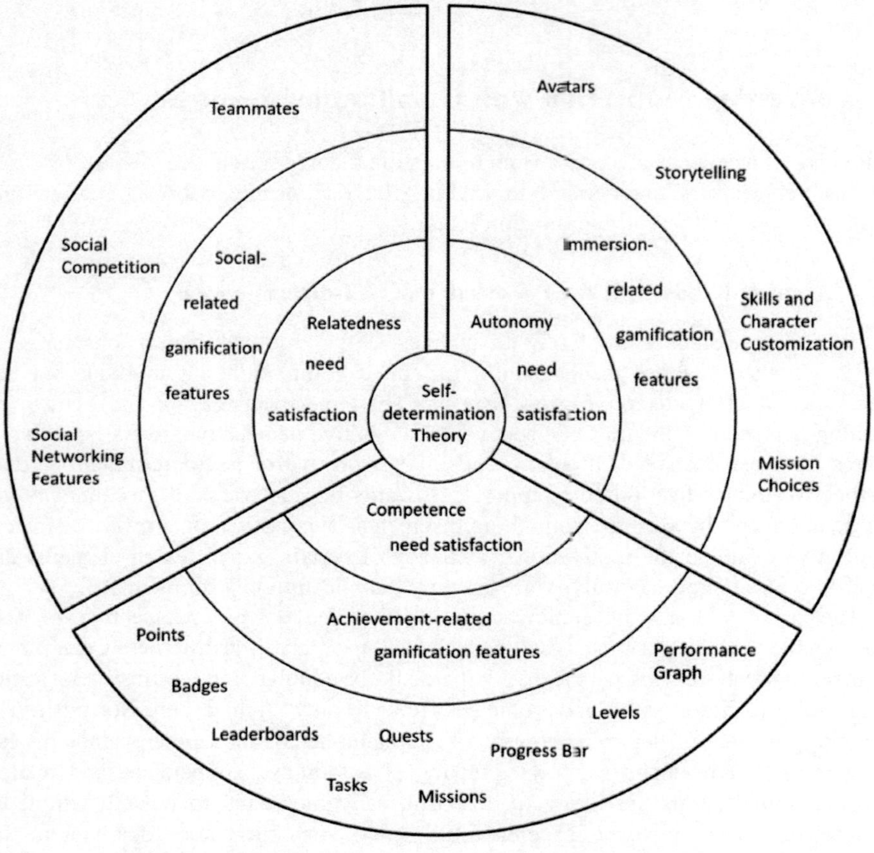

Fig. 1. Matching intrinsic needs with gamification features

3.3 Satisfying Needs with Gamification in Practice

Although the relationship between intrinsic needs and gamification features seems to be conclusive, there is a serious lack of systematic (empirical) evaluation that confirms the theoretical concept in general. Furthermore, existing empirical studies in this research area focus on their own restricted set of gamification features, which address only a subset of intrinsic needs. Consequently, the present state of empirical research represents just a fraction of the whole conception. This fraction, however, reports mainly positive effects on intrinsic motivation. Table 1 provides an overview of existing evaluations towards need satisfaction with gamification [10].

Due to the lack of evaluation that covers the full spectrum of intrinsic needs and features of gamification, Xi and Hamari's recent study [10] deserves closer attention. The purpose of this work was to confirm the relationship between intrinsic needs (autonomy, competence, relatedness) and gamification features (immersion, achievement and social-related) in particular. According to this study, immersion-related features only affected autonomy need satisfaction positively. Achievement-related features addressed all three kinds of need satisfaction, but predominantly autonomy and competence. Social-related features had a positive effect on all needs as well, but the main impact was on relatedness. Therefore, the findings of this study indicate a relationship between intrinsic needs and game features. Moreover, features of one category can have positive effects on multiple needs as well [10].

Table 1. Empirical studies towards need satisfaction with gamification [10]

Source	Features	Results	Motivation
Bormann and Greitemeyer (2015)	Storytelling	Positive	Autonomy, competence, Relatedness
Hanus and Fox (2015)	Badges, leaderboard	Negative	Intrinsic motivation
Kim et al. (2015)	Character customization	Positive	Autonomy
Peng et al. (2012)	Mission choices, skills & character customization	Positive	Autonomy
Peng et al. (2012)	Badges, dynamic difficulty, heroism meter	Positive	Competence
Roy and Zaman (2018)	Badges, social competition, challenges	Positive	Autonomy, competence, relatedness
Sailer et al. (2017)	Badges, leaderboard, performance graph	Mixed	Autonomy, competence
Sailer et al. (2017)	Avatar, storytelling, teammates	Mixed	Relatedness
Thom, Millen and DiMicco (2012)	Badges, points	Positive	Intrinsic motivation

4 Suitable Gamification Features for a Programming Course

This chapter reviews the choice of game design elements for an introductory programming course deployed in Moodle, a widespread free open-source learning management system (LMS).

4.1 Evaluation Methodology

As stated in the introduction, this paper aims at answering the following question: Which game design elements tend to be suitable for an introductory programming course? In order to answer this question, it needs to be clarified what "suitable" actually means with respect to the given platform and programming context. To do so, game elements need to match predefined selection criteria with different priorities to be suitable for the course.

This includes a high priority criterion that encompasses all the technical requirements for an implementation. Moodle already incorporates some fundamental principles of gamification innately although the platform was not explicitly designed for that purpose. Nevetheless, the possibilities to include additional (gamification) elements directly in th LMS are very limited, but can only be realized through plug-ins. Moreover, this highly prioritized criterion excludes game features unconditionally if they cannot be implemented technically.

The evaluation also includes a criterion with medium priority that filters all game elements that affect competence need satisfaction. Since the course for introductory programming follows a competence-oriented approach, e.g. the ability to develop own programs autonomously instead of memorizing specific source code without understanding what they mean, competence-oriented game design elements are preferred. This criterion may exclude features that may be implemented technically, but do not address the satisfaction of competence needs directly. It should be noted that non-competence-oriented features could affect the satisfaction of competence needs after all.

Additionally, a low prioritized criterion is included that targets elements with already conceivable use cases within the programming course. This criterion excludes features that do not have a meaningful use within the course until now, i.e. the feature could technically be implemented and addresses competence needs as well, but it is still unclear how these elements might be integrated in the course reasonably.

4.2 Analysis of Game Design Elements

The game design elements considered for the evaluation are presented in Fig. 1. This compilation represents just a subset of all possible features. Furthermore, an overview of the described selection criteria is given in Fig. 2.

Figure 2 illustrates the continuous process of identifying suitable elements for the course. In the first pass, the elements are analyzed to meet the technical requirements for an implementation in Moodle. Within this pass, all social-related game features got already excluded due to the fact that there is no technical support for them. There are also no plugins that could provide these game features explicitly.

The second pass of the evaluation excluded all remaining non-competence-oriented gamification features. Therefore, all immersion-related gamification features are discarded, although there may be game elements involved that affect satisfaction of competence needs indirectly (see Sect. 3.3). Nevertheless, for this evaluation, only game elements were considered that address the satisfaction of competence needs directly.

The last pass of the evaluation excluded all game features with an unclear course integration. Tasks, quests, and missions could be implemented as special assignments within the course, but it is still unclear when and how to do this and what the benefit really would be. The same applies to the performance graph that could be applied but it is still unclear to which performance the graph could refer precisely.

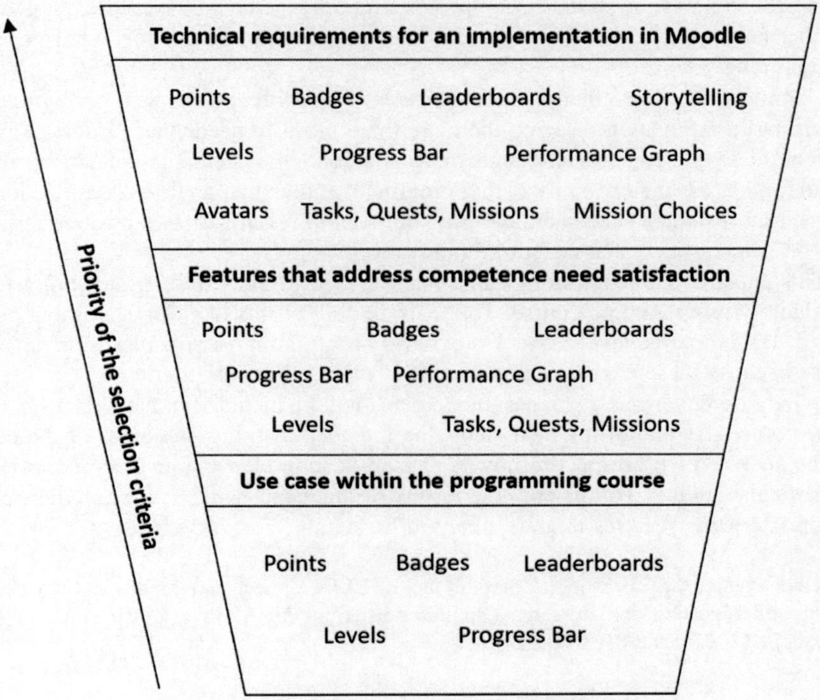

Fig. 2. Selection process for suitable game design elements

4.3 Discussion of the Analysis

In Sect. 4, various game design elements have been analyzed towards their suitability within an introductory programming course in Moodle.

The evaluation indicates that only a small selection of gamification elements fulfils all necessary criteria, namely technical feasibility, competence-orientation, and availability of a reasonable usage scenario. In particular, only points, badges, leaderboards, levels, and progress bars are suitable features for an introductory programming course since they match all selection criteria reasonably well.

5 Summary

Learning to program is a widely known problem in many technical study programs. Reasons for this may be found in a considerable technical complexity, but also in a lack of (intrinsic) motivation to get involved in the subject. Gamification is a promising approach for tackling the latter issue. This concept may transfer game characteristics to real-world contexts in order to improve user engagement. Therefore, a gamified learning environment could potentially increase students' motivation with respect to programming.

Gamification offers many game features with different impact on motivation. According to motivation research there are three intrinsic needs that can be addressed by game design elements. Although there is a lack of systematic evaluations in this research area, the majority of existing empirical studies that analyzed the relationship between intrinsic need satisfaction with gamification features shows positive results at least on some aspects that constitute motivation.

This paper examined various game features with respect to their suitability for an introductory programming course. The main finding of this work indicates that points, badges, leaderboards, levels, and a progress bar are the most suitable features for the course because these elements match each of the used selection criteria.

In the future, various game design elements will be included in the didactic concept of the course. As one of the next steps, the intended learning outcomes of the course will be analyzed with respect to how well gamification elements may support them in the particular setting. To this end, the results of this paper will serve as a guidance for appropriate game features that are likely to be suitable for the course.

Acknowledgement. This work is part of the EVELIN project and funded by the German Ministry of Education and Research (Bundesministerium für Bildung und Forschung) under grants 01PL12022A and 01PL17022A.

References

1. Bennedsen, J., Caspersen, M.: Inroads: paving the way towards excellence in computing education. Commun. ACM **10**(2), 30–35 (2019)
2. Bennedsen, J., Caspersen, M.E.: Failure rates in introductory programming. SIGCSE Bull. **39**(2), 32 (2007)
3. Deterding, S., Dixon, D., Khaled, R., Nacke, L.: From game design elements to gamefulness: defining "gamification". In: Proceedings of the 15th International Academic MindTrek Conference: Envisioning Future Media Environments. MindTrek 2011. Association for Computing Machinery, New York, NY, USA, pp. 9–15 (2011). https://doi.org/10.1145/2181037.2181040.
4. Dicheva, D., Dichev, C., Agre, G., Angelova, G.: Gamification in education: a systematic mapping study. Educ. Technol. Soc. **18**, 75–88 (2015)
5. McGonigal, J.: Reality is Broken: Why Games Make us Better and How They Can Change the World. Penguin Press, New York (2011)
6. Sailer, M., Hense, J., Mandl, H., Klevers, M.: Psychological perspectives on motivation through gamification. Inter. Des. Archit. J. **19**, 18–37 (2013)

7. Sailer, M., Hense, J.U., Mayr, S.K., Mandl, H.: How gamification motivates: an experimental study of the effects of specific game design elements on psychological need satisfaction. Comput. Hum. Behav. **69**, 371–380 (2017)
8. Sleeman, D.: The challenges of teaching computer programming. Commun. ACM **29**(9), 840–841 (1986)
9. Werbach, K., Hunter, D.: For the Win. How Game Thinking Can Revolutionize Your Business. Wharton Digital Press, Chicago (2012)
10. Xi, N., Hamari, J.: Does gamification satisfy needs? A study on the relationship between gamification features and intrinsic need satisfaction. Int. J. Inf. Manage. **46**, 210–221 (2019)
11. Zichermann, G., Cunningham, C.: Gamification by Design: Implementing Game Mechanics in Web and Mobile Apps. O'Reilly Media (2011)

Development and Implementation of Gamified Technologies in the Life Long Learning System (Based on a Multidisciplinary University)

Anastasia Tabolina[1(✉)], Sergey Salkutsan[2], Pavel Kozlovskii[3], Dmitrii Popov[4], Olga Kunina[1], Inna Yudina[5], and Anastasia Ryushenkova[6]

[1] Institute of Humanities, Department of Engineering Education and Psychology, Peter the Great St. Petersburg Polytechnic University, St. Petersburg, Russia
asyatabolina@gmail.com, okun96@mail.ru

[2] "SPbPU Advanced Manufacturing Technologies Center, National Technology Initiative" for Education, Peter the Great St. Petersburg Polytechnic University, St. Petersburg, Russia
salkutsan@spbstu.ru

[3] Advanced Manufacturing Technologies Center (National Technology Initiative), Scientific Laboratory of Strategic Development of Engineering Markets, Peter the Great St. Petersburg Polytechnic University, St. Petersburg, Russia
pavelspb@mail.ru

[4] Institute of Humanities, Higher School of Mediacommunications and PR, Peter the Great St. Petersburg Polytechnic University, St. Petersburg, Russia
popov_dg@spbstu.ru

[5] Research Laboratory "Systems of Data Streaming Processing", Peter the Great St. Petersburg Polytechnic University, St. Petersburg, Russia
inna.yudina@spbpu.com

[6] Career Center and Pre-university Training, Peter the Great St. Petersburg Polytechnic University, St. Petersburg, Russia
ryushenkova_ai@spbstu.ru

Abstract. Active use of gamification in training, development and socialization has established itself among the best educational and managerial practices. Specialists note that this phenomenon, apart from being a practical implementation of theoretical educational technologies, also reflects a change in the approach to socialization of a new generation of students and specialists. The key objective of the educational process is to shape skills and competence of students as they prepare for their future professional activities. The potential of gamification is primarily based on the basic need of individuals for development. In the conditions of transition to the economy of knowledge and even of entertainment, educational institutions are facing increasingly more requirements in the field of developing Life Long Learning technologies. In response to this challenge, universities start creating educational programs that combine gamification opportunities and requirements to continuity of the educational process for target audiences.

M. E. Auer and T. Rüütmann (Eds.): ICL 2020, AISC 1328, pp. 510–516, 2021.
https://doi.org/10.1007/978-3-030-68198-2_47

Keywords: Gamification · Life Long Learning · Multidisciplinary university

1 Context

The number of educational services that use gamification in the training process keeps growing year on year. Gamification of education is one of the best ways to use new technologies in transferring knowledge and skills more effectively.

Gamification of education implies shaping a community whose members engage in mutual help, challenge, and motivation. Gamification has become an inalienable trend in development of the information society; it encompasses all spheres of the economy, governance, science, education, and social development.

Johan Huizinga postulated that games often transcend the borders of purely biological or physical activities. He viewed gaming as a function rich in meaning.

Eric Berne considered games to be a sequence of meta-level hidden interactions that lead to a definite predictable outcome [1].

N. Almazova wrote that "a game is a reasonable and goal-oriented, systematic, socially coordinated system of behavior that is subject to certain rules" [2, 4, 8].

According to M. Olennikova, the game, due to its inherent characteristics, simultaneously puts a player in several positions being a two-dimensional process. In a game, a person finds himself or herself concurrently in two dimensions: real and conventional (player) space; this situation creates unique opportunities for development [6, 7].

P. Kozlovsky notes that today the use of gamification in university education is a promising area of focus. Gamification should be viewed as a tool for increasing the effectiveness of the educational process within the Life-Long Learning System [9, 10].

D. Popov writes that the question of role and place of gamification technologies in the educational process has long progressed from scientific discussion to practical implementation. Games are recognized as one of the most effective ways of presenting information in an accessible format. Therefore, efficient use of gamification technologies is one of the drivers of innovation in the educational system [9].

According to T. Baranova, gamification elements introduced into learning materials and training process improve interactivity of education and thus help implement and support new types of academic activities [3, 5].

E. Gulk notes that the main areas of focus of the Life-Long Learning System at a technical university are: humanization and intellectualization of popular professions, digitization of information databases and systems, customization of educational products, introduction of gaming technologies (gamification) as a tool for professional testing, arrangement of cultural awareness raising and developing classes, arrangement of scientific research in FabLabs, facilitating personal growth of young scholars as a part of Factories of the Future, and introducing active dialog forms of learning [3, 10].

In the authors' opinion, use of gamification technologies as an interactivity tool in the educational process is currently one of the most promising areas of improvement of the student training system.

2 Purpose

Purpose: to develop and implement gamification technologies in the Life Long Learning system (on the basis of a multidisciplinary university).

To solve the tasks, the following research methods and activities were used:

1. Theoretical - analysis of philosophical, sociological, psychological, educational, scientific and methodological literature on the topic under study; analysis of the curriculum content to determine the organizational and meaningful possibilities of using social and humanitarian design in the technological education of students;
2. Experimental - pedagogical experiment, experimental work, questioning, observation, testing, expert assessment method, modeling and development of teaching materials, mathematical methods for interpreting data from a pedagogical experiment, ascertaining and teaching experiment, design training seminars, organizational activity games, psychological and pedagogical testing, conversation, questioning, Socio-engineering technologies, game-, role-playing and training technologies, development of game simulators that immerse students into real activities through modeling of situations, arrangement of small- and large-scale educational projects.

The research-and-experiment approach is based on insider observation of experimental and associated projects, including creation of engagement concepts and scenarios, development and testing of educational games and their trial integration into the system of educational processes.

Research objectives:

1. A review of literature on the topic of gamification in education
2. Development of key stages of remote gamification (with SPbPU as a case study)
3. Definition of main elements of the project activity
4. Recommendations for introducing gamification technologies into the Life-Long Learning (LLL) system
5. Summary.

3 Approach

3.1 Stages of Gamification Technology Implementation

Development and integration of gamification technologies into the Life-Long Learning System (LLL) in a multidisciplinary university helps create an information-based learning environment that encourages students to actively and independently pursue knowledge acquisition, develop their professional skills, such as critical thinking, decision-making, teamwork and cooperative skills. Therefore, gamification helps uncover creative abilities in students and motivates them to pursue self-education. Gamification is an effective tool for improving the professional Life-Long Learning (LLL) in a higher education institution, as it improves the quality of learning and involves students into the educational process as it becomes more appealing, provides for systemic, continuous, and profound study of the discipline, builds a stronger personality through the experience of mistakes and failures, refines behavior, and forms systemic knowledge, skills and competencies applicable in everyday life. Training through gamification becomes innovative, agile, personalized, and autonomous. It provides fair assessment of achievements and motivation, develops communication skills, fosters the students' independence and proactivity, and facilitates their self-development.

The main stages of remote gamification (with SPbPU as a case study) are presented in Table 1:

Table 1. Main stages of gamification technology implementation

Introduction	Game design
Main part	Selection of participants, distribution of roles: 1. Participants use the quest game model to autonomously work through the academic materials and develop new knowledge, skills and competencies 2. Moderator/instructor becomes a consultant, organizer and coordinator of educational, developmental, problem-solving and research activities of learners 3. The moderator creates conditions for independent intellectual and creative activities of students and supports their initiative
Projecting part	Obtaining instructions, fulfilment of assignments. The gamification process at SPbPU uses team activities that are crucial for achieving the game objectives. As a result, the participants become a team, learn to share responsibilities, and, in ideal models, show the skills necessary for their specialization
Reflective part	Summary. Gamification technologies help develop soft digital skills in a game format, give students a chance to manage the pace, time, route and place of learning. They help develop self-regulation, planning and control skills, distribution of responsibilities and assessment of performance in group and individual activities

3.2 Project Training Activities

We use project training methods as the organizational form of remote work. The project activities are aimed at creating subjectively and objectively new phenomena of the surrounding reality which differ from those already known by their attributes and characteristics. SPbPU has developed a course entitled "Introduction to Dramaturgy and Creation of Media Content".

Objective: to teach students how to create scripts for remote training programs.

Task: to create a gamified version of the course in Philosophy at SPbPU, write a script and come up with the framework concept in six key topics. Tools: online remote education technologies and traditional offline technologies, such as:

1. Information and communication technology
2. Critical thinking development technology
3. Project technology

4. Developing education technology
5. Health management technologies
6. Gaming technology
7. Modular technology
8. Workshop technology
9. Case technology
10. Integrated learning technology
11. Cooperation pedagogy
12. Level differentiation technologies
13. Group technologies.

The main elements of project activities are as follows:

1. Didactic orienteering – choice of a topic, analysis of significance of the problem, definition of the type of project, study of the period, historical, cultural and architectural aspects, coming up with the name, studying the rules of project activities, verifying the number of project participants and allocating necessary resources;
2. Concept planning – development of the methodological tools for the research, formulating goals and objectives, creating a hypothesis, defining the necessary result, planning of presentation and reporting forms, distribution of responsibilities among the participants, creation of the information base, graphic representation of materials (Adobe Photoshop, Illustrator, 3S Max);
3. Instrumental design – analysis of project implementation tools, ways of obtaining information, peculiarities of interpretation and presenting the results;
4. Practical implementation – implementation of the project, execution of the planned types of activities, implementation of interaction, personification of the epoch (character, epoch, cultural and interior context);
5. Productive reflection – assessment of the process of project execution and obtained results, investigating the causes of deviations, analysis of each participant's contribution, checking if conclusions are well-justified;
6. Preparation of the presentation and defense of the project.

Instruction: two Polytech students find themselves in the past. All characters must complete the quest on their own. They have to fulfil a task of the epoch, meet a character from that epoch, and get introduced to the philosopher.

The students were tasked with developing unique content which would engage the participants in completing their quest. Detailed graphic design and thorough impersonation were deemed to be important. The material is mastered via identification with the character and reflection on the choice. Vivid graphic context improves retention of the material. Gamification technologies are important. Through developing a game, students become involved in the educational process more effectively. This generation is used to playing, so it is important to teach them to play while learning and developing. Transition from the role of the observer to the role of the creator.

Roles:

1. Observer – a passive figure with pronounced extrinsic motivation ("must" position);
2. Researcher – intrinsic motivation ("want" position);

3. Creator – top level of engagement (“transform” position).

Remote forms of teaching Philosophy using the student-authored gamification technologies increase the quality of resultant knowledge by 27%. Remote teaching methods through gamification help design professional competencies of the learners and incorporate emotional aspects into training activities.

4 Conclusions

A number of conclusions can be drawn from successful application of gamification technologies in the educational process of St. Petersburg Polytechnic University.

Firstly, development of the game, communication with proactive, creative students favorably affects educational motivation, stimulates reflection, increases satisfaction with the process and the results of students' work, builds a sense of community, stimulates creative thinking, drives the learners' progress, activates professional and personal self-development.

Secondly, educational courses and disciplines, as well as educational materials created in the game format or amplifying the game world, not only contribute to acquisition of certain knowledge by target audiences, but also have a great potential in terms of scientific communication, scientific volunteering and complementary professional education.

Thirdly, gamification elements and game-based engagement methods improve the extent of academic materials retention and reproduction during testing. Gamification as an independent or complementary academic subject is of interest to students specializing in pedagogy, public relations, and as a basis for complementary education programs.

Fourthly, in the current conditions gamification can be used not only as a way or tool of training and testing but also as an independent educational discipline which focuses on creating such educational content. This process can involve students working on specialization-based course projects, as well as teachers taking their further training and proficiency improvement courses.

References

1. Berne, E.: Games People Play – The Basic Hand Book of Transactional Analysis. Ballantine Books, New York (1964)
2. Baranova, T., Almazova, N., Tabolina, A., Kunina O., Yudina, I.: The study on psychological constitutions of comprehensive university students with different levels of academic procrastination. In: Anikina, Z., (eds) Integrating Engineering Education and Humanities for Global Intercultural Perspectives. IEEHGIP 2020. Lecture Notes in Networks and Systems, vol. 131 (2020)
3. Baranova, T.A., Gulk, E.B., Tabolina, A.V., Zakharov, K.P.: Significance of psychological and pedagogical training in developing professional competence of engineers. In: Auer, M., Tsiatsos, T., (eds) The Challenges of the Digital Transformation in Education. ICL 2018. Advances in Intelligent Systems and Computing, vol. 917 (2019)

4. Khalyapina, L.P., Baranova, T.A., Almazova, N.I.: Models of interdisciplinary coordination in higher education area of Russia. In: International Conference on Education, Research and innovation. – 16–18.11.2017. Saville, Spain. – C.0780-0785
5. Kruglikov, V.N., Kasyanik, P.M.: The role of active learning in the concept of global engineering education. Scientific and technical statements of the St. Petersburg State Polytechnic University. Hum. Soc. Sci. **3** (227), 159–168 (2015)
6. Olennikova, M.V., Tabolina, A.V.: Psycho-pedagogical support of students project activities in multi-functional production laboratories (Fab Lab) on the basis of technical university. Adv. Intell. Syst. Comput. **917**, 732–740 (2019)
7. Posokhova, S.T., Olennikova, M.V., Tabolina, A.V., Khalyapina, L.P.: Professional self-concept of students future psychologists. In: The European Proceedings of Social & Behavioural Sciences EpSBS 19th PCSF Professional Culture of the Specialist of the Futur 28–29 November 2019, pp. 926–933 (2019)
8. Rubtsova, A.V., Almazova, N.I.: Productive model of foreign languages learning: realities and prospects. international conference communicative strategies of information society (CSIS 2018). Adv. Soc. Sci. Educ. Hum. Res. **289**, 319–324 (2018). https://doi.org/10.2991/csis-18.2019.65
9. Tabolina, A., Kozlovskii, P., Popov, D., Yudina, I., Snegirev, N., Tretyakov, D.: Socio-psychological program for the selection of students in the adapters public institute. In: Anikina Z. (eds) Integrating Engineering Education and Humanities for Global Intercultural Perspectives. IEEHGIP 2020. Lecture Notes in Networks and Systems, vol, 131 (2020)
10. Tabolina, A.V., Olennikova, M.V., Tikhonov, D.V., Kozlovskii, P., Baranova, T.A., Gulk, E.B.: Project activities in technical institutes as a mean of preparing students for life and professional self-determination. In: Auer, M., Hortsch, H., Sethakul, P., (eds) The Impact of the 4th Industrial Revolution on Engineering Education. ICL 2019. Advances in Intelligent Systems and Computing, vol. 1134 (2020)

Teaching Environmental Entrepreneurship to Students with a Serious Gaming Approach

Development, Implementation, and Evaluation of *Bikorama@TUM*

Ferdinand Shuyu Xiong[1(✉)], Florian Bajraktari[1], Benedikt Pirmin Ströbl[1], Victor Karl Möslein[1], Nilüfer Faizan[1], Robert Heininger[1], Matthias Christoph Utesch[1,2], and Helmut Krcmar[1]

[1] Technical University of Munich (TUM), Arcisstraße 21, 80333 Munich, Germany
ferdinand@xiong.de

[2] Staatliche FOS BOS Technik München, Orleansstraße 44, 81667 Munich, Germany

Abstract. As the climate debate engages ever more high school students to take to the streets, the objective of this research is to combine environmentally conscious thinking with entrepreneurial ambition. By teaching business concepts in a way that is both captivating and immediately applicable, the serious gaming approach is implemented. The game developed during this project follows the story of a young student who initially rents his own bike and later identifies a business opportunity for building a large bike rental company. This paper explains the game and the corresponding lesson, arguing that positive learning outcomes can be achieved through a playful IT tool.

Keywords: Green mobility · Pedagogy · Serious gaming · Entrepreneurship

1 Introduction

The aim of the game *Bikorama@TUM* is to teach the fundamentals of entrepreneurship and business mathematics, e.g. to students of the upper vocational school in Bavaria (FOS, BOS). At the same time, it promotes the development of ecologically sustainable businesses in the sector of green mobility. The importance of conveying these skills to young people has become ever more important in a world of uncertainty [1], especially considering the challenges humanity will face over the course of the next decades, like climate change [2]. In order to tackle these challenges, there is a need for inventive new ideas and businesses. This requires early training as an entrepreneur, since dealing with the topic at an early stage increases the likelihood of starting a company later [3]. At the same time, students must be made aware of the importance of sustainable companies that take the environment into account when making their business decisions [4]. With its combination of educational purposes and playful presentation, *Bikorama@TUM*

M. E. Auer and T. Rüütmann (Eds.): ICL 2020, AISC 1328, pp. 517–529, 2021.
https://doi.org/10.1007/978-3-030-68198-2_48

therefore fits squarely into serious gaming and is designed for 14- to 20-year-old students. It is part of a lesson thoroughly structured according to pedagogical concepts.

The goal of serious gaming in general is to achieve better learning outcomes by taking a playful approach [6]. At the same time, games shall strengthen the students' capabilities in areas such as teamwork and personal study skills [5]. Therefore, by using *Bikorama@TUM* at the FOS and BOS, the game is meant to prepare the school students for university. However, the Bavarian upper vocational schools are particularly well suited for projects such as *Bikorama@TUM*, since they place a strong focus on practical application and are less focused on teaching theory [5]. In addition, the serious gaming concept is not only aimed at the students who ultimately play the game, but also takes into account the goals of teachers and scholars. Teachers gain a new role in being "a supporting entity, rather than [a] lecturer" [7] and scholars are given the opportunity to connect theory to practice [7]. In addition to the basic possibility of combining theory and practice, computer games that adapt to the player make it possible to adapt the learning experience of the students to their individual skills. This positive impact of personalization on student motivation is another benefit of serious games [8, 9]. Research shows that the effects of serious gaming can consequently far exceed the realm of school or university, and in fact be pivotal for the development of personal capabilities and contribute to the vision of life-long learning [1]. An example is self-regulated learning, whereby "students select and use self-regulated learning strategies to achieve desired academic outcomes on the basis of feedback about learning effectiveness and skill" [10]. Thus, the concept of self-regulated learning is especially well epitomized by serious games [10, 11]. Students have to master tasks like selecting a team partner, setting the right goals, and choosing the smartest strategy by themselves under the supervision of the teacher. Even the methods for solving complex problem statements have to be applied independently. But, there are other key capabilities which are required for success at the university and beyond, such as initiative, intrinsic motivation, and personal responsibility, which could be acquired by the students via *Bikorama@TUM*. Because new developments in the educational system are rare, technology makes it possible to "engage students in deep, meaningful, realistic, and relevant problems, the kinds of complex collaborative problems that education reformers have been clamoring for many years" [12]. Another key benefit of this playful approach to learning is that it creates a dynamic within the classroom that bolsters the cooperation among students, which creates interaction and leads to students contributing to each other's game progress [13]. This approach, that was developed by Aronson in the jigsaw experiment, is pronounced distinctively in games like *Bikorama@TUM*, as collaboration between participants is allowed and enabled by the supervisors [6]. At the same time, the student teams compete with their colleagues and this competition is actively reinforced by comparing the game parameters. The competitive nature helps to improve the motivation of the students and to participate enthusiastically in the class. In addition, this leads to better learning outcomes [14].

Previous literature specifically mentioned that metacognitive skills such as the capabilities required for successful teamwork are seen as the key assets that are strengthened through serious gaming [5, 15]. Since these results are difficult to quantify, another work also measured whether serious games have an impact on the final grade of the school students at German upper vocational schools. The results show that the overall grades of the students participating in the project were actually better than those of the control group [5]. With this in mind, the main goal of this research is to evaluate the learning outcomes of *Bikorama@TUM*. To achieve this, the following research questions are posed:

RQ_1) To what extend does *Bikorama@TUM* achieve the learning objectives designed according to Bloom's Revised Taxonomy?
RQ_2) To what extend does *Bikorama@TUM* improve the students' understanding of the fundamentals of economics and entrepreneurship covered in the lesson?

To answer the research questions, preliminary and follow-up tests were carried out to assess whether the game and the lessons had a positive impact on the knowledge of the students [16]. This paper gives a detailed explanation of both the game and the lesson of which it is a part in Sects. 2 and 3, and about the methodology of the test setting in Sect. 4. Section 5 presents the results of the test. Section 6 concludes with comments on the data collected during the lesson as well as possible future uses of *Bikorama@TUM*.

2 Game Description

According to [17], a game is a structured form of play, usually undertaken for enjoyment and sometimes used as an educational tool. *Bikorama@TUM* has a clearly structured storyline that consists of four different stages as stated in the learning objectives part. Whilst interacting with the software, the player acts within a fictional and playful world that transmits the joyful emotions of a game in the definition of [17]. *Bikorama@TUM* has been designed to create an enjoyable game experience, but at the same time, its main purpose is to serve as an educational tool that helps the student to achieve the learning objectives. A simulation, on the other hand, is defined as an attempt to model a real-life or hypothetical situation on a computer so that it can be studied to see how the system works [18]. By changing variables in the simulation, predictions may be made about the behavior of the system. In *Bikorama@TUM* the real-life scenario of founding a company is modelled and the player changes variables to modify the outcome and behavior of the economic system. While observing the impact of the decisions made by the player, the game allows for investigating the behavior of the economic system in which *Bikorama@TUM* is acting. Based on that, *Bikorama@TUM* is a game that has similarities with a computer-based discrete system simulation. For the purpose of assessing the students learning success, it is described as a game and is based on a simulated storyline that is explained in the next section.

2.1 Key Game Options and Objectives

The story of *Bikorama@TUM* begins with an environmentally conscious student who owns a bike and was asked by a friend to lend it to him. As a thank you, he receives some money from the friend and comes up with the idea to rent the bike regularly for money while using the subway himself. With the income generated in this way, further bicycles are purchased. Building on this story, *Bikorama@TUM* teaches entrepreneurship and the commitment to environmental sustainability to school students. Consequently, the two key metrics that the player should maximize are money and CO_2 savings. At the end of the game, these two indicators are compared between all players to determine the final winner in both categories. Therefore, the game is designed in a way that demonstrates that it is possible to perform well in both of these categories, without sacrificing one or the other. However, various strategies can be followed in order to reach maximum values. The strategies manifest themselves mainly in the choice of how many bikes to purchase at which stage of the game and from where to buy these bikes.

Four categories of bikes exist in the game, which differ in price, CO_2 savings per rented day, demand, and additional wear off per rented day. The categories are road bikes, e-bikes, mountain bikes, and city bikes. For each day a bike is rented, the additional wear is added to the current status of the bike. Once a degree of wear of 100% is reached, the bike can no longer be used for rentals and disappears from the player´s assets. Since there is a specific demand for each bike category, better bikes can be offered for a higher rental charge and thereby generate more revenue. At the same time, those bikes naturally have higher purchase prices. At various specific points in the game, those bikes can be bought at two facilities. The bicycle shop offers new bikes, while the recycling depot offers bikes at half the price of the shop and with a random wear degree between 30% and 70%. Consequently, there is a risk associated with buying bikes at the recycling depot, since they may not be available to generate revenue and CO_2 savings for a long time. However, there is also the chance of benefiting by paying less than the market value for a bicycle that is still in fairly good condition.

Another important decision that the player has to make at the start of the game is choosing a suitable warehouse. Each option has a different added demand and monthly rental price. Since the demand drives the price that can be charged for rentals, which in turn drives the revenue, it is necessary for a successful game strategy to create and satisfy high demand. The same concept also applies to the choice of which bike lane to build later in the game. Again, the player needs to focus on maximizing the demand while keeping in mind whether he can fulfill the demand at all.

Selecting one out of three possible employees also requires conscious entrepreneurial decision-making. Depending on whether the company's biggest challenge is that the bicycles are subject to high wear or that the bicycles are not generating enough money, a specialist in mechanics or a sales professional should be employed. The third option is an all-rounder who has some expertise in both of these areas. Each candidate has a different monthly cost that should also be considered when making the hiring decision.

Overall, it is important to state that there is not one optimal game strategy but rather a manifold of possible ways to reach a very good result. Thereby, the game reflects the real circumstances that entrepreneurs are exposed to, such as uncertainty, risk, lack of human capital and the need for some luck to be successful.

2.2 Theoretical Content Taught in the Game

The overarching context of *Bikorama@TUM* and the accompanying lesson are to teach theoretical concepts of entrepreneurship and financial mathematics. The knowledge acquired must be applied directly in the game to achieve a good score, which creates an incentive for participants to learn and understand the content. Following the general theme of the game, the presented concepts are those necessary when starting a company and help to convey a more profound picture of economic coherences.

In stage I, the **Survival Stage**, the theory of supply and demand and reservation price are presented and have to be applied when making the decision of how much to charge for each bike depending on its specific demand and the maximum willingness to pay of the customers. The opportunity costs also play a role here, since they make it possible to calculate the best alternative when deciding on how many days the subway should be taken in order to make money by offering the bike for rentals.

In stage II, the **Growth Stage**, the balance sheet is introduced as well as the differences between variable and fixed costs. This is necessary to understand that some costs, like for example for the storage facility, will always arise each month, independent of whether any bikes are actually rented or not. The balance sheet is something that all entrepreneurs need to know of, since it forms the basic administrative structure underlying all assets of a company.

In stage III, the **Expansion Stage**, concepts that become increasingly relevant with a growing size of a company are being introduced. These include loans and interest rates, which the bank will offer depending on the trust it can put in an entrepreneur to be able to pay it back. The break-even point method can be used to calculate the exact date on which an investment paid off. In the game, the loan and its interest rate are used as a cost function, while the number of bicycles available and the rental price determined indicate the revenue function. In this way, students can calculate when and whether it makes sense to take out a loan at the interest rate offered by the bank. In addition, debt financing and equity are introduced in connection with the bank loan and the investor wishing to acquire a stake in the company. Finally, market capitalization and the initial public offering are highlighted as outlooks and to convey a fundamental knowledge on capital markets.

It was an additional focus to introduce the students into the world of business and finance using commonly used vocabulary and English terms. While the main content and process of the game is in German, some parts are purposely implemented in English in order to increase the students′ sensibility in a globalized business world that largely depends on English as the predominant language.

3 The Learning Objectives

According to the research by [19] and [20], the structure of the lesson is documented in the didactic framework of *Bikorama@TUM*, shown in Fig. 1. It equals a process flow of three repetitive stages (see Fig. 2) similarly to the didactic framework introduced in [5]. To learn the process of entrepreneurship and company development, it is recommended that students learn the underlying basic concepts of finance as well as market mechanisms. The meaning of these concepts is part of almost every introductory lecture in economics, e.g. in [21], a compulsory lecture for every student of economics at a technical university in Europe. Only with this basic knowledge can the best results be achieved in the game, whereby various milestones for learning progress must be reached, which are formulated as Learning Objectives (LO). To avoid an overflow of theoretical content, the LOs are grouped by stages of company growth [22]. Hence, the three learning stages Survival Stage (I), Growth Stage (II), and Expansion Stage (III) are conducted. Every stage consists of three steps A, B, and C and occasionally additional content.

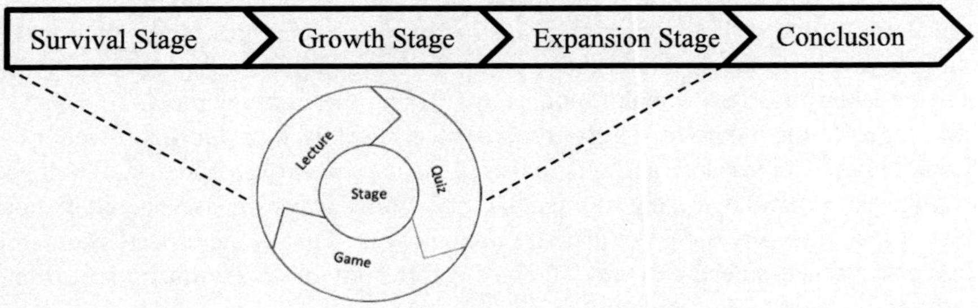

Fig. 1. Didactic framework of *Bikorama@TUM*

A. Lecture: fundamental structures about concepts of finance and entrepreneurship
B. Quiz: Understanding phase
C. Game: Application phase

A common approach to teach contents systematically is basing the lecture on Bloom's Revised Taxonomy. Each learning stage consists of several LOs which can be represented in a table based on Bloom's Revised Taxonomy [23]. This Taxonomy provides a structure for classifying learning contents by two dimensions, the "Knowledge Dimension" and the "Cognitive Process Dimension". The combination of these dimensions implies the type as well as the complexity of a learning content. Every step of *Bikorama@TUM* targets both specific dimensions as shown in Fig. 3. As each phase of each stage contains at least one LO, there are 17 LOs in total, of which the most important ones will be explained.

In the **Survival Stage** (Fig. 2), the students learn the early processes of a company. The game shows them how the idea of founding a company arises and leads them to decisions where knowledge about concepts of economy are required. This includes the understanding of the terms "demand" and "supply". Hence, LO_3_IB2: "The students understand the concept of price formation." is the most important LO of stage I, as it is one of the key concepts of economics as referred to in [21] and is crucial for every student to understand.

Stage I		The Cognitive Process Dimension					
		1. Remember	2. Understand	3. Apply	4. Analyze	5. Evaluate	6. Create
The Knowledge Dimension	A. Factual knowledge	LO_1_IA1	LO_2_IA2				
	B. Conceptual Knowledge		LO_3_IB2				
	C. Procedual Knowledge			LO_4_IC3			
	D. Metacognitive knowledge						

Fig. 2. Bloom's Revised Taxonomy stage I

Stage II		The Cognitive Process Dimension					
		1. Remember	2. Understand	3. Apply	4. Analyze	5. Evaluate	6. Create
The Knowledge Dimension	A. Factual knowledge	LO_5_IIA1	LO_7_IIA2				
	B. Conceptual Knowledge	LO_6_IIB1	LO_8_IIB2				
	C. Procedual Knowledge			LO_9_IIC3			
	D. Metacognitive knowledge						

Fig. 3. Bloom's Revised Taxonomy stage II

In the **Growth Stage** (Fig. 3), students start realizing that the initial idea of bike renting is able to generate money, thus they can think about how to conduct it on a slightly larger scale and test the initial idea in a more realistic scenario with an anonymous market. As the students rent their bikes, they approach the key element of the game. Finding the best pricing scheme enables LO_9_IIC3: "The student can set a price under consideration of demand and supply.", as this proves their understanding of the learning contents of the previous stage.

As the students accumulate more money, it is hinted that the idea of renting bikes seems to work even at a large scale. To achieve a good result, students must learn to deal with more possibilities of how to invest their money during the **Expansion Stage** (Fig. 4). While playing, they learn the mathematics behind taking a loan by calculate a break-even point. This is the first step to LO_15_IIIC4: "The students can analyze the impacts of taking a loan on a company given the financial situation."

Stage III		The Cognitive Process Dimension					
		1. Remember	2. Understand	3. Apply	4. Analyze	5. Evaluate	6. Create
The Knowledge Dimension	A. Factual knowledge	LO_10_IIIA1	LO_12_IIIA2				
	B. Conceptual Knowledge	LO_11_IIIB1	LO_13_IIIB2				
	C. Procedual Knowledge			LO_14_IIIC3	LO_15_IIIC4		
	D. Metacognitive knowledge						

Fig. 4. Bloom's Revised Taxonomy stage III

Conclusion Stage		The Cognitive Process Dimension					
		1. Remember	2. Understand	3. Apply	4. Analyze	5. Evaluate	6. Create
The Knowledge Dimension	A. Factual knowledge	Company founding (Stage I - III)					
	B. Conceptual Knowledge						
	C. Procedual Knowledge						
	D. Metacognitive knowledge				LO_16_D4	LO_17_D5	

Fig. 5. Bloom's Revised Taxonomy stage IV

When playing the final stage, the **Conclusion Stage** (Fig. 5), the students have already learned and applied the main concepts of company foundation. As the students finish the game, they are asked individually, which decisions resulted in which impacts on their company. Figure 7 shows the highest LOs the students can obtain by playing *Bikorama@TUM*. As the students summarize their decisions, they start realizing the connection between their decisions, which is listed as LO_16_D4. In the last step, the students reconsider their decisions and think of possible improvement of their strategy resulting in their final Learning Objective LO_17_D5: "I am able to analyze the economic process of founding a venture."

4 Test Setting

4.1 Sample Group

For testing the LOs, a partnership with the FOS BOS Weilheim in southern Bavaria, Germany, was formed. The goal was to define a sample group of 14 to 20 years old school students to participate in the 90 min lecture unit, which involves playing the game and engaging with the intermediate lecture parts that contain explanatory content about the discussed concepts. In order to measure and validate the outcome of the LOs, a test was created, consisting of two parts that every student of the sample group had to answer both before and after the lesson. A sample size of 151 students was collected together with the partner school, the vast majority of whom were between 17 and 19 years old. Every school student in the sample attended the lecture unit and answered the pre- and post-test. The pre- and post-test were created by using a five-point Likert scale, "which is a common way to record and evaluate the knowledge of the target group regarding a specific topic" [24]. The answers of the Likert scale ranged from *1 - Strongly Disagree* over *2 - Disagree*, *3 - Neither Agree nor Disagree*, and *4 - Agree* to the most positive answer *5 - Strongly Agree* [24].

As mentioned earlier, *Bikorama@TUM* is focusing on defined LOs, classified in Bloom's Revised Taxonomy, as previously illustrated. Based on that, Test Statements (TS) with a focus on entrepreneurship and the general content of the lecture unit, as described in Sect. 3, were developed in the form of questions for the pre- and post-test, which are explained in the next subsection.

4.2 Test Statements

In the following, the most important TSs are explained, based on which the performance of the students is evaluated in relation to the previously defined LOs. To clarify the relations of TSs and LOs, a cross-reference table was created for each stage of *Bikorama@TUM* that indicates which TS is assessing which LOs. Table 1 shows exemplary the cross-reference table for the Survival Stage.

Table 1. Cross reference table for the Survival Stage

Test statement (TS)	Classification in bloom's revised taxonomy	Learning objective (LO)	
TS_1_A2	Remember/Factual knowledge	LO_1_IA1	Survival stage
TS_2_A2	Understand/Factual knowledge	LO_1_IA2	
TS_3_C3	Apply/Procedural Knowledge	LO_1_IC3	

To give an example, the first TS of the Survival Stage is briefly described in the following: *TS_1_A2: I understand the concept of the reservation price* tests knowledge of what a reservation price is and why it is useful to understand customers' buying decisions. It refers to the first stage of the lecture unit where the concepts of reservation prices and opportunity costs are introduced. Testing *LO_2_IA2*, this TS is classified as Understand/Factual Knowledge in Bloom's Revised Taxonomy.

5 Results

Two questionnaires were filled out to track the success of RQ_1 and RQ_2. The students filled out their pre-test before the beginning of the 90-min lesson, which uses eleven TSs to gauge the students' pre-existing knowledge around entrepreneurship. After class, the students completed their post-test, which contained the same TS, but the answers from the pre-test were not visible. The pre-test helps to observe if students were already knowledgeable in some of the topics *Bikorama@TUM* wants to teach them. The test monitored the overall progression of the nine different courses containing between 9 and 23 students. There is no differentiation between courses. The tests were checked for invalid answers. These included unchecked boxes, multiple checked boxes for the same question or crosses outside the five boxes on the Likert scale. Sheets containing one or two invalid answers were not included in the result of the corresponding TSs. Two questionnaire sheets containing over five invalid answers were completely discarded and not used in this evaluation. Excluding these two questionnaires the evaluation observed the progress of 151 students using *Bikorama@TUM*. Based on this data both research questions can be answered.

5.1 Evaluation of the Pre-test

The students' ages ranged between 14 and 27. The median age was 18 years, with the 25th percentile being 19 years old and the 75th percentile being 17 years old. When being asked about their general knowledge in entrepreneurship, 48.7% chose a 1 or 2 on the Likert scale. 8.0% chose a 3, 22.7% a 4 and only 5.3% strongly agreed with the TS stating they already have previous knowledge in entrepreneurship. Overall, students tended to not be confident in their knowledge of key LOs in stage I, with 90% of answers giving a negative self-assessment (*1 - Strongly Disagree* and *2 - Disagree*) of the knowledge and skills they will learn in stage I. Therefore, the first three TSs are fitting to find out whether *Bikorama@TUM* is able to teach students unfamiliar topics.

Stage II involved knowledge and concepts more students were already familiar with and therefore were more confident about in their self-assessment. This is interesting to further test if *Bikorama@TUM* can not only help with learning the basic concepts but deepen students' knowledge further.

Stage III has had on average over the five TSs 53% disagreeing answers (*1 - Strongly Disagree* and *2 - Disagree*) with the TSs, 18.9% answering with *3- Neither Agree nor Disagree* and 38.2% agreeing (*5 - Strongly Agree* and *4 - Agree*) with them (Fig. 6).

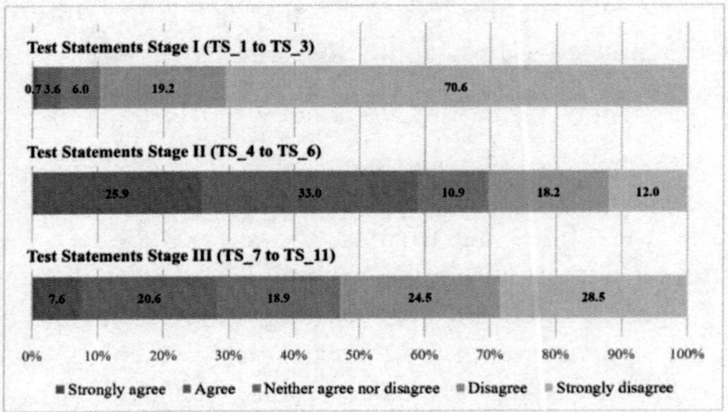

Fig. 6. Combined pre-test Test Statements of the stages I to III

With these different distributions the evaluation can indicate whether *Bikorama@TUM* excels in either teaching the basics by looking at stage I or whether it is a great teaching tool to deepen students' preexisting knowledge by looking at stage II.

The focus of the evaluation will be on stage I, since only a combined 4.3% of the pre-test answers were an agreement with the corresponding TSs. This allows to answer RQ_1 by looking at to what extend the game *Bikoroma@TUM* achieves the LOs designed according to Bloom's Taxonomy. Furthermore, stage I includes TS_3_C3, which can answer RQ_2 because it checks to what extend the game *Bikoroma@TUM* improves the students' understanding of the fundamentals of economics and entrepreneurship covered in the lesson.

5.2 Comparison of the Pre- and Post-test

To measure the impact of a lesson driven by *Bikorama@TUM*, the results of the post-test were compared to the results of the pre-test. TSs 1 to 3 were overall mostly negative in the pre-test but got one of the best overall scores in the post-test. Students were generally optimistic about their own knowledge and abilities to apply that knowledge asked about in the TSs of stage I after the 90-min lessons. TS_1_A2 asked the students if they understand the term reservation price. It went from 62.7% of students answering with *1 - Strongly Disagree* and 24.7% *2 - Disagree* to 55.7% *5 - Strongly Agree* and 33.6% *4 - Agree* with it. A similar shift can be observed for TS_2_A2 and TS_3_C3, asking the students if can explain opportunity costs and apply the knowledge of the previous two TSs. Both went from over 76% disagreement (*1 - Strongly Disagree* and *2 - Disagree*) to over 87% agreement (*5 - Strongly Agree* and *4 - Agree*) from students. A significant shift for every of the three TSs of stage I can be observed. RQ_1 can be answered by the first two TSs, with the students themselves in vast majority stating they achieved the Learning Objectives. By looking at TS_3_C3, RQ_2 can be answered. A vast majority of students agreed that *Bikorama@TUM* improved the students' understanding of the fundamentals of economics and entrepreneurship covered in the lesson.

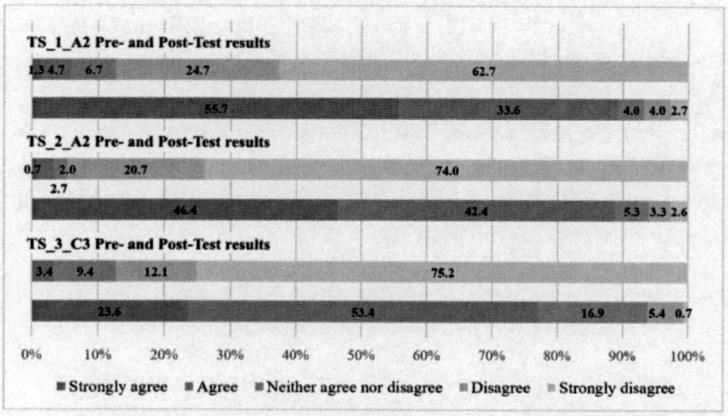

Fig. 7. Comparison pre- and post-test Test Statements 1−3

6 Conclusion and Discussion

By interweaving lessons with the usage of *Bikorama@TUM*, students can apply their recently gained knowledge and solidify it long term. Therefore, the research goal was to create *Bikorama@TUM* which can be used for teaching the basics of entrepreneurship to students aged between 14 and 20 years old without much pre-existing knowledge. Eleven Test Statements (TS) were created with the intent of having a broad spectrum of knowledge and methods for an insight into entrepreneurship. These TSs could be divided into the three stages of *Bikorama@TUM*: Survival (I), Growth (II), and Expansion Stage (III). Furthermore, questions were diversified enough to cover both research questions. *Bikorama@TUM* was tested by more than 150 school students in the grades 11 to 13 of an upper vocational school in Weilheim, Bavaria, Germany. To measure the teaching success, students had to fill out a pre-test before lecture and post-test after the 90-min lecture containing 11 identical TSs. For each of these, students had to fill out a 5-point Likert [24] scale on how much they agree or disagree with these TSs. By comparing these, the success of the lecture unit *Bikorama@TUM* could be measured. The evaluation of the progress proved that the 151 participants did improve their knowledge with regards to the eleven TSs. In the pre-test, the knowledge around all the three stages had a negative or neutral median in nine TSs. In the post-test, negative answers to questions ranged from 1.3% to a maximum of 12.7%, with an overwhelming majority of the answers being positive. Over 95% of the test group agreed that they learned a lot due to the game and that the game brings in an additional value.

Bikorama@TUM can help teachers in developing a higher participation rate and involving more students in their lessons compared to traditional teaching methods. This success of a playful didactic approach has been observed before [6]. With serious games covering the success factors, including audience integration, audience centered focus, and instant feedback, *Bikorama@TUM* proves again the success IT-based learning games can have. With more focus shifting towards online classrooms and autonomous learning, serious games like *Bikorama@TUM* can help improve the learning success of future generations and have once again proven to be a valuable addition to the future of teaching.

Acknowledgements. The authors would like to thank Mrs. Andrea Jochner-Weiß, Chief District Administrator of Weilheim-Schongau in Bavaria, Germany, and Mr. Christian Dick and the teaching staff of the upper vocational school FOS BOS Weilheim in Bavaria, Germany, for supporting the evaluation of Bikorama@TUM.

References

1. Castro, M., Sancristobal, E.: From technology enhanced learning to ethics and critical thinking as part of the engineering education: skill driven with humanities comprehension editorial. iJEP **10**(1), 4 (2020)
2. Utesch, M.: Five theses on a renaissance of engineering education: skill-driven leaning and teaching SDLnT. iJEP **9**(5), 4 (2019)
3. Elert, N., Andersson, F., Wennberg, K.: The impact of entrepreneurship education in high school on long-term entrepreneurial performance. J. Econ. Behav. Organ. **111**, 209–223 (2015)
4. Lans, T., Blok, V., Wesselink, R.: Learning apart and together: towards an integrated competence framework for sustainable entrepreneurship in higher education. J. Cleaner Prod. **62**, 37–47 (2014)
5. Utesch, M.: A successful approach to study skills: Go4C´s projects strengthen teamwork. iJEP **6**(1), 35 (2016)
6. Heininger, R., Prifti, L., Seifert, V., Utesch, M., Krcmar, H.: Teaching how to program with a playful approach: a review of success factors. In: 2017 IEEE Global Engineering Education Conference (EDUCON), Athens, Greece, pp. 189–198 (2017)
7. Heininger, R., Seifert, V., Prifti, L., Utesch, M., Krcmar, H.: The playful learning approach for learning how to program: a structured lesson plan. In: 30th Bled Econference: Digital Transormation – From Connecting Things To Transforming Our Lives, Bled, Slovenia, pp. 215–230 (2017)
8. Peirce, N., Conlan, O., Wade, V.: Adaptive educational games: providing non-invasive personalised learning experiences. In: Second IEEE International Conference on Digital Game and Intelligent Toy Enhanced Learning, Banff (2008)
9. Tinto, V.: Research and practice of student retention: what next?. J. College Student Retention Bd. **8**(1), 1–19 (2006–2007)
10. Zimmermann, B.J.: Self-Regulated learning and academic achievement: an overview. Educ. Psychol. **25**(1), 3–17 (1990)
11. Zimmerman, B.J.: Becoming a self-regulated learner: an overview. Theory Pract. **41**, 64–70 (2002)
12. Klopfer, E.: Augmented Learning: Research and Design of Mobile Educational Games, Education next (2008)
13. Aronson, J.: Improving Academic Achievement. Elsevier Science, New York, pp. 209–278 (2002)
14. Cagiltay, N., Ozcelik, E., Ozcelik, N.: The effect of competition on learning in games. Comput. Educ. **87**, 35–41 (2015)
15. Löffler, A., Prifti, L., Levkovskyi, B., Utesch, M., Krcmar, H.: Simulation games for the digital transformation of business processes. In: 2018 IEEE Global Engineering Education

Conference (EDUCON), Santa Cruz de Tenerife, Canary Islands, Spain, pp. 1475–1483 (2018)
16. Faizan, N., Löffler, A., Heininger, R., Utesch, M., Krcmar, H.: Classification of evaluation methods for the effective assessment of simulation games: results from a literature review. Int. J. Eng. Pedagogy (iJEP) **9**(1), 19 (2019)
17. Definition of GAME, Merriam-webster.com, 2020. https://www.merriam-webster.com/dictionary/game. Accessed 10 Mar 2020
18. Banks, J.: Discrete-event System Simulation. Prentice Hall, Upper Saddle River (2001)
19. Pivec, P.: Game-based Learning or Game-Based Teaching, in Becta (2009)
20. Garris, R., Ahlers, R., Driskell, J.E.: Games, Motivation, and Learning: a Research and Practice Model," Simulation & Gaming, pp. 441–467 December 2002
21. Kurschilgen, M.: Volkswirtschaftslehre 1 (2020). https://campus.tum.de/tumonline/wbLv.wbShowLVDetail?pStpSpNr=950429349&pSpracheNr=1
22. Mel, S., Bruce, R.: Five stages of Survival in small business. Long Range Plan. **20**(3), 45–52 (1987)
23. Anderson, L., Krathwohl, D.: A Taxonomy for Learning, Teaching, and Assessing, A Revision of Bloom's Taxonomy of Educational Objectives. Addison Wesley Longman, New York (2001)
24. Likert, R.: A technique for the measurement of attitudes. Archives of psychology (1932)

Towards a Framework for Adaptive Gameplay in Serious Games that Teach Programming: Association Between Computational Thinking and Cognitive Style

Anastasios Theodoropoulos[1](✉), Vassilis Poulopoulos[2], and George Lepouras[1]

[1] HCI-VR Laboratory, University of the Peloponnese, Tripoli, Greece
ttheodor@uop.gr
[2] Knowledge and Uncertainty Research Laboratory, University of the Peloponnese, Tripoli, Greece

Abstract. Learning basic programming principles is the main process to acquire computational thinking skills, which comprise an essential capability for education worldwide. The game-based learning approach is a very promising learning strategy towards that direction. Digital games help learners become active and engaging in the acquisition of knowledge. Though, the learning process is not always clear through serious games that teach programming. Personality characteristics like cognitive style, seem to affect how efficiently and effectively learners perceive the knowledge. Cognitive style deals with the way learners perceive information and is a significant feature in improving programming learning. Thus, in order learners to achieve better performances through serious games that teach programming concepts, we need to investigate how cognitive style can be incorporated into the design of game-based learning environments. This article proposes a framework for adaptive gameplay that can support and encourage researchers and game designers to properly identify the concept of cognitive style when designing game-based learning environments for programming learning. The framework models an ontology that searches personality traits and enables personalised decision-making learning activities.

Keywords: Serious games · Programming learning · Cognitive style · Adaptive gameplay · Conceptual framework

1 Introduction

Basic programming knowledge and Computational Thinking (CT) are core capabilities for the 21st century. By learning programming principles, one learns to solve problems, to design systems and to understand the behavior of machines, in other words to develop CT skills [1]. This attitude is important to almost every engineering task. It can help young people develop technological skills and a deeper understanding of the new digital reality, but even to make them able to intervene in it.

M. E. Auer and T. Rüütmann (Eds.): ICL 2020, AISC 1328, pp. 530–541, 2021.
https://doi.org/10.1007/978-3-030-68198-2_49

However, teaching and learning basic programming principles is a great challenge for educators worldwide. New approaches, such as digital games, are used in innovative education, both in formal and informal learning, offering an unprecedented opportunity for teachers to adopt immersive and progressive methods. The basic elements that dominate through the digital Game-Based Learning (GBL) method are: the fun, the playful mood, the sense of joy, imagination and natural curiosity, thus achieving high levels of participation and involvement of the learner in the learning process [2]. Despite the continuous and increasing production of digital game-based material for learning purposes, the efforts to gather and evaluate overall data that affect programming learning, based on the principles and material of the games, have not progressed to the same extent. In addition, due to the nature of the subject matter of programming, the factors that seem to affect its learning are several [3, 4].

Personality traits and particularly cognitive style seem to relate with achievement in programming [5]. Cognitive style is a concept used in cognitive psychology to describe the way individuals make choices, gather, organize and process information [6]. Previous research have found significant relationships between cognitive style and programming related skills [5, 6]. When the educational content is adapted to the student's cognitive style through games, learning becomes more effective [4]. Therefore, it is important to study personality traits in regards to CT skills and programming learning from serious games.

To this end we introduce a new conceptual framework to create personalized content in serious games, exclusively and uniquely for a learner. There are several game design frameworks and models about system design, UX, narrative, or new monetization methods [7–10]. Moreover, there are games that use strategies to adapt their gameplay according to the needs of the player in order to provide a better experience [11, 12]. By expanding this idea into programming education, a framework towards an adaptive gameplay depending on the personality of the learner and the actions that trigger him, is a valuable addition.

1.1 Outline - Research Question

The aim of this research is to highlight the relationship between digital programming games and the player's cognitive style. In this mapping, the following research question is posed: RQ: *Can digital games that teach programming adopt users' personality traits effectively in the learning process and affect both motivation and learning outcomes?* Rationale: Present a framework that creates personalized content based on user's cognitive style.

2 Related Work

2.1 Programming Learning Through Playing Games

Digital games can be used as a vehicle to train students to acquire CT skills in a fun and engaging environment and can have very important gains in the educational settings [13]. Integrating them into the learning process can enable teachers to systematize learning in an adaptive way [11]. Moreover, the fact that every student encompasses a totally different style of learning or processing data has been recognized by educationalists. Digital games can encourage STEM participation and learning and support identity exploration [14]. They involve a dynamic individual-environment interaction for learners. From a situative perspective on learning, games promote knowledge transformation through participation in activities that put players in a dynamic interaction between the individual and the environment [14].

Learners need to apply knowledge in different situations while playing computer games to learn programming (coding games). They have to solve coding problems and develop essential skills such as strategies for abstraction, generalization, decomposition and problem-solving. Mental approaches have to be turned into coding [15]. In order to understand the role of coding games, it is very important to identify insights from previous researches. As found through literature, the largest number of empirical works with students and games in the field, target to enhance CT, algorithmic thinking and basic programming skills [16]. Most games that aim to teach basic programming concepts belong to the puzzle games genre. Such games are based on basic computational logic and pattern recognition and thus are easier to follow [10, 11]. Other popular genres are simulations and strategy games. The main coding activities are: write code, program an agent, debug the code or fill the code.

Considering the need to enhance the value of learning programming through electronic games, previous work [17] presents a taxonomy which includes four categories for serious games classification: Modality, Interaction style, Environment and Learning approach. Based on this classification, the most common modality fields used in games for programming learning, are visual and auditory, but sometimes it may be haptic or tangible interactive environments [18]. As for the interaction style, most games use combinations of styles like graphical direct manipulation, questions and answers, avatars etc. The basic feature that differentiates the learning environment, is whether they use a block-based approach (e.g. Blockly) or a text-based programming language (e.g. Python). Most studies use the block-based approach, since there is no typing and syntax errors are eliminated which makes the game suitable for novice programming.

2.2 Personality Traits in Programming Learning

Within the GBL domain, studying issues relevant to individual learning and particularly linked to learning with the use of technology (like digital games) is very important. Post-modern educational theorists like Hlynka [19], considered that educational technology is not neutral in the sense that it does not simply promote a learning process by making it more efficient, but it changes the nature of learning itself. The nature of learning really changes with the use of technology in the learning process [20]. There are several factors that can significantly influence an individual's learning. Numerous theories have evolved around the factors that influence individual's learning and they can be divided in two main categories: 1) situation independent and 2) situation dependent. Cognitive style is one of the main concepts within the situation independent category the one most used in technology related educational research. It has a strong relationship to the personality of the individual and it is not easily affected by the different learning circumstances.

Cognitive style is a research construct that assists the study of cognitive learning and refers to the preference and habitual approach of a person to organizing and representing information [6]. It describes deeper cognitive structure than learning style and students whose learning methods and material matches their cognitive style perform better [4]. It can be easily revealed with the use of tools like the Myers-Briggs Type Indicator (MBTI) [21], or the Big Five Inventory (BFI) [22]. The MBTI describes learners on four dimensions: Extraversion-Introversion, Sensing-Intuition, Thinking-Feeling and Judging-Perceiving, while the BFI considers five factors to classify individual differences of personality characteristics: Openness, Conscientiousness, Extroversion, Agreeableness and Neuroticism.

Researchers studied GBL and programming learning and found that personal learning characteristics are important. For example, Lau and Yuen [23] found that students of a particular learning style (sequential learners) performed better in programming tasks than students of a different learning style (random learners). Milovanović et al. [24] showed that GBL is very effective when it is correlated on specific cognitive styles. Authors researched the cognitive style in higher education students who engaged in game design for understanding computer networks. Another study, by Theodoropoulos et al. [4], measured the learning achievement and motivational appeal of games for learning basic programming concepts, in school settings with secondary education students. Authors correlated cognitive style with learning outcomes from code.org' s games and introverts achieved better outcomes than extroverts. Concluding, personality traits and particularly cognitive style seems to be a significant characteristic that should be taken into consideration when using serious games to study programming concepts.

3 Method

Our approach takes into account past experience from widely accepted game design frameworks, but also incorporates the concept of cognitive style which serves as a strong organizing factor for programming learning. The proposed modelling framework, features cognitive style in terms of both learning objectives as well as learning activities and material. To ensure the validity of the framework, we also present guidelines that can be applied in game design process and research and help the learner/player to think analytically and widely in solving different CT tasks through games.

3.1 Adapting the Learning Experience

Our methodology is to provide support for learners with different personality traits like Introverts, Extraverts, Neuroticism, Openness to Experience, etc. The goal is to increase motivation and performance of participants while they play a coding game. To adapt the learning experience, it is crucial to consider impact on both students and teachers. At the student level, we want to find and present ways to help the them think analytically and comprehensively in solving problems through games. At the teacher level, we want to focus on learning goals, needs and evaluation of the gaming experience. Therefore, before shaping the framework we must take into account several factors for designing serious coding games (see Fig. 1).

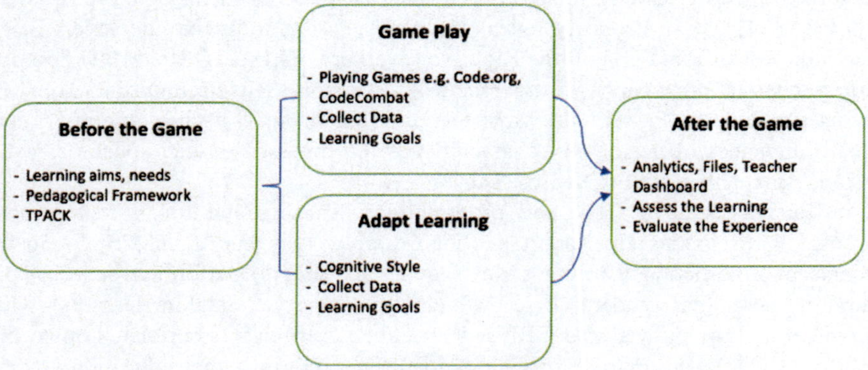

Fig. 1. Using digital games to learn programming.

The steps presented above, describe the elements that should be taken into account in three main stages: before using a game, during the playtime and after the gaming activities. This outline stands as the basic idea for the proposed adaptive framework. It shows the interactions of teaching, knowledge and technology that should be taken into account when using digital games to learn programming. The adaptation process can lead to success when including the context of specific learning aims and needs, pedagogical frameworks [3] and technological models [25]. Furthermore, the levels for the adaptation process can vary in specific programming tasks or even re-consider after the gaming experience. Game designers or educators should address this knowledge to develop more effectively and engaging games.

3.2 Game Design Frameworks

Game design frameworks can help us to understand how games are designed and how they affect the player experience. Prensky [7, 26] presented a detailed model of the learning levels that a game develops, i.e. topics that players learn in their efforts to achieve the goals of a game. This model can be a first guide on how to incorporate learning elements into a game. The model distinguishes the following: How (gameplay), What (game rules), Why (strategy), Where (conditions) and When/If (decisions/ethics).

In a successful game, learning experiences should not give the feeling of interruption or arbitrary course of the game. A widely accepted tool for engaging games is the Mechanics-Dynamics-Aesthetics (MDA) framework [8]. MDA breaks down games into three main components: Mechanics, Dynamics, and Esthetics (Fig. 2).

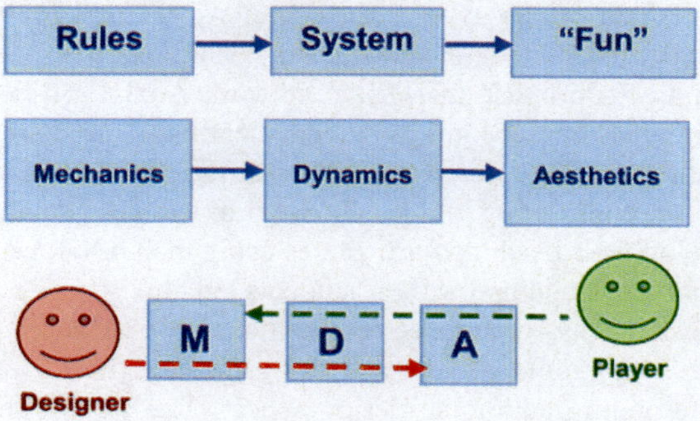

Fig. 2. The MDA game design framework by [8].

Mechanics refer to every basic action that the player can make during the gameplay, which comes from specific algorithms and data structures developed in the game engine. In other words mechanics are the games' rules [8]. Dynamics refer to the actions on the player input by the run-time behavior of the mechanics [8]. They basically describe how the game is played. Finally, aesthetics stand for the emotional reactions (feelings) of the participant during the game [8].

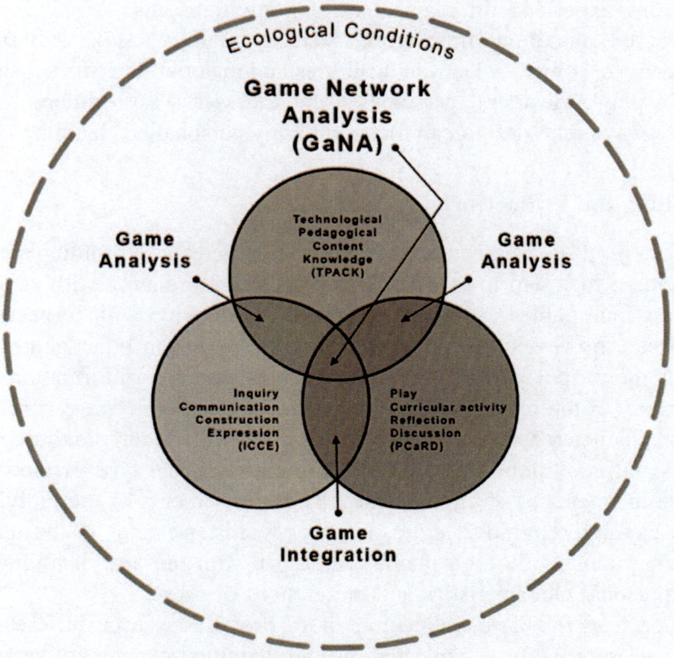

Fig. 3. The Game Network Analysis Model by [9].

Since the aim of this study is to form a framework for learning purposes, it is important to include works that meet educators learning goals during the gaming process. Foster et al. [9] propose the Game Network Analysis (GaNA), a systematic way to teach with games and encourage students learning and interest, by analyzing a game and integrating it within context. GaNA comprises of some other models that form a network for considering games in several learning environments (Fig. 3). Teachers that develop their teaching with games using the GaNA, can get achieve both knowledge and motivation gains for their students [9].

Finally, the Design-Play-Experience (DPE) framework by Winn [10] describes the relationship between the game players and the game designer. It expands the MDA framework, by interpreting additional design aspects like storytelling and user experience into layers.

4 A Framework for Adaptive Gameplay

The works mentioned in the previous section stand as the base for a new adaptive framework, particular for programming learning through a GBL environment. Since the main addition of this work refers to the personality traits of the player, we seek ways to include or exclude learning objectives and adapt the learning context and related activities based on learners' cognitive style. The need to model the aforementioned aspects in correlation with personality traits, can be helpful for enabling efficient learning experience in adaptive gaming environments.

The proposed modelling framework, features cognitive style in terms of both learning objectives as well as learning activities and material. It defines a structure that can point out players/learners' personality and access it within different stages of a game. By this way game design can focus in highly personalised learning experiences.

4.1 Modelling the Framework

To this end, we propose a framework architecture that creates a profiling service which aims to provide a first step towards personal traits in correlation with serious games. The present serious games are not designed to be adaptive with respect to learners personality traits, however they adapt the progress based on other game mechanics. Towards this, the proposed framework collects and processes information concerning to the personality of the players, with gamified techniques. The framework will collect the individual characteristics that affect learning and divide them into those that depend on environmental conditions (for example learning style and type of motivation) and those that are independent of circumstances and are elements of the individual's personality (for example cognitive style). In other words, we refer to the use or incorporation of mechanisms through playful activities, with the aim of finding solutions through the personal characteristics and preferences of users.

Figure 4 presents the basic architecture of the proposed user profile creation model. The appropriate information is collected through gamified personality test techniques, which can be expressed in different ways. The application for the creation of profiles (profiling server) focuses on storing, maintaining and analyzing these elements which are strictly related to the personality of the user. This information can be treated separately at each gaming session.

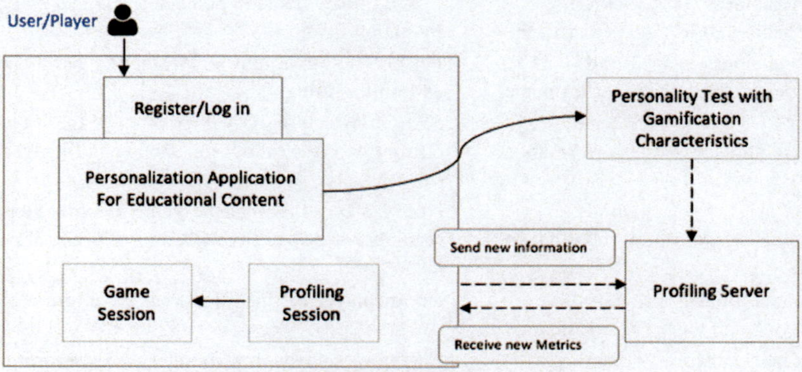

Fig. 4. The architecture of the proposed framework.

The constituent parts of this architecture are as follows:

– Game registration, is the fragment that encompasses and provides personal information of the player such as age, gender, previous programming experience etc. and can be available for usage in the gaming session.
– The "Personalization Application", is the main dimension of this framework. It uses the "Personality Test with Gamification Characteristics", which refers to a gamified process that will occur as part of the game and will reveal the cognitive style of the player. This will be a part of the initial stages of the game and will store the data for the gaming sessions.
– The "Profiling Server", stores and transfer new information to the games, but will also collect metrics from the gameplay and the performance of the player, in order to (a) analyze computational metrics and (b) adapt the player's profile. Furthermore, the procedures for creating a user's profile will be anonymized.
– The "Game Session" represents the programming activities. Although this refers to new games, it could also be used to existing games as an additional element, or for research purposes.

Table 1 presents examples of guidelines that lead to a response from the game, depending on the personality of the learner and his performance during the gaming activities.

This will allow us to adapt the game based on different personality traits and actions that trigger players and affect their motivation and achievement towards the game. Digital games adapt the game's progress based on the user's progress and are not designed to be customized in relation to their personal characteristics. In order to create

Table 1. Examples of adaptation rules for different personality types.

Learner/Player personality	Trigger/Action	Game response/Adaptation
Low openness to experience (BFI)	Good performance	Ask if player wants to proceed to another stage/level by encouraging him
Low openness to experience (BFI)	Bad performance	Perform a simple tutorial, presenting basic elements as familiar things
Low conscientiousness (BFI)	Good performance	Show player how he will advance if the he keeps doing good (e.g. show milestones and badges)
Low conscientiousness (BFI)	Bad performance	Notify player that that the game has no strict regulations, however low scores will not assist him
High neuroticism (BFI)	Good performance	Honor player for his effort (e.g. give him a badge)
High neuroticism (BFI)	Bad performance	Encourage player in a friendly and supportive way to try again
High intuition (MBTI)	Bad performance	Re-explain by using symbols and abstraction (e.g. we need to add one more block in order to achieve an action)
High sensing (MBTI)	Bad performance	Support by presenting facts and examples (e.g. what happens if he performs actions right or wrong)
High feeling (MBTI)	Bad performance	Present stories and/or real life problems to give opportunities for emotional learning

a user profile and enable the required educational content adaptation, we need to collect as fully as possible the characteristics that affect a person's learning about programming.

5 Discussion

This research highlights and presents a systematic way to help learners think analytically and comprehensively in playing games to learn programming. The proposed framework encourages the use of digital game activities in the educational process by providing the appropriate conditions for this to be done systematically for every learner. Games can be very powerful tools in the learning process by adopting personality traits of students that affect both motivation and learning outcomes. The framework focuses on dynamic adaptation by promoting learner interaction with pervasive manner in a smart learning environment and incorporating and sharing knowledge about personality traits and pedagogical experiences through actuating agents. Therefore it provides a new personalized approach in programming learning that could represent a wide range of educational settings like Engineering Education, STEM or Computer Science education. In addition, through thorough research and appropriate experiments, this framework will gain valuable new knowledge about the nature and way in which learners achieve better through games, which will then be used to highlight all the positive practices that can be offered by GBL.

This work is a first step towards a holistic framework for adaptive gameplay in serious games that teach programming, by highlighting the need to find associations between CT and cognitive style. Although currently, there are not concrete games that encompass the proposed framework, game designers should have this in mind and incorporate clear mechanics and rules that will keep the gaming boundaries narrow enough and construct a clear vision of what we want, what we need to do, and when to complete it, in order to achieve the learning goals. The framework can help game designers and researchers to specify the requirements of gaming environments from the perspective of personality traits.

Finally, although this work is about games and game design, it could help curriculum designers to better understand the impact of digital games on learning programming and increase the range of innovative products in education that reflect the real needs of modern society.

6 Conclusions

The goal of this article was to highlight the need to associate CT and cognitive style when learning from playing programing games and for this need to introduce a game design framework. As a first step, we focused on identifying relevant works and propose a profiling model. Furthermore, we have provided examples of adaptation rules for different personality types.

The proposed framework has possible limitations. Cultural differences and other personal problems may be causing conflicts. Therefore, to minimize these disputes, the data-gathering process should be carried out properly. Cognitive style variables are collected through gamified activities and questions and some users may not provide trustworthy responses. To check if the adaptation has any effect on learners, we may imply various aspects from the profiling process, such as the duration or a recurrence of the process. In this paper we have not explicitly discussed the role of instructional support to enhance learning. However, metacognitive interventions can be accommodated by the framework.

Currently, the framework is built on insights from literature, both previous works on the field and game design models. However, we have not yet verified the usefulness of the presented dimensions in the analysis, comparison and design of game-based learning environments. The next step in developing this framework is to invite experts from different fields to discuss the proposed framework and its application to design and research. Concerning the development of this framework, a future step would be to identify how different personalities affect learning trough experiments with more existing coding games. Further examining how CT is associated with different cognitive styles. In the current study, we only focused on presenting basic design elements and the model's complexity. We plan our future research to be more comprehensive, and to cover as much range of cognitive skills needs to be covered.

The work presented in this article serves the purpose of incorporating the characteristic of cognitive style into game design, about programming learning. A final framework about this kind of serious games may undergo future changes. Some elements may receive additional consideration and there are still parts which deserve being explored deeper. In summary, we consider that this framework contributes to shaping advances in game-based learning within this new area and eventually will benefit researchers and game designers in incorporating cognitive style in coding serious games.

References

1. Wing, J.M.: Computational thinking. Commun. ACM **49**(3), 33–35 (2006)
2. Tan, W.H.: Design, Motivation, and Frameworks in Game-Based Learning. IGI Global, Hershey (2018)
3. Theodoropoulos, A., Vassilakis, C., Antoniou, A., Wallace, M., Lepouras, G.: ATMF: A Student-Centered Framework for the Effective Implementation of Alternative Teaching Methods for CSEd. In: Conference on e-Business, e-Services and e-Society, pp. 116–127 (2019)
4. Theodoropoulos, A., Antoniou, A., Lepouras, G.: How do different cognitive styles affect learning programming? Insights from a game-based approach in Greek schools. ACM Trans. Comput. Educ. (TOCE) **17**(1), 3 (2017)
5. Bishop-Clark, C.: Cognitive style, personality, and computer programming. Comput. Hum. Behav. **11**(2), 241–260 (1995)
6. Riding R., Rayner, S.: Cognitive Styles and Learning Strategies: Understanding Style Differences in Learning and Behavior. Routledge, Abingdon (2013)
7. Prensky, M.: Digital Natives, Digital Immigrants. On Horizon **9**(5), 1–6 (2001)
8. Hunicke, R., LeBlanc, M., Zubek, R.: MDA: A formal approach to game design and game research. In: Proceedings of the AAAI Workshop on Challenges in Game AI, vol. 4, no. 1, p. 1722 (2004)
9. Foster, A.N., Shah, M., Duvall, M.: Game network analysis: for teaching with games. In: Teacher Education: Concepts, Methodologies, Tools, and Applications, IGI Global, pp. 371–403 (2016)
10. Winn, B.: The Design, Play, and Experience Framework in Handbook of Research on Effective Electronic Gaming in Education. IGI Global Hershey, PA (2008)
11. Pinto, J.F., Carvalho, H.R., Chambel, G.R., Ramiro, J., Goncalves, A.: Adaptive gameplay and difficulty adjustment in a gamified upper-limb rehabilitation. In: 2018 IEEE 6th International Conference on Serious Games and Applications for Health (SeGAH), pp. 1–8 (2018)
12. Goršič, M., Darzi, A., Novak, D.: Comparison of two difficulty adaptation strategies for competitive arm rehabilitation exercises. In: 2017 international conference on rehabilitation robotics (ICORR), pp. 640–645 (2017)
13. Gros, B.: Digital games in education: The design of games-based learning environments. J. Res. Technol. Educ. **40**(1), 23–38 (2007)
14. Shah, M., Foster, A., Barany, A.: Facilitating learning as identity change through game-based learning, Game-Based Learning: Theory, Strategies and Performance Outcomes, pp-257–278. New York, Nova Publishers (2017)
15. Winslow, L.E.: Programming pedagogy—a psychological overview. ACM Sigcse Bulletin **28**(3), 17–22 (1996)

16. Theodoropoulos, A., Lepouras, G.: Digital game-based learning and computational thinking in P-12 education: a systematic literature review on playing games for learning programming. In: Handbook of Research on Tools for Teaching Computational Thinking in P-12 Education, IGI Global, pp. 158–179 (2020)
17. Combefis, S., Beresnevičius, G., Dagienė, V.: Learning programming through games and contests: overview, characterisation and discussion. Olympiads Inf. **10**(1), 39–60 (2016)
18. Marco, J., Bonillo, C., Cerezo, E.: A tangible interactive space odyssey to support children learning of computer programming. In: Proceedings of the 2017 ACM International Conference on Interactive Surfaces and Spaces, pp. 300–305 (2017)
19. Hlynka, D.: Postmodernism in educational technology: update: 1996-present. Handbook Res. Educ. Commun. Technol. **2**, 243–246 (2004)
20. Leblanc, S., Saury, J., Sève, C., Durand, M., Theureau, J.: An analysis of a user's exploration and learning of a multimedia instruction system. Comput. Educ. **36**(1), 59–82 (2001)
21. Myers, I.B., McCaulley, M.H., Most, R.: Manual, a Guide to the Development and use of the Myers-Briggs type Indicator. Consulting psychologists press, Palo Alto (1985)
22. Judge, T.A., Higgins, C.A., Thoresen, C.J., Barrick, M.R.: The big five personality traits, general mental ability, and career success across the life span. Pers. Psychol. **52**(3), 621–652 (1999)
23. Lau, W.W., Yuen, A.H.: Exploring the effects of gender and learning styles on computer programming performance: implications for programming pedagogy. British J. Educ. Technol. **40**(4), 696–712 (2009)
24. Milovanović, M., Minović, M., Kovačević, I., Minović, J., Starčević, D.: Effectiveness of game-based learning: Influence of cognitive style. In: World Summit on Knowledge Society, pp. 87–96 (2009)
25. Archambault, L.M., Barnett, J.H.: Revisiting technological pedagogical content knowledge: Exploring the TPACK framework. Comput. Educ. **55**(4), 1656–1662 (2010)
26. Prensky, M.: Digital game-based learning. Comput. Entertainment (CIE) **1**(1), 21 (2003)

The Counterintuitive Concept of Ergodicity in the Context of a Business Simulation Game

Chrysa Bika[1(✉)], Sarah Koblitz[1], Nilüfer Faizan[1], Robert Heininger[1], Matthias Christoph Utesch[1,2], and Helmut Krcmar[1]

[1] Technical University of Munich (TUM), Arcisstraße 21, 80333 Munich, Germany
chrysa.bika@tum.de
[2] Staatliche Fachober- und Berufsoberschule Technik München, Orleansstraße 44, 81667 Munich, Germany

Abstract. The concept of ergodicity as described in the *London Mathematical Laboratory* seems very counterintuitive. Something is ergodic when its time average equals to its expectation value. Transferring this sentence into a more suitable real-life situation, one can say pooling and sharing earnings will increase one's wealth steadily over time. One may wonder why should they cooperate with each other when they all have bad exterior conditions? Will cooperation not make everything worse? We developed a Business Simulation Game, *New Rising of the Ylsung* concept, which tests different parameters to visualize the concept of ergodicity. We distinguished that in most of the cases it brings a greater advantage to collaborate than to work by oneself. Moreover, the simulation helps high school students to understand the research of the *London Mathematical Laboratory* in a playful manner. Hence, they do not have to read research papers, but simply play through our Business Simulation Game while adjusting the parameter sets and analyzing the graphs. Our main aim is to find out to what extend the application of different parameters between cooperation and non-cooperation differ from each other and how the application of those looks in real-life scenarios.

Keywords: Ergodicity · Business simulation · Game · Playful learning · Economic concept · Ergodicity economics

1 Introduction and Motivation

Marc Elsberg's book "Gier" published in 2019 focuses on the works of the *London Mathematical Laboratory* (LML) scientists. The thriller builds on inequality, injustice, polarization, division of society and the great dissatisfaction with how politics and businesses deal with those aspects. Furthermore, the thriller presents a scientific approach to the discomfort of bringing the LML's approach closer to people [4]. The LML has several research areas from economics to inference, from models to limits and to learning. The concept of the *Farmer's Fable*, hence the concept of ergodicity, can be found within the economics research [16]. The *Farmer's Fable* within "Gier" is a fictional story describing the economic concept of ergodicity. In this

M. E. Auer and T. Rüütmann (Eds.): ICL 2020, AISC 1328, pp. 542–553, 2021.
https://doi.org/10.1007/978-3-030-68198-2_50

story, two farmers cooperate and share the earnings of their farming, whereas the other two farmers do not want to cooperate and consequently keep their whole earnings [4]. The advantage of cooperation lies in the concept of not risking everything you own. You always have somebody to rely on in difficult times. Even though you might have to give away most of your earnings, over time the profit will be larger. Whereas if you never share, you might make high profits for a couple of years, considering if any unexpectedly happens, like a natural disaster, you are at risk of losing everything you worked for.

We created the Business Simulation Game (BSG) *New Rising of the Ylsung* based on the *Farmer's Fable*. This project is a part of the *Business Simulation Design* module at the *Technical University of Munich*, with the target group of teenagers and young adults in middle and high schools. For this purpose, we constructed a single player game in Unity with C# and integrated Python. The BSG presents valuable information about the economic phenomenon of ergodicity.

Moreover, we conducted a user study and developed a quiz, which are not part of this paper, to get a better understanding of how the simulation is perceived by out standers and which parts we can improve during the development process. The group of play testers consisted of pupils at the *Staatliche Fachober- und Berufsoberschule Technik München*, a technical college in Bavaria, and of students at the *Technical University of Munich*. This quiz is designed to function as a pre- and post-test, to examine how well the participants understand the concept of ergodicity before and after playing the BSG and whether they are able to apply it to real-life situations.

It is important to consider that the *Farmer's Fable* takes up a span of several years [4]. When taking only a couple of years into account, they might not discover any significant advantages towards neither cooperating nor working alone. For this reason, we implemented an adjustable timeline. The user can regulate it between a span of 10 up to 250 years and observe the different outcomes.

2 Background

In the following sub-sections we define a simulation and the concept of ergodicity.

2.1 Simulation

The process of reproducing systems in various application fields is called "Simulation" [11]. These simulations may analyze possible issues within the system, without putting anyone or anything at risk [11]. A business simulation facilitates a faster growing of the learning curve, by creating an immersive feeling and with no risk of damaging a working system [3, 8, 9]. It allows to gain an overview over the project, while highlighting bottlenecks and raising questions for improvement of the overall product [3, 8, 9]. Furthermore, taking a closer look onto our target group, teenagers and young adults between the ages 14 to 25, it is of high importance for us to create a simulation that is easily understandable as well as enjoyable. Compared to only reading about a system or a process, the user has a first-hand experience and gains some hands-on experience. Since games, in particular serious games and business simulation games are able to create a link between entertainment and learning content, the BSG *New Rising of Ylsung* aims to be an interactive and playful way of studying economic principles and learning to adapt them quickly.

2.2 The Concept of Ergodicity

Ergodicity is a part of probability theory and related to statistics [16]. Both are difficult topics that students in German schools usually have to study during their mathematics classes at high school. We created the BSG *New Rising of the Ylsung* to make the concept of ergodicity more intuitive for this target group.

2.2.1 A Definition of Ergodicity

Ergodicity can be found in equilibrium statistical mechanics. For a system to be ergodic, it is assumed that the time average of an observable equals its expectation value. Even though, this concept is applied a lot in economics, it does not hold in most cases, since wealth is not ergodic [16].

We have to take a closer look on dynamical systems to find ergodicity. Now, a dynamical system is used as an object in physics to model certain phenomena. These objects can have states, which will develop over time. Three main themes can be distinguished within dynamical systems, namely the predictive ones, diagnostic ones and ones that aim to explain physical phenomena by theories [1]. Moreover, attention on the systems that are not only dynamical, but also indecomposable dynamical, has to be paid. Indecomposability means that something cannot be partitioned into several parts [13]. Our aim is to reach the points in a set of another orbit. Allowing the system to be broken down into smaller sets would make this impossible [1].

2.2.2 New Rising of the Ylsung

In this subsection, we give an overview of our BSG. At the start of the Game (see Fig. 1), the user has the option to choose between four and up to 20 farmers that can either cooperate or farm by themselves. Moreover the user can apply conditions to the farmer's fields. These conditions vary between good, bad and random (see Sect. 4.1). The adjustable timeline of the BSG ranges from 10 to 250 years.

While the BSG runs through, one can watch the farmers running around their fields, as shown in Fig. 1. Meanwhile a graph is depicted on the top left corner, which shows how much profit each farmer respectively each collaboration of farmers makes. This is useful to grasp the idea of the business concept in real time. To get a more elaborate explanation, the user receives more graphs when the simulation is completed. Those graphs are ordered in the shape of spider webs to explain different parameter sets that act upon the profits of the farmers, simultaneously. The impact of various parameter sets will be discussed in Sect. 4.1.

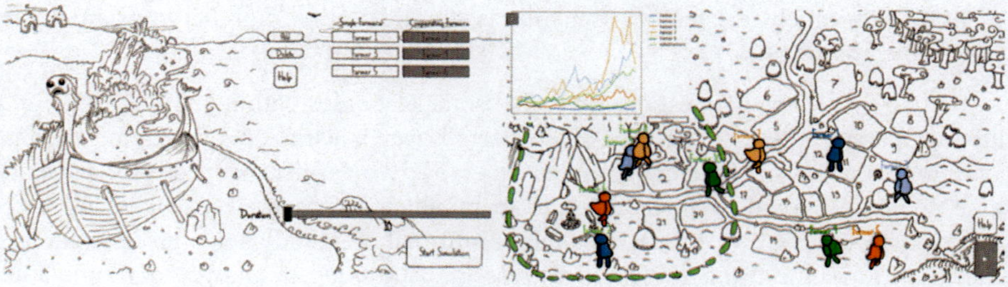

Fig. 1. Start screen and game screen of the BSG *New Rising of the Ylsung*.

2.2.3 Ergodicity in *New Rising of the Ylsung*

According to the philosopher Thomas Hobbes' conception of a human being, a human being can be understood as selfish by nature [7]. Moreover, interpreting John Locke's philosophies, man is altruistic [12]. At the beginning of the BSG, the user is asked which and how many farmers should collaborate for a certain amount of time. A first intuition makes us wonder why the farmers should cooperate in the first place. It is a competitive world that we live in, hence, would not it make more sense for the farmers to work on their own and keep the whole amount of profit they make? Nowak argues in his paper on *Five rules for the evolution of cooperation*: "[…] cooperation is the decisive organizing principle of human society […]" [1] and that "Cooperation is needed for evolution to construct new levels of organization" [1]. So in contrary to the first intuitive belief, it should be beneficial, not only for "genes [to] cooperate in genomes" [14] and for "chromosomes [to] cooperate in eukaryotic cells" [14], but for farmers to cooperate in *New Rising of the Ylsung*, so that they can increase their profits.

This paper explores how in every cooperation one *entity* has more of a certain aspect, such as profits in our case, than the other cooperating *entity* can provide. This means that one *entity* will decrease their previous value when they share [14, 18]. However, considering cooperation in a long-run, it is quite possible that giving and receiving will balance out between the entities. In research of the LML, Adamou and Peters [18] describe this idea as reciprocity and relatedness. Reciprocity describes the balancing of giving and receiving [18]. In some cases, *entity_1*, in our game called a farmer, will share their profits and in other cases *entity_1* will gain from their cooperation partner, *entity_2*, as well a farmer. The latter, relatedness, aims to spread genetic material that *entity_2* is carrying [14, 18]. This case is of lower importance for the BSG. We will not explicitly look at the biological aspects, as the farmers are not related to each other, neither are their crops.

3 Research Questions and Methodology

The LML provides a great knowledge base, which we try to support with our BSG. For our research, we posed the following two research questions:

RQ1: How does a parameter set have to look like to support the *London Mathematical Laboratory* concept of ergodicity by using a business simulation?

RQ2: For which real human scenarios is the concept of ergodicity a particularly suitable solution?

For the first research question, we define some test cases within our BSG and apply different parameter sets. These parameter sets cover a variation of extreme weather conditions among six different farmers, whereby two farmers respectively have the same conditions for their fields in order to be able to compare harvest yields retrospectively. Moreover, we conducted a minor informal user study and a presentation at a Games Symposium at the *Technical University of Munich* to analyze our simulation and the parameter outcomes.

However, in real-life one cannot influence the weather as they like, neither do farmers have unlimited field sizes. These reasons lead us to the next question, RQ2. More situations than farming exist where one encounters the concept of ergodicity. To answer RQ2, a literature research according to [19] was done by using the following keywords on the website of the LML and in Google Scholar: "ergodicity, applied ergodicity, ergodicity in nature, ergodicity in real-life". According to Peters [17] ergodicity economics predict functional forms, given the wealth dynamics. However, wealth dynamics is not a well-known or well predictable factor. Therefore, ergodicity economics is not a trivial concept in mathematics. Apart from the economics factor, the question leads to other real-life scenarios where ergodicity has been applied, can be applied or will be applied in future – even if only partially.

4 Results

In the following sub-sections, we will answer RQ1 with the findings of our BSG and RQ2 with literature research.

4.1 Applying Parameter Sets

To analyze and answer RQ1, various sets of parameters were applied and the results were compared, accordingly. Not fixating the BSG to the original story of Marc Elsberg [4], where he explains the concept with the use of four farmers but giving the user the opportunity to change the parameters, the BSG becomes more interactive. This interaction helps users to learn more thoroughly, since they can manipulate the game and observe the different outcomes, instead of following a strict presentation of the outcomes as they would do in frontal education. As Zimmermann and Miliband explain in their paper on self-regulated personalized learning (SRPL), the student, in our case the user, holds the responsibility for learning themselves [20].

While playing through the BSG, an intuition might lead to the Prisoner's Dilemma [10], where the most common strategy for winning is tit-for-tat. Both parties rely on each other's answer, where one will cooperate only if the other does the same. However, they cannot know each other's reaction, and might fall back into egoistical thinking due to mistrust in the other party. While changing the strategy in the midst of the game is not possible in the BSG, the next simulation cycle offers the opportunity to change the arrangement of the cooperating farmers, such that adequate results of the various strategies can be drawn.

Finally, to answer RQ1 the BSG was tested multiple times with different parameter sets. Two interesting sets are explained in further detail in the following. For our examples we use six farmers. Thus, *Farmer 1* and *2* have the same weather conditions, as *Farmer 3* and *4* and *Farmer 5* and *6* do. While *Farmer 1*, *Farmer 3* and *Farmer 5* do not cooperate, *Farmer 2*, *Farmer 4* and *Farmer 6* share their earnings after each year.

An important factor of the BSG lies in the adjustable timeline, since the most significant results are achieved after a huge time span. It is important to consider several generations of farmers, rather than just one generation. To conclude our results, we sat the timeline to 50 years, 100 years, and 250 years and compared the results under the same weather conditions. For our exemplary cases, we are setting the timeline to 100 years.

Please note that farmers might be able to predict on average how the climate is going to change in the future. However, they are not able to make accurate predictions about the weather [6]. It can stay the same over a period of time or change rapidly. The decision of whether the farmers want to collaborate for the next certain amount of time gets tougher when they are not able to distinguish how the exterior factors will act upon their fields.

In the following table (see Table 1), we distinguish the influence of the weather conditions on the fields by their multiplying factors displayed as float values.

Table 1. Weather conditions and their multiplying factor

Weather Conditions	Multiplying Factor
Sunny	1.33
Cloudy	1.15
Rainy	0.97
Stormy	0.79
Deer attacks	0.6

These values only apply if the user influences the weather. Otherwise, a range between 0.6 and 1.5 is given for the random weather condition. Nevertheless, these results, which might be adaptable to real-life situations, could be derived within the prefixed range.

Table 2 is an exemplary observation for six different simulations with different parameters that run with an annual time span of 100 years. The same parameters are also used for simulations lasting 50 and 250 years, but they are not shown here.

Table 2. Weather conditions for each simulation and harvest amount after 100 years.

Farmers	Weather Conditions											
	1st simulation		2nd simulation		3rd simulation		4th simulation		5th simulation		6th simulation (divided in even sections)	
Farmer 1	Random		Random		Sunny		Deer Attack		Sunny		Sunny -> Cloudy -> Rainy -> Stormy -> Deer Attack	
	Harvest kg	0,62	Harvest kg	37,66	Harvest kg	2,43 * 10^12	Harvest kg	6,53 * 10^(-23)	Harvest kg	2,43 * 10^12	Harvest kg	<0,01
Farmer 3	Random		Random		Cloudy		Stormy		Rainy		Deer Attack -> Stormy -> Rainy -> Cloudy -> Sunny	
	Harvest kg	0,17	Harvest kg	5,11	Harvest kg	1,1 * 10^6	Harvest kg	5,79 * 10^(- 1)	Harvest kg	0,05	Harvest kg	<0,01
Farmer 5	Random		Random		Rainy		Rainy		Deer Attack		Rainy -> Cloudy -> Stormy -> Sunny -> Deer Attack	
	Harvest kg	0,42	Harvest kg	967,01	Harvest kg	0,05	Harvest kg	0,048	Harvest kg	6,53 * 10^(-23)	Harvest kg	<0,01
Collab. Farmer	Mix of above		Mix of above		Mix of above		Mix of above		Mix of above		Mix of above each period	
	Harvest kg	3,63	Harvest kg	477,23	Harvest kg	47,87	Harvest kg	100,05	Harvest kg	2,37	Harvest kg	2,62

4.1.1 Cooperation Under Random Exterior Conditions

Since random weather conditions are closest to reality, we will outline an exemplary case in this section. For all simulations, we used 1.00 kg as initial value. In a real-life situation, farmers would use tons of crops and not only 1.00 kg. We can extract the values of profit with a total of ten decimal digits, but decided to use only two decimal digits in this paper.

For our first observation (see Fig. 2) with random weather conditions, we consider a lifetime of 100 years for the simulation cycle. In the first 40 years, the growth curves of all farmers look quite similar. Only *Farmer 1*, who is not collaborating, cannot expand their profits and is staying stable at a very low level. After 80 years, they finally start to grow, only to reach their peak of 15.81 kg after a time span of 100 years.

Farmer 3 has a highly fluctuating growth curve. They have several peaks, for instance, at 43 years and again at 60 years. However, every few years they must deal with great losses. Their curve is not stable at all and is continuously decreasing, down to 4.00 kg after 100 years. This value is very close to their initial starting value.

Farmer 5 can never reach their opponent *Farmer 3*. We observe that after the major loss from year 45 to year 55 *Farmer 5* manages to develop a more stable growth. They are even able to outperform *Farmer 3* and reach their top performance with 21.99 kg after 100 years.

The most remarkable are the collaborating *Farmers 2, 4* and *6*. Even though they are outperformed by *Farmer 3* between year 40 and 65, they never have to deal with great losses. Their growth curve is continuously increasing. We observe the huge differences between the respective farmers. Those who do not cooperate have to deal with great losses and keep their profits quite low, whereas the cooperating farmers are steadily increasing their value. They even reach 118.49 kg after 100 years.

Hence, we conclude that under random influences it is of advantage to cooperate. This might seem counterintuitive in today's society, since everybody wants to grow their own profits without sharing. However, as the LML observes with their research: "[...] sharing increases the long-time growth rate for cooperating entities, meaning that cooperators outgrow similar non-cooperators" [18].

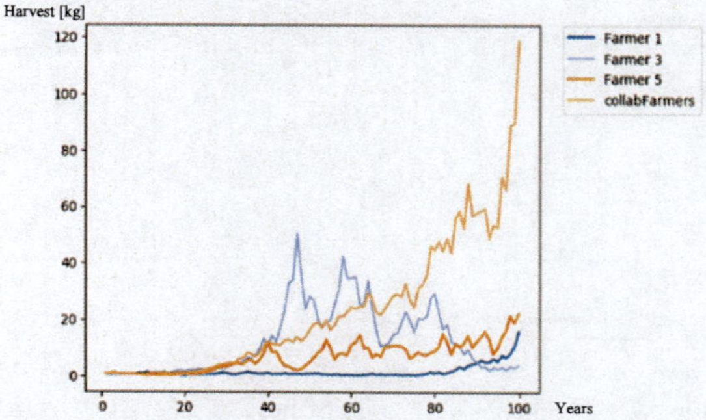

Fig. 2. Harvest after 100 years with random weather conditions.

4.1.2 Cooperation Under Bad Exterior Conditions

Since we heavily influenced the parameters in this case, the outcomes are observed in more detail. We set the parameters such that the farmers only have bad weather conditions through the whole iteration. The time frame of the simulation is set to 100 years. Rapidly we can distinguish a trend: it is of advantage to cooperate.

We fixed following exterior conditions for the fields of our farmers:

Farmer 1 and *2*: only deer attacks
Farmer 3 and *4*: only storm
Farmer 5 and *6*: only rain

That explains why the weather influences for each variant are depicted as straight lines in the spider web graphs (see Fig. 3).

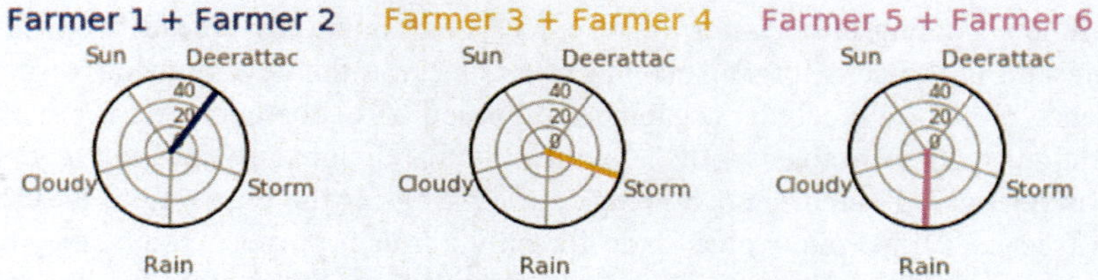

Fig. 3. Bad weather conditions acting on the variants of fields.

In Fig. 4, *Farmer 1*, *3* and *5* are only depicted as straight lines, which means they have a constant value of zero. Even if, within the 100 years of farming, they might have raised their profits a little bit, they dropped rather quickly to zero again. Those minor fluctuations are so insignificant that they are depicted as the straight lines we see at the bottom. Moreover, the farmers even lost their initial value and could not raise it again. On the contrary, the collaborating *Farmers 2, 4* and *6* who had bad weather conditions, as well, managed to make profits as high as 120 kg.

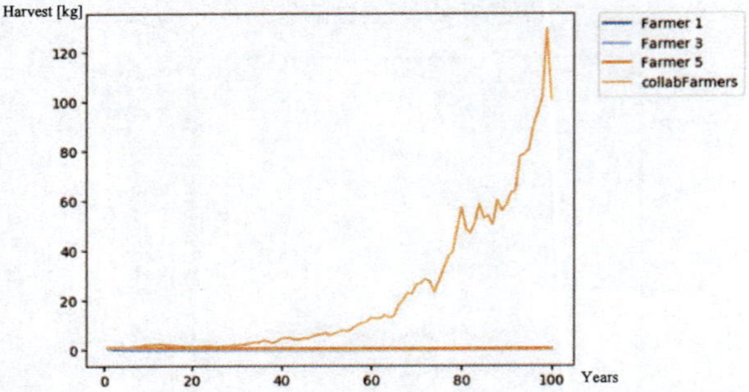

Fig. 4. Second cycle. Profits of harvest after 100 years.

We distinguish that if the farmers work together, they can share even the little amount of harvest they make and hence, grow their profits. About 30 years into the simulation, it already shows the advantage of collaborating farmers. One might think it would be better to hold on to the little profit they make and to try to expand it on their own. Nevertheless, the simulation shows the exact opposite. Ergodicity might seem very counterintuitive, especially when the exterior conditions are bad for all participating teams. However, it certainly improves the situation for everybody when they share. Again, we conclude that it is of advantage to cooperate when we know that we are going to encounter consecutive years of bad exterior conditions. To answer RQ1, a parameter set should consist of randomized weather conditions, which means uncertainty for the future of the farmers, or of bad weather conditions for all farmers, where they depend on each other to cooperate and share their profits.

4.2 Ergodicity as a Solution for Real-Life Scenarios

In evolution, pooling and sharing of resources is a commonly spread behavior. It is omnipresent, in the cells, organisms and more, because the development respectively emergence of cells and cellular organisms are based on cooperation and for evolution, cooperation is indispensable [14, 18]. On the other hand, humanity is intuitively seen as profit-driven, as individual human beings are looking for the best way to ensure their wealth (see 2.2.2). No one would share for only altruism purposes, since they cannot gain anything in return. Apart from this counterintuitive wealth factor, as asked in RQ2, can we apply ergodicity also on further human made situations? As defined by [15] ergodicity means: "In an ergodic scenario, the average outcome of a group is the same as the average outcome of the individual over time." For instance, coin toss is an ergodic situation as if 100 people flip a coin once or one-person flips a coin 100 times the outcome is the same. Even though consequences of these outcomes are typically not ergodic. To find real-life scenarios we must compare two outcomes, the individual's and the group of individuals', and when they have the same outcome, it is an ergodic scenario. Since almost every human scenario is non-ergodic, we must have a closer look at how non-ergodic situations can be made more ergodic. Thus, diversification plays an important role, since it is defined as "the process of starting to include more different types or things", according to the Cambridge Dictionary [2]. There are various human scenarios that we could take on, but in order not to go beyond the scope of this paper, we use an example of an investment strategy that we create with the help of diversification ergodic. A businessman called Bill wants to invest his money in a secure way but also with profit. So, he could combine two different investing approaches for example treasury bonds with stocks. What matters is not the average correlation, but correlation during a crisis, e.g. the actual Covid-19 crisis. It is suggested that Bill should put, during a timespan of his life, around 80 percent into something less risky, such as real estates, and 20 percent into something much riskier, such as starting another business [15]. This strategy of diversification is called "Barbell Approach" and it can be inferred that the outcome of Bill's investment is more ergodic when he splits the risk management into smaller risk factors instead of taking one big risk [15].

As an answer to RQ2, one can say, there is not *that one* scenario where the concept of ergodicity is a suitable solution but with diversification, you can make almost every scenario more ergodic than it was before.

5 Conclusion

While answering RQ1 we found out that the concept of ergodicity from the *London Mathematical Laboratory* can be supported with the random and bad weather condition parameter sets inside our simulation. However, there are much more factors which are not considered for today's farmers, such as limited fields, planting space, and investing the profit into better agricultural vehicles and crops. With these factors in mind, we can relate to RQ2, which explains that there is no real ergodic scenario, but it is possible to improve a scenario and make it more ergodic than it was before. With this research one can say, that a world ruled by the concept of the *Farmer's Fable* is a utopia. However, with the concept of ergodicity partially applied to real-life scenarios one can raise their outcome over time. This might be especially helpful for teenagers and young adults. By applying different parameter sets themselves, they gain hands-on experiences. Furthermore, the classroom should provide the students with a safe space to discuss their findings and dive deep into the topic of ergodicity economics. Our aim is not to impose a fixed strategy on them, but we rather aim at provoking their thoughts to develop suitable strategies themselves and to be able to think further of how they can exceed the scope of the Business Simulation Game to real-life scenarios. Moreover, the students gain familiarity with the concept of not being always right. They have to try out several parameter sets to find the best suitable one, sometimes there might be more and sometimes they might not find even a single one, with their respective strategy. As mentioned previously we conducted a case study during the development of the Business Simulation Game. The outcome of further improvement of the game, such as game balancing, and discussion of the case study combined with bringing statistics and stochastic closer to students will be a topic of another paper. For instance, the students should learn the fundamental concept of ergodicity economics and be able to scrutinize its value for real-life scenarios. As a famous (disputed) quote from Richard Feynman says: "I would rather have questions that can't be answered than answers that can't be questioned" [5].

Acknowledgments. The authors like to express their gratitude and appreciation to the thought-provoking impulse by Prof. André Thomas and Prof. Dr. François Bry during the Game Based Symposium at the *Technical University of Munich*. Furthermore, they thank the group of students (Leonard Jungkunz, Cynthia Drews, Alexander Williams, Simon Stolz, and Timotej Svitec) with whom they developed the Business Simulation Game and all the participants of their informal user studies who gave them the constructive feedback they needed to improve their simulation. Moreover, they would like to thank the reviewers and editors for their helpful and supportive comments. Last but not least, the authors would like to express their gratitude towards their families and friends.

References

1. Brown University: Learning Dynamical Systems: A Tutorial (1998). http://cs.brown.edu/research/ai/dynamics/tutorial/home.html
2. Cambridge Dictionary (ed.): diversification: Cambridge University Press. Available online at 23. https://dictionary.cambridge.org/de/worterbuch/englisch/diversification. Accessed 27 May 2020
3. Celemi - the power of learning: What is a Business Simulation? https://celemi.com/business-simulations/what-is-a-business-simulation/. Accessed 7 May 2020
4. Elsberg, M.: Gier. Wie Weit Würdest du Gehen?: Roman, 1st edn. Blanvalet, München (2019)
5. Goodreads: Quote: "I would rather have questions that can't be answered than answers that can't be questioned". https://www.goodreads.com/quotes/1134331-i-would-rather-have-questions-that-can-t-be-answered-than. Accessed 7 May 2020
6. Harper, L.: What Are Climate Models and How Accurate Are They? Edited by State of the Planet. Earth Institute Columbia University (2018). https://blogs.ei.columbia.edu/2018/05/18/climate-models-accuracy/. updated on 5/1/2018, checked on 5/27/2020
7. Hobbes, T.: Leviathan oder Stoff, Form und Gewalt eines bürgerlichen und kirchlichen Staates. With assistance of Iring Fetscher, Walter Euchner. Neuwied, Berlin: Luchterhand (Politica, Bd. 22) (1966)
8. Blažič, A.J., Novak, F.: Challenges of business simulation games — a new approach of teaching business. In: Gradinarova, B. (ed.) E-Learning - Instructional Design, Organizational Strategy and Management, InTech, Rijeka (2015)
9. Keys, B., Wolfe, J.: The role of management games and simulations in education and research. J. Manage. **16**(2), 307–336 (1990)
10. Kuhn, S.: (ed.) Prisoner's Dilemma. With assistance of Edward N. Zalta. Winter 2019: Metaphysics Research Lab, Stanford University (The Stanford Encyclopedia of Philosophy) (2019). https://plato.stanford.edu/archives/win2019/entries/prisoner-dilemma/

11. Lackes, R., Siepermann, M.; Rottmann, H., Benjamin R.A.: Lübbecke, Marco: Simulation. Edited by Springer Gabler. Gabler Wissenschaftslexikon. https://wirtschaftslexikon.gabler.de/definition/simulation-43833/version-267158, updated on 2/19/2018, checked on 7/15/2020
12. Locke, J.: Two treatises of government. A critical ed. with an introd. and apparatus criticus/by Peter Laslett. Hg. v. Peter Laslett. Cambridge: Cambridge University Press (1960)
13. Miriam-Webster (ed.): indecomposable: Miriam-Webster.com Dictionary. https://www.merriam-webster.com/dictionary/indecomposable. Accessed 27 May 2020
14. Martin, A.N.: Five rules for the evolution of cooperation. In Science vol. 314 (5805), pp. 1560–1563 New York (2006)
15. Taylor , P.: A Big Little Idea Called Ergodicity. Or The Ultimate Guide to Russian Roulette (2019). https://taylorpearson.me/ergodicity/. updated on 10/10/2019, checked on 5/27/2020
16. Peters, O.: The ergodicity problem in economics. Nat. Phys. **15**(12), 1216–1221 (2019)
17. Peters, O.: The Copenhagen Experiment (2018). https://ergodicityeconomics.com/2018/05/29/the-copenhagen-experiment/#more-3248, updated on 5/29/2018, checked on 5/27/2020
18. Peters, O.: Adamou, A.: An evolutionary advantage of cooperation (2015). http://arxiv.org/pdf/1506.03414v2
19. vom Brocke, J., Simons, A., Niehaves, B., Niehaves, B., Reimer, K., Plattfaut, R., Cleven, A.: Reconstructing The Giant: On The Importance of Rigour in Documenting The Literature Search Process. Ecis 2009 Proceedings (2009). https://aisel.aisnet.org/ecis2009/161
20. Zimmerman, B.J.: A social cognitive view of self-regulated academic learning. J. Educ. Psychol. **81**(3), 329–339 (1989)

A Guide for the Development of Game-Based Evacuation Simulators

George Kougioumtzoglou(✉), Anastasios Theodoropoulos, and George Lepouras

HCI-VR Laboratory, University of the Peloponnese, Tripoli, Greece
gkoug80@gmail.com

Abstract. The evacuation process of complex scenarios like ships and buildings evacuation is a challenge under many conditions. Simulators are a worthy method to experiment in the field. However, it is very difficult to simulate evacuations and engage participants like in real conditions. This article presents a guide based on previous works about 3D evacuation simulators. The guide incorporates (i) hazardous elements such as fire sources, smoke, poisonous gasses, falling/flying objects; (ii) spatial elements such as bottleneck areas, congestion areas, stairs, elevators; (iii) human and the crowd behavior such as the behavior under psychological pressure and (iv) movement features such as the mimic behavior, the initial reaction time, the family-follow-up behavior, and the group-leader-follow-up behavior. Moreover, some evacuation scenarios such as ships and underground structures require the implementation of specific elements. The proposed guide, suggests all the aforementioned features, with gaming characteristics and with a game engine software for the development of evacuation simulators. The game simulators should support features such as gravity, physics, AI, multiplayer ability, programming ability etc., in order to adapt in different scenarios effectively.

Keywords: Evacuation simulation · Game-based simulator · Guide for simulators

1 Introduction

The development of evacuation models and computer-based evacuation simulators arose from the inadequacy of evacuation drills because of their high cost, their poor repetitive capability and their easiness to cause accidents [1]. Although an evacuation drill may provide useful information it lacks some basic elements. The most notable omissions are the psychological influence on a human being, while he/she is acting under stressful conditions, the social forces between the evacuees, the influence of panic reactions by other evacuees, and finally the unexpected incidents that may happen during an emergency evacuation [2]. On top of that, the fire alarm systems (fire is the most common evacuation cause) have often failed to work "as planned" when a real fire breaks up because they were put in place with false expectations regarding how occupants actually behave during fires [3]. Therefore, the simulation of the human

M. E. Auer and T. Rüütmann (Eds.): ICL 2020, AISC 1328, pp. 554–566, 2021.
https://doi.org/10.1007/978-3-030-68198-2_51

behavior using a computer program is a challenging task. Moreover, the evacuation drills of simulators can rarely provide trustworthy and useful results.

To this end, come the game-based evacuation simulators, programs that aim to simulate evacuations using an artificial environment, based on realistic conditions and represent the emergency egress from an area or a structure. An evacuation simulation can be conducted as many times as it needs to and under every possible condition. The main division of computer-based evacuation simulators is based on the users' perspective. Two-dimensional simulators use dots or artifacts for the representation of the evacuees or any other elements such as a fire and smoke and viewers perspective is usually a top-down one. A 2D approach is not inferior than a 3D one, but its usability is restricted to specific functions and specifications. The evacuation models are functionally implemented, while the evacuees are rendered symbolically, while the simulation can be accelerated or decelerated according to the user settings. On the other hand, a 2D simulator cannot support the direct participation of a player impersonating an evacuee and therefore their unexpected behavior. The computer-controlled bots do not always act as a human being, rendering the collected data less reliable [4]. Improved and well-designed algorithms may control a bot towards realistic reactions and behaviors but can only simulate a small portion of the qualities and the unpredictable behavior of a human being or a group of human beings. Three-dimensional simulators usually use 3D human models for the evacuees or any other roles (rescue team staff and crew members etc.). The structures (e.g. buildings) are usually 3D models of existing structures further enhancing the realism and are more or less in a gaming environment. The importance of this approach, is based on the easiness and improvement of the understanding of human behavior in evacuation situations, as the subjects are monitored while they are playing the game and some performance measures are logged to be further analyzed later on [5]. Nygren [6] proposes that a 3D evacuation simulator can combine the focus on the behavior of individuals and the behavior of crowd, as it allows the participation of real players simultaneously with computer-controlled bots.

This article presents a guide based on previous works about 3D evacuation simulators. The guide incorporates the most important elements from previous simulators and human behavior studies and movement features. Moreover, it incorporates gaming characteristics and a game engine software for the development of evacuation simulators.

2 Background

Users' immersion is one of the key features of virtual environments, such as computer games or simulators, especially if the technology of Virtual Reality is implemented. Brown and Cairns [7] argue that in the stage of the "total immersion" the user's senses are cut off from the real world and only the end of the computer game is all that matters. The level of immersion is defined by the interest that a virtual world causes to the player, the permanent existence of tasks to complete and continuous update of challenges) and the feeling of flow, where people are completely absorbed in the activity [8]. The main features of 3D evacuation simulators, developed with immersion characteristics, are briefly reviewed below.

Ren et al. [9] proposed the application of computer simulation and information technology for the disaster prevention and mitigation, under the name, VR system. This simulator is purposed to simulate fire emergency evacuation drills, while provides emergency evacuation training. The main concept is that a single player controls a virtual avatar, while performing firefighting tasks. The simulator was developed combining, the Vega Engine for the virtual environment, the Multigen Creator for the rendering of the 3D building models, and the Fire Dynamics Simulator for the simulation of the fire sources and Visual C++ 6.0 (Fig. 1).

Fig. 1. Simulation of fire, screenshots from the VR system [9].

Ginis et al. [10] developed the VELOS Simulator, a multi-user Virtual Reality (VR) system that aims to support designers on the assessment of the activities and movement of the passengers and the crew members of a ship, early in the design process, for normal and hectic conditions of operations in order to improve the design of passenger ship (Fig. 2).

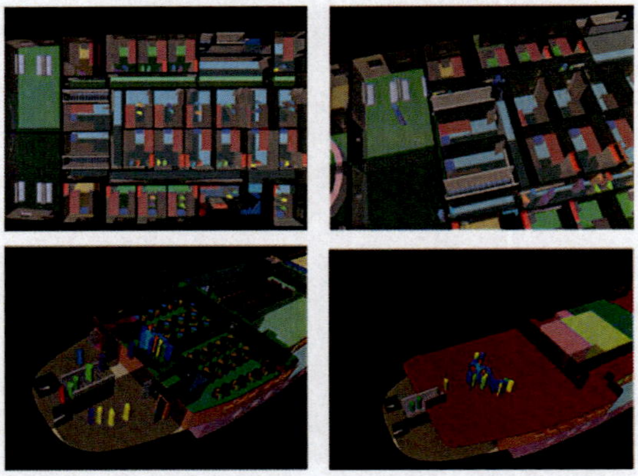

Fig. 2. Different views and levels from Velos [10].

The VELOS assesses various ship features, such as the evacuation process, the ergonomics and the comfortability. This work is in line with IMO Interim Guidelines for Evacuation Analysis of Passenger Ships and aims to develop an integrated environment for the rational analysis and assessment of real emergency conditions, removing this way restrictive assumptions and omissions of IMO, which lead to the need of safety factors.

Finally, Serious Game [11] is a 3D game-based evacuation simulator and is a combination of a simulation and a serious game. Serious Game simulates the evacuation of buildings by a crowd of computer-controlled bots with the participation of only one human player (the developers claim that the next step is the participation of more human-players). Ribeiro et al. [5] state, that the Serious Game extends a popular game engine to implement a pedestrian simulator to study evacuation dynamics and to provide an appropriate environment for testing the influence behavior of egresses of a building in hazardous situations, such as fires (Fig. 3).

Fig. 3. Crowd trying to find an exit, screenshot from Serious Game [11].

3 Evacuation Simulators Models

Evacuation Simulators Models simulate the movement and behavior of crowds [12]. Kuligowski et al. [13, 14] propose a categorization of evacuation simulators into three main types: Behavioral-based, Movement-based and Partial-Behavioral. These three categories imply the most prominently applied element of each model. A Behavioral model is focused on the simulation of evacuee's behavior while a Movement on the simulation of evacuee's movement. Partial-Behavioral models are primarily Movement-based but Behavioral elements are implemented during the evacuation sessions.

The Behavioral models are categorized further to No Behavior, Implicit Behavior, Conditional (or Rule), Artificial Intelligence (AI) and Probabilistic Models. The main Movement Models are the Density Correlation, the User's choice, the Inter-person Distance, the Potential, the Emptiness of next grid cell, the Conditional, the Functional analogy, the Other model link, the Acquiring Knowledge, the Unimpeded Flow and the Cellular automata [14]. Summarizing, Ribeiro et al. [11] ascertain, that the most commonly used models are:

- **Cellular Automata Models**; these models represent the floor as a cells grid. The occupants move from cell to cell by a simulated throw of a weighted dice.
- **Forces-based Models**; these models apply the behavior of natural forces such as Magnetic Forces, Fluid Spreading or Social Forces, onto evacuees' movement.

- **Artificial Intelligence-based Models**; these models attempt to simulate the real human behavior, based on AI algorithms.

The following sections present the most used elements on evacuation models and describe their effects on the simulation process.

3.1 Hazardous Elements

The developers of an evacuation simulator should consider to implement elements, such as fire or smoke, which significantly increase the injuries and the fatalities.

A fire is one of the most common evacuation causes. Thus, an evacuation simulator should include fire as a hazardous element and more specifically a feature that (i) spawns fire sources of various sizes and intensities and (ii) controls their expansion randomly or controllably. Moreover, the flames should inflict damage on humans and/or bots. If the exposure is prolonged, the subject should be, virtually killed. The implementation of the fire not only affects the emotional and behavioral characteristics of an evacuee but also affects his/her physical constitution [15]. If one of the avatars' qualities is fatigue, an overexposure to high temperature should decrease it.

The outbreak of a fire is usually accompanied by smoke. If a simulator includes the presence of fire should also include the presence of smoke. Attention should be paid to (i) match fire and smoke, (ii) of the smoke, while, it is filling the rooms and the corridors restricting the movement and vision of the occupants and (iii) the harmful consequences of smoke inhalation.

3.2 Spatial Elements

The evacuated areas include various spatial elements such as stairs, elevators etc. that may alter the speed and the locomotion of crowds or individuals. These elements should be implemented to the most evacuated areas and should affect the pedestrian movement. A bottleneck situation occurs, when many pedestrians rush to an evacuation exit causing a drastic breakdown of outflow flux from the exit owing to human jam effect, sometimes referred to as the human arch effect [16]. A bottleneck denominates a limited area of reduced capacity or increased demand. For pedestrian movement bottlenecks are usually formed by direct capacity reduction (door or corridor). Bottleneck areas are crucial, as they affect (i) the evacuation time, (ii) the moving speed and the locomotion of the human crowds and (iii) the behavior of the evacuees inside and the evacuees in the immediate proximity of the crowd.

Wang et al. [17] describe this phenomenon as congestion. In normal evacuation process, all the agents only need to arrive at the exit as soon as possible and there is no overtaking. In congestion evacuation process, some evacuees want to move to the exit quickly by crowding others. Since under intense stress people try to move faster than normal, the overtaking and casualty phenomenon usually arise. The congestion phenomenon is very common in emergency situations.

Moreover, after a disaster, a number of people tend to moving against the main stream of evacuees toward their cabins or seats in order to search their relatives and gather their belongings or their life jackets. This flow of passenger is called Opposite-Flow or

Counter-Flow significantly reducing the overall moving speed, because of the resulted overcrowding and counter flow [18]. It has been observed, that during a fire, even the firefighters might create an Opposite-Flow phenomenon, while they move towards the fire source and the evacuees attempt to move away from it [19].

Another spatial element is the staircase movement, a complicated three-dimensional movement and its modeling is a challenging task. Several studies are focused on the effect of the crowd or individual's movement on stairs and stairwells and the use of elevators during an evacuation. Previous work state particular issues, that should be taken into consideration and applied on evacuation simulators [20].

Finally, an evacuation that uses elevators has many advantages such as the vertical opposed to smoke's motion movement of the elevators, the easiness of elderly, children and physically impaired evacuation, the familiarity of people because of their daily use and the minimum physical effort that their use demands [14]. On the other hand, the disadvantages are the poor reliability, the piston effect that can induce fire spread, the difficulties on evacuation organization, especially when a crowd of people attempts to enter the limited space of an elevator [21].

3.3 Human Behavior

Human evacuation demands the implementation of the behavioral characteristics on the evacuees, in evacuation simulators. These characteristics aim to realistically predict the human behavior during an egress.

Table 1. Phases that the simulation of a crowd should follow.

Pre-movement or initial response time phase	Movement phase	Rescue phase
• Recognizable signs forecast a disaster • Initial Awareness of the Incident • Perception of Seriousness • Gathering of personal belongings • Alerting other people/discussion with other people	• Beginning of immediate evacuation based on specific behaviors: Overactive, noneffective behavior, out of emotional control (15%). Apathetic and nervous behavior, lack of initiative (75%). Calm, with overall picture, vigorously, potential leadership (10%) • Searching/Gathering of family/friendly persons • Location recognition • Follow or Lead a group • Some help other evacuees/Firefighting (a percentage of the evacuees) • Move toward to an exit • Search for an alternative exit if others are blocked	• Survivors are more or less safe • Evacuation is sometimes acute, because of injuries and physical damage • Survivors are ready to evacuate or to be evacuated

There are three main approaches for evacuation simulation models; the macroscopic, mesoscopic and microscopic. The macroscopic considers the evacuees as a whole while the microscopic handles each evacuee, as an individual. Mesoscopic approach combines the macroscopic and microscopic approach. Although a microscopic approach is computational expensive, it provides more reliable results, than a macroscopic, as each individual acts according to the behavior of a single person. The movement and behavior of a crowd of evacuees may harm members of the crowd and/or for other evacuees or may lead to disastrous incidents. Analyzing studies on hypothetical scenarios, real evacuations and evacuations drills, it is suggested that the simulation of a crowd or individual evacuees should follow basic phases (Table 1).

Moreover, a stressful condition may cause panic reactions. Panicked evacuees sometimes transfer the control to others but contrary to the Leader-following behavior, this behavior is irrational and leads to harmful decisions such as the ignoration of the side doors and eventually to an increase of the overall fatalities. Helbing et al. [4] list several reaction of panicked evacuees (the elements of this list should be randomly microscopically and macroscopically implemented on an evacuation simulator as it is proposed in brackets by the authors of this article, Table 2.)

Table 2. Reaction of panicked evacuees.

Microscopic	In escape situations, panicked individuals get nervous
	People try to move considerably faster than normally
	Individuals start pushing and interactions among people become physically in nature
	Escape is slowed down by fallen or injured people turning into obstacles
	People tend to show herding behavior, i.e., to do what other people do
	Alternative exits are often overlooked or not efficiently used in escape situations
Macroscopic	The physical interactions in jammed crowds add up and can cause dangerous pressures up to 4,500 N/m, which can bend steel barriers or tear down brick walls
	Moving and in particular passing of a bottleneck frequently becomes in coordinated
	At exits jams build up. Sometimes, arching and clogging are observed
	People tend to show herding behavior, i.e., to do what other people do

Furthermore, people tend to follow leaders during a disaster. A leader could be an unknown person, a relative or a friendly person that takes this role after the outbreak of a disaster. Alternatively, a leader could be a crew member, a rescue team member or the staff of a store. The followers usually accompany one or more leaders, composing groups of evacuees, towards an exit or a safe place. A Leader-Following Behavior should be implemented on an evacuation simulator microscopically measuring on the one hand the scale of social status, the influence, the prestige of the leader and on the other (i) the familiarity between the leader and every follower, (ii) the alleged family relationship and (iii) the age and the social status of the followers.

A special group of evacuees are the children especially from 2 to 7 years old. A child is motivated by different incentives compared to an adult. A characteristic feature of children's behavior is acting without thinking, under the influence of momentarily occurring feelings and wishes [22]. The children's' pre-movement time (time before their initial movement after the outbreak of the disaster) is longer compared to an adult because they tend to stay in place even after a vocal sign waiting for their educators (many experiments are conducted at schools) to take them by the hand. Therefore, the immediate intervention of adults is critical for the evacuation of children, because of their tendency to remain in the same place or to search their parents or teachers in random locations.

Finally, physically impaired people compose a separate category of evacuees. There are divided to two main categories; the ones who can move on their own and the ones who cannot. Disabled people on a wheelchair can move on their own but they cannot use stairwells (the use of a motion platforms is an alternative for stairwells, if available) and their speed is generally slower compared to the other evacuees. Physically impaired people who cannot move on their own, ought to be carried by others (relatives, staff or random individuals).

3.4 Special Elements

The elements in a simulated area are not always harmful nor the scenery of an evacuation simulator is always a building. Evacuations may occur in ships, airplanes, oil platforms, underground metro and train stations and in every natural or artificial human inhabited area.

After analyzing the data from previously conducted simulations [18, 19] we concluded that a ship demands a specialized approach on evacuation research, because of its unique evacuation environment. The characteristics, that should be taken into consideration in a ship evacuation simulation are (i) the list of a ship under natural forces such as the inlet of water, the shift of the cargo, a strong storm, a collision with other ships or large objects (the probable rotation cases are Listing, Rolling and Pitching in both directions) (ii) the motion of a ship by ship's engines or by the waves, (iii) the varieties of types of ships (High Speed Passenger Ships, Ro-Ro, Cruise Ships), (iv) the difficulty to escape from a ship because of the large body of water around it, (v) the presence of crew members who usually contribute positively during an evacuation (the degree of contribution is defined by their training level), (vi) the high number of families and children.

Finally, the most special characteristic of a ship evacuation are ship's list and water inlet. Cold, frozen or hot water may damage the health of humans and if one is submerged begins to drown. In contrary, a long stay inside a volume of cold water, gradually, damages the victim and eventually guide them to faint and/or die. On the other hand, a douche of hot water may result in serious burns. The total immersion in a volume of water causes suffocation to the victim and after a period of time he/she faints and dies.

4 Simulators and Gaming Characteristics

A well-designed virtual environment with improved graphics and proper sound effects can immerse the users in a virtual world distracting temporarily their senses from the real world and focusing their thoughts and attention in the artificial environment. A game-based evacuation simulator which allows the participation of users via virtual avatars should immerse the users into the virtual world, so that they could act such as being in a real accident.

The development of 3D virtual evacuation simulators requires the use of game engines, for the design of the virtual environment. The most popular game engines so far, are the Unreal Engine 4[1] and the Unity Game Engine[2]. Selecting the proper Game Engine is fundamental as the essential features and qualities of the Simulator are defined by that. Apart from the main simulated area all simulations require to be part of a broader environment in order to represent a realistic environment. The sky, the ground, the static items such as the buildings and the trees, the water volumes, the lighting are developed with the game engine's built-in tools. Some of these elements can also be developed by 3D graphic rendering software but the implementation in the game engine's built-in tools simplifies the development process. A proper game engine should be:

- **Effective;** since the importer models should be intact and well-shaped models.
- **Fast;** the importation should not last for a long time, because this can disrupt the normal function of the computer system due to the high computational cost of a virtual development procedure.
- **Simple;** since a complex import procedure can result to repeated failed importation attempts and non-properly imported models importation.
- **Flexible**; has to be able to import and handle numerous 3D models' types and recognize the discrimination of a Skeletal and a Mesh Model.

Moreover, a standalone simulation, assigns the role of evacuees to AI-controlled bots, as human players do not directly participate. A single-player simulator allows the participation of a single player (as evacuee) and the control of other evacuees is assigned to AI-controlled bots. A useful addition to an evacuation simulation should be the multiplayer option, as most game engines support this type of game architecture. If the developers of a simulator are interested in enhancing their software with multi-player, this addition will provide reliable data such as (i) the results human-players of the behavior, (ii) the measurement of the scale of immersion, (iii) the motion and the actions of single evacuees or of crowds of evacuees (consisted of real humans) can be recorded or observed and be applied on AI-bots movement in order to increase the reliability of the locomotion of AI-controlled bots. Another application would be the assignment of rescue team members, crew member or stuff member to real players for training purposes.

---

[1] Unreal Engine suite of creation tools for game development, https://www.unrealengine.com/en-US/, last accessed 5th July 2020.

[2] Unity game development platform, https://unity.com/, last accessed 5th July 2020.

Furthermore, a crucial architecture setup is the support of two modes: Administrative mode and Player mode. An Administrator should (i) create and set up the options of simulation sessions, (ii) spawn events during a session, (iii) observe and assess the overall procedure, (iv) assure the normal execution of every session, (v) collect and store the output of the evacuation sessions. On the other hand, a player should participate on several evacuation sessions (created by an administrator) following the instructions and notes in order to assure the validity of the result of his involvement.

In addition, the evacuation simulators usually assign the control of the evacuees (bots) to the AI (with the exception of the simulator that support the participation of real players). The game engines usually contain a build-in AI control system. Some of them are based on behavior tree and others on programming. Every evacuee should be enhanced with some instructions that control his behavior and implement a decision-making mechanism. E.g. the selection of the escape door or the avoidance of a flame is implemented on bots by the AI system. The AI algorithm of a bot who mindlessly runs through a flame toward a door is not properly developed. A useful feature is the crowd management systems that some engines provide. These systems allow the definition of AI-controlled crowds in order to simulate the locomotion of huge groups of evacuees (apart from the AI implementation on every evacuee).

When an object such as a ball, is pushed by a force (or the hand of a human), will possibly roll on the floor and after a period of time it will stop moving, under the effect of rubbing. A game engine should implement and modify the parameters of physics on every object or creature. The most critical feature to be implemented on every object or creature are gravity, movement, rubbing, determination of mass and shadow casting. Even if the objects and the creatures in a simulation act according to the laws of physics, it is very important to interact with other objects. The effect of the interaction of two or more objects or creatures is called Collision and is partially related to Physics.

Moreover, the selection of the material and textures of objects is crucial for the realism of the simulator. A game engine should include the physical material application option. A proper physical material application enhances the realistic locomotion of objects and creatures, as it defines the physical abilities of an object. A rubber ball should bounce on a wall but an iron one should fall against the floor. Light types have their own qualities and are used on different occasions.

Designing a virtual environment is only one aspect of a game engine. A game engine, thought, implements various elements of the virtual world, using programming. Game engines use three methods in order to code a simulator.; existing programming languages (C++, C# etc.), integrated on the engine programming languages (Boo) and visual scripting (Unreal Engine Blueprints).

Although the creation of menus is not crucial for the evacuation simulation itself it may increase the software's usability. On the other hand, a HUD (Heads-Up Display) provides the user with an in-game menu which informs him of his avatar's health condition, position or other information. Elements such as the total evacuation time, the casualties' number, the door selection etc. should be stored for further reference. Additionally, non-numerical data such as the bottleneck detection should be overviewed by the administrators or the spectators of a simulation.

Another important element for the development of a virtual simulation is the 3D graphics and rendering software (e.g. Studio Max, Maya, Blender). The main requirements that such a 3D graphics software should meet are: Object Creation; Colliders application; Pivot Point application; Materials application capability; Animation application capability and Exportation Format.

Finally, in order to demonstrate, assess and evaluate the results of a virtual evacuation session we propose the measurement and observation of the elements as are shown in Table 3.

Table 3. Evacuation Output Assessment.

Numerical outputs	Evacuee's Crowd Composition, Total Evacuation Time, Total Percentage and Numerical output of Safely Evacuated, Percentage and number of deceased divided by death cause, Percentage and number of injured divided by injury cause, Percentage and number of deceased divided by category, Percentage and number of injured divided by category, Exit selection
Empirical outputs	Recognition of Congestion Areas, Recognition of Bottleneck Areas, Counter-Flow Effect, Behavior

However, the reliability of the results of an evacuation simulation cannot be solely certified, because the simulations are conducted in a virtual environment and the lack of a crucial element, such as a problematic collider of a locked door, which allows objects and human to pass through it, can ruin their credibility.

5 Conclusion and Discussion

This article is a guide for game-based evacuation simulators based on existing evacuation simulators. The guide reviews the elements of human behavior and movement and how those features can be implemented to Game-based simulators. The impersonation of evacuees by real players aims to simulate assess the realistic human behaviors under stressful evacuation conditions and the impact of such heterogeneous crowd on evacuation simulations. Real players can also be trained in evacuation tasks, participating in evacuation scenarios.

Concluding, the basic criteria for the development of a Game-based Simulator are (i) a reliable Game Engine for the development of the virtual world, the physics, the AI and the locomotion, (ii) a 3D Graphics and Rendering Software for the rendering of the objects, the hazardous elements (fire, smoke etc.) and (iii) a Human Modeling and Animation Software for the development of the human models.

Although this work is based on the literature, the proposed guidelines on evacuation simulators misses an evaluation. However, an evacuation simulator based on these guidelines is under development and evaluation. Our future work will focus on the improvement by designing a VR evacuation simulator that realistically simulates a virtual world towards a higher user's immersion.

References

1. Rozo, K.R., Arellana, J., Santander-Mercado, A., Jubiz-Diaz, M.: Modelling building emergency evacuation plans considering the dynamic behaviour of pedestrians using agent-based simulation. Saf. Sci. **113**, 276–284 (2019)
2. Vorst, H.C.: Evacuation models and disaster psychology. Procedia Eng. **3**, 15–21 (2010)
3. Sun, Q., Turkan, Y.: A BIM Based simulation framework for fire evacuation planning. In: Mutis, I., Hartmann, T. (eds.) Advances in Informatics and Computing in Civil and Construction Engineering, Springer, Cham, pp. 431–438 (2019)
4. Helbing, D., Farkas, I.J., Molnar, P., Vicsek, T.: Simulation of pedestrian crowds in normal and evacuation situations. Pedestrian Evacuation Dyn. **21**(2), 21–58 (2002)
5. Ribeiro, J., Almeida, J.E., Rossetti, R.J., Coelho, A., Coelho, A.L.: Using serious games to train evacuation behaviour. In: 7th Iberian Conference on Information Systems and Technologies (CISTI 2012), pp. 1–6 (2012)
6. Nygren, M.: Simulation of human behaviour in stressful crowd situations. Numerisk analys och datalogi, Kungliga Tekniska högskolan (2007)
7. Brown, E., Cairns, P.: A grounded investigation of immersion in games. In: Extended abstracts of the 2004 conference on Human factors and computing systems CHI, vol. 4 (2015)
8. IJsselsteijn, W., De Kort, Y., Poels, K., Jurgelionis, A., Bellotti, F.: Characterising and measuring user experiences in digital games. In: International Conference on Advances in Computer Entertainment Technology, vol. 2, p. 27 (2007)
9. Ren, A., Chen, C., Shi, J., Zou, L.: Application of virtual reality technology to evacuation simulation in fire disaster. In: CGVR, pp. 15–21 (2006)
10. Ginnis, A.I., Kostas, K.V., Politis, C.G., Kaklis, P.D.: VELOS: a VR platform for ship-evacuation analysis. Comput. Aided Design **42**(11), 1045–1058 (2010)
11. Ribeiro, J., Almeida, J.E., Rossetti, R.J., Coelho, A., Coelho, A.L.: Towards a serious games evacuation simulator. *arXiv preprint* arXiv:1303.3827 (2013)
12. Li, Y., Chen, M., Dou, Z., Zheng, X., Cheng, Y., Mebarki, A.: A review of cellular automata models for crowd evacuation. Physica a: Stat. Mech. Appl. **526**, 120752 (2019)
13. Kuligowski, E.D.: Computer evacuation models for buildings. In: Hurley, M.J. et al. (eds.) SFPE Handbook of Fire Protection Engineering, Springer, New York, pp. 2152–2180 (2016)
14. Kuligowski, E.D., Peacock, R.D., Hoskins, B.L.: A review of building evacuation models. US Department of Commerce, National Institute of Standards and Technology, Gaithersburg (2005)
15. Ding, Y., Yang, L., Weng, F., Fu, Z., Rao, P.: Investigation of combined stairs elevators evacuation strategies for high rise buildings based on simulation. Simul. Modell. Pract. Theory **53**, 60–73 (2015)
16. Sticco, I.M., Frank, G.A., Dorso, C.O.: Effects of the body force on the pedestrian and the evacuation dynamics. Saf. Sci. **129**, 104829 (2020)
17. Wang, J., Zhang, L., Shi, Q., Yang, P., Hu, X.: Modeling and simulating for congestion pedestrian evacuation with panic. Phys. a Stat. Mech. Appl. **428**, 396–409 (2015)
18. Yoshida, K., Murayama, M., Itakaki, T.: Study on evaluation of escape route in passenger ships by evacuation simulation and full-scale trials, Research Institute of Marine Engineering, Japan (2001)
19. Łozowicka, D.: Problems of opposite flow of people during evacuation from passenger ships. Zeszyty Naukowe/Akademia Morska w Szczecinie, pp. 82–86 (2010)
20. Sano, T., Ronchi, E., Minegishi, Y., Nilsson, D.: A pedestrian merging flow model for stair evacuation. Fire Saf. J. **89**, 77–89 (2017)

21. Chen, J., Wang, X., Fang, Z.: Collaborative evacuation strategy of ultra-tall towers among stairs and elevators. Procedia Eng. **135**, 170–174 (2016)
22. Chen, L., Tang, T.-Q., Song, Z., Huang, H.-J., Guo, R.-Y.: Child behavior during evacuation under non-emergency situations: Experimental and simulation results. Simul. Model. Pract. Theor. **90**, 31–44 (2019)

Educational Virtual Environments

Virtual Environments for Smart House System Studying

Anzhelika Parkhomenko[1(✉)], Olga Gladkova[1], Yaroslav Zalyubovskiy[1], Andriy Parkhomenko[1], Artem Tulenkov[1], Marina Kalinina[1], Karsten Henke[2], and Heinz-Dietrich Wuttke[2]

[1] National University Zaporizhzhia Polytechnic, Zhukovskogo 64, Zaporizhzhia 69063, Ukraine
parhom@zntu.edu.ua

[2] Ilmenau University of Technology, Helmholtzplatz 5, 98693 Ilmenau, Germany
karsten.henke@tu-ilmenau.de

Abstract. Today, virtual worlds are used not only in the gaming industry. Virtual, augmented and cross-reality also allow to organize effectively 3D environments that provide the effect of immersion and user interaction with objects and processes of the learning environment. The combination of physical objects and virtual models is a modern trend in the online lab development. Based on this approach, online experiment becomes more visual and interesting for students. In addition, it reduces queues for experiments on the real equipment. The paper presents the results of the development of an interactive web-oriented virtual model that expands the Smart House & IoT remote laboratory functionality as well as 3D virtual environment of Smart House for the virtual reality helmet. The proposed solutions will motivate students to study home automation technologies and the features of Smart Home systems realization.

Keywords: Remote and virtual laboratories · Smart House system · Interactive web-oriented model · Virtual reality environment · Virtual reality helmet

1 Introduction

Nowadays Ambient Assisted Living is a relevant and promising area [1]. The activities of today's IT professionals are directly related to the realization of the concepts of Smart Cities and Smart Houses. Thus, the practical studying of Internet of Things (IoT) technologies and Smart House systems (SHS) development is an urgent educational task.

The showrooms and demonstration stands from well-known manufacturers in the branch of SHS development are usually created for advertising purposes. They are not oriented to students' training and their cost is quite high for universities [2].

Remote laboratories (RL) are Internet resources that have already become popular in the world, especially in the conditions of COVID-19 pandemic quarantine. They are free and available to all interested users who have access to the Internet. Existing RLs

M. E. Auer and T. Rüütmann (Eds.): ICL 2020, AISC 1328, pp. 569–576, 2021.
https://doi.org/10.1007/978-3-030-68198-2_52

are used for training in various fields of knowledge - from image processing [3] to the design of embedded and control systems [4, 5]. However, there are many open issues regarding the improvement of RL services, in particular user interfaces, and the development of unified software and hardware of RL [6–9].

For example, in the Smart House & IoT remote laboratory [10] the user works in an interactive mode with a real demonstration stand, which can be viewed online with a webcam [10]. Multiple users can simultaneously perform certain experiments that correspond to specific Smart House subsystems (Illumination control, Access control, Safety control and others). This leads to the appearance of the users queues. To give complete system control to one user (for example, to study the interconnected operation of different subsystems of the house (Presence-Lighting, Climate-Ventilation-Heating, etc.), the administrative module needs to block access to experiments for all other users. Therefore, the remote laboratories for the study of SHS have their specifics that developers must take into account.

Users queue is not a problem for virtual labs, in which 2D or 3D models are used to create virtual reality environments (VRE). To increase the attractiveness of online labs, a virtual and real parts of functionality can be combined for better operation. Due to this, laboratories move into the hybrid laboratories category. A well-known example of such a laboratory is GOLDi (Grid of Online Lab Devices), which was created at the Technical University of Ilmenau, Germany [11].

Implementation of an online educational laboratory based on modern virtual, augmented and cross reality environments expands the frames of the learning process as it makes experimenting faster, safer, more visual and fun as well as it provides the possibility of simultaneously work of many students [12].

Thus, the goals of this work are research and development of specialized VRE for studying issues of Smart House systems development and features of home automation technologies realization.

2 State-of-the-Art

Investigations have shown that in the works [13–18] the results of the development of simulators and virtual environment for SHS rapid prototyping are presented. All of them have specific features of realization, advantages and disadvantages.

A multi-purpose scenario-based simulator [13] gives 2D virtual environment for house plan development and home appliances placement in it. It can be used as a tool for real system prototyping at the beginning stages of SHS design. However, the lack of 3D environment for character moving and interacting with the model reduces the visibility of the proposed system.

The virtual environment for rapid prototyping of the intelligent environment [14] allows to design the interior of the house and to place the sensors. The user can work in the 3D environment and two modes of the operation are available for him. In the first mode, the user can control the inhabitant and interact with different objects in the house. The second mode is based on scenarios usage to control the inhabitant`s behavior. However, with this approach, the user can`t get practical experience with actuators and real home automation equipment.

OpenSHS is an open Smart Home simulator [15] that is intended for forming a Smart Home data set by implementing several stages: design of the initial virtual environment, events modeling and aggregated data set generating. However, the presented tool is an outline that makes it not so interesting for the user.

In work [16] the authors describe tools for virtual design of different configurations for SHS to find out the best variant of topology without the application of real equipment.

The paper [17] presents an environment for modeling typical actions of smart home residents based on uncontrolled teaching methods and associative rules. This approach can be used to detect changes in the behavior of the house residents, which may indicate, in particular, health problems. At the same time, this approach requires the generation of big data sets by their long-term collection from sensors, and it can be problematic for students because of the limited time of the learning process.

The authors of the paper [18] propose tools for modeling the operation of the software on virtual devices in a virtual 3D environment to control changes of input parameters and to format the corresponding aftereffect. Nevertheless, the application doesn`t provide an online working mode.

It is well known that virtual reality (VR) has entered the mass market with appropriative gadgets. Accordingly, various applications were created for the SHS studying, based on the usage of VR glasses and helmets. We have analyzed the possibilities of several virtual reality environments for SHS modeling [19–21].

Electronic House 3D [19] is the software for a smart home virtual environment that allows a person to move freely around the virtual house. The Electronic House works with Oculus Rift on a Windows PC platform. The application has a user-friendly interface. The Electronic House operates at speeds of up to 30 fps. This application gives the possibility to interact with objects in the VE to study their features.

Virtual Home [20] is a software of the house virtual environment, which allows a person to move around the house using teleportation at predetermined points. Live Home works with HTC Vive on the Windows PC platform. The application does not provide a very user-friendly interface. Live Home operates at up to 60 fps. This application allows a person only to inspect the house without interacting with it.

Planner 5D [21] is a virtual home environment software that gives the possibility of a person moving around the house. GearVR from Samsung, which runs on the Android platform, is used as the hardware. This program has a low level of graphics and it is also poorly optimized. However, the application has a user-friendly interface, which provides convenient interaction with the environment. The virtual environment has no smart home sensors. Planner 5D runs at up to 20 fps. This program allows a person to plan and inspect a future housing place.

The experience of simulators and virtual reality environments implementation for studying the SHS that is presented in other authors' works, in particular, their possibilities and approaches to VRE realization is useful for the further improvement of our RL Smart House & IoT.

Thus, we can conclude that the development of several types of VRE can expand the functionality of our RL and make it more interesting for students. Further integration of VREs with real equipment of RL will help to make it more accessible and clear for users and move it to the hybrid laboratory category.

3 SHS Virtual Environments Implementation

Based on the research of solutions proposed by other authors, we have decided to realize two different approaches to VRE implementation.

The first approach is based on the development of interactive web-oriented models for VRE. This can be done, for example, using Mozilla A-Frame markup language using JavaSctipt, HTML5, WebGL. Such a web application provides the ability to play VR content on users' computers or smartphones and they don`t need expensive additional equipment in this case.

The second approach to the development of VRE is based on the usage of game engines. The most popular game engines for VR development are Unreal Engine 4 (EU4) [23] and Unity 3D [24]. Both have a wide range of capabilities and they are reliable tools for 3D environment control, importing content (3D models, images, sound, video), as well as programming of interactivity and gameplay. The Unreal Engine 4 is more optimized in terms of calculations, it gives a more reliable image, but it is more difficult in studying. Unity provides wide opportunities for commercial games creation, as well as it remains more intuitive and effective for novice developers.

The created interactive web-oriented VRE for our RL Smart House & IoT (Fig. 1) is based on a WebGL platform in combination with HTML5 and JavaScript. Unity 3D game engine and Autodesk 3Ds Max 3D environment were used for its realization [22].

Fig. 1. Web-oriented VRE for RL Smart House & IoT

The developed web-oriented VRE allows to:

- Simulate the activities of the house resident and his interaction with the home environment.
- Get information about the equipment and functions of SHS.
- Perform virtual experiments (Security control, Zone control, Access control, Climate control, Solar station, Lighting control, Irrigation control) instead of the usage of real equipment in the case of queues or technical problems with Smart House & IoT remote lab.
- Motivate students to learn and create SHS in the future professional activity based on the interesting learning experience.

The application runs on the Android mobile platform and provides functions of character control using a helmet and a controller, calibration of the helmet, as well as interaction with the interface using the controller (Fig. 2).

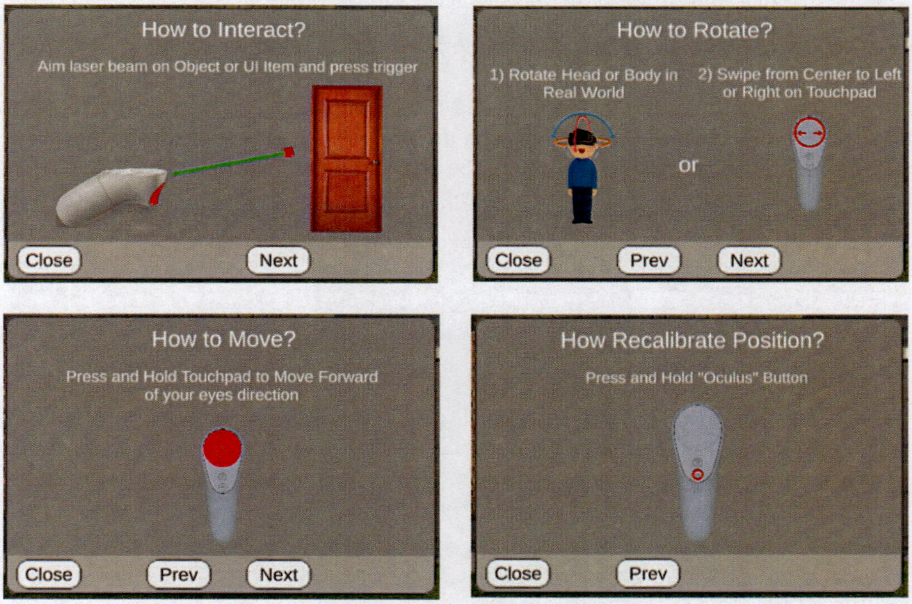

Fig. 2. Interaction with the VRE interface using a controller

The product was developed on the Unity3D platform which features were used for the VRE optimization:

- Texture compression was set at Adaptive Scalable Texture Compression (ASTC), that provides the best balance between quality and file size. The textures, that are used for the user interface, were not compressed to ensure the quality of the visualization.

- The automatic API of graphics has been disabled and installed on the OpenGLES 3 API, which is supported with all mobile devices that are currently running Oculus VR.
- Multi-threaded rendering has been activated. It moves graphical API calls from the main thread to a separate worker thread.
- Static batching was turned on, and all immovable objects were marked as static. Because static batching merges meshes that are marked as static into one large mesh if they have the same material. When several meshes are connected, they can be drawn with a single draw call. It works due to the extra memory and storage space.
- Stereo rendering method. Single-pass rendering has been enabled. Instead of duplicating calls for each eye buffer, objects are transferred once to the left eye buffer and after that they are duplicated to the right buffer automatically. When an object is duplicated in the buffer of the right eye, the corresponding vertex position changes and viewing-dependent variables, such as mapping, are applied.
- Backend scripting. The backend script was exposed to IL2CPP, which works faster on the device, but at the same time increases the assembling time.

Using these optimization techniques, the indicator FPS was increased from 10 to 25 frames per second.

The developed VRE (Fig. 3) provides interaction with subsystems of Lighting control, Security control, Climate control, Irrigation control, etc. It gives the possibilities for students to gain knowledge of the structural and functional features of SHS, as well as the principles of systems` control.

Fig. 3. Smart House VRE for VR helmet

4 Conclusion

Today the developed web-oriented VRE of Smart House is available on our RL website and it allows students to get useful information about SHS components and to simulate control of different SHS subsystems. The VRE Smart House for virtual reality helmet also works stably, it gives students useful experience with scenarios of Smart House inhabitants' activity as well as motivates them to study home automation technologies and systems.

The practical works, based on the application of created VREs allow students to gain experience with SHS design technologies, human behavior simulating as well as creating control scenarios for SHS. The proposed practically-oriented teaching method, based on the usage of Smart House VRE, allows lecturers to give students the knowledge and practical skills for successful application in future professional activities, valuable practical experience in the field of home automation systems and technologies.

Future work will be aimed at the realization of extra functionality for VRE, in particular at the creation of multiple characters (or several Smart House inhabitants) for web-oriented VRE.

References

1. Rucinski, A., Garbos, R., Jeffords, J., Chowdbury, S.: Disruptive innovation in the era of global cyber-society: with focus on Smart City efforts. In: The 9th International Conference on Intelligent Data Acquisition and Advanced Computing Systems: Technology and Applications, Bucharest, Romania, vol. 2, pp. 1102–1104. IEEE (2017)
2. Tulenkov, A., Parkhomenko, A., Yaremchenko, Y., Sokolyanskii, A., Zalyubovskiy, Ya., Parkhomenko, A., Kalinina, M., Stepanenko, A., Andreiev, M.: Investigation and development of demonstration system for training in the field of home automation technologies. In: The 2020 IEEE European Technology & Engineering Management Summit, Dortmund, Germany. IEEE (2020, in publishing)
3. Niederstaetter, M., Klinger, Th., Zutin, G.D.: An image processing online laboratory within the iLab shared architecture. JOE **6**(2), 37–40 (2010)
4. Parkhomenko, A., Gladkova, O., Ivanov, E., Sokolyanskii, A., Kurson, S.: Internet-based technologies for design of embedded systems. In: The XIII International Conference on the Experience of Designing and Application of CAD Systems in Microelectronics, Lviv, Ukraine, pp. 167–171. IEEE (2015)
5. Poliakov, M., Larionova, T., Tabunshchyk, G., Parkhomenko, A., Henke, K.: Remote laboratory for teaching of control systems design as an integrated system. In: The XIII International Conference on Remote Engineering and Virtual Instrumentation, Madrid, Spain, pp. 339–346. IEEE (2016)
6. Tawfik, M., Salzmann, C., Gillet, D., Lowe, D., Saliah-Hassane, H., Sancristobal, E., Castro, M.: Laboratory as a Service (LaaS): a novel paradigm for developing and implementing modular remote laboratories. iJOE **10**(4), 13–21 (2014)
7. Zutin, G.D., Auer, M., Orduna, P., Kreiter, Ch.: Online lab infrastructure as a service: a new paradigm to simplify the development and deployment of online labs. In: The 13th International Conference on Remote Engineering and Virtual Instrumentation, Madrid, Spain, pp. 202–208. IEEE (2016)

8. Orduna, P., Rodriguez-Gil, L., Angulo, I., Dziabenko, O. Towards a microRLMS approach for shared development of remote laboratories. In: The 11th International Conference on Remote Engineering and Virtual Instrumentation, Porto, Portugal, pp. 375–381. IEEE (2014)
9. Parkhomenko, A., Gladkova, O., Sokolyanskii, A., Shepelenko, V., Zalyubovskiy, Y.: Implementation of reusable solutions for remote laboratory development. iJOE **12**(7), 24–29 (2016)
10. Parkhomenko, A., Tulenkov, A., Sokolyanskii, A., Zalyubovskiy, Y., Parkhomenko, A.: Integrated complex for IoT technologies study. In: Auer, M.E., Zutin, D.G. (eds.) Online Engineering & Internet of Things. LNNS, vol. 22(31), pp. 322–330. Springer, Cham (2017)
11. Henke, K., Vietzke, T., Hutschenreuter, R., Wuttke, H.-D.: The remote lab cloud "GOLDi-labs.net". In: The 13th International Conference on Remote Engineering and Virtual Instrumentation, Madrid, Spain, pp. 31–36. IEEE (2016)
12. May, D.: Cross reality spaces in engineering education. Online laboratories for supporting international student collaboration in merging realities. iJOE **16**(3), 4–26 (2020)
13. Jahromi, Z.F., Rajabzadeh, A., Manashty, A.R.: A multi-purpose scenario-based simulator for Smart House environments. Int. J. Comput. Sci. Inf. Secur. **9**(1), 13–18 (2011)
14. Francillette, Y., Boucher, E., Bouzouane, A., Gaboury, S.: The virtual environment for rapid prototyping of the intelligent environment. Sensors **17**(11), 2562 (2017)
15. Alshammari, N., Alshammari, T., Sedky, M., Champion, J., Bauer, C.: OpenSHS: open smart home simulator. Sensors **17**(5), 19 (2017)
16. Vasilateanu, A., Popescu, I.A., Cergan, A.S., Goga, N.: Smart home simulation system. In: IEEE International Symposium on Systems Engineering, Edinburgh, UK, pp. 96–101. IEEE (2016)
17. Rodner, T., Litz, L.: Data-driven generation of rule-based behavior models for an ambient assisted living system. In: IEEE Third International Conference on Consumer Electronics, Berlin, Germany, pp. 35–38. IEEE (2013)
18. Nishikawa, H., Yamamoto, S., Tamai, M., Nishigaki, K., Kitani, T., Shibata, N., Yasumoto, K., Ito, M.: UbiREAL: Realistic Smartspace simulator for systematic testing. In: 8th International Conference on Ubiquitous Computing, CA, USA. LNCS, vol. 4206, pp. 459–476. Springer (2006
19. Electronic House 3D for Windows PC. https://www.livehome3d.com/win/live-home-3d. Accessed 30 May 2020
20. Virtual Home free software (3D or VR). https://store.steampowered.com/app/703060/connect__Virtual_Home_3D_or_VR. Accessed 30 May 2020
21. Planner 5D Google Play page. https://play.google.com/store/apps/dev?id=4677036829894589804&hl=ru. Accessed 30 May 2020
22. Parkhomenko, A., Bilov, O., Tulenkov, A., Sokolyanskii, A., Zalyubovskiy, Ya., Henke, K., Wuttke, H.-D.: Virtual model for remote laboratory Smart House & IoT. In: 10th IEEE International Conference on Intelligent Data Acquisition and Advanced Computing Systems: Technology and Applications, Metz, France, pp. 985–990. IEEE (2019)
23. Ferro, L.: Unreal Engine Blueprints Visual Scripting Projects: Learn Blueprints Visual Scripting in UE4 by Building Three Captivating 3D Games. Packt, Birmingham (2017). 528 p.
24. B-de Byl, P.: Holistic Game Development with Unity: An all-in-One Guide to Implementing Game Mechanics, Art, Design and Programming. A K Peters/CRC Press, Natick (2019). 504 p.

Experiences from Maritime Logistics Distance Learning Course

Olli-Pekka Hilmola[1,2(✉)], Riina Palu[2], and Andres Tolli[2]

[1] LUT University, Kouvola Unit, Prikaatintie 9, 45100 Kouvola, Finland
olli-pekka.hilmola@lut.fi

[2] Estonian Maritime Academy, Tallinn University of Technology (Taltech), Kopli 101, 11712 Tallinn, Estonia
{riina.palu,andres.tolli}@taltech.ee

Abstract. For the purposes of project, "Taltech in the Caspian Sea Region", was in the end of 2019 initiated distance learning course. Topic was maritime logistics and lectures were recorded as well as edited either at the site of Taltech (lecturing room) or at office. Lectures combined interactive parts with students (examples and assignments were part of lecturing module). Purpose of distance course was to deliver it in prompt fashion and with high quality for distant student audience. We report this as a qualitative case study. Research illustrates what kind of tools and approaches were felt best suited to deliver course. It is essential that both technical tools of distance courses as well as teachers and materials are ready and serving demanding digital lecturing task. Collaboration between members of project group is also important, and the professional level as well as experience of all actors. Giving distance education in short notice, it is required that organization is having basic readiness. Technical issues are always something, which require attention (software and purchased hardware become old, while new options arise). Given course had implications on in-service program development, which is also described in here.

Keywords: Distance learning · Implementation · Technology · Case study

1 Introduction

Distance education has been coming increasingly more mainstream in the past decades, and Covid-19 pandemic brought it eventually to masses, and this at all levels of education. Although, changes in 2020 were rather disruptive, the development of "distance education", or more largely "open learning" has been at the development agenda and implementation since 1990's [1]. In the beginning "open learning" was seen as flexible way of bringing education and courses for students, and it was thought to be centralized, with "learning centers" or "open learning offices" supporting all the activities [1, 2]. As early analyses showed, with this sort of structures, distance education was seldom "cheap" option as compared to traditional courses as also preparation of courses, recording lectures and checking exercises, seminar works and exams required a lot of time and resources [2]. This is still considered as one Achilles heel [3], however, keen desire to centralize learning has changed as people are more familiar of using distributed

M. E. Auer and T. Rüütmann (Eds.): ICL 2020, AISC 1328, pp. 577–585, 2021.
https://doi.org/10.1007/978-3-030-68198-2_53

and networked learning devices and software [4–7] – we do not need any longer coordinators and "centers" in between lecturers and students, and it is rather direct contact between parties. Together with positive development in technology and its adaption as well as use, another issue is also providing long-term support for distance learning, and that is lower carbon emissions. It is amazing how much lower emissions distance and online learning can bring – based on studies it is more than 80% lower CO_2 emitted as compared to traditional teaching courses [8–10]. Travel (domestic and international), housing and campus site operations produce quite substantial emissions. It is of course so that information and communications technology (ICT) will consume more energy in distance/online mode, however, its increase is very limited in CO_2 as compared what is actually saved in others. Although, there exist a number of positive issues and drivers of distance learning, it needs push to bring it to main-stream and masses. Due to Covid-19 outbreak, and fear of public health, distance learning has exploded. Early experiences highlight that it is important to use instructors/lecturers voice in different levels and forms that gestures and maneuvers of on-site and classroom teaching could be replaced in learning experience [11]. It is also very important to control that students do work with subjects outside of classes as online participation of courses is more difficult to control. It is known from earlier experiences that adding student interaction e.g. through quizzes, increases the likelihood of providing successful distance learning course [12]. We as well have experiences out of distance/online courses [5] and digital (mobile) platforms [13] used in more traditional and older population tasks, and they have been reported to work rather well.

In this research work we report results of project *"Taltech in the Caspian Sea Region"* concerning its distance learning course. This course was given in the early parts of 2020 for Kazakh audience, and lectures as well as overall distance learning package was managed by Tallinn University of Technology and Estonian Maritime Academy. Lecturers and instructors of the course came from this institution as well. Target audience of this distance learning course was project affiliated Kazakh people, dealing somehow with international trade and logistics. Participants were from business and public sector. Partner in Kazakhstan helped to promote course for local residents. Production of lectures and online material started in the late 2019 (November), and was announced and put online (available for students) in February 2020. This research reports our experiments and experiences of producing this distance education course, and how we were able to have it available in limited amount of time. Research is a case study, and it is qualitative [14, 15]. Research problem could be stated in form of question as follows: "How to produce distance education course in short amount of time within higher education?".

This research is structured as follows: In the following Sect. 2 we introduce process of producing distance education course, and also highlight what kind of software and hardware configurations were used. In Sect. 3 distance education course is introduced through used Internet platform, and content of lectures. As given distance education course had influence and implications on the development of one year in-service program of "Smart Ports & Global Logistics", its plans and actions are described in Sect. 4. Research is concluded in Sect. 5, where we also propose further avenues for the studies in this area.

2 Process of Producing Distance Education Course Content

Idea of producing distance education package for project *"TalTech in the Caspian Sea Region"* arose from the needs of the project to provide information for Kazakh audience from the contemporary maritime logistics themes, which would contain possibly novel Information Technology (IT) applications. This is rather typical information dissemination task of any scientific project from one country to another. Participation in the task was on voluntary basis, and was not specifically paid. Lecturers based their contribution on current lecturing materials, and recent acquisitions of material to university (like simulators, books, reports, computers and software). Without voluntary and charity based lectures, it would have been difficult to reach tight schedule deadlines of this course implementation.

Project started with informal queries from teaching and research staff about the interest to participate. In the beginning of course pool of potential lecturers was much larger than what was the end result (those who actually contributed to this task). After getting idea of interested people, were meetings arranged about the production of teaching material. It was known that some years ago at Taltech and Estonian Maritime Academy, there was investment made to one teaching class, where distance lectures could be recorded. This was at first needed to be checked. However, it was learnt that distance lecturing class was a bit of out of date, and proper recording of lectures with it was difficult task. It is simply so, that software changes so rapidly and licenses expire (at larger university level – leaving some older software abandoned), which eventually leads to malfunctioning of the entire distance education recording class. Of course this issue was tackled by using another class in other location, but it gave good reminder about checking functionality of systems (and how distance education had declined before Covid-19 crisis – it was pretty much popular decade earlier). Together with class based recording, one lecture was recorded using MS PowerPoint screen recorder (it had functionality to record entire screen, if needed and could incorporate also use of other programs, [16]) – this recording was completed in smaller parts, and after editing with free and open source editor program called Shotcut [17], it was incorporated to be part of delivered package. In the end all teaching videos were checked and edited by competent people of Multimedia Center of Taltech.

3 Distance Education Course in Used Platform

As all lectures were properly edited as viable teaching units, they were all combined together in platform called Easygenerator [18], where there was opening page, which simply accessed to five lectures (Fig. 1). Lectures were uploaded to Youtube, however, they were not (and still are not) publicly searchable and only visible via link. These lectures were linked to Easygenerator course platform page. Duration of lectures varied quite much, starting from few minutes, and in the other end being more than one hour long. Two lectures were given in English (Palu & Hilmola), while three remaining were in turn given in Russian (however, many slides were still in English). This simply to serve the target audience as Russian is well spread in old Soviet bloc countries still. Specifics of lecturers, topics, and duration are as follows:

- Riina Palu (duration: 2:20): Smart Ports and Global Logistics (introduction to in-service training program)
- Dan Heering (duration: 17:27): Cybersecurity for seafarers
- Inga Zaitseva-Pärnaste (duration: 11:04): The basics of spatial data analysis for smart cities
- Yrjo Saarinen (duration: 39:50): New INCOTERMS® 2020 Rules
- Olli-Pekka Hilmola (duration: 1:12:18): Using Linear Integer Programming to Solve Physical Distribution Problems

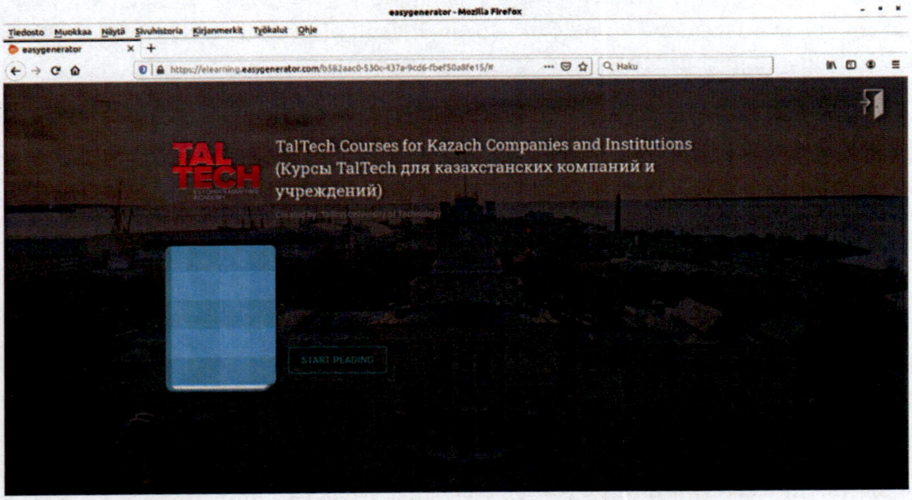

Fig. 1. Opening page of maritime logistics course at course platform.

Fig. 2. Lectures page of maritime logistics course at course platform.

In general, these lectures were acting as an introduction to their specific topics, and did not contain that many exercises or student interaction. Typically, lectures contained lecturer's upper body part, and then edited items from slides (or documents; see Fig. 2). This of course required recording lectures at site of Taltech, and also video editing was needed. Some news of topic were also incorporated in the lectures, if they were relevant (like from cybersecurity and using specific laboratory and simulation of sea vessel operations at Taltech). Only in the last lecture on the list had one online exercise and interactive task being offered for the students (Fig. 3). This required working with distance matrix and spreadsheet optimization package using evolutionary algorithm (Fig. 4).

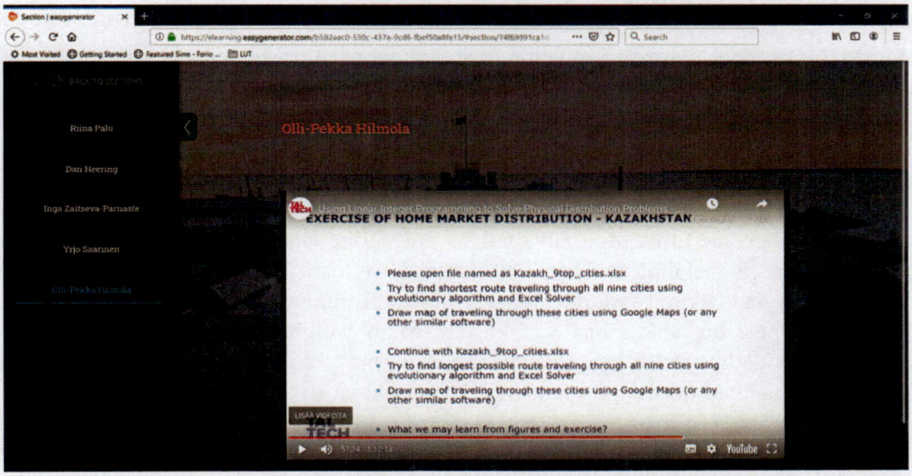

Fig. 3. Lectures page of maritime logistics course at course platform.

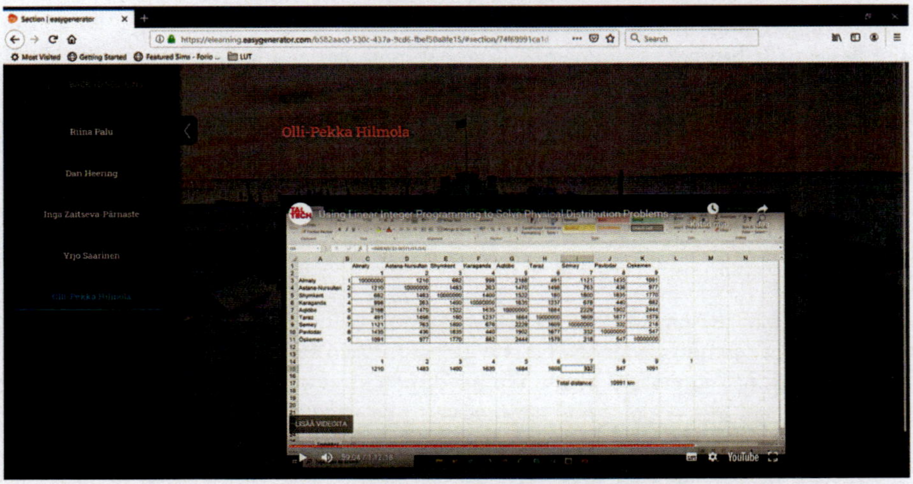

Fig. 4. Lectures page of maritime logistics course at course platform.

Due to Covid-19 pandemic, we were not able to achieve high enough interaction with Kazakh partner of the project, and student interaction (intended questions and answers of lecturers from students in live mode) as well as student feedback are still not completed (at the time of writing this research). However, these are still at the agenda during autumn of 2020, and are possibly to be completed in one form or another. Interest so far from Kazakh audience on this course has been moderate. Somewhat above 20 persons have watched uploaded videos.

4 Implications on Further Development of In-Service Training Program

During this distance course completion time authors of this research were also involved with the internal development of in-service training program, called "Smart Ports & Global Logistics". This program is intended to be a fruitful addition to the students of other subjects and also to the field professionals as in-service program of one year under "open university" (further education center) of Taltech. From this distance learning course execution reasons as well as due to Covid-19 situation throughout the world, it was decided that program shall be put online and within distance education mode entirely. Development has been changed to such that program will be more networked style, having experts giving lectures from Estonia as well as Northern Europe and possibly also from Kazakhstan. Main weight is on high quality video material provided, and having good lecturers within program. Following topics and lectures are planned on the program (this is preliminary planned list):

- Hinterland container transport and the relevance of information exchange; optimizing the planning and control of container logistics processes; developing customized ICT solutions for hinterland transportation
- Global supply chains
- Financing of global infrastructure development: risks and opportunities
- Special planning in water environments
- Effects of (Chinese) e-commerce on global, regional and local logistics
- Maritime cyber security
- Smart Ports case study of Port of Tallinn
- Maritime law
- Global politics
- Business development in new era
- Global financial systems
- Innovative solutions for global logistics

Due to the reason that in-service program is now going to be completely online/distance learning mode, and this will last for two semesters, we have completed different kinds of actions and plans. We have spotted new application called "Mastery Team" [19], and this is one good alternative to be used in toolbox for program execution. This application has been developed in collaboration of one of the leading YouTubers of Estonia. Key idea of in-service program is that it is based on "easy learning" (from student perspective), however, this in other side means that it is very demanding for lecturers as well as on used technology. However, due to the reason that students of this in-service program are intended to be arising also from Central Asia besides Estonians, program development in current world environment leaves very limited amount of options in other regards. This in-service package is in the future also be offered for M.Sc. Program students as an option of specialization of M.Sc. Degree. In the new mode it should fit well in working life being students. Table 1 provides preliminary details of program structure.

Table 1. Intended in-service program structure called "Smart Ports & Global Logistics".

Subjects	EAP	I semester EAP	II semester EAP
	50	**25**	**25**
General subjects	**16**	**7**	**9**
Transmission of (business) messages in digital era	3	3	0
Writing of scientific texts	3	0	3
International and Maritime Law	4	4	0
Practical research methods	6	0	6
Global logistics	**20**	**12**	**8**
Management and financing of global logistics and infrastructure development projects	6	6	0
Global supply chains and international strategy	4	0	4
Global politics, economics and finance	6	6	0
Practical research	4	0	4
Smart ports	**14**	**6**	**8**
Smart Ports case study of Port of Tallinn and Port of Helsinki	4	0	4
New technologies and innovative solutions in logistics	6	6	0
Port community systems/e-Governance	4	0	4

Making in-service program successful, it is not only required to have technology and lecturers there. Program will have some sort of learning center (advising students, if they have problems), but this in limited form. Institutional partners from Estonia, other universities and branch companies are also required together with Program Council. Funding for program development and initial execution is also one area to be solved, and in there we have as alternatives different European Union funding tools (like Erasmus+, Horizon, Green Deal etc.) as well as local state funds together with other country-level funds.

5 Conclusions

Due to numerous reason, and lately driven mostly by Covid-19 epidemic, distance and online courses and education has been brought high at agenda of different schools, institutes and universities [11]. Distance learning has its roots deep in the early days of ICT and Internet development – it was firstly seen as an opportunity in the late 1990's and early 2000s [1, 2, 4]. In that time distance learning courses using video format and some interaction were introduced. However, as it was introduced to the old structures and world did have back then distributed solutions of IT (like mobile and fast Internet, tablets, Internet capable mobiles and applications), it was most often regarded as costly exercise and having higher price tag than conventional teaching of courses. Cost is of course only one factor in decision-making of shifting more on distance education. We have used so much on traditional form of courses at site that this habit is difficult to overcome. Now due to virus epidemic, and also in parts due to environmental reasons [8–10], world and institutions have "push" for new forms of teaching. It is of course not problem free or easy task to transform from traditional courses to distance courses. Most often this increases the work-load and stress of lecturers, and in turn creates difficulties to control whether students actually learn something and do things [3, 7, 12].

Providing maritime logistics course for Kazakh students using distance learning format was one awakening event for Taltech and Estonian Maritime Academy to consider other programs developed for future students. As institute learnt that producing distance learning material is not that difficult, and staff is even at voluntary base willing to contribute in the development of it, this should be workable format in later on. Therefore, as reported earlier in this research that one in-service program lasting for one year is intended to be given in distance learning format. It was also learnt during the process that new software applications are needed and IT applications to aid in one year program, and these have been searched further. As bigger changes always come outside of bigger institutions and structures, it would be good to suggest that universities take some learning points from YouTubers as well. It is worthwhile to think about their approach for videos (often providing instructions or learning material) – typically these videos are short, they contain interaction (comments or live-streaming chat), are made with very limited resources and are from contemporary topics. In the end YouTubers nowadays (best ones) attract audiences larger than national or international TV networks (having millions of subscribers and individual video views). We feel that in here there is a learning point for even university distance education courses and programs. Rigid use of technology as well as low and direct structures are way to go forward.

As a further research in the topic area, we would be interested to continue with student interaction and also gathering modes of student feedback. In many research works interaction is seen as vital in distance learning courses, and even in there tone of voice of lecturer is seen as vital. These should be researched further in order to understand how successful distance learning courses and programs are built in the future.

References

1. Davis, H.J.: A review of open and distance learning within management development. J. Manag. Dev. **15**(4), 20–34 (1996)
2. Lawton, S., Barnes, R.: Developing distance learning courses in a "traditional" university. Qual. Assur. Educ. **6**(2), 106–111 (1998)

3. Major, A.E., Chaudhury, S.R., Gilbertson, B.M., King, D.T.: An integrated science course modes online: four concurrent perspectives. J. Appl. Res. High. Educ. **6**(2), 248–268 (2014)
4. Guha, A.S., Subhashish, M.: E-learning: the latest spectrum in open and distance learning. Soc. Responsib. J. **4**(3), 297–305 (2008)
5. Tung, M.-C., Huang, J., Keh, H.-C., Wai, S.: Distance learning in advanced military education: analysis of joint operations course in the Taiwan military. Comput. Educ. **53**, 653–666 (2009)
6. Herodotou, C., Rienties, B., Hlosta, M., Boroowa, A., Mangafa, C., Zdrahal, Z.: The scalable implementation of predictive learning analytics at a distance learning university: insights from a longitudinal case study. Internet High. Educ. **45**, 100725 (2020)
7. Scarinci, A.L.: Evaluating resources for scientific modelling in a distance education course. Procedia Soc. Behav. Sci. **167**, 238–244 (2015)
8. Roy, R., Potter, S., Yarrow, K.: Designing low carbon higher education systems: environmental impacts of campus and distance learning systems. Int. J. Sustain. High. Educ. **9**(2), 116–130 (2008)
9. Caird, S., Lane, A., Swithenby, E., Roy, R., Potter, S.: Design of higher education teaching models and carbon impacts. Int. J. Sustain. High. Educ. **16**(1), 96–111 (2015)
10. Oliveira, J.H., Giannetti, B.F., Agostinho, F., Almeida, C.M.V.B.: Decision making under the environmental perspective: choosing between traditional and distance teaching courses. J. Clean. Prod. **172**, 4303–4313 (2018)
11. Bao, W.: COVID-19 and online teaching in higher education: a case study of Peking University. Hum Behav. Emerg. Tech. **2**(2), 113–115 (2020)
12. Raes, A., Vanneste, P., Pieters, M., Windey, I., Noortgate, W.V.D., Depaepe, F.: Learning and instruction in the hybrid virtual classroom: an investigation of students' engagement and the effect of quizzes. Comput. Educ. **143**, 103682 (2020)
13. Nuanmeesri, S.: Mobile application for the purpose of marketing, product distribution and location-based logistics for elderly farmers. Appl. Comput. Inform. (2020, accepted)
14. Eisenhardt, K.M.: Building theories from case study research. Acad. Manag. Rev. **14**(4), 532–550 (1989)
15. Eisenhardt, K.M., Graebner, M.E.: Theory building from cases: opportunities and challenges. Acad. Manag. J. **50**(1), 25–32 (2007)
16. MS Office homepage advise to screen recording. https://support.office.com/en-us/article/record-your-screen-in-powerpoint-0b4c3f65-534c-4cf1-9c59-402b6e9d79d0. Accessed 13 May 2020
17. Shotcut program homepage. https://shotcut.org/. Accessed 13 May 2020
18. Easygenerator homepage. https://www.easygenerator.com/en/. Accessed 12 May 2020
19. Mastery Team homepage. https://masteryapp.eu/. Accessed 18 May 2020

MOOCs in Logistics – Preliminary Data on University Curricula Coverage

Tarvo Niine[1(✉)], Franca Cantoni[2], and Miguel Córdova[3]

[1] TalTech School of Business and Governance, Tallinn, Estonia
tarvo.niine@taltech.ee
[2] Department of Social Sciences and Economics, Università Cattolica del Sacro Cuore, Milan, Italy
[3] Department of Management Sciences, Pontificia Universidad Católica del Perú, Lima, Peru

Abstract. After a decade of MOOC and open education development there is an abundance of available online content. The aim of this study is to find out whether the MOOC landscape in logistics has grown to a point of topically covering entire university curricula worth of topics. Provided the affirmative outcome, this would mean greater competition but also greater opportunities for universities teaching logistics programs to apply blended learning. We present an overview of logistics-related material on three major platforms totaling 95 courses and compare a sample of five logistics curricula against this list to demonstrate the extent of coverage by online material as well as to point out the gaps. The data suggests that the current status of logistics MOOCs can mostly cover more introductory and broader managerial-type programs but not material on logistics operations in-depth. Also, MOOCs tend to struggle with more interdisciplinary topic approaches. The findings allow to discuss on the nature of identified gaps as well as to encourage and foresee continuous growth of blended learning.

Keywords: MOOCs · Open education · Logistics curricula

1 Introduction

Online courses, a technological novelty only a decade ago, are quite commonplace today with global accessibility and multiple competing platforms offering broad array of choices. The focus of this paper is to observe whether the MOOC landscape in logistics has already grown to a point of topically covering entire university curricula. Provided the affirmative outcome, this would mean both a greater level of threat but also greater level of opportunity for universities teaching logistics programs. Provided, however, that coverage gaps are observed, it begs the question whether they are mostly random by nature and therefore perhaps filled in near future, or, alternatively, there are some systematic obstacles that would make the MOOC landscape to be relatively better equipped in some topic areas and much less so in other areas, which would indicate at least a longer process of bridging with perhaps more targeted actions assumed. Logistics and Supply Chains organizational fields represent a proper opportunity for

M. E. Auer and T. Rüütmann (Eds.): ICL 2020, AISC 1328, pp. 586–597, 2021.
https://doi.org/10.1007/978-3-030-68198-2_54

this study due to the long history of conceptual growth and evolution, starting from a transport-centric view to advanced logistics operations in global supply chains [1], and from supply chain management to newest sustainable supply chain management discipline towards sustainability goals and ethical behavior [2].

The study is organized as follows. The next section provides an overview of the literature related to MOOCs and how various treatments have seen MOOCs interacting with formal higher education. The third section describes the method applied in the study. The fourth section presents findings from analyzing courses of five examples of logistics curricula from universities in Estonia, Italy and Peru. Finally, the fifth section concludes the paper, highlighting implications for academia and practitioners as well as raising further questions for researchers in the field.

2 Studies on MOOCs and Formal Higher Education

In the period of 2010–2015, where the MOOC landscape took form of platforms with international appeal, MOOCs were seen as a disruptive force to education. Disruptive to the extent that it was described as a direct threat to education institutions with lesser resource base [3]. Clearly, this has not realized to groundbreaking extent and so we see university courses and MOOCs co-existing – sometimes in a more competitive setting, but also often in a collaborative setting. This section provides a brief look to the linkages between the two worlds.

As the prime stakeholder of MOOCs is the student, it is important to understand what motivates the students of MOOCs to take the courses. According to Shapiro *et al.*, in a study about MOOCs of chemistry and statistics, the main motivators were personal (knowledge, interest, convenience) as well as career (work, career change). While about 30% of students also mentioned "certificate" as a reason, it was still a less significant argument even than "fun" (about 50%). [4] Other studies have concluded similarly [5, 6]. This suggests that there is a relatively low linkage between MOOCs and formal higher education pursuits, which can be interpreted as notable growth potential for MOOCs to further boost motivation, participation, contribution and success.

A broad coverage of relationships between MOOC movement and higher education was presented by Schuwer *et al.*, which charted broadly opportunities and threats. Notable opportunities treated were collaboration potential, acceleration of online learning, existing policies and European ECTS framework, marketing potential and using MOOCs as innovation test-bed, while lack of recognition, quality concerns, innovation-hindering regulation, missing evidence and low completion rates were pointed as threats. In addition, "*availability of multiple platforms for adopting MOOCs is tangled with a threat of too much fragmentation*". [7] On a similar theme, Nortvig and Christiansen have analyzed institutional collaboration on MOOCs and pointed out advantages in terms of ensuring quality and innovation in the common learning designs. According to their study, noteworthy barriers to institutional collaboration are reluctance of institutions to engage due to fear of competitive concerns and the teachers' hesitancy or even resistance to new educational platforms and formats [8].

Inamorato dos Santos *et al.* propose a similar tally of open education motivators and growth barriers. Some key motivators listed are increased access, development of teaching strategies and learning technologies. The obstacles are mostly focusing on the academia side: lack of shared vision, passive resistance and time constraints, additional training requirements and financial sustainability. Still, the authors point out that almost half of the institutions in their study are involved in open education and find that open education practices are increasingly significant in higher education in Europe. [9] Sandeen has noted that a growing number of colleges and universities are experimenting with various MOOC integration and credit-recognition models, but with approaches as diverse as the institutions themselves. The physical exam option by edX and certificates involving identity identification by Coursera are examples how MOOCs can be appealing to an institution considering acceptance of open education for credit transfer [10]. A common theme through these studies appears to be cautious optimism.

A specific type of MOOC in which locally produced and third-party MOOCs are integrated into traditional courses are known as hybrid MOOCs (H-MOOCs). Perez-Sanagustin *et al.* have proposed an H-MOOC framework with two key dimensions of MOOC variety: curricular content alignment and institutional support [11]. This approach emphasizes ways how MOOCs can be utilized by higher education institutions (HEIs). This further supports the idea and trend of unbundling in education – that, by and large, learning ought not to be based on a single fixed block of content input. Moskal *et al.* have pointed out that blended approaches tend to have higher student satisfaction than both face-to-face teaching and fully online teaching as well as lower dropout rates. However, the authors caution that to implement blended learning properly is not a course-level consideration but an institution-wide strategy, facilitated with proper resources and clear vision, in order to become a transformational force [12].

Rothe *et al.* have provided a study on the business models and development trajectories of platforms of MOOC provision. The authors categorize three business models: Premium Services (free access, pay for certificate), Professional Degrees (packages providing entry to a professional career) and Corporate Training (packages aimed at corporate employee training rather than individuals). [13] While this approach shows that MOOC platforms can also be viable businesses without credit recognition by HEIs, it can also be proposed that recognition is a significant hurdle to broader acceptance of MOOCs. In one recent study of 50 institutions providing MOOCs in China, only a small proportion of providers were found to provide quality management and monitoring [14]. This indicates that recognition can sometimes be a non-trivial process and there is development room for both MOOC providers as well as HEIs.

According to Witthaus, recognition is a key aspect of opening up education by transition from non-formal to formal education. The authors provided an open learning recognition traffic light model, which lists six elements for MOOCs to facilitate recognition by HEIs – learner identity verification, supervised assessment, quality assurance, informative credentials, credit points and collaboration with recognizing institutions. The model came with a dedicated recommendation to "ensure policies on validation and recognition of non-formal learning embrace open education and MOOCs". [15] The challenge of recognition is further reflected by the issue that there does not appear to be a universally accepted quality model for online and open education. Ossiannilsson *et al.* have presented a comparison of 15 models, concluding that all reviewed quality systems suffer certain deficiencies [16].

The legal framework of e-learning has been extensively covered through a timeline view by Salajan and Roumell. The authors point to emergence of an increasingly coherent and formal approach to supporting e-learning initiatives from initially cautious and tentative steps to progressively formalized policy documents. The authors suggest that the policy focus of 2010s has shifted away from continued emphasis on ICT, with relegating it to the status of supporting structures [17]. A possible way to interpret this would be to see e-learning not so much as a stand-alone entity but that the domains of e-learning and traditional learning are fusing both de jure as de facto.

A strong counterpoint to MOOC-related optimism has been made by Reich and Ruiperez-Valiente, who analyzed Harvard and MIT courses on edX from 2012–2018, pointing out that the numbers of new unique users to courses turned to decline after 2016. Their study notes that "*rather than creating new pathways at the margins of global higher education, MOOCs are primarily a complementary asset for learners within existing systems*" [18]. From the perspective of academia, this suggests that the worlds of MOOCs and HEIs are somewhat closer than traditionally perceived. According to Belleflamme and Jacqmin, MOOCs are not replacing incumbent institutions but are pushing teaching practices to evolve [19].

3 Methodology

As pointed out by various studies, most of MOOC-related research has been concentrated at the student level [20–22]. In contrast, this study focuses on the viewpoint of university as an external stakeholder and potential beneficiary – here not as a direct provider of MOOCs but as an agent being able to recognize MOOCs as part of studies as well as integrating MOOCs in classroom in various blended ways. In this study, the research goal is to observe the MOOC landscape in the area of logistics with the intent to chart the topical coverage of MOOCs by comparing them against a small sample selection of university curricula in logistics across the world. The assumptions of willingness and ability are important constraints of the study. They are both practical concerns that a university, faculty or a local program head might be not able to leverage MOOCs in credit recognition process due to administrative hurdles or that a program head or teacher might mistrust MOOCs, perhaps for quality reasons, or is unwilling to blend the materials for other considerations – such constraints fall outside the scope of this work. We are asking the question – if a university is both able and willing to embrace MOOCs, what is the maximum application potential of it as of writing this research (in April 2020) across particular selected curricula in logistics.

In this study, we are approaching MOOCs and curricula quantitatively, estimating the topical coverage and linkages without taking into account quality considerations. It is acknowledged that the input data to this study is limited and that our conclusions can only be as reliable as our input data.

This study was conducted following three consecutive phases – mapping, selection of sample curricula and comparison. In mapping phase, a list of courses and relevant video material was compiled. In this study, the mapping process was limited to three platforms: edX, Udemy and Coursera. It is acknowledged there are more relevant platforms out there), but we believe that the current sample pool allows to identify sufficient patterns. Our work on extending and updating the database is continuing. The three platforms were selected based on personal experiences as well as student suggestions as they host online material in a wide range of disciplines to a worldwide student body. As the study was carried by a global perspective, only material available in English language was considered. Search was made with the following keywords: *logistics, transport, supply chain, operations, inventory, vehicle, traffic, mobility*. It is curious that in this mapping process, we could not directly rely on platforms' own search tools. The typical case was that a keyword search outputted manifold more results than what were actually topically relevant. Such design, it can be speculated, is the result of platform marketing efforts – to push discovering more courses and support the perception of abundance. This meant that the mapping process was both iterative (keyword by keyword), as well as browsing through piles of "false positives".

Altogether, 95 material packages were documented. The term package here suggests that it is difficult to draw the line, what is the smallest size (as well as the presence of other defining elements) to really call it a proper MOOC. Here, we adopt a pragmatic view: all available titles were included that were topically suitable and had a minimum length of the video material of 3 h. Clearly, a 3 h package would not carry the title "course" in a formal education setting. However, the smaller packages are also relevant for blended learning purposes, especially in the areas of this total coverage.

The typical organizational model varies between platforms. On edX, the courses are mostly free to audit and a charge is applied only for a certificate. There are both time-limited MOOCs and courses available all the time. We found 31 topical courses hosted by edX for a total amount of 1825 h of estimated student effort. On Coursera, pricing is for certificate and applies either lump-sum or a monthly fee principle. 29 logistics-themed courses were found for a total of 480 h' worth of videos. On Udemy, 35 elements were listed totaling 267 video hours. While on average they are significantly smaller to an average title on other two platforms, this allows also more niche approaches and less generic titles. On Udemy, typical offerings are from private experts, which is aligned with a somewhat lower access fee.

In terms of most active institutions, our mapping identified MIT (8), Rutgers (7), Delft (6) and Illinois (5). The list of all 95 MOOCs is extensive and was therefore omitted from this paper but the authors are happy to share it on demand, as well as the specifics of which course connected to which topic area in the evaluation phase.

After the mapping phase, the following concerned selecting sample curricula. At this preliminary stage of research, two considerations were prominent. One was focusing on programs that the authors have first-hand experience with. The second stemmed from the approach of "four types of logistics curricula" that were categorized in a cluster analysis approach in a study in 2015. This study proposed the four distinct types of logistics curricula as: business administration with logistics branch, interdisciplinary logistics management, transport management and logistics engineering [23]. The ambition was to cover all four types, however, in this study due to practical limitations, the study is limited to samples of first three types.

We present an analysis of five logistics curricula in this paper to explore different educational contexts: two from Estonia, two from Italy, one from Peru – three programs on undergraduate and two on graduate level.

The comparison of programs and MOOCs first focused on each individual logistics-themed course in the program and listing the MOOCs with a topical linkage, judging MOOCs by title, the description and content headings and formal courses by their stated learning outcomes and expected content. This was followed by evaluating the outcome to see, if there were topics intended to be treated in courses, which were apparently left uncovered by the linked MOOCs. In evaluating the overall coverage, it seemed appropriate to use a 4-step scale: no coverage (0%), scarce supply (less than 1/3 of topics covered), moderate supply (1/3 to 2/3 coverage), strong supply (over 2/3 coverage), proposing that this is as detailed as such observation can be for practical purposes.

There are a few more acknowledged limitations. The first being that in time, both the intended material in formal courses as well MOOC material is aging and therefore some currently defined linkages might not stand the test of time (especially concerning technology topics). Secondly, it is possible that some MOOC descriptions do not necessarily represent the content 100%. The authors did not meet such significant cases during the study. We are relying here on the assumption that while there might be occasional discrepancies between MOOC content and intended content, we position that this discrepancy is not greater than the occasional discrepancy between ILOs of a formal course and the topics what the teacher is actually covering in class. In other words, this study carries the assumption that an average academic teacher in the classroom is as reliable as an average MOOC teacher (as far as our sample platforms are concerned).

4 Findings

Here we present findings from evaluating the MOOC database from the topical connections perspective against five logistics curricula:

- Curriculum #1 – Business administration with logistics and supply chain specialization, undergraduate level, TalTech School of Business and Governance, Estonia;
- Curriculum #2 – Logistics, undergraduate level, Estonian Enterpreneurship University of Applied Sciences;
- Curriculum #3 – Supply Chain Management and Digital Innovation (SCHMIDT), graduate level, Università Cattolica del Sacro Cuore, Italy;
- Curriculum #4 – Global Executive Master in Operations and Supply Chain (GEMOS), Polimi, Italy;
- Curriculum #5 – Management and Top Direction, Pontificia Universidad Católica del Perú, Peru.

Curriculum #1 is a 3-year full-time program in TalTech, one of the largest universities in Estonia. The students split between five majors, one of which is "logistics and supply chain". As the program is built upon broad business foundation, it dedicates ~20% of attention to logistics topics with 6 courses, all 6 ECTS and including 40–64 h of classroom activity. How these courses link to MOOC database is shown in Table 1.

Table 1. Linkages between curriculum #1 and MOOCs

Curriculum course	Connected MOOCs	Evaluated coverage
Logistics and inventory management	15 courses	Strong supply
Transportation and forwarding	4 courses	Scarce supply
Purchasing management	16 courses	Strong supply
Operations management	7 courses	Strong supply
Logistics technologies	17 courses	Strong supply
Management of logistics enterprise	16 courses	Strong supply

As the intended material coverage on this curriculum is rather introductory compared to more specialized logistics programs, the overall linkage to MOOCs is strong and approaching full coverage. The main topical gap appears to be in transport arrangements – out of 4 courses linked in Table 1, only one (#35) is dedicatedly about transport regulations and the downside of that particular specimen is that this MOOC is too specific (for the general audience in this context) and more appropriate for graduate level.

Curriculum #2 is a 3-year program in Estonian Enterpreneurship University of Applied Sciences, a private school located in Tallinn, Estonia. This program is also built upon general business foundation, but includes logistics-specific courses more prominently and has a dedicated transport focus – to an extent that the students are eligible to take "FIATA diploma in Freight Forwarding" (a recognized vocational certificate) exam. We separated 52 ECTS of logistics courses (Table 2) from the foundational courses.

Table 2. Linkages between curriculum #2 and MOOCs

Curriculum course	Connected MOOCs	Evaluated coverage
Foundations to logistics	7 courses	Strong supply
Logistics systems and cost management	27 courses	Strong supply
Land transport	6 courses	Moderate supply
Rail transport	1 course	Scarce supply
Water transport	0 courses	No supply
Air transport	0 courses	No supply
Multimodal transport	2 courses	Scarce supply
Foundations of forwarding	3 courses	Scarce supply
Risks in logistics	3 courses	Moderate supply
Manufacturing logistics	8 courses	Strong supply
Purchasing and inventory management	13 courses	Strong supply
Material handling and warehousing	2 courses	Scarce supply
Customs regulations	2 courses	Scarce supply
Passenger travel	1 course	Scarce supply

It is clear that current MOOCs are much less able to cover more specific and more transport-focused topics than more general managerial and "big picture" views. Table 2 still lists a number of courses applicable, but often, the linkages are weak and contribution remains secondary. The gaps are notable in transport mode specific topics both concerning physical world of operations (capabilities, performance characteristics) as well as administrative, regulative and direct managerial implications of transport operations (such as costs, efficiency, analytics). Another shortcoming appears to be in the field of warehousing and material handling. It appears like the MOOC landscape of logistics is tilted towards broad perspective topics than actual physical operations.

Curriculum #3 is a master level program at the Faculty of Economics and Law, Università Cattolica del Sacro Cuore in Italy. The program is structured into four interdisciplinary teaching modules in Table 3.

Table 3. Linkages between curriculum #3 and MOOCs

Curriculum course	Connected MOOCs	Evaluated coverage
Supply chain management and digital innovation	7 courses	Moderate supply
Supply chain planning	42 courses	Strong supply
Market, marketing and supply chain management	10 courses	Moderate supply
Supply chain innovation: regulation and policies	15 courses	Moderate supply

The core courses are well covered by MOOCs. Topics such as the managerial approach for supply chain in Industry 4.0, leadership and team management, projects and decision support systems for supply management, technologies for the digitalization and innovation are all areas of interdisciplinary approach. Such perspectives are, at least currently, not widespread on the MOOC landscape. While one can find good material to mix and match to an extent, there is usually no single ready-made solution.

Curriculum #4 is a 12-month part-time program taught on campus at two European schools, MIP Politecnico di Milano in Milan, Italy, and EADA Business School in Barcelona, Spain. The program focuses on delivering in a highly active and hands-on manner, emphasizing "learning by doing". Teaching includes business cases, and simulations, with students constantly working on real-life business cases and strategic decisions. The program has four relevant modules: operation management 4.0, digital operations management, sustainable supply chain management and innovative supply chain management. A module of leadership skills was omitted from our analysis due to being more foundational than logistics-specific (Table 4).

Table 4. Linkages between curriculum #4 and MOOCs

Curriculum course	Connected MOOCs	Evaluated coverage
Operations management 4.0	28 courses	Strong supply
Digital operations management	4 courses	Moderate supply
Sustainable supply chain management	9 courses	Moderate supply
Innovative supply chain management	6 courses	Moderate supply

The keywords of content are mostly present in MOOCs. Similarly to Table 3, viewing process management topics from perspectives of sustainability and innovation are good examples of areas where online content is not extensive. It is also relevant to note that the more a program is designed to be hands-on and applicable in learner's own practical context, the more difficult it would be to rely on a MOOC format. Still, there is a lot of available material to recommend as additional optional sources.

Curriculum #5 is a 5-year undergraduate program in Pontificia Universidad Católica del Perú. The program is built upon organization management foundation and includes three logistics and operations specific courses: operations management, logistics management, and international commerce management (the latter is an optional course). The 3 logistics courses are shown in Table 5.

Table 5. Linkages between curriculum #5 and MOOCs

Curriculum course	Connected MOOCs	Evaluated coverage
Operations management	9 courses	Moderate supply
Logistics management	31 courses	Strong supply
International commerce management	3 courses	Scarce supply

Topics uncovered by MOOCs in Operations Management course are: processes management, processes flowcharts, SIPOC methodology, stochastic simulation of discrete processes, material requirement planning, and linear programming optimization. Even when SIPOC and stochastic simulation, and linear programming are specific topics that could be outside of our key words used for searching, we expected to find coverage on topics such as processes management, processes flowcharts, and material requirement planning, because these appear closely related to our search patterns.

Regarding the Logistics management course, it results as strongly supplied with MOOCs. The only topics uncovered were megatrend in logistics and SCOR methodology, and we argue that the first one is an introductory subject, and the second one is specific methodology that could have been outside of our search range.

Finally, the International commerce management course was scarcely supplied, being partially covered just by three MOOCs from our list. One of the missing topics was local regulations for international trade, which we argue is also due to local focus and language barrier – if such course would exist, it probably would not be in English.

5 Conclusion and Discussion

It appears that while the MOOC landscape is much heterogeneous, the landscape of logistics curricula is still notably more heterogeneous. From competitive view between MOOCs and HEIs, this makes sense – the more faculty programs have local unique characteristics, the less they link with MOOCs, where M stands for "massive", suggesting more standardized approach. Broadly, our study echoes that MOOC development has turned out slower than initially anticipated. The situation on the MOOC landscape in logistics topics shows some notable gaps, such as in terms of freight forwarding and international trade regulations, specific handling and warehousing topics but also interdisciplinary perspectives, where process and tech views (automation and digitalization) are linked to human elements of leadership, teamwork and change.

For the former, we propose two explanations, what really are sides of the same coin – demand and supply. The first obstacle is smaller market demand. It appears there is an order of magnitude more "supply chain management"-focused topics than directly transport-focused topics. Perhaps it can be suggested that "it is cool to be a generalist rather than a specialist" and that many MOOCs are much driven by popular demand (and more than the desire to fill knowledge gaps). For the latter, it is interesting that interdisciplinary topic treatments are not "mainstream" on MOOC landscape. Perhaps it is a matter of time – as competition is expected to increase between MOOCs, topical differentiation seems logically a potent strategy.

Let's turn to future. It can be assumed that MOOC landscape will keep growing and the main reason is that the motivation of students is probably growing and MOOCs will be there to follow the demand. Watted and Barak have categorized MOOC-related student motivators as personal, career and educational benefits [24]. Today, the situation is that most students are taking MOOCs because of personal interests or due to perceived career impact, but not for the credits to be recognized in formal learning. Why? Partially because a notable proportion in academia is not recognizing it. There are clearly systematic benefits and synergies if more of academia would give proper credit to online learning. Should there be formal educational benefit added to the mix, all the involved stakeholders (students, universities, MOOC providers, employers and society) would benefit.

Of course, this line of thought has been around for some time and efforts have been made while progress is still slow to come. However, it could be proposed that after 2020 the changes might speed up. This study was conceived way before the COVID-19 pandemic. However, writing this in the middle of pandemic, it appears that the idea of HEIs working together with open education is more relevant than ever. Many universities involved in pandemic areas, which also covers the homes of all three authors of this paper around the globe, were pushed into distance teaching. A vast number of teachers needed to handle delivering content and achieving ILOs in a non-contact situation. This could impact the average mindset and boost the understanding that if an average teacher could handle the situation (and our experiences are a positive indication), then surely it is easier to give credit to MOOC teachers who have been doing pretty much the same thing, only that in some cases with years more preparation and experience. Also, not only students but many teachers probably turned to MOOCs for aid in crisis, getting to know the MOOC landscape better and discovering ways how to integrate this into their own courses. While not all these experiences might have been positive, surely many of them were a much-needed help. All in all, awareness would, over time, increase experiences, good experiences boost acceptance, which in turn allows to draw the formal education world and open education closer together and kick-start the synergies. This is the "optimistic progress track" guiding this study. Finally, with more recognition, more MOOCs will appear. The teachers with successful online experimentation, might be interested in "scaling up" and create their own MOOCs (while others would leave their content locked on university platforms).

If MOOC recognition would become mainstream, it would boost the demand also for the current gap areas which we speculated is a partial result of a lack of demand-pull.

In terms of future, we suggest that universities will not disappear, but will adapt. Imagine this line of thought: a 100% face-to-face program would always see value added in adding online elements, due to arguments of superior accessibility, flexibility, efficiency and perhaps also broader networking reasons. Similarly, a 100% online program would always see value added elements in also involving face-to-face elements. The optimal balance point needs to be somewhere between those two extremes. This leads us to conclude by lending support to blended learning. From university viewpoint, the current MOOC situation, even with the coverage shortcomings, still offers a potent boost to enrich the learning experiences for logistics students. Therefore we propose that a key message of this study is to recommend raising awareness and encouraging teachers and directors of logistics curricula to seek ways on how to integrate the relative abundant availability. The digitalization of education ought not to be seen as a threat lowering intake to classical universities, but as an opportunity to embrace tech not only remain relevant but to further their own advantages. Perhaps the pandemic has also served as an innovation boost for universities, which would be a silver lining.

Our findings shed light on which could be the gaps between universities curricula and MOOCs content regarding logistics. However, we believe that the number of curricula examples and analysed MOOCs are an important limitation. Hence, we encourage other researchers to leverage our findings for further research in the field of MOOCs and online learning as a whole, using a larger sample of curricula, different geographical areas of study, and other organizational fields, in order to provide additional insights of which are those gaps, how universities are bridging for them, and how a broader digital revolution would fill or expand them.

References

1. Southern, N.: Historical perspective of the logistics and supply chain management discipline. Transp. J. **50**(1), 52–64 (2011)
2. Fritz, M.: Sustainable supply chain management. In: Leal Filho, W., et al. (eds.) Responsible Consumption and Production. Encyclopedia of the UN Sustainable Development Goals, Springer, Cham (2019)
3. Ng'ambi, D., Bozalek, V.: Massive open online courses (MOOCs): disrupting teaching and learning practices in higher education. Br. J. Educ. Technol. **46**(3), 451–454 (2015)
4. Shapiro, H.B., Lee, C.H., Wyman Roth, N.E., Li, K., Çetinkaya-Rundel, M., Canelas, D.A.: Understanding the massive open online course (MOOC) student experience: an examination of attitudes, motivations, and barriers. Comput. Educ. **110**, 35–50 (2017)
5. Milligan, C., Littlejohn, A.: Why study on a MOOC? The motives of students and professionals. Int. Rev. Res. Open Distrib. Learn. **18**(2), 92–102 (2017)
6. Loizzo, J., Ertmer, P.A., Watson, W.R., Watson, S.L.: Adult MOOC learners as self-directed: perceptions of MOOC motivation, success, and completion. Online Learn. **21**(2), 1–24 (2017)

7. Schuwer, R., Gil-Jaurena, I., Aydin, C., Costello, E., Dalsgaard, C., Brown, M., Jansen, D., Teixeira, A.: Opportunities and threats of the MOOC movement for higher education: the European perspective. Int. Rev. Res. Open Distance Learn. **16**, 20–38 (2015)
8. Nortvig, A.-M., Christiansen, R.: Institutional collaboration on MOOCs in education—a literature review. Int. Rev. Res. Open Distrib. Learn. **18**, 306–316 (2017)
9. Inamorato dos Santos, A., Punie, Y., Muñoz, J.C.: Opportunities and challenges for the future of MOOCs and open education in Europe. In: From Books to MOOCs? Emerging Models of Learning and Teaching in Higher Education, pp. 81–92. Portland Press, London (2016)
10. Sandeen, C.: Integrating MOOCs into traditional higher education: the emerging "MOOC 3.0" era. Change Mag. High. Learn. **45**(6), 34–39 (2013)
11. Pérez-Sanagustín, M., Hilliger, I., Alario-Hoyos, C., Kloos, C.D., Rayyan, S.: H-MOOC framework: reusing MOOCs for hybrid education. J. Comput. High. Educ. **29**, 47–64 (2017)
12. Moskal, P., Dziuban, C., Hartman, J.: Blended learning: a dangerous idea? Internet High. Educ. **18**, 15–23 (2013)
13. Rothe, H., Täuscher, K., Basole, R.C.: Competition between platform ecosystems: a longitudinal study of MOOC platforms. In: ECIS 2018 Proceedings, 134. AISeL, Portsmouth (2018)
14. Liu, M., Zha, S., He, W.: Digital transformation challenges: a case study regarding the MOOC development and operations at higher education institutions in China. TechTrends **63**(5), 621–630 (2019)
15. Witthaus, G., Inamorato dos Santos, A., Childs, M., Tannhäuser, A., Conole, G., Nkuyubwatsi, B., Punie, Y.: Validation of non-formal MOOC-based learning: an analysis of assessment and recognition practices in Europe (OpenCred). Joint Research Centre Science for Policy Report. EUR 27660 EN (2016)
16. Ossiannilsson, E., Williams, K., Camilleri, A.F., Brown, M.: Quality models in online and open education around the globe: state of the art and recommendations. ICDE Reports Series. International Council for Open and Distance Education, Oslo (2015)
17. Salajan, F.D., Roumell, E.A.: Two decades of e-learning policy evolution at EU level: motivations, institutions and instruments. Eur. J. Educ. **51**(3), 391–407 (2016)
18. Reich, J., Ruipérez-Valiente, J.A.: The MOOC pivot. Science **363**, 130–131 (2019)
19. Belleflamme, P., Jacqmin, J.: An economic appraisal of MOOC platforms: business models and impacts on higher education. CESifo Econ. Stud. **62**(1), 148–169 (2016)
20. Ebben, M., Murphy, J.: Unpacking MOOC scholarly discourse: a review of nascent MOOC scholarship. Learning **39**(3), 328–345 (2014)
21. Veletsianos, G., Shepherdson, P.: A systematic analysis and synthesis of the empirical MOOC literature published in 2013–2015. Int. Rev. Res. Open Distrib. Learn. **17**(2), 198–221 (2016)
22. Zhu, M., Sari, A., Lee, M.: A systematic review of research methods and topics of the empirical MOOC literature (2014–2016). Internet High. Educ. **37**, 31–39 (2018)
23. Niine, T., Koppel, O.: Typology of logistics curricula – four categories of logistics undergraduate education in Europe. Int. J. Eng. Pedag. **5**(2), 4–11 (2015)
24. Watted, A., Barak, M.: Motivating factors of MOOC completers: comparing between university-affiliated students and general participants. Internet High. Educ. **37**, 11–20 (2018)

Real and Virtual Lab Activities and the Effect of the Switching of Their Order in Teaching Science Concepts to Students with Learning Difficulties – A Case Study

Charilaos Tsihouridis[1(✉)], Marianthi Batsila[2], and Denis Vavougios[1]

[1] University of Thessaly, Volos, Greece
{hatsihour, dvavou}@uth.gr
[2] Directorate of Secondary Education, Ministry of Education, Larissa, Greece
marbatsila@gmail.com

Abstract. Concepts of physics, such as those of electricity or heat, present difficulties in their understanding, for students in general and for students with learning difficulties in particular. The present study investigates the perceptions of a high school student with Attention Deficit Hyperactivity Disorder (ADHD) in concepts of electricity and heat and observes his responses to experimental personalized activities in real and virtual environments in order to evaluate the cognitive results that emerge after the interventions. The initial interview that preceded the interventions showed that the student was not actively involved in the classroom and had similar alternative ideas about electricity and heat concepts. However, the results of the instructive virtual and real experimental interventions revealed the learner's active participation in the learning process, the enhancement of the conceptual change and the satisfactory apprehension of the science concepts. The findings also indicate that the switching from virtual to real labs lead to better cognitive results, lessening the stress and anxiety of the unknown.

Keywords: Science · Experiments · Learning difficulties · ADHD first section

1 Introduction

The study of the natural sciences has been the center of research interest for many decades now, with experiments being an important part of scientists' focus. Comparing experimental teaching to traditional, it can be seen that the first reinforces and supports learning [1]. So researchers all over the world are trying to improve their experimental teaching methods for the benefit of the students. Experiments are important for science as they facilitate the understanding of science concepts. They are considered one of the most important teaching strategies, especially for the teaching of difficult or abstract concepts [2]. This is because experiments enhance the development of technical skills (using devices, understanding and performing instructions), which will be particularly useful to students, both for school and social purposes. Experiments aim to connect theory with practice and promote scientific thinking [3]. In addition, experiments aim to strengthen students in both a cognitive and a scientific level. Experiments also display a

M. E. Auer and T. Rüütmann (Eds.): ICL 2020, AISC 1328, pp. 598–607, 2021.
https://doi.org/10.1007/978-3-030-68198-2_55

significant pedagogical value in science teaching because they involve students in an active educational process as opposed to passive learning [4]. Through experimental procedures students learn to collaborate, become more social and address their alternative perceptions. What is more, experiments cultivate skills such as patience, initiative taking, responsibility and equipment operating skills. They enhance creativity, objectivity, and scientific thinking, and if they are designed properly and accompanied by relevant questions and activities, they can have positive results for learning [5].

Experiments can be performed either by the teacher or by the students in collaboration with the teacher. Usually, they are laboratory activities included in curricula or syllabi and are performed based on available lab equipment emphasizing however the safety of the students. Their purpose is mainly to represent a phenomenon, in order to validate the corresponding theory that children have already learned. These experiments are called demonstration experiments where students are mainly spectators and passive participants. These are conducted by teachers to validate some theory, and/or to avoid the risk of materials misuse by students. However, there are also experiments in which students are active participants in the studying of a phenomenon, thus, gaining authentic research experiences. Their active participation helps them become familiar with scientific methodology and scientific thinking. Similarly, virtual experiments are simulations of phenomena that are studied through software. They simulate real objects using sound and image, elements that motivate students in the learning process and arouse their interest [6]. Virtual experiments allow the study of a phenomenon that otherwise might had been impossible or very difficult to do, either because of their dangerous characteristics or due to extensive time that would be required to implement in real laboratory conditions [7].

Experiments are suggested as an effective approach to deal with students' alternative ideas. These are ideas about concepts and phenomena of the natural world that children have from an early age and continue to have throughout their lives. They are created almost automatically and are usually quite difficult to change. Research shows that students have important alternative ideas in concepts such as heat or electricity [8]. For example, and in relation to heat, it has been found that young children believe that heat is something that "lives" in objects [9], a "substance" that can flow in or out of bodies such as air [10], that flows from one point to another [11] or that could be added or removed from an object [12]. It has also been shown that students are not able to perceive that different materials could be classified as cold, hot or in the between [13] or that some students consider temperature to be an instrument for measuring heat [14]. Regarding electricity concepts, research reveals that primary school children do not understand the concept of open and closed circuits, nor do they have representations of the connection of electrical appliances or the structure of an electrical installation [15]. But even in Secondary Education, students cannot separate electricity from energy [16] and their main alternative ideas relate to the nature of the electric current, that the current depends on the direction and order of the data or that it is consumed in the circuit [17].

Experiments are also suggested for teaching science to students with learning difficulties, that is, students who have a difficulty to learn something [18]. People with learning difficulties can face many challenges that often remain throughout their lives. Based on the type and severity of the disorder, teaching interventions and digital technology can be used effectively to lead such people to learning success or at least to reducing difficulties. Some interventions can be very simple, while others are more complicated. Educating and familiarizing students with digital technology can prove to be a useful teaching tool in the classroom. Within the broader category of learning difficulties we discern ADHD (Attention Deficit Hyperactivity Disorder), a situation characterized by hyperactivity and impulsivity. School students with ADHD find it difficult to comprehend the learning process in the same way that their classmates can do it, and at the same time, their attention is distracted, a fact that does not allow them to focus on what is happening in the classroom, thus, failing to participate. It is believed that students with learning difficulties benefit from specific guidelines which address their personal needs in a systematic way. For this reason, personalized educational programs are considered very helpful [18]. When it comes to students with ADHD, targeted educational interventions, encouragement and emphasis on the students' strengths/talents are suggested to assist the teaching process.

2 Rationale of the Research

The importance of laboratory practice in Science studies at all levels of education is acknowledged by the educational research community, while the pedagogical value of the experiments in science teaching has been acknowledged for years. Experiments are considered the most important educational tools and strategies in the classroom, especially for the teaching of difficult or abstract concepts. The ultimate aim of experiments is to connect theory to practice, help students develop experimental skills, and become acquainted with scientific thinking. In addition, experiments aim to enhance students' cognitive level by helping them observe, interpret, and express natural phenomena in order to gain a positive attitude towards natural sciences, develop self-efficacy, and boost their self-esteem and social skills. Teachers opt for experiments in science teaching, be these real, virtual or a combination of both, trying to lead their learners into scientific thinking and achieve efficient learning results. Teachers' choice lies on the fact that both real and virtual experiments display their own features and advantages and are therefore used extensively in science classes.

In particular, real experiments offer learners the possibility to experience phenomena as these evolve in real-life authentic situations, become aware of the experimental errors, and familiarize themselves with the use of real objects and equipment and their operation. Additionally, virtual laboratories display advantages such as attractive environment, simulation of concepts and phenomena studied in less time than the time needed for real experiments, minimizing the need for costly or specialized apparatuses, and offering an easy and with minimum risk frequent repetition of the experiments. Over the years, a great interest has been shown by researchers throughout the world regarding the effectiveness of real or virtual laboratories in different age groups, with sometimes-controversial conclusions. However, there is not extensive information regarding this matter with students that appear to have difficulties in learning and/or ADHD. Focusing on this issue and on the need to detect the effectiveness of laboratory work to students with learning difficulties, the authors of this study decided to investigate the effect of experimental lab work and the switching of the activities (real to virtual and vice versa) to a student with Attention Deficit Hyperactivity Disorder (ADHD).

3 Research Methodology

3.1 Research Purpose and Research Questions

The purpose of the research is to investigate the perceptions of an ADHD student regarding the heat and electric circuits phenomena through real and virtual experimental tasks. For this purpose, an experimental intervention took place aiming to examine the impact virtual and real lab environments have on the learning of the above concepts. The main objectives for the teaching of the heat phenomena are for the student to understand the differences between heat and temperature concepts, to allow him experiment with different quantities and be able to compare and understand the resulting changes, to familiarize himself with temperature and time measurements, to practise creating and using time temperature charts. Regarding the learning of electric circuits the aim is for the student to recognize the different types of electric sources, to experiment with batteries and measure them with a voltmeter, to create a simple electric element, to identify short circuit cases and determine their location, to understand the dangers, the operation and usage of electric circuits, to recognize the characteristics of resistor connections and use the measuring instruments correctly. What is more, the study also aimed to investigate the effect of the switching of the experimental activities (real to virtual or vice versa). Based on the above, the research questions are the following: 1. What is the impact of virtual and real lab environments on the ADHD student's learning in relation to heat and electricity phenomena? 2. Which of the two approaches –when switching from real to virtual or vice versa - gives better results in the ADHD student's learning in relation to the teaching of heat and electricity phenomena?

3.2 Research Participant Profile

The participant student in this research is a 13 year old male learner attending the first grade of Junior High School. He has been diagnosed with ADHD, language and math deficiencies and speech problems. It is therefore suggested by the state educational diagnosis educational committee to attend the conventional class but with emphasis on differentiated instructions and appropriately adapted activities based on his needs and difficulties. Based on the diagnosis committee (permission was taken to discuss the student profile by parents, committee and the school) the student a) displays difficulties in reading, b) makes many spelling mistakes, c) his level of understanding oral questions is satisfactory whereas his difficulty in understanding written questions is major. In order to support the student, his teachers need to make a special effort to help him understand and formulate orally any written questions (it is suggested by the diagnosis committee that the student is assessed orally), d) presents a relatively poor vocabulary and has difficulty in formulating sentences (they are relatively small and semantically problematic), e) displays lack of cognitive background and his answers are often repetitions of other people's opinions f) displays distraction, hyperactivity and impulsiveness, g) has difficulty concentrating and focusing on a particular activity or even a game for a long time h) shows significant memory impairment, i) his spelling is very slow and presents a very slow reading rate, j) has difficulty to organize time but has a good understanding of space.

What is more, generally, and according to his teachers, the student generally finds it difficult to understand relatively simple questions, does not understand (and usually misinterprets) the pronunciation of physics-chemistry-mathematics problems given to him and therefore does not show interest in solving them. He asks for multiple repetitions of questions asked to him, and does not fully complete his thought. He is relatively good at mathematics and mainly has practical skills in the laboratory courses in which he participates. He has very good knowledge of computer use, which he handles with relative easiness. Based on the DSM-IV diagnosis tool, the student has attention span, impulsivity and hyperactivity. In particular, the student cannot concentrate, is easily distracted by irrelevant stimuli, does not seem to listen carefully, does not pay attention to details, makes careless mistakes, finds it difficult to follow instructions, avoids tasks that require systematic effort, forgets schoolwork, loses things and is generally disorganized. These problems may be due to difficulties in decoding, lack of cognitive background and the way students with such characteristics approach the study process, i.e. the strategies they use [19].

3.3 Research Method, Tools and Process

Before the interventions, and for ethical reasons, permission was asked and received by the parents, and the school (teachers and principal). In particular, permission was asked to read and refer to the student profile ADHD characteristics. It was explained and reassured that this information would be used only for the purposes of this research, and it would remain anonymous. What is more, the student was also informed and was explained that he could withdraw from the process anytime he felt uncomfortable. For the purposes of this research, a few things were kept in mind: one, an individualized teaching intervention would be chosen mainly due to the student's difficulty in participating in group work in the classroom. Based on his profile, this difficulty relates to the fact that the student does not cooperate with his classmates and in cases that group work tasks are assigned in class, the student remains silent and inactive; two, due to the students' ability to use the computer with relevant easiness virtual experiments were chosen; three, the student enjoys creating things with his hands and therefore, experiments with real objects were also implemented. Additionally, the particular intervention experimental activities - based on constructivist and inquiry process theories - were selected, based on the fact that they constitute an effective means to bring about conceptual change. The intervention took place in five phases of ten teaching hours for the heat phenomena and another five phases of ten teaching hours for the teaching of electric phenomena where the student was guided into implementing relevant experimental activities. The student implemented first the real experiments and then the virtual for the heat phenomena, and vice versa for the electric phenomena. The student was observed throughout the whole process and detailed notes were taken in each phase and step. In the end, a thorough conversation took place with the student to examine his inner thoughts regarding the process and have a clearer understanding of the research aims. The data received were analyzed with the content analysis method.

3.4 Heat and Electric Circuit Phenomena Intervention Phases

Before the actual intervention of the heat phenomena the student was informed and familiarized with the process. Thus, during the first phase, which lasted for two teaching hours the student was introduced to the concept of heat and temperature through an interesting and playful way (using rich visual stimuli and many images on the computer). During the second phase of two hours, the student was asked to give his written opinion on stimuli, which were not displayed on the computer this time but in printed form, with corresponding space to write under the pictures. The third and fourth phases of the interventions concerned the implementation of real experiments. More specifically, in these two phases the student was able to work with real objects in tasks that were designed for the research purposes and were actually modified activities derived from the class textbook. Their aim was: a) to provoke an extensive discussion with the student about the difference but also the relationship between temperature and heat, b) to help the student practice with (and without) help and guidance in order to recognize the processes that lead bodies to thermal equilibrium, c) to help him practice temperature increase/decrease, d) to allow him take and record a series of temperature and time measurements and to create temperature-time charts as well as to process and interpret them e) to recognize temperature up to how it is measured, and how it is related to hot and cold, f) to help him be confronted with visual-experimental dilemmas and to cause possible changes in the final results of an experiment. The duration of these phases was four teaching hours. In the fifth phase of two hours a particular type of software was used regarding heat phenomena. The main goal was for the student to work with the main concepts in a virtual environment and to interact with the software to some extent with appropriate guidance and then independently.

In relation to the electricity phenomena the order was the opposite. The student work first on virtual and then with real objects. There was a first pre-intervention phase of two teaching hours. A six minute video was displayed to introduce the student to the topic. The choice of the video was based on his profile analysis based on which he enjoyed watching cartoons and/or children TV series. Therefore, this was considered an appropriate motive to attract his attention, which was achieved. Then, a worksheet was given to the student with activities with graded difficulty from easier to harder. Before the answers were provided the student was explained the questions of the tasks that were accompanied with relevant pictures and symbolisms. The tasks had to do with questions such as: What do you think is a battery? Can you describe why the bulb lights up? They were also tasks with pictures that he had to observe and predict whether the bulb would light up or not in a simple or not simple circuit, justifying his answer in a relevant space provided or predict the brightness of bulbs. He was also given a circuit with three bulbs in series and then in parallel and was asked to predict what would happen in relation to the brightness of the bulbs, if the middle bulbs were taken off. During the actual second and third intervention phases of four hours altogether, the software EDISON 4 was used. The aim was the student's familiarization with the software, the design of virtual electric circuits and their measurement. In particular, the objectives of the activities were for the student a) to experiment with batteries and measure them with a voltmeter, b) to design a simple electrical element, c) to design electrical circuits, d) to identify cases of short-circuits and their positions, e) to predict short circuits in everyday life and to prevent them, to be aware of their risks, to understand the mode of operation and the usefulness of "safety" in electrical circuits. Finally, the purpose of the last two phases of two hours each was to introduce the student to the real aspects of experimental physics, through a game based approach, using real objects. In particular, the aim of the activities was to arouse the student's interest to move from simple observation to systematic observation, to challenge his thinking and to promote learning, to highlight his ideas about electricity and address his alternative ideas.

4 Results

For the heat phenomena interventions, the specific student was given appropriately designed pre-experimental tasks on the concepts to be studied, in order to detect his alternative ideas in combination with his already acquired knowledge at the cognitive level. It is noteworthy that the student had not been actively involved in experimental procedures and tasks before, which played a very important role both in his response to them and in their performance. Initially, the student was very hesitant and cautious when he was explained the personalized teaching process and was asking questions like "What are we going to do?", "Is it difficult?". He seemed quite upset and kept asking what to write and whether his written reply would be read by others. He displayed a low self-esteem and confidence in his abilities, and he seemed very anxious. He had a negative attitude towards the written recording of his ideas and he felt quite embarrassed. However, his embarrassment was overcome when it was explained to him that these actions were carried out in order to implement a number of creative experiments which would not be evaluated, examined or graded by anyone. Although he had already been explained exactly what would happen, he was clarified again that he was not obliged to know anything particular in advance and that he could express his opinion freely, without any anxiety or fear of being ridiculed by his peers, as he would work on his own. This seemed to calm him down and he looked more confident and relaxed. Additionally, he seemed eager to start after he had been explained that the whole process would not be an examination but a different teaching approach tailored to his needs, in order to help him learn interesting and useful things for his daily life and to fill some of his gaps in Physics. Thus, he gradually became more interested to work and participate in the process. He was also told from the beginning that he would deal with both real experiments, that use real objects, and virtual experiments through a computer, which he particularly liked.

Throughout the experimental phases, the student appreciated constant encouragement, guidance, and confirmation to be able to render his thoughts in writing. There were noticeable differences between the first and last stages both in the answers and in the justifications chosen by him. For example, in the beginning, when the student was asked to render in writing the correct answer, he gave his opinion more recklessly and superficially, with many spelling mistakes, problems with the use of small and large letters (confusion of oral speech attributed to the written word) and inability to find the right words to convey the desired meaning. However, in the last stages the student was more active and participant and the justifications were more thorough and careful.

In the meantime, and in the phase with the real experiments, the student had generally ambiguous feelings. There were moments when he was very excited and not very anxious, as he stated that the real experiments reminded him of real life situations (e.g. filling the jars with water from the tap and using it), and moments when he said he was scared and believed that the experiment would affect his safety (e.g. by touching a hot surface or having the thermometer ready to use in his hand). In the experimental process with the virtual environments, the student already had an idea of his contact with concepts related to heat, he was quite willing to deal with the tasks, given his great interest in computers, but because of the slow speed of the experimental phases of the software used, he lost interest. He expressed his discomfort and said that he would have been more relaxed had he worked on the virtual tasks first because he would know what to expect. However, he completed all the activities following the steps of the inquiry research model based on which, he formulated hypotheses, predicted, observed, experimented and managed to formulate his own conclusions on the topics under study.

In relation to the electric circuits phenomena, the student initially was again somewhat reluctant to work. However, he soon adapted himself to the process and participated actively in the steps that followed. He showed interest and enthusiasm and tried to implement all the activities. In relation to the order of the activities it seemed that the sequence of virtual to real experiments helped him to adapt better as he was able to work with the real objects without the fear of the unknown or possible damages and he seemed more confident this time. This was because he felt comfortable having worked first on the virtual tasks, knowing in a way the process, and then continued to implement it with the real objects. This way, it seemed that the tasks ran smoothly and he was able to follow the whole task implementation with less anxiety, more interest and curiosity.

5 Discussion

In this research we have tried to look into the impact of virtual and real experiments on an ADHD learners' cognitive level and particularly, towards science learning and the extent to which they can enhance his interest, motivation, creativity and participation in the educational process. What is more, the aim was to detect the effect of the switching of the order of the experimental activities (virtual to real or vice versa). According to Siemens [13] learning is a complicated and demanding process which is best fulfilled when individual needs are addressed in order to reach the desired goals. The multi-faceted process of teaching is enhanced with a variety of tools both virtual and real that can be used to promote learning and achieve the desired teaching and learning goals. It is argued that the concepts related to Heat, Temperature and Electricity are among the most important and complex science topics for school students and among the ones with the most difficult and abstract concepts for students of all levels. Based on the results of this research, in relation to the study of heat and electricity concepts, the ADHD student managed to respond very well in both cognitive and metacognitive level as through the experimental procedures he understood the above concepts. It can be then argued, and according to the results, that, for learners with learning difficulties, such as the target student, the use of properly designed experimental activities, with individualized characteristics, can possibly offer better learning outcomes, enhanced with positive feedback and encouragement.

For the purposes of the research, specific tasks were assigned to the student regarding the concepts to be studied, before the implementation of each experimental intervention, in order to assess his alternative ideas in combination with his cognitive level as well as during the interventions. It is noted that the particular student had not been actively involved with experimental procedures before the interventions. Initially, the student was quite hesitant and cautious as he was characterized by low self-esteem. He then adapted to the research processes and actively participated in the experimental phases that followed. Throughout the process, he gradually showed interest and enthusiasm and tried to implement his experiments and "new inventions" as he called them. It is important to mention that after the implementation of the teaching interventions, significant conceptual changes took place, as well as quite a satisfactory degree of understanding of the concepts taught. Satisfactory improvement of the specific student with ADHD was observed in terms of his performance in the target concepts, taking into account the previous situation. Concepts that came to the forefront of the experimental process could be recalled by him satisfactorily even after a few days, based on his class teachers. However, an encouraging fact was a change in the student's attitude as, according to his teachers' comments, after the interventions, he participated in the class lessons, together with his fellow students, more actively now and with more interest than in the past.

In general, it seems that both real and virtual environments had a positive effect in this direction, as through the experimental process the student was able to comprehend the new concepts and be more active in learning. He developed experimental data collection skills (e.g. temperature or time measurements) and experimental data processing and interpretation (e.g. creating and interpreting time temperature charts or electric circuits). In conclusion, the student's positive response to the subject matter, throughout the intervention processes, should be emphasized as well as the effective learning outcomes, in terms of cognitive, socio-emotional and cognitive skills. Therefore, it is noteworthy to mention that when teaching science concepts, it is important on the one hand, to design appropriate teaching interventions and on the other hand, to select an appropriate teaching method based on students' needs and capabilities and especially on students with special needs and learning difficulties. Finally, it seems that the order of virtual experiments to real ones lessens the burden and fear of the unknown and allows learners and especially learners with learning difficulties to concentrate on the study of a phenomenon with more interest and active participation.

References

1. Tsihouridis, Ch., Vavougios, D., Ioannidis, G.S.: The timeless controversy between virtual and real laboratories in science education - "And the winner is…". In: International Conference of Interactive Collaborative Learning and Engineering Pedagogy (ICL), Kos, Greece, 25–28 September 2018 (2018)

2. Etkina, E., Van Heuvelen, A., Brookes, D.T., Mills, D.: Role of experiments in physics instruction–a process approach. Phys. Teach. **40**(6), 351–355 (2002)
3. Psillos, D., Niedderer, H.: Issues and questions regarding the effectiveness of labwork. In: Psillos, D., Niedderer, H. (eds.) Teaching and Learning in the Science Laboratory, pp. 21–30. Kluwer Academic Publishers, Dordrecht (2002)
4. Marshall, J.A., Young, E.S.: Pre-service teachers' theory development in physical and simulated environments. J. Res. Sci. Teach. **43**(9), 907–937 (2006)
5. Soares, B.C., Castelhano de Campos, M.E., Thomaz, J.R., Cruz Pereira, G., Roehrs, R.: The importance of experimentation in the teaching of sciences to elementary school. Revista Monografias Ambientas, REMOA **15**(2), 1–7 (2016)
6. Abdulwahed, M., Nagy, Z.K.: The impact of the virtual lab on the hands-on lab learning outcomes, a two years empirical study. In: 20th Australasian Association for Engineering Education Conference, pp. 255–260 (2009)
7. Tsihouridis, C., Vavougios, D., Ioannidis, G.: The effectiveness of virtual laboratories as a contemporary teaching tool in the teaching of electric circuits in Upper High School as compared to that of real labs. In: Proceedings of 2013 International Conference on Interactive Collaborative Learning (ICL), Kazan National Research Technological University, Kazan, Russia, 25–27 September 2013, pp. 816–820. IEEE (2013). ISBN 978-1-4799-0152-4/13/$31.00 ©2013
8. Teixeira Dorneles, P.F., Veit, E.A., Moreira, M.A.: A study about the learning of students who worked with computational modeling and simulation in the study of simple electric circuits. Revista Electronica de Ensenanza de las Ciencias **9**(3), 569–595 (2010)
9. Albert, E.: Development of the concept of heat in children. Sci. Educ. **62**(3), 389–399 (1978)
10. Erickson, G.L.: Children's conceptions of heat and temperature. Sci. Educ. **63**, 221–230 (1979)
11. Harris, W.F.: Heat in undergraduate education, or isn't it time we abandoned the theory of caloric? Int. J. Mech. Eng. Educ. **9**, 317–321 (1981)
12. Erickson, G.L.: Children's viewpoints of heat: a second look. Sci. Educ. **64**, 323–336 (1980)
13. Choi, H., Kim, E., Paik, S., Lee, K., Chung, W.: Investigating elementary students' understanding levels and alternative conceptions of heat and temperature. Elem. Sci. Educ. **20**, 123–138 (2001)
14. Harrison, A.G., Grayson, D.J., Treagust, D.F.: Investigating a grade 11 student's evolving conceptions of heat and temperature. J. Res. Sci. Teach. **36**, 55–87 (1999)
15. Hamid, R., Widodo, A., Sopandi, W.: Students' conceptual change in electricity, advances in social science. Educ. Humanit. Res. (ASSEHR) **57**, 48–52 (2016)
16. Madu, B.: Exploring senior secondary school two students' alternative conceptions of current electricity in physics in Nigeria. SAUSSUREA **6**(4), 257–274 (2016)
17. Kallunki, V.: A historical approach to children's physics education: modelling of DC-circuit phenomena in a small group. Dissertation, Department of Physics, University of Helsinki, Finland (2009)
18. Lyon, G.R., Shaywitz, S.E., Shaywitz, B.A.: Defining dyslexia, comorbidity, teachers' knowledge of language and reading: a definition of dyslexia. Ann. Dyslexia **53**, 1–4 (2003). https://doi.org/10.1007/s11881-003-0001-9
19. Brigham, F.J., Scruggs, T.E., Mastropieri, M.A.: Science education and students with learning disabilities. Learn. Disabil. Res. Pract. **26**(4), 223–232 (2011)

Online Learning as a Necessary Measure During a Pandemic and as an Opportunity to Increase the Engineering Education Efficiency

Irina Makarova[1(✉)], Anton Pashkevich[2], Polina Buyvol[1], Eduard Mukhametdinov[1], and Vadim Mavrin[1]

[1] Kazan Federal University, Syuyumbike prosp., 10a, 423812 Naberezhnye Chelny, Russia
kamIVM@mail.ru, skyeyes@mail.ru, funte@mail.ru, vadim_mmite@mail.ru

[2] Politechnika Krakowska, Warszawska 24, 31-155 Krakow, Poland
apashkevich@pk.edu.pl

Abstract. This research aim was to analyze problems, which educational systems are facing in the global upheaval era, statement of tasks, which must be solved, as well as search for solutions. The engineering education system has been improved in recent years through the use of new features, which are provided by e-learning software. Future engineers will require such skills as creative thinking, ability to make informed, effective decisions, as well as social responsibility and eco-thinking. The research was carried out on the tools used to organize online learning during the COVID-19 pandemic in several universities. It gave a possibility to compare the effectiveness of using different virtual environments to organize the training of different profiles' engineers, as well as to analyze the educational content and its difference from that used in traditional classroom teaching. To make a correct assessment, evaluation criteria were identified, which can be formalized, and also motivational characteristics were classified, which can be influenced on the quality improvement. For a case study, different student's groups were selected to compare them involvement in the process with traditional and online training forms as well as their performance. Additionally, was done an analysis how a training form impacts on their projects quality.

Keywords: Engineering education · Online learning · Virtual educational environments · Educational content quality

1 Introduction

Modern society faces unprecedented challenges arising from urbanization and mobility needs, which associate with the globalization of all life spheres. One of these challenges is pandemics, similar to COVID-19, leading to a collapse in many activities areas and the economic systems' imbalance. In such conditions, the educational system

M. E. Auer and T. Rüütmann (Eds.): ICL 2020, AISC 1328, pp. 608–620, 2021.
https://doi.org/10.1007/978-3-030-68198-2_56

gets increased requirements associated with the need to specialists training by various economy sectors, who could work in critical situations. The educational system should maintain ability to overcome inertia, in the same time it must be enough flexible both from the viewpoint of educational methods and technologies and from the process control viewpoint. In recent years, the engineering education system has been improved through the use of new opportunities e-learning software [1]. In the same time, the growing complexity of engineering and technology requires considerable efforts from students during both training and subsequent professional activities. Future engineers be able to make informed effective decisions, should be have creative thinking and eco-thinking and social responsibility. Currently, the emergence of new tools and techniques for engineering education is associated with processes in all spheres of public life and activities. These are such trends as Industry 4.0, Smart City, smart transport and logistics, industry digitalization, digital twins, etc. Smart education, virtual and augmented reality technologies allow improving the educational environment, increasing the students' motivation and their competitiveness, as they prepare them for life in the digital age.

University professors are often committed to traditional lecturing and laboratory teaching methods, which is why many of them experience stress under unusual external circumstances. To avoid stress, it is necessary to pre-test various resources, combining traditional and online learning. A similar stress is experienced by students who have in ordinary conditions a motivating factor by the conscious choice of the form and method of receiving education, but in critical situations there is no such choice and the student is faced with a dilemma: either to submit to the conditions accepted due to external circumstances, or to refuse training. Based on the above-mentioned facts, the following challenges were formulated: (1) What changes in the management system of universities, departments as well as educational processes should be implemented in order to increase efficiency during critical situations? (2) What changes in educational contents are necessary to ensure the necessary quality of education as well as involvement and motivation of students? (3) How will it be possible to use the gained experience after the crisis? In the authors' opinion, the most effective tool for the reactive management of learning process, as a complex system, is feedback. Due to the duration of learning process, the educational system is very inert. That is why the effective feedback is a tool to overcome this inertia. The way to implement feedback between teachers and students is a key factor determined the quality of the educational process and student involvement. To give a teaching staff the real possibility to monitor and evaluate the quality of education, it is necessary to identify objective criteria for such evaluation in the context how motivational features are differentiated. In addition, in such way an academic teacher will be able to assess the effectiveness how a particular method influences on the motivational criterion. Such a system, which were developed using formalized criteria, will be more objective and reveal problems for an individual student or group of students, as well as within the framework of a training course or a separate section of a course unit.

2 Problem Status: E-learning and Distance Study Tools

2.1 Concepts and Methods to Realize Distance Study

Distance study is an important tool of a modern education. The article [2] presents how the students' experience was tested as well as the methodology of teaching in remote format was assessed by students in real time. The results show that the flexibility and usefulness of the course format are important elements for students' experience. The article identifies a number of questions and starting points for the next study, focused on the perception of gender or the real impact of course content.

The author of the article [3] proposes a new approach for learning and examination of students in the discipline "Theory of Electrical Engineering". It is recommended to duplicate theoretical materials from lectures, uploading them to Moodle, where the student can study the theory while solving examples, then to check the acquired knowledge with the help of tests uploaded to the system. Thus, students could prepare for the final exam. The final grade is made up of two parts: the first one is the sum of the points obtained when studying the course and passing the control tests, the second part is the grade for the exam. The article [4] presents a learning model, which uses the YouTube video repository for engineering courses. Some results show students' preference for this model over the more traditional learning models, which use a live video. The results can be applied to develop hybrid models for engineering courses, which can have a positive impact on student satisfaction and, in general, on the course where the model is used. The article [5] describes an experimental study of the Online Flipped Classroom Learning Model (OFCLM) in relation to open and distance learning for adults in a training center. The main results showed that students liked the flexibility and convenience of the proposed model. Smart Classroom is defined as a concept, which needs to adapt to synchronous and asynchronous learning using artificial intelligence technology, that teachers and students can approach different learning styles, engage interactively, and share content. The study [6] provides a system of "intelligent classes", which covers the process of distance learning within the framework of the Smart Learning System (SLS) as well as learning in a classroom using the Smart Learning Environment (STE). The proposed Smart Classroom allows to facilitate both classroom learning and distance learning processes.

2.2 Interaction and Its Realization During Distance Education

Distance learning is one way to achieve equal quality education. The study [7] proposes a new technological solution for video conferencing using webRTC technology as an alternative for applications, which are still based on the flash memory technology. The technology allows users to conduct video conferences, use additional functions, such as a function, which allows to share a recording screen.

Distance learning courses are becoming more and more attractive for students and educational institutions, and it is necessary to assess the extent, to which online education can provide opportunities to promote the personality of researchers. The researcher's identity can be defined as a combination of such features as confidence, logical thinking and the ability to plan experiments, interpret the results and the desire to seek and succeed in genuine research. But as the authors of the article [8] point out, real research should only be partially replaced by online activities, although the activities of outbreak case study—group project are an economical and attractive tool. Students at a distance often experience social exclusion and lack of communication with their peers. A survey was conducted to better understand the nature of the relationships established between distance students and peers. The analysis performed by the authors [9] shows how connections are formed and tested in the online learning environment in various social contexts and technologies, which strategies and practices promote or impede connections that can satisfy students' social needs.

The article [10] discusses the level of the joint learning strategy that students of distance learning programs have. The reason is so that team work allows people to realize the common goal and objectives, as well as to understand what everyone in the team must accomplish to achieve a common goal. The authors believe that one of the ways to achieve this positive interdependence is to motivate students and that the development of teamwork requires the intervention of both the student and the teacher.

2.3 Remote Laboratories

E-learning should include not only web courses and virtual classes, but also remote laboratories with an effective recognition system, which will allow students to interact with real-life experiments conducted at a distance. The work [11] analyzes the influence of students' emotions on the improvement of the educational process in two aspects: (i) filling in existing gaps among students and (ii) improving the usability and readability of the platform. The authors create a new intelligent learning system called "LabTutor", which is adapted to the profile of each student, providing students with access to experimentation on real laboratory equipment in the field of engineering education. Social requirements promote an educational approach based on the premise of "anywhere, anytime". Remote laboratories start to be a response to the requirements of technical fields of education to adapt to this scenario. The result [12] not only benefited students of distance learning, but also provided new learning scenarios for both teachers and students, and also allowed a flexible approach to experimental topics.

Web courses have achieved a high level of recognition in all areas of education. In this sense, many web-based platforms for online courses offer resources as a complement to theoretical studies [13]. In the fields of STEM (science, technology, engineering and mathematics), online resources can include on-line labs, which are typically needed in an automated learning process. Remote laboratories [14] can serve as the basis for the development of physical experiments for training in the field of STEM. Thus, one of the VISIR + project goals is to disseminate best practices in the field of teaching electrical and electronic circuits. The article [15] authors proposes a model for the informational interaction integration in projects to create training courses, which are based on virtual and complementary reality technologies. A sequence of technological stages to create virtual learning systems and algorithm for its creation is proposed.

2.4 Issues Connecting with the Organization Process of Distance Study

In the field of engineering education, the possibility of using a real management system and virtual laboratories in the educational process motivates teachers to improve students' knowledge and skills. That also can reduce costs and increase the variety of experiments. The authors of the study [16] suggest that distance experimentation is an alternative didactic methodology. The WebLabs use will improve the engineering education quality, as it will allow direct information analysis in real time and bring industrial scopes problems closer to the academic space. Student feedback indicates that this is the engineering education's future.

The article [17] shows how tests were conducted such as training, which is becoming more common in the educational environment. This hybrid teaching method combines a massive open online course (MOOC) with students' personal work. As a new form of distance learning, the Blended Synchronous Learning Environment (BSLE) is a new learning environment that allows students to participate in the learning process synchronously. In China, BSLE has a more unique practical meaning - the Blended Synchronous Classroom (BSC) - to solve the problem of educational equilibrium. The purpose of the study [18] is to compare and analyze the preliminary results of the BSC project, then investigate the key problems of the BSC and give appropriate countermeasures in combination with specific situations to lay the foundation for the future development of BSC and increase its reliability. In recent years, the successful integration of virtual reality in the context of distance learning has led to the development of various structures related to virtual learning approaches, which are used to improve interaction with students and improve their overall learning experience. The purpose of the study [19] is a thorough study and a deep understanding of the impact that embodied pedagogical agents have on the students' learning experience, as well as on their academic performance. In a future work, the authors of the article plan to create agents that resemble well-known scientists (for example, Albert Einstein), and evaluate how the appearance of agents affects the learning experience of students.

Feedback is an integral part of the distance learning system. This helps students to recognize strengths and areas for improvement, as well as identify actions to increase their effectiveness. In addition, it helps teachers focus on educational policies to improve their content. The authors of [20] propose a model for analyzing student traces in the cloud. Teaching styles in the field of natural and technical sciences were compared by analyzing the totality of lecture transcripts in English and Japanese at leading universities in the USA and Japan, respectively [21]. Science and engineering education in the Japanese educational context tends to reflect and reinforce a teacher-centered personalized knowledge-style transfer that characterizes the field style of the Japanese cultural context and can reinforce this style among students, while teaching in the American context seeks to match and strengthen learning styles characterized by impersonal, inductive thinking, typical of the mainstream American culture.

3 Case Study: Experience of Distance Study Realization During COVID-19 Pandemic

3.1 Features of the Educational System and Its Readiness for Global Challenges

It makes sense to consider the educational system from the point of view of both the processes and the participants. The impact on the process is carried out through organizational and managerial structures, which include different departments, all of which is responsible for its own process and, accordingly, controls it. For these purposes, there are various regulatory documents and instructions, as well as feedback. As a rule, management is carried out in accordance with the activities types. Traditionally, the learning process was carried out by transferring educational content in the classroom: in the lectures, laboratory and practical exercises. With the advent of computers and the information technology's development, new types of educational content have appeared, as well as new teaching methods. To expand the possibilities in implementing the concept of "life-long learning", technologies for the distance education implementation have appeared. As can be seen from the above review, these technologies currently provide the opportunity to receive a quality education, although they have problems.

The modern university implements both traditional and distance learning. However, it should be borne in mind that, firstly, these are two completely autonomous processes with their own management systems, and secondly, the participants in the process (administrators, teachers and students) consciously chose one form or another and, therefore, are mentally prepared to the features of the learning implementation.

3.2 Problems Encountered in the Countries of the World: The View of Teachers, Students and Administrators

Enough time has now passed since the pandemic began, for which some experience has been accumulated, a number of problems have emerged, including those that were resolved to some extent and those that could not be resolved promptly. Interviews appeared in the public domain, in which teachers, students and administrators share experiences, problems and ways to solve them. As shown in [22], Covid-19 forced Indian universities, as in the world rest, to switch to online classes, suspending physical classes. Online training is conducted both using recorded classes, including MOOC, and through online classes conducted in the sessions form with zoom or webinars. To do this, professors need high-speed Internet and platforms to provide education or a learning management system, as well as a stable IT infrastructure and professors who are ready to teach online. High-speed Internet and computers or mobile phones are also needed by students to attend or review previously recorded classes.

In India, diverse online education platforms are supported by the Ministry of Human Resources Development (MHRD), the National Council for Educational Research and Training (NCERT), and the Department of Technical Education. There are also initiatives such as SWAYAM (online courses for teachers), e-PG Pathshala (electronic content) and NEAT (enhancing employability). Other online platforms are aimed at improving communication with institutions and access to content. They are used for training materials and classes, as well as for working with online modules. These include the National Knowledge Network (NKN), the National Project on Technology Enhanced Learning (NPTEL), the National Academic Depository (NAD), and others.

For provide online education, to have web and video courses in engineering, science and management, NPTEL and project of MHRD, initiated by seven Indian Institute of Technology (IIT), along with the Indian Institute of Science Bangalore was created in 2003. During Covid-19, more than ever before, the need for online learning increased for both institutions and teachers and students. However, despite the availability of technology, there is a problem of access, since not every student has a fast Internet and a computer. A survey conducted by an IIT teacher showed that 9.3% of its 2789 students were unable to download materials sent by the institute or study online. Only 34.1% of them had a reasonably good internet connection for streaming lectures in real time. Another study conducted by LocalCircles among 25,000 respondents found that only 57% of students at home had the necessary equipment — a computer, router, and printer — to attend online classes. Students note difficulties in obtaining materials due to lack of internet-traffic and connection problems, as well as the stress caused by the need to be in touch all the time and watch the whole day to a computer or phone screen. In addition, both students and teachers note the insufficiency of direct contact, discussions, debates, etc. According to the director of Takshashila University, the model of online education cannot replace the physical class, but can only be an addition.

Similar problems were encountered in other countries. For example, on the World Economic Forum platform has accent the gap in digital learning; both between countries and within many countries [23]. Thus, according to the OECD, while 95% of students in Switzerland, Norway and Austria have a computer for study, in Indonesia this index is only 34%. In the US, many are also concerned that the pandemic will gap to a digital technology use, as there is a significant divide: while almost all 15-year-olds from privileged families have their own computer, and almost 25% of children from disadvantaged families do not have such opportunities.

There is evidence that online learning can be more effective for those who have access to the right technology, since according to research, on average, students retain 25–60% more material when online learning compared to only 8 –10% in class. E-learning requires 40–60% less time for learning, as students can learn at their own pace, skipping part of the lecture or speeding up the pace at their discretion. But it should be understood that the effectiveness of online learning varies among age groups. But transition to online learning can be a catalyst for creating a more effective method of teaching students. The problems identified during the pandemic confirm the book [24] author opinion that schools continue to focus on traditional academic skills and rhythm-based learning, rather than on skills such as critical thinking and adaptability, which will be more important for future success. However, there is a positive point in the fact that the though haste of the transition to online learning may have reduced its quality, but, using the advantages of such training, can be done its part of the traditional.

3.3 Features of the Educational Process Implementation in KFU During COVID 19

In the forced "switch" case, problems arise due to both the participants' internal unreadiness for changes and the need for a quick transition to another process implementation way, which may be not prepared both from a technical-organizational and normative-legal points of view. In order to increase the educational system sustainability and its readiness to work under stress, it is necessary to identify the risks causes, ways to its minimize and measures to manage the system in a risk situation. Since the transition to distance learning was carried out in accordance with the order of the Ministry of Russia Education and Science, the Ministry, together with the Institute for Social Analysis and Forecasting of the Russian Academy of National Economy and Public Administration at the end of the academic year, conducted a mass survey of the universities' teaching staff on the development of the online environment in the context of coronavirus infection. It summarizes the attitude of teachers to distance education in the context of a sharp change in their professional activities and the organization of their personal lives. 33987 teachers, or about 15% of the total of the teaching staff, were taken part in the survey. The results of the study showed that teachers are organizationally ready for the transition to distance learning formats, but psychologically they do not accept such a sharp break with traditional full-time education. A skeptical attitude to what is happening is due to both the peculiarities of the disciplines taught (for example, technical and experimental), and conservative views on the nature of teaching. Although teachers formally accept the transition to distance education, they have an internal, latent rejection of it in all areas of training, regardless of gender, age, social and professional status. The survey recorded two positive social characteristics of the professional community of university teachers: the adoption of a state policy on countering coronavirus infection and the availability of skills and abilities to work in an online environment.

At the same time, the breakdown of the usual life way, the destruction of the existing day race-order, in which work and rest are distributed not only in time, but also in space, led to stress and, as a result, rejection of distance education: 66% of teachers indicate that they don't like working at home; 34% of teachers have no place at home for comfortable teaching; 85.7% of teachers have less free time, they have an idea of an increase in the workload; 87.8% believe that it is better to conduct their classes in full-time format. The main request of teachers comes down to three components: material (providing with computer equipment and software); communicative (an environment for communication, necessary and sufficient to support distance learning, inclusion in the team and maintaining a high level of learning); organizational (reducing bureaucratic pressure and providing more freedom in the choice of teaching means and methods).

46.1% of the respondents disagree with the change in education towards individualization and adjustment for each student. The main threats associated with the impossibility of liberalizing education, the transition to a distance format, are called: the decline in students' motivation to learn; lack of students' skills and abilities to maintain discipline and diligence in distance learning; emotional breakdowns of both students and teachers; increased workload on teachers; lack of an individual approach in the education system, impersonality; inability to control the level of knowledge; restriction in some spheres (for example, in engineering) on the remote transfer of knowledge; formalization of educational processes, a tendency to stereotyped, unified solutions. In general, the results of the survey in Russia coincide with the opinion of the teachers in our university (Fig. 1). In our research, we studied the following issues: (1) What virtual environments have shown the greatest efficiency in terms of process control, education quality, student involvement, increasing their motivation; (2) How the transition to online learning has affected the quality of educational content, its structure, the employment of teachers and students, class attendance; (3) What resources should be developed to control learning quality and the process organization. The university implements both the traditional form of education and the distance learning. For these purposes, a special "virtual audience" has been created. Each teacher can access to it and use this resource as an addition to traditional teaching. At the beginning of the transition to online mode, courses were held for teachers on the use of this resource. But when the mass connection began, it turned out that the technical capabilities of mass simultaneous use of this resource are clearly not enough since access is through the university's website. It was decided to use Microsoft Teams for these purposes.

Fig. 1. Results of the teacher survey

This solution turned out to be effective for most teachers, since almost all of them have high-speed Internet and a computer at home. For most students, this option of training also did not cause problems, most of them quickly adapted to the new form of training, as evidenced by the increase in attendance at classes by 10–15% at different courses. From an organizational point of view, there were no problems with the control of classes, since the classes were conducted according to the schedule approved at the beginning of the semester, however, the load on the controlling staff, who had to analyze and summarize the data, increased. Assessment of students' knowledge was carried out in different ways - by testing, checking the work sent to the teacher, protecting online projects during their presentation. An online conference of students was also held, where they shared the results of their research.

To study the quality of teaching academic disciplines and the entire educational process, two questionnaires "Teacher through the eyes of a student" and "The educational process through the eyes of a student" were developed. The research was carried out with the help of the information-analytical system "Electronic University", where access to the questionnaires was opened to students in their Personal Accounts. The research results were accumulated in a single final base, and then processed.

In the course of the questionnaire "Teacher through the eyes of a student" for each student, according to the disciplines of the curriculum, a list of teachers was generated and 10 criteria were formulated. Each criteria could be evaluated on a 10-point scale (minimum - 1 point, maximum - 10 points): I always understand the lesson purpose; The teacher presents the material in a clear and creative way; The teacher emphasizes the importance of the subject for the future profession; I actively use teaching aids developed by the teacher; The teacher uses visual aids and technical teaching aids during classes; The teacher demonstrates general erudition; The teacher is objective in assessing the work of students; The teacher is friendly and tactful, treats students with respect; I am impressed by the appearance and demeanor, organization and discipline of the teacher, his culture of speech; The teacher is available for consultation.

During the questionnaire "The educational process through the eyes of students" the following questions were asked: (1) Evaluate the quality of the knowledge obtained in the disciplines of the basic part (compulsory disciplines); (2) Evaluate the quality of the knowledge obtained in the disciplines of the variable part (discipline of choice); (3) Are you satisfied with the list of variable disciplines for your specialty/direction? (4) Are you satisfied with the quality of the organization of laboratory and practical classes? (5) Are you satisfied with the quality of teaching foreign languages? (6) Are you satisfied with the quality of the industrial practice organization? (7) Do teachers help you to effectively organize your independent work? (8) Do teachers help you find the information you need for independent work? (9) Are you satisfied with the control over the performance of independent work by teachers? (10) What difficulties do you experience when working independently? (11) Are you satisfied with the procedure for providing topics for coursework? (12) Appreciate the help of supervisors in writing term papers; (13) Are you satisfied with the current knowledge assessment system.

On the whole, students were satisfied with the organization of the educational process. At the same time, technical difficulties arose, as in other countries. So, during the online broadcasts there were interruptions in communication, students complained about the difficulty of listening to online lectures. In order to minimize this shortcoming, many teachers gave students self-study materials. The greatest difficulties arose with laboratory experiments in which students were unable to participate. Only some of them became possible to implement as online broadcasts (Fig. 2).

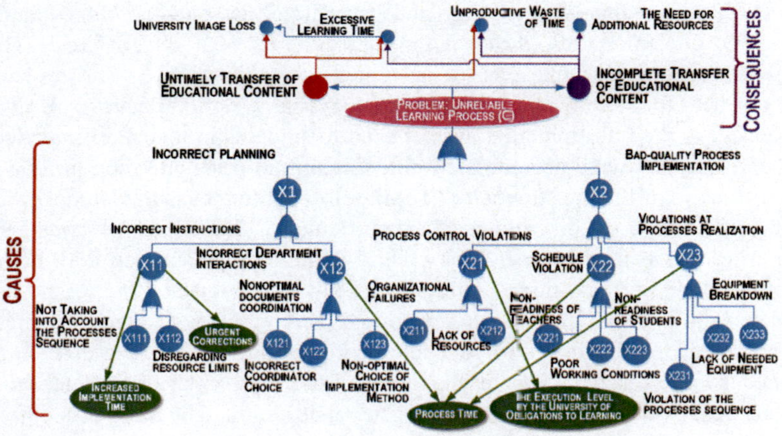

Fig. 2. The tree of problems in the educational system when critical situations arise

We have summarized organizational problems that must be resolved before crisis situations arise. According to the results obtained during the educational process during the COVID-19 pandemic, it is possible to obtain the probabilities of their occurrence and develop measures to overcome it.

4 Conclusion

Major global events are often a watershed for rapid innovation, as SARS has seen an increase in e-commerce. Although it is too early to make conclusions whether COVID-19 will affect the development of e-learning, it is already clear that it is important to spread knowledge across borders, include all companies and sectors of society in this process, fully revealing the potential of online learning technology.

The researches revealed the advantages of online education for teachers: the possibility of innovative teaching methods using online tools; wider coverage of students from different regions; development of distance learning technologies. At the same time, there are drawbacks: such a transition takes time and practice; there are doubts about the fairness of student assessment; the impossibility of personal communication with students; the inability to reach all students due to technological limitations. The advantages of online learning for students: the ability to use various online tools and methods; observance of the education terms; the ability to re-listen to classes and study at your own speed. Disadvantages: lack of personal communication; technological difficulties associated with weak devices or Internet access; the need for online communication and evaluation skills; learning in conditions not adapted to study - off campus.

References

1. Makarova, I., Shubenkova, K., Tikhonov, D., Buyvol, P.: Improving the quality of engineering education by developing the system of increasing students' motivation. In: Advances in Intelligent Systems and Computing, vol. 716, pp. 150–161. Springer, Cham (2018)
2. Núñez, M.E., Rojas, J., Rodriguez-Paz, M.X.: Real-time distance courses to improve satisfaction and competence - a case study on the performance of students observing their grades. In: IEEE Global Engineering Education Conference, Dubai, pp. 519–525 (2019)
3. Filipova-Petrakieva, S.K.: Some problems in distance learning on theoretical electrical engineering in Technical University of Sofia. In: 16th Conference on Electrical Machines, Drives and Power Systems, Varna, Bulgaria, pp. 1–5 (2019)
4. Rodriguez-Paz, M.X., Gonzalez-Mendivil, J.A., Rojas, J., Núñez, M.E.: Use of an offline video repository as a tool to improve students' performance in engineering courses versus real-time long distance courses. In: IEEE Global Engineering Education Conference (EDUCON), Dubai, United Arab Emirates, pp. 544–551 (2019)
5. Shu, F., Zhao, C., Wang, Q., Huang, Y., Li, H., Wu, D.: Distance learners' learning experience and perceptions on the design and implementation of an online flipped classroom learning model. In: Eighth International Conference on Educational Innovation through Technology (EITT), Biloxi, MS, USA, pp. 7–11 (2019)
6. Yuliantoputri, A.R., Muhamad, W., Suhardi, S.: Smart classroom services system design based on services computing system. In: International Conference on ICT for Smart Society (ICISS), Bandung, Indonesia, pp. 1–6 (2019)
7. Alimudin, A., Muhammad, A.F.: Online video conference system using WebRTC technology for distance learning support. In: International Electronics Symposium on Knowledge Creation and Intelligent Computing, Bali, Indonesia, pp. 384–387 (2018)
8. Massimiliano, M., Galindo, S., Silva-Lugo, J.L.: Fostering researcher identity in STEM distance education: impact of a student-led on-line case study. FEMS Microbiol. Lett. **366**(6), fnz068 (2019)
9. Na, S., Xiying, W., Rosson, M.B.: How do distance learners connect? In: Proceedings of the CHI Conference on Human Factors in Computing Systems, Paper 432, p. 12. Association for Computing Machinery, New York (2019)
10. Barreda Ramírez, C., González Cárdenas, C.E., Constain Moreno, G.E., Mora Pedreros, P. A.: Determination of the appropriation level in the collaborative work, a challenge in distance education focused on E-learning. In: Agredo-Delgado, V., Ruiz, P. (eds.) Human-Computer Interaction, HCI-COLLAB 2018. Communications in Computer and Information Science, vol 847. Springer, Cham (2019)
11. Khalfallah, J, Ben Hadj Slama, J.: The effect of emotional analysis on the improvement of experimental e-learning systems. Comput. Appl. Eng. Educ. **27**, 303–318 (2019)
12. Garcia-Loro, F., et al.: PILAR: sharing VISIR remote labs through a federation. In: IEEE Global Engineering Education Conference (EDUCON), Dubai, pp. 102–106 (2019)
13. Sáenz, J., de la Torre, L., Chacón, J., Dormido, S.: A new architecture for the design of virtual/remote labs: the coupled drives system as a case of study. In: 2019 24th IEEE International Conference on Emerging Technologies and Factory Automation (ETFA), Zaragoza, Spain, pp. 769–775 (2019)
14. Lerro, F., et al.: Improving the use of remote laboratories. The case of VISIR at Universidad Nacional de Rosario. In: 2019 5th Experiment International Conference (exp.at'19), Funchal (Madeira Island), Portugal, pp. 183–188 (2019)

15. Morozov, V., Shelest, T., Proskurin, M.: Create the model for development of virtual reality E-learning. In: 2019 IEEE 2nd Ukraine Conference on Electrical and Computer Engineering (UKRCON), Lviv, Ukraine, pp. 1265–1270 (2019)
16. Kagami, R.M., Kovalski da Costa, G., Uhlmann, T.S., Mendes, L.A, Freire, R.Z.: A generic WebLab control tuning experience using the ball and beam process and multiobjective optimization approach. Information **11**(3), 132 (2020)
17. Plaza, P., et al.: Portable blended MOOC laboratory. In: 2019 IEEE Learning With MOOCS (LWMOOCS), Milwaukee, WI, USA, pp. 15–20 (2019)
18. Sun, J., et al.: Preliminary outcomes, problems and countermeasures of blended synchronous learning environment: a case study in China. In: 2019 Eighth International Conference on Educational Innovation through Technology, Biloxi, MS, USA, pp. 210–215 (2019)
19. Grivokostopoulou, F., Kovas, K., Perikos, I.: The effectiveness of embodied pedagogical agents and their impact on students learning in virtual worlds. Appl. Sci. **10**(5), 1739 (2020)
20. Lahbi, Z., Sabbane, M.: U-Edu: multimodal learning activities analytics model for learner feedback in ubiquitous education system. Int. J. Adv. Trends Comput. Sci. Eng. **8**(5), 2551–2555 (2019)
21. Kunioshi, N., Noguchi, J., Tojo, K.: Evidence of cultural differences between American and Japanese mainstream science and engineering contexts from analysis of classroom discourse. Eur. J. Eng. Educ. **44**(4), 535–544 (2019)
22. Education in the time of Covid-19: how institutions and students are coping. https://www.business-standard.com/article/education/education-in-the-time-of-covid-19-how-institutions-and-students-are-coping-120043001575_1.html. Accessed 21 May 2020
23. The COVID-19 pandemic has changed education forever. This is how. https://www.weforum.org/agenda/2020/04/coronavirus-education-global-covid19-online-digital-learning/. Accessed 21 May 2020
24. Harari, Y.N.: 21 Lessons for the 21st Century. Jonathan Cape, London (2018)

Virtual Reality for Developing Intercultural Communication Skills of Engineering Students

Julia Ziyatdinova(✉) and Artem Bezrukov

Kazan National Research Technological University, Karl Marx Street 68, 420015 Kazan, Russian Federation
uliziatd@gmail.com

Abstract. Nowadays, virtual reality (VR) systems gain popularity in various applications ranging from space to entertainment. Education took up this challenge thus promoting virtual reality into the learning environment, first of all, in teaching science and engineering. Courses in humanities, however, are often ignored, though they have a great potential of applying VR techniques to develop students' soft skills, intercultural communication among them. To bridge this gap, the paper aims at analyzing the demand for VR, the experience in virtual reality and readiness of Master's degree engineering students for taking courses with immersion into virtual reality, using the example of intercultural communication courses. The authors analyzed recent publications in research journals and conference proceedings on the use of VR tools in higher education paying special attention to engineering curricula. The authors also conducted a survey among 87 MSc students focusing on the degree to which students are familiar with VR and ready to study through immersion into VR. The overall analysis of the survey results revealed that the students have a limited experience of using VR in any applications but most of them are ready for this experience in education. Based on these interpretations of the survey results, the authors made an attempt to develop a draft course to introduce elements of VR into EFL teaching for engineering students. Further research can be done to disseminate the initial experience of VR applications and to update the course developed.

Keywords: Virtual reality · Intercultural communication · Engineering students

1 Introduction

The modern globalized world made communications between the nations and countries easy. Higher education as part of this global world is following the same internationalization trends with the changing roles of professors and students [1] and multiple academic mobility programs [2, 3] within the European Union and beyond [4].

The recent COVID-19 outbreak, however, changed many of the academic mobility practices, urging students and faculty members to stay home. These circumstances lead to new scenarios of using digital technologies, and virtual reality can substitute mobilities inside or outside the country [5].

M. E. Auer and T. Rüütmann (Eds.): ICL 2020, AISC 1328, pp. 621–628, 2021.
https://doi.org/10.1007/978-3-030-68198-2_57

Any immersive virtual reality is a communication process, either between a human and artificial intelligence or between humans mediated by computers through additional visual and sensual effects. We are already facing VR under different circumstances such as 3D movies and google watch. Virtual reality systems gain popularity in various applications ranging from space to entertainment. Therefore, university students should be aware of the use of VR tools in order to be competitive in the globalized world [6] and ready for cross-cultural communications [7, 8].

Education took up this challenge thus promoting virtual reality into the learning environment. Some studies propose that VR can be very effective as an instructional media. There are great prospects for the use of VR which are nowadays limited due to a number of technical problems, including high requirements to computer hardware and software.

At the same time, many researchers state that these problems will be solved in the nearest future. This will lead to a paradigm shift in the whole educational process, where VR will be an indispensable component of teaching and learning. VR technologies can be further used for intensification of professional training [9]. Nowadays, however, there is still insufficient empirical data as in regard of VR efficiency in teaching and learning.

Secondary schools are actively testing VR tools, both in primary and secondary education and in different subjects. Being digitally native, schoolchildren very quickly adopt these new technologies and show a growing motivation to study subjects with the use of VR [10, 11].

VR is also making its way to higher education through different courses. These technologies are used, first of all, in teaching courses related to science and engineering, where experimental practicums can be demonstrated, e.g. space engineering [12], architectural practice [13], or environmental engineering [14]. In these cases, VR is considered as an active learning method involving students with the course material through virtual practice [15].

VR tools, however, are often ignored in courses in humanities. There have recently been very few publications in research journals regarding VR in humanities, these discussions mainly occur during conferences or round tables.

To bridge this gap, the paper aims at analyzing the demand for VR, the experience in virtual reality and readiness of MSc students for taking courses with immersion into virtual reality to study humanities and languages.

2 Methods

The authors started with theoretical analysis of recent publications in research journals and conference proceedings on the use of VR tools in higher education paying special attention to engineering curricula. Numerous recent years publications have supported the authors' ideas on the fast dissemination of VR tools and their application in teaching.

The authors decided to introduce a course in intercultural communication into an engineering MSc curriculum in the academic year of 2020/21, and received seed financing from one of the national foundations to buy a VR headset, a 3D camera, computer hardware and software.

In order to have a better understanding of the MSs students' expectations in regard of VR use in education, the authors developed an online questionnaire using SurveyMonkey free tools and conducted a survey among 87 MSc students. The survey focused on the degree to which students are familiar with VR and ready to study through immersion into VR. They survey included 10 questions.

The questions were of two types: close-ended and open-ended. The close-ended questions included multiple choice questions concerning the definition of virtual reality, the presence of experience of using VR, the plans to own a VR headset, the importance of a free English language command in order to use VR tools, the readiness to use VR in teaching and learning, and the expected effect of VR use on academic progress.

The survey included two close-ended rating scale questions. In the first question, the respondents had to rate the following spheres of VR application: education, computer games, leisure time, retail, and work. In the second question, the respondents set their priorities for the use of VR tools in the following university curriculum components: natural sciences, engineering sciences, humanities, foreign languages, experimental practicums.

The two open-ended questions included the description of the respondents' experience of using VR tools (if any), and additional spheres of human activities where VR tools could be useful.

Through summarizing the results for close-ended questions and content analysis for open-ended questions, the authors categorized the survey results thus arriving at conclusions regarding the prospects of VR in developing soft skills of engineering students [16].

Based on the results of the theoretical analysis of recent research and publications and the empirical survey results, the authors drafted elements of a course in intercultural communication with the use of VR tools for MSc students majoring in Innovation Management [17] and Smart Materials [18] degree programs. The authors bear in mind that in the future, the course can be also used in other degree programs [19, 20].

3 Results and Discussion

3.1 Survey Results

Recent publications in research journals and conference proceedings show a great interest toward VR tools in education [21].

In order to reveal the factors contributing to the use of VR in higher education environment, the authors conducted a survey on VR among 87 MSc students. The survey gave the following results.

The majority of respondents, 73 survey participants (84%) chose a correct definition of virtual reality as the world created by technical tools communicated to a person through his or her perceptions. At the same time, 6 survey participants (7%) thought that VR is the same as a computer play station, and 8 respondents (9%) considered VR to be an imaginary world of a person's emotions and relationships.

Out of 84 respondents, only 29 (33%) had an experience of using a VR headset. They listed the following popular locations where they had this experience: game clubs (10 respondents, 11%), retail shopping centers (7 respondents, 8%), home (6 respondents, 7%). Only 48 respondents (55%) want to own a VR headset. These answers suggest that VR tools and technologies are still available for a limited number of people, even those in the higher education, and, there is a great potential for the growth of interest in this sphere.

Table 1. Ranks of potential VR application spheres in % of the responses from 87 respondents.

Spheres of VR application	Rank 1	Rank 2	Rank 3	Rank 4	Rank 5
Leisure time	25.61%	28.05%	19.51%	17.07%	9.76%
Computer games	35.71%	14.29%	11.90%	10.71%	27.38%
Education	18.07%	22.89%	37.35%	16.87%	4.82%
Retail services	4.94%	13.58%	24.69%	30.86%	25.93%
Work	14.46%	21.69%	6.02%	24.10%	33.73%

In rating the spheres of potential VR applications (see Table 1), 30 respondents (36%) topped computer games which were followed by leisure time chosen number two by 23 respondents (28%), education selected by 31 respondents (37%) as priority number 3, retail services ranked number 4 by 25 survey participants (30%), and work rated number 5 by 28 respondents (34%). Among other spheres of VR applications, 10 respondents suggested travelling, 7 respondents voted for medicine, and 5 respondents offered sports. Only 63 respondents (72%) think that a free English language command is necessary for an efficient use of VR tools. These answers show the interest to VR in different spheres, education, however, not being in the top position.

Table 2. Ranks of courses to be taught using VR in % of the responses from 87 respondents.

Spheres of VR application	Rank 1	Rank 2	Rank 3	Rank 4	Rank 5
Natural sciences	28.05%	18.29%	28.05%	14.63%	10.98%
Engineering courses	25.93%	29.63%	17.28%	14.81%	12.35%
Humanities	9.76%	10.98%	21.95%	24.39%	32.93%
Foreign languages	15.85%	14.63%	12.20%	32.93%	24.39%
Experimental practicums	21.69%	26.51%	19.28%	12.05%	20.48%

At the same time, 57 respondents (66%) are ready to study at the university with the use of VR tools with the following priorities as for the course taught through VR technologies (see Table 2): engineering courses top the list followed by natural sciences, experimental practicums, foreign languages, and humanities. 59 respondents

(67%) think that VR tools will improve the efficiency of their education. Thus, the majority of students are ready for infusing VR tools into their university education.

The overall analysis of the survey results revealed that the students have a limited experience of using VR in any applications. Only half of them are willing to own VR devices at home, but over 65% of the students are ready to infuse VR into their university classes. Students are more likely to assume that VR can be useful in engineering courses rather than in humanities. Nevertheless, as much as 72% of the respondents understand the importance of the English language in using VR tools, although only 16% of the survey participants consider foreign languages as the top priority for studies with the use of VR.

3.2 Draft Course Development

Based on these interpretations of the survey results, the authors proceeded to development a draft course to introduce elements of VR into EFL teaching for engineering students. Alongside with VR, the course uses the recent approaches to development of communicative competences in a foreign language [22–24]. This VR based course will be introduced into the curriculum in the next academic year.

The VR course was inspired by similar experience of the university international partner in the USA. This university opened the VR learning center for students who study English, Spanish, Chines, Japanese and Russian as foreign languages.

The course content is based on several technologies, which can be used to create a VR learning environment. First, a VR station was purchased: a VR headset with 4K video capabilities and a computer that supports a VR environment. The VR station is also equipped with a projector and a projection screen. The basic idea behind this VR station is that every student in a language learning group (10–15 students) can experience VR immersion during a class, which usually lasts two academic hours. Other students stay also involved into this student's current VR experience by watching the respective video on the screen.

The software selected for the introduction of EFL students to VR is Google Earth VR. The advantage of this free software is that students can get a truly immersive experience of virtual visiting almost any space in the world and feel the language environment around them. Such virtual visits to London or New York can be accompanied with teacher's comments in English or publicly available tour guide audio materials.

Another important consideration behind this draft course is to provide opportunities for students themselves to be involved into the course content development. The course equipment also includes a 360 VR video camera, which students can use to make their own immersive videos with comments in English or conversations with native speakers. Such videos were first shot and tested by the project team and turned out to be efficient tools for EFL learning with VR.

The course, therefore, includes 3D videos made by the students in the potential work setting environment. For example, such videos include visits to innovative parks with chemical laboratories, production facilities and administrative offices. With a VR headset on, a student can take a tour of this work setting, listen to the employees describing their responsibilities, or practice his English to describe the environment.

The third technology is based on modern software that can generate virtual environment. The project team processed 360 panoramic photos and videos with Adobe Creative Cloud software (Adobe Photoshop for images and Adobe Premiere Pro for videos) to add learning components such as vocabularies into a VR environment.

The course will be taught as a semester-long discipline in a special room, which has been modified to accommodate a VR station for a group of EFL students. The course consists of several consecutive steps. First, the teacher introduces VR learning materials to students and provides basic instruction of how to use VR for learning (2 academic hours). Then, students get familiar with VR first with Google Earth VR (4 academic hours). The further step is watching VR-videos with comments of native speakers and developing skills of listening, comprehension, and interpreting (10 academic hours). The next stage is focused on VR visits to professional business or research facilities, such as innovative parks and/or chemical laboratories (12 academic hours). The final stage is the student EFL project: students make a VR-video presentation in English and show it to their groupmates with the VR station (8 academic hours for preparation and presenting by all students in the group).

Students can also use the VR content (videos and images) privately with their smartphones and compatible VR headsets as part of their extracurricular activities. The project team, however, preferred modern high-quality desktop-based VR headset as it provides a truly immersive experience, so we can feel the advantages of the VR-based learning.

4 Conclusions

The authors draw on a conclusion that VR is inevitable in higher education, engineering degrees in particular. The popularity of distance learning programs contributes to the need of using virtual reality techniques in both engineering courses and humanities.

These techniques create a new game like learning environment for the digitally native students and encourage them to invest more time and effort into developing soft skills. The English language virtual reality resources develop intercultural communication skills of the students in any courses learned.

Survey results and the experience of university partners in implementing similar projects resulted in development of a VR EFL course for engineering students. This course uses several key VR technologies available for universities, such as VR headsets, panoramic VR cameras and software for creating and processing the VR content.

An important component of an EFL course with VR technologies is involvement of students in development and/or upgrading the course content. It is recommended for VR course developers to consider adding student projects to the course curriculum, so students can develop their own VR content with available technologies. For EFL, such an approach provides additional motivation for students to learn English in an immersive environment offered by virtual reality.

Acknowledgements. The research was done with the support of Vladimir Potanin Foundation, Agreement № GSGK-25/20.

References

1. Valeeva, E.E., Kraysman, N.V.: The impact of globalization on changing roles of university professors. In: Proceedings of 2014 International Conference on Interactive Collaborative Learning, ICL 2014, pp. 934–935 (2014)
2. Kraysman, N.V., Shageeva, F.T., Mullakhmetova, G.R., Pichugin, A.B.: Poster: preparation of engineering university students for academic mobility to French universities. In: Advances in Intelligent Systems and Computing (AISC), vol. 1135, pp. 713–718 (2020)
3. Shageeva, F.T., Erova, D.R., Kraysman, N.V.: Poster: social-psychological readiness of engineering university students for academic mobility to European countries. In: Advances in Intelligent Systems and Computing (AISC), vol. 1135, pp. 719–724 (2020)
4. Valeyeva, N.S., Kupriyanov, R.V., Valeyeva, E.R.: Results and challenges of Russia's integration into Bologna Process. In: Proceedings of 2015 International Conference on Interactive Collaborative Learning, ICL 2015, pp. 404–406 (2015)
5. Khasanova, G.F., Shageeva, F.T.: Poster: variable scenarios of the VR use in training specialists for chemical industry. In: Advances in Intelligent Systems and Computing (AISC), vol. 1135, pp. 808–813 (2020)
6. Osipov, P.N.: Training competitive specialists as the priority of modern education. In: 2013 International Conference on Interactive Collaborative Learning, ICL 2013, pp. 251–254 (2013)
7. Panteleeva, M., Sanger, P.A., Bezrukov, A.: International approaches to the development of cross-cultural education at high school. In: ASEE Annual Conference and Exposition, Conference Proceedings (2016)
8. Ziyatdinova, J., Bezrukov, A., Sanger, P.A., Osipov, P.: Cross cultural diversity in engineering professionals - Russia, India, America. In: ASEE 2016 International Forum (2016)
9. Osipov, P.N.: Intensification of professional training as a pedagogical problem. In: 2012 15th International Conference on Interactive Collaborative Learning, ICL 2012 (2012)
10. Cooper, G., Park, H., Nasr, Z., Thong, L.P., Johnson, R.: Using virtual reality in the classroom: preservice teachers' perceptions of its use as a teaching and learning tool. Educ. Media Int. **56**(1), 1–13 (2019). https://doi.org/10.1080/09523987.2019.1583461
11. Zhang, H., Yu, L., Ji, M., Cui, Y., Liu, D., Li, Y., Liu, H., Wang, Y.: Investigating high school students' perceptions and presences under VR learning environment. Interact. Learn. Environ., 1–21 (2020). https://doi.org/10.1080/10494820.2019.1709211
12. Aoki, H., Oman, C.M., Natapoff, A.: Virtual-reality-based 3D navigation training for emergency egress from spacecraft. Aviat. Space Environ. Med. **78**(8), 774–783 (2007)
13. Spaeth, A.B., Khali, R.: The place of VR technologies in UK architectural practice. Archit. Eng. Des. Manag. **14**(6), 470–487 (2018). https://doi.org/10.1080/17452007.2018.1502654
14. Taylor, G.L., Disinger, J.F.: The potential role of virtual reality in environmental education. J. Environ. Educ. **28**(3), 38–43 (1997). https://doi.org/10.1080/00958964.1997.9942828
15. Barabanova, S.V., Nikonova, N.V., Pavlova, I.V., Shagieva, R.V., Suntsova, M.S.: Using active learning methods within the andragogical paradigm. In: Advances in Intelligent Systems and Computing (AISC), vol. 1134, pp. 566–577 (2020)
16. Valeeva, E., Ziyatdinova, J., Galeeva, F.: Development of soft skills by doctoral students. In: Advances in Intelligent Systems and Computing (AISC), vol. 1135, pp. 159–168 (2020)

17. Sultanova, D., Maliashova, A., Bezrukov, A.: Consistent development of the training program "Innovation Management". In: Advances in Intelligent Systems and Computing (AISC), vol. 1135, pp. 234–243 (2020)
18. Bezrukov, A., Sultanova, D.: Development of a "Smart Materials" Master's degree module for chemical engineering students. In: Advances in Intelligent Systems and Computing (AISC), vol. 1135, pp. 169–180 (2020)
19. Bezrukov, A., Sultanova, D.: Application of microfluidic tools for training chemical engineers. In: Advances in Intelligent Systems and Computing (AISC), vol. 1135, pp. 496–504 (2020)
20. Osipov, P., Ziyatdinova, J., Girfanova, E.: Factors and barriers in training financial management professionals. In: Advances in Intelligent Systems and Computing, vol. 916. pp. 167–175 (2020)
21. Chen, N.-S., Hwang, W.-Y., Chen, G.-D.: The disruptive power of virtual reality (VR) and serious games for education. Interact. Learn. Environ. **21**(2), 101–103 (2013). https://doi.org/10.1080/10494820.2012.704249
22. Khusainova, G.R., Astafeva, A.E., Gazizulina, L.R., Fakhretdinova, G., Yakimova, J.Y.: Poster: development of communication skills for future engineers: organizational forms, methods and tools for their communication skills development at foreign language classes. In: Advances in Intelligent Systems and Computing (AISC), vol. 1134, pp. 119–124 (2020)
23. Tsareva, E., Bogoudinova, R., Khafisova, L., Fakhretdinova, G.: Poster: multilingualism as a means of students' technocommunicational competence forming at engineering university. In: Advances in Intelligent Systems and Computing (AISC), vol. 1134, pp. 137–142 (2020)
24. Semushina, E.Y., Ziyatdinova, J.N.: Studying English on-line as a part of a course "English for Special Purpose" in technological university. In: Advances in Intelligent Systems and Computing (AISC), vol. 1135, pp. 21–29 (2020)

Emotion Analysis in Distance Learning

Dalila Durães[1,3(✉)], Rámon Toala[1,2], and Paulo Novais[1]

[1] Algoritmi Research Centre, Department of Informatics, University of Minho, Braga, Portugal
dad@estg.ipp.pt, id7410@alunos.uminho.pt, pjon@di.uminho.pt
[2] Technical University of Manabí, Portoviejo, Manabí, Ecuador
[3] CIICESI, ESTG, Politécnico do Porto, Felgueiras, Portugal

Abstract. The COVID-19 pandemic has changed education forever because schools, universities, teachers, and students had to adapt to distance learning. Multiple differences are identified with online learning compared to face-to-face education. First, students must be more responsible. Second, users' familiarity with using computers varies significantly. Third, the traditional interaction between teacher, student and content are made more complicated by the introduction of technology. The application of new tools related to the student, teacher, content, technology, software, and communication results in the improvement of teaching methods in online learning. When new tools are applied and there is an improvement in the results in online education, the student, teacher, and educational institutions benefit from it. Emotion plays an important role in the knowledge, acquisition, and decision process of an individual. Consequently, they directly influence perception, learning process, and the way people communicate. There is also significant evidence that rational learning in humans is dependent on emotions. In this paper, we presented a solution with a new Intelligent Tutoring Framework, that analyzed emotions in a non-intrusive and non-invasive way.

Keywords: Intelligent system · Sentiment analysis · Distance learning

1 Introduction

Todays, we live in a world where the student do not go to the school, because of pandemic situation. Distance learning is a teaching modality that, due to the Covid-19 pandemic, has now become the alternative. This distance learning was only possible, due to the integration of information and communication technologies (ICT) in the teaching and learning processes. However, when the teacher explains the contents, he does not know if the student is or not motivated to learn.

Emotion plays an important role in the knowledge, acquisition, and decision process of an individual. Consequently, they directly influence perception, learning process, and the way people communicate. There is also significant evidence that rational learning in humans is dependent on emotions.

M. E. Auer and T. Rüütmann (Eds.): ICL 2020, AISC 1328, pp. 629–639, 2021.
https://doi.org/10.1007/978-3-030-68198-2_58

To analyses emotion, there are several theories, which attempted to specify the interrelationships of all the components involving an emotion and the causes, the reasons, and the function of an emotional response. Some research intended to relate emotion and computer, so for Ortony, Clore and Collins [1] emotion identification is generally used in the field of cognitive science has a connection to affective computing enabling computers to recognize and express emotions.

According to the literature [2], there is a connection between emotion and learning process. Although, this process is not simple or direct, it is accepted that positive and negative emotional states can cause different kinds of thinking and can influence the learning perspective.

Intelligent Tutoring System (ITS) is a learning environment that allows students to acquire knowledge and skills in a targeted manner and adapted to their own pace. ITS contains intelligent algorithms that adapt to users and allow the application of complex learning principles. Some basic activities of this system must incorporate the student's active learning, interactivity, adaptability, and feedback [3].

In this paper, we proposed a new ITS framework to obtain data from behavior biometric, specially user's emotional state during e-learning activities.

So, this paper is organized as follows. After this introduction, Sect. 2 introduces the state of the art of ITS and student emotion analysis. Then, Sect. 3 presents a proposed ITS framework. Next, Sect. 4 presents some application methods and results. Finally, Sect. 5 concludes the study by performing a global analysis of the presented research.

2 State of Art

In the last decades, the rapid development of ICT has benefited all areas of knowledge. However, and according to [4], these technologies were applied in education very late. Even so, from these technologies emerged the Virtual Learning Environments (e-learning systems), in which students interact as if they were in a real environment. These environments are combined with other applications that provide intelligent tutoring and are called Intelligent Virtual Environments or ITS. These ITS aim to adapt to the student's profile, applying techniques that best suit each one, to obtain better learning results.

Currently, there are several such tutors, however, these tutors have not completely achieved the desired goals, as they do not consider an important element that affects students' learning: their emotional state. There are some of these tutors who assess the student's emotional state only at the end of the work sessions, which is not enough to improve the learning environment. For better functioning of the ITS, it should perform an analysis of the data presented in Fig. 1, so that, from there, the best teaching mechanism could be applied [5].

2.1 ITS

The typical architecture of an ITS has the following four basic components: the Expert Model, the Student Model, the Tutor Model and the Interface [5, 6].

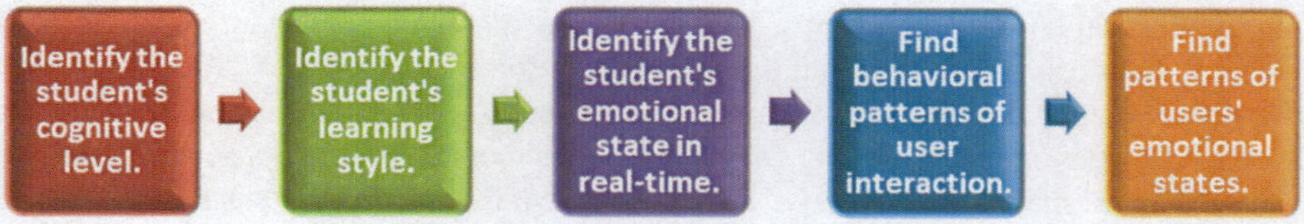

Fig. 1. Student analysis of an ITS.

The Domain Model, also known as the Expert Model, contains all the concepts, facts, rules, and problem-solving strategies for a given domain. This model serves as a source of specialized knowledge, a standard for assessing student performance and diagnosing errors [5, 6]. This model also performs data analysis and can also make predictions about the knowledge of a given student, since it observes the actions performed by that student.

The Student Model is an overlay of the Expert Model. This model contains the cognitive and affective states of the student in association with their evolution as the learning process progresses. As the student works step by step in the problem-solving process, the system analyzes the student's interaction with the system [5, 6]. This model contains the dynamic monitoring of the student's emerging knowledge and skills.

The Tutor Model is the part of the ITS that designs and regulates interactions with the student. This model accepts information from the Student Model and the Expert Model. In addition, it is closely linked to the Student Model, since it makes use of knowledge about the student and its own structure of tutorial objectives, to design the pedagogical activity to be introduced. It also monitors student progress, creating a profile of strengths and weaknesses in relation to production rules [5, 6].

The Interface is the front-end interaction with ITS. This system integrates all types of information necessary to interact with the student, through graphics, text, multimedia, video, menus, etc. The Interface is the communication component of the ITS that controls the interaction between the student and the system. The interface is translated between the internal representation of the system and an interface language understandable to the student [5, 6].

2.2 Student Emotion Analysis

There are several types of learning. In 1956 Benjamin Bloom [7], identified three domains of educational activities cognitive: mental skills (Knowledge); affective: growth in feelings or emotional areas (Attitude); psychomotor: physical or manual skills (Skills). The combination of all domains influences the way one learns, and the way rational decisions are made.

Behavioral analysis is the scientific area that tries to understand the behavior of individuals, and how it has been affected by the surrounding environment. There are several ways to monitor and obtain information about the user, for example, through the mouse, keyboard, and even cameras and sensors.

Some research indicated that a slight positive mood could produce an effect on memory, well-organized open-minded, flexible problem-solving and thinking as well as more efficiency and thoroughness in decision-making. This can be found in groups of different ages and professions [8, 9]. The effect on cognition is not restricted to positive states of mind. Negative affective states like anger, sadness or fear can influence the brain activity affecting the thought process [9].

Todays, we live in a world where the student do not go to the school, because of pandemic situation. Distance learning is a teaching modality that is a quality alternative for students unable to attend a school in person, based on the integration of ICT technologies in the teaching and learning processes as a means for all have access to education. However, teacher do not know if the students are or not motivate for learning.

Sentimental analysis allows the use of tools and techniques to identify the opinion and feeling(s) that a person may have about various entities, namely the learning. With the emergence of social networks in distance learning, there is now a large volume of opinion data that can be analyzed, increasing interest in this area. These techniques are used by organizations to market their products, identify new opportunities, and monitor their online emotions to improve learning.

Once information about the individual's text exists in these terms, it is possible to start monitoring emotion in real-time [9]. This makes this approach especially suited to be used in learning activities in which students use computers, as it requires no change in their working routines. This is the main advantage because present an emotional analysis module that is more accurate and that can provide and determine the emotions present in a text and calculate the impact of the emotions present in the text on a person. Besides, the system will send the professor indications of the results obtained. It is possible to collect data that describes the text interaction of the students.

Affective Norms for English Words (ANEW) [10] was developed to offer a set of normative emotional ratings for many words in the English language. The objective is to construct a series of verbal materials that were rated in terms of pleasure, arousal, and dominance. ANEW complements the International Affective Picture System (IAPS) [10] and the International Affective Digitized Sounds (IADS), which are collections of photographs and sounds stimuli, respectively, that also contain affective ratings.

Pang and Lee [11] reviewed the concept of sentiment and opinion analysis, which refers to the application of natural language processing, computational linguistics, and text analytics to recognize subjective information, like emotion, in the text.

3 Proposed Framework

Based on the state-of-the-art section, the idea is to create an ITS system adapted to each student. In this first phase, a general structure of an ITS was developed, which is shown in Fig. 2.

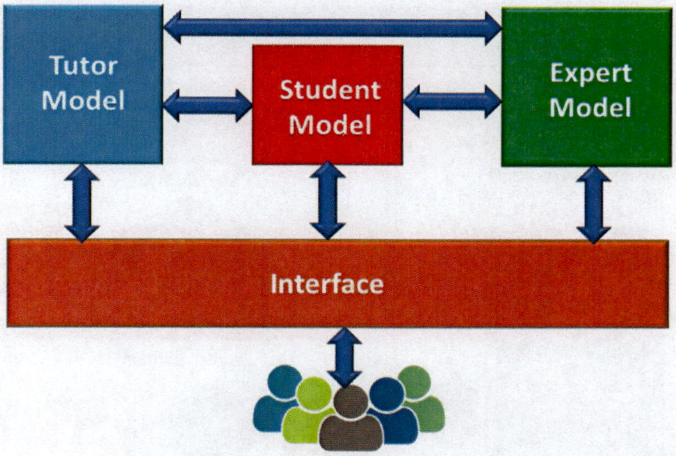

Fig. 2. General structure of an ITS.

The idea is to have an Interface that communicates directly with users and captures the data necessary to create a student profile. From there, the system, based on the content it has to address and the student profile, applies the tools necessary for the student to acquire the necessary knowledge.

Making a more complete description of the system, Fig. 3 presents the ITS system in more detail. In this figure, you can see in more detail the architecture of the Student Model and the Tutor Model.

The Student Model is divided into three levels: student style, student emotion and state of knowledge. In the first level, student style, the student's learning style and the interaction of the student's standard behavior are defined. In the second level, student emotion, the student's emotional profile is created, based on his emotional state. Finally, at the third level, the state of Knowledge, the student's profile is created regarding their learning evolution. All this information is stored in its database.

The Tutor Model, based on the Student model, should adjust, and adapt the level of learning difficulty to each student, depending on their parameters. In this way, it applies to learn strategies and rules for a given content.

The Interface captures data from the student's interaction with the ITS. Data capture is done using a non-invasive and non-intrusive approach. There is a log application that runs in the background, saving the student's necessary events with ITS. This application has a device that generates raw data that describes the student's interaction with the mouse and keyboard. There are also flexible sensors that use the information available from other measurements and process parameters to calculate and estimate the amount of raw data. The raw data generated is stored locally until it is synchronized with the web server in the cloud at regular intervals, usually every 5 min. In this layer, each event is coded with the necessary information (for example, timestamp, coordinates of the mouse movement, type of click, keypress, etc.).

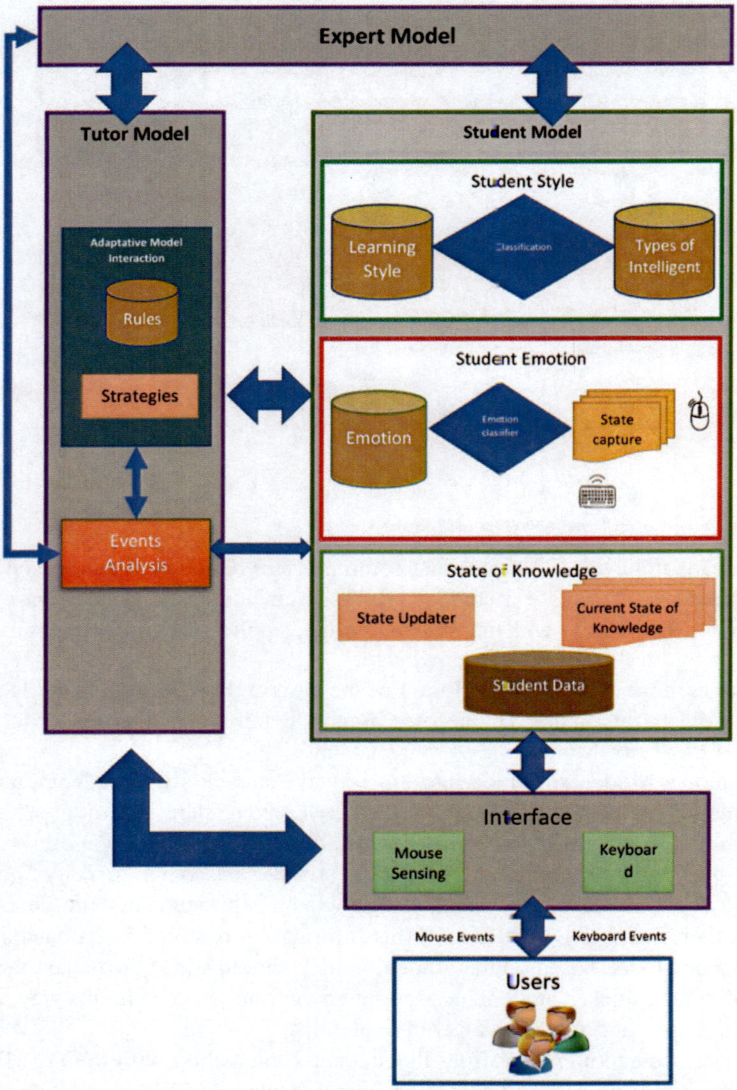

Fig. 3. General architecture of an ITS.

4 Methodology and Results

To implement the proposed system, it was decided to create an algorithm that based on the student's profile, on the Emotion Classifier and on the contents to be taught, MOOCs are used to explain the contents and, from there, a set would be applied questions, based on levels of difficulty. In Fig. 4, the ITS operating structure is shown.

4.1 ITS Application

Based on the framework presented in Sect. 3 and Fig. 4, we have created an ITS Application that is presented in Fig. 5. This ITS Application it will be applied at the Technical University of Manabí, Portoviejo, Manabí, Ecuador.

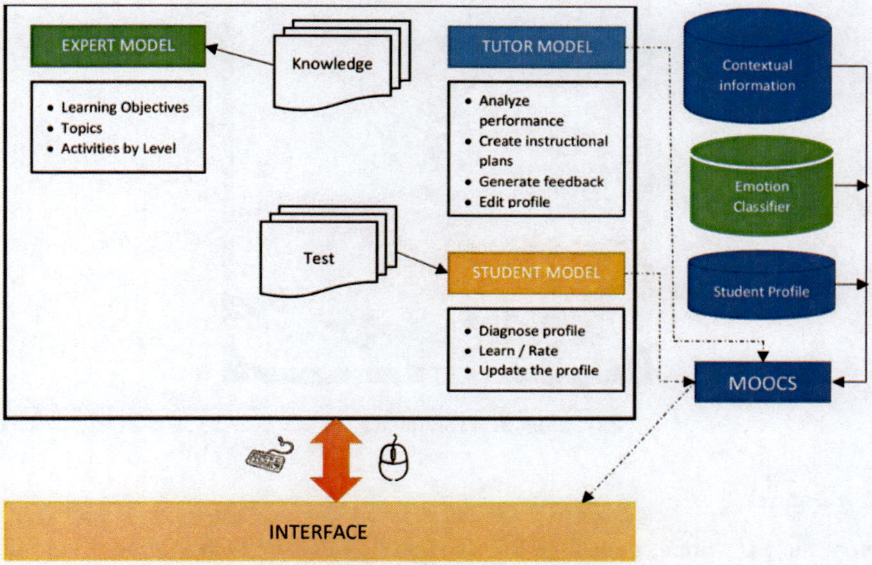

Fig. 4. Proposal for operation of ITS.

When the student opens the application for the first time, we need to register. To register, it is necessary to indicate the following data: date of birth, gender; course, and address. The system creates a student profile from this data. When the student makes a test, we can indicate the difficulty level of the test to be performed. In Fig. 5 it is presented the second page of a register page and the difficulty level test to be performed.

4.2 Population

To capture the data, the approach followed is based on the dynamics of the mouse and the keyboard, to propose a completely non-intrusive method for evaluating student-computer interaction. To do this analysis, an activity to be carried out on the computer was applied to a class of Statistics subject, from the Technical University of Manabí, Ecuador. Each computer has a keyboard, a mouse, and a monitor. The assessment activity starts at the same time for all students and they log in to the standard software using their credentials and the activity begins.

Fig. 5. ITS Application.

4.3 Dataset

Before the test started, students were asked a questionnaire with the following information: "How prepared are you for the evaluation?", "Have you studied for this evaluation?". All the answer has only the possibility of "Yes" or "No".

During the evaluation if the studying as a wrong answer, the answer will be at red color. If the answer is correct it will be at green.

At the end of the evaluation, another questionnaire will be applied with the following three questions: "Was the test easy?"; "Will you get a good grade?"; "Write your opinion about the test". The first two questions only have the possibility of answer of "Yes" or "No". The third question is a free text write with limit of 70 characters.

Figure 6 presented question example of the ITS application, where the student answer corrected to a question (green mode).

4.4 Results

Figure 7 depicts the type of information that these features provide. It shows the evolution of the performance of a specific student during the lesson through two features: Click Duration and Mouse Velocity. The velocity of the mouse increases until approximately the same point in time and then it starts decreasing. The duration of each

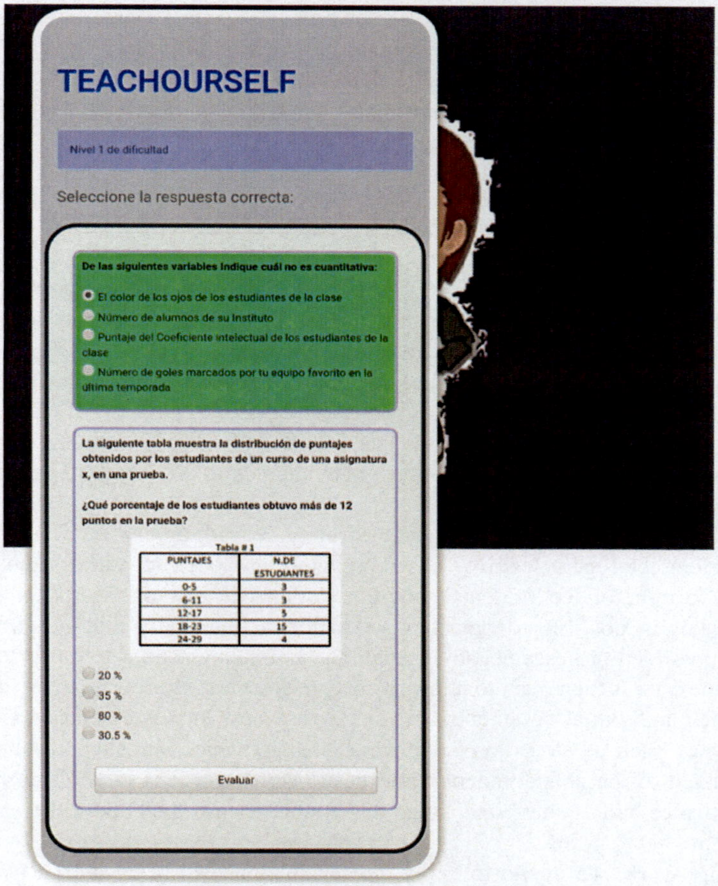

Fig. 6. Question example of the ITS application.

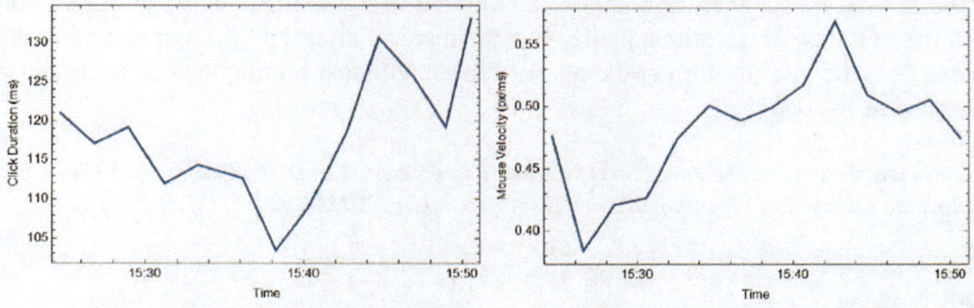

Fig. 7. - Real-time performance: evolution of one student's interaction performance during a video Class. Left: Click Duration. Right: Mouse Velocity.

click decreases until roughly the middle of the exam and then increases up to a global maximum. Both features point out an initial improvement of performance, followed by a degrading. Figure 7 reveals a classical effect of stress: performance tends to improve for some time after the beginning of the stressor stimulus, with a drop off in performance after some time performing above average.

5 Conclusions and Future Work

This paper presents a first approach to the model of an ITS. The approach is non-invasive and non-intrusive for an ITS. It is proposed based on the biometric analysis of work behavior in different students with different emotions styles. The system monitors and analyzes the dynamics of the mouse, keyboard, and tasks to determine the student's interaction with the computer.

These results are crucial to improve learning systems in an e-learning environment and to predict student behavior based on their interaction with mobile devices or the computer.

The ITS makes possible the enhanced learning/teaching processes. The architecture of an ambient intelligent learning environment is proposed to address these issues, especially to monitor the students' emotion students in distance learning. With this architecture it is possible to detect those factors dynamically and non-intrusively, making it possible to foresee negative situations, and taking actions to mitigate them. In this case, the door is then open to allow to analyze students' emotion profile, taking into account their individual comments and to propose new strategies and actions, minimizing issues such as stress, anxiety, which can influence students' results and are closely related to the abandonment occurrence. Moreover, it is possible to maximize the performance and attentiveness since the teacher is informed about the emotion of each student.

In future work, we intend to apply two evaluation tests. One with a non-limited time to answer with the possibility to change the answer, and another with time limited for each question with no possibility to change the answer of the question, and when the answer is wrong, the answer will change to a red color. The idea is to improve stress in the second type of test. Another improvement in the work is to make correlations with the two type of questionnaire (before the test and after) and the test results. It also be analyses the last question of the questionnaire, with text mining techniques to obtain a sentiment analysis.

Acknowledgements. "This work has been supported by national funds through FCT – Fundação para a Ciência e Tecnologia through project UIDB/04728/2020."

References

1. Ortony, A., Clore, G.L., Collins, A.: The Cognitive Structure of Emotions. Cambridge University Press (1990).

2. Durães, D., Carneiro, D., Jiménez, A., Novais, P.: Characterizing attentive behavior in intelligent environments. Neurocomputing **272**, 46–54 (2018)
3. Durães, D., Toala, R., Gonçalves, F., Novais, P.: Intelligent tutoring system to improve learning outcomes. AI Commun., 1–14 (2019). (Preprint)
4. Méndez Pozo, G., Una arquitectura software basada en agentes y recomendaciones metodológicas para el desarrollo de entornos virtuales de entrenamiento con tutoría inteligente (Doctoral dissertation, Informatica) (2008)
5. Carneiro, D., Novais, P., Durães, D., Pego, J.M., Sousa, N.: Predicting completion time in high-stakes exams. Fut. Gener. Comput. Syst. **92**, 549–559 (2019)
6. Ahuja, N., Sille, R.: A critical review of development of intelligent tutoring systems: retrospect, present and prospect. Int. J. Comput. Sci. Issues (IJCSI) **10**(4), 39 (2013)
7. Best, R.M., Floyd, R.G., Mcnamara, D.S.: Differential competencies contributing to children's comprehension of narrative and expository texts. Read. Psychol. **29**(2), 137–164 (2008)
8. Picard, R.W., Papert, S., Bender, W., Blumberg, B., Breazeal, C., Cavallo, D., Machover, T., Resnick, M., Roy, D., Strohecker, C.: Affective learning—a manifesto. BT Technol. J. **22**(4), 253–269 (2004)
9. Isen, A.M.: An influence of positive affect on decision making in complex situations: theoretical issues with practical implications. J. Consum. Psychol. **11**(2), 75–85 (2001)
10. Bradley, M., Lang, P.: Affective Norms for English Words (ANEW): Instruction Manual and Affective Ratings. University of Florida: Psychology (Vol. Technical). The Center for Research in Psychophysiology (1999)
11. Lang, P., Bradley, M., Cuthbert, B.: International Affective Picture System (IAPS): Technical Manual and Affective Ratings. Center for the Study of Emotion and Attention 1997. Psychology (1997)

Computer Aided Language Learning (Call)

Work in Progress: Web-Delivered Reading Improvement Battery of Tasks

Aikaterini Striftou[1], Nikolaos C. Zygouris[1](✉), Georgios I. Stamoulis[2], and Denis Vavougios[3]

[1] Department of Computer Science and Telecommunications, University of Thessaly, Lamia, Greece
nzygouris@uth.gr
[2] Electrical and Computer Engineering Department, University of Thessaly, Volos, Greece
[3] Department of Special Education, University of Thessaly, Volos, Greece

Abstract. Learning to read requires the development of highly organized brain systems that are capable of incorporating spelling, phonological and lexical-semantic features of written words. The main aim of the present research protocol is to strengthen the speed and accuracy of phonological awareness for children with reading impairments by designing and implementing a web delivered set of tasks. "Poke the Reading Ability" is a web-delivered application, that helps readers increase their reading speed, accuracy and comprehension by training them to avoid subvocalization (saying words in their head while reading), backtracking (going back to re-read words or sentences), and fixations, factors that have negative effects on reading speed and accuracy. "Poke the Reading Ability" includes a graphical environment that encourages learners to complete activities, while "playing an internet game". HTML5, CSS, JavaScript were used in order to design all tasks included in "Poke the Reading Ability" web application. There are software applications that are used as intervention programs. However, "Poke the Reading Ability" is an online program designed in order to improve the reading capacity of children and adolescents. Its tasks are not only targeting at building phonological awareness but also, include tasks that improve visual and auditory memory, visual discrimination ability and text comprehension. The web application that is presented is carefully designed both in its pedagogical and computer programming aspects in order to offer a solution to tutors at everyday school practice.

Keywords: Reading intervention · Web application · Reading speed · Reading accuracy

1 Context

Learning to read requires the development of highly organized brain systems that are capable of incorporating spelling, phonological and lexical-semantic features of written words. Sets of reading components provide useful frameworks on which to base course design, teaching, and test development [1]. Learning to read is vital to full participation in modern society. Reading is the basis for gaining knowledge, developing culture,

M. E. Auer and T. Rüütmann (Eds.): ICL 2020, AISC 1328, pp. 643–654, 2021.
https://doi.org/10.1007/978-3-030-68198-2_59

democracy and success in the workplace [2]. The word recognition literature converges to suggest that reading involves the common activation of spelling, phonology, and semantics [3, 4].

Reading achievement is stable over time [5, 6]. Students who are the poorest readers in the first years of primary school tend to remain poor readers for the rest of their academic life [7]. Letter knowledge and phonemic awareness are key contributors to children's understanding of the alphabetic principle and its application to the decoding and spelling of single words [8, 9].

Greek is considered an orthographically transparent language, but while its alphabetic system has high feed forward (reading) consistency, it has substantial feedback (spelling) inconsistencies [10]. Reading as a skill permeates all areas of the curriculum. Reading offers pleasure and enlightenment only when reading becomes a pleasant activity [11].

2 Purpose or Goal

Reading skills are described as the cognitive ability a person is able to use when interacting with texts [12]. Learning to read is a critical but often challenging academic task for young children [13]. The main aim of the present research protocol is to strengthen the speed and accuracy of phonological awareness for children with reading impairments by designing and implementing a web delivered set of tasks. Reading is a complex linguistic-cognitive ability that is uniquely human [14]. It represents the translation of abstract graphs into the corresponding phonemes (decoding) and an intransitive and automatic word recognition [Frith, 1985 in 15]. Reading involves visual word recognition, phonological awareness, syntactic processing, and efficient processing speed between them [16].

Reading is generally considered to be the fundamental ability for academic success [17]. During the reading process, information from visual, semantic, conceptual, and linguistic sources is combined to understand sentences and phrases [18].

The ability to perceive and understand speech sounds that make up syllables and words are subject to the term "phonological processing". This term is divided into "phonological awareness" (the ability to hear separate words in a sentence, listening to separate syllables in a word, hearing separate sounds in a word) and phonetic awareness (the ability to listen and handle the sounds in syllables and words). Sound control in phonological processing includes segmentation, mixing, sound deletion and rotation [19].

There is a growing interest in deepening reading comprehension skills and developing strategies to improve reading development in young readers. Opportunity has long been seen as the ability to read quickly and accurately, especially through automatic word recognition [20].

This work presents a tool that is most accessible to users according to their needs. The present work is an attempt to research the child's relationship to reading. The aim is to improve the speed and accuracy of phonological awareness by enhancing children's reading skills.

3 Approach

Learning to decode written text in any type of language entails discovering what the written symbols represent and is important skill for learning to read an alphabetic script [21]. A learner must be instructed on the principles of how a certain written symbol (e.g., a letter) connects to something that the learner already knows (i.e., a speech sound) [22].

Understanding is the essence of reading and the active process of "constructing" the meaning of the text. Understanding reading involves the interaction of language processing and conceptual processing, resulting in the interpretation or conceptual representation of a text in memory. As key requirements for understanding, words must be recognized, semantic sentences must be derived from sentences, and ideas from different sentences must be linked together [23].

Effective reading takes place when the reader is able to understand the words of the reading material, understand their declarative and symbolic meanings and assimilate them [11]. There are two components that are considered critical to successful reading: decoding and comprehension [24].

These two main components are clearly separable [25] but are associated with individuals without reading difficulties [26]. However, there are children who have difficulty in understanding but have normal decoding skills [27]. This inequality leads to the assumption that understanding and decoding are based on different underlying abilities [28].

Phonological analysis skills in particular contribute to the development of phonological reconstruction skills used to decode unknown words and are evaluated by pseudo-reading tests [29, 30]. Working memory allows individuals to temporarily remember information while competing for information [31]. Working memory should be considered a factor that can affect reading comprehension [32, 33]. There is a strong relation between working memory, visual discrimination and reading comprehension [34]. Comprehension does not correlate with tasks that simply involve the passive storage of information [35].

The relation between working memory and comprehension skills have been presented with tasks that require the processing and storage of words [36], sentences [37] and numbers [38]. Working memory serves as a buffer for the most recently read propositions in a text, enabling their integration to establish coherence, and holds information retrieved from long-term memory to facilitate its integration with the currently active text [39].

Teaching reading based on computer software is a relatively new and very promising approach to improving children's phonological awareness with reading difficulties [40]. "Poke the Reading Ability" – PtRA, is a web-delivered application, that helps readers increase their reading speed, accuracy and comprehension by training them to avoid subvocalization (saying words in their head while reading), backtracking (going back to re-read words or sentences), and fixations, factors that have negative effects on reading speed and accuracy. PtRA includes a graphical environment that encourages learners to complete activities, while "playing an internet game". This web application can be useful as it has the ability to lure "players" into performing, and if the user enjoys the "web game", he/she will probably continue to "play", putting more effort in order to complete all needed tasks that are included.

The software, which focuses on repetitive decoding skills and adapts to the child's level, can lead to beneficial results for young readers [41]. The genius of PtRA is that it is so easy to use and does not require installation as it is available to those who have an internet connection.

HTML5, CSS, JavaScript were used in order to design all tasks included in "Poke the Reading Ability" web application. Specially, HTML5 is the latest version of Hypertext Markup Language, the code that describes web pages. It's actually three kinds of code: HTML, which provides the structure, Cascading Style Sheets (CSS), which take care of presentation and JavaScript, which makes things happen. For graphics Adobe Photoshop is a software application for image editing and to creating intricate digital paintings and drawings that mimic those done by hand.

More specifically, we used:

(1) A word-reading task that includes words that are missing the first syllable and children have to choose the correct syllable to complete the word (see Fig. 1).
(2) A word-reading task that includes words in 5 different ways of spelling and children have to choose the correct spelling of the word (see Fig. 2).
(3) A visual discrimination combined with a computational task. This assessment consists of images of the same design with similar color combinations, and children have to identify the images with the same colors, count them, and select the number of times that each image appears with a specific color combination (see Fig. 3).
(4) A word-reading task that includes words that are missing the middle syllable and the children have to choose the correct syllable to complete the word.
(5) A word-reading task that includes a word that appears with pseudowords 5 times and children have to choose the right word (see Fig. 4).
(6) A visual discrimination combined with a working memory task. This assessment consists of a series of diagrams or patterns and the children have to memorize the pattern and the order in which they appear. The children are then asked to choose the patterns that appeared in the correct order in which they appeared.
(7) A word-reading task that includes words that are missing the middle syllable. The syllable "happens" to be anagrammatically and the children have to distinguish and choose the correct syllable to complete the word.
(8) A word-reading task that includes a word that appears in anagram 3 times. Children need to stand out and choose the right word.
(9) A working memory task. The essence of the job is to find the same letters/numbers. Children have to find all the pairs of letters/numbers hidden in the closed huge ones.
(10) Kryptolex is a painting framed by scattered letters of the Greek alphabet and children have to sweep the painting vertically and horizontally in order to discover the hidden word.
(11) A reading discrimination combined with working memory task. Children need to read and understand a text that has words in it. The words are given at the top of

the text mixed up and the children have to put the right word in the right place in the text in order to make sense (see Fig. 5).

(12) A working memory task. Children need to read the questions and choose the right answer. The questions are related to the previous task and the children have to remember the text they read (see Fig. 6).

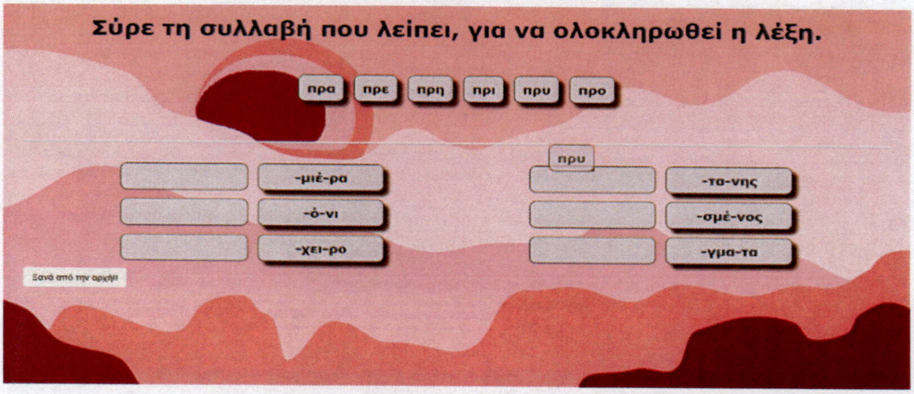

Fig. 1. Children must identify the appropriate syllable and drag it to the appropriate word. If they do not place the syllables correctly, they have the option by pressing "Again from the beginning!!" to resume the placement of the syllables. Example in English language: we suppose that we have at least the syllables: pra, pre, pri, pro, pru, pry and the child has to form the word (pra)ctice

All activities are based on improving phonetic correspondence, phonological decoding, and visual working memory. The software is constantly tested and modified immediately with new versions/upgrades. PtRA web app aims to improve people's reading ability. By constantly working on the same application, some moves are now made mechanically without noticing why we chose that particular answer. In PtRA, the answers are constantly mixed without giving the operator the ability to memorize the order and position of the correct answers.

The presentation of the tests is done with austerity and visualization of all the possibilities of the application. A simple and attractive graphical interface is provided without the user losing focus on important points. The application includes a series of activities that support the development of reading and vocabulary skills.

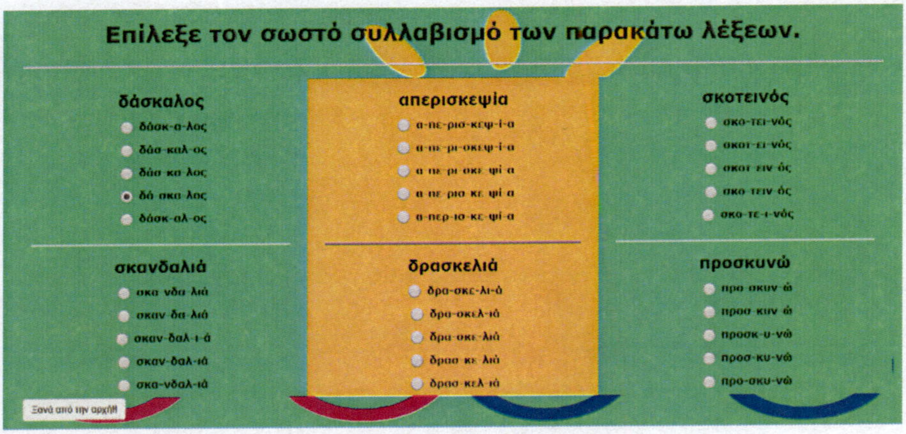

Fig. 2. Children need to identify the correct spelling of the original word. If they do not succeed, they have the option by clicking "Again from the beginning!!" to begin locating the syllables. Example in English language: we suppose that the child has to spell the word teacher. It must identify the correct between five options: Te-a-ch-er, Te-a-ch-er, Tea-cher, Tea-che-r, Te-ac-her.

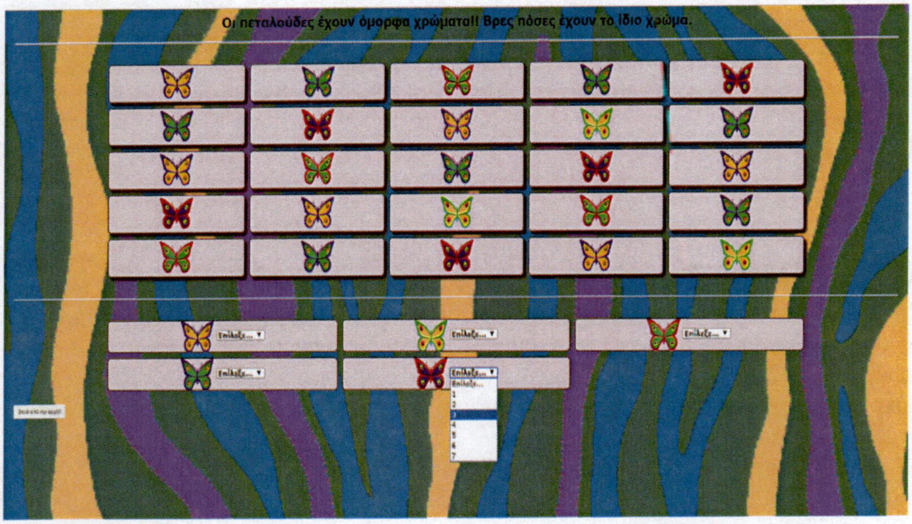

Fig. 3. Children have to locate and count the number of similar colored butterflies and then state it by selecting the number that corresponds to their number. If they do not manage to count correctly, they have the option by clicking "Again from the beginning!!" to start counting from the beginning.

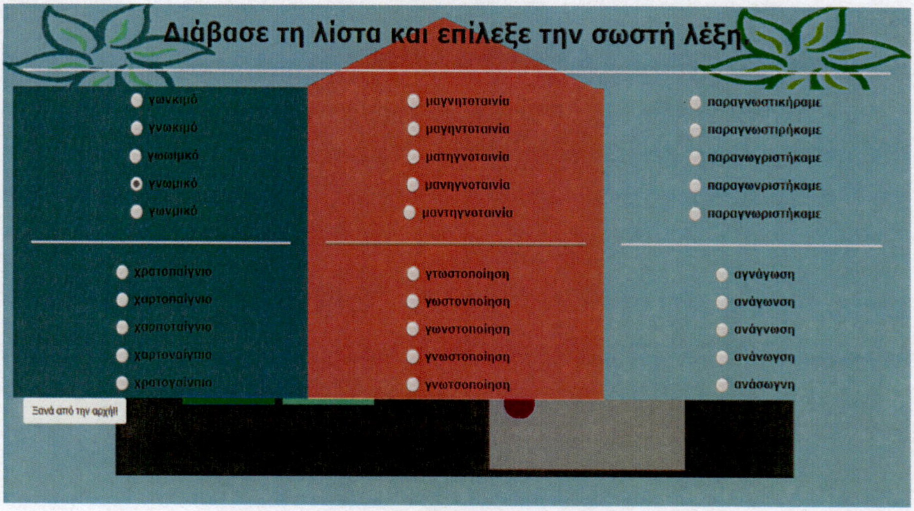

Fig. 4. Children are asked to read and determine which of the 5 options the right word is. If they don't succeed, they have the option by clicking "Again from the beginning!!" to start locating the words. Example in English: We have the words butterfly buffertly, (butterfly), bufterfly butfertly and children have to read and detect the correct word between five options.

Fig. 5. Children are asked to read carefully and find the right word that fits each gap in order to make sense of the text. The missing words are given at the top of the text. If they fail, they have the option to click "Again from the beginning!!" to start the effort again

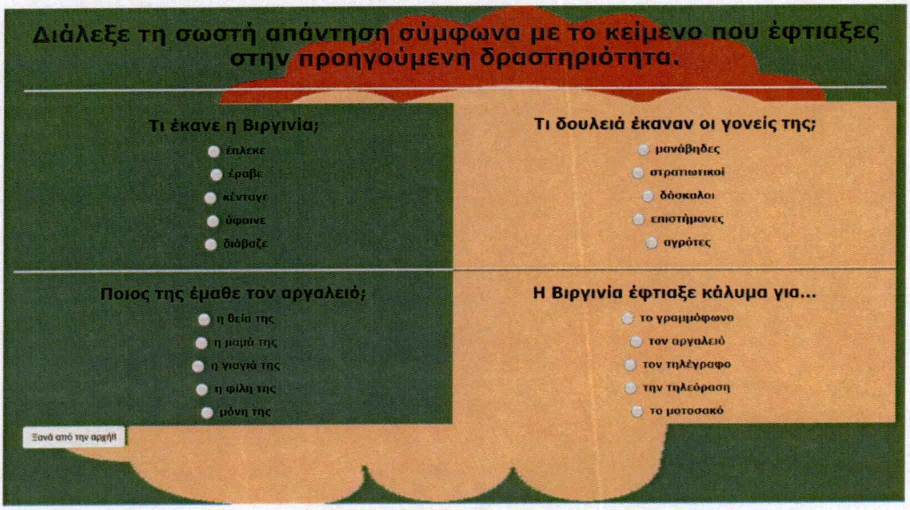

Fig. 6. Children are asked to read the questions carefully and check their memory by choosing the correct answer according to the text they read in the previous task. If they fail, they have the option to click "Again from the beginning!!" to start the effort again.

4 Actual or Anticipated Outcomes

Speed and accuracy in reading and comprehension is a determining factor of reading efficiency and plays a critical role in academic achievement [42]. Lack of reading skills or even a reading ability lower than normal can lead to fugitives, low-paying jobs or perpetuate the illiteracy cycle for the next generation [43].

The ability to translate a printed word into its oral form is a fundamental skill that readers need to acquire. There are several ways in which this can be achieved. For example, the correlation of the shape of a written word with its phonological (oral) form. Another strategy could be to associate smaller sections of text (letter by letter) with their phonological form and create words by decoding each section in order. However, in all alphabetical languages, learning the graph-phoneme correspondence remains the most profitable way to read all words [44].

Furthermore, the present study lends support to studies that have used paper-pencil tasks in order to strengthen the reading abilities of children with reading impairments. However, none of them, to the best of our knowledge, has used the internet in order to deliver the tasks to the child.

Acquiring reading is an important milestone in children's development. Learning to read is intended to train the visual system in decoding letter combinations and to provide a new entry into already developed cortical areas for speech comprehension [45].

5 Conclusion

The WinABC software developed by Tressoldi [46], the Fast ForWord Language intervention program developed by Tallal and Merzenich [47] and the GraphoGame educational tool designed by researchers at Jyväskylä University in Finland [48, 49] are some of the software applications that are already used in several languages.

Speed reading is one of reading skills that must be possessed by a student. This reading technique is done at high speed by not ignoring the comprehension of the reading [50]. However, "Poke the Reading Ability" is an online program designed in order to improve the reading capacity of children and adolescents. Its tasks are not only targeting at building phonological awareness but also, include tasks that improve visual memory, visual discrimination ability and text comprehension.

Software can enhance children's learning, with positive effects on reading skills development [51, 52]. The web application that is presented is carefully designed both in its pedagogical and computer programming aspects in order to offer a solution to tutors at everyday school practice.

Its activities are based on improved graphophysical correspondence, phonological decoding and visual working memory. An advantage is that children are usually very willing to use digital "learning toys" [53].

References

1. Sandak, R., Mencl, W.E., Frost, S.J., Pugh, K.R.: The neurobiological basis of skilled and impaired reading: recent findings and new directions. Sci. Stud. Read. **8**(3), 273–292 (2004)
2. Castles, A., Rastle, K., Nation, K.: Ending the reading wars: reading acquisition from novice to expert. Psychol. Sci. Public Interest **19**(1), 5–1 (2018)
3. Grainger, J., Jacobs, A.M.: Orthographic processing in visual word recognition: A multiple read-out model. Psychol. Rev. **103**(3), 518–565 (1996)
4. Seidenberg, M.S., McClelland, J.L.: A distributed, developmental model of word recognition and naming. Psychol. Rev. **96**(4), 523–568 (1989)
5. Spira, E.G., Bracken, S.S., Fischel, J.E.: Predicting improvement after first-grade reading difficulties: the effects of oral language, emergent literacy, and behavior skills. Dev. Psychol. **41**(1), 225 (2005)
6. Cunningham, A.E., Stanovich, K.E.: Early reading acquisition and its relation to reading experience and ability 10 years later. Dev. Psychol. **33**(6), 934 (1997)
7. Kotaman, H.: Impacts of dialogical storybook reading on young children's reading attitudes and vocabulary development. Read. Improv. **57**(1), 40–45 (2020)
8. Ehri, L. C.: Development of sight word reading: phases and findings. In: The Science of Reading: A Handbook, pp. 135–154 (2005)
9. Vellutino, F.R., Fletcher, J.M., Snowling, M.J., Scanlon, D.M.: Specific reading disability (dyslexia): what have we learned in the past four decades? J. Child Psychol. Psychiatry **45** (1), 2–40 (2004)
10. Protopapas, A.: Learning To Read Greek. Learning to Read Across Languages and Writing Systems, pp. 155–180 (2017)
11. Nestor, M.T., Ange, A.M.: The effectiveness of reading in the development of a learner education. Int. J. Res. Soc. Sci. Humanit. (IJRSS) **1**(1), 1–9 (2020)

12. Liu, F.: Reading abilities and strategies: a short introduction. Int. Educ. Stud. **3**(3), 153–157 (2010)
13. Ramirez, G., Fries, L., Gunderson, E., Schaeffer, M.W., Maloney, E.A., Beilock, S.L., Levine, S.C.: Reading anxiety: an early affective impediment to children's success in reading. J. Cogn. Dev. **20**(1), 15–34 (2019)
14. Grainger, J., Dufau, S., Montant, M., Ziegler, J.C., Fagot, J.: Orthographic processing in baboons (Papio papio). Science **336**(6078), 245–248 (2012)
15. Horowitz-Kraus, T., Vannest, J.J., Holland, S.K.: Overlapping neural circuitry for narrative comprehension and proficient reading in children and adolescents. Neuropsychologia **51**(13), 2651–2662 (2013)
16. Breznitz, Z., Misra, M.: Speed of processing of the visual–orthographic and auditory–phonological systems in adult dyslexics: the contribution of "asynchrony" to word recognition deficits. Brain Lang. **85**(3), 486–502 (2003)
17. Carrell, P. L., Grabe, W.: Reading. In: Schmitt, N. (ed.) An Introduction to Applied Linguistics, pp. 215–231 (2010)
18. Pokharel, P.K.: Learning to read and reading to learn in English. J. NELTA Surkhet **5**, 75–81 (2018)
19. Breznitz, Z., Rubinsten, O., Molfese, V.J., Molfese, D.L.: Reading, Writing, Mathematics and the Developing Brain: Listening to Many Voices. Springer (2012)
20. Godde, E., Bosse, M.-L., Bailly, G.: A review of reading prosody acquisition and development. Read. Writ. **33**(2), 399–426 (2020)
21. Torppa, M., Lyytinen, P., Erskine, J., Eklund, K., Lyytinen, H.: Language development, literacy skills, and predictive connections to reading in Finnish children with and without familial risk for dyslexia. J. Learn. Disabil. **43**(4), 308–321 (2010)
22. Larsen, L., Schauber, S.K., Kohnen, S., Nickels, L., McArthur, G.: Children's knowledge of single-and multiple-letter grapheme-phoneme correspondences: an exploratory study. Early Child. Res. Q. **51**, 379–391 (2020)
23. Horiba, Y., Fukaya, K.: Reading and learning from L2 text: effects of reading goal, topic familiarity, and language proficiency. Read. Foreign Lang. **27**(1), 22–46 (2015)
24. Gough, P.B., Tunmer, W.E.: Decoding, reading, and reading disability. Remedial Spec. Educ. **7**(1), 6–10 (1986)
25. Pazzaglia, F., Cornoldi, C., Tressoldi, P.E.: Learning to read: evidence on the distinction between decoding and comprehension skills. Eur. J. Psychol. Educ. **8**(3), 247–258 (1993)
26. Klauda, S.L., Guthrie, J.T.: Relationships of three components of reading fluency to reading comprehension. J. Educ. Psychol. **100**(2), 310–321 (2008)
27. Oakhill, J.V., Cain, K., Bryant, P.E.: The dissociation of word reading and text comprehension: evidence from component skills. Lang. Cogn. Process. **18**(4), 443–468 (2003)
28. Johann, V., Könen, T., Karbach, J.: The unique contribution of working memory, inhibition, cognitive flexibility, and intelligence to reading comprehension and reading speed. Child Neuropsychol. **26**(3), 324–344 (2020)
29. Stuart, M., Masterson, J.: Patterns of reading and spelling in 10-year-old children related to prereading phonological abilities. J. Exp. Child Psychol. **54**(2), 168–187 (1992)
30. Wagner, R.K., Torgesen, J.K.: The nature of phonological processing and its causal role in the acquisition of reading skills. Psychol. Bull. **101**(2), 192 (1987)
31. Miyake, A., Shah, P.: Models of Working Memory: Mechanisms of Active Maintenance and Executive Control. Cambridge University Press (1999)
32. Titz, C., Karbach, J.: Working memory and executive functions: effects of training on academic achievement. Psychol. Res. **78**(6), 852–868 (2014)

33. Cain, K., Oakhill, J., Bryant, P.: Children's reading comprehension ability: concurrent prediction by working memory, verbal ability, and component skills. J. Educ. Psychol. **96** (1), 31–42 (2004)
34. Cain, K.: Children's reading comprehension: the role of working memory in normal and impaired development. In: Working Memory and Education, pp. 61–91. Academic Press (2006)
35. Swanson, H.L., Berninger, V.: The role of working memory in skilled and less skilled readers' comprehension. Intelligence **21**(1), 83–108 (1995)
36. De Beni, R., Palladino, P., Pazzaglia, F., Cornoldi, C.: Increases in intrusion errors and working memory deficit of poor comprehenders. Q. J. Exp. Psychol. Sect. A **51**(2), 305–320 (1998)
37. Seigneuric, A., Ehrlich, M.F., Oakhill, J.V., Yuill, N.M.: Working memory resources and children's reading comprehension. Read. Writ. **13**(1–2), 81–103 (2000)
38. Yuill, N., Oakhill, J., Parkin, A.: Working memory, comprehension ability and the resolution of text anomaly. Br. J. Psychol. **80**(3), 351–361 (1989)
39. Cook, A.E., Halleran, J.G., O'Brien, E.J.: What is readily available during reading? A memory-based view of text processing. Discourse Process. **26**(2–3), 109–129 (1998)
40. Torgesen, J.K., Barker, T.A.: Computers as aids in the prevention and remediation of reading disabilities. Learn. Disabil. Q. **18**(2), 76–87 (1995)
41. Watkins, C. P., Caporal, J., Merville, C., Kouider, S., Dehaene, S.: Accelerating reading acquisition and boosting comprehension with a cognitive science-based tablet training. J. Comput. Educ., 1–30 (2020).
42. Anderson, R.C.: Becoming a Nation of Readers: The Report of the Commission on Reading. National Academy of Education, Champaign (1985)
43. Kutner, M., Greenberg, E., Jin, Y., Boyle, B., Hsu, Y.C., Dunleavy, E.: Literacy in Everyday Life: Results from the 2003 National Assessment of Adult Literacy. NCES 2007-490. National Center for Education Statistics (2007)
44. Vousden, J.I., Ellefson, M.R., Solity, J., Chater, N.: Simplifying reading: applying the simplicity principle to reading. Cogn. Sci. **35**(1), 34–78 (2011)
45. Dehaene, S., Pegado, F., Braga, L.W., Ventura, P., Filho, G.N., Jobert, A., Dehaene-Lambertz, G., Kolinsky, R., Morais, J., Cohen, L.: How learning to read changes the cortical networks for vision and language. Science **330**(6009), 1359–1364 (2010)
46. Tressoldi, P.E.: ImparareGiocando, Winabc - Programma per il trattamento della Dislessia. https://www.impararegiocando.it/WinABC50.htm. Accessed Dec 2018
47. Tallal, P.: Improving language and literacy is a matter of time. Nat. Rev. Neurosci. **5**(9), 721 (2004)
48. Lyytinen, H., Erskine, J., Kujala, J., Ojanen, E., Richardson, U.: In search of a science-based application: a learning tool for reading acquisition. Scand. J. Psychol. **50**(6), 668–675 (2009)
49. Lyytinen, H., Ronimus, M., Alanko, A., Poikkeus, A.M., Taanila, M.: Early identification of dyslexia and the use of computer game-based practice to support reading acquisition. Nordic Psychol. **59**(2), 109–126 (2007)
50. Walipah, E., Rahman, R., Sopandi, W., Sujana, A.: School literacy movement in the industrial revolution era 4.0 in building speed reading skills. In: International Conference on Elementary Education, vol. 2, no. 1, pp. 1763–1770 (2020)
51. Heikkilä, R., Aro, M., Närhi, V., Westerholm, J., Ahonen, T.: Does training in syllable recognition improve reading speed? A computer-based trial with poor readers from second and third grade. Sci. Stud. Read. **17**(6), 398–414 (2013)

52. Magnan, A., Ecalle, J.: Audio-visual training in children with reading disabilities. Comput. Educ. **46**(4), 407–425 (2006)
53. Zygouris, N.C., Avramidis, E., Karapetsas, A.V., Stamoulis, G.I.: Differences in dyslexic students before and after a remediation program: a clinical neuropsychological and event related potential study. Appl. Neuropsychol. Child **7**(3), 235–244 (2018)

Learners' Preferences in ESP Instruction for Higher Medical Staff

Ludmila Faltynkova(✉), Ivana Simonova, Katerina Kostolanyova, and Tomas Barot

University of Ostrava, F. Sramka 3, Ostrava, Czech Republic
faltynkova@epol.cz, iivana.simonova@gmail.com, {katerina.kostolanyova, tomas.barot}@osu.cz

Abstract. The blended model of foreign language instruction in professional higher education has increased rapidly and amounts of research studies have proved that it can enhance the quality of learning English for Specific Purposes (ESP). However, not much has been discovered on how the blended model affects learning outcomes in higher education, of medical staff in particular. Therefore, the main objective of this pilot study is to analyze and consider students' perception of blended model compared to the face-to-face course delivery, and their learning outcomes achievement.

The research was conducted at the Tertiary school for medical staff in Olomouc, Czech Republic, within the ESP course. Totally, the research sample included 86 first year students of various medical study programmes. Both forms of instruction, face-to-face and blended model, were implemented for 16 weeks, i.e. one semester. Data were collected via two written didactic tests. Students' feedback on the blended learning mode was monitored by the questionnaire. They expressed higher motivation and appreciated traditional strengths, i.e. the possibility of individual pace and independence of study. They also demonstrated better learning outcomes in the test scores, despite no significant differences were discovered. However, the students with lower level of English performed slightly better in the group taught in the blended mode.

Keywords: English for Specific Purposes · ESP · Higher medical staff · Blended learning

1 Introduction

The implementation of blended model in foreign language instruction in higher education has increased rapidly and amounts of research studies have proved that it can also enhance the quality of learning English for Specific Purposes (ESP), e.g. [1, 2]. However, not much has been discovered on how the blended model affects learning outcomes in higher education, of medical staff in particular.

According to Shorey et al. research [3], in educating nurses and healthcare providers, integrating blended learning within the course enhances communication skills and improves self-efficacy among nursing students Moreover, Halasa et al. [4] suggest

M. E. Auer and T. Rüütmann (Eds.): ICL 2020, AISC 1328, pp. 655–662, 2021.
https://doi.org/10.1007/978-3-030-68198-2_60

that the best teaching practices that create students taking the lead in the knowledge of nursing, arise from other than the traditional model of learning.

So, the main objective of this paper is to present results of the research dealing with learning ESP in higher medical staff from the view of students' knowledge and feedback on the blended learning model.

2 Theoretical Background

The process of acquiring new knowledge, skills, attitudes through appropriate learning content and feedback, and reflecting learner's individual needs and preferences, must not be necessarily directly managed by the teacher. As generally accepted, the ICT-supported learning is an educational technique characterized by self-paced, self-administered instruction presented in logical sequence and with much repetition of concepts Learner's activities depend on how precisely defined learning objectives are, on appropriate learning conditions, teacher's proper knowledge of the learning process, and last but not least, on the learning content [5].

The blended approach provides more flexibility for students and instructors. It facilitates specific kinds of learning activities that might not be possible without the technology, as presented e.g. by the SAMR (Substitution – Augmentation – Modification – Redefinition) model [6], students activate different abilities and skills to build new knowledge, and consequently, demonstrate what they learned. Last but not least, both teachers and students have the opportunity to develop their skills in using latest devices and applications. Considering a wide range of delivery methods, blended learning is appreciated for the possibility to study using individually preferred learning style and pace, as researched e.g. by Poon [7]. The experience in blended learning proved that well-designed blended courses not only enhanced students' learning towards acquiring new knowledge but also increase the retention, even in large classes [8].

3 Research Methodology

3.1 Research Problem, Question, Objective

For numerous higher institutions, the blended model of instruction represents a connecting link between face-to-face and distance online courses. In the field of *higher medical staff* preparation, hardly any research study has been conducted. The question is whether blended approach can enhance and support the process of ESP learning for medical professionals. Therefore, the main objective of this pilot study is to analyze and consider students' perception of blended model compared to face-to-face course delivery, as well as their learning outcomes achievement.

3.2 Hypotheses

Six hypotheses were set to reach the research objectives.

$1H_1$: There are the statistically significant paired differences between results of pretest and posttest1 in group E.
$2H_1$: There are statistically significant paired differences between results of pretest and posttest2 in group E
$3H_1$: There are statistically significant paired differences between results of pretest and posttest1 in group C.
$4H_1$: There are statistically significant paired differences between results of the pretest and posttest2 in group C.
$5H_1$: There are statistically significant paired differences between results of posttests1 in groups E and C
$6H_1$: There are statistically significant paired differences between results of posttests2 in groups E and C.

To get p values, the exact or non-parametric tests should be applied [9], from two reasons: (1) a small number of samples in the data file ($N = 46$ in experimental group, $N = 40$ in control group), (2) the appearance of zero values in some data columns, which do not show the normality distribution (expressed by e.g. Shapiro-Wilk or Kolmogorov-Smirnov tests and required for the exploitation of parametric tests). The Wilcoxon paired exact test belongs to non-parametric tests and is used for processing the paired based comparisons. It works as an alternative to the parametric paired T-test in case of the normality of data distribution [10].

For testing hypotheses, a pair of a zero hypothesis H_0 and an alternative hypothesis H_1 were used. According to p value with regards to the significance level α, the zero hypothesis can be failed to reject (for $p > \alpha$), or the zero hypothesis can be rejected in favour of the alternative hypothesis (in case of $p < \alpha$).

Besides the p value, the achieved statistic results are supported by the effect size that confirms the strength of rejecting the zero hypothesis.

3.3 Methods, Tools, Sample

The comparative method was applied when focusing on data collected via didactic tests. All students stated they had not studied ESP for higher medical staff before, so, the entrance knowledge was not monitored but considered zero. The process of instruction took 16 weeks (one semester). The research was conducted at the Tertiary school for medical staff in Olomouc, Czech Republic, within the ESP course in the first year of study. Students were divided into two groups; group 1 (experimental group, E) consisted of 46 students and the blended model of instruction was applied. It means that students attended face-to-face lessons and received study materials and tasks for via LMS Moodle. Group 2 (control group, C), comprising 40 students, followed the traditional face-to face-lessons and LMS Moodle was not exploited at all in this group. All materials which were available in LMS for group 1 were provided in the printed form to them. In the middle of the semester (i.e. after eight weeks of instruction in the blended, or face-to-face manner) both groups sat for posttest1 (posttest1 in E group, posttest1 in C group) which monitored students' knowledge. After another eight-week period, i.e. at the end of semester, posttest2 was administered (posttest2 in

E group, posttest2 in C group). Both the posttest1 and posttest2 consist of 70 tasks providing maximum score of 100 each. Active replies were required from the respondents; the tasks were not of multiple-choice type. Tests were assessed according to the following criteria: grade A: 100%–90% of maximum test score, B: 89%–79%, C: 78%–60%, F: 59%–0%.

Moreover, students of the experimental group reflected their opinions on the blended learning in the questionnaire. It included five areas of interest to provide the feedback, particularly focusing on students' preference to face-to-face, or blended approach, other (which ones) subjects taught through blended learning would students appreciate, which type/s of study materials students consider the most useful, and which approach was more motivating for their learning were under the focus.

4 Research Results

Following the research design, results of testing six hypotheses are presented first, followed by the results of questionnaire collecting the feedback from students of the experimental group.

Using the techniques based on the quantitative research, the determined research problems, questions and objectives are confirmed in this contribution. Testing the hypotheses was generally and frequently used in this type of research complementing with descriptive statistics, e.g. boxplot. As one of the advantages of testing hypotheses, the guarantee of the statistical significance can be considered. Particularly, the paired comparisons of obtained results of the students' improvement can be observed. For purposes of comparisons of appeared pretests and posttests, the paired tests are the appropriate option [10].

With regards to a normal probability distribution of data, parametrical or non-parametrical statistical paired test were exploited. In each testing the hypothesis or normality distribution of data, the obtained p value is compared to the declared statistical significance level. In the field of education, the applied significance level is 0.05. For purposes of the paired comparisons, the parametrical test Paired T-test should be used (for p greater than significance level) or the non-parametrical Wilcoxon paired test should be applied (for p lower than significance level) [10].

Data collected via didactic tests were processed by software PAST Statistics version 2.17 [11] and IBM SPSS Statistics, version 26.

4.1 Hypotheses $1H$ and $2H$

The zero and alternative hypotheses were tested to compare the pretest and posttest1 scores and the pretest and posttest2 scores in the experimental group. Results of statistic testing are displayed in Table 1.

According to $p = 3.51 \times 10^{-9} < \alpha = 0.05$, the zero hypothesis $1H_0$ is rejected in favour of the alternative hypothesis $1H_1$. Therefore, there are the statistical significant paired differences between results of pretest and posttest1 in group E at the significance level $\alpha = 0.05$. Due to effect size $r = 1$, the rejecting the zero hypothesis was significantly proved.

Table 1. Results of testing hypothesis $1H$ – Paired Comparison between Pretest and Posttest1 in Group E and $2H$ – Paired Comparison between Pretest and Posttest2 in Group E.

	*1*H: E Pre	*1*H: E Post1	2H: E Pre	2H: E Post2
Means:	0	73.15	0	76.09
Medians:	0	74.50	0	75.00
p value:	3.51×10^{-9}		3.49×10^{-9}	
Testing criterion W:	1081		1081	
Effect size - r:	1		1	

According to $p = 3.49 \times 10^{-9} < \alpha = 0.05$, the zero hypothesis $2H_0$ is rejected in favour of the alternative hypothesis $2H_1$. Therefore, there are the statistical significant paired differences between results of the pretest and posttest2 in group E at the significance level $\alpha = 0.05$. Due to effect size $r = 1$, the rejecting the zero hypothesis was significantly proved.

4.2 Hypotheses 3*H* and 4*H*

The zero and alternative hypotheses were tested to compare the pretest and posttest1 scores and the pretest and posttest2 scores in the control group. Results of statistic testing are displayed in Table 2.

Table 2. Results of testing hypothesis $3H$ – Paired Comparison between Pretest and Posttest1 in Group C and $4H$ – Paired Comparison between Pretest and Posttest2 in Group C.

	*3*H: C Pre	*3*H: C Post1	*4*H: C Pre	*4*H: C Post2
Means:	0	66.09	0	68
Medians:	0	69.50	0	69
p value:	3.49×10^{-9}		3.49×10^{-9}	
Testing criterion W:	1081		1081	
Effect size - r:	1		1	

According to $p = 3.49 \times 10^{-9} < \alpha = 0.05$, the zero hypothesis $3H_0$ is rejected in favour of the alternative hypothesis $3H_1$. Therefore, there are statistically significant paired differences between results of the pretest and posttest1 in group C at the significance level $\alpha = 0.05$. Due to effect size $r = 1$, the rejecting the zero hypothesis was significantly proved.

According to $p = 3.49 \times 10^{-9} < \alpha = 0.05$, the zero hypothesis $4H_0$ is rejected in favour of the alternative hypothesis $4H_1$. Therefore, there are the statistically significant paired differences between results of the pretest and posttest2 in group C at the significance level $\alpha = 0.05$. Due to effect size $r = 1$, the rejecting the zero hypothesis was significantly proved.

4.3 Hypotheses 5*H* and 6*H*

The zero and alternative hypotheses were tested to compare the pretest and posttest1 scores and the pretest and posttest2 scores in the experimental and control group. Results of statistic testing are displayed in Table 3.

Table 3. Results of testing hypothesis 5*H* – Paired Comparison between Posttests1 in Groups E and C and 6*H* – Paired Comparison between Posttests2 in Groups E and C.

	5H: E Post1	5H: C Post1	6H: E Post2	6H: C Post2
Means:	73.15	66.09	76.09	68
Medians:	74.50	69.50	75.00	69
p value:	7.41×10^{-9}		3.18×10^{-9}	
Testing criterion *W:*	1029		1081	
Effect size - *r*:	0.95		1	

According to $p = 7.41 \times 10^{-9} < \alpha = 0.05$, the zero hypothesis $5H_0$ is rejected in favour of the alternative hypothesis $5H_1$. Therefore, there are the statistical significant paired differences between results of the posttests1 in groups E and C at the significance level $\alpha = 0.05$. Due to effect size $r = 0.95$, the rejecting the zero hypothesis was significantly proved.

According to $p = 3.18 \times 10^{-9} < \alpha = 0.05$, the zero hypothesis $6H_0$ is rejected in favour of the alternative hypothesis $6H_1$. Therefore, there are the statistically significant paired differences between results of the posttests2 in groups E and C on the significance level $\alpha = 0.05$. Due to effect size $r = 1$, the rejecting the zero hypothesis was significantly proved.

4.4 Students' Feedback on Blended Learning

Students expressed their opinions on blended learning through the questionnaire. The results show that.

- 96% of 46 students preferred the blended approach;
- 79% would appreciate to have blended learning in other subjects as well, particularly in Anatomy and Physiology (21%), Pharmacology (10%), Chemistry (9%), Pharmacognosy (4%), Latin language (2%) and others;
- some types of study materials were considered very useful for learning: video-recordings (21%), grammar exercises (18%), vocabulary exercises (14%), extra study materials providing texts with extra or more detailed information about the related topic, links to such materials, sometimes also extra exercises, labelling pictures, reading comprehension etc. (7%), topic revision in Czech language (6%), links to dictionaries (2%);
- blended approach (62%) was more motivating to self-study for the students compared to the face-to-face learning (21%), both modes were also appreciated (17%).

5 Summary and Conclusion

To sum up, students reported significantly higher preferences in the blended model of the course in comparison to the face-to-face lessons. After studying in the blended model, they expressed higher motivation and appreciated traditional strengths, i.e. the possibility of individual pace and independence of study. They also demonstrated better learning outcomes in the test scores, despite no significant differences between the groups were discovered. However, the students with lower level of English performed slightly better in the group, which was presented with the blended model.

Both course delivery approaches examined in this pilot study proved themselves efficient in today's higher education. The authors believe that the traditional course delivery will continue to offer benefits that cannot be fully obtained in any other formats. Nevertheless, blended instruction does not appear to impair students' performance and it might enhance their appreciation of the concepts in some cases. Students in the blended learning group indicated they would definitely take another course using this method. Moreover, this delivery format offers increased flexibility to both students and teachers, who can operate from home not only in cases like the corona virus crisis, which we are experiencing right now.

Research results were analysed using the quantitative research including the testing the hypotheses and descriptive boxplot (displayed in Fig. 1). The data show that starting from no knowledge at the beginning of the research period, they significantly improved in posttest1 which was administered in the middle of the semester. However, hardly any increase can be detected at the end of the period (+2.94). Nevertheless, the increase in pair scores in the experimental group compared to the control one in posttest1 is statistically significant (+7.06, Table 3), as well as in posttest2 (+7.541, Table 3).

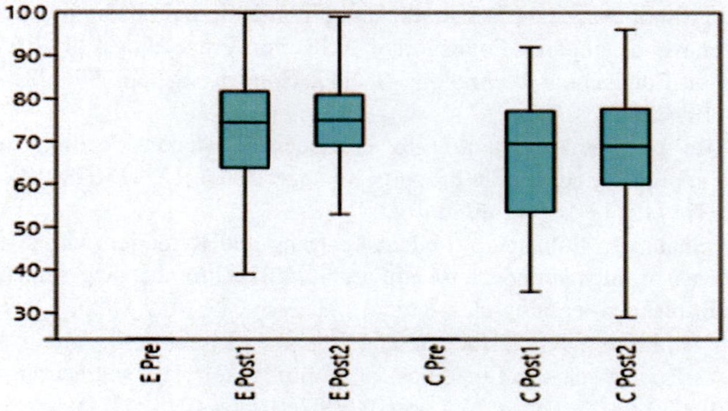

Fig. 1. Boxplot summarizing the test results.

The fact that students reach higher cognitive increase when the process of instruction is supported by information and communication technologies, i.e. in blended learning, is also proved by these research results.

Moreover, as clearly seen from the feedback questionnaire, test scores and numerous studied conducted in the past, higher medical students also appreciated traditional positive characteristics of blended learning, i.e. the face-to-face contact with teachers and at the same time the possibility to study any time, any place, any pace, as displayed in results of questionnaire. These features refer to learning general English and ESP. However, as ESP for higher medical staff does not appear frequently in the study programmes, the contribution of this study is high. As the principles of ESP teaching/learning are identical, the discovered findings can be applied in engineering study programmes.

Acknowledgements. The paper is supported by the SGS project N. SGS03/PdF/2019-2020 ICT-Enhanced Teaching English.

References

1. Simonova, I.: Blended approach to learning and practising English grammar with technical and foreign language university students: comparative study. J. Comput. High. Educ. **31**, 249–272 (2019). https://doi.org/10.1007/s12528-019-09219-w
2. Kostolanyova, K., Simonova, I.: Adaptive e-learning model for learning English as a second/foreign language. Int. J. Innov. Learn. **27**(3), 274 (2020). https://doi.org/10.1504/IJIL.2020.106811
3. Shorey, S., Kowitlawakul, Y., Devi, M.K., Chen, H.C., Soong, S.K.A., Ang, E.: Blended learning pedagogy designed for communication module among undergraduate nursing students: a quasi-experimental study. Nurse Educ. Today **61**, 120–126 (2018)
4. Halasa, S., et al.: Comparing student achievement in traditional learning with a combination of blended and flipped learning. Nurs. Open (2020). https://doi.org/10.1002/nop2.492
5. Malach, A.: Programované vyučování [Programmed learning], 1st edn. Akademie, Praha (1977)
6. Netolicka, J., Simonova, I.: SAMR model and bloom's digital taxonomy applied in blended learning/teaching of general English and ESP. In: Proceedings of 2017 International Symposium on Educational Technology (ISET), Hong Kong, pp. 277–281 (2017). https://doi.org/10.1109/ISET.2017.68
7. Poon, J.: Use of blended learning to enhance the student learning experience and engagement in property education. Property Manage. **30**(2), 129–156 (2012). https://doi.org/10.1108/02637471211213398/full/html
8. Amaral, K., Shank, J.: Enhancing Student Learning and Retention with Blended Learning Class Guides. https://er.educause.edu/articles/2010/12/enhancing-student-learning-and-retention-with-blended-learning-class-guides. Accessed 12 May 2020
9. Kitchenham, B., Madeyski, L., Budgen, D., Keung, J., Brereton, P., Charters, S., Gibbs, S., Pohthong, A.: Robust statistical methods for empirical software engineering. Empir. Softw. Eng. **22**, 579–630 (2017). https://doi.org/10.1007/s10664-016-9437-5
10. Barot, T., Burgsteiner, H., Kolleritsch, W.: Comparison of discrete autocorrelation functions with regards to statistical significance. In: Advances in Intelligent Systems and Computing. Springer (2020, in Print)
11. Hammer, Ø., Harper, D.A.T., Ryan, P.D.: Past: paleontological statistics software package for education and data analysis. Palaeontologia Electronica **4**(1), 1–9 (2001)

Training Professional Vocabulary On-line when Studying "English for Special Purpose" in Technological University

Elena Yurievna Semushina(✉) and Elena Volkova

Kazan National Research Technological University, Kazan, Russian Federation
epospelova12@yandex.ru, wolfkova@mail.ru

Abstract. Development of information technologies makes it possible to carry out distance support of the course "English for special purpose" studied at Technological University. The purpose of the study is to analyze the system of mastering lexical skills, based on combination of online study and face-to-face classes which allows to take into account the individual trajectory of each student. The study examines the organization of work with professional vocabulary in the framework of distance support of courses "English for special purpose" for Masters (major "Chemical technology") of Kazan national Research Technological university. Introduction and training of vocabulary is based on texts which are analyzed using the Flash Kincaid scale and processed later using the methods analyzed in the research. Introduction of new vocabulary online is carried out through visualization, word formation analysis and translation. Training lexical skills online is carried out through translation and training exercises (multiple choice, finishing a sentence, inserting a word), which are organized through the "test function", as the possibilities of MOODLE are limited. Automated control of three types (diagnostic, intermediate and final) is used for assessment. The system of lexical exercises allows not only to understand the meaning of the unit in the context, but also allows students with a low level of English not to be behind the group. As a result, the average percentage of completion of the online final test raised from 70% to 80–85%.

Keywords: On-line study · Distance support · Professional vocabulary

1 Introduction

Today, knowledge of professional vocabulary is a necessary component of competence of a young scientist. A special role here is played by the possession of professional vocabulary of the English language, since working with scientific literature in a foreign language is an important part of research of a competitive specialist [1]. Knowledge of terminology in English allows them to keep up with the trends of the modern world within this specialty. In addition, as a result of globalization, a significant part of professional terms is international, which significantly simplifies understanding and improves international interaction [2].

In the same time, the high level of development of information technologies makes it possible to carry out distance (on-line) support of the course studied. Distance

M. E. Auer and T. Rüütmann (Eds.): ICL 2020, AISC 1328, pp. 663–670, 2021.
https://doi.org/10.1007/978-3-030-68198-2_61

support is used for the students to study the lexical material independently after being absent at the lesson, to conduct automated control, to master lexical skills if a student has a low level of initial knowledge of the English language [3].

In general, the study of English as an academic subject at Kazan National Research Technological University (KNRTU) is characterized by the following:

- Inter-subject character – professional vocabulary is studied because adequate translation and understanding of the text is necessary to understand the processes described in the text. It can determine the correct choice of lexical units or synonyms.
- Multilevel structure – three stages of studying are defined. For Bachelors, it is necessary to form a business foreign language competence for written and oral communication in everyday and professional spheres of communication; for Masters, it is necessary to form a foreign language competence for solving professional tasks; for Postgraduates, the foreign language competence consists of the formation of academic communication skills [4].

At the moment, on-line learning is integrated into the educational process in higher education institutions in two forms:

- An autonomous learning system where students work online independently.
- A mixed learning model that involves the simultaneous use of face-to-face and distance learning, for example, a number of course elements, such as additional training materials, video materials, or automated control, are provided for distance support and speaking is operated in traditional form [5].

2 Purpose of the Research

The purpose of the study is to analyze the system of mastering lexical skills, based on combination of online independent study and face-to face classes. Using distance (on-line) support for training lexical skills is based on the possibility of additional development of lexical skills using training tasks, video material and automated control. The work is carried out taking into account the individual trajectory of each student, as it is necessary to constantly monitor the knowledge, level of motivation and self-discipline of students who study online. Professional vocabulary to study is selected according to the principles of consistency, frequency, compatibility, accessibility and communicative expediency.

Teaching a foreign language in the master's program has certain peculiarities. Firstly, most Masters have to work to earn their living and that may coincide in time with some classes. Distance support allows these students to do certain tasks by themselves and send them for the teacher to check. Secondly, every year the number of hours allocated to the subject "English for special purpose" is reduced, that's why, a number of aspects must be worked out in distant form to save time in the classroom for practicing other skills, such as speaking. Thirdly, the level of foreign language proficiency of most students does not meet the requirements for graduates of secondary educational institutions. In addition, foreign citizens who study in the same groups as

Russian-speaking students sometimes do not speak a foreign language at all. These students need additional training to master the language according to the requirements of the Federal State Educational Standard.

3 Approach

The study examines the organization of work with professional vocabulary in the framework of distance support of courses "English for special purpose" for Masters of the 1st year of study. Distance support in Kazan National Research Technological University (KNRTU) has been used since 2016, and by now more than 300 students have been trained on the platform moodle.kstu.ru. In the 2019/20, 102 students of the major "Chemical technology" and 45 students of the major "Information technologies" used the on-line course. The discipline "English for special purpose" is focused on mastering the language, speech and socio-cultural competence of students, which allows them to communicate with foreign specialists, as well as to do research in a foreign language. So, distance support of the subject "English for special purpose" has the following peculiarities:

- Individualization of studying. The student studies the material at an individual pace, the teacher can select elements of the on-line part of the course according to the level and capabilities of a particular student.
- Modularity. The training course consists of several blocks (modules), which makes it possible not only to choose a specific module to study, but also to combine them in a certain sequence.
- Strengthening the role of independent work of the student, since most tasks involve independent work with on-line sources and dictionaries.
- Development of self-discipline of students as well as teachers. The student should organize their time so that they can complete tasks according to the plan, and the teacher should check the tasks on time. These actions require a high level of motivation.

The ability to communicate using a foreign language directly depends on how well lexical skills are formed. To know a lexical unit means to know its form, meaning and usage. Accordingly, additional practical training of lexical skills and control are provided for distance (on-line) support of the course. The results of training lexical skills online depend first of all on the quality and systematic character of electronic material presented.

When organizing work with professional vocabulary online, the following methodological issues are taken into account:

- Communicative approach, which is used at all stages of studying vocabulary online: introduction, training and actualization.
- Relying on the native language. Translation exercises are compulsory to train lexical skills. Ignoring the reliance on the native language leads to disastrous consequences, since the student does not live in the language environment of the English language and the material studied is quickly forgotten [6].

- Visibility of material. Video material is actively used, and it should have inscriptions, since introducing and fixing lexical material using video and text simultaneously ensures high level of language learning.
- Accessibility of material. The problem is that students with advanced level of knowledge and graduates of schools with low level of language teaching can study in the same group, and the material studied should be accessible to all students.
- Differentiated approach, which is an inevitable consequence of teaching students with different levels of language proficiency in the same group. The vocabulary is selected taking into account individual semantic fields of students. This fact requires a special system for organizing tasks within each module, in other words, three levels of tasks – basic (tasks 1, 2), intermediate (tasks 3 and 5) and advanced (tasks 5–7), which are selected later taking into account the individual trajectory of the student [7].
- Gradual formation of lexical skills which becomes important because the student have to work a lot with the texts on their own without a constant control of the teacher.

4 Outcomes

To make mastering lexical skills online effective the following rules are followed:

1) Introduction and development of vocabulary is based on text material that is selected from authentic sources using the Flash Kincaid scale (recommended readability indicators are 15–40 points on sites https://readability.io/ for Russian and https://app.readable.com/text/?demo for English). If the selected texts do not meet the requirement of gradual increase in the complexity of the texts, they are processed contextually and linguistically using the methods analyzed in the research.

The Flash Kincaid scale is based on the average sentence length and the average number of syllables in a word. The higher the readability index is, the easier it is for the reader to understand the text. To increase the index, the following methods are traditionally used: reducing the average sentence length, minimizing punctuation density, reducing the size of paragraphs or increasing their number, getting rid of unnecessary words and sentences, simplifying sentence structures, and getting rid of unnecessary sentences [8].

The texts selected for introduction and training lexical skills online for the course "English for special purpose" (major "Chemical technology") must correspond to the curriculum of the students. When the online part of the course was created the following indicators of text complexity were obtained when the authentic texts were chosen: Chemistry (Introduction) – 33,3, Chemical elements – 42, Properties of chemical elements – 49, Nanotechnology (Introduction) - 30,2, Nanoengineering – 30.1, Nanomaterials – 27.5, Nano systems – 20.3, Nanotechnology and other sciences - 13.1, Microscope – 28.1, Polymers – 45, Polymerization – 28, Oil refinery – 35. It turned out that this arrangement of topics and the complexity of texts did not meet the requirements of a gradual increase in the complexity of the material studied. As a result, text processing was performed to increase or decrease the readability of the text using the following methods: simplification - complication of the vocabulary; division – unification; reduction - lengthening of sentences; changing the voice and number of passages; complete restructuring of sentences.

As an example of changing the complexity of a text, the text "What is Chemistry?" (Module 1) is given. Initially its index was 33 and it had to be increased to 36. The following simplifications were processed:

- Simplification of vocabulary (3 cases): *qualitative and quantitative – different.*
- Articulation of sentences (5 cases): *Chemistry is the study of matter, its properties, how and why substances combine or separate to form other substances, and how substances interact with energy. - Chemistry is the study of matter and its properties. It studies how and why substances combine or separate to form other substances, and how substances interact with energy.*
- Shortening of the sentences (5 cases): *All of the molecules that make up living tissue have carbon as part of their makeup. - All of the molecules that make up living tissue include carbon.*
- Change of the voice (4 cases): *Agricultural chemistry is concerned with the substances and chemical reactions that are involved with the production, protection and use of crops and livestock. - Agricultural chemistry deals with the substances and chemical reactions in producing crops and livestock.*
- A number of passages increased from 5 to 7.
- Complete restructuring of sentences (2 cases): *Industries require chemical engineers to devise new ways to make the manufacturing of their products easier and more cost effective. - Chemical engineers manufacture more cost-effective products.*

Processing of all texts to make it more complex or simple, depending on the place of the text in the course structure, allowed us to put the texts in the course according to increasing complexity of texts from 36 to 18.5: Chemistry (introduction) – 36, Chemical elements – 35.5, Properties of chemical elements – 32.2, Polymers – 31.7, Polymerization – 31.2, Nanotechnology (introduction) – 30.9, Nanoengineering – 30.1, Microscopes – 28.1, Nanomaterials – 27.5, Nano systems – 20.3, Nanotechnology and other sciences - 19.1, Oil refinery – 18.5.

Both course structures (original and modified) were offered to the students. The students with approximately the same level of English language proficiency, but different year of study, were selected to analyze the results. After analyzing the test results and taking into account the feedback from listeners, an improvement in the level of assimilation of the material was observed.

2) The course is structured according to the following principles: each module corresponds to a specific topic and has a semantic completeness; the structure of modules is identical; it is necessary to use three levels of vocabulary in each module; the study of lexical material is carried out step by step within each module [9].

3) Planning of the group's work is carried out taking into account the differentiated approach and individual trajectory of the student, which, despite its obvious effectiveness, takes a lot of time of the teacher. The teacher determines which lexical exercises are performed face-to-face and which are performed online.

4) The material is selected according to the major of the students. Special attention is paid to international words, that is, positive interference of the native and English languages is used [10].

5) Three stages of working with vocabulary (introduction, the training itself and actualization) are preserved, but they have certain peculiarities comparing to traditional study.

– Introduction of new vocabulary

New vocabulary is selected based on the principles of consistency (professional language material should be linked to each other, both by belonging to the same topic and word formation models), frequency, compatibility (lexical units that can give the greatest number of combinations with other words), accessibility and communicative expediency.

The following ways of introducing lexical material are used:

- visualization (task "Introductory video"),
- word formation analysis (task "Learn the words", for example, *technology, technological*)
- translation (task "Learn the words" - 30–40 words for further training)
- work with international terms (task "Learn the words", for example, *profession, engineer, engineering*).

The use of such methods of introduction as use of synonyms/antonyms and language guesswork through context is not effective here, as it requires careful face-to-face control by the teacher which is not possible if the work is processed online.

– Training vocabulary

The purpose of these exercises is to make students memorize the material. When activating vocabulary, students perform a variety of exercises in which the vocabulary should be included in system connections: paradigmatic, syntagmatic, semantic, associative, word-forming and typological. The following ways of vocabulary training are offered:

- Translation (tasks "Translate the text from English into Russian", "Translate the text from Russian into English") – tasks can be performed on-line or in a traditional form, depending on the trajectory of the student, developed by a teacher.
- Use of new vocabulary answering the questions (task "Answer the questions") – the task is done during full-time classes.
- Revision of vocabulary with the help of video material (task "Revision") - links to open sources are given.
- Training exercises (task "Revise the words" – to choose synonym/ antonym, to do multiple choice task, to finish a sentence, to insert a word into the text, and to play a game). These exercises are used online, and the teacher can only see the result, but cannot control the process. To compensate this disadvantage, the restrictions on the time and number of attempts to complete the task are removed. In other words, the students can repeat the tasks until they get the result desired.

– Actualization (generation of sentences)

To update the new vocabulary, speech practice is used, which contributes to the final formation of lexical skills. This aspect is fully transferred to full-time classes, and only the tasks themselves are provided as part of distance support:

- Make up a text (for example, tasks "Provide a brief summary of aims, products, advantages and disadvantages of chemical engineering using the words and phrases from the unit", "Complete the summary about your university education and career plans", "Present yourself (CV)")
- Make up a dialogue (for example, tasks "Practice in pairs", "Make a list of questions about different subjects you study at the university including mathematics, physics and chemistry for your groupmate. Then take turns to ask each other").
- Make a report about your research.

6) Lexical skills control has two forms in the framework of the on-line support:

- Synchronous (test). Automated control of three types (diagnostic, intermediate and final) in form of tests is used to evaluate the knowledge of the students. Assessment of knowledge online must meet the following criteria: validity, reliability, rating, personality, ease of calculating the result, cost-effectiveness, ease of management, compliance with the goals of testing, the relationship between content and form, representativeness, balance, variability, unambiguity of interpretation of results, correctness of formulation [8]. Each module presents a test that contains two types of questions: multiple choice and short answer.
- Asynchronous. After completion, the task is sent to the teacher (the task typically consists of 10–20 sentences for translation from Russian to English).

5 Conclusions

According to feedback of the students, introduction of vocabulary by combining the use of video and text material significantly facilitates the assimilation of a new vocabulary online. Despite certain limitations that arise due to insufficient capabilities of the MOODLE system, a step-by-step study of lexical material is implemented within each module (introduction, training, generation). If the first two stages can be conducted on-line, the third stage, aimed at training oral speech, is conducted during face-to-face classes. The system of lexical exercises created on the basis of working with authentic text with gradual complexity allows not only to update the meaning of the unit in the context, but also allows students with a low level of English not to be behind the group. Vocabulary exercises are used according to individual trajectory of students and manage to reduce the gap in proficiency levels of students in the group, which allowed to avoid a situation where the teacher is forced to work more with students with a low level of English in face-to-face classes. As a result, the average percentage of completion of the final test was brought to 80–85%, mainly due to an increase in the level of training of students who initially had a low level of foreign language proficiency.

References

1. Shageeva, F.T., Erova, D.R., Gorodetskaya, I.M., Kraysman, N.V., Prikhodko, L.V.: Training the achievement-oriented engineers for the global business environment. Adv. Intell. Syst. Comput. **716**, 343–348 (2018)
2. Kraysman, N.V., Shageeva, F.T., Mullakhmetova, G.R., Pichugin, A.B.: Poster: preparation of engineering university students for academic mobility to French universities. In: Advances in Intelligent Systems and Computing (AISC), vol. 1135, pp. 713–718 (2020)
3. Semushina, E.Y., Ziyatdinova, J.N.: On-line testing of engineering students as a form of assessment when studying English in distant form. In: Advances in Intelligent Systems and Computing, vol. 716, pp. 475–480 (2018)
4. Valeeva, E., Ziyatdinova, J. Galeeva, F.: Development of soft skills by doctoral students. In: Advances in Intelligent Systems and Computing (AICS), vol. 1135, pp. 159–168 (2020)
5. Semushina, E.Y., Ziyatdinova, J.N.: Studying English on-line as a part of a course "English for special purpose" in technological university. In: Advances in Intelligent Systems and Computing (AISC), vol. 1135, pp. 21–29 (2020)
6. Lefterova, O., Giliazova, D., Valeeva, E., Ziyatdinova, J.: Computer-aided translation course for students majoring in engineering. In: Advances in Intelligent Systems and Computing (AISC), vol. 1135, pp. 154–158 (2020)
7. Amurskaa, O.Y., Gimaletdinova, G.K., Khalitova, L.K.: Multimedia Sanako study 1200 for TEFL in institution of higher education. Xlinguae **10**(3), 229–236 (2017)
8. Semushina, E.Y.: Osobennosti Otbora i Ispol'zovaniya Tekstov pri Izuchenii Inostrannogo Yazyka v Distantsionnoy Forme (na Materiale Kursa "Professional English"). Vysshee Obrazovanie Segodnya **5**, 63–66 (2020)
9. Volkova, E.V.: Different approaches to the problems of intercultural communicative competence. In: Proceedings of the 16th International Conference on Interactive Collaborative Learning and 42nd International IGIP Symposium on Engineering Pedagogy, pp. 456–457 (2013)
10. Sultanova, D., Maliashova, A., Bezrukov, A.: Consistent development of the training program "innovation management". In: Advances in Intelligent Systems and Computing (AISC), vol. 1135, pp. 234–243 (2020)

Utilizing NLP Tools for the Creation of School Educational Games

Aristides Vagelatos[1(✉)], Monica Gavrielidou[1], Maria Fountana[1], and Christos Tsalidis[2]

[1] Computer Technology Institute and Press, Athens, Greece
vagelat@cti.gr
[2] Neurocom S.A., Athens, Greece

Abstract. The use of digital games to support learning (game-based learning) through an alternative, more attractive way is rapidly growing in both European and worldwide educational sector. Digital games are a rapidly developing field, as they are amongst the most popular technologies that young people use for their entertainment. Within this framework, project "Lexipaignio" was initiated in order to develop an innovative and state-of-the-art NLP (Natural Language Processing) environment for the creation of digital educational games for Greek students. These games will be dynamically generated by the educator for his/her students, in order to improve various vocabulary and linguistic skills, while understanding the context of specific school subject areas. In this paper we present the NLP environment for the subject of Geography.

Keywords: Educational games · NLP · Game-based learning · Digital games

1 Introduction

Lately, with the integration of new technological achievements into both educational and everyday life of students, important changes are under development in the educational and learning processes, where the Information and Communications Technology (ICT) plays a major role. From this perspective, the use of digital games to support learning (game based learning) in a more attractive way is rapidly emerging. Digital games are a rapidly developing field, as they are amongst the most popular technologies that young people use to amuse themselves [7]. The educational potential of digital games is correlated to motivation, amusement and the trigger of interest, which are considered consistent with positive learning results. According to relevant research, computer games provide a quick and interesting learning pace in contrast to the conventional teaching methods and in this perspective, they can evolve into a challenging way as far as digital learning is concerned [3].

In the "Lexipaignio" project, an innovative and state-of-the-art computational environment is under development, through the creation of digital educational games for students of upper primary and lower secondary education in order to: a) improve their language competency level and overall linguistic abilities, b) develop various vocabulary and linguistic skills, while deepening in the context of specific school subjects (biology, geography etc.).

M. E. Auer and T. Rüütmann (Eds.): ICL 2020, AISC 1328, pp. 671–681, 2021.
https://doi.org/10.1007/978-3-030-68198-2_62

At the point of convergence of computer science and linguistics, Natural Language Processing constitutes a challenging area for numerous applications in our everyday life. Focusing on Education, the "Lexipaignio" project aims at the use and further development of a series of Natural Language Processing infrastructure tools (Morphological Lexicon, Lemmatizer, Mnemosyne language editing system, corpus of Greek school subjects, etc.), for the implementation of dynamically created gamified educational material. This paper emphasizes on the utilization of NLP tools for the creation of minigames regarding the subject of Geography in Greek secondary schools. As part of an ongoing project, the development of Geography minigames aims at providing useful feedback regarding the use of Natural Language Processing for the development of dynamic gamified materials in other school subjects.

In the rest of the paper, the developed NLP infrastructure is presented, then, the selected test bed (Geography) is described, and finally the drawn conclusions are being quoted.

2 NLP Infrastructure

Natural Language Processing (NLP) is a research area that gains extreme interest in the post-information age. The ability to process information and transform it to knowledge is considered essential in today's "information jungle" [2]. In the context of education it seems that NLP has great benefits to offer [1]. In this section, the necessary NLP infrastructure is examined in order to support the dynamic creation of educational games within the project's scope. Most of these tools are based on earlier research work which has been further developed to constitute the basis of the "Lexipaignio" project.

2.1 Greek Language

In specific, statistical language models that select "word terms" do not work well with Greek text due to its rich morphology, returning poor results for both "precision" and "recall" metrics [9]. For single word terms, we must handle the morphology of words along with the part of speech. With more than 20 word forms for adjectives and more than 200 word forms for verbs, counting frequencies of words does not provide good results. A combination of POS (Part Of Speech) filtering, lemmatization (normalization) and TF/IDF scoring give the best results for single word terms. Equally, multi word terms have similar problems: e.g., a 2-tuple sequence of the form |ADJECTIVE NOUN| can have more than 100 different forms and only a few of them can be considered as valid terms. In this case, lemmatization and POS filtering does not help. For n-tuples with $n > 2$ the challenge remains. Thus, we use "KANON" formalism [4] to recognize valid patterns of candidate terms, e.g. |ADJECTIVE_GCN NOUN_GCN| where _GCN suffix codifies the G = Gender, C = Case, N = Number, i.e. the adjective and noun must agree in gender, case and number in order to be considered as candidate terms. Recognizing a candidate term is the first step of filtering. The second step is to normalize the expression by lemmatizing both words. These filtered expressions then feed the n-tuple language model for the statistical evaluation of candidate terms.

Another challenge is the "compound words" which are common in the scientific terminology of the Greek language. Together with morphology and extensive vocabulary they increase the difficulty of classifying the texts and finding common meanings. To handle this phenomenon, we split the words in q-gram characters and take these q-grams as units instead of entire words. For example, the compound words "geology" and "geography" have the word "geo" common and if we want to utilize this fact in a classification process we must use 3-grams ("geo") or 4-grams ("_geo") as units.

2.2 Corpus Processing

Within the projects' framework, a geographical corpus was compiled (see next section) to serve as a basis for the rest of the tools as well as input for the educational games, after being annotated. The annotation process of the LEXIPAIGNIO corpus involved almost all NLP components that already existed and included: a tokeniser, a sentence splitter, a morphosyntactic tagger, a toponym gazetteer and a single and multi-word term recognizer.

The collection of documents used includes more than 180 chapters from the educational books used in upper primary and lower secondary education in Greece, cleared from problematic typographical and layout related elements. Then a number of NLP tasks was applied in order to find ways to utilize the content in the games. Specifically:

Candidate Terms Extraction by Applying a Set of Morphosyntactic Patterns. We used 20 morphosyntactic patterns expressed with "KANON" rules that recognise candidate terms. The distribution of terms in these 20 rules along with the frequency for each candidate term as well as the frequency of the prefixes, provide valuable information about a) the most common morphosyntactic patterns of terms b) confidence level and c) patterns for synthetic (fake) terms that can be used in games. Table 1 shows the number of candidate terms per pattern as well as the lengths (number of words) recognized by each pattern.

Table 1. Candidate terms

Pattern	#Words	#Candidate terms	#Instances
1	2	4845	8272
2	2–8	1200	1484
3	2–12	912	1199
4	3–11	215	239
5	3–8	410	481
6	3	383	433
7	3–7	76	81
8	3	1042	1185
9	4–7	62	62
10	4	79	80
11	4	17	17

(*continued*)

Table 1. (*continued*)

Pattern	#Words	#Candidate terms	#Instances
12	4–10	56	60
13	4–10	118	118
14	4–8	72	77
15	4	91	99
16	5–7	12	12
17	5–9	14	14
18	5	8	8
19	5	7	7
22	3–10	250	256

Classification of Chapters. Two unsupervised classification algorithms were applied in order to discover the thematic categories of the chapters: a) *K-Means* with *Euclidean distance* as metric and b) *Hierarchical clustering* with *cosine similarity* as metric [8]. To minimize the problem of rich morphology and to consider common stems of compound words we used three types of elements: words, 3-gram characters and 4-gram characters. In Table 2 we present the results of the different clustering methods.

Table 2. Clustering method results

Method	#Clusters	Unit
K-Means (K = 10)	6 (4 empty)	Word
Hierarchical	24	Word
K-Means (K = 10)	7 (3 empty)	3-gram
Hierarchical	18	3-gram
K-Means (K = 10)	5 (5 empty)	4-gram
Hierarchical	18	4-gram

Discovery of Keywords Per Chapter. Keywords are important in the classification and characterization of a document/chapter. In order to select the appropriate words that distinguish a chapter from the rest of the chapters, the TF-IDF [6] (Term Frequency-Inverse Document Frequency) algorithm has been used in words, lemmas, and truncated words. For each word and chapter, we compute its TF-IDF metric which is the weight of the word for the chapter. It is computed by the product of TF * 1/DF where TF is the number of times a word is presented in a chapter and DF is the number of documents in the collection that contain this word. This means that a word increases its weight for the specific chapter in case it appears many times (frequency) but it loses (minimize) its weight if it also appears in many chapters in the collection (common words). Later on, the Okapi-BM25 [11] algorithm was applied, which is an alternative of TF-IDF, with two approaches: In the first approach we used the lemmas of the words for every chapter and for unknown words, the words themselves have been used. In the second case for unknown words we truncated the words removing the suffix in case it matched a common suffix of the Greek morphology. The words with higher scores were extracted and used as keywords for the corresponding chapters.

2.3 Augmenting Morphological Lexicon

The Greek morphological lexicon (~90.000 words, ~ 1.200.000 word forms) that has been implemented in a previous project [9, 10] was enriched with the geographical terminology extracted from the corresponding Corpus. The process of collecting corpus-based geography terms was based on the hypothesis that if a word-form found in the corpus is either unknown to the lexicon or has got high TF/IDF score in the corpus, then there are good chances for it to be a geographical word-form.

2.4 Tools

Several tools were utilized in the process of term extraction. Among these, a concordance was particularly helpful to the linguists that had to manually inspect and correct the output of the automatic terms selection tool. As can be seen in Fig. 1, the concordancer, allowed the linguists to inspect the usage of any term within the corpus in order for them to decide whether or not it constitutes a valid geographical term.

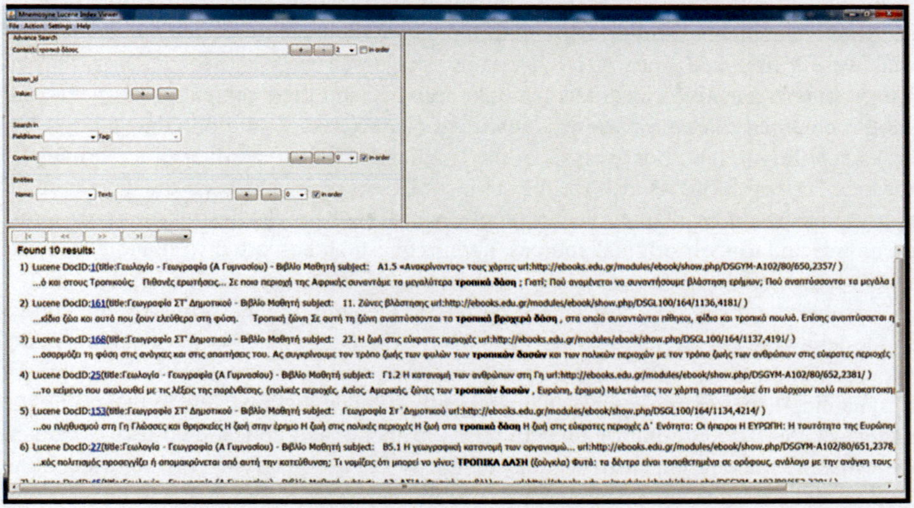

Fig. 1. The concordancer (a screenshot presenting the use of the term "τροπικό δάσος" (tropical forest)).

3 Test Bed: Geography

Among several other candidate school subjects (i.e. Biology, Home Econnomics, Music, Religious Education, etc.), with Modern Greek Language holding the main focus of the project for full deployment of the developed infrastructure, Geography was selected as the subject to be used for the project hypothesis test bed.

Although it was initially estimated that schoolbooks of the first two grades of secondary school would be sufficient to compile a substantial corpus for the subject of Geography, it was decided by the partnership to also explore educational activities from online Greek collections and repositories such as 'Ifigeneia' and 'Photodentro' as well as other relevant educational resources. As a result, the corpus of Geography was additionally compiled by full educational packages of teacher and student books taught both in upper Primary and lower Secondary education in Greece, escorting curricula and complementary educational material which embraced Geology and other related sciences complying with the interdisciplinary structure of the relevant schoolbooks.

Developing the process for the automatic extraction of thematic vocabulary and other structured information from the collected educational material and proceeding with the generation of taxonomies in the field of Geography has been a challenging task. It involved the identification of patterns both in terms of text format and layout, but also of linguistic morphosyntactic structure and resulted in an additional, on the side and beyond the scope of this project, interesting list of suggestions on the structure of future interactive books.

The corpus of Geography was, then, automatically annotated through the use of the NLP tools to automatically extract structured information from the semi-structured electronic documents. Additionally, with the use of the built-in "KANON" formalism, rules were formulated since KANON, as already described (see Sect. 2.2.), is able to define context sensitive constructs that take into account their contextual elements and has the potential to exploit words' individual characteristics (morphological, stylistic, etc.), according to the prior analysis of the language and the recording of its formalistic imprint. The result of this task for the subject of Geography has been the development of a unique resource of monolectic (single word) and polylectic (multi word) terms interconnected to their original sources, upon request, and escorted with additional text related information.

For the "Lexipaignio" project, the corpus of Geography, and primarily the corpus of Modern Greek Language to be developed at a later stage, is a valuable system resource out of which structured linguistic material will be drawn for the development of dynamic, thematic mini games. Consequently, teachers will be able to determine the content of the dynamic mini games in terms of theme and educational level / student age whereas they might also be able to go as far as specifying the relevant book chapters to be used as the primary source for the mini games that they will be releasing to their students.

3.1 Taxonomy of Geographical – Geological Terms

To facilitate the production of a large number of questions suitable for thematic games, a taxonomy of geographical and geological terms was developed based on the thematic vocabulary extracted from the relevant corpus. Expanding the possibilities of the "KANON" formalism by utilizing the syntactic forms which result from the co-occurrence of terms, will allow the automatic creation of classifications and will establish the functionality of formalism for further use in texts and other thematic fields.

For the compilation of the taxonomy, a step towards the formation of an ontology that will constitute the standard description of the geographical – geological knowledge, the following steps were followed:

- Four school textbooks were selected and isolated from the corpus of Geography: two textbooks taught in the senior classes of the Primary School and two textbooks taught in the junior classes of the High School.
- The contents of the above books were merged (overlapping sections among the books were grouped together).
- A first version of the distribution of the material throughout the textbooks was developed and mapped based on the prior grouping and the categorization of their chapters.
- Verification of the functionality of the categorization was attempted to additionally ascertain any overlaps and repetitions.
- Corrections were made to the initial categorisation and the taxonomy was finalized (part of the upper level can be seen in Fig. 2).
- Code numbers were assigned to each category of the classification as well as to each section of all school textbooks.
- The taxonomy was embedded in the project's portal, in the following address: https://lexipaignio.cti.gr/apotelesmata/ypodomes/taksinomia.

The categories of concepts that compile the initial classification concern a wide range of Geography-Geology study areas (e.g. the natural environment, the anthropogenic environment, continents, etc.), which serve the needs of the project effectively. In addition, the range of the categories ensures coherence in the knowledge of the relevant subject area, which can be generalized and reused in a variety of other subfields.

3.2 The Concept of Mini Games

Designing scenarios for educational games is a challenging process that requires creativity, perseverance, tough negotiations, continuous reviews and adjustments.

What has been acknowledged as absolutely critical and innovative in Lexipaignio was not the game per se but the mechanism behind the game that would enable teachers/users to deepen in (i) various grammatical, morphological and vocabulary related phenomena taught at school (ii) by using ready-made, customized or even created from scratch simple and popular mini games, (iii) drawing exemplary sections directly from school text books and (iii) automatically connecting them to specific online tools such as online spellers, grammar checkers and various dictionaries developed and maintained by members of the project partnership.

In the case of geography, the development of the taxonomy of geographical – geological terms, as described above, is seen as the liaison between the mini games and the school text books. The 'game mechanism' under development foresees that the

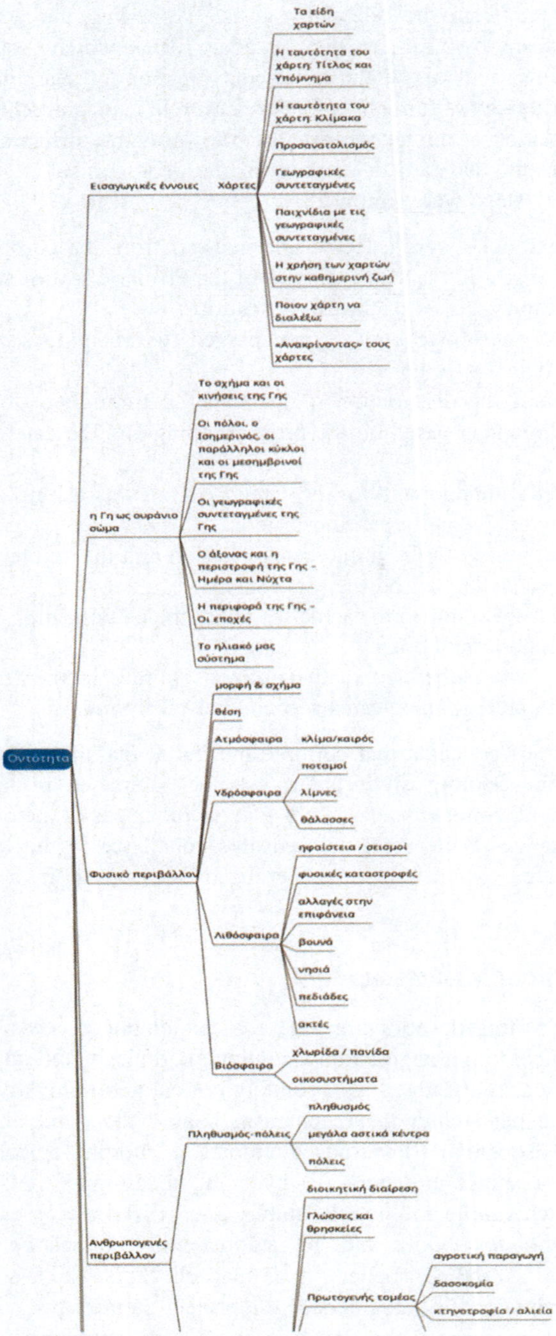

Fig. 2. Part of the upper levels of the taxonomy of Geographical – Geological Terms.

teacher/user will be able to connect sections of the taxonomy to specific, available mini games through a set of predefined, specific rules. Activities approached through the mini games may extend from the exploration of different meanings of specific vocabulary in geography to the comparison of specific geographical phenomena presented through different school text books addressing students of different age.

Game features such as 'levels', 'scoring', 'time limit' and 'help' aim to turn the mini games more appealing to end users/students. Online 'help', in specific, will constitute an additional way for the mini games to connect to the available, online tools (one simple example of an online help could be to check the meaning of a 'geographical term' on the online dictionary) with the primary connection being made through the set of rules that will be used in order for specific content from the database of the online tools to be 'fed' into a specific mini game (one simple example of such a rule could result in 'feeding' a mini game with all or specific phrases from all geography school text books that contain the word 'water').

3.3 The Role of the Teacher

Initial shift of the emphasis from one exhausting stand-alone, ready to play, educational game to a collection of multiple, simple, model mini games also changed the perspective of the role of the teacher, who can select to be in charge of game play by specifying the areas or the phenomena (grammatical, morphological, lexical) of focus.

Thus, it is estimated that teachers will be able to either select the suitable mini games for their students from an extended collection that will be available or create a new version of one of the available mini games by specifying the content and/or one (grammatical, morphological, lexical) phenomenon from the school books or the taxonomy, in the case of geography.

Furthermore, mini games will be equipped with extra features in order for teachers to be able to monitor their students' performance and most importantly to be able to provide feedback and assist the learning process whenever deemed necessary.

Finally, teachers will be able to load new material to a mini game (e.g. questions – answers) through ready-made templates whereas the possibility of defining additional grammatical, morphological or lexical phenomena of interest through tagging/annotations is still being explored.

3.4 Intentions of Using NLP Techniques and Pedagogical Objectives

In recent years, the presence of NLP technologies in digital games has been eminent. Similarly, NLP is being used in educational digital games. However, there is little research on NLP methodologies in this educational context. Specifically, Picca et al. [5] argue that this is related to the fact that NLP techniques are widely used to facilitate the game pedagogical objectives and are not treated as distinct game features. They stress on the importance of clarifying the objectives of using NLP and the pedagogical objectives of an educational digital game. Effectively, these two objective categories should be aligned in order to serve the overall game concept.

NLP technologies form a useful tool for the creation of language-based activities par excellence. In Lexipaignio project NLP technologies are linked to the dynamic creation of a variety of language mini games based on structured linguistic material. Thus, the Geography test bed serves as an example of how NLP technologies can assist in creating learning tools for the teaching of subjects other than Greek Language. The NLP techniques in the Geography test bed share the same objectives with the development of the Greek language games. At the same time, the pedagogical objectives in the Geography test bed focus on helping teachers combine educational resources on the basis of the geography taxonomy to dynamically create mini games according to their students' needs. Furthermore, the Geography test bed aims at improving students' knowledge on Geography and Geology through the study of terminology and the construction of geographical knowledge (information on the earth natural and human environment) according to the Geography syllabus for the Greek secondary school. The question types used in the geography test bed mini games are multiple choices, gap filling, true/false, put in the right order, sentence synthesis and riddles.

4 Conclusions

With the aim to deploy Natural Language Processing infrastructure for the creation of educational games in a variety of school subjects (Geography, Modern Greek Language, Biology), so far, the language processing techniques applied in "Lexipaignio" project provide encouraging results. With the infrastructure almost completed, the next step is to implement the mini games designed, in order for them to be tested and approved by the educators. Though this is not a straightforward task since it needs a lot of interdisciplinary co-operation as well as extensive "on site" testing, the research so far is more than promising as far as the successful implementation of the project is concerned. This will enable educators to easily create minigames according to their students' needs, by regulating the game content.

Acknowledgment. This research has been co-financed by the European Regional Development Fund of the European Union and Greek national funds through the Operational Program Competitiveness, Entrepreneurship and Innovation, under the call RESEARCH – CREATE – INNOVATE (project code: T1EDK-05094).

References

1. Alhawiti, K.M.: Natural language processing and its use in education. Int. J. Adv. Comput. Sci. Appl. **5**(12), 72–76 (2014)
2. Al-Rfou, R., Perozzi, B., Skiena, S.: Polyglot: distributed word representations for multilingual NLP. In: Proceedings of CoNLL (2013)
3. Gregory, S., Torsten, R., Wood, L., Henderson, M.: Gamification and digital games-based learning in the classroom. In: Teaching and Digital Technologies: Big Issues and Critical Questions. In: Henderson, M., Romeo, G. (eds.), Cambridge University Press (2015)
4. Kokkinos, T., Gakis, P., Iordanidou, A., Tsalidis, C.: Utilising grammar checking software within the framework of differentiated language teaching. In: Proceedings of the 2020 9th International Conference on Educational and Information Technology, Oxford, United Kingdom (2020)

5. Picca, D., Jaccard, D., Eberle, G.: Natural language processing in serious games: a state of the art. Int. J. Serious Games **2**(3), 77–97 (2015)
6. Ramos, J.: Using tf-idf to determine word relevance in document queries. In: Proceedings of the first instructional conference on machine learning, vol. 242, pp. 133–142 (2003)
7. Reinders, H. (ed.): Digital Games in Language Learning and Teaching. Palgrave Macmillan UK, London (2012)
8. Singla, A., Karambir, M.: Comparative analysis & evaluation of euclidean distance function and manhattan distance function using k-means algorithm. Int. J. **2**(7), 298–300 (2012)
9. Tsalidis, C., Vagelatos, A., Orphanos, G.: An electronic dictionary as a basis for NLP tools: the Greek case. In Proceedings of 11th Conference on Natural Language Processing, Fez, Morocco (2004)
10. Vagelatos, A., Mantzari, E., Pantazara, M., Tsalidis, C., Kalamara, C.: Developing tools and resources for the biomedical domain of the Greek language. Health Inf. J. **17**(2), 127–139 (2011)
11. Whissell, J.S., Clarke, C.L.: Improving document clustering using okapi BM25 feature weighting. Inf. Retrieval **14**(5), 466–487 (2011)

Work in Progress: Designing an Academical Online Course for Technical Students: Structure, Content, Assessment

Elena Makeeva[1](✉), Julia Lopukhova[2], and Ekaterina Gorlova[2]

[1] Samara State University of Social Sciences and Education, Samara, Russian Federation
helenmckey2205@gmail.com

[2] Samara State Technical University, Samara, Russian Federation
j.v.lopukhova@mail.ru, eagorloval3@rambler.ru

Abstract. Distant education and online courses have recently become a part of a new reality, especially in the last four months as many institutes and universities all over the world had to shift most of their educational content online because of the ongoing COVID-19 pandemic. It appeared that some higher educational institutions and a certain part of their students were almost ready to go online as these institutions had been developing their online platforms thus integrating traditional face-to-face classes with computer-assisted learning as well as integrating massive open or shared online courses into their curricular. At the same time, many other universities and institutes proved to be absolutely unprepared to shift online as they and their students had neither their own educational environment and online resources developed by their teaching staff nor any experience of working in an external online environment and choosing courses appropriate for their educational purposes. It turned out that most existing online courses could not be successfully used for implementing into most universities curricula and for teaching a majority of technical students as they simply reproduced the traditional classroom learning process and missed all advantages and opportunities afforded by online learning environments. The authors objective for this research in progress is to identify the key features and describe structure, content and assessment specifics of an academical online course for technical students. This paper provides tips on how to organize the course content into a proper structure stressing what types of teaching materials could be uploaded to the educational platform and goes into its assessment specifics. In a long-term perspective the authors aim to design, develop and introduce a modern and progressive constructivist learning environment for teaching business English in technical universities.

Keywords: Engineering education · Online course · Online course design · Teaching materials · Online tests and quizzes · Formative assessment · Summative assessment · Peer reviewing

M. E. Auer and T. Rüütmann (Eds.): ICL 2020, AISC 1328, pp. 682–689, 2021.
https://doi.org/10.1007/978-3-030-68198-2_63

1 Introduction

Recent events as well as official instructions and guidelines suggest that online education has come to stay and when most university classes come back to a traditional face-to-face format there will be a certain amount of educational university courses which stay online. For technical students it is mostly academical courses in humanities, such as Philosophy, History, Theoretical Economics, Fundamentals of State and Law, National Languages, Intercultural Communication, Foreign Languages & Business English, etc., covering a significant amount of content to be delivered online to a large number of learners to save money and work expenditure. At the same time, most existing online courses in social and humanitarian disciplines cannot be successfully used for implementing into most universities curricula and adjusted for teaching a majority of technical students as they simply reproduce the traditional classroom learning process and miss all advantages and opportunities afforded by online learning environments.

2 Project Description

2.1 Research Background

In spring 2020, when all universities in Russia had to switch to computer-assisted distance learning, it turned out that most of them were not ready to successfully go on with it. At first, they had to resort to e-mail correspondence and social services' chat-groups which turned out to be unproductive. Though limited either in time or in the number of participants, web-based video conferencing sessions via Zoom, MS Teams, etc. proved themselves better. Still, it became clear that this unexpected online educational process required careful planning and managing.

Universities were actively encouraged to search for ready-made open or shared online courses which content was similar to the subjects they taught and involve these courses into their studies [12]. Some teachers even managed to hastily elaborate and refine or even develop their own online courses in either their universities official learning management systems (such as Moodle, WebTutor, Blackboard, iSpring Learn, etc.) or while using such free resources as Edmodo, Sakai, etc.

The authors of this paper studied 34 online courses in social disciplines (among them 18 massive open online courses (MOOCS) on both internationally recognized platforms such as Coursera, EdX & etc., as well as 16 local online courses developed by the teachers of Samara State University of Social Sciences and Education (SSUSSE) and Samara State Technical University (SSTU) in LMS Moodle. The research was qualitative, data collection continued for over ten weeks and is still in progress. Two mechanisms for collecting data were employed: observations of structure, content and assessments tools employed in these courses & short interviews with course participants (both teachers and students) from SSUSSE and SSTU. Both mechanisms considered the categories that are often used to evaluate online courses in terms of course organization: Course Overview and Introduction; Learning Objectives (Competences); Assessment and Measurement; and so on [11, 19].

2.2 Purpose

The objective for this research in progress is to identify the key features and briefly describe structure, content and assessment specifics required of an academical online course for technical students as many courses (mostly presented on Russian educational resources) simply reproduce the traditional classroom learning process, lack most advantages and opportunities afforded by online learning environments and expect your students to consume knowledge rather than create it. In a long-term perspective the authors aim to consider the principles behind good online design, go deeper into the features of a successful course, outline the features of an unsuccessful course design and content, and then develop and introduce a constructivist learning environment starting with a course on teaching business English in technical universities.

2.3 Approach

When dealing with online education, it is necessary to rethink the methodologies, criteria, and approaches normally adopted for conventional educational design [8]. The authors resort to constructivism as a philosophical and pedagogical approach and consider its fourth (lower) technological methodology level to give a description of methods and techniques which can be fruitful for developing constructivist learning environments to enhance online courses. The research is based on the assumption that analysis and design stages as well as the course content, materials and assessment tools are essential to ensure course effectiveness and learners' motivation and participation; learning activities offered by the course should be both interesting, inspiring and well-matched to the level of the participants and their learning objectives.

3 Literature Review

From the first offering of a fully online course in 1981 [7] it was clear that this new model of education had much potential to impact the design and delivery of education at all levels [9]. Soon after online learning emerged as a learning approach, certain advantages became apparent – such as flexibility, alleviation of overcrowded classrooms, increased enrollment, reduced cost, and increased profit [4, 5]. In almost forty years that passed after this memorable event, online education has changed greatly. As long ago as in 2001, Taylor identified five generations of distance education development and assumed that the latter, fifth, generation, 'the intelligent flexible learning model', that had just begun to take shape would include Internet-based access to resources, easily accessible online interactive multimedia, and computer-mediated communication using automated response systems …" [17]. In a few years other researchers expressed hopes that online learning would soon transform education from instructor-centered to student-centered, where students had more responsibility for their learning [14]. Now we still hope that it would really happen one of these days as most online courses still aim at pouring content into student minds rather than supporting students in making that knowledge their own through practice and experience [16].

The reasons for that are as follows. Institutional learning environment is still mainly used as a repository for lecture materials and reproduces the traditional paradigms of teaching [2]. Thus, academics remain largely teacher-centered in their use of technology and are very reluctant to adopt more learner-centered ideas [15]. One of the great potentials of online learning that students can take more responsibility for their learning is obviously not harnessed in most cases [16].

At the same time, successful online course developers share a 'social constructivist' view of learning which means that the teacher should primarily act as a facilitator and guide rather than a deliverer of information, and that peers will influence an individual's approach to learning [18]. Constructive approach – an outcomes-based teaching and learning framework, proposed by Biggs and Tang [1], in which teaching activities and assessment tasks are systematically aligned to the intended learning outcomes – is based on the idea that teaching by telling doesn't really work, and that the most effective learning is achieved by students doing something active. Active learning, defined as any teaching method that engages students in the learning process, requires students to complete meaningful learning activities and to think about what they are doing [14]. Online education, in its turn, is a different medium for teaching and learning, and therefore requires a different approach and even a different pedagogy [8].

4 Main Body

In this part of the research the factors that contribute to constructing a high quality and meaningful course for learners are considered. It is also important to stress that while studying online courses during the preliminary stage of the research, the authors initially concentrated on the drawbacks and pitfalls evident in the courses' structure, content, adopted approaches, etc. These observations also helped to establish a framework for the researchers' schematic overview of an academical online course for technical students, its structure, content and assessment tools.

The importance of providing students with a structure for learning is one the most important steps towards quality teaching and learning. In this case, it is evident, that being aimed for technical students of Russian universities, an online course in social or humanitarian disciplines should be fitted within an autumn or spring term starting and ending at the same time with face-to-face courses [13]. As technical students are in the habit of going to class at the same time every week, these students need to form the same habits to maintain consistent performance across the term while having an online course. Making sure that assignments are always due on the same day of the week and modules always begin on the same day of the week also helps. Thus, the course should be organized over a specific time scale, with modules and lessons for each week of the program, and the number of modules aligned with the number of academical weeks.

Creating course content is one of the most challenging steps of creating an online course [10]. While investigating existing academical courses created on the Russian platforms the authors couldn't help but notice that these courses instructors are often tempted to pack too much content into each module pouring on students all materials they have. It might be long texts to and even whole books to summarize or 90-minute-long recorded video lectures as the other extreme. Most instructors ignore the assumption that an online course should include a variety of materials, both audio, visuals, video, texts, etc.

Briefly characterizing the types of teaching materials that could be uploaded to the educational platform, it is vital to stress that video lectures are a widely-used kind of resource for online learning. But simply recording a lecture in front of a live classroom and using that video in an online course has many drawbacks and is not as effective when put online. Live classroom lectures are much longer than the short videos designed for online learning experiences; and, when delivering a live classroom lecture, the primary audience are the learners in the live classroom, which can make online learners feel like mere bystanders. Effective video lectures need to engage directly with the online learner. Besides, online lecture videos need to be short (their length varying from 6 to 12 min), engaging, and tailored specifically to the online audience [6].

The next type of educational content is a variety of text materials, often referred to as 'text-only documents' [3]. It can be additional reading materials, video subtitles, notes, manuals, etc. This is the type of content that is easiest to edit and update.

The third type of content is a variety of interactive elements: slide shows, simulators and simulation facilities, education games of different types. For example, on the Coursera platform, you can integrate your own external plugins both inside the lesson and inside the test connecting your model with Quizlet or Kahoot and the like.

Assessment tools and test formats in online courses and special aspects of their use stand aside. In online courses, assessment is used for two purposes: formative and summative assessment. Formative assessment allows students to test understanding of the material before passing to summative assessment, but summative assessment allows authors of a course to verify understanding of the material by the listener.

We especially stress such principles for effective online assessment as designing learner-centred assessments that include self-reflection; rubrics for the assessment of contributions to discussions, as well as collaborative assessments through public posting of papers; encouraging students to develop skills in providing feedback by modelling what is expected; designing assessments that are clear, easy to understand, and likely to work in the online environment.

As for the types of assessment tools, an academic course for technical students should include:

1. Tests integrated inside video lectures or the so-called 'In-Video Quizzes'. Tests integrated inside video lectures are usually simple non-timed 'true/false' questions, multiple choice and matching questions with no grades that the listener should undergo right while watching video fragments. In-Video Quizzes are usually checked automatically. This test format is only used for formative assessment, meaning that the results of such tasks do not contribute to summative assessment.

Questions integrated inside video lectures should not include difficult tasks or complicated tasks. They can even be even humorous in cases where the joke fits the style of presentation of the material. In-Video Quizzes serve as a way to interact with the listener. They compensate for the absence of face-to-face contact in the process of online education. The recommended number of questions in these quizzes is 1–2 questions per 5–10 min of video, that is, it is about 1–2 questions per average online course video.

2. The second test type is standard tests. Tests in an online course consist, firstly, of multiple-choice questions where students choose one or several correct answers. Online course designers can also include such question types as calculated simple or calculated multi-choice questions, drag and drop into text/onto image tasks, short answer questions when the student types in a word or phrase in response to a question, flexible questions consisting of a passage of text that has various answers embedded within it, etc. Answers to these questions are also checked automatically. This test format can be used for both formative and summative assessment.

As a rule, a typical quiz for formative assessment includes 5–10 questions for 30 min of video. In turn, a typical quiz for summative assessment is about 10–20 questions per hour of video.

The number of attempts also differs. The number of attempts for formative assessment is not limited. At the same time, the number of attempts for summative assessment is usually limited to 3 staggered attempts.

3. The next form of checking-up is peer-reviewing. Peer-reviewing assumes that at the first stage students independently construct their responses according to a certain pattern, then upload their responses and finally check and grade each other's works. Such tasks are obviously not automatic in terms of verification. Using peer reviews in online courses enhances student learning and allows students the opportunity to read a variety of student writing, learn from each experience, learn by providing feedback, and develop their writing skills in the process. Peer-reviewing is one of the keyways to test the achievement of complex learning outcomes, such as critical thinking or academic writing skills. Reviewees here benefit from the reviewers' comments, which improves their understanding of the subject and vice versa. This task format is suitable in situations where listeners have to construct a complex response or complete a creative task, and their response could not be checked automatically.

This, of course, is only a brief outline of an academical online course for technical students design features.

5 Recommendations

One of the most significant requirements for further adoption of online learning is the development of well-designed courses with interactive and engaging content, structured collaboration between peers, flexible deadlines to allow students to pace their learning, continuous monitoring of student progress, and the provision of formative feedback when it is needed.

It is important to keep in mind that online learning can be as effective as face-to-face learning, but it is not the same and requires different planning, one should resort to constructive alignment to design and develop an online course. Decisions need to be made about the content, structure, timing, pedagogical strategies, sequence of learning activities, and type and frequency of assessment in the course, as well as the nature of technology used to support learning. Technology, when it is used appropriately, can play a very important role in this process, particularly where students locate material, undertake formative assessment and, critically, where effective collaborative working and critical discussion with peers is to occur.

Course content should be presented in organized, manageable segments for students to access and learn content in an organized and manageable format appropriate for the students' prime goal which is the consumption of knowledge generated by the teacher. At the same time, the concept of active learning should not be overlooked as in an online environment students are supposed to create knowledge rather than consume it.

The consideration of appropriate assessment is central to course planning. most importantly, developers need to remember the importance of constructive alignment, and ensuring that they plan their assessment to support the intended learning objectives or intended learning outcomes.

6 Conclusion

The analysis of existing courses suggested by such international platforms as Coursera and EdX and National Open Education Platform created specially for Russian-speaking learners has produced a set of recommendations. Some of these tips are as follows. Online courses should be divided into logical units, each of which is further divided into a series of modules. All materials must be carefully designed and must embed adequate instructional support to allow learners to function throughout the course. Types of teaching materials that could be uploaded to the educational platform include 6–12 min long video lectures; a variety of text materials (additional reading materials, video subtitles, notes, manuals, etc.); a variety of interactive elements: slide shows, simulators and simulation facilities, games of different types. Designers can integrate their own external plugins both inside the lesson and inside the test or attach a survey to their texts that was made using external services such as Survey Monkey, SurveyGizmo or Microsoft Forms.

As for assessment, in online courses, assessment is used for two purposes: formative and summative assessment. Formative assessment allows students to test understanding of the material before passing to summative assessment, but summative assessment allows authors of a course to verify understanding of the material by the listener. Formative assessment does not contribute to summative assessment, while summative assessment results is the student's overall score for the course and so on…

As the authors are still working on giving a more detailed description of a highly efficient online course on humanitarian subjects for technical students, during the next term they plan to go deeper into their analysis of existing courses, point out their main drawbacks and thus make more steps forward to their ambitious objective.

References

1. Biggs, J., Tang, C.: Teaching for Quality Learning at University: What the Student Does. Society for Research into Higher Education & Open University Press, Maidenhead (2007)
2. Browne, T., Jenkins, M.: Achieving academic engagement? The landscape for educational technology support in two UK institutions. In: Proceedings ascilite Melbourne 2008 (2008). http://www.ascilite.org.au/conferences/melbourne08/procs/browne.pdf. Accessed 22 May 2020

3. Chao, L.: Utilizing Open Source Tools for Online Teaching and Learning: Applying Linux Technologies. Information Science Reference, New York (2009)
4. Clardy, A.: Distant, on-line education: effects, principles and practices. Distance, On-line Education, 1 (2009). https://files.eric.ed.gov/fulltext/ED506182.pdf. Accessed 20 May 2020
5. Grandzol, C.J., Grandzol, J.R.: Best Practices for Online Business education. The International Review of Research in Open and Distance Learning, 7(1), (2006). http://www.irrodl.org/index.php/irrodl/article/view/246. Accessed 22 May 2020
6. Guo, P., Kim, J., Rubin, R.: How video production affects student engagement: an empirical study of MOOC videos. In: Proceedings of the first ACM conference on Learning @ scale conference. https://doi.org/10.1145/2556325.2566239 (2014)
7. Harasim, L.: Shift happens: online education as a new paradigm in learning. Int. Higher Educ. **3**(1–2), 4–61 (2000). https://doi.org/10.1016/S1096-7516(00)00032-4
8. Huang, H.: Toward constructivism for adult learners in online learning environments. British J. Educ. Technol. **33**(1), 27–37 (2002)
9. Joksimovic, S, Gasevic, D., Kovanovic, V.: The history and state of online learning. In: Preparing for the digital university: a review of the history and current state of distance, blended, and online learning. Chapter 4. Athabasca University, pp. 93–131 (2015)
10. Lynch, M.M.: The Online Educator: A Guide to Creating the Virtual Classroom. Routledge Falmer, London (2002)
11. Martín-García, A.V.: Blended Learning: Convergence between Technology and Pedagogy. Springer, Cham (2020)
12. Order# 397 issued by the Ministry of Science and Higher Education of the Russian Federation (March 14, 2020). https://www.minobrnauki.gov.ru/ru/documents/card/?id_4=1064&cat=/ru/documents/docs/ Accessed 18 May 2020
13. Osipovskaya, E., Miakotnikova, S.: Using gamification in teaching public relations students. In: Auer, M., Tsiatsos, T. (eds.) The Challenges of the Digital Transformation in Education. ICL 2018. Advances in Intelligent Systems and Computing, vol. 916, pp. 685–696. Springer, Cham (2020). https://doi.org/10.1007/978-3-030-11932-4_64
14. Prince, M.: Does Active learning work? a review of the research. J. Eng. Educ. **93**, 223–231 (2004)
15. Sharpe, R., Benfield, G., Francis, R.: Implementing a university e-learning strategy: Levers for change within academic schools. Res. Learn. Technol. **14**(2), 135–151 (2006). https://doi.org/10.3402/rlt.v14i2.10952
16. Germain, E, S.: 'Five Common Pitfalls of Online Course Design', in Faculty Focus [website] (2011). www.facultyfocus.com/articles/online-education/five-common-pitfalls-of-online-course-design. Accessed 15 May 2020
17. Taylor, J.C.: Fifth generation distance education. Presented at the keynote address delivered at the iCDe 20th World Conference, Dusseldorf, Germany (2001). http://www.usq.edu.au/users/taylorj/conferences.htm. Accessed 22 May 2020
18. Trentin, G.: Designing online courses. In: Maddux, C.D., Johnson, D.L. (eds.) The Web in Higher Education: Assessing the Impact and Fulfilling the Potential, The Haworth Press Inc., New York, London, Oxford, pp. 47–66. (2001)
19. Tripathi, P., Mukerji, S. (eds.): Handbook of Research on Technology-Centric Strategies for Higher Education (Advances in Educational Marketing, Administration, and Leadership). IGI Global, Hershey (2017)

The OpenLang Network Platform: Building a European Community of Language Learners and Teachers

Alexander Mikroyannidis[1(✉)], Maria Perifanou[2], Anastasios Economides[2], and Antonio Giordano[3]

[1] Knowledge Media Institute, The Open University, Milton Keynes MK7 6AA, UK
alexander.mikroyannidis@open.ac.uk
[2] University of Macedonia, Thessaloniki, Greece
[3] Pixel, Florence, Italy

Abstract. This paper presents the OpenLang Network platform, an open and collaborative online environment for networking between language learners and teachers across Europe. The OpenLang Network platform brings together educators wanting to discover and share open language learning resources, as well as Erasmus+ mobility participants that wish to improve their language skills and cultural knowledge. This initiative aims at raising language awareness of mobility European languages and fostering the Open Education European multicultural and multilingual vision via Open Educational Resources (OERs) and Massive Open Online Courses (MOOCs).

Keywords: Language learning · Mobility · Networking · Open Educational Resources · Massive Open Online Courses

1 Introduction

The free movement of people among European Union (EU) member states has been recognised as one of the cornerstones of the European society [1]. With 4 million participants by 2020, the Erasmus+ programme offers unique opportunities to study, train, gain work experience or volunteer across the EU [2]. However, the lack of language competences is still one of the main barriers to participation in European education, training and youth programmes. As languages are the heart of mutual understanding and comprehension, it is essential to promote and facilitate language learning for Erasmus+ mobility participants.

The OpenLang Network is an Erasmus+ project,[1] addressing the needs for linguistic skills and culture awareness of Erasmus+ mobility participants and the training needs of language teachers. In particular, the project is targeting the following 2 groups of stakeholders:

[1] https://www.openlangnet.eu/.

M. E. Auer and T. Rüütmann (Eds.): ICL 2020, AISC 1328, pp. 690–700, 2021.
https://doi.org/10.1007/978-3-030-68198-2_64

1. Erasmus+ KA1 mobility participants (Higher Education students & staff, Vocational education and training, Adult & School education staff, Youth learners, Youth Workers, Youth Entrepreneurs) that need to boost their language skills and cultural awareness.
2. Volunteer language teachers who will support the Erasmus+ KA1 mobility participants offering their professional experience while receiving professional certified training on the creation, sharing and use of language Open Educational Resources (OERs).

The OpenLang Network project envisages to:

- Connect the above 2 groups of stakeholders in an interactive collaborative environment (web-based and mobile-based) that will support more efficiently their effort to raise language awareness of the target mobility European languages and to develop European intercultural knowledge covering all European cultures.
- Foster the Open Education European multicultural and multilingual vision to all OpenLang Network members.

The OpenLang Network project is promoting the wider use and adoption of OERs for addressing the linguistic needs of Erasmus+ mobility participants. OERs can be described as "teaching, learning and research resources that reside in the public domain or have been released under an intellectual property license that permits their free use or repurposing by others depending on which Creative Commons license is used" [3]. The emergence of OERs has greatly facilitated online education through the use and sharing of open and reusable learning resources on the Web. Learners and educators have been allowed to access, download, remix, and republish a wide variety of quality learning materials for use within different levels of education [4–11].

The OpenLang Network platform[2] is a core output of the OpenLang Network project, as it offers the technological foundations for the success and sustainability of the project's language learning community. To this end, the OpenLang Network platform is providing a variety of services to language learners and teachers, including:

- A personal member's dashboard.
- An e-Tandem language learning matching service for 24 European languages.
- An open & highly interactive Forum.
- A European database of Language Open Educational Resources (OERs) of high quality.
- Training MOOC materials on OERs for language teachers.
- A responsive design accessible through different devices (tablets, smartphones, etc.).

The remainder of this paper is organised as follows. First, we introduce the requirements that have driven the design of the OpenLang Network platform. We then proceed with presenting the implementation of the platform and the services it offers to the community of language learners and teachers. Finally, we conclude the paper and outline the next steps of this work.

[2] https://openlang-network.kmi.open.ac.uk/.

2 Platform Design

In this section, we outline the requirements that have driven the design of the OpenLang Network platform. These requirements have been largely based on the language needs of Erasmus+ KA1 mobility participants, which have been captured via an online questionnaire and a series of interviews [12].

2.1 Openness

As the OpenLang Network project promotes an open approach for language learning, teaching and community support, it is essential that the project's platform, which is at the very core of this approach, is also open to the community of language teachers and learners, as well as to the general public. We have therefore opened the OpenLang Network platform to the public, making it available to anyone who wishes to join the community of language learners and teachers.

2.2 Customisation

As the OpenLang Network platform offers customised services to the community, it is important that the platform is dynamic and flexible and offers different ways for customisation of both its front-end and back-end. One popular way for customisation is through plugins, which are software components that extend the codebase of a platform. By developing custom plugins, we aim to address the specific needs of our users and offer them the services outlined in the Introduction section of this paper.

2.3 Interoperability

Another important requirement for the design the OpenLang Network platform is interoperability, i.e. the ability to exchange data with other systems. This can be achieved by supporting standard specifications for importing and exporting educational data, such as SCORM[3] and xAPI.[4] These are standard specifications widely used by educational platforms and facilitate the exchange of data between these platforms. For example, learning materials can be exported as a SCORM packages from one platform and be imported into another. Learning Analytics data can be exported as xAPI statements from one system and imported into another for further analysis. Interoperability is essential for the OpenLang Network platform, as we want to make it as easy as possible to import learning materials, as well as export Learning Analytics data from our system.

[3] https://scorm.com/scorm-explained/one-minute-scorm-overview/.

[4] https://xapi.com/overview/.

2.4 Mobile Interface

As we our primary target group is mobility participants across Europe, it is safe to assume that they will not always have access to a desktop or laptop computer device and will need to access the OpenLang Network platform from different European countries and while travelling. It is therefore important that the OpenLang Network platform has a responsive web design so that it can be accessed by the widest possible range of devices, including tablets and smartphones that have different screen sizes. Responsive web design provides a website with the flexibility to adapt to any of these devices, by dynamically adapting its interface to the resolution of a given device [13].

2.5 Community Support

As the project has limited resources, it is important that we design a platform supported by the community, in order to receive regular updates of the codebase and bug fixes. This usually goes hand in hand with openness, i.e. open-source platforms have the advantage of receiving support from the community of developers that regularly update the codebase, thus fixing bugs and security holes, plus delivering new features.

2.6 Monitoring

An important aspect of online teaching and learning is the monitoring of student progress and tools utilisation in online courses. Educational research shows that monitoring the students' learning, for example via Learning Analytics, is an essential component of high-quality education [14]. Feedback about the status and the history of the activities in online courses can be useful to teachers, students, study program managers and administrators. For example, it can help in better understanding whether the courses provide a sound learning environment (availability and use of discussion forums, etc.) or show to which extent best practices in online learning are implemented (students provide timely responses, teachers are visible and active, etc.).

2.7 Storytelling

The most effective way to convey information in eLearning is to get the audience to care about the content, connect with the content, and to commemorate the content. This can be done through storytelling, the most fundamental form of communication. Storytelling is how we tap into our common humanity, evoke emotional connection, and facility recall to anything we are learning [15]. When we are told something in the form of story, we remember it better than if we are simply exposed educational snippets, lists, or disconnected piles of information. The first step in creating learning solutions through storytelling is to get learners to care about the content: content creators can do this by simulating the learner's reality by adding characters and/or other real-life and work-life situations and dialogues. The second step is to get learners to connect with the content, and to goal is to elicit emotion: creating an emotional connection with learners helps them to believe in the value of the content and can drive them to want to learn even more. The third step is to get learners to commemorate the content — to get the learners to talk about it, repeat it, and think about it and how it affected them.

2.8 Interactivity

Although eLearning platforms are able to overcome the geographical constraints that arise in the conventional class, they also create uncertainty amongst the learners about the quality and frequency of the interaction they enjoy. In distance learning, feedback is categorized into feedback among the learner towards the content, the trainer/educator towards the learner, and finally the learners. Regular interaction among the teachers, the students, and the content is a vital factor in achieving learning objectives [16]. On one hand, educators play a key role in engaging students with other participants in a modern and asynchronous way, encouraging them to respond to their questions by creating a learning community that will provide learners with equal opportunities for communication such as live lessons and interaction in the virtual environment. On the other, thanks to interactivity, the student will overcome the fear of insecurity due to the distance barrier.

2.9 Gamification

Gamification has had a significant impact on education and learning [17]. As a teaching tool, it has proven both its effectiveness and popularity time and again. As such, e-learning programs with gamification functionality are in high demand among learners and educators alike. Gamification has many benefits for (e-) learners. Amongst them, it: boosts engagement, provides motivation through goal tracking, improves knowledge retention, promotes team building and offers valuable feedback. Of course, it does all this for a reason: it is fun and therefore engages learners in a more effective way than other more conventional pedagogical approaches.

3 Platform Implementation

In order to best address the design requirements of the OpenLang Network platform, we have implemented our platform on top of a Learning Management System. A Learning Management System (LMS) is a software system for the administration, documentation, tracking, reporting and delivery of educational courses, training programs, or learning and development programs. An LMS essentially allows learners and educators to manage the learning process. The core features of an LMS are in line with the core features of the OpenLang Network platform, especially regarding the provision of learning materials in the form of OERs or MOOCs, the availability of a personal dashboard, as well as features for community building and interactions among the community members.

Several LMS options are currently available, both proprietary and free ones, such as Moodle,[5] Sakai,[6] Canvas,[7] Blackboard,[8] and more. We have selected Moodle, an open-source LMS, used by millions of learners and teachers worldwide. Moodle offers a powerful and modular environment for online and blended learning [18]. Additionally, Moodle fulfils the design requirements of the OpenLang Network platform as follows:

- **Openness:** Moodle is completely free and open source.
- **Customisation:** Moodle supports custom plugins to extend its codebase.
- **Interoperability:** Moodle supports the most widely used standard specifications for educational data, including SCORM and xAPI.
- **Mobile interface:** Moodle offers a dedicated app for mobile devices and supports the development of responsive interfaces.
- **Community support:** Moodle has a vast community supporting it, currently consisting of more than 900 developers.[9]
- **Monitoring:** Moodle allows students to monitor their progress when studying a course. Moodle also offers detailed reports to teachers about the progress of their students.
- **Storytelling:** Moodle supports storytelling via its tools for course authoring and via third-party plugins.
- **Interactivity:** Moodle offers various tools for facilitating interactivity, such as discussion forums, private messaging, live chats, and more.
- **Gamification:** Moodle supports gamification via tools such as badges and also via third-party plugins.

Figure 1 shows a snapshot of the front page of the OpenLang Network platform. First of all, the front page of the platform features two pathways, which provide guidance to language learners and teachers via recommended sequences of activities to be performed on the platform. In particular, the pathway for language learners starts with determining the language proficiency level of the learner via language placement tests and subsequently finding a tandem learning partner based on their language proficiency level. The pathway for language teachers starts with teachers registering their language proficiency level, followed by finding and sharing language learning resources.

As it can be seen in Fig. 1, the OpenLang Network platform offers the following services and activities to the community of language learners and teachers:

- **OpenLang MOOC:** The OpenLang MOOC, which will be launched in autumn of 2020, will aim at the creation, the sharing and the use of multilingual and interactive language OERs.
- **Language placement tests:** This activity offers 24 language placement tests, covering all European languages. Members of the platform can take these tests in

[5] https://moodle.org/.
[6] https://www.sakailms.org/.
[7] https://www.canvasvle.co.uk/.
[8] https://www.blackboard.com.
[9] https://moodle.org/plugins/.

Fig. 1. The front page of the OpenLang Network platform.

order to assess their level of expertise in a particular language. Each test is available via a link and a PDF file. Members may visit the link to take the test online or download the PDF file to take it offline.

- **Tandem language learning:** This activity offers a matching service between members of the OpenLang platform. Through this service, a member of the platform is able to discover other members, with which they can communicate synchronously and asynchronously, in order to be taught a language or learn a language together. Figure 2 shows the search interface for finding tandem learning partners. Through this interface, members of the platform can find other members that have a

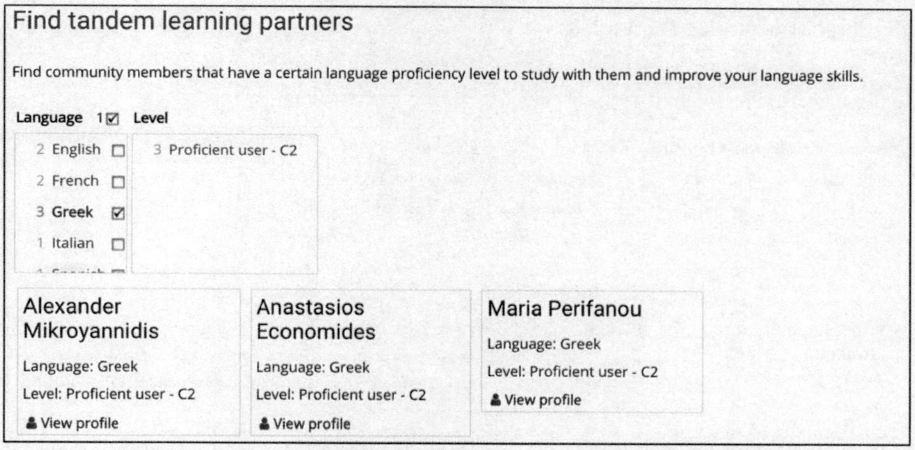

Fig. 2. Finding tandem language learning partners on the OpenLang Network platform.

certain language proficiency level and view their profiles in order to connect with them and start studying a language together.

- **Recommended language resources:** This activity offers language learning resources recommended by the OpenLang Network project. These are high-quality OERs or MOOCs that are offered by renowned educational institutions for learning a European language. Members of the platform can search these resources using a variety of filters in order to customise their search results, for example by specifying the language of the resource, its target language, its level, as well as the institution offering the resource, as shown in Fig. 3.
- **Shared language resources:** In this activity, members of the platform can share language learning resources with the community and find resources shared by other members of the community. As in the previous activity, the search for shared language learning resources can be filtered based on a set of metadata.
- **Community discussions:** This activity offers a forum dedicated for discussion between all members of the platform. Through this forum, members can ask questions, search for answers, exchange experiences, seek advice, etc. Members of the platform can also subscribe to the whole forum or certain discussion threads that interest them, so that they receive notifications when a new message is posted in the forum.

Additionally, the OpenLang Network platform offers a personal dashboard for its members. This dashboard provides an overview of a member's personal data stored in the platform, including the courses that a member has registered for, the messages they have sent and received, their network of contacts in the platform, their private files, and more.

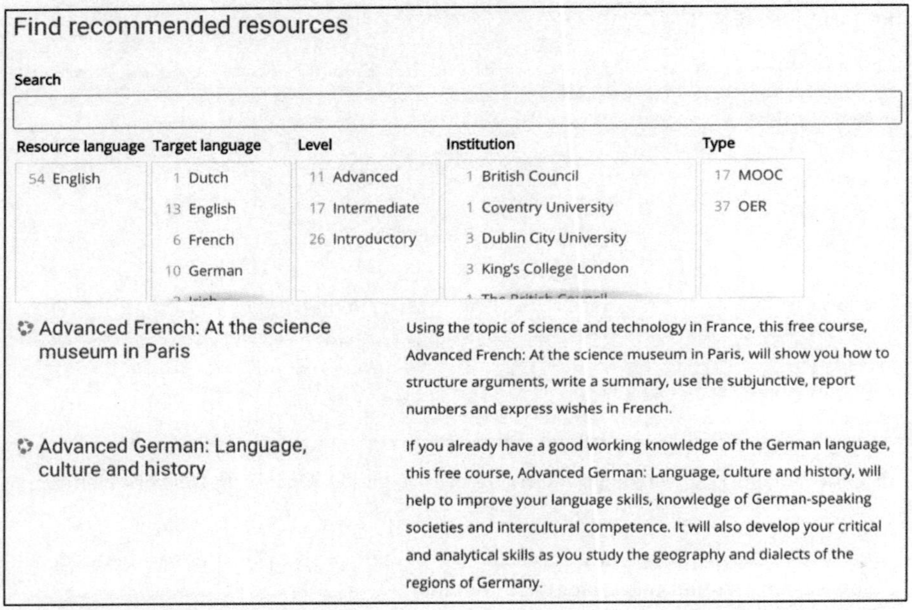

Fig. 3. Searching for language OERs and MOOCs on the OpenLang Network platform.

4 Conclusions and Next Steps

The OpenLang Network platform aims at building a community for educators that want to discover and share open language learning resources, as well as for Erasmus+ mobility participants that want to improve their language skills and cultural knowledge. The design of the platform has been driven by a number of requirements, including openness, customisation, interoperability, monitoring, storytelling and interactivity. The platform is an open and collaborative online environment, offering a variety of services to the community of language learners and teachers, such as language placement tests, tandem language learning matching, a wide range of recommended language OERs, as well as the ability for community members to share their language OERs.

The next steps of this work will be focused on engaging the community of language learners and teachers via a dedicated MOOC, which will be delivered via the OpenLang Network platform. The MOOC will offer opportunities for learners and teachers to discuss their needs and exchange experiences and best practices regarding the creation, sharing and use of multilingual and interactive OERs. MOOC participants will be invited to use the various services offered by the platform and provide their feedback, which will be used to further improve the platform and align it with the needs and preferences of the community of language learners and teachers.

Acknowledgement. This work has received funding from the European Union's Erasmus + programme under grant agreement 2018-1-EL01-KA203-047967 (OpenLang Network).

References

1. Ritzen, J., Haas, J., Kahanec, M.: EU Mobility. IZA Institute of Labor Economics (2017)
2. European Commission: Erasmus+ Annual Report (2019)
3. Atkins, D.E., Brown, J.S., Hammond, A.L.: A Review of the Open Educational Resources (OER) Movement: Achievements, Challenges, and New Opportunities. The William and Flora Hewlett Foundation (2007)
4. Lane, A., Connolly, T., Ferreira, G., McAndrew, P., Wilson, T.: Reusing, reworking and remixing open educational resources. In: Marshall, S., Kinuthia, W. (eds.) Cases 'n' Places: Global Cases in Educational and Performance Technology. IAP - Information Age Publishing, Charlotte (2010)
5. Wilson, T., Schuwer, R., McAndrew, P.: Collating global evidence of the design, use, reuse and redesign of open educational content. In: Open Educational Resources Conference (OER10) (2010)
6. de los Arcos, B., Farrow, R., Pitt, R., Weller, M., McAndrew, P.: Personalising learning through adaptation: evidence from a global survey of K-12 teachers' perceptions of their use of open educational resources. J. Online Learn. Res. **2**, 23–40 (2016)
7. Mishra, S.: Open educational resources: removing barriers from within. Distance Educ. **38**, 369–380 (2017)
8. Baas, M., Admiraal, W., van den Berg, E.: Teachers' Adoption of Open Educational Resources in Higher Education. J. Interact. Media in Educ. **2019**, 1–11 (2019)
9. Hilton, J.: Open educational resources, student efficacy, and user perceptions: a synthesis of research published between 2015 and 2018. Educational Technology Research and Development, pp. 1–24 (2019)
10. Mikroyannidis, A.: Collaborative Authoring of Open Courseware with SlideWiki: A Case Study in Open Education. In: 10th Annual International Conference on Education and New Learning Technologies (EDULEARN18). International Academy of Technology, Education and Development (IATED) (2018)
11. Mikroyannidis, A., Pallonetto, F., Mangina, E., Pyrini, N., Sadauskas, M., Trepule, E., Volungevičienė, A., Panagiotakopoulos, C., Karatrantou, A., Armakolas, S., Mauro, A.N., Cheniti, L., Korbaa, O.: Lessons learned from the use of the SlideWiki OpenCourseWare platform in different learning contexts. In: 11th annual International Conference of Education, Research and Innovation (ICERI 2018). International Academy of Technology, Education and Development (IATED) (2018)
12. Kosmas, P., Parmaxi, A., Perifanou, M., Economides, A.A., Zaphiris, P.: Creating the Profile of Participants in Mobility Activities in the Context of Erasmus+ : Motivations, Perceptions, and Linguistic Needs. In: International Conference on Human-Computer Interaction, pp. 499–511. Springer (2020)
13. Mohorovičić, S.: Implementing responsive web design for enhanced web presence. In: 2013 36th International Convention on Information and Communication Technology, Electronics and Microelectronics (MIPRO), pp. 1206–1210. IEEE (2013)
14. Ferguson, R.: Learning analytics: a firm basis for the future. In: Sheehy, K., Holliman, A. (eds.) Education and New Technologies: Perils and Promises for Learners, pp. 162–176. Routledge, Abingdon (2017)
15. Schmoelz, A.: Enabling co-creativity through digital storytelling in education. Thinking Skills Creativity **28**, 1–13 (2018)
16. Dailey-Hebert, A.: Maximizing interactivity in online learning: Moving beyond discussion boards. J. Educ. Online **15**, 1–26 (2018)

17. Majuri, J., Koivisto, J., Hamari, J.: Gamification of education and learning: A review of empirical literature. In: 2nd international GamiFIN conference (GamiFIN 2018). CEUR-WS (2018)
18. Sulisworo, D., Agustin, S.P., Sudarmiyati, E.: Cooperative-blended learning using Moodle as an open source learning platform. Int. J. Technol. Enhanced Learn. **8**, 187–198 (2016)

Teaching Best Practices

Challenges of Teaching Programming in StackOverflow Era

Jaanus Pöial(✉)

School of Information Technologies, Tallinn University of Technology, Raja 4C, 12616 Tallinn, Estonia
jaanus.poial@taltech.ee

Abstract. Teaching computer programming has changed a lot during the last 15 years. Instead of visiting lectures, reading books, documentation and manuals, solving individual programming tasks, etc., students prefer to "find" solutions to all problems in Google, StackOverflow and similar forums, YouTube videos and other helpful tools. It could be enjoyed, if there were no negative side effects: programs are created without proper understanding using "copy and paste" from different sources and brute force hacks to comply them to the tests. Often these internet resources are incomplete, contain errors and do not fit to solve the given problem.

This article covers some techniques used by the author in the following areas:

1. Choice of assignments - full scale projects vs. small (but meaningful and complex enough) exercises.
2. Using version control to track students' progress.
3. Using learning environments, e.g. Moodle.
4. Role of automated tests.
5. Cheating issues.
6. Providing enough feedback to massive amount of homework.
7. Using "flipped classroom" and "work in pairs" approaches.

Daily teaching in a rapidly changing environment necessitates constant updating of teaching methods using peer experience, continuous student feedback, professional publications and common sense. Author has tried "flipped classroom" and "work in pairs" approaches combined with automated assessment and version control for a medium level programming course ("Algorithms and Data Structures"). The results were impressive - dropout decreased from 54% to 17% in two years.

Keywords: Teaching best practices · Collaborative learning · Flipped classrooms

1 General Teaching Practices for Courses of Programming

1.1 Potential Weaknesses of Traditional Methods of Teaching

Teaching computer programming is an easy and inspiring task in case of a small class of dedicated, focused, active and co-operative students – instructor can give immediate

M. E. Auer and T. Rüütmann (Eds.): ICL 2020, AISC 1328, pp. 703–710, 2021.
https://doi.org/10.1007/978-3-030-68198-2_65

feedback to everybody, monitor students' progress, advise improvements to solutions, make flexible changes to the teaching process according to the audience, give unique individual tasks depending on interests of a learner, act as a mentor. Unfortunately, this ideal situation is impossible and in practice the drop rates from the programming courses are high (up to or even greater than 50%) for different reasons:

- number of students per instructor is too big for individual supervision,
- students are not interested in the subject, just want to pass it,
- methods and tools used to teach the subject are not fully suitable for the particular audience (or for the particular subject),
- the choice of assignments and the way in which they are checked entices the use of outside assistance or other issues of cheating,
- amount of homework and other tasks does not meet students' expectations (or official working hours),
- group atmosphere in a large heterogeneous class does not support active learning.

In this paper we discuss some possibilities to improve the situation using ideas and methods of extreme programming [3, 7] tools of automated assessment, version control and e-learning.

1.2 Choice of Technical Tools

Suppose the level of a programming course ("for dummies", beginners. intermediate, advanced, professional) and the language/framework to be used (Python, Java, C, ...) are predetermined. Even then there are a lot of technical decisions that a lecturer must make in order to teach the subject successfully.

Command Line vs. Integrated Development Environment. Most of the programming languages have integrated development environments (IDEs) that support editing, formatting and following the code style, documentation, debugging, testing, version control, etc. all in one. Some IDEs maintain multiple programming languages (e.g. Eclipse, https://www.eclipse.org/). IDE serves as a powerful aid to make programming more efficient if it is used professionally. On the other hand, tight integration of so many tools and aspects turns IDEs hard to exploit for beginners, who tend to use only a small fraction of the functionality provided. Even worse, if hints and suggestions given by IDE are not understood, but still applied, the code may easily get out of hand.

There are several choices possible:

- Not to use IDE at all. Every tool is used separately on command line. There are code editors available for text input, stand-alone debuggers, etc. This approach (surprisingly) works for beginners, who study to become software developers, because they need to understand the command line interface anyway and each aspect of development is explicitly exposed. For professionals this stage is inevitable and afterwards they can move on to make use of IDE features. In case of "dummies" command line is not generally a good choice.
- Use one specific simple IDE. It is applicable for beginners, who typically take one course of programming (e.g. Thonny for Python, https://thonny.org/). Advanced

features for big projects are usually not needed, code formatting and debugging tools are introduced.
- Use one specific professional IDE. This approach is reasonable for intermediate classes (e.g. IntelliJ IDEA for Java, https://www.jetbrains.com/idea/), where instructor demonstrates the basic techniques during lab sessions. All questions and answers are limited to this specific IDE. Teacher can also explain more advanced features (refactoring, version management, etc.). Careful monitoring is needed to avoid anti-patterns of IDE usage (e.g. IDE is used as an editor only, program is compiled and run from command line – strange practice that author has seen several times, probably caused by misconfiguration of software).
- Let student use any IDE of her choice. All advanced and professional courses expect that students are familiar with development tools and are free to choose any. On the other hand, in case of difficulties the student is on her own with this choice. Teacher must monitor the situation to be sure that IDE is used efficiently.

For teaching purposes there is no need to purchase professional full versions of IDEs (beginners do not need all the "bells and whistles"). Community editions work fine and are freely available to all students. The other aspect of choice of tools for the massive courses concerns administrator privileges – whether it is possible to install the chosen software as a non-privileged user (e.g. for Python non-privileged user can install Thonny, but not PyCharm).

Unit Testing. Testing is an integral part of programming that must be taught even (and especially) on beginner courses. Student has to understand basics of unit testing ([5]). On the other hand, unit tests may serve as a pre-filter for submission of student assignments. Unit tests for teaching purposes have some specific features – they should reflect most typical student errors, be resistant to malicious attacks and give maximum informative feedback targeted to learning.

Sometimes students misuse the concept of test-driven development [4] and twist their solutions found from suspicious sources to comply to the tests formally. It is better to avoid such a possibility or at least monitor all attempts of cheating. Fully automated assessment is not a good idea in case of intermediate (and higher level) courses but is imminent for massive beginners' courses with easy tasks and the low risk to use the above-cited method of cheating.

We have provided unit tests for Java and Python exercises in public for homework as study material but hidden for tests and exams to avoid "test driven" approach (massive number of tests in short time are hard to check by hand).

Code Style and Documentation. The often-debated question is whether and how much to teach documentation and following a code style. It used to be an important topic before the era of agile methods. As professional IDEs take care of formatting and warn about violation of formal code style, we only teach basic documentation requirements. At the same time, for bigger Java projects with many classes and methods involved, we demand that Javadoc is written in correct manner. At courses for "dummies" it is enough to give a few examples and not to go into details.

Version Control. Processes are very important in learning in opposite to the "results", that might be easy to find in internet, forums, from classmates etc. in case assignments are not carefully designed. In software industry version-control systems (like git [9]) allow to track changes in source code - who changed what and when. As version control is compulsory for software developers it makes sense to also introduce it to the learning process of programming courses. In case of massive courses automated version control helps the teacher to track students' progress and also supports the work in pairs that we are going to discuss later. But it is possible to forge, for example, submission deadlines, so version-control systems are not absolutely safe against cheating and must be used only where appropriate.

E-learning Environment. E-learning environments (like Moodle [11]) help a lot in case of massive courses, but they nearly lack of generally accepted tools for checking the advanced programming exercises. If a course resides in e-learning environment, it is not convenient (especially on courses "for dummies") to force the students to use other external systems. Some attempts are made to integrate unit testing with Moodle (e.g. [1]), but these are not generally available. Even if the teacher must check programming exercises partially by hand, it is still reasonable to use the e-learning environment that takes care of bookkeeping issues, checks the deadlines of the assignments, tracks the progress of every student, supports delivery of materials, conducts automated assessment, etc.

For advanced courses that have bigger projects, group work and variety of developer tools involved the e-learning environment is less inevitable (but still appreciated if it does not introduce unnecessary difficulty level).

1.3 Choice of Assignments

On programming courses, we usually concentrate on coding craftsmanship [10], not so much on software industry processes. Instead of writing nice essays about modern agile methods students are forced to write code. For the same reason we write "programming courses" instead of "software development courses" in this article.

In case of massive beginner courses e-learning environment allows to prepare a great variety of small exercises that can be chosen randomly and checked fully automatically. As these tasks are really small the cheating is not usually a risk. Learning by a lot of tiny steps is a good method to drill "mechanical" skills.

In case of professional advanced courses one can use the real-life tasks. But the risk is to exceed an official number of working hours a lot (even up to ten times). Dedicated students are happy with getting the full-scale experience, average students complain about time. Teacher has to be able to reduce the scope of the project if such problems arise and check the uniqueness of the task to avoid plagiarism. Group work with 3–5 students per group has been resultful for advanced courses. Less than three students are not a group, work in pairs is very different. Group of more than five students raises the risk of unequal workload. It is reasonable to grade all group members equally, leaving distribution of tasks to the group leader elected by group members.

The hardest situation arises with mid-level courses where tasks are not trivial, yet it is not easy to generate a big number of essentially different variations (for example, how many variations of a singly linked list implementations can one think of). We have used the elements of a flipped classroom (see [8, 12], to decrease the damage caused by potential plagiarism. The main difference from classical approach is that students' initial solution serves only as a starting point to a number of enhancements made in classroom and by the instructor. Of course, the initial solution has to be functionally correct (unit controlled tests can check this) and all the sources of information must be referenced. Students work on their solutions to add new features (like proper error handling) suggested during the lab.

1.4 Assessment and Feedback

Timely and adequate feedback to student assignments is an integral part of teaching. General observation is that beginners need more attention, one has to give immediate feedback even to small exercises to keep students motivated. Automated assessment is inevitable in case of massive courses, it does not make sense to include essay-type (or other subjective) questions that teacher has to grade by hand.

For professionals the feedback has to be more profound – not only about what is needed to be done and how, but also why, what if …, etc. Automated assessment is partially possible as a pre-filter (e.g. unit tests for programs). Discussion of the solution is important, and it is hard to make this part computer aided.

1.5 Flipped Classroom

Contact time is a valuable resource, it doesn't make sense to waste it on the tasks that students can do on their own. In case of a flipped classroom [12], students attend the lab prepared. During the lab they discuss their solutions, improve the code and elaborate the details of the topic. To guarantee that student comes prepared, her solution has to pass all automated tests before the lab.

1.6 Work in Pairs

If the number of students per instructor is too big for individual supervision/feedback and automated assessment is not fully possible (or reasonable) one can use "work in pairs" approach (similar to "pair programming", see [6–8]) to use the contact time productively. Work in pairs is effective in case its objectives are well-defined. For example, students explain their solutions to each other and improve them in pairs. If teacher asks a question about some work, the partner of the author has to answer it. Technically, it is possible to apply a version control system to document the work in pairs giving the teacher a quick overview of the results.

2 Experiments and Results

Author has been teaching a course "Algorithms and Data Structures" for second year students for more than 10 years. Dropout rate of this course used to be nearly 50%, methods described above allowed to reduce it significantly, the exact numbers for last five semesters are provided in Table 1 (to make the numbers comparable we only chose official results of compulsory course taught in Estonian over all these semesters).

Table 1. Dropout rate in "Algorithms and Data Structures".

Semester	Total number of students/not passed	Rate
Fall 2017	87/47	54%
Spring 2018	67/35	52%
Fall 2018	99/24	24%
Spring 2019	60/24	40%
Fall 2019	115/20	17%

2.1 Introducing Automated Testing

At early period all student programs were presented as free-text answers to e-learning environment. Teacher had to download, run, test and mark all attempts by hand, also write feedback comments and make these available to student. It was not much better than marking paper submissions last century. Integration of unit-testing and e-learning environment [1] allowed to shift teachers' workload from tedious technical tasks to giving more qualified feedback. It is very important to treat all students equally - in case of massive number of students everybody has to get her attention and the feedback she needs. It is not a good idea to only work with top 20% and leave others without help. There has to be sufficient number of assistants to work with everybody (including low 20%). If there are no assistants available, the planning of all processes becomes crucial. Automated assessment as a front-end to live discussion with the lecturer helps to decrease the lecturer's workload a lot. On the other hand, student's motivation gets concentrated to passing these automated tests, sometimes leaving other aspects out of focus. If the teacher only relies on automated tests and does not deal with students more thoroughly, then problems with cheating may arise. Fighting plagiarism with strict measures only (applied by the author during fall 2017, see Table 1) does not improve productivity of students and creates negative atmosphere. Extreme example of those measures was "Recognize your solution" test in case of plagiarism attempts, but irony and punishment caused some negative feedback to the whole subject.

2.2 From Plagiarism Detection to Flipped Classroom

Students tend to start problem solving from Google search (for programming exercises from StackOverflow forum [2]). In the context of algorithms there is a joke definition for undecidability: "Problem is said to be undecidable if one cannot find its solution in internet". As rigid methods to avoid plagiarism turned out to be counterproductive, we flipped the classroom [12]: students are allowed to modify solutions inspired by sources from internet, as long as they provide correct references and are able to develop the solution further. Nevertheless, a lot of students developed their original solutions as they used to do before this relaxation. Teacher has to find meaningful additions to well-known solutions and check that students perform these tasks singly. We have used ideas from the clean code methodology book [10] and adding proper error handling to students' initial solutions. At the beginning (spring 2018) we used only "hidden tests" that failed for well known (but not fully correct) solutions and students had to improve their solutions to pass these tests. As these tests also became well known, a more permanent solution to motivate and activate students was needed.

2.3 Introducing Work in Pairs

Students have to write a technical report about one individual task during the course. The main drawback of these reports used to be readability; the reports were often completely incomprehensible to the external reader. We started work in pairs (fall 2018) from the reports. Student had to explain her report to the partner and partner explained her understanding to the teacher who graded it. As this is still a time-consuming task from the teacher's viewpoint, the results with distance learning students (spring 2019) were less impressive. Real breakthrough came with expanding work in pairs to all contact labs (fall 2019). Automated assessment before the lab and version control during the lab kept teachers' workload tolerable. Pairs are formed dynamically during the lab (it is important to avoid unwanted cooperation before the lab) using some simple rule, e.g. pick the nearest student in the classroom. Some example tasks for work in pairs:

- Make at least two suggestions to the partner of how to improve her code based on clean code methodology [10] (starting from style issues up to full refactorization of code). Both partners create git commits [9] with the improvements agreed during the discussion and present these to the teacher (it is possible to check them quickly).
- Find a new solution to a given problem together so that it would be different from initial solutions of both partners. This exercise provides enough variations and is convenient to be checked using version control. We applied it to the binary insertion sort method where there are at least four different solutions.
- Defend the homework of your partner (used to check quality of technical reports since fall 2018).
- Arrange a competition of solutions (used for comparison of speed of sorting tasks).

Experience from work in pairs demonstrates that thorough feedback from the student partner motivates a learner more than brief (but still profound) feedback from the teacher and also keeps the plagiarism issues under control.

3 Conclusion

The methods described above and introduced to teaching of one programming course caused a measurable decrease in the dropout rate (possibly there were more factors involved that were not considered). The student satisfaction also increased, but there are no comparable numbers for the time period of current investigation, so this estimation is subjective. The largest impact was achieved with "work in pairs" approach. No changes in the teaching of the course were made for experimental purposes but only to improve the quality of teaching. Flipped classroom and work in pairs (supported by technical tools) proved to be two useful techniques for teaching computer programming and author is going to use these in the future.

References

1. Andreimann, A.: Integrating Software Testing Frameworks with the Moodle Course Management System. MSc Thesis, Tallinn University of Technology School of Information Technologies, Department of Informatics, Chair of Information Systems. Estonian Research Information System (2011). https://www.etis.ee/Portal/Mentorships/Display/bee4f37f-430f-4b71-b2e9-5be53dc9c02e?lang=ENG. Accessed 21 Apr 2020
2. Atwood, J., Spolsky, J.: Stack Overflow. Website, Creative Commons license (2008). https://stackoverflow.com. Accessed 21 Apr 2020
3. Beck, K., Andres, C.: Extreme Programming Explained: Embrace Change, 2nd edn, Addison-Wesley Professional, Boston (2004). ISBN: 978-0321278654
4. Beck, K.: Test Driven Development: By Example. Addison Wesley Professional, Boston (2002). ISBN: 978-0-3211-4653-3
5. Hunt A., Thomas, D.: Pragmatic Unit Testing in Java with Junit (Pragmatic Programmers). Pragmatic Bookshelf, US (2003). ISBN: 978-0974514017
6. Kafilongo, K.W.M.: The Use of Pair-programming to Enhance the Academic Performance. LAP LAMBERT Academic Publishing, Germany (2017). ISBN: 978-3-330-03565-2
7. Kessler, R., Williams, L.: Pair Programming Illuminated. Pearson Education, London (1990). ISBN: 9780201745764
8. Lepp, M., Tõnisson, E.: Integrating flipped classroom approach and work in pairs into workshops in programming course. In: Hamid, R.A., Bahrami, A., Deligiannidis, L., Jandieri, G., (eds.): Proceedings of the 2015 International Conference on Frontiers in Education: Computer Science and Computer Engineering, pp. 220–226. CSREA Press (2015)
9. Loeliger, J., McCullough, M.: Version Control with Git: Powerful Tools and Techniques for Collaborative Software Development. 2nd edn. O'Reilly Media, Inc, Newton (2012). ISBN: 978-1449316389
10. Martin, R.C.: Clean Code: A Handbook of Agile Software Craftsmanship. Prentice Hall, Saddle River (2008). ISBN: 9780136083238
11. Rice, W.: Moodle E-Learning Course Development: A Complete Guide to Successful Learning Using Moodle. Packt Publishing, Birmingham (2006). ISBN: 9781904811299
12. Talbert, R.: Flipped Learning: A Guide for Higher Education Faculty. Stylus Publishing, LLC, Sterling (2017). ISBN: 9781620364321

A Digital Services Course to Promote Teaching of Informatics in Estonia

Kaido Kikkas(✉) and Birgy Lorenz(✉)

Tallinn University of Technology, Tallinn, Estonia
{Kaido.kikkas,Birgy.lorenz}@taltech.ee

Abstract. The question of how and what to teach at schools as Informatics has been a hot topic for years. Different countries have taken different roads - some have focused on teaching basic digital competencies, some have opted for in-depth IT teaching, and some have not decided yet. In Estonia, ways to reform the teaching of Informatics (which is still an elective) has been sought for years. In 2018, it was finally decided to start compiling the Informatics curriculum for gymnasium (secondary school) level, and in 2019 the largest three Estonian universities co-operated in creating the study materials. One of the courses was named Digital Services. Students and teachers of gymnasiums (7 schools, 50 students) were used as the focus group in the pilot. The textbook developed for the course is used by various Estonian gymnasiums and complemented by various training events for teachers. The planned further development includes the compilation of a repository for practical exercises (inviting contributions from various practitioners from business, government, local authority, and education). This article seeks to have a critical look at the situation.

Keywords: Teaching of Informatics · Best practices · New learning models · K-12 programs · Flipped classroom

1 Background

1.1 The K-12 Informatics in Europe

Teaching Informatics at secondary schools has been a recurring topic for discussion in both local and international level for some time now [9]. The economy needs people with at least basic digital skills, the same skillset is sought by universities and vocational schools for their students as well. Students are expected to know various tools and methods, as well as have an idea about their future working life - already well before finishing K-12. Yet, all this is not guaranteed by their education as Informatics is often still a complementary subject. Among the various reasons for this, probably the most quoted ones are the lack of official curriculum, materials and teachers, inadequate skills of existing teachers, and as always, lack of time. In this, the problem seems to regenerate itself.

The report [9] by Informatics Europe makes the following recommendations:

- All students must have access to ongoing education in Informatics in the school system;

M. E. Auer and T. Rüütmann (Eds.): ICL 2020, AISC 1328, pp. 711–720, 2021.
https://doi.org/10.1007/978-3-030-68198-2_66

- Informatics teaching should preferably start in primary school, and at the latest at the beginning of secondary school;
- Informatics courses must be recognized by each country's educational system as being on par with courses in other STEM disciplines (also credit-wise);
- Teaching Informatics must be undertaken only by teachers with formal education, qualification in Informatics and appropriate methodological training.

At the same time, we see the following happening in Europe (see Fig. 1):

- In Finland, computational thinking and programming are mentioned as part of the curriculum, but there is very little content-wise [1];
- Meanwhile, Sweden has bundled ICT with media and engineering studies [1];
- In Germany, schools have extensive freedom about teaching Informatics – some schools go very deep, others do not teach it at all [13];
- In Latvia, Informatics is compulsory, schools with STEM focus also add programming (starting in primary school) [3];
- Croatia has based its curriculum on the DigComp framework [16] extending it further towards programming [15];
- In France, vocational programmes like the Initiative of School 42 (https://www.42.fr/) offer many country students real life/university-like experience in working in IT sector [17].

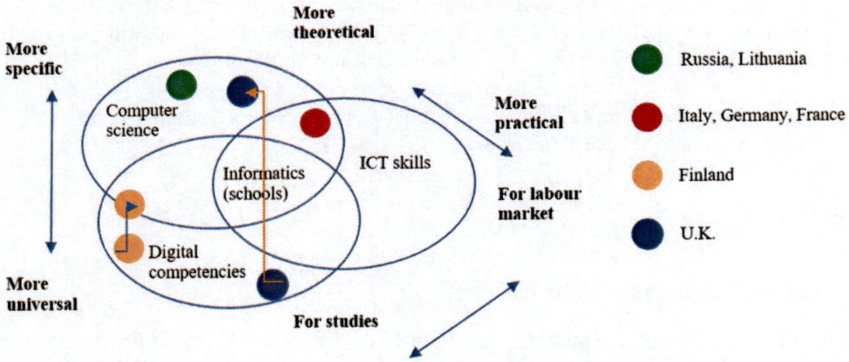

Fig. 1. Recent trends in teaching Informatics in various European countries (adapted from Laanpere [11] and [7]).

A point of discussion is the potential goal (or 'production') for Informatics at schools: should it aim for tertiary education (produce university students) or rather labour market (produce workers)? In Europe, universities have traditionally aimed strategically to produce independent thinkers capable of choosing their further career. At the same time, the labour market has had a more tactical insight, the companies need the employees 'right now' and need them to have the current state-of-the-art skillset. This discrepancy has been especially visible in more dynamic fields of IT for some time – even if the university heeds the companies and starts to teach the skill X, by the time it takes from students to graduate (3–4 years) the demand for the skill may already be fading. This is exacerbated further by the increasing perspective of AI 'taking over' and threatening many lower-profile professions [10].

Note: The cooperation of higher education with enterprises is notably stronger in the U.S. where the success of a university is largely determined by the success of its alumni in the labour market. However, the size of the U.S. leaves room for both high-level research universities and practically-oriented vocational institutions catering to the specific needs of the enterprise. This model does not fit Europe with its many countries and therefore fragmented education [2, 12].

1.2 The Gymnasium Informatics Plan in Estonia

Mart Laanpere (often considered the *grand old man* of Informatics in Estonia) has compared the changes in Estonian teaching of Informatics to a pendulum. In the early years (1980s) it was considered a specialization, thus the focus was set on programming and hardware handling. With computers becoming mainstream, the 1990s brought along a change – Informatics started to be seen as everyman's business, ultimately envisioning that Informatics as a subject would become obsolete as all other subjects will adopt the IT tools relevant to them. By 2016, it was clear that this vision did not materialize – while other teachers had adopted the teaching of digital competencies, they had failed to grasp the principles of Informatics that are needed to raise new generations of IT experts and to direct young people towards IT careers. The pendulum has therefore moved back towards the need for a separate subject, one that has specific learning outcomes and better connection to computer science [11]. The changes in Estonia over time are shown in Fig. 2 below.

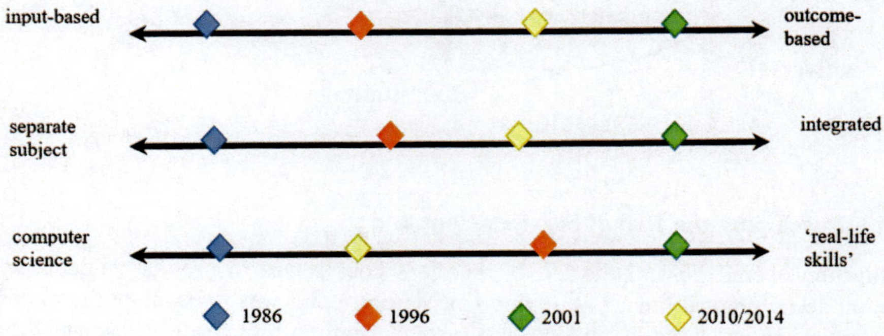

Fig. 2. The 'pendulum' in teaching of Informatics in Estonia 1986–2014 (adapted from Laanpere [11]).

Currently, Informatics is taught in parallel streams – several new courses have been added to the previous ones from earlier periods that are still offered. Examples of the new ones include Programming, Software Development, User-Centred Design and Prototyping, Software Analysis and Testing, Digital Technologies, Cybersecurity etc. Of earlier courses, 76% of Estonian gymnasiums [8] offer any of the following: Using

Computers in Research (81%), Technical Drawing (44%), Informatics/Computer Science (39%), Programming and Applications (39%), Robotics (21%), 3D Modelling (16%), Geoinformatics (15%), Media Studies (8%) etc. As schools are free to add electives, there have been others like Animation, Web Media etc.

In 2018, it was finally decided to start compiling the Informatics curriculum for gymnasium (secondary school) level, and in 2019, study materials were created for it. Various topics were distributed among the three main universities - the University of Tartu prepared the programming and software development courses, Tallinn University had the topics of user-centered design and prototyping as well as software analysis and testing, and Tallinn University of Technology compiled the Digital Services course reflected in more detail in this article. Additionally, a final project called DigiHive (Digitaru) [4] was proposed to bring all the different courses together in a startup-like environment. The possible options for Gymnasium Informatics Plan in Estonia (GINF) as defined by Laanpere are shown in Fig. 3 below.

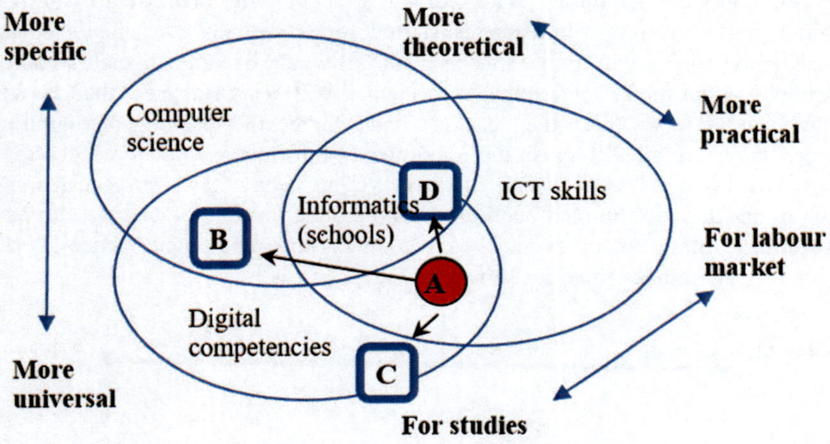

Fig. 3. The GINF options in Estonia (adapted from Laanpere [11] and [7]).

1.3 Developing the Digital Services Course

While the overall goal of the Digital Services course was to provide students with generic technological literacy, an important starting point was to promote the mindset that allows approaching IT projects and systems from both the project manager's and administrator's view (essentially what is often called DevOps) and building a strong base for subsequent courses [5]. The project manager would determine the content and duration of the project, compile the budget, and distribute the tasks. The administrator would set up and maintain systems, ensure security, and manage users.

According to the learning outcomes for the course, the student will be expected to: name different digital services for different target groups (including students); describe principles, components, and functionalities of a given service; compare governmental or local information systems based on their documentation and user guides, outlining similarities and differences; analyze his/her behavior as a user of a digital service targeting youth; name common security risks at using and developing digital services; compare the life cycles of two start-up companies, studying their beginning, development, reputation and target groups; explain the choices and steps made in the project from different points of view (entrepreneur, developer, project manager, customer/user).

The course has the following main topic groups [6]: Estonian e-government, e-identity, and nationwide digital services in Estonia; the public e-services (on the school, local and national level) that impact students the most; startups in Estonia and elsewhere; basics of information systems (e.g. architecture) and core principles of systems integration (the Estonian X-Road, the 'once-only' principle of data collection, open data, security, etc.); the development/control cycle of services (planning, design, implementation, launch, improvements); laws and regulations addressing digital services (GDPR, the law of personal data protection, eIDAS, SLAs); development of digital services (vision, planning, implementation, integration); risks involved in digital services (types, evaluation, countermeasures); cybersecurity in the context of digital services; incident handling and data protection (the 'CIA triad' of confidentiality, integrity and availability, the Estonian ISKE standard, data ownership).

2 Methodology

The design method was used to develop the course, the starting point being the curriculum of digital services [5, 6]. The development of study materials and the pilot phase were held in cooperation with the Estonian Information Technology Foundation for Education. Each chapter went through several cycles of feedback and related modifications, adding external reviewers where necessary. The team of developers included researchers and managers of various IT-related programs at Tallinn University of Technology.

Students and teachers of gymnasiums (7 schools, 50 students) were used as the focus group in the pilot of the Digital Services course. The general feedback was collected twice during the pilot, plus specific feedback from students and teachers about each chapter and additional feedback from teachers and external reviewers. The pilot schools were expected to offer supervision to students, use the flipped classroom method (using home for most of the studies and school predominantly for reflection and exchange of ideas) and allow the participation of 6–12 students at the pilot (8 weeks, 2 × 45 min of lessons per week).

The main points that were agreed upon:

- Technical aspects - the platform (Pressbooks, WordPress), list of main components (abstract, content, pictures and videos, terminology, guides, H5P tests, tasks, additional reading), feedback methods and additional materials needed (Moodle course support, the book of methodology for teachers);
- Pedagogical principles – the flipped classroom model supported by assignments, tests, text, and additional reading was used. Teachers were directed to be mentors

rather than lecturers. Students had to work in pairs or teams to complete some of the assignments, forcing them to develop cooperation skills;
- Schedule - 6 months from the beginning of content development to the end of the pilot and completion of study materials.

Both theoretical study and practical sessions were involved. A set of electronic study materials was made available (study texts, exercises, examples, links etc.). Grading is based on the student's independent work (performed both in-class and at home). The main goal is to get familiar with project management – to achieve this, constant feedback and hints/recommendations are provided (either by teachers or external experts), and the practice project is chosen as lifelike as possible. Continuous, direct participation of students is vital – the project is a complex group task that has the desired effect only on active participants.

The timetable for development was:

- January – content creation;
- February - piloting the first part, feedback from students and teachers;
- March – piloting the second part, feedback from students and teachers; up-dates to the first part based on the feedback;
- April – updates to the second part;
- May - feedback collection from teachers and external evaluators;
- June – content updates based on the feedback.

3 Results

3.1 Examples from the Workbook Exercises

The content of the workbook helps students to develop an understanding of e-governance in Estonia and basic principles of startups (it is available at https://web.htk.tlu.ee/digitaru/digiteenused).

The tasks include:

- Map local and national e-services and connect them to actual focus groups: elderly, students, newborn parents, industry representatives, volunteers, etc. The goal is to learn to from the best practices – what is available and where are the gaps in eservices;
- Study the Startup Estonia database (available from https://startupestonia.ee/startupdatabase) to find out what the companies are up to, what kind of services are provided, how much funding they have secured, who they hire. The goal is to understand how the ideas get funded and what topics are trending at the moment in service development;
- Analyze five services you use the most as a student. Look into user and service agreements to understand the GDPR and security issues related to developing services and acquiring user's data. The goal is to explain the responsibilities and knowledge of the law needed to deal with sensitive and personal data.

Developing an e-service (available at https://web.htk.tlu.ee/digitaru/digiteenused/):

- Test your skills and interests as a worker. The Pathfinder portal (http://ametid.rajaleidja.ee/ikt-ametid provides more than ten different sample careers in IT to explain what kind of occupations are available in the industry. By testing their skills, students learn about the needs and values of different occupations and their relations. GINF provides the possibility to role-play at least eight of them (from project manager to programmer to security specialist);
- Develop a small project about an e-service. Study the current situation (where the market gap is) and draft a plan complete with the needed workforce and development schedule. It is a complex exercise that is divided up to 4 different tasks, but in the end, it will be one project the students will be working on as a group for 1–2 weeks;
- Develop a prototype of the service. The exercises branch out for three difficulty levels: advanced level would install a Linux server, intermediate would install a Web hosting service (e.g. XAMPP) and beginners would use an online service (e.g. WordPress). The goal of these 2–3 week tasks is to teach necessary IT skills and competences to work as a project manager or content developer, advanced students learn also the roles of administrator and security manager.

3.2 Results from Feedback and Changes to the Course

Concerning the project management topics (e-services, Estonian e-government, startups, etc.), the feedback from students was positive (informative texts and videos, thought-provoking yet not too difficult tasks). The only notable critical remark concerned the group work-only approach ("There should be a way to work alone, too").

Teachers pointed out that the students were surprisingly good at adapting the tasks to local circumstances, especially in the areas where local community initiative could be more effective than the national, 'umbrella' approach. The main criticism addressed the volume of exercises, wishing for more guidance in determining which tasks were 'must-have' (as opposed to 'nice to have').

The administration topics were considered difficult. Both students and teachers mentioned that technical tasks like setting up a virtual machine, installing a Linux-based web-server, and setting up web services (e.g. a WordPress website) proved too complicated due to no previous experience. Notably, the students were more positive in their feedback ("you had to think – it felt good when you finally succeeded") but pointed out that teachers were often unable to provide any support. The latter was affirmed by teachers' feedback as well.

The lessons learned for redesigning the course included:

- The current version focused on two positions/professions: the IT project manager and the administrator. The specifics of the latter should be explained better (also differentiating between different administrative profiles), the positions of IT analyst and security manager should be added;
- Due to the lack of technical skills (among both students and teachers) in most schools, the materials should have three skill levels (beginner, intermediate, advanced), and so should the final project.

4 Discussion

In general, the wheels turn slowly in education. At many schools as well as universities, the few yearly changes often depend on specific persons as well as resources (time, funds, tools, and staff). At the same time, much of the IT world moves in a hectic pace comparable to a startup or a Garage48 or School 42 projects.

The positive aspect is that the understanding that old ways do not suffice anymore is there. While the needs and possibilities are somewhat uneven, it has been understood that the success of Estonian information society depends on the new generation grasping both the technology itself and its application possibilities (startup mentality and financial aptitude also help – the PISA 2018 study [14] shows that Estonians manage rather well there). The Digital Services aims to contribute here as well, also considering the lessons from the GINF about the reinclusion of teaching of programming.

While the vision of the future direction seems to be clearing, there are practical challenges:

- Even if the practical industry needs have been more recognized lately, there is still a gap. The GINF does not comply with professional standards (also, neither do many university IT curricula) – the problem is exacerbated further in that it is also recommended for vocational schools that are supposed to output directly to the labor market. On the one hand, the enterprises complain about a lack of systematic approach and inability to serve the labor market. On the other hand, universities point out that the more American way of focusing on specific tools and languages is short-sighted as well, given the development pace of Estonian information society. Ideally, both parties should come together and try to envision the situation in 15–20 years, to allow these predictions to take root at today's schools;
- Informatics, just as other subjects but perhaps more so, needs to strive to strike the balance point between the increasing student-centeredness of modern schools and the somewhat more rigid business environment of IT companies (universities are somewhere in between these two). On the one hand, school focusing on the student (and not the teacher as a 'sage on the stage') promotes initiative, creativity and problem-solving attitude – the qualities much needed at future IT professionals (this was also witnessed at the Digital Services pilot). On the other hand, 'lowering the bar' should not be allowed, and the students should ideally be prepared to manage in a variety of environment and cultures (e.g. a large multinational vs a local startup).
- The way of teaching is up to school and thus eclectic as every institution chooses its own based on its interests and resources. This results in inequality in both education and skills (even if different skills can also have a positive aspect in the case of well-organized cooperation), there is no 'glue' to keep it together. The Digital Services course has the potential to become the 'glue', building on various technical subjects in GINF and adding the problem-solving, management and administration layers;
- The government should signal their expectations and preferences more clearly – as long as the activities are non-compulsory, it is difficult to point out the problems and set standards, instead, the work is still left on the shoulders of volunteers. In a

longer perspective, it will harm the image of Estonia as a digital nation, and make it infeasible to invite large international companies to the country due to the lack of qualified personnel.

Thus, while Estonia (being small) is good for piloting, the serious problem is the lack of a systematic approach. The latter results in diminishing numbers of students, which in turn leads to a smaller talent pool. And those who still manage to 'find their own way' will be tempted to go abroad where the career-building is easier.

5 Conclusion

The development of the Digital Services course helped to sort out the topics that are more difficult for schools and students. The later course design also added three levels of difficulty to allow beginners, intermediate and advanced participants all benefit from it. The digital services textbook is used as the base of studies at various Estonian gymnasiums and complemented by various training events for teachers. The planned further development includes a compilation of a repository for practical exercises (inviting contributions from various practitioners from business, government, local authority, and education).

Estonia should make up its mind – whether it is a digital nation that affords to take larger steps towards a universal IT curriculum, or it remains an enthusiastic but small startup dependent on acquiring the next project funding for a couple of years. Building a sustainable IT education with a wide base that also produces a sufficient number of top talents is not possible on the enthusiasm of a few activists only.

Acknowledgements. This research was supported by the IT Didactics project at Tallinn University of Technology.

References

1. Bocconi, S., Chioccariello, A., Earp, J.: The Nordic approach to introducing Computtional Thinking and programming in compulsory education. Report prepared for the Nordic@ BETT2018 Steering Group (2018). https://doi.org/10.17471/54007
2. Chadha, D., Toner, J.: Focusing in on employability: using content analysis to explore the employability discourse in UK and USA universities. Int. J. Educ. Technol. High. Educ. **14** (1), 33 (2017)
3. Dagienė, V., Jevsikova, T., Stupurienė, G.: Introducing informatics in primary education: curriculum and teachers' perspectives. In: International Conference on Informatics in Schools: Situation, Evolution, and Perspectives, pp. 83–94. Springer, Cham (2019)
4. Gümnaasiumi informaatika ainekava: digilahenduse arendusprojekt. HITSA (2020). https://www.hitsa.ee/ikt-haridus/progetiiger/gumnaasiumi-informaatika-ainekava/digilahenduse-arendusprojekt
5. Gümnaasiumi informaatika valikkursuste loomine. Gümnaasiumi informaatika ainekava tööversioon; Riigihange Lisa 1–2 (2018). https://riigihanked.riik.ee/rhr-web/#/procurement/728628/documents?group=B

6. Gümnaasiumi informaatika valikkursuste loomine. Gümnaasiumi informaatika kur-suste ja õppeprotsesside kirjeldused; Riigihange Lisa 1–3 (2018). https://riigihanked.riik.ee/rhr-web/#/procurement/728628/documents?group=B
7. Gümnaasiumi informaatika valikkursuste loomine. Gümnaasiumi informaatikaõppe kontseptsioon; Riigihange Lisa 1–1 (2018). https://riigihanked.riik.ee/rhr-web/#/procurement/728628/documents?group=B
8. IKT-haridus: digioskuste õpetamine, hoiakud ja võimalused üldhariduskoolis ja lasteaias Hariduse Infotehnoloogia Sihtasutus/Praxis (2017). http://www.praxis.ee/wp-content/uploads/2016/08/IKT-hariduse-uuring_aruanne_mai2017.pdf
9. Informatics Education in Europe: Are We All In The Same Boat? Informatics Europe (2017). http://www.informatics-europe.org/component/phocadownload/category/10-reports.html?download=60:cece-report
10. Krusell, S., Rosenblad, Y., Michelson, L., Lambing, M.: Eesti Tööturg täna ja homme 20192027, Tööjõuvajaduse seire ja prognoosisüsteem OSKA, Mjandus- ja Kommunikatsiooniministeerium/ SA Kutsekoda (2020). https://oska.kutsekoda.ee/wp-content/uploads/2018/11/IKTterviktekst.pdf
11. Laanpere, M.: Changes in Schoollevel Informatics/Muutused Eesti kooliinformaatikas - Mart Laanpere/ Gümnaasiumi in-formaatika, HITSA. Youtube presentation (2019). https://youtu.be/n_OkCetom4E
12. Lo Presti, A., Pluviano, S.: Looking for a route in turbulent waters: employability as a compass for career success. Organ. Psychol. Rev. **6**(2), 192–211 (2016)
13. Passey, D.: Computer science (CS) in the compulsory education curriculum: implications for future research. Educ. Inf. Technol. **22**(2), 421–443 (2017)
14. PISA 2018. Haridus ja Teadusministeerium/Inniove (2019). https://www.hm.ee/sites/default/files/pisa_2018–19_raportweb.pdf
15. Žiljak, T., Baketa, N.: Education policy in Croatia. In: Policy-Making at the European Periphery, pp. 265–283. Palgrave Macmillan, Cham (2019)
16. Vuorikari, R., Punie, Y., Gomez, S.C., Van Den Brande, G.: DigComp 2.0: The digital competence framework for citizens. Update phase 1: The conceptual reference model (No. JRC101254). Joint Research Centre (Seville site) (2016)
17. Walt, V.: Radical Programming School'42' Still Solving for the Skills Gap. Fortune (2019): https://fortune.com/2019/03/16/42-computer-programming-school-skills-gap/

Poster: Technique of Active Online Training: Lessons Learnt from EngiMath Project

Oksana Labanova[1], Elena Safiulina[1], Marina Latõnina[1], Anne Uukkivi[1(✉)], Vlad Bocanet[2], Cristina Feniser[2], Florina Serdean[2], Ana Paula Lopes[3], Filomena Soares[3], Ken Brown[4], Gerald Kelly[4], Errol Martin[4], Anna Cellmer[5], Joanna Cymerman[5], Volodymyr Sushch[5], Igor Kierkosz[5], Javier Bilbao[6], Eugenio Bravo[6], Olatz Garcia[6], Concepción Varela[6], and Carolina Rebollar[6]

[1] TTK University of Applied Sciences, Pärnu mnt 62, 10135 Tallinn, Estonia
uukkivi@tktk.ee
[2] Technical University of Cluj-Napoca, Memorandum Street 28, 400114 Cluj-Napoca, Romania
[3] Polytechnic Institute of Porto, Rua dr Roberto Frias 712, 4200 465 Porto, Portugal
[4] Letterkenny Institute of Technology, Port Road, F92 FC93, 6539078V, Letterkenny, County Donegal, Ireland
[5] The Koszalin University of Technology, Śniadeckich 2, 75453 Koszalin, Poland
[6] University of the Basque Country, Barrio Sarriena s/n, 48940 Leioa, Bizkaia, Spain

Abstract. The goal of this paper is to introduce a technique of creating self-tests that has allowed to actively incorporate university students into the learning process. The study was conducted within the framework of the Erasmus + Project EngiMath. Partners' peer reviews, the survey results and the students' comments in forums and test results were used to conduct the research. The students' overall satisfaction was in a high level. However, opportunities for some technical improvement has been emerged like the formulation of the tasks needs to be very clear and the time required to perform the tests must be limited. The following conclusions can be drawn from the study. The use of self-tests at all stages of training has intensified the assimilation of the material, i.e. increased understanding of theoretical material and developed computational skills. By completing a series of such assignments on each topic of the course, students had mastered the methodology of studying the topic and mastered specific teaching material on this topic. Feedback made, taking into account typical errors, has allowed the students to analyse their knowledge. A large number of variations for such tasks has allowed students to be involved in the process of active independent and individualized self-study.

Keywords: Online self-assessment · Teaching of mathematics · Test design

M. E. Auer and T. Rüütmann (Eds.): ICL 2020, AISC 1328, pp. 721–729, 2021.
https://doi.org/10.1007/978-3-030-68198-2_67

1 Introduction

As far back as in 1968, it was shown by Bloom that the quality of knowledge acquired by students directly depends on the amount of time personally spent by students on independent assimilation of material [1]. A learning approach called Learning for Mastery (LFM) was later developed [2]. According to LFM students regularly are tested using formative tests, and they are required to demonstrate a correct answer to 90% of the test tasks, that is, demonstrate "mastery". When student fails to master the subject teaching and testing are further repeated and several times if necessary. The development of tasks for self-monitoring in all studied modules is one of the stages of the system. Students believe that self-tests are a useful learning strategy [3]. Various adapted versions of this system are now being actively implemented both in developed and developing countries higher and secondary schools [4]. In Estonia, this learning approach is presented in the works of Krull [5]. At the same time, organizing assignments for students who are potentially at different levels of knowledge is difficult. Online opportunities come to the rescue. Bloom's LFM model was adapted using online exercises and formative tests [6, 7]. Self-assessment through online exercises with feedback plays an important role in getting students to reflect on their own learning and to self-manage their learning process, which can increase their autonomy and intrinsic motivation [8]. According to Sumantri and Satriani [9] a student's higher level of independence affects the results of that student's mathematics learning positively.

Current study introduces the technique of developing self-tests according to Bloom's LFM model that has allowed to actively incorporate students into learning process. According to the authors, the strong side of the present article is the experience and knowledge gained during this research, which can be further used to create, develop and conduct online self-tests, and can be also useful to other educators in organizing the learning process.

2 Background

2.1 Project and Course Overview

The role of technology is to facilitate teaching and promote learning in an authentic situational manner and this role is now central to support of "student-centred" learning. Prompt feedback provides reinforcement, guidance, confirmation of learning and encouragement; as all is necessary to favour and promote an active learning ethos. The EngiMath project – a mathematics online learning model in Engineering education was launched to implement this approach. The project started in 2018. It includes teachers from TTK University of Applied Sciences/TTK UAS from Estonia (project coordinator), Letterkenny Institute of Technology/LYIT from Ireland, Polytechnic Institute of Porto/P.PORTO from Portugal, University of the Basque Country/UPV/EHU from Spain, Technical University of Cluj-Napoca/UTC from Romania and Koszalin University of Technology/PK TUK from Poland. One of the main outputs of this project is a 3 ECTS on-line course on engineering mathematics, which is created in the

Moodle TTK UAS learning environment. The main goal of this course is to develop basic and structured knowledge and practical skills in mathematics sub-topic of Matrices, Determinants and Linear Equations Systems related to Engineering. After identifying common problems that exist for students of the first-year of study at the universities of partner countries, and conducting a comparative analysis of their curriculum, a common core of mathematical topics was identified [10]. Using general basic agreement, 22 lessons with theoretical materials and a question bank of practical and assessment tests were developed [11].

2.2 Tests and Questions

The Estonian partner with years of experience started to create practical activities in the form of tests. A corresponding practice test has been created for each theory lesson, i.e. 22 self-tests have been created. Each test consists of 8−12 questions, depending on the theme and the order of questions followed from lightest to complex principle. Self-tests are characterized by a large number of questions related to specific topics in order to build knowledge and skills [12]. These tests focus on issues with learning difficulties and common mistakes, so the point system is not used. Since the test is educational in nature, the number of attempts to solve it is unlimited, moreover, the duration of the tests is also unlimited. This gives students an opportunity to analyse the mistakes made, assess the depth of knowledge gained, identify gaps in learning, more carefully repeat and study the training material, and then pass a new exam. On the other hand, an unlimited number of attempts allows students with different levels of knowledge to pass testing at an individual pace. A testing model was compiled based on the objectives of tests: for each lesson a technological matrix containing competencies selected for testing. For each competency, questions with variability in average of 10 versions were created. Some versions were created by using the STACK question type which generates new initial data each time a new test run.

The question bank consists of 200 questions. Therefore, in total when considering the number of variations of questions there are at least 2000 items in question bank. There are 50 close-ended and 150 open-ended type of questions. Exhaustive feedback has been provided for each question. Test developers created the questions following the guidelines for creating tests while writing questions [13–15]. An important point in creating test questions is their wording, which provides brevity, simplicity and unambiguity. The authors controlled the correspondence of the terms in theory and in questions. The wording of the questions assumes that students have different pre-school backgrounds so complex mathematical constructions were avoided. When creating the tests, it was taken into account that in one task it is recommended to ask only one question or statement.

Further are described the types of questions used in the practice tests. For each question a question type was selected. The authors also substantiate why this or that type of question was used.

The following types of closed-ended questions were used: “Multiple Choice”, “Word Select”, “Matching”, “Drug and Drop” and “True/False”. The questions “Multiple Choice” and “Word Select” pursued diagnostic goals, that is, incorrect answers were formulated on the basis of “close to true” taking into account expected and typical errors. “Word Select” and “Multiple Choice” have a similar idea, but the authors used these questions purposefully: the first one was used to avoid a long list of possible answers, to make visually compact (see Fig. 1).

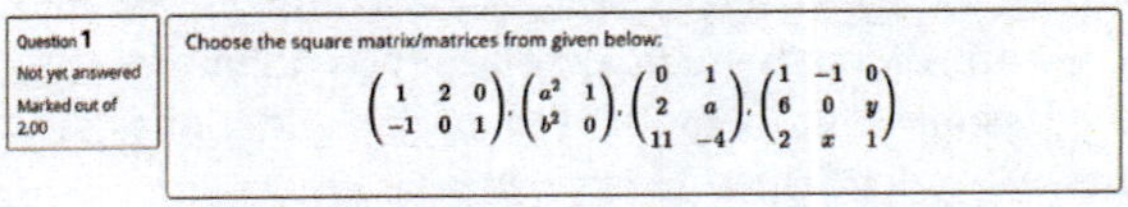

Fig. 1. Word select type of question

“Matching” and “Drag and Drop” were used to assess homogenous knowledge. Training to remember factual information is also important. “True/False” questions were used to get quick feedback on student’s understanding of theoretical material to measure their ability to recognize a correct statement or definition.

Open-ended questions were used to avoid random answers for students. For such purpose questions like “Close Embedded”, “Gapfill” and STACK were used. When filling the gap student can either enter the final result or some part of the solution or add the whole solution in a step-by-step manner (see Fig. 2).

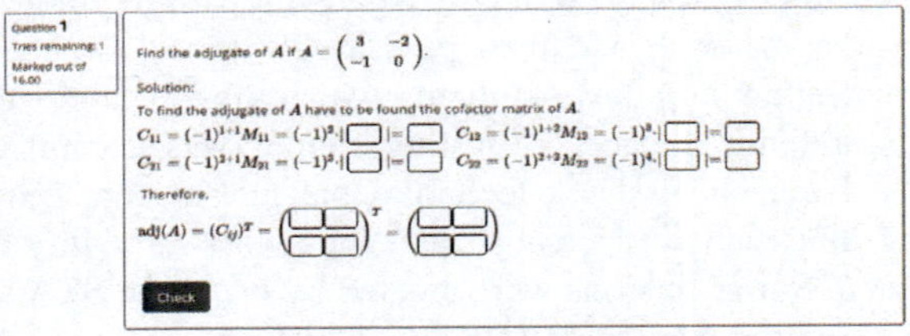

Fig. 2. Step-by-step question using Gap Fill type

To be able to answer the questions presented in a step-by-step question and solve the problem student should study relevant sections of the theory. STACK type of question was chosen for possibility of generating initial data and feedback for each question [6]. Unfortunately, open-ended assignments often generate several correct, but logically disproportionate answers, which is undesirable from the point of view of an unambiguous assessment and technological control. For preventing unwanted interpretations by students, we use constraints as “calculate to the nearest hundredth” or technical hint as “abs (x)”, “sqrt (x)”, also tooltips added to questions using HTML code in Moodle.

3 Approach

3.1 Collaborative Peer Review

A review of all question for content quality and clarity by external subject-matter experts (collaborative peer review) is a necessary step in the test development process [16]. The analysis of expert assessments was carried out by the brainstorming method and suggested that the experts fill out a joint document in which a comprehensive analysis (linguistic, didactic) of test questions is made and possible ways to improve the situation are identified [17]. In our case, colleagues and educational institutions of partner countries acted as experts, which allowed us to avoid a superficial study of the material. Constant comparative analysis was used to analyse the data.

3.2 Approbation

Approbation of Lessons 1–7 was conducted in 2020 to determine the quality and effectiveness of the self-tests, in which 40 first-year Building Construction and Facilities Management daytime students of TTK University of Applied Sciences participated alongside students from other countries. The current study is based only on the results and feedback of Estonian students. Each lesson includes a forum for sharing ideas and concerns by posting comments and questions. Content comparative analysis was also used to analyze these data. Statistical processing of test results was carried out by calculating difficulty index and average test time. For difficulty index calculating a dichotomous grading scale was used: for each test a matrix of answers was formed (rows are student's serial number, columns are questions' number) with repeated attempts and entries 0 (incorrect or partially correct answer) and 1 (correct answer). Each lesson includes a forum to exchange ideas and problems by posting comments. Additionally, students' forums comments were analysed.

3.3 Students' Survey

At the end of approbation an online questionnaire was sent to students to assess their perception of the online materials. The questions were measured on Likert scales of varying lengths (either 6 or 7 points depending on the question). Some scales had an even number of options to force the respondent to more thoroughly think through their response. Others had an odd number of options as they were ranging from too little to too much, the middle value being the optimal. For six-point scales, the first two values (0, 1) were considered as reflecting a low level of agreement, the next (3, 4) were a moderate level, and the last two (4, 5) a high level of agreement. Open-ended questions were also asked. The questionnaire was written in Estonian and was disseminated by using Google Forms in February 2020.

4 Results

4.1 Partners' Peer Review

There were 55 expert comments in the joint document. Six categories emerged through gathered expert comments and suggestions: guidelines for question names, test setup, test adequacy, question comprehensibility, question correctness, question feedback.

Guidelines for question names- comments on the description of general format of a given question name.

Entry 3: "Suggestion – we feel that it is important to distinguish all questions –… – question name begins with section number to make it distinguishable."

Test Setup – test's administering settings block with timing, grade, layout, question behaviour and review options.

Entry 18: "Test 2 – it does not allow to repeat as in Test 1 – we felt this was a plus for formative/practical purposes."

Test adequacy - the content and duration of the test are enough to test knowledge, skills etc. on given topic.

Entry 5: "Test 1: better to remove the question (Matrix I is not defined at this point)."

Question comprehensibility - the question is clear, unambiguous.

Entry 43: "We propose to change the text to something like": "Without the application of Sarrus' rule or any other computations point out the values of $$x$$ and $$y$$."

Question correctness - the question is logical, linguistically and technically correct.

Entry 34: "Please see what happens in some trials – Not every time…We went to html-code and changed "ENG" to "ENG-US" … "

Question feedback - the feedback on the question is correct.

Entry 13: "Questions in 2.2 – general feedback – insert spaces" "A matrix is said to be square if the (SPACE) number of rows (SPACE) is equal to the number of columns".

4.2 Results of the Approbation

Results of Students' Forum Postings. There were 12 posts in the online course's forums. As a result of forums' posts analysis, three categories appeared: presence of typos, ambiguous formulation and content questions. Usually the posts were provided with a corresponding image of error.

Presence of typos. Via forums' posts students pointed to test questions' portions where they had found typos. Typos can be divided into two groups: technical errors and errors made when translating from English into Estonian.

Post 8: "There is an error in the test, the first element of the matrix is incorrect."

Ambiguous formulation. In posts students cited parts of test questions in which the question to their opinion was ambiguous.

Post 2: "Confusing terms? A matrix row is a matrix whose number of rows is equal to the number of columns."

Content's questions. Students used forums to ask advice from teacher. They also entered their solutions into forum in order to receive feedback from the teacher.

Post 7: "Question about the calculation procedure in question/item number 6: is it correct that x = −1? What is the calculation procedure?"

Lecturers of TTK UAS systematically responded to posts, corrected the wording and typos and gave feedback to students if necessary.

Results of Tests Analysis. The results of the analysis of tests that have passed approbation are presented in the Table 1. The question analysis was conducted for each test by computing the difficulty index using the classical test theory, and test duration.

Table 1. Tests analysis.

Criteria	Test 1	Test 2	Test 3	Test 4	Test 5	Test 6	Test 7
Number of questions	9	8	9	10	8	11	9
Number of questions with variations	90	81	91	80	53	40	40
Number of students	40	40	39	38	37	32	27
Number of attempts	62	59	73	47	46	47	30
Mean difficulty index (%)	13.08	9.11	55.10	38.53	39.67	65.96	37.5
Easy questions (%)	88.90	87.50	0	20	12.5	0	11.11
Difficult questions (%)	0	0	11.11	0	0	45.45	0
Average time (min)	~7	~3	~21	~27	~23	~48	~26

The target range for difficulty index was defined from 20 to 80 and the percentage of items considered too easy (less or equal to 20) or too difficult (more or equal to 80) were calculated. The reliability was not calculated, because the testing purpose is to increase the level of students' knowledge i.e. with each new attempt the number of tasks that students cope with increases. The total number of questions was 64 (range: 8–11 questions per test) and total number of questions with variations was 475 (range: 40–90 questions per test), while the number of examinees was 40 (range: 27–40 examinees per test). The number of students' attempts were counted in compiling the table, each student has used more than one attempt. Mean difficulty index ranged from 9.11 to 65.96. Easier questions appear in Test 1, Test 2, Test 4, Test 5 and Test 7. More difficult questions appear in Test 3 and Test 6. Average time to pass the test ranged from 3 to 48 min.

4.3 Results of Students' Survey

Most students (83%) considered that practical material was highly related to the theory and they had no difficulty with self-tests. About 42% of students thought that they had enough time for self-tests, and 50% even had too much time for this. Students considered that all course materials will help with final exam (83%) but were less convinced the material is helpful in real life. Slightly more than 30% strongly agreed while about half were moderate in their answer to this question.

5 Conclusions

From peer review, in accordance with the experts' suggestions outlined above, a system for numbering test questions was developed and posts' analysis from the forums the confusing questions and tests were redid. The authors found forums to be a good tool for getting quick feedback from students during passing the tests. The authors conclude that it takes a student for an average of one hour to pass one test. Based on the data shown in Table 1, can be seen that Test 1 and Test 2 turned out to be the least difficult, also it took least time to complete them. We can conclude that in the future these tests can be unified. The complexity of Test 3 can be explained by the presence of mathematical symbols in this test, which may interfere with the visual perception of the question. The table shows that Test 6 was the most difficult. When examining the test in more detail and relying on students' feedback, it was decided to pay attention to the corresponding theoretical part, whether everything is understandable and are there enough examples. Was planned to split Test 6 into two tests. In future work, all tests are planned to be created according to difficulty levels (easy, medium, difficult) in accordance with the development of the course.

Feedback from students shows that the number of self-tests was sufficient to master the subject and that the tests are logically structured. The formulation of tasks sometimes raised questions; and needs to be reviewed. Students emphasized that independent learning increases sense of duty and allows active participation in learning process.

In the future, it is planned to study the number of attempts made to pass the test and time spent taking the test with grades achieved. Moreover, correlation between the results of self-tests, assessment tests and course assessments will be studied.

References

1. Bloom, B.S.: Learning for mastery. instruction and curriculum. regional education laboratory for the carolinas and virginia, topical papers and reprints, number 1. Eval. Comment **1**(2), 1–11 (1968)
2. Bloom, B.S.: The 2 sigma problem. Educ. Res. **13**(6), 4–16 (1984)
3. Kulik, J.A., Kulik, C.C., Bangert, R.L: Effects of practice on aptitude and achievement test scores. Am. Educ. Res. J. **21**(2), 435–447 (1984)
4. Anderson, L.W., Block, J.H.: Mastery Learning Models. Dunkin, M.J. (ed.) The International Encyclopedia of Teaching and Teacher Education. Pergamon Press, Oxford (1987)
5. Krull, E.: Mastery learning experience in Estonian schools. In: Lehtinen, E. et al., (eds.) Fourth European Conference for Research on Learning and Instruction, p. 338. Turku University Press, Turku (1991)
6. Pelkola, T., Rasila, A., Sangwin, C.J.: Blended Mastery Learning in Mathematics. arXiv: History and Overview (2017)
7. Pelkola, T., Rasila, A., Sangwin, C.J.: Investigating bloom's learning for mastery in mathematics with online assessment. Inform. Educ. **17**(2), 363–380 (2018). https://doi.org/10.15388/infedu.2018.19

8. Ibabe, I., Jauregizar, J.: Online self-assessment with feedback and metacognitive knowledge. High. Educ. (00181560), **59**(2), 243–258 (2010). https://doi.org/10.1007/s10734-009-9245-6
9. Sumantri, M.S., Satriani, R.: The effect of formative testing and self-directed learning on mathematics learning outcomes. Int. Electron. J. Elementary Educ. **8**(3), 507–524 (2016)
10. Brown, K., Uukkivi, A., Soares, F., Lopes, A.P., Cellmer, A., Feniser, C., Rebollar, C., Varela, C., Bravo, E., Safiulina, E., Kelly, G., Bilbao, J., Cymerman, J., Latõnina, M., García, O., Labanova, O., Bocanet, V.: A European Educational Math Project – Comparative Needs Analysis in Partner Institutions. In: EDULEARN19 Proceedings, pp. 742–749. IATED, Palma (2019). https://doi.org/10.21125/edulearn.2019.0248
11. Soares, F., Lopes, A.P., Cellmer, A., Uukkivi, A., Rebollar, C., Varela, C., Feniser, C., Safiulina, E., Bravo, E., Kelly, G., Bilbao, J., Cymerman, J., Brown, K., Latõnina, M., Labanova, O., Garcia, O., Bocanet, V.: Development of a mathematics on-line project in engineering education. Open Educ. Stud. **1**, 257–261 (2019)
12. Labanova, O., Uukkivi, A., Safiulina, E., Latõnina, M.: Engaging Engineering Students in the Educational Process Using Moodle Learning Environment. In: Soares, F., Lopes, A.P., Brown, K., Uukkivi, A. (eds.) Developing technology mediation in learning environments, pp. 239–263. IGI Global (2020). http://10.4018/978-1-7998-1591-4
13. Beilmann, M.: Küsimustiku koostamine. http://samm.ut.ee/k%C3%BCsimustiku-koostamine. Accessed 04 June 2021
14. Testide moodustamise ning testidele vastamise programmisüsteemi APSTest ver. 2.0 kasutajajuhend. https://study.risk.ee/wp-content/uploads/sites/3/2012/04/apstest2-1.pdf. Accessed 04 June 2021
15. Ruul, L.: Testimine e-õppes. https://www.hariduskeskus.ee/opiobjektid/test/ksimuste_konstrueerimine.html. Accessed 04 June 2021
16. Downing, S.M., Haladyna, T.M. (eds.): Handbook of test development. Erlbaum, Mahwah, NJ (2006)
17. Putman, V., Paulus, P.: Brainstorming, Brainstorming Rules and Decision Making. The Journal of Creative Behavior 43(23/39) (2009). https://doi.org/10.1002/j.2162-6057.2009.tb01304

A New English Course for the Program "International Transport Policy"

Tatiana Polyakova(✉)

Moscow Automobile and Road Construction State Technical University,
64, Leningradskiy Pr., 125319 Moscow, Russia
kafedral01@mail.ru

Abstract. Due to radical changes in technologies, there appear new professions and functions of engineers. Engineering universities respond designing and introducing new educational programs. Moscow State Institute of International Relations (MGIMO) run by the Russian Federation Ministry of Foreign Affairs and Moscow Automobile and Road Construction State Technical University (MADI) with the support of the United Nations Association of Russia launched a new additional educational program "International Transport Policy". The aim of the Program is the development of professional competences of students necessary for the professional activities in the sphere of international transportation processes, including cooperation of organizations in the system of the United Nations Organizations. MADI Foreign Languages Department was responsible for implementation of the English course. The paper describes the course design, the content of compiled teaching materials, learning outcomes and highlights the importance of learner-centered approach. Special attention is given to the results of the survey revealing students' motivation, their language background, self-assessment of their English communicative competence level, the difficulties they have studying the foreign language, evaluation of the course, the skills they have acquired, their attitude towards teaching materials and recommendations for the course improvement.

Keywords: International relations · Transport · Additional educational program · English course design · Survey

1 Context

Fast radical changes in technologies cause the appearance of new professions and new functions of engineers. In order to respond to these changes, engineering universities design and introduce new educational programs.

At present in international relations, there is a demand for synchronization and harmonization of transportation processes in the sphere of multimodal and automobile transport, cooperation of international transport organizations, providing traffic safety and rights of all stakeholders. As a result, the specialists require professional competences both in the field of international relations and multimodal transportation.

For that purpose, Moscow State Institute of International Relations (MGIMO) run by the Russian Federation Ministry of Foreign Affairs and Moscow Automobile and

M. E. Auer and T. Rüütmann (Eds.): ICL 2020, AISC 1328, pp. 730–736, 2021.
https://doi.org/10.1007/978-3-030-68198-2_68

Road Construction State Technical University (MADI) launched a new additional educational program “International Transport Policy”. The United Nations Association of Russia has supported the initiative.

The Program is aimed at the development of professional competences necessary for the professional activities in the sphere of international transportation processes, including cooperation of organizations in the system of the United Nations Organizations (UNO). The curriculum and syllabus were designed on the base of the Federal State Educational Standard “International Relations” [1]. The total load is equal to 396 h, or 11 credit points. At the end of the Program, the students who pass final examinations and defend their final project successfully receive Bachelor’s Degree. The Program is intended for those who already hold either Bachelor’s or Master’s Degrees in Transport and for those who are still students of MADI.

2 Goal

The Program “International Transport Policy” includes three modules: General, General Theoretical and Specialized modules. The first two modules are common and at the level of the third module, the students have the opportunity to elect one of the three specializations:

- “Global international transport policy”;
- “Traffic safety and respect for the rights of all traffic participants”;
- “The role of road and transport sector for the sustainable development of regions”.

Each module of the Program provides one English course. The first module contains a course of General English, the second module includes a course of English for Specific Purposes and in the third module there are three elective courses of English for Specified Purposes. The election of the course depends on one of the three specializations mentioned above. All the foreign language courses are 36 contact hours each, or 1credit point.

According to the Program, the course of General English is aimed at the development of students’ communicative competence the level of which allows them to discuss general problems of international policy and participate in Moscow International UN Model. The course of English for Specific Purposes is mainly supposed to train students to be able to use terminology of international policy, transport and law both in oral and written forms. The English for Specified Purposes is aimed at developing students’ communicative competence necessary for oral and written communication in the professional activities according to one of the three specializations they have chosen.

MADI Foreign Languages Department is responsible for linguistic training of students. The goal of the Department was to design and implement the General English course of the first module. According to the Program, the content of teaching materials should cover the issues connected with the UN, the principal organs of the UN, the system and structure of the main UN organizations, the economic and social policy of the UN including international conventions on traffic safety.

With the knowledge and the skills acquired, at the end of the course students should be ready to participate in Model United Nations. It is also known as "Model UN" or "MUN". Model UN is a conference organized as an extra-curricular activity of students. This conference is a role play in which students have the roles of UN delegates to the United Nations and simulate UN committees. After the debate, the participants adopt a resolution aimed at solving some problem.

With the necessity to achieve such results, it was evident that there was no chance to find an English textbook already published corresponding to such content and aims. Therefore, the second goal of the Department was to compile teaching materials for the curriculum developed.

The Program aroused interest and 45 students applied for the Program. They were MADI first-year students of the Departments of Economics, Logistics and Management. They enrolled at the Program without any entrance examinations in a foreign language, which made the task of the Foreign Languages Department even more difficult.

3 Approach

The General English course of the first module for the Program "International Transport Policy" was designed on the base of the concept of long-life foreign language training diversification in engineering education [1] that provides the corresponding technology [2]. The area of the language application and the language activity in this sphere were already determined in the Program as synchronization of international multimodal and automotive transportation and participation in international conferences and negotiations. The organization where the language will be applied was defined as a multinational team.

It was necessary to determine communicative needs' of learners and specify educational aims and objectives, to select the content of teaching and to describe learning outcomes. That will give the opportunity to develop teaching materials for the English course. For that purpose, a survey of all the students of the Program "International Transport Policy" was undertaken which is of vital importance with the learner-centered approach. For the survey, two questionnaires were prepared: a preliminary one and a final one.

At the first stage, it was necessary to work out a preliminary questionnaire and to analyze its results. The questionnaire was supposed to find out motivation of learners, their language background, the level of their English communicative competence and the difficulties they have when studying English. Later the results of the students' self-assessment were compared with the outcomes of the placement test.

At the second stage, according to the description of the learning outcomes in the Federal State Educational Standard "International Relations" [3], the data collected in the survey and the outcomes of the placement test, we worked out teaching materials.

The third stage provided the teaching process. There were four groups of students and two English teachers. Both of them were involved in compiling teaching materials.

At the fourth stage, the students were tested and participated in Model UN conference to assess the outcomes of learning.

At the fifth stage, after the course the final questionnaire was prepared and analyzed. It was supposed to reveal the learners' evaluation of the course, the opinion on the skills they acquired, their attitude towards teaching materials and recommendations for the course improvement. The results were compared with the outcomes of learning.

4 Actual Outcomes

At the first stage, the preliminary questionnaire gave the data concerning learners' language background, the course expectations, self-assessment of their English communicative competence, and the main difficulties they have while studying the foreign language. The placement test gave objective assessment of the learners' level of communicative competence.

The preliminary questionnaire showed that before entering the University the majority of students studied in ordinary secondary schools (93.4%) and only 6.6% of students left specialized language schools. However, they turned out to be highly motivated to master English. Before the beginning of the Course 95% of them had studied English additionally, 48.8% of students had had one-to-one lessons with a private teacher, 31.1% of them had had the experience of self-study, 11.1% of them had attended different English courses, 8.8% of them had used other means of improving their English. The students' interest to studying English is also confirmed by the fact that almost half of the students (46.6%) elected All-Russian State Examination in a Foreign Language as one of their final secondary school examinations and passed it with good marks.

These results of the survey correspond to the self-assessment of the students' level of English communicative competence. According to the students' opinion, their average level is Intermediate but they feel more confidence in written speech. In reading, 35.5% of them assess their level as Upper Intermediate and 35.5% of them as Intermediate. In writing, the level of 33.3% is Upper Intermediate and of 44.4% is Intermediate. The self-assessment of oral speech is lower. In speaking, 53.3% of the students assess their level as Intermediate and 20% as Pre-Intermediate. In listening, the results are even still lower: the level of 37.7% is Intermediate and of 26.6% is Pre-Intermediate. As we see the students feel more self-confidence in reading and writing than in listening and speaking.

Consequently, the students give the priority to acquiring skills in oral speech. 84.4% of the students want to acquire skills to understand oral speech, 82.2% – to participate in formal oral communication, 77.7% – to participate in oral informal communication, 60% – to read professional literature, 64.4% – to write texts, letters, 53.3% – to compile documents, 51.1% – to interpret from English into Russian, 44.4% – to translate written texts, and only13.3% – to make presentations. (The sum is more than 100% in cases when it was possible to choose more than one answer; 6.6% of the students chose some other activities).

According to the students' opinion, the main difficulties they have when studying English are also connected mainly with oral speech. They have problems with the usage of grammar structures for expression of their own ideas both orally and in writing (53.3%); understanding oral texts (46.6%); maintaining a conversation with a partner

(44.4%); defining the content of their own oral speech (37.7%); writing the text of a presentation (33.3%); learning new words (31.1%); understanding grammar structures in reading or translation (26.6%); translating written texts from English into Russian (11.1%); understanding written texts (9%); translating written texts orally (6.6%).

The results of the survey were confirmed by the outcomes of the placement test of the four groups of the students that showed that the average level of the students in reading was Upper Intermediate, in writing it was Upper Intermediate, in listening it was Intermediate, and in speaking it was Intermediate.

At the second stage, the teaching materials were compiled according to the requirements of the Program, the data of the preliminary questionnaire and the workload of 36 contact hours. The materials are grouped in four units; each of them takes eight contact hours of classwork. The program defined the UN as the content of English texts and Model UN as a final event for the assessment of learning outcomes. Thus, the main topics of the units are the following: Unit 1 "The history of the UN", Unit 2 "The structure of the UN", Unit 3 "The transportation policy of the UN", Unit 4 "Model UN".

The structure of each Unit was traditional and contained the following sections: "Introduction", "Active Vocabulary", "Listening", "Reading", "Speaking", "Writing", and "Self-study Work". The content of each section was determined according to the results of the preliminary questionnaire. In order to help students to overcome the difficulties they have the section "Introduction" uses tasks in oral speech. The section "Active vocabulary" contains about 900 words and word combinations, work with synonyms and antonyms, definitions of terms, names of the countries, nationalities and languages, etc. Special attention is given to the section "Listening" that includes two or three video films and oral texts. The section "Speaking" contains various tasks preparing students for participating in a monologue, a dialogue, and a discussion. The section "Reading" includes two or three texts on the topic of the Unit for reading and translation. In "Writing", the students learn to compile formal documents and letters. The section "Self-study Work" provides a variety of tasks to overcome the students' difficulties revealed, for example, writing scripts of video films or oral texts. The project work is connected with the students' preparation for Model UN conference.

At the third stage, four groups of the students had English classes with the teaching materials developed. However, because of coronavirus pandemic each group had only two face-to-face classes. After that, the situation changed radically and the teachers had to switch to distance education. MADI immediately provided the access to Microsoft Teams. All the English classes in the form of webinars were organized with the help of this communication tool. 100% of the students concluded that Microsoft Teams were effective for the English course. Only 11% of the students think that it is necessary to increase the number of contact hours. All the groups studied three Units and participated in Model UN as a final lesson.

The fourth stage provided the organization of remote Model UN. The topic of the conference was "Coronavirus Pandemic: Different Countries and Strategies" and provided team project work. The students were divided into groups of four or five and collected the materials for the presentation of various countries. At the same time, all of them participated in compiling the final resolution. The role-play "Model UN" integrated all the language skills developed and showed the improvements of the students' skills in reading, writing, speaking, and listening. These results were confirmed by the learning outcomes of the final tests.

At the fifth stage, the students filled in the final questionnaire and the information received was analyzed. According to the data of the final questionnaire, the majority of the students of the Program "International Transport Policy" evaluate the English course as effective. 33% of the students consider the course to be extremely effective, 59.4% – effective, and 6.6% – more likely to be effective than not effective.

The course improved the students' level of English communicative competence. 22% of the students are sure that they have improved their level considerably, 55% of them are sure that they have improved it but they would like to achieve even a higher level, 22% - have improved it but not considerably.

The students think, that by the end of the course they have acquired various skills in English and they are able: to translate written texts (44.4%); to interpret from English into Russian (44.4%); to understand oral speech (41.8%); to read professional literature (37.4%); to participate in formal oral communication (35.2%); to participate in oral informal communication (33%); to write texts, letters (24.2%); to make presentations (15.4%); to compile documents (8.8%).

Due to the course, the majority of the students have overcome the difficulties they had: 50.6% of them have overcome them to some extent, 35.2% - considerably, 6.6% - completely and only 6.6% have failed to do it.

Answering the open question concerning the efficiency of the English course especially for them, the majority of the students wrote that the information about the UN and other international organizations was relevant to them (two students commented that they even got interested in international political events). Most part of the students concluded that it was useful for them to learn a large number of new words and expressions, to improve their pronunciation, to begin using synonyms of the words they had already known, to use actively the names of the countries, nationalities, etc. Many students are sure that they have improved their listening skills that are most difficult for them. Some of them find that it was watching a lot of video films and especially writing their scripts that contributed to this improvement. Many students appreciate that at the lessons they had the opportunity to communicate with teachers and groupmates in English.

As far as the students' recommendations on the improvement of the English course are concerned, the typical answer is that the course was effective and does not need any changes. At the same time, individual suggestions were quite different. Among the things that the students suggested are the following: increasing the number of tests, video films, presentations, project work and pair work, simulations for speaking, etc.

5 Conclusions

The additional educational program "International Transport Policy" is the response of the two universities, MGIMO and MADI, to a new demand for the specialists having professional competences both in the sphere of international relations and multimodal transport. The Program aroused interest of MADI students.

The concept of long-time foreign language training diversification in engineering education allowed designing and implementing a new course of General English for the Program. The survey gave the opportunity to reveal the students' motivation, their language background, self-assessment of their English communicative competence level, the difficulties they have when studying the foreign language. The data of the survey helped to design a tailor-made course for the language needs of the particular target audience. As a result, the students have acquired the necessary skills and the students evaluate the course of General English as effective. The final leaners' outcomes show that the structure of the course and the teaching materials compiled are adequate to the General English course objectives. The recommendations of the students will be used to improve the course.

References

1. Polyakova, T.Y.: Variety of engineers' needs in the foreign language usage as a basis for their training diversification. Soc. Behav. Sci. Procedia **214**, 86–94 (2015)
2. Polyakova, T.: Designing methods-and-curriculum basic for foreign language training courses in modern non-linguistics universities. Methods-and-curriculum basis of foreign language vocational training at a non-linguistics institution for higher professional education, issue **14** (725), Moscow, FSFEI HPE MSLU, pp. 64–76 (2015)
3. FSES HE training program in the field of international relations 41.03.05 (Bachelor's degree) (2016). http://fgosvo.ru/410305. Accessed 25 May 2020

Interdisciplinary Sustainable Development Module for Engineering Education

Marina Zhuravleva, Natalia Bashkirtceva, Galina Klimentova, Nina Kotova, and Elvira Valeeva(✉)

Kazan National Research Technological University, K. Marx Street 68, 420015 Kazan, Russia
elviravaleeva75@yandex.ru

Abstract. Global priorities of the new industrial revolution aim at intensive developing advanced technologies for industrial clusters promoting success of regional and national economies. Chemistry and petrochemistry are among the key industries in Russia. The existing Russian petrochemical industrial clusters, however, do not contribute to resilient society development due to complex scientific content, inherent engineering and ecological hazards. The EU is the leader in resilient society development policy at global scale. The EU has initiated the strategy 'Industrie 4.0' as a start of the fourth industrial revolution. Therefore, the EU approaches can be widely used in petrochemical engineering for employers, researchers, university professors and students. It is important to improve and upgrade the educational programs for training highly qualified specialists for the chemical industry. The article presents the interdisciplinary module developed to teach Master's degree students majoring in chemical engineering at Kazan National Research Technological University. The module consists of four interdisciplinary teaching courses aimed at developing professional skills in the field of sustainable development of the chemical industry based on the EU industrial principles, tools and technologies.

Keywords: Engineering education · University · Sustainability

1 Introduction

The EU counties have over 50 years' experience in developing and operating industrial clusters. The EU policy of sustainable development indicates the main principles of a continuous long-term improvement of life quality and implements them successfully to create resilient communities. The EU platform 'Industrie 4.0' promotes global principles for a new high-tech strategy [1]. Therefore, introduction and dissemination of EU approaches in economy of other countries will give better results.

The success of Russian development depends on industrial modernization of petrochemical and petroleum refining clusters which are one of the most promising sectors of National economy [2]. Therefore, the sustainable development of petrochemical cluster facilities is the priority for the National policy.

Global trends in ensuring natural and human security leads to the transformation of industrial technologies which includes:

M. E. Auer and T. Rüütmann (Eds.): ICL 2020, AISC 1328, pp. 737–743, 2021.
https://doi.org/10.1007/978-3-030-68198-2_69

- conducting fundamental research;
- designing and implementing new advanced technological equipment;
- developing environmentally friendly processes and materials.

Under these conditions, a modern engineer must be able to make decisions according to the principles of sustainable development [3], based on the existing technological limitations for social, economic and environmental consequences of engineering operations. Therefore, the development of engineering education focusing on the principles of sustainable development is of great importance. The purpose of the study is to identify the features and conditions for the implementation of advanced engineering training in the Master's degree program, to develop and evaluate the effectiveness of an interdisciplinary teaching module aimed at training students in accordance with the trends of future engineering.

2 Interdisciplinary Teaching Module

Innovative and multifunctional aspects of engineering activity require the development of combined competencies in forecasting, developing and managing industrial technologies, taking into account global technological priorities. The global trend for modern industrial companies is sustainable development based on the strategy 'Industrie 4.0' [1], the key goals of which include:

- development of eco-friendly technologies;
- digitalization of production processes;
- application of renewable power sources;
- reduction of production time and cost.

The European Union has long experience in an integrated approach to the development of industrial sectors [4]. The implementation of European strategy of sustainable development may lead to the deep modernization of industries based on "breakthrough" technologies and the creation of new rapidly growing markets for the high-tech economy. Therefore, it is important to foster interdisciplinary approach to developing the engineering education content focusing on the integrated knowledge of natural, technical, economic, social and humanitarian sciences [5–8]. Nowadays, engineering education in Russia aims at to train highly qualified specialists who are able to provide safe and effective operation of new and existing industrial processes using advanced research methods and information technologies.

The leading university in the field of chemistry and petrochemistry of the Republic of Tatarstan is Kazan National Research Technological University (KNRTU). It is the largest educational center in Russia, which provides advanced training of specialists in the field of chemical technology [9].

At KNRTU, the teaching interdisciplinary module focused on studying EU principles of sustainable development based on the strategy 'Industrie 4.0' for developing Russian chemical industry has been developed.

The module 'Sustainable Development of Petrochemical Industry' has been integrated into the existing MSc program in chemical engineering.

The module consists of 4 teaching courses:

- Chemical Process Reliability;
- Lean Production in Oil and Gas Industry;
- Advance Regulation and Ecological Safety;
- Breakthrough Technologies and Resources for Sustainable Development of Chemical Industry.

The course 'Chemical Process Reliability' trains students to apply managerial and engineering approaches to enhancing chemical process reliability to provide smooth and safe service of equipment. The course content includes EU principles of operational reliability based on sustainable development of the chemical industry. These principles are used by students to estimate the technological ways for increasing the reliability of industrial processes in Russian chemical companies.

The course 'Lean Production in Oil and Gas Industry' trains students to apply principles, tools and methods of lean production for reducing production time and cost at all stages of a petrochemical company lifecycle and improving the quality of final products. The course is developed on the basis of the EU experience in the sustainable development and lean production.

The course 'Advance Regulation and Ecological Safety' aims at disseminating the principles and approaches of the EU environmental policy, improving regional environment and training highly qualified chemical specialists who are able to solve ecological problems and increase the quality of final products through engineering principles. The course contributes to the development of environmentally friendly technologies of the region. The course trains students to work with Russian, European and international standards for the chemical industry.

The course 'Breakthrough Technologies and Resources for Sustainable Development of Chemical Industry' trains students to estimate and characterize hydrocarbon feedstock, the processing depth of hydrocarbons and their effective use. Students study the activity of Russian and international oil and gas companies, aimed at sustainable development of chemistry and petrochemistry. The course provides knowledge on technological methods related to catalytic processes and environmental catalysts to solve ecological problems and safe production of chemical industries based on EU experience in this field.

The module is an interdisciplinary course including lectures, discussions, seminars, multipurpose research project, and business role-plays and management decision simulation. The module develops integrated competences to apply the EU criteria for operation and management of chemical complex as a sustainable system that leads to the technological progress of an industrial company. These teaching methods foster the practical orientation of the courses and expand the range of theoretical training of students. Such interactive activities not only involve students in the educational process, but also contribute to the development of their soft skills which are necessary for critical thinking, teamwork, creativity, sociability, and an active life position [10].

The educational process is based on the concept of the zone of proximal development originally developed by Russian psychologist L. Vygotsky [11], and fosters new student knowledge and skills to predict, organize, operate, and control the hazardous production according to the objectives of resilient society and EU principles of sustainable development.

The module allows students to develop their engineering skills in real petrochemical companies. Students are to propose the innovative technological solution for waste treatment and recycling. This task is based on the analysis of existing technological systems, the best practices of the EU countries and Russia in this field. Based on the analysis, students select the optimal system for cleaning emissions from the petrochemical facility. The choice of the system is to be confirmed by technological and economic calculations of equipment. The final results of this work are presented at the students' conference and published in conference proceedings. At the end of the conference the best technological system proposed for waste treatment is selected.

To increase the efficiency of the educational process, the lectures are held in the form of conferences. The purpose of such lectures is to intensify the educational and cognitive activities of students. The topic of the lecture is outlined, and each student within a few minutes formulates the most interesting questions that are sorted by their content. In response to students' questions, the lecturer conveys the basic information of the lesson. In conclusion, an assessment of students' questions as a level of their knowledge and interests is carried out.

The method of group project training for solving problems is used during the module. Students in groups are to solve a specific problem for a real chemical or petrochemical company. This project work includes several stages that may improve the skills of logical thinking, maximize the creative abilities of students and foster their research activities.

3 Interdisciplinary Research Projects of MSc Students

MSc students will pursue interdisciplinary research project 'Sustainability Prospects of Petrochemical Production, Process, Company'. The research project is aimed at applying the EU mechanisms and tools of sustainable development to a regional commercial petrochemical company. The research project includes the impact of a technological process on the cluster, a process test, estimation of ecological parameters, process engineering, parameters of feedstock and final products, a fire threat level and explosive hazards of a company. During the research, MSc students conduct SWOT analysis and define environmental risks and bottlenecks of a company. The EU experience, tools and mechanisms of sustainable development are used to find the new ways of increasing safety and reliability of equipment, optimizing technological processes, solving ecological problems, and developing energy and resource saving technologies for a commercial petrochemical company. The research project is held gradually during the internship programs of MSc students. The program includes the following types of internship:

- Theoretical research internship (selection of the subject of research project; analysis and study the EU principles of sustainable development of chemical industry).
- Practical research internship (patent analysis; study and analysis of scientific publications; development of project implementation plan; development of mechanisms

for implementing the EU principles of sustainable development in a commercial petrochemical company).

- Industrial research internship (detailed study of a commercial company, its industrial technologies and processes, generation of proposals for sustainable development of a company, economic assessment of the proposals).

The research project results are presented at the conferences and published in scientific journals. Industrial experts participate in conferences, public presentations and review the MSc theses. The close cooperation between the University and industrial companies accelerates the implementation of the results of research projects in industry. The new engineering solutions are implemented in regional chemical and petrochemical industries. The research project develops interdisciplinary competences and trains students to solve managerial, economical, ecological and engineering problems through the EU principles of resilient society.

4 Results

The teaching module 'Sustainable Development of Petrochemical Industry' trains MSc students to apply the principles, methods and tools of sustainable development to reduce production costs, increase productivity and quality of final products based on the experience and best practices of the EU countries. The module develops soft and hard skills of students which are necessary for solving engineering, technological, economic, environmental, organizational problems. These skills allow students to quickly adapt to new conditions of a real industrial company and transfer their knowledge, skills, and individual abilities to objects of professional activity.

Graduates of our program (over 90%) work in chemical and petrochemical companies with innovative technologies. Examples of Russian companies are Gazprom, SIBUR holding, Tatneft, Rosneft, and LUKOIL. These enterprises are characterized by a high level of technology, as well as they are the leaders in the digitalization of industrial processes. Our graduates work in the international companies such as Haldor Topsoe, British Petroleum. The development programs of these Russian and international companies aim at sustainable development of their industries. In this regard, the environmental issues are of great importance. Therefore, they continue to need highly qualified chemical and petrochemical engineers to develop and implement their advanced technologies and processes meeting the requirements of industrial and environmental safety.

The industrial companies assess the competences of graduates. The special tests developed by the companies show the high level of technology, economic, managerial and ecological skills of our graduates. These skills allow students to be employed as engineers and technologists after their graduation and become competitive in the labor market.

Innovative research projects of our Master's degree students (up to 15%) are being implemented in the technological processes of industrial companies. The most promising projects are [12–15]:

- Synthesis, regeneration and utilization of catalytic systems using supercritical fluid media.
- Development of organized polyfunctional molecular materials for optoelectronics and biomedicine.
- Self-organizing crystalline structures in disperse oil systems.
- Spontaneous supramolecular organization in oil disperse systems under the influence of external factors.

For university professors, the development and implementation of an interdisciplinary module may be considered as an additional opportunity for professional development, since it requires the knowledge of the latest materials for the educational process, ensuring the interconnection of different branches of science and technology. Interdisciplinary module content requires the development and implementation of new teaching forms and methods, which increases the effectiveness of engineering training.

5 Conclusions

The global industrial processes lead to changes in the employer requirements for engineers. Education focusing on training chemical and petrochemical engineers of new generation is characterized by the principles of advanced training system. These principles include integration, interdisciplinarity, variability, intensification of education, social partnership, motivation, and personal development.

Multifunctional engineering activity requires the development of interdisciplinary training. Therefore, the most important task is to create interdisciplinary teaching courses based on new training methods and approaches. The proposed teaching module allows us to develop the following engineering skills of Master's degree students:

- the ability to adapt quickly to changing conditions of modern industry;
- ability to predict industrial changes;
- ability to develop, introduce and analyze innovations as well as the social and professional consequences of their use;
- ability to operate modern industrial equipment and instruments according to the standards of production and environmental safety;
- ability of creative thinking, critical analysis and organizational socialization;
- ability of personal and career development.

Such professionals possess world-class competencies, who are capable of organizing, coordinating, and managing complex scientific and technical projects that form the basis of sustainable development and growth of industrial complexes. Our graduates is ready to use the EU experience, tools and mechanisms of sustainable development and find the new ways of increasing safety and reliability of equipment, optimizing technological processes, solving ecological problems, and developing energy and resource saving technologies for a chemical company.

References

1. Buyanova, M.E., Mikhaylova, N.A.: Industrial revolution 4.0: tendencies and risks of social and economic changes in the regions of Russia. In: Inshakova, A., Inshakova, E. (eds) Competitive Russia: Foresight Model of Economic and Legal Development in the Digital Age 2019, CRFMELD, Lecture Notes in Networks and Systems, vol 110. pp. 95–102. Springer, Cham (2019)
2. Dmitrievskii, A.N.: The fundamental basis of innovative development of the oil and gas industry in Russia. Her. Russ. Acad. Sci. **80**, 7–21 (2010)
3. Adibi, N., Ataee-pour, M.: Consideration of Sustainable development principles in ultimate pit limit design. Environ. Earth Sci. **74**, 4699–4718 (2015)
4. de Sadeleer, N.: Sustainable development in EU law: still a long way to go. Jindal Glob. Law Rev. **6**, 39–60 (2015)
5. Zhuravleva, M.V.: Outstripping professional training in internationalization of higher education conditions. In: 2013 International Conference on Interactive Collaborative Learning, ICL 2013, pp. 367–368 (2013)
6. Shageeva, F.T., Bogoudinova, R.Z., Kraysman, N.V.: Poster: teachers-researchers training at technological university. Adv. Intell. Syst. Comput. **917**, 977–980 (2019)
7. Sultanova, D., Maliashova, A., Bezrukov, A.: Consistent development of the training program "innovation management". In: Advances in Intelligent Systems and Computing, vol. 1135 AISC, pp. 234–243 (2020)
8. Ziyatdinova, J., Bezrukov, A., Sanger, P.A., Osipov, P.: Cross cultural diversity in engineering professionals - Russia, India, America. In: ASEE 2016 International Forum (2016)
9. Zhuravleva, M.V., Bashkirceva, N.Y.: Cluster system of specialist training for petrochemical industry. J. Chem. Technol. Metal. **50**(3), 321–324 (2015)
10. Valeeva, E., Ziyatdinova, J. Galeeva, F.: Development of soft skills by doctoral students. In: Advances in Intelligent Systems and Computing, vol. 1135, pp. 159–168 (2020)
11. Gielen, U.P., Jeshmaridian, S.S., Lev, S., Vygotsky, S.: Vygotsky: the man and the Era. Int. J. Group Tensions **28**, 273–301 (1999)
12. Bakirova, I.N., Zenitova, L.A., Romanov, D.A., Rozental, N.A., Kirpichnikov, P.A.: Waste-free process for production of castable polyurethanes. Russ. J. Appl. Chem. **70**(9), 1483–1487 (1997)
13. Bakirova, I.N., Romanov, D.A., Gubanov, E.F., Zenitova, A.A.: Thermomechanical analysis of poly(urethanes) produced from recycled raw materials. Poly. Sci. Series **40**(9–10), 323–326 (1998)
14. Lakhova, A.I., Valieva, G.R., Valieva, A.A., Karalin, E.A., Petrov, S.M., Bashkirtseva, N.Y: Catalytic activity of an Al-Cl-Re system for N-hexane cracking. Chem. Technol. Fuels Oils. **55**(1), 3–7 (2019)
15. Sladovskaya, O.Y., Tsyganov, D.G., Bashkirtseva, N.Y., Mukhametzyanova, A.A.: Peculiarities of the process of destruction of stable water-oil emulsions in intermediate layers. J. Chem. Technol. Metal. **53**, 191–201 (2018)

STEM4Girls-Workshop on Machine Learning

Veronika Thurner(✉) and Johanna Sickendiek

Munich University of Applied Sciences, Munich, Germany
veronika.thurner@hm.edu , johanna.sickendiek@gmail.com

Abstract. Although female specialists in the STEM area are urgently needed in the work force, the percentage of women in STEM study programs in higher education is still rather low. One reason for this are gender-related stereotypes that still persist in many modern cultures, which associate STEM as being a "male" domain. To actively draw young women towards STEM and Computer Science, we design a workshop for girls in the age group of 10th grade in secondary education. This workshop introduces the girls to machine learning in the application context of fashion and shopping. We define both cognitive and affective learning objectives and corresponding observable outcomes that indicate whether the learning objectives were reached or not. Based on these learning objectives, we develop a didactic concept for our workshop, as well as a web application as a front end for the machine learning application. Finally, we execute a pilot run of our workshop, which we evaluate with respect to participant satisfaction and the achievement of the learning objectives.

Keywords: Gender gap · Girls in CS · K12 education · STEM education · Machine learning

1 Motivation

The fifth of the United Nations Sustainable Development Goals addresses gender equality. At least in the western hemisphere, much has been accomplished towards equal rights of men and women. Nevertheless, a vast amount of gender-related stereotypes still persist, influencing both men and women in many areas of their lives, including their choice of studies and profession. As a result, the percentage of women in STEM courses is very low [9], as many women regard this domain as being "male". But it is precisely in these areas that female specialists are urgently needed.

To attract more women to study STEM, we must circumvent existing stereotypes and reach out to girls and young women, help them get into contact with STEM and support them in finding an emotional access to this domain, free from prejudice and traditional dos and don'ts. In order to achieve this, it is important to chose a setting and examples that allow the girls and young women to bond with the topic; i.e., the STEM content must relate to their existing world and their interests.

M. E. Auer and T. Rüütmann (Eds.): ICL 2020, AISC 1328, pp. 744–755, 2021.
https://doi.org/10.1007/978-3-030-68198-2_70

In the following, we provide a brief overview of related work (Sect. 2) before introducing the goals of this contribution in Sect. 3. Section 4 describes the approach we took in the development of the STEM4Girls-Workshop. Learning objectives are specified in Sect. 5, distinguishing between cognitive and affective learning objectives. In Sect. 6, we illustrate the course concept that we employ to achieve these learning objectives. First experiences are provided in Sect. 7, before summing up our conclusions in Sect. 8.

2 Related Work

There already exists a variety of offerings that try to attract young people to technology at an early age. A popular one is LEGO mindstorms (www.lego.com/mindstorms), a programmable robotic device whose physique can be custom designed using LEGO building blocks [6]. However, most prepackaged lego boxes and project instructions focus on action figures, space ships and the likes, which still appear to be more appealing to boys than to girls.

Similarly, the majority of the course offerings of the local initiative FabLab (www.fablabkids.de) focusses on robotics and electronics in an optical presentation that turns out to be still more appealing to boys than to girls.

In contrast to this, the offerings on the website MachineLearningForKids.co.uk comprise projects whose topic and optical presentation is of great appeal to both sexes. However, most of these projects focus on younger children rather than teenagers.

A popular program that specifically focusses on attracting young women to tech-topics and professions is the Girls' Day, which is regularly organizied in more than 20 countries world wide (see e.g. www.girls-day.de). At the Girls' Day, businesses and universities open their doors to female students of 5th grade or higher, to give them insights into professions that traditionally are male dominated, get them into contact with female role models and thus widen their perception of possible career paths [4].

3 Goals

To actively draw young women towards STEM and Computer Science, we design a workshop for girls in the age group of 10th grade in secondary education. This workshop introduces the girls to machine learning in the application context of fashion and shopping – which is a typical favourite topic of interest of the addressed age group in our geographic area.

This workshop aims at getting the girls interested in machine learning, and to make them aware of the many application areas in which they already encounter machine learning in their everyday life, usually without noticing it. As well, it attempts to help overcome existing stereotypes that machine learning is "just for boys", can only be applied to "nerdy stuff" or is generally "boring", as well as the gender stereotype that "girls are intellectually unfit to study something technical". Finally, it contributes to triggering a process in which girls reflect on the educational and career path they intend to enter after finishing their secondary school education.

4 Approach

To start with, we analysed existing offerings that aim at attracting girls and young women to STEM. Realizing that there are quite a few offerings for younger kids, but comparatively few for teenage girls that are about to finish school, we decided to specifically address this age group of girls between 15 and 17, which is also on the verge of choosing their future career paths they want to pursue after school.

As a next step, based on a literature analysis on the life contexts and living realities of today's youth, we identified possible areas of interest which we hope to be attractive to the girls in our target group. We backed up our choice by executing interviews with selected representatives of our target group. On this basis, we decided on using fashion and shopping as application domain for our workshop.

Then, for the content and skills we intend to address within the workshop, we define both cognitive and affective learning objectives and corresponding observable outcomes that indicate whether the learning objectives are reached or not. Cognitive learning objectives address the intended increase of knowledge and mental skills. In our definition of cognitive learning objectives, we distinguish different skill levels according to Bloom's taxonomy [3] in its revised form [1]. Achieving these cognitive learning objectives requires some sort of affective motivation. This relates to affective learning objectives, which address a development of values, attitude and feelings. Here again, we distinguish different levels, following the taxonomy of [7].

Based on these learning objectives, we develop a didactic concept for our workshop. We structure the workshop into phases, starting from bonding with the role model that moderates the workshop, via establishing a learning setting and awakening motivation, activating previous knowledge (and making the girls aware of the contacts most of them already had to machine learning applications, without being aware of it). On this basis, we enter into the more technical part, providing an introduction into some machine learning essentials, intertwined with their immediate application and an analysis of the results that the girls achieve with their models.

To support the didactical concept, we develop a web application as a front end for the machine learning application, which the girls use throughout the workshop to create, train and then use their own model. Furthermore, we develop a questionnaire to assess participant satisfaction.

Finally, we execute a pilot run of our workshop, which we evaluate with respect to participant satisfaction and the achievement of the learning objectives.

5 Learning Objectives

To ensure that the workshop has a fair chance of achieving the desired effect, we clarify the desired outcome in terms of learning objectives. For each of these learning objectives, we specify a corresponding observable behaviour or effect that indicates whether the learning outcome fulfills the intended learning objective.

5.1 Cognitive Learning Objectives

Cognitive learning objectives address the intended increase of knowledge and mental skills. In our definition of cognitive learning objectives, we distinguish different skill levels according to Bloom's taxonomy [3] in its revised form [1].

As our workshop intends to *attract* girls to the STEM area and specifically to machine learning (and not repel them), we must ensure to address the right level of complexity. I.e., we need to take care that the girls will be neither bored nor pushed too hard. Since the target group of our workshop possesses little or no previous knowledge on machine learning and the time of the workshop is limited to two hours, we restrict ourselves to learning objectives on the lower levels of the taxonomy, C1: Remember, C2: Understand and C3: Apply.

Table 1 shows a selection of the cognitive learning objectives for our workshops, and their corresponding fulfillment indicators.

Table 1. Selection of cognitive learning objectives and their corresponding fulfillment indicators

Level	Number	Learning objective: students ...
Remember	C1.1	... name examples for applications of machine learning that they encounter in their everyday life
	IC1.1	> discuss with their neighbours and point out several examples of occurences of machine learning in their everyday life
Understand	C2.1	... explain why the results of machine learning can be incorrect
	IC2.1	> explain the relevance of the amount and the quality of the training data, and the significance of the threshold value
Apply	C3.1	... train their own machine learning model using the tool Clarifai
	IC3.1	> work on the exercise and stay focused on the training process

5.2 Affective Learning Objectives

Achieving these cognitive learning objectives requires some sort of affective motivation. This relates to affective learning objectives, which address a development of values, attitude and feelings. Here again, we distinguish different levels, following the taxonomy of [7].

For our workshop, we focus on the lower levels, A1: Receive, A2: Respond and A3: Value. (Due to the comparative shortness of our workshop and as each set of girls participates only once, we deem levels A4: Organize and A5: Internalize to be out of our reach.)

Table 2 depicts a selection of the affective learning objectives for our workshops, and their corresponding fulfillment indicators.

Table 2. Selection of affective learning objectives and their corresponding fulfillment indicators

Level	Number	Learning objective: students ...
Receive	A1.1	... realize that gender stereotypes inhibit their studying in the STEM area
	IA1.1	> select the check box "prejudice" (instead of "reality") at questions 8.1–8.4 of the provided questionnaire. (8.1: STEM is just for men. 8.2: For studying STEM, I need top grades in math. 8.3: Studying STEM is only for nerds. 8.4: If I study CS, I will sit in front of my computer all by myself all day long.)
Respond	A2.1	... want to know more about studying in the STEM area
	IA2.1	> ask for more information on study programs or job opportunities in the STEM area
Value	A3.1	... form an opinion about studying in the STEM area
	IA3.1	> appraise the pros and cons of their studying in the STEM area and explain their opinion

6 Course Concept

To make reasonably sure that our workshop will be successful, we carefully developed a didactic concept.

6.1 Role Model as Instructor

According to [5], it is the person (and personality) of the teacher that has the greatest impact on students and on the learning outcomes they achieve.

To increase the chance that the young women in our target group bond emotionally to the domain we address in our workshop, the workshop should be supervised by a young female tech professional that can serve as a role model. In addition to her technical expertise, this role model must be well skilled in a variety of non-technical competences. For example, she should be highly motivated, enthusiastic about her professional domain, self-ensured as well as empathic and sensitive, self-reflected, open to share her own experiences (both good and bad) along her path into her current position, and, generally, approachable.

Furthermore, the role model becomes easier to identify with, if she has been somewhat "similar" to the girls in the workshop when she was their age, i.e. "just a normal girl", rather than some technical Einstein or Ada right from the cradle. As well, being authentic and adding personal information to her messages and information makes the role model more convincing and thus helps the participating girls to bond with her, and to take her and the experience she shares as some kind of inspiration for their own steps into the future.

6.2 Combine Theoretical Input with Acitivity and Self-reflection

The vast majority of teenagers in the age group adressed by the workshop, possesses an attention span of approximately 10 to 15 min, with attention rapidly decreasing after spending this time span on a single acitivity. However, changes in activity or the applied teaching and learning methods help to keep the students' attention high throughout a longer period of time [2]. Therefore, the workshop was designed to incorporate a variety of didactic methods, emphasizing on involving the participants in an active way as much as possible and in every phase of the workshop.

Table 3 provides an overview of the workshop structure, the applied didactic methods and the tools employed.

Table 3. Overview of the workshop structure, didactic methods and tools

Phase	Duration	Content	Method	Tools and Media
1	10 min	Provide overview, get connected	- Presentation - Self reflection	- PowerPoint - VoxVote
2	15 min	Introduce STEM study programs	- Presentation - Personal anecdotes - Q & A	- PowerPoint
3	5 min	Activate prior knowledge	- Test style questions - Buzz group	- VoxVote
4	10 min	Introduce Machine Learning	- Advance organizers - Didactic reduction	- PowerPoint - VoxVote
5	50 min	Apply ML	- Exercise - Team work	- "Shop the Look"-app - Laptops, internet access
6	20 min	Analyse results	- group discussion - Peer feedback	- "Shop the Look"-app - Laptops, internet access
7	10 min	Evaluate workshop	- Questionnaire - Flashlight	- Pen & Paper

At the beginning of the workshop, the instructor (i.e., the role model) introduces herself briefly, in order give students a chance to emotionally bond with her. As well, she provides an overview of the goals and the structure of the workshop and encourages the participants to ask questions and to interact. To actively involve the students as early as possible within the workshop, they are asked a variety of questions that induce them to reflect on their own career plans as well an on the mental images that they associate with different professional domains. Results are made visible via the online voting system VoxVote, which participants adress via their personal mobile devices.

Picking up on these mental images (and some underlying social stereotypes), the role model then names some Computer Science related study programs that are offered at the local universities, and presents typical job profiles that these study programs prepare for. Based on this information, participants are asked to reflect on some of the core competences that they deem to be necessary to successfully study or work in this area.

After this brief and very general introduction to the professional domain of Computer Science, the workshop narrows in on Machine Learning. As a first step, students activate their prior knowledge by answering some initial test questions.

On this basis, the role model provides a quick and very high level introduction to some key ideas of Machine Learning, using the recommendation system of a popular streaming service as an example that most participating students already know from personal experience. As well, students are sensitized to risks, limitations and ethical issues of Machine Learning. As a next step, students collect in buzz groups any application of Machine Learning that they may have encountered so far in their personal life.

Core part of the workshop is the exercise, in which students execute themselves some basic steps of Machine Learning. More precisely, they provide a system (which has been set up by the instructor) with some training data, label these data, train the system and then challenge their system by offering it new data that the system then has to classify.

After that, they critically analyze the results that the system achieved, discuss their shortcomings and finally derive typical pitfalls that occur when dealing with machine learning. Thus, students gather a first impression of aspects such as the amount and appropriateness of training data, the significance of proper labelling of the training data, and the implications of the threshold value on the quality of the decisions that the Machine Learning system makes.

In the closing phase, the instructor collects feedback of the participants, to evaluate both the students' learning outcomes and the quality of the workshop. Based on these insights, the workshop concept can be improved, before applying it again in the future.

6.3 Machine Learning Tool and Front-End

To facilitate the introduction to machine learning in the setting of our example topic of fashion and shopping, we designed a web application called "Shop the look" as an appealing front end. Through this web application, participants add their training data, train their model and finally test it on new data.

Note that this web application serves merely as a front end to the machine learning logic, which is incorporated using the web services of Clarifai (www.clarifai.com), a start-up specialized on image and video recognition.

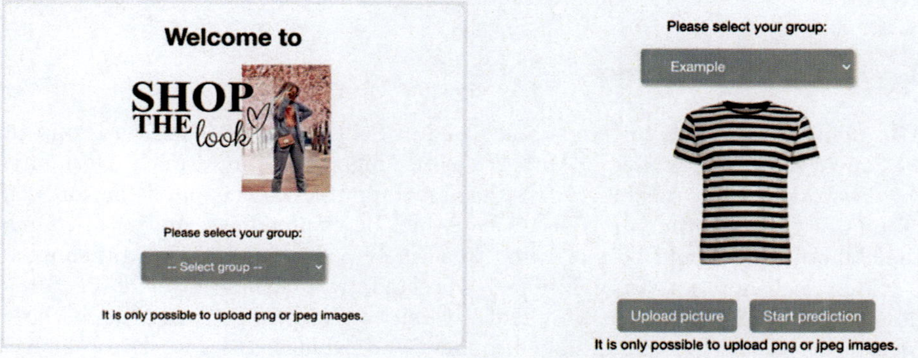

(a) Landing page, select team

(b) Upload image, start prediction

Fig. 1. Impressions of the "Shop the Look" web application [8]

Figure 1 provides an impression of the look and feel of the "Shop the Look" web application. The left part (*a*) depicts the landing page, on which students select the team they are in. This is somewhat important, as each team trains a different instance of the system, so that student will be able to observe the effects of what they did. The right part (*b*) of Fig. 1 depicts the screen that allows students to upload an image of some piece of clothing, with which they then can challenge the predictive quality of the system. Here, they can challenge both their own instance of the system that they trained themselves, as well as the system instances of the other groups.

Fig. 2. Example of classification results by the "Shop the Look" web application [8]

Figure 2 shows the classification results that the system achieved. On the right side, the different possible classification labels are listed, together with the probability that each label achieved. In the middle of the picture, the threshold value that has been previously specified is documented. Below, a button invites participants to upload a new image and challenge the system again.

7 First Experiences

The pilot run of the workshop took place in September 2019 with a group of 14 female students enrolled in 10th grade of a high school in Munich, Germany. As the workshop replaced a regular class on computer science, participation was mandatory for the students. Thus, it could not be assumed as a matter of course that all students would be motivated to actively participate in the workshop.

The workshop was organized and directed by a female student of Business Information Systems at Munich University of Applied Sciences (Hochschule München), Germany, who designed, piloted and evaluated the workshop as her master's thesis project. Being just a few years older than the participants and having successfully studied in a male-dominated area, she was a great role model for the participating girls.

In spite of the previous communication between the instructor and the school, some things did not turn out as expected. For one thing, instead of the 120 min that were indicated beforehand, the instructor was allowed merely 90 min for her workshop. As a consequence, she had to cut short the introductory information and instructions on the exercise. For another thing, the school computers allowed only limited access to the internet. Thus, for training their machine learning models, participants had to use the instructor's Laptop that did not need to rely on the school's network.

For assessing the achievement of affective learning objectives, we observed the participant's behaviour during the workshop. All the girls participated actively in the discussions, answered self reflection and quiz questions and tackled the exercise in groups. Moreover, they were highly attentive during the presentation parts of the workshop as well, not being distracted even in those phases when students had a less active role.

As well, we incorporated several questions on the students' beliefs and attitudes into the workshop. At the beginning, we asked students what they associate with STEM study programs (Fig. 3), and to give some reasons for why girls only rarely choose a technical career (Fig. 4). The answers to both questions indicate that the participating girls are higly influenced by traditional stereotypes, with respect to both the STEM domain and to gender roles.

After the workshop, we evaluated the benefit that the workshop created for the participants with respect to their choices of their future career paths (Fig. 5). The answers indicate that at least temporarily, students were able to distance themselves from the stereotypes they had voiced initially, and to look on the STEM domain and related study programs in a more open or even positive manner.

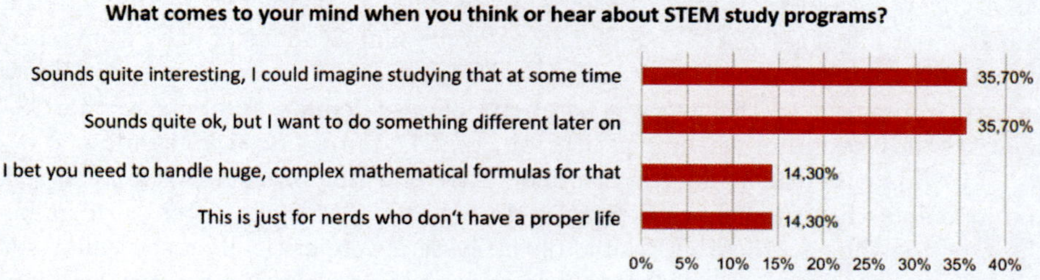

Fig. 3. Participating girls' associations with STEM study programs

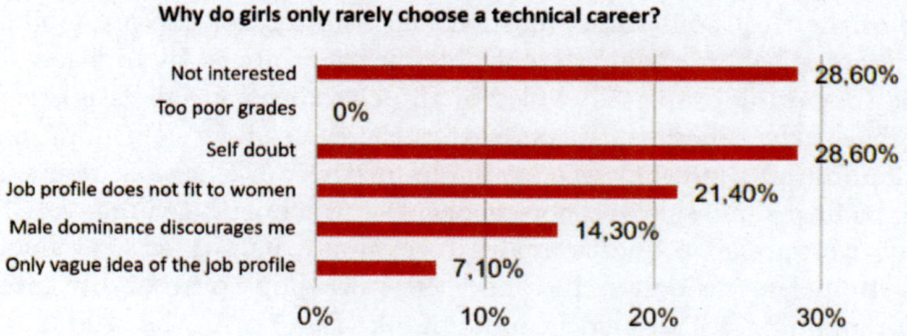

Fig. 4. Reasons the students give for why few girls aspire to a technical career

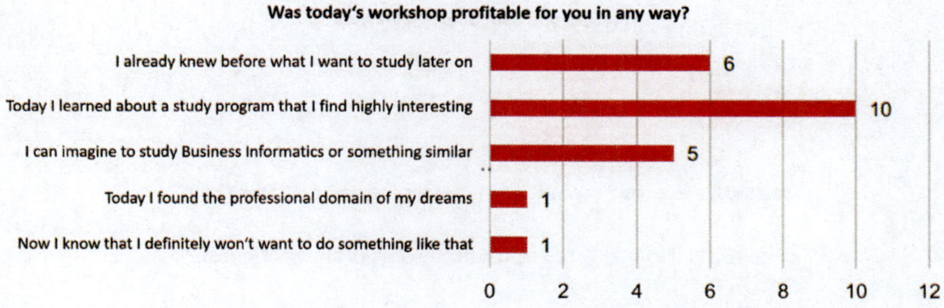

Fig. 5. Benefit of the workshop for the participants

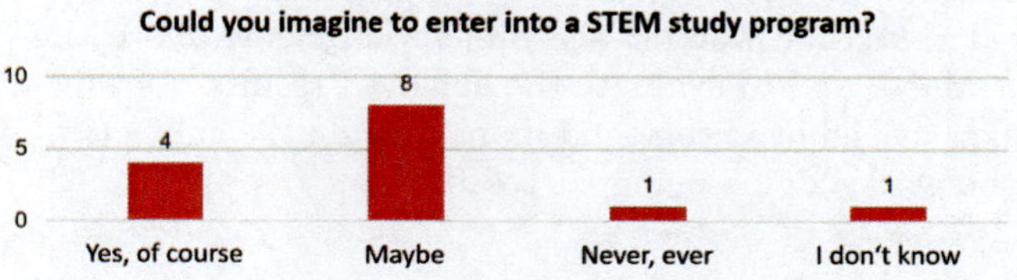

Fig. 6. Attitude of the participants towards STEM study programs

Figure 6 visualizes that 4 students are certain that they will study in the STEM area and that another 8 can imagine it. One student is absolutely convinced that she definitely *won't* go into STEM, and one student does not know yet. As compared to the average of about 25% of female students in the CS-related study programs at our institution, that is quite a positive result.

For evaluating the achievement of cognitive learning objectives, we analyzed the artefacts that participants created during the workshop. After performing the exercise, the classification results of the system were critically discussed. As a result of the discussion, students realized and explained in their own words that a vast amount of high quality data is necessary to properly train a Machine Learning system. If the training data is somewhat selective, the system's classification will be biased correspondingly. Furthermore, students realized the meaning and relevance of the threshold value, and learned to interpret the probabilities that the system computed for the different classification labels. In addition, students found out that an appropriate choice of the classification labels is essential, to properly reflect the different aspects within the data under consideration. Thus, even in the limited time that was available for the workshop, students gained a good insight into some selected core aspects of Machine Learning.

Finally, we applied a questionnaire to evaluate participant satisfaction. Of the 14 participating students, 10 found the workshop to be highly interesting, while the remaining 4 classified it as being ok (Fig. 7). As the students did not participate voluntarily, but because they had to, we celebrate this result as a great success.

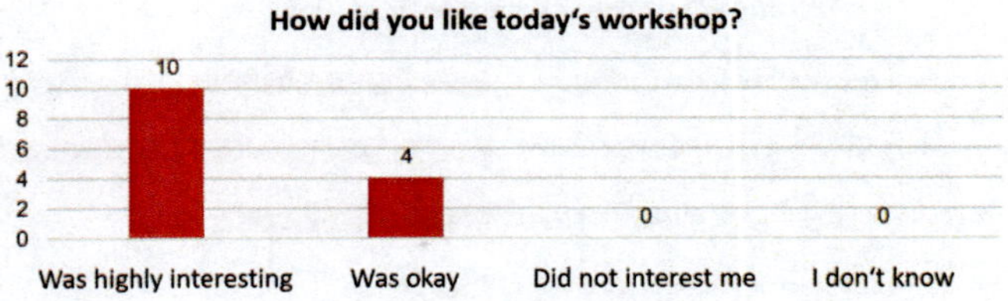

Fig. 7. How the participants enjoyed the workshop

In the free text fields of the evaluation questionnaire, students stated among others that they greatly liked the insight into job profiles of the STEM area, and the information on the related study programs. As improvement potential, they expressed that they would have preferred to have more time to test the system they trained, and to evaluate the achieved results. As this indicates that students would have enjoyed to spend more time on the subject, we perceive our pilot run of the workshop as being successful.

8 Conclusions

All in all, the pilot run of the workshop went well and produced the desired effect, as the majority of the participants found the topics adressed to be interesting, and realized that they were well able to cope with the tasks at hand. Some of the participating girls even stated that the workshop motivated them to enquire into the programs on computer science offered by the local universities – which they did not have in their focus before.

How long this effect will hold needs to be observed, when in a years time the participants of the workshop graduate from high school and decide on their future professional domain.

References

1. Anderson, L.W., Krathwohl, D.R., Airasian, P.W., Cruikshank, K.A., Mayer, R.E., Pintrich, P.R., Raths, J., Wittrock, M.C.: A Taxonomy for Learning, Teaching, and Assessing, 1st edn. A Revision of Bloom's Taxonomy of Educational Objectives. Longman, New York (2001)
2. Biggs, J.B., Tang, C.S.: Teaching for Quality Learning at University: What the Student Does, 4th edn. McGraw-Hill Education, Maidenhead (2011)
3. Bloom, B.S., Engelhart, M.B., Furst, E.J., Hill, W.H., Krathwohl, D.R.: Taxonomy of Educational Objectives: The Classification of Educational Goals. David McKay Company, New York (1956)
4. Funk, L. and Stühmeier, R., Wentzel, W.: Girls' Day und Boys' Day — Klischeefreie Berufsorientierung, die wirkt!, Kompetenzzentrum Technik-Diversity-Chancengleichheit e.V. (2019)
5. Hattie, J.: Visible Learning: A Synthesis of over 800 Meta-Analyses Relating to Achievement. Routledge, London (2010)
6. Klassner, F., Anderson, S.D.: LEGO mindstorms: not just for K-12 anymore. IEEE Robot. Autom. Mag. **10**(2), 12–18 (2003)
7. Krathwohl, D., Bloom, B., Masia, B.B.: Taxonomy of educational objectives, book ii. affective domain (1964)
8. Sickendiek, J.: Development of a didactic concept and a web training environment in the area "application of machine learning" for a MINT4Girls workshop. Master Thesis, Hochschule München, Germany (2019)
9. So viele Frauen wie noch nie (2019). https://www.zeit.de/campus/2019-10/geschlechterverhaeltnis-studiengaenge-frauen-maenner-studium-universitaet

Remote Training for Firefighter Group Commanders

Thomas Klinger[1](✉), Klaus Tschabuschnig[2], Marvin Hoffland[1], Vera Ratheiser[1], and Christian Kreiter[1]

[1] Carinthia University of Applied Sciences, Villach, Austria
t.klinger@fh-kaernten.at
[2] Carinthian State Firefighting School, Villach, Austria
https://www.fh-kaernten.at/en/en

Abstract. This paper describes the results of a project funded by the Austrian government that has set a starting point for the digitalisation of the Austrian fire brigade training. Within the scope of this project, the theoretical contents of the group commander training were transferred to an online learning platform (Moodle). This has led to the great advantage that the trainees are not absent from their workplace for an entire week, but only two to three days. In addition, the time gained can be used to offer the important practical training more intensively and in better quality. A Moodle course was set up to teach the theoretical content in advance of the shortened but more intensive practical training. At the end of the course, an exam is held which is the entitlement for the practical part. The participants receive a certificate which is automatically generated by the learning platform.

Additionally, Citizen Science methods were used to ensure the quality of the online education. Approximately one hundred people were interviewed on the design of the course, including participants, future participants, teachers at the fire brigade academy, and citizens affected by firefighting operations. For this purpose, teaching content and videos of courses held were made available on a public web platform. The results were used to further improve the course and increase its quality.

Keywords: Firefighter training · Distant education · Citizen science

1 Introduction

The Austrian fire brigades are on the threshold of digitalisation. The regional fire brigade academies are in the process of transferring teaching and learning content to online platforms in order to be able to offer modern and up-to-date training in the future. In addition to members of the professional fire brigades, training is provided above all for members of the voluntary fire brigades who have a main occupation [1] and therefore must spend additional time on these training courses. For example, training to become a group commander currently

M. E. Auer and T. Rüütmann (Eds.): ICL 2020, AISC 1328, pp. 756–763, 2021.
https://doi.org/10.1007/978-3-030-68198-2_71

requires that participants are not available to their main employer for a whole week.

The main objective of this project (it was named GriSu, after a cartoon dragon created in Italy [2]; see Fig. 1) was to facilitate access to training, especially for volunteer firefighters. Moreover, by bringing forward the theoretical training online, the practical training should be more intensive and of better quality. The project should start with the group commander course, which usually keeps volunteer firefighters away from their main job for a whole week, but with on-line training, the attendance times should be halved.

Education in this form is an absolute novelty in Austrian fire brigade training. Starting in our province of Carinthia and with the group commander course, the interactive online training will be extended to the rest of Austria as well as other courses.

Fig. 1. Grisu, the firefighting dragon [2].

2 Implementation

The developed online course has an interactive structure. Intermediate checks ensure that content that has already been heard or read is also understood. Only after a positive intermediate check will further content be released.

The course contents are offered in the following formats:

2.1 Multiple-Choice Questions

This question type is mainly used for the check of text-based content, such as law and organisation topics. An example (translated) is:

Which terms are covered (for example) by the fire brigade's task of preventing "other dangers of a local nature"?
Select one or more answers:
- *Flooding*
- *Vehicular accident involving trapped passengers*
- *Fire prevention*
- *Earthquake*

2.2 Scenarios

Figure 2 shows a tactical example scenario. The (translated) assignment is as follows (TLFA2000 is the name for a special fire brigade unit [3]):

Fig. 2. Tactical example.

General situation: office building on the outskirts of the city, surface hydrant (sufficient water) in approx. 80 m distance, 2.00 pm, little to no traffic. Own situation: TLFA2000 (1:6) incl. breathing apparatus, further forces are on their way. Damage situation: room fire on the ground floor, probably several people in the building (it could not be confirmed that nobody is in the building!), intensive fire (see picture).
What actions (note: max. 4 actions due to team size) do you take with the TLFA2000?

2.3 Videos

Two video types were recorded and made available to the participants:

- Recordings of demonstrations in front of an audience (of a previous course, see Fig. 3);
- Teaching content in a studio atmosphere, without any audience (Fig. 4).

The efficiency of the two video types was tested by interviewing some of the participants. Both types have their advantages and disadvantages: The “studio videos” convey the content in a clear and structured way, whereas the other videos are perceived more vividly when the audience is involved.

Fig. 3. Video type 1: demonstrations.

Fig. 4. Video type 2: studio.

Again, the learning flow is interactive. For example, if a set of videos describes the fire classes A to D (solids, liquids, gases, and metals), the video is stopped after the description of fire class A and the knowledge of the trainee is tested with quiz questions. Trainees can answer the questions as often as they want; only after correct answers the next part (fire class B) is activated.

2.4 Calculation Exercises

Additionally, calculation exercises need to be completed by the trainees. Figure 5 shows an example, where the resulting force absorbed by the fire truck needs to be calculated. The (translated) assignment provided together with the graphic is:

> *An accident vehicle (see figure) must be secured with force F. The constellation of loose and/or fixed rollers can be seen in the figure.*
> *What force must be absorbed by the winch of the fire truck?*

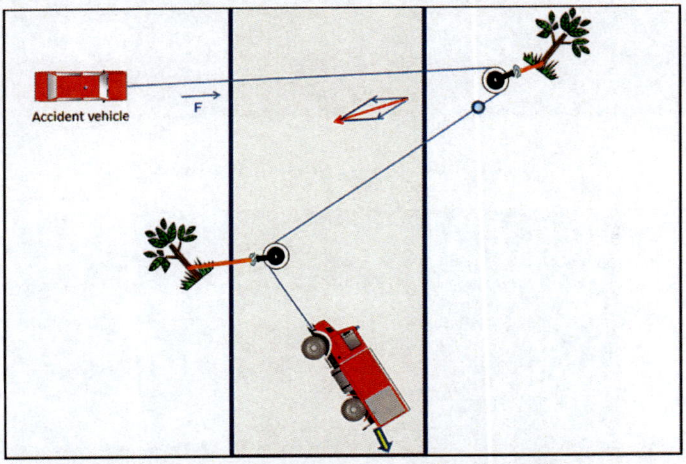

Fig. 5. Calculation exercise: force on a fire truck.

2.5 Final Exam

Participation in the final examination is activated at a previously defined time and can be taken online. The exam content consists of a selection of questions that have already been answered during the interactive course. In contrast to the previous procedure, there is only one possibility to answer the questions.

After passing the final examination, participants receive a certificate. This certificate entitles them to participate in the practical training, which can now be more intensive due to the time saved by the online course.

3 First Results

In order to implement the e-Learning product and further on to transfer the new concept, the Carinthian State Firefighting school started an evaluation phase in April 2020 with 20 firefighters, which were selected by parameters such as age, gender, experience and duty-time in the fire brigade. Those candidates were given the task to work through all the phases of the digitalized part of the group commander curriculum, do the intermediate checks and maintain their progress until the final exam. The participants had the possibility to deliver feedback by an internet platform, which was directly evaluated by the firefighting school in order to fix problems and/or to improve the contents in general. Additionally, the participants were guided by the trainers of the firefighting school using the communication possibilities of the Moodle platform. Finally, an all-hands meeting with the participants and the involved trainers will lead to the first released version.

Due to the Covid-19 crisis, all presence courses had to be cancelled until autumn; which means, that also the planned presence part (3 days) of the "Firefighting Group Commanders" course had to be transferred to September. The participants of the evaluation are now been led to this part and will be finishing the whole education with the final exam in presence by the end of September.

Until now, the concept has surpassed the expected results as the participants have expressed a very positive view to the new technology. Moreover, the trainers of the firefighting school improved themselves and developed a new concept to design courses using the "inverted classroom" concept. In the meantime new learning-products, for instance in the segment of disaster release relief, were developed on this basis.

4 Citizen Science Approach

To ensure the quality of the online education Citizen Science methods [4–6] in combination with tools of qualitative and quantitative empirical social research were used. On the one hand, the fire brigade was interested to find out from people who needed a fire brigade intervention in the last three months (January to March 2019) whether there were still points regarding the implementation and the procedure that should be considered even more in the training of fire fighters in the future. To this end, an online survey was carried out with 100 people from the target group in Carinthia and eleven people personally affected were interviewed.

67% of the survey participants, of which 58% were male and 42% female, were in need of a firefighting mission for the first time. The most common type of mission was a fire, followed by vehicle recovery, water damage and a traffic accident. Most people in this situation were excited (29%) and at the same time grateful (35%) for the help they received. The most stressed factor for the respondents was their own awkwardness (41%). The most important aspects for a good firefighting operation are, from the point of view of the affected persons, the immediate readiness for action, the professional and quick handling as well as the friendly support and information on site (see Fig. 6).

All respondents rated the fire brigade team after the experienced mission as professional, experienced, friendly, reassuring, well-organised, competent and sufficiently trained. The personal interviews also confirmed the thoroughly positive image of the fire brigade, which was confirmed by the deployment in the

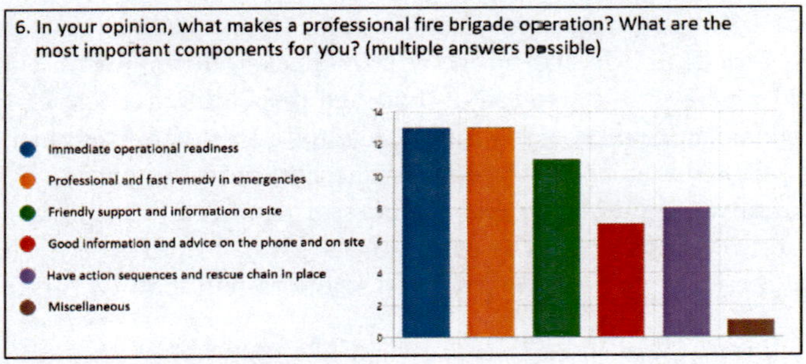

Fig. 6. Question about the professionalism of a fire brigade mission.

personal environment of the persons concerned (quick appearance, good equipment, helpfulness, professional and calm action). In the areas of professionalism and personal crisis management, the expectations of the persons concerned were exceeded.

All respondents were also asked for their opinion on the planned, additionally offered combination course (online and presence). They were also presented with a sample instructional video in order to obtain their assessment of its quality and comprehensibility. The presented instructional video was rated by the oral and digital interviewees as instructive as well as very good in terms of quality and comprehensibility. During the personal interviews, the importance of group work, the possibility to ask questions and the time of presence was emphasized. This was therefore given particular attention in the design of the combination course. All participants in the survey are of the opinion that there should be more training or information opportunities for citizens and pupils (see Fig. 7).

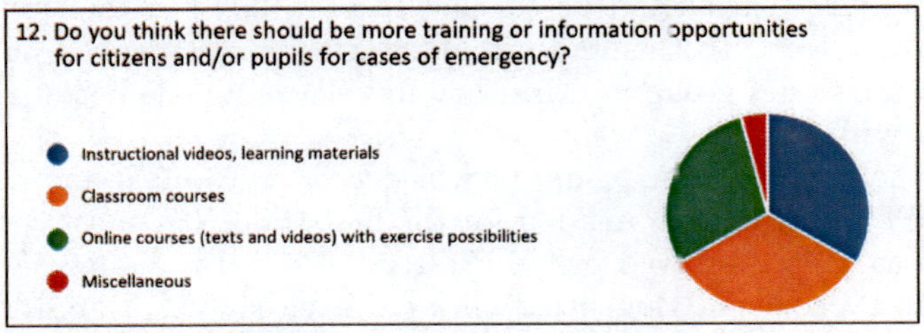

Fig. 7. Question on information opportunities for citizens and pupils.

The eleven oral interviews led to further results. It was mentioned several times that it is becoming more and more difficult to get enough firefighters for an operation in the countryside, especially during the week. There is a fear of reduced budgets and insufficient funding. As a result, not all fire brigades in a county are getting all the equipment they need. Many companies no longer release firefighters, which makes it impossible for many to get involved. The Citizens Science survey also included two group discussions with seven former course participants, four guideline-based interviews with instructors and a survey of all instructors at the Carinthian State Firefighting School. All of them were asked about their experiences with the previous course design and about their opinion on the newly planned course preparation and the digital teaching and learning setting presented.

5 Conclusions and Outlook

Started in 2020, the group commander course will be conducted alternately in the online form and in the original form, so that the findings can be implemented, and the quality further improved. First evaluation results are expected in June 2020.

The next step in the cooperation between our university and the fire brigade academy school will be to also offer practical elements of training online or remote. As an example, a remote-controlled extinguishing system is planned, which will be set up as a remote laboratory [7].

References

1. Feuerwehr in Österreich 2019. Österreichischer Bundesfeuerwehrverband (2020). https://www.bundesfeuerwehrverband.at/wp-content/uploads/2020/02/Statistik_2019.pdf. Accessed 15 May 2020
2. Grisù il draghetto. It.wikipedia.org (2020). https://it.wikipedia.org/wiki/Gris'u_il_draghetto. Accessed 18 May 2020
3. Medium-sized fire engines. Magirus Group (2020). https://www.magirusgroup.com/de/en/products/fire-engines/medium-pumpers/. Accessed 18 May 2020
4. What is Citizen Science? Center for Citizen Science at the OeAD (2020). https://www.zentrumfuercitizenscience.at/en/citizen-science-en. Accessed 27 May 2020
5. Citizen science. Austrian Federal Ministry of Education, Science and Research (2019). https://www.bmbwf.gv.at/en/Topics/Research/Research-and-Public/Citizen-science.html. Accessed 27 May 2020
6. Citizen science. En.wikipedia.org (2020). https://en.wikipedia.org/wiki/Citizen_science. Accessed 27 May 2020
7. Klinger, T., Garbi Zutin, D., Madritsch, C.: Parallel use of remote labs and pocket labs in engineering education. In: International Conference on Remote Engineering and Virtual Instrumentation (REV), New York City, USA, March 2017

Examining Technology and Teaching Gaps in Russian Universities Amid Coronavirus Outbreak

Elizaveta Osipovskaya[1(✉)], Svetlana Dmitrieva[2], and Vadim Grinshkun[3]

[1] Peoples' Friendship University of Russia (RUDN University), 6 Miklukho-Maklaya Street, Moscow 117198, Russian Federation
e.osipovskaya@gmail.com
[2] Perm National Research Polytechnic University, Komsomolsky Avenue 29, Perm, Russia
[3] Moscow City University, 2-Y Sel'skokhozyaystvennyy Proyezd, 4, Moscow, Russia

Abstract. The current study facilitates an evidence-based discussion on teaching during the period of an abrupt shift to virtual classrooms caused by coronavirus spread. There has been a massive investment to bring technology to universities; however, there are some significant gaps that must be addressed. The first is the availability of university online platforms and the second is the digitalization of the curriculum combined with teachers' confusion and anxiety around the ability to adjust to the new digital environment. The study is aimed to research the current situation at two Russian universities: RUDN (People's Friendship University of Russia) and PNRPU (Perm National Research Polytechnic University) and describe what changes have been implemented into teaching by the university administration and professors. The research questions include two groups: 1) Learning management system (LMS): Did the university offer a teaching platform to be used by all teachers? Or was it an individual choice of every teacher? 2) Curriculum: Did teachers have to modify the content of their course? In what way? Our study contributes several meaningful findings regarding understanding the advantages, disadvantages, and challenges in changing the education landscape.

Keywords: Coronavirus · Curriculum · Learning management system · Massive Open Online Course

1 Introduction

Every time a crisis hits the economy in general or industry in particular it has no choice than to adapt to a new environment and go ahead. The spring of 2020 was an extremely challenging time for higher education worldwide when universities had to make rapid and hard decisions to shut down in-person classes and transit to online teaching to fulfill commitments to learners.

Lockdowns increased the number of online learners up to 89.5% of the world learners including those enrolled at universities, according to UNESCO [1]. On the

M. E. Auer and T. Rüütmann (Eds.): ICL 2020, AISC 1328, pp. 764–774, 2021.
https://doi.org/10.1007/978-3-030-68198-2_72

other hand, the research made by the Monitoring of Educational Economics of Higher School of Economics before the coronavirus crisis revealed that digital technologies accounted for 4% of all educational technologies offered by higher education in Russia in late 2019. Only 1% of all students received their diplomas on MOOC (Massive Open Online Course); 73% of students and 41% of teachers never heard about the possibility of getting higher education on MOOC; 11% of teachers had their training on online platforms and 40% planned to do it in the future [2].

As a result, almost all Russian universities struggled and continue to struggle with accommodating online teaching due to different reasons. Among them, one can name technical reasons and considerations with online education technologies which include familiarity with online technologies and tools and teaching approaches, and professors' and students' easy access to Wi-Fi. Universities have to train academic staff to use online educational technologies and adopt a new methodology of teaching content online while involving students in online discussions. Professors should be ready to embrace online teaching and motivate students to get and stay engaged in learning.

The virus has not only revealed vulnerabilities in the higher education system but emphasized that it has to be flexible and resilient to respond to unpredictable changes in the future.

2 Literature Review

Classroom management during the COVID-19 pandemic when educators use various tools to maximize collaboration, creativity, communication, and students' engagement is often used interchangeably in educational circles now (e-learning, online learning, digital learning, technology-based learning, distance learning, remote learning, synchronous learning). Typically, they refer to the educator's capacity to construct curricula, design student assessment by taking advantage of synchronous and asynchronous learning which works best in digital formats [3, 4]. COVID-19 is the greatest challenge for teachers to harness the power of technology to extend learning.

Nowadays, rearranging the landscape of education is a vital element of institutional response. Therefore, it is extremely crucial to follow certain rules and standards. The internationally recognized ISTE (International Society for Technology in Education) Standards provide "clear guidance for the skills, knowledge, and approaches students need to succeed in the digital age" [5]. We would like to list the standards for students and teachers as we believe they are highly essential elements of any technology-rich learning environment, especially amid coronavirus outbreak. ISTE standards for students: 1) Empowered Learner, 2) Digital Citizen, 3) Knowledge Constructor, 4) Innovative Designer, 5) Computational Thinker, 6) Creative Communicator, 7) Global Collaborator [6]. ISTE standards for teachers: 1) facilitate and inspire student learning and creativity; 2) design and develop digital age learning experiences and assessments; 3) model digital age work and learning; 4) promote and model digital citizenship and responsibility; 5) engage in professional growth and leadership [7].

A huge array of scientific papers has been published about how to ramp up teachers' ability to teach remotely in the context of the COVID-19. These publications cover several major topics. For example, a researcher Petar Jandrić classifies current reality as post-digital viral modernity with viral education, viral post-truth, and viral open science [8]. Sir John Daniel proposes a selection of professional development resources that educators can adopt during remote learning: The Commonwealth of Learning is an intergovernmental agency of the Commonwealth tasked with helping developing countries use technology effectively in education and training at all levels; UNESCO's list of educational applications (digital learning management systems), platforms (collaboration platforms that support live-video communication) and resources (self-directed learning content), Massive Open Online Course (MOOC) platforms; collected resources on digital tools and general tutorials for online teaching. The tutorials were prepared by David Gaertner, an Assistant Professor and a settler scholar in the First Nations and Indigenous Studies Program at the University of British Columbia [9]. Some scientists raise the issues about how the pandemic has amplified the existing societal and educational inequities [10].

There is an earlier study of academic teachers' experience in Norway conducted by the Centre of Experimental Legal Learning [11]. The survey is based on 172 responses and identifies the needs and challenges of the academic staff, significant changes in teaching methods, and approaches to conveying the curriculum. The study reports that the majority of respondents are positive about innovations. Based on the findings and their interpretations, the study provides a set of recommendations to academics and teachers in order to make short-term adjustments in the subsequent rounds of online teaching.

A scholar from Indonesia has designed a primary school student worksheet based on the scientific literacy of indicators in the COVID [12]. Other authors have designed the action plan that could be implemented at the time of COVID-19 to provide high-quality distance education at Beijing Normal University and to train academic staff to be more resilient to similar crises. This guideline embeds the following actions: 1) immediate dissemination of information via email, Wechat and school website; 2) development of open educational platforms which allow access to the high quality of learning resources; 3) quantitative and qualitative research and evaluation of current e-learning models; 4) cooperation between universities, international organizations [13]. Moreover, there are publications about challenges and opportunities for the education system in Morocco [14], India [15], Mexico [16], and Turkey [17] in the context of coronavirus.

There is a considerable amount of literature on the implications of the COVID-19 outbreak on medical education [18, 19], in particular the examination process and the way to assess student's performance and competence [20, 21]. There have been analyzed peculiarities of pediatric [22], vascular surgery [23], urology [24], pharmacy [25] education, pre-clerkship, and clerkship learning environments, how to organize virtual classes and what educators could do to create experiences for students who are usually assigned to inpatient or outpatient rotations [26].

Besides, there are some crucial findings regarding COVID-19′s implications on an educator's role. Experts believe that teachers are becoming curators for virtual learning environments. They create problem-solving activities for students and let them get to the bottom of complex situations on their own. In such an education ecosystem, students can self-regulate their learning switching to video tutorials or guidance from teachers when they need to learn new skills to proceed [27].

3 Methodology

The survey was conducted in the context of the Russian higher education system to respond to the measures that were imposed by the government amid the coronavirus outbreak. Our knowledge of synchronous and asynchronous teaching during total lockdown is largely based on responses from two universities: the Russian University of People's Friendship (RUDN) and Perm National Research Polytechnic University (PNRPU).

The objectives of the survey were as follows:

1.1. To identify which online services and pedagogical techniques were employed.

2.2. To determine obstacles and challenges in digital learning.

3.3. To evaluate teachers' workload in delivering online courses.

The content of the survey can be synthesized as follows:

1. Learning management system (LMS): Did the university offer a teaching platform to be used by all teachers? Or was it an individual choice of every teacher? Did students have a voice in decision making? What platforms were used? What was the choice based on? What advantages and disadvantages did teachers find in a new educational system?

2. Curriculum: Did teachers find any suitable MOOC courses? Did teachers have to modify the content of their course? In what way? Did teachers use interactive forms of digital teaching? Did teachers plan to increase the online part of their courses in the future after the crisis? Or did they plan to return to the before-the-crisis way of teaching? Did teachers receive technical support from IT-staff?

Two following research methods were employed in the study: descriptive statistics and qualitative analysis of free responses to the survey questions.

4 Results and Discussion

The research has also a specific aim to describe and compare the experiences of the academic staff from two Russian universities. The paragraph focuses on how online education was organized with specific reference to two divisions of the Universities: Mass Communication Department (42 members) of RUDN University, Moscow, Russia, and Foreign Languages and Public Relations Department (34 members) of Perm National Research Polytechnic University, Perm, Russia (see Fig. 1).

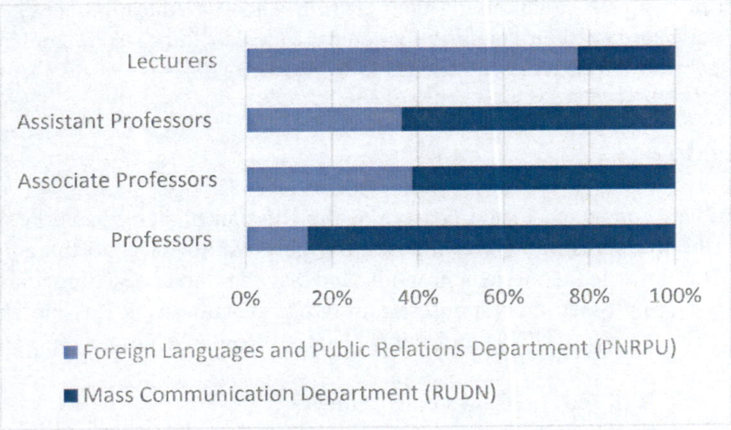

Fig. 1. Distribution of respondents by position.

4.1 RUDN University

RUDN University had to cancel face-to-face lectures and switch to online teaching to try to halt the spread of the coronavirus. On the one hand, the University is technology-equipped for computer-based learning. There are two centralized online collaboration workspaces that professors and lecturers can use. They are Microsoft Teams and a Telecommunication Education and Information System (TEIS) based on Moodle platform.

TEIS – is an educational environment, an ecosystem where a teacher can manage all learning activities such as creating an online course, sending messages, sharing class materials, assessing students, enhancing collaboration, or tracking achievement and make learning accessible from anywhere. It encompasses a collection of links to useful resources, photos, inspirational videos, and interactive assignments and tests. Students need to enter corporate e-mail to access TEIS, so it is quite a secure environment that helps students to stay safe online.

RUDN University has a commitment-based Microsoft Volume Licensing agreement (Microsoft Enrollment for Education Solutions) that allows academic staff and students to utilize all Microsoft applications. To shift to remote learning teachers were suggested to use Microsoft Teams. It is a shared space that provides group chat, channeled conversations, instant messaging, live document collaboration, audio or video calls, and meetings. That is why the majority of lecturers of the Mass Communication Department used MS Teams for online seminars or virtual meetings (see Fig. 2).

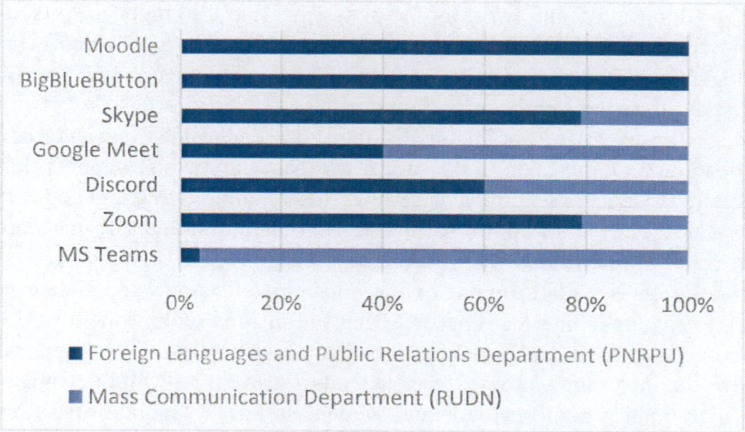

Fig. 2. Distribution of online platforms in RUDN University and PNRPU.

Our survey revealed various challenges in online education. The most frequently identified obstacles are struggling with a lack of experience in online programs (63%) and a lack of knowledge about converting paper-based exercises into digital forms (59%). The administration of RUDN University organized training courses for academic staff to reduce techno-anxiety. But the question of how to adapt the style of teaching to digital space fell on teachers' shoulders and intuition. The next obstacle is an attempt to balance online teaching with parenting (23%). According to educators, there were a lot of confusing situations when teachers' kids' heads popped into a virtual lesson. Respondents also said about zoombombing (14%), where uninvited participants interrupted meetings and distracted students' attention. So many teachers decided to switch from Zoom to Microsoft Teams. Moreover, the study demonstrated a lack of participation in online discussions (41%), especially compared with face-to-face interaction during traditional lessons. In the professors' opinion, students seemed to experience some sort of anxiety and lack of confidence in an unfamiliar environment.

The survey also found out that teaching at a distance requires much more of an educator's time. Academic staff had to design courses from scratch to be taken fully online. Most of the time was spent on creating PowerPoint presentations (73%) and giving feedback on students' projects (54%). The new digital environment has greatly intensified and increased the teachers' workload. Finally, the study showcased the ideal class size for an online course. The ideal traditional practical class size was 18 students while for digital learning it decreased to 11 students.

4.2 PNRPU

Perm National Research Polytechnic University had to transit to online teaching within a week. That was a challenging week for the professors and lecturers. On the one hand, the University has the technical resources to rapidly change the mode of teaching. There are three centralized university services that professors and lecturers can use.

They are BigBlueButton and two Moodle platforms: one for engineering faculties and another one for the Faculty of Humanities. Each of the platforms provides technical and methodological assistance in creating content and using it; one can find the contacts of technical staff when registering on the platform.

Moodle is an excellent option for distant and blended learning and has a lot of practical features: adding texts, links, images, videos, uploading support documents, working with students' assignments: setting the deadlines, making comments, and grading works. There is an option to link to Bigbluebutton and Zoom video conferences and teach in real-time.

BigBlueButton is a platform for online education. It is an open-source web conferencing system for online learning. BigBlueButton runs within one's web browser. There is no need to download a plugin, or to install the software. It is supported by all major browsers, including Chrome, FireFox, Safari, and Safari Mobile. BigBlueButton provides high-quality audio, video, and screen. It has a "mobile-first" design. Its interface can be run on a mobile device with no need to download or install a mobile app. BigBlueButton has a useful feature that enables an instructor to recall, demonstrate, apply, or ask questions about the teaching material. The platform has tutoring/virtual office hours, flipped classrooms, group collaboration, and online classes. Instructors can use BigBlueButton for engaging students and bringing virtual classrooms closer to a traditional classroom. They can exploit multi-user whiteboard, break out rooms, public and private chat, polling, shared notes, emojis.

Thus, the University has all the necessary technical resources to meet the educators' demands and organize online teaching most efficiently. So, why then transition to online teaching became a formidable problem for a part of teaching staff.

At the end of March 2020, when the COVID-19 forced the University to move on to online, many thought that the lockdown would only last for a couple of weeks. The university administration issued the order to organize online teaching using Moodle, MOOC, Coursera, and other available resources. There were no guidelines or step-by-step instructions about the platform, mode of teaching (distant/online), or recommendations on how to organize the process. All the aspects were given to the consideration and responsibility of the Chairs and teaching staff. They had to work out feasible solutions to the questions. They also had to consider the technical possibilities of students who could have limited access to online technologies.

All the above mentioned revealed that the University was not ready for a rapid transition to online education. The academic staff was expecting clear guidelines and unified policies on the recommended online platform, webinars on how to accommodate it in teaching different subjects, how to make online lesson plans, and many other things. On the other hand, many professors enjoyed the freedom of making their own decisions about what online platforms to use; or decision was made due to technical characteristics of the electronic devices at their disposal. However, such variety sometimes posed obstacles to students who had to adapt to different resources utilized by subject teachers.

A survey conducted 2 months after the lockdown revealed ambivalent attitudes of the academic staff to the enforced online teaching. Younger and middle-aged teachers considered it as a challenge and were ready to seize the opportunity, improve their technical skills and master new tools, while teachers "over 60" believed that the situation was temporary and were rather reluctant to change the way they were teaching. If they had had any choice and had been asked for their opinion, they would probably have said that the best teaching is when one teaches in front of the classroom through the textbooks [28]. The group of technologically competent teachers and those who had already had their materials on online platforms were ready to assist and consult those who wanted to overcome technical barriers.

The survey unveiled that only 4 teachers used Moodle before the closure of the University for uploading their lectures or presentations. Then the respondents were asked about the type of software they used to teach digitally and if they taught synchronously or asynchronously. About 70% taught in real-time, other 30% asynchronously via social networks, email, and messengers. The most popular platforms were Zoom, Google Classroom, Discord, Moodle, Skype, and BigBlueButton (see Fig. 2).

Some respondents used more than one platform; it depended on the aims of the course and the group of students they were teaching. The majority of teachers used Skype which had proved itself as reliable software for business conferences, despite the Zoom's boom; rather behind them were Google Classroom and Discord. Google Classroom is a very convenient free service for those teachers who want to have all the material in one place, streamline assignments, give feedback to students, and interact with them individually. Classroom keeps students' works; automatically creates a chart with their grades which tremendously facilitates teachers' work. Discord is also free software, a messenger for easy communication which is not aimed at the academic environment, in our opinion. Its choice was likely based on students' preferences and familiarity with this program. BigBlueButton had only 1% of users which is surprising as it is the official platform of the University. Other programs included Hangouts, VKontakte, Jitsi, and email. Hangouts is a service for video conferences and communication which has a significant drawback as it limits the number of a group to only10 people. VKontakte is a Russian social network similar to Facebook. Jitsi is a free video-conferencing service that is available on laptops and mobile phones. The email was used by the group "over 60" for giving assignments and feedback on them.

The survey exposed some positive and negative attitudes toward teaching exclusively online. Table 1 presents what teachers wrote about the advantages and challenges in free-text answers.

The teachers were also asked to evaluate their experience of using MOOCs. The majority of the teachers (71%) could not find the course suitable for their program; others (10%) said that they found the course, but it needed adaptation; some could not find the time to browse MOOCs (6%). However, there were a few lecturers (13%) who adapted a free English language course and developed evaluation criteria to monitor students' progress.

To conclude, the study revealed that not all academic staff demonstrated a favorable attitude to online teaching; there were considerable challenges to respond to. 4% of the teachers said that they did not see any advantages of changing the mode of education from in-class to online. However, about 70% of the teachers turned to synchronous delivery of their material and embraced new challenges – they self-educated in technical programs and online teaching methodology and inspired their colleagues to do the same. The remained 26% were more focused on challenges and looked forward to returning to the classrooms.

Table 1. Positive aspects and challenges of teaching exclusively online

Positive aspects	Challenges
1. Possibility to improve teachers' technical skills 2. Better visualization of material; 3. Mobility (one can teach from any place provided they have internet access); 4. Students' self-development; 5. Individual approach to every student; 6. A way to study during epidemics; 7. Time-saving - no commuting; 8. Easier to correct home assignments; 9. New formats of teaching; 10. Good for students to learn new skills; 11. Excellent for giving creative group assignments	1. Technical reasons: the absence of Internet coverage, poor Internet connection; technical characteristics of computers and mobile phones; 2. The increased amount of work: to find materials and plan lessons 3. Adapting existing materials to online mode 4. Absenteeism of students in virtual classrooms 5. Students' reluctance to participate in online discussions 6. Online communication cannot replace personal contacts in-class 7. Spending too much time in front of a computer screen 8. Physically and mentally demanding for teachers

5 Conclusion

The study was aimed at examining the experience of the academic staff at two big Russian Universities (RUDN and PNRPU) within two months of teaching online. The specific aim was to compare the experiences of the academic staff and to decide if teachers faced the same or different challenges.

We understand that the survey data was rather small, limited to 76 respondents (42 from RUDN and 34 from PNRPU); so, it does not allow making general conclusions or recommendations. The study gives the first insight into the situation when the Russian higher education system was forced to change almost overnight and how it responded to that challenge.

No strong evidence was found that RUDN as the university situated in Moscow met a rapid transition to teaching exclusively online head-on. The advantage was that the University provided the academic staff with Microsoft platform and gave a webinar on how to use it. It enabled over 40% of the staff to deliver their teaching on MS Teams. However, due to the absence of prior experience, the teachers needed a more comprehensive training program about the features of the platform and pedagogic support on how to transform their programs into online mode. The academic staff at PNRPU found themselves in the situation when they had to rely on their technical expertise and intuition about which online service to choose and how to use it. Not surprisingly, only 1% of the teachers used BigBlueButton - an official university platform, others resorted to Skype (59%) and Zoom (50%).

This highlights that if a university wants to gain a competitive edge in the future, it should organize sustained training programs that will help its academic staff acquire digital competence and confidence; develop a strategy to train the staff in pedagogical online expertise; organize in-house (online) workshops or chatrooms where professors and lecturers can share experience, exchange ideas and educate each other on best practices.

Another finding of the survey was about teachers' preparedness for online delivery which resulted in consequential positive or negative attitudes to it. A geographical location of a university does not have any influence on one's attitude to the teaching mode. Even for those with a positive feeling, it was challenging and demanding. It is quite understandable; in such an emergency they did not have a choice but had to adjust to online methods in haste. However, the silver linen is that the transition triggered creativity, experiment, and desire to master new technical and pedagogical skills. Now all teachers have to revise their teaching methods, give honest feedback on their strengths and weaknesses, and bridge the gap in digital knowledge at webinars and teachers' forums. "Learning anywhere, anytime" concept should become a habit integrated into daily life – a true lifestyle.

We are aware that our research may have limitations. The first is the rather small scope of data. It would be beneficial to conduct a further study on a bigger number of participants, e.g. include professors from technical departments and discover their attitudes to online teaching. The second, new research should dig deeper into professors' negative experiences and feedback on current teaching. The results should be reported to the university administration that has power and financial resources to address the problems and solve them. The third is the absence of students' opinions; the research did not include a questionnaire of students' impressions about online learning.

References

1. Neil, B., Asha, K., Stamenka, U.-T.: A Basic guide to open educational resources (OER). Commonwealth of Learning. United Nations Educational, Scientific and Cultural Organization (2015). https://unesdoc.unesco.org/ark:/48223/pf0000215804. Accessed 11 May 2020
2. 2020 Online learning market research. EdMarket Research. https://research.edmarket.ru/. Accessed 11 May 2020
3. Boguslavskaya, V., Budnik, E., Azizulova, A., Sharakhina, L.V.: Cybersport community: social structures transformation as a basis for intercultural dialogue. In: Bodrunova, S. (eds.) Internet Science. INSCI 2018. Lecture Notes in Computer Science, vol. 11193. Springer, Cham (2018). https://doi.org/10.1007/978-3-030-01437-7_23
4. Galikhanov, M.F., Guzhova, A.A.: Complex approach for preparation and implementation of continuous professional education programs in technological university. In: 2013 International Conference on Interactive Collaborative Learning, ICL 2013, pp. 54–55 (2013). https://doi.org/10.1109/ICL.2013.6644535. Article no. 6644535
5. Dowd, H., Green, P.: Classroom Management in the Digital Age, p. 70. EDTechTeam Press, California (2016)
6. ISTE students. https://www.iste.org/standards/for-students. Accessed 11 May 2020
7. ISTE teachers. https://id.iste.org/docs/pdfs/20-14_ISTE_Standards-T_PDF.pdf. Accessed 11 May 2020
8. Jandrić, P.: Postdigital research in the time of Covid-19. Postdigit. Sci. Educ. **2**, 233–238 (2020). https://doi.org/10.1007/s42438-020-00113-8
9. Daniel, S.J.: Education and the COVID-19 pandemic. Prospects (2020). https://doi.org/10.1007/s11125-020-09464-3

10. Verma, G., Campbell, T., Melville, W., Park, B.: Science teacher education in the times of the COVID-19 pandemic. J. Sci. Teach. Educ. (2020). https://doi.org/10.1080/1046560X.2020.1771514
11. Langford, M., Damsa, C.: Online Teaching in the Time of COVID-19. https://www.jus.uio.no/cell/digitaldugnad/nasjonal_evaluering_laerere.html. Accessed 11 May 2020
12. Setiawan, A.R.: Scientific Literacy Worksheets for Distance Learning in the Topic of Coronavirus 2019 (COVID-19), 23 March 2020. https://doi.org/10.35542/osf.io/swjmk
13. Zhu, X., Liu, J.: Education in and after Covid-19: immediate responses and long-term visions. Postdigit. Sci. Educ. (2020). https://doi.org/10.1007/s42438-020-00126-3
14. Naji, A.: The repercussions of Covid-19 on the field of education (2020). https://doi.org/10.37870/naji.1.2020
15. Gupta, A., Goplani, M.M.: Impact of COVID-19 on educational institutions in India, March 2020. https://doi.org/10.13140/RG.2.2.32141.36321
16. Ramos Santillan, V.O., Andrade-Romo, J.S., Barajas-Ochoa, A.: Challenges for medical education in Mexico in the time of COVID-19. Gaceta medica de Mexico **156**(4) (2020). https://doi.org/10.24875/GMM.M20000376
17. Varol, G., Tokuc, B.: Medical education in Turkey in the time of COVID-19. Med. J. (2020). https://doi.org/10.4274/balkanmedj.galenos.2020.2020.4.003
18. Lattouf, O.: Covid and a call for reinventing medical education (2020). https://doi.org/10.22541/au.158938585.59758681. Authorea
19. Ferrel, M.N., Ryan, J.J.: The impact of COVID-19 on medical education (2020). https://doi.org/10.7759/cureus.7492
20. Ahmed, H., Allaf, M., Elghazaly, H.: COVID-19 and medical education. Lancet Infect. Dis. (2020). https://doi.org/10.1016/S1473-3099(20)30226-7
21. Lopukhova, J., Makeeva, E., Rudneva, T.: Using flipped classroom in foreign language teaching: implementation of interactive educational technologies. In: Auer, M., Hortsch, H., Sethakul, P. (eds.) The Impact of the 4th Industrial Revolution on Engineering Education. ICL 2019. Advances in Intelligent Systems and Computing, vol. 1135. Springer, Cham (2020)
22. Winn, A.S., Winthrop, Z., Chiel, L.: The COVID-19 pandemic and pediatric graduate medical education. Pediatrics (2020). https://doi.org/10.1542/peds.2020-1057
23. Tinelli, G., Sica, S., Minelli, F., Tshomba, Y.: Vascular surgery education during COVID-19 pandemic. J. Vasc. Surg. (2020). https://doi.org/10.1016/j.jvs.2020.04.496
24. Ding, M., Wang, Y., Braga, L., Matsumoto, E.: Urology education in the time of COVID-19. Can. Urol. Assoc. J. Journal de l'Association des urologues du Canada **14**(6) (2020). https://doi.org/10.1007/978-3-030-40271-6_61
25. Adebayo, A.Y., Obaloluwa, A.P., Okereke, M.: COVID-19 pandemic: medical and pharmacy education in Nigeria. Int. J. Med. Students (2020). https://doi.org/10.5195/ijms.2020.559
26. Rose, S.: Medical student education in the time of COVID-19. JAMA (2020). https://doi.org/10.1001/jama.2020.5227
27. Thomas, M.S.C., Rogers, C.: Education, the science of learning, and the COVID-19 crisis. Prospects (2020). https://doi.org/10.1007/s11125-020-09468-z
28. Kebritchi, M., Lipschutz, A., Santiague, L.: Issues and challenges for teaching successful online courses in higher education: a literature review. J. Educ. Technol. Syst. **46**(1), 4–29 (2017). https://doi.org/10.1177/0047239516661713

Applying Strategies of Higher Order Thinking in Pre-gradual Preparation of Technical Subject Teachers

Monika Valentová, Peter Brečka, and Alena Hašková(✉)

Faculty of Education, Constantine the Philosopher University in Nitra, Nitra, Slovakia
{mvalentova2, pbrecka, ahaskova}@ukf.sk

Abstract. Now-a-days in the professional community of teachers, broad discussions on efficiency of teaching are led. At the same time teacher trainees are more and more also involved into these discussions, a key point of which are the teachers' skills to choose appropriate education strategies to develop students' higher order thinking, i.e. thinking at a level that is higher than remembering, memorizing and understanding facts. These discussions follow Bloom's taxonomy known as the taxonomy of educational objectives. A question still remains, to what extent is the theory incorporated in teacher trainee practical training and as well to what extent this theory is used and applied in the educational and teaching practice itself. The paper presents some results of research focused on the current state of implementation of development strategies of higher order thinking in pre-gradual preparation of teacher trainees majoring in technical subjects teaching.

Keywords: Abilities of higher levels of thinking · Bloom's taxonomy of educational goals · Development strategies · Higher order thinking · Teacher training · Technical subjects teaching

1 Introduction

Every teacher should be interested in thinking process and possibilities to develop this area of human personality through different strategies, as teachers should be able to provide their pupils and students by adequate activities to help them to develop higher order thinking. Bloom's taxonomy has been for several decades the most widely used strategy on base of which pupils and students' thinking at schools is developed. For teachers, it is a practical tool for planning learning objectives and outcomes of lessons and applying adequate teaching tasks to ensure maximal pupils and students' progress.

Bloom's taxonomy covers three domains: cognitive (mental development), affective (attitudes) and psychomotor (physical skills). However, it is most commonly used to define cognitive skills [1]. Bloom taxonomy is an impetus for different concepts introduction, including educational concepts, such as ability to solve problems, creative and critical thinking or thinking at a higher level [2]. It has been revised over the years as several shortcomings and practical limitations have occurred [3]. The revised taxonomy consists of 6 levels arranged from the lowest thinking abilities to the highest [4]:

M. E. Auer and T. Rüütmann (Eds.): ICL 2020, AISC 1328, pp. 775–786, 2021.
https://doi.org/10.1007/978-3-030-68198-2_73

- remember - consists in student's ability to remember new facts, memorize knowledge and, when necessary, to recall it in memory and repeat it or reproduce it.
- understanding - is based on the student's ability to recall the remembered, memorized facts, however at this level the student can present the recalled knowledge also in a different form than s/he remembered and memorized. For example, s/he can explain the information, say it in his/her own words.
- application - at this level of thinking, the student is expected to be able to use the acquired knowledge in practice, in solving problems.
- analysis - consists in student's ability to analyse information, phenomena and activities into parts and determine relations among the analysed parts themselves, as well as between the analysed parts and the whole. The analysis consists in finding causes, connections and consequences.
- evaluation - at this level the student is able to assess facts, things and phenomena according to the criteria or standards, to evaluate their advantages, disadvantages, consequences, and is also able to argue.
- creation - is understood as the ability to create new original ideas, procedures, proposals, concepts, wholes.

In comparison to Bloom's taxonomy Vygotskij distinguishes lower and higher order thinking based of a child's cultural development (cultural-historical psychology). According to Vygotskij [5] lower order thinking involves unconscious biological functions (memory, attention and intelligence, language acquisition, writing, painting, calculating skills development) which are linked by biological, social, affective and cultural factors, are not value-free and are influenced by man' feelings and emotions. In contrast, higher order functions involve conscious deliberate intentions and actions (e.g. problem solving, logical reasoning, abstract thinking). Central is the idea that the child's naturally given mental processes become transformed by the acquisition of speech and meanings. Through speech the child acquires a worldview that reflects reality in a more adequate way. In this point Vygotskij overemphasized the role of speech to the detriment of the child's concrete operations with reality (see also Rubinštejn and Leontiev's activity theory, [6, 7]).

The particular levels of Bloom's taxonomy are arranged in a cumulative hierarchical order, what means that to achieve a higher thinking ability requires to achieve the previous lower ones [8]. The lower levels of thinking include remember, understanding and application, and to the higher levels belong analysis of arguments, statements or evidence [9, 10], evaluation and judgment [11, 12] and inference, synthesis by inductive or deductive arguments [13, 14]. At present, the need to develop higher level thinking is emphasized, so it is necessary at schools to apply educational strategies that lead to deeper learning and transfer of knowledge and skills [15, 16]. Higher thinking skills are expected to better prepare students for the challenges of the labour market and everyday life. They help students to think critically, which is thinking that involves logical thinking and reasoning, including such skills as comparison, classification, analogy, induction, deduction, planning, assumption, and criticism [17–19]. However, different analyses have shown that learning objectives in many training programs [20, 21] and school curricula [22] are focused mainly on lower order thinking, i.e. on remember and understanding. But at the same time the analyses have shown that teachers' knowledge of higher order thinking, their ability to develop and upgrade higher levels of students' thinking are still low, despite the fact that teachers understand its importance [23]. In some authors' opinion [24] teachers mostly use the traditional style of education, i.e. education focused on teaching, not on learning and thinking. Higher order thinking cannot be manifested itself, if students do not have any opportunity to react to specific situations, if they are provided only with passive learning possibilities.

2 Research Methodology

Our aim was to find out current state of applying strategies of the development of higher order thinking in the subject of Technology taught in Slovakia at lower level of secondary education (ISCED2), as well as to find out whether teacher trainees are prepared for this task during their pre-service training (pre-gradual preparation). These our research activities were a part of the national project *Practice in the Centre of Branch Didactics, Branch Didactics in the Centre of Practical Training*, main goal of which was "to identify adaptive teaching strategies applying a cognitively oriented approach to the development of students' critical and creative thinking and other key psychological and didactic topics and strategies in specific professional (branch) didactics and their implementation in undergraduate practical training of secondary education teachers through an excellent centre of practical training" [25].

Research sample of the whole project consisted of 221 teachers of Slovak primary and secondary schools (ISCED 2 – ISCED 3), of which 25 were mentor teachers, and 31 teacher trainees studying Technology as their major subject [26]. Within the project two online questionnaires were administrated, one for the teachers and the other one for the teacher trainees, purpose of which was to find out frequency of applying higher order thinking development strategies by the respondents. The scope of the addressed respondents was limited by the number of education institutions preparing teacher trainees with Technology as their major subject.

To respond to the questionnaire items, the respondents were asked to use a 6-point scale with following meanings of the particular points:

- 6–5 for *applied in each lesson,*
- 4–3 for *not applied in each lesson,*
- 2–1 for *not applied at all.*

In this paper, we present only those project results which are directly related to applying education strategies aimed at the development of higher order thinking in teaching technical subjects.

Purpose of the presented survey was to find out which levels of thinking (according to Bloom's taxonomy) are subject of the teaching intervention dominantly and how often. Within the survey:

- the mentor teachers were asked to assess the teacher trainees from the point of view of the way how they implement strategies of higher order thinking development in their teaching practice (teaching training),

- the teacher trainees were asked to assess themselves,
- the teacher trainees were given the task to assess their mentor teachers from the same point of view as they were assessed.

The recorded research results were processed by means of the descriptive statistics and analysed based on the achieved frequencies of the recorded responses to the particular scales.

Overviews of the survey results are summarised in a graphical form in Figs. 1, 2 and 3.

3 Results and Their Discussion

Figure 1 shows how the teacher trainees applied strategies of the development of the particular thinking levels (following to Bloom's taxonomy) in their teaching practice according to their mentor teacher assessments.

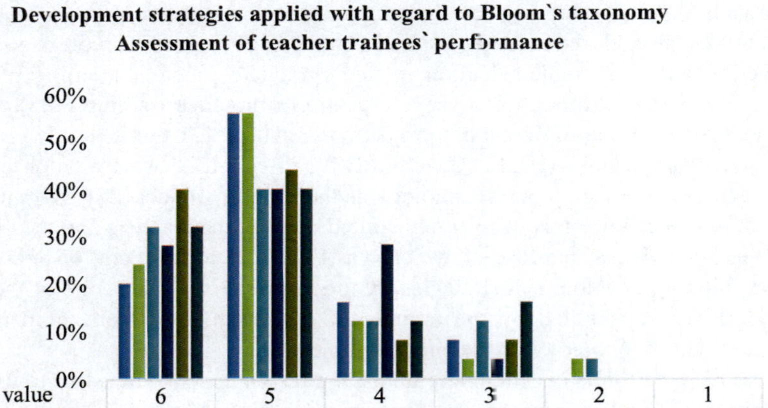

Scale value	6	5	4	3	2	1
Remember	20%	56%	16%	8%	0%	0%
Understanding	24%	56%	12%	4%	4%	0%
Application	32%	40%	12%	12%	4%	0%
Analysis	28%	40%	28%	4%	0%	0%
Evaluation	40%	44%	8%	8%	0%	0%
Creation	32%	40%	12%	16%	0%	0%

Fig. 1. Intervention of development strategies applied by the teacher trainees during their teaching practice – results of the evaluation done by the teacher trainees' mentor teachers.

As can be seen from Fig. 1, evaluations of the teacher trainees done by their mentor teachers have been very positive. However, in the analysis of the presented results we paid attention to the records related to the intervention of the development strategies in regard to the higher order thinking. Results of the analysis can be summarised in following points:

- In opinion of almost 70% of the mentor teachers the teacher trainees created during their teaching practice situations aimed at the development of students' ability to analyse in each lesson (scale range 6–5), and 32% of the mentor teachers had the opinion that the teacher trainees did not create relevant situations in every lesson (scale range 4–3). As the thinking level of analysis is very closely related to evaluation, therefore the results of applying the analysis strategy (see in the next point) were very positive.
- Up to 84% of the mentor teachers rated applying development of thinking at the level of assessment during the subject Technology lessons taught by the teacher trainees by the highest scale values (6–5). From this fact it can be deduced that the teacher trainees during the lessons taught by them created situations which gave their students opportunities to assess, evaluate, criticize, etc. In opinion only of 16% of the mentor teachers, the teacher trainees did not create such possibilities for their students in each lesson (scale range 4–3).
- In a similarly positive way the mentor teachers evaluated the teacher trainees with regard to applying development strategies of the thinking level of creation. Even 72% of the mentor teachers reported applying this strategy by the teacher trainees on the highest scale values (6–5) and only 28% of the mentor teachers reported it on the scale values meaning that the teacher trainees did not apply the strategy at every lesson of Technology taught by them.

So as the teacher trainees involved in the presented survey were asked to make a self-assessment of their teaching performance from the same point of view, so also their mentor teachers did it.

In the literature there is mentioned a number of benefits that self-reflexion (self-assessment) brings, for example development of critical thinking, communication, and increasing self-confidence. For teacher trainees this is also a good way to re-evaluate their teaching methods, strategies or activities they have learned [27], and so it is an integral part of their practical training and preparation for practice. Therefore, the teacher trainees' self-evaluation was also used within the carried out survey to find out how the students themselves evaluate their skills in the implementation of strategies to develop higher thinking skills of students. Results of the teacher trainees' self-assessment in a graphical form are presented in Fig. 2.

Results of the analysis of the results presented in Fig. 2 can be again summarised in following points:

- According to the teacher trainees 51% of them was convinced that they created during each of the lesson of Technology they taught appropriate situations giving their students possibilities to develop their ability of analysis (scale range 6–5). This number is lower than the number recorded in the previous case of the assessments of the teacher trainees done by their mentor teachers. So a general conclusion can be done that at least a half of the teacher trainees really created in each lesson of Technology taught by them situations in which their students could analyse or classify arguments.

- In case of the assessment of the development strategies applied with regard to the thinking level of evaluation there has occurred a significant discrepancy. This strategy was reported by the teacher trainees within the scale range of 5–3 scales, but mostly on the scale value of 3 (58%). This means that only 23% of the teacher trainees were convinced that they created during each of the lesson of Technology they taught appropriate situations giving their students possibilities to develop their ability of evaluation. On the other hand, the mentor teachers reported that even 84% of the teacher trainees applied this strategy at each of the taught lesson. However, 77% of the teacher trainees was convinced that they applied this strategy at least in some of the taught lessons (scale values 4–3). This discrepancy can be caused by the fact, we have already mentioned above, that analysis is very closely related to evaluation, and so it is very hard to distinguish whether the created situation supports the development only of analysis or even of evaluation. Acceptation of the given 77% cases of the reported scale values 4–3 would mean that 77% of the teacher trainees gave very rare their students a possibility e.g. to present results of their work or to express satisfaction with their own activities or work progress.

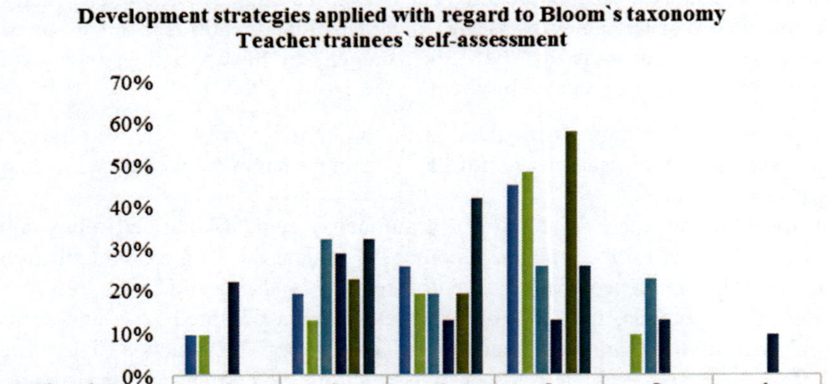

Scale value	6	5	4	3	2	1
Remember	10%	19%	26%	45%	0%	0%
Understanding	10%	13%	19%	48%	10%	0%
Application	0%	32%	19%	26%	23%	0%
Analysis	22%	29%	13%	13%	13%	10%
Evaluation	0%	23%	19%	58%	0%	0%
Creation	0%	32%	42%	26%	0%	0%

Fig. 2. Intervention of development strategies applied by the teacher trainees during their teaching practice – results of the teacher trainees' self-evaluation.

- The teacher trainees similarly evaluated themselves also in relation to creation of situations supporting development of creation ability of their students. The scale values stated by them differed from 5 to 3, most often stated value was 4. Analogically to the previous case, also here is a significant discrepancy between the assessment of the teacher trainees done by the mentor teachers and by the teacher trainees themselves (i.e. between the evaluation of teacher trainees done by their mentors and the teacher trainees' self-evaluations). While according the mentor teachers even 72% of the teacher trainees applied development strategies for creation in each Technology lesson taught by them, according to statements of the teacher trainees only 32% did so knowingly.

As it is clear, the mentor teachers were more positive in their assessments of the teacher trainees' performance. However, in our opinion the assessments done by the teacher trainees, whether in relation to their self-assessment or their assessment of their mentor teachers' performance (see hereinafter), are objective as they were more receptive to assessing the particular levels of thinking and their views were more heterogeneous, expressed in broader scale values range.

In the last part of the presented survey the teacher trainees assessed the mentor teachers. Results of this part of the carried survey are presented in Fig. 3.

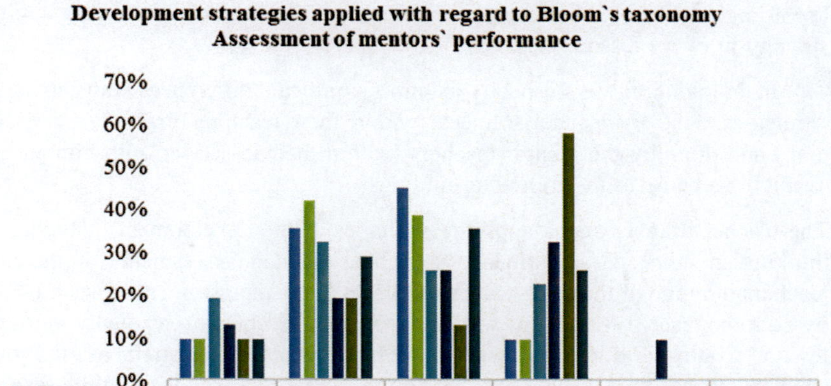

Scale value	6	5	4	3	2	1
Remember	10%	35%	45%	10%	0%	0%
Understanding	10%	42%	39%	10%	0%	0%
Application	19%	32%	26%	23%	0%	0%
Analysis	13%	19%	26%	32%	10%	0%
Evaluation	10%	19%	13%	58%	0%	0%
Creation	10%	29%	35%	26%	0%	0%

Fig. 3. Intervention of development strategies applied by mentor teachers in teaching the subject of Technology – results of the evaluation done by the teacher trainees.

As it is clear from Fig. 3, most of the teacher trainees assessed the mentor teachers within the scale values interval from 3 to 5, from which we can deduce that the mentor teachers apply the development strategies of thinking often, but not in every lesson. Also in this case our attention was paid to the records related to the intervention of the development strategies in regard to the higher order thinking and the findings can be summarized as follows:

- According to 32% of the teacher trainees the mentor teachers applied development strategies of analysis in each lesson (scale values 6–5) and according to almost 60% of the teacher trainees the mentor teachers applied this strategy occasionally, i.e. not at every lesson (scale values 4–3). According to 10% of the teacher trainees the mentor teachers did not even lead their students to analyses of facts and things, to classifications of knowledge and information and so on.
- Neither evaluation the teacher trainees perceived as a frequent strategy applied in mentor teachers' teaching, in consequence of what only 29% of them assessed the mentors' performance by the highest scale values (6–5). According the assessments done by the other teacher trainees, the mentor teachers applied situations leading their students to evaluation rarely, as up to 71% of the teacher trainees ranked the mentors at the scale level of 4–3.
- In a similar way the mentor teachers were assessed by the teacher trainees as to applying creation development strategy. According to 38% of the teacher trainees this strategy was applied at each lesson taught by the mentor teachers, and according to 61% of the teacher trainees the mentor teachers did not apply this strategy in every lesson.

Our main intention was to specify the most significant differences between applying the strategies done by teacher trainees (within their teaching practice, i.e. teaching training) and done by the mentor teachers (within their in-service). By comparing all the results, we came to the following findings:

- The teacher trainees stated applying strategies of the development of higher order thinking in both self-assessments as well as in their assessments of the mentor teachers in range of the scale values from 5 to 3, i.e. the given strategies are applied by both the teacher trainees as well as the mentor teachers occasionally, not at each lesson. Contrary to the assessments done by the teacher trainees, the mentor teachers assessed the teacher trainees very positively, ranked them at the scale level of 6–5.
- In comparison to the assessments done by the mentor teachers, the teacher trainees were more critical in their self-assessments, as they stated applying of comprehension, application, and analysis strategies also at the scale level 2–1, i.e. they did not apply them at their lessons at all.
- Through a deeper analysis of the particular strategies, we came to the conclusion that the level of the use of analysis is perceived by some teachers (32%) may be as insufficient, as it was stated in the assessments of the teacher trainees at the scale level 4–3. Also in case of the self-assessments done by the teacher trainees we see ranking of this strategy at the lower levels of the used scale, even at the levels 2–1.

So even the teacher trainees assessed themselves that they did not apply this strategy at each lesson taught by them, or even that they did not apply it at all. Also in the assessments of mentor teachers almost 60% of the teacher trainees stated this strategy at the scale levels 4–3 and 10% at a scale level of 2. It can therefore be assumed that so as the mentor teachers do not apply the given strategy, neither the teacher trainees do it. We often face the opinions that teacher trainees during their teaching practice (teaching training) cannot be creative and have to imitate teaching styles of their mentor teachers, what can lead to a lack of the use of the given strategy.

- Analysis creates a smooth transition to the next level of thinking which is evaluation. As some shortcomings in applying the analysis strategy occurred, it could be assumed that the same would be also in applying the evaluation strategy. This our assumption was confirmed through comparison of the assessments done by the mentor teachers and the teacher trainees. So it can be stated that in Technology lessons there is almost no evaluation strategy applied. In case of the mentor teachers this strategy was assessed very positively, so as also the other strategies were. But the teacher trainees ranked (almost equally in the self-assessments as well as in the assessments of their mentor teachers) applying this strategy in the scale range of 6–3, even with the highest percentage almost of 60% at the scale value 3. From the above we can conclude that with the current one lesson allocation of the subject Technology, which is very insufficient due to the scope of the subject matter included in its curriculum as well as the practical activities, teachers have very limited time to evaluate activities. At the same time, these findings confirm our assumption that teacher trainees imitate the mentor teachers' learning styles and do not try to plan teaching activities in a different way, so that they would have time also for reviewing and evaluation.

As mentioned above, from our point of view, self-reflective evaluations done by the teacher trainees are more objective and also more significant than their evaluations done by their mentor teachers, mainly in relation to the development of higher order thinking. If we are aware that in teaching such situations can arise, especially when introducing the topic for the first time, when it is necessary to give more time for tasks at lower levels of thinking, i.e. when it takes more time until the students remember the presented knowledge and understand it and only then (may be at the next lesson) they can transform it, analyse or suggest new solutions, creations [28]. This also underlines the main principle of Bloom's taxonomy, which is that students cannot pass to creation of new and more complex products and ideas at the higher level if they have not mastered some skills at a lower level [4]. If students do not remember or understand the information (subject matter) they are learning, they are unlikely to be able to apply it to a new situations or analyse or evaluate it correctly. The problem we see, however, is that some levels of thinking neither teachers nor teacher trainees include in teaching at all. From this reason in the following part we present recommendations in a form of strategies for the development of higher order thinking based on the principle of Bloom's taxonomy, which are suitable for the technical subjects.

4 Recommendations to Strategies Aimed at Development of Higher Order Thinking Skills in Technical Subjects Teaching

To the teaching strategies through which the development of higher order thinking skills in teaching technical subjects the following ones are proposed:

- Goal setting - it is important, first of all, clearly to set the goal to be achieved, i.e. what ability regarding thinking the student is expected to use, achieve or develop in the field of thinking. A quick way for a teacher to check a goal at a given level of thinking is to answer such questions as: What will students do in this topic? Do they name or describe something in their own words? - remember, understanding - lower thinking; Do they analyse or evaluate something? Will they design or create something? - higher thinking. If the aim of the topic is that students to understand its content, then lower level activities are appropriate. On the other hand, if the goal is to broaden basic knowledge and apply it in a new task in order to create a new product or idea, then higher level activities are appropriate.
- Experimentation - educational standards of technical subjects are designed so that teachers can create such situations as: searching, verifying, constructing. Through them students can look for and find solutions, produce new ideas, so these activities almost always involve students in analytical and evaluative thinking.
- Exploration, discovery - teachers of technical subjects tend to include practical activities in the spirit of imitating a model product designed by them. However, such activities do not lead to the development of thinking or creation, only to imitation. Teachers should therefore create situations in which they allow students to manipulate with tools and think about their function, so that they have the opportunity to choose tools and their own technological process. But even the imitation of the finished product itself can lead to the development of thinking at a higher level, if the teacher does not inform the students in advance of the progress of the work, but creates a space for them to think about the procedure and the necessary tools.
- Asking questions - not include suggestive questions or closed questions with a one-word positive or negative answer (yes-no), but rather questions supporting the ability to observe, classify, think, e.g.: Why? What if?
- Problem solving - teachers should create, retell, illustrate or demonstrate problem situations which activate students to solve them. It is important to make students to think about the problem, to analyse it in parts, to look for relations among them on the basis of which they will suggest ways how to solve the problem and thus will work towards its solving. Thinking about problems presupposes the development of analytical, critical and creative thinking.
- Applying creative and own ideas - the best way how to develop critical thinking and creativity is to create situations for the development of self-activity and self-responsibility, when students learn by means of problem solving, through particular activity in which they can plan, implement, search and verify solutions where they can formulate their own proposals, look for non-traditional solutions or several

possible solutions, where they can critically evaluate result of their work and correct their mistakes.

5 Conclusion

Our aim was to analyse applying development strategies of higher-level thinking in pre-service training of future teachers of technical subjects and to specify the most significant differences in applying these strategies by both teacher trainees and their in-service mentor teachers. The results showed that both teacher trainees and mentor teachers pointed out shortcomings in development of students' abilities of higher order thinking, such as analysis and evaluation. Based on these findings, there were formulated the above-mentioned recommendations in the form of strategies for the development of higher order thinking to improve quality of the teacher trainee pre-gradual preparation in this area. So as in the other areas also here new research questions arise to improve pupils and students' learning achievements. As examples of topics for further research can be stated assessment of the levels of higher order thinking achieved by pupils and students, investigation of the relation between the achieved levels of higher order thinking and school specializations (or in the schools by teachers applied teaching strategies).

Acknowledgement. This work has been supported by the Cultural and Educational Grant Agency of the Ministry of Education, Science, Research and Sport of the Slovak Republic under the project No. KEGA 021UKF-4/2018.

References

1. Adams, N.E.: Bloom's taxonomy of cognitive learning objectives. J. Med. Libr. **103**, 152–153 (2015)
2. Forehand, M.: Emerging Perspectives on Learning, Teaching, and Technology. Global Text, Michael Orey (2010). https://textbookequity.org/Textbooks/Orey_Emergin_Perspectives_Learning.pdf
3. Amer, A.: Reflections on Bloom's revised taxonomy. Electron. J. Res. Educ. Psychol. **4**(1), 213–230 (2006)
4. Bloom, B.S.: Taxonomy of Educational Objectives: The Classification of Educational Goals. Longmans, Green, New York (1956)
5. Vygotskij, L.S.: Vývoj vyšších psychických funkcí. SPN, Praha (1976)
6. Rubinštejn, S.L.: Základy obecné psychologie. Praha, SNP (1964)
7. Leontiev, A.N.: Činnosť, vedomie, osobnosť. Bratislava, Veda, SAV (1979)
8. Krathwohl, D.R.: A revision of Bloom's taxonomy: an overview. Theory Pract. **41**(4), 212–218 (2002). https://doi.org/10.1207/s15430421tip4104_2
9. Facione, P.A.: Critical Thinking: A Statement of Expert Consensus for Purposes of Educational Assessment and Instruction. The California Academic Press, Millbrae (1990)
10. Halpern, D.F.: Assessing the effectiveness of critical thinking instruction. J. Gen. Educ. **50** (4), 270–286 (2001)
11. Lipman, M.: Critical thinking - what can it be? Educ. Leadersh. **46**(1), 38–43 (1998)

12. Tindal, G., Nolet, V.: Curriculum-based measurement in middle and high schools: critical thinking skills in content areas. Focus Except. Child. **27**(7), 1–22 (1995)
13. Paul, R.W.: Critical thinking: what, why, and how? New Dir. Community Coll. **1992**(77), 3–24 (1992)
14. Willingham, D.T.: Critical thinking: Why is it so hard to teach? Arts Educ. Policy Rev. **109**(4), 21–32 (2008). https://doi.org/10.3200/AEPR.109.4.21-3
15. Mayorov, A.N.: Osnovy pedagogicheskih izmereni. The Fundamentals of Pedagogical Measurements. Logos, Moscow (2010)
16. Musina, V., Peresypkin, A., Makotrova, G., Shumakova, I., Shekhovskaya, N.: The problem of setting educational goals: from Socrates to b. Bloom. Amazonia Investiga **7**(14), 119–127 (2018). https://amazoniainvestiga.info/index.php/amazonia/article/view/477
17. Shaw, A., Lydia, L.O., Gu, L., Kardonova, E., Chirikov, I., Li, G., Hu, S., Yu, N., Ma, L., Guo, F., Su, Q., Shi, J., Shi, H., Loyal-ka, P.: Thinking critically about critical thinking: validating the Russian HEIghten® critical thinking assessment. Stud. High. Educ. **45**, 1933–1948 (2019)
18. Kudrinskaia, L.A., Kubarev, V.S.: Characteristics of the learning motivation of students in a higher technical educational institution. Russ. Educ. Soc. **55**(4), 25–37 (2013)
19. Johnson, L., Lamb, A.: Teacher tap (2011). https://eduscapes.com/tap/topic69.htm
20. Blanco, M.A., Capello, C.F., Dorsch, J.L., Perry, G.J., Zanetti, M.L.: A survey study of evidence-based medicine training in US and Canadian medical schools. J. Med. Libr. Assoc. **102**(3), 160–168 (2014)
21. Freitas, A.F., Thompson-Leduc, P., Borduas, F., Luconi, F., Boucher, A., Witteman, H.O., Jacques, A.: The majority of accredited continuing professional development activities do not target clinical behavior change. Acad. Med. **90**(2), 197–202 (2015)
22. Wei, B., Ou, Y.A.: Comparative analysis of junior high school science curriculum standards in Mainland China, Taiwan, Hong Kong, and Macao: based on revised Bloom's taxonomy. Int. J. Sci. Math Educ. **17**, 1459–1474 (2019)
23. Retnawati, H., Djidu, H., Apino, E., Anazifa, R.D.: Teachers' knowledge about higher-order thinking skills and its learning strategy. In: Problems of Education in the 21st Century, vol. 76, no. 2 (2018)
24. Ackoff, R.L., Greenberg, D.A.: Turning Learning Right Side Up: Putting Education Back on Track. Prentice Hall, New Jersey (2008)
25. Duchovičová, J., Tomšik, R.: Critical and creative thinking strategies in teaching internal consistency of the research tool: Stratégie kritického a tvorivého myslenia vo vyučovaní vnútorná konzistencia výskumného nástroja. Slavonic Pedagogical Stud. J. **6**(2), 375–394 (2017)
26. Vítečková, M., Gadušová, Z.: Teacher professional induction in the Czech Republic and Slovakia. Am. J. Educ. Res. **2**(12), 1131–1137 (2014)
27 Klimova, B.F.: Self-reflection in the course evaluation. Procedia – Soc. Behav. Sci. **141**, 119–123 (2014)
28. Fastigi, W. (2020). https://technologyforlearners.com/applying-blooms-taxonomy-to-the-classroom/

Developing Engineering Students Writing Competence: An Intervention Based on Formative and Peer Assessment

Tom O'Mahony(✉)

Cork Institute of Technology, Cork, Ireland
tom.omahony@cit.ie

Abstract. The ability to communicate is recognised as an important undergraduate engineering competence. While acknowledging its importance, employers often express dissatisfaction with engineering graduates ability to communicate. In one course that the author teachers, learners are asked to document project outcomes via an engineering-style conference paper. Over the years, many learners consistently struggle to write high quality abstracts, introductions, conclusions and to apply referencing conventions appropriately. This year the author designed and implemented a writing intervention, based on formative and peer-assessment, to address this issue. This article explore the impact of this intervention. Impact was evaluated by comparing writing artifacts produced in this academic year with artifacts produced in two other years prior to the intervention. The results indicate that the intervention was impactful with the quality of these elements improving from an average score of 35% (2017 & 2018) to 70% this academic year. An implication is that specific, targeted supports are likely to be more impactful than repeated practice – especially when practice is not supported with feedback. More generally, the article adds to the extensive literature evidencing the impact of formative assessment on student learning.

Keywords: Writing · Graduate attributes · Formative assessment · Peer review

1 Introduction

Writing is important. It is one of those "generic" competences that crosses disciplines. Most employers expect that those who graduate from higher education will exhibit communication skills, including writing, to a high professional standard. Regardless of whether we agree with this neo-liberal view on the purpose of higher education, graduates that do not possess those skills are likely to be disadvantaged in a competitive labor market. In that sense we, educators, have an ethical responsibility to try and develop those skills. In addition to its importance as a graduate attribute, writing is also one of those really effective learning strategies that we should be encouraging all learners to adopt [1]. For these reasons, writing features strongly in all of the courses that I teach.

One such undergraduate engineering course, the focus of this paper, asked learners to document project outcomes in a conference style engineering paper. This task represents the main assignment within the course and has been consistently so for many

M. E. Auer and T. Rüütmann (Eds.): ICL 2020, AISC 1328, pp. 787–796, 2021.
https://doi.org/10.1007/978-3-030-68198-2_74

years. Over those years I have mostly used exemplars – examples of papers from past students and others – to help clarify the requirements and expectations associated with this assessment task. This, along with feedback on drafts, has had a positive impact on writing. Despite this general positive impact, year after year I see that my students struggle to write good abstracts, introductions, conclusions and their referencing practice is really poor. This paper summarizes how I attempted to address this problem and evaluates the impact of the intervention that I developed. The next section provides a more general background and justification for the approach adopted and subsequent sections document the research approach, including the intervention itself, and evaluates its impact on student's writing performance.

2 Background and Justification for This Approach

2.1 Writing in the Engineering Curriculum

Writing is important as a professional competence within engineering, as a graduate attribute that can enhance employability and as a really effective learning strategy [1]. However, despite its importance, and indeed the regularity with which engineering undergraduates produce laboratory and other types of written reports, there is ample evidence that graduate writing skills are underdeveloped [2, 3]. One reason why writing remains underdeveloped is attributed to the lack of formal writing instruction within the engineering curricula. We (engineering educators) seem to assume that our undergraduates have been taught everything that they need to know about writing from the secondary-school curriculum and students just need more practice in order to develop writing proficiency. This is in stark contrast with our attitude to mathematics. As a specific example, on the B.Eng. in Electronic Engineering programme in my home department, there is a component of one first-year, semester 1 module that is dedicated to writing instruction. There are two mandatory 5-credit modules dedicated to mathematics in first year, second year and third year with an optional 5-credit mathematics module in four year. Hence, through our curriculum, our undergraduates can easily see that mathematics is important but that writing is something "other". This "otherness of writing practices in the engineering curriculum" is developed in [4] where the authors argue that engineering identity centres around technical knowledge (exemplified by science and maths) and that writing is not what engineers do. Undergraduate students then adopt this identity. Hence, there is a tension between the demand from employers for graduate engineers that can write and engineering curricula's willingness to meet that demand.

As this tension is not new, different approaches have been explored to redress it. A common approach is to introduce one or more writing or communication modules into the engineering curriculum. These usually appear in the early years of the programme. The obvious advantage of this approach is that writing instruction is delivered by experts and a mandatory module that focuses on writing means that learners cannot sidestep this learning if they want to progress. On the other hand, if this writing instruction is provided by experts outside of the engineering department it can reinforce the "otherness" of engineering writing. It is also possible that the writing assignments that are posed are less meaningful or relevant to engineering students, making engagement harder. The alternative approach, and the one that the Dept. of Electrical & Electronic Eng. at CIT has adopted in our programmes, is to integrate writing instruction and practice throughout the curriculum. This provides multiple opportunities to develop and practice the skill. Moreover, by embedding writing within technical courses that are delivered by engineering staff we may change engineering student's perception of writing and they may come to see writing as being important to the engineering profession [4, 5]. However, it does now become possible for students to avoid this skill and successfully complete courses by just focusing on the remaining technical parts. It also, arguably, more challenging to develop an integrated curriculum as the progression of writing development from year to year needs to be defined to create a spiral curriculum [6]. Furthermore, engineering staff may need to be convinced to make the changes to individual course and then be supported as they implement those changes. Despite these challenges, our Department's approach has been to integrate writing into the curriculum as we feel that perceptions of writing is a big part of the problem and, like any skill, providing multiple opportunities to practice and develop is an important part of the solution.

2.2 Solution Components

Within individual modules, our Department's solution integrated a number of teaching strategies - specifically formative assessment and peer review in a blended learning environment. The assessment literature presents persistent, strong evidence that formative assessment has a positive impact on learning [7]. However, for this impact to be realized it is critical the learners engage with and apply the feedback that is provided. To enable this [8] recommend a "social constructivist" model whereby learners are encouraged to interact with the criteria and standard associated with assessment components, apply those to assessment examples and discuss both the process and outcome with their peers. Throughout this literature the use of two stage assignments that consist of drafts, an opportunity to re-work based on feedback and then the submission of a final assignment is recommended. This model naturally incorporates many of the principles of good feedback practice advocated in [9].

The second strategy that formed part of our solution was to incorporate peer-reviews of writing assignments. Traditional feedback practices are heavily criticized for encouraging learners to be heavily dependent on a more knowledgeable others for validation and feedback [10]. As a consequence, peer-reviews are gaining in currency as one strategy that enables more independent learners. The peer-review process is also advocated as a means of encouraging learners to develop the capacity to make critical and evaluative judgements [11]. With peer-reviews, learning can happen from the feedback received by peers. It is also accepted that the process of creating feedback can have a powerful learning impact. This happens because the process requires learners to actively engage with the assessment criteria, apply the criteria to examples and justify the application [12, 13]. This can result in a more explicit or concrete understanding of the assessment requirements and as a result better enable individuals to meet those requirements. Challenges associated with the implementation of peer-reviews relate to the need to provide training around the process, the need to develop a shared understanding of the assessment criteria and students holding mixed feelings towards the strategy [14]. However, despite these challenges the potential of peer-reviews to enhance critical thinking and to develop academic writing [13, 15] convinced the author to pilot this approach.

The contribution of this article then is to describe how these strategies – formative assessment and peer-review were combined in a blended learning environment to create a writing intervention for final year engineering students in a particular module. The impact of this intervention is then analyzed by comparing the quality of written papers from two years prior to the intervention and this current academic year when the intervention was piloted. Data analytics from the Learning Management System used to host the resources will also be used to assess student engagement with the developed interventions. In the next section, the research method, including the intervention, will be described.

3 Research Method

3.1 Research Question

The research question explored in this article is *what is the impact of a specific formative and peer-assessment writing intervention on the quality of specific writing components in a particular undergraduate engineering course?* The specific writing components are: title, abstract, introduction, referencing and conclusion.

3.2 Problem Context

The context that this intervention is situated in is a 5 ECTS credit course that forms part of year 4 of the B.Eng. in Electronic Engineering at that author's institution. The module timetable specifies two hours of lecture content and two hours of laboratory content per week. The laboratory content consists of a semester-long group mini-project in which learners are expected to engage in design, test and evaluation work. Content is primarily delivered through lectures that incorporate a range of individual and group-based active learning techniques. The module is assessed via coursework, which includes an individual research-style paper (worth 70%) that is typically submitted in two phases as a draft, on which learners receive instructor feedback and then a final submission. This two stage draft/feedback/final assignment has previously been shown to positively impact on learning [16]. However, because learners have limited time between the draft and final submissions there is only so much feedback that can be acted on. This often means that the focus (both mine and theirs) is on important technical aspects of the project that need to be addressed. Consequently, for a number of years final assignments have consistently displayed a range of common problems associated with weak titles, incomplete abstracts, inadequate introductions, poor or no conclusions, few if any references and poor formatting of references. One aim of the intervention was to address those issues. The second aim of the intervention was to try and spread the writing workload by requesting submissions of sections of the paper throughout the semester rather than leaving this large writing task to the busy end of the semester. Learners then received feedback on these submissions either from their peers or the instructor.

3.3 The Intervention

The intervention consisted of a range of short focused writing activities to address the common problems identified in the previous subsection. So for example to address the referencing problem, learners were firstly asked to respond to a general question on what purpose(s) referencing serves and then given a paragraph of text which referenced a number of sources and these sources were then listed after the text. Common errors were deliberately built into this reference list. Learners were then asked to identify which of the sources were correctly referenced according to the IEEE referencing format. This, and all the interventions, were hosted on the institute's Learning Management System (LMS). To enable fast, efficient feedback the activity was structured as a multiple-choice question where learners had to pick one from a range of options. Feedback was automatically built into this so once learners engaged and selected an option the feedback explained which of the references were incorrectly formatted and the formatting issue with each. For the abstract, introduction, conclusion learners were pointed to an on-line resource that discussed the purpose and nature of these elements. They were then given a previously published paper that was similar in content to their projects [17] and asked to critique these sections of the paper. With the introduction, for example, they were asked to identify the topic sentence of each paragraph in the introduction so that they would see how that was structured. They were then asked to create topic sentences for a three-paragraph introduction for their own papers. In general the format of these small interventions was similar.

a) learners were provided with a general resource on the topic
b) learners were asked to engage with a specific task e.g. critique an example
c) learners were then asked to apply that to their own paper

Over the course of the semester 10% of the course grade was assigned to these activities to encourage engagement.

3.4 Evaluation

The impact of this intervention on writing was evaluated by comparing the quality of reports produced at the end of this current academic year with those produced during the 2016–17 and the 2017–18 academic years. The criteria summarised in Table 1 was developed for this purpose. This criteria is quite subjective in nature. Given the authors "insider" role and personal involvement this raises concerns related to the objectivity of this evaluation and the role that biases e.g. confirmation bias might play. In an attempt to mitigate this some more objective measures were also considered. These included engagement data from the institutes LMS, word counts for the introduction, number of sources referenced in the text and the marks awarded for the quality of the writing. While the marks associated for writing are arguably a subjective judgement, over the period in question (2016–2020) the same assessment rubric has been used and these marking decisions are available for scrutiny by two external examiners. These processes help to ensure a degree of consistency and objectivity in the grading process. Grading sheets for the academic year 2018–19 could not be located and that is the only reason for not including that year in this analysis.

Table 1. Criteria used to evaluate the impact of the intervention

Criteria	Points	Explanation/Example
Title	0	Generic title e.g. "Report for Control Engineering Module"
	1	Title describes the content of the report e.g. "Controller Design and Analysis of a Balance Beam"
Abstract	0	Missing
	+1	For including a specific aim or objective of the work; a description of the method; specific results/conclusion is included in the abstract. A maximum of 3 points if all three are included
Introduction	1	The aim/objective of the work and an outline of the tasks involved is described
	2	In addition to the criteria for (1) above some more general background context is included
References	1	More than two sources are cited in the text
	2	More than two sources are cited plus accurately referenced
	3	More than two sources are cited, referenced and used to justify an argument or support an explanation
Conclusion	0	Missing
	1	Focuses on summarizing the work
	2	Uses key results to emphasize findings and draw conclusions

3.5 Participants

Participants were all enrolled on the same final year course. They are predominantly male. To be included in this study they must have submitted a valid paper for the final assignment. Valid in this case means that the assignment was complete, judged to represent the student's own work, be free from plagiarism or other infringements and not be accompanied by requests for a deferral (e.g. on account of the impact of COVID-19). Table 2 lists the participant numbers by academic year.

Table 2. Summary of participant numbers per academic year

2016–17	2017–18	2019–20
12	9	14

4 Results

Report quality for each of the three academic years was evaluated using the criteria reported in Table 1. Scores for each year were then averaged and converted into a percentage of the maximum possible score (1, 2, or 3) depending on the criteria. These average percentages for each of the three years are displayed in Fig. 1. Along with these percentages, the actual grade that was assigned for the quality of the written report is also presented in Fig. 1, labelled as *Writing Mark*. Again the average is recorded as a percentage. This *Writing Mark* represents the overall quality of the written communication and the technical details are graded separately. So the *Writing Mark* includes things like the quality of the writing, the logical structure of the paper, the quality of graphs, tables and equations etc. in addition to the aspects (abstract, introduction, conclusion and referencing) that are the focus of this article. Comparing the 2017 and 2018 academic years the results show a small overall improvement in most of these elements and a significant jump in quality for the 2020 academic year. The five specific criteria (Title, Abstract, Introduction, Referencing, Conclusion) increased from an average grade of 31% in 2017 to 39% in 2018 to 70% in 2020. This is reflected in a modest increase in the overall writing mark which remained stable at 63% (2017) and 64% (2018) but increased to 73% in 2020.

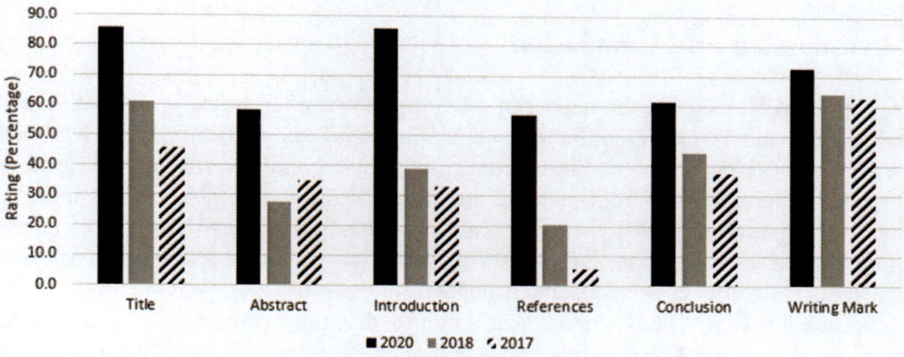

Fig. 1. Evaluation of report quality based on the criteria reported in Table 1.

A more objective lens is provided by Table 3 and Table 4. Table 3 presents some quantitative data from the reports which count the average number of articles referenced in each year and also the average word count in the introduction. There is clear evidence that in both cases these numbers have increased. Table 4 presents the organization of the intervention over the semester and student engagement with the intervention. Student engagement data was extracted from the institute's LMS. Section 3.3 outlined how learners were expected to engage or "do something" as part of each activity and in each case the "doing" was submitted to the LMS. In some cases (week 3 for example) the engagement was to answer multiple choice questions while in other cases (week 5 for example) it was to submit a draft of a section of the paper. So the engagement data records the number of learners that engaged with the requested activity. Throughout the semester the engagement data remained high and helps to explain or justify the results presented in Fig. 1. COVID-19 did have an impact. Our institution closed to students on week 7 and the average student engagement after that is, understandably, lower than it was before. Of course, the second half of the semester is always a busier one for students so this drop in engagement may not exclusively be caused by the impact COVID-19.

Table 3. Quantitative evaluation of paper quality

	2016–17	2017–18	2019–20
Average no of references per paper	1.3	2.6	4.6
Average word count introduction	116	160	399

Table 4. Engagement with formative and peer assessment activities

Intervention or activity	Referencing	Writing an introduction	Peer review	Draft Sect. 2	Draft Sect. 3	Writing a conclusion	Title & abstract	Draft Sect. 4	Peer review
Timing	Week 3	Week 5	Week 6	Week 7	Week 11	Week 12	Week 12	Week 13	Week 14
Engagement (%)	93	93	93	86	79	64	64	71	57

5 Discussion and Conclusion

This article has evaluated the impact of a writing intervention based on formative and peer assessment. The intervention focused on addressing specific writing issues around the quality of titles, abstracts, introductions, referencing and conclusions. These were identified as problematic from previous iterations of the course. The results indicate that the intervention was largely successful in that the quality of these elements improved, on average, from 35% (2017 & 2018) to 70% this academic year. This represents a significant and substantial impact.

Examining Fig. 1, it is evident that some of the more substantial improvements were linked to the abstracts, introductions and references. In 2017 there was on average 1.3 sources referenced per student paper and in almost all cases these references were not formatted in accordance with the recommended style (IEEE). While the number of references increased in 2018 the quality of the formatting did not. This year, the intervention was successful in increasing both the number of sources and the quality of the referencing. The most common use of these references is to identify components or software that was used, rather than using references to support an academic argument. This is perhaps a consequence of the intervention that I designed which did focus on the technical aspects of referencing rather than the more abstract purpose and use of references. A future iteration will address this. Likewise, while the quality of the abstract have improved, most abstracts still do not summarize the entire paper. They generally do not report findings or conclusions/implications. Additional emphasis on these aspects may help to develop this aspect in the future. Conclusions tend to be similarly limited and focus on summarizing in a quite generic way and not emphasize findings by linking them to specific results. In contrast, introductions were developed to a high standard. In previous iterations of the module, the introduction was very short (Table 3) and did little more than provide an outline of the paper or the technical tasks that were completed. This intervention resulted in the development of significantly longer (Table 3) introductions that included a general background (importance of control engineering), explicitly included the aim/objective of the work in addition to an outline of the tasks that needed to be completed.

A somewhat unexpected outcome, evident from both Fig. 1 and Table 3 is that the quality of these writing elements improved somewhat between 2017 and 2018. Since then the institute has migrated to a new LMS and I no longer have access to the details of the courses as they were implemented in 2017 and 2018. Hence it is not possible to explore or explain what features of the course or cohort that might have resulted in this small improvement.

There are a number of limitations associated with this research. The small sample size in each year raises questions about the generalizability of the findings. There is also a question as to whether the intervention is feasible with large cohorts. Some components of the intervention involved individual instructor feedback and these are certainly problematic for larger groups. Other aspects involved the use of multiple choice questions and peer-reviews which should be easy to scale. A significant issue is the potential for bias to influence the outcome due to the "insider" nature of the research. As both the instructor who teaches the course and designed the intervention, I have a vested interest in the outcome. In addition, my expectation at the start is that it would be successful as the literature on formative assessment generally indicates positive outcomes [7, 8, 12]. Hence there is real potential for confirmation bias and sources to contaminate the findings. I tried to manage that by developing a simple and relatively explicit rubric to evaluate the impact of the intervention prior to evaluating the written papers themselves. Again, in an attempt to manage the subjective, qualitative nature of the evaluation, I considered some more objective measures such as the number of references and the length of introductions and student engagement with the intervention. These more objective measures generally support the qualitative evaluation.

It is not uncommon to hear academics in higher education bemoaning the generic skills levels, especially those linked to writing, mathematics and evaluation, possessed by current students. Faculty are often reluctant to address these as they believe they do not have the time or background to teach these skills [4]. An implication of this small-scale piece of research is that specific aspects of writing can be developed with no formal in-class teaching and without reducing the normal technical content of the course. In this case, the intervention was exclusively hosted on the institutes LMS and a small portion of the course grade was used to encourage learners to actively engage with this writing intervention. While the focus of this intervention was on abstracts, introductions etc., there is no reason why a similar type of intervention could not be used to develop other writing elements if they were to be found wanting. In a more general context, the article contributes to the extensive literature evidencing the impact of formative assessment on student learning.

Acknowledgements. The author gratefully acknowledges the support of the National Forum for the Enhancement of Teaching and Learning in Higher Education under their Strategic Alignment of Learning and Teaching Enhancement Fund and the Faculty of Engineering and Science at CIT for their continued support and encouragement.

References

1. Kuh, G.D.: High-impact educational practices : what they are, who has access to them, and why they matter. Association of American Colleges and Universities (2008)
2. ASEE (American Society for Engineering Education): Transforming Undergraduate Education in Engineering Phase II: Insights from Tomorrow's Engineers (2017)
3. Brunhaver, S.R., Korte, R.F., Barley, S.R., Sheppard, S.D.: Bridging the gaps between technology and engineering education. In: Freeman, R.B., Salzman, H. (eds.) U.S. Engineering in a Global Economy, pp. 129–163. National Bureau of Economic Research, University of Chicago Press (2018)
4. Goldsmith, R., Willey, K.: The otherness of writing in the engineering curriculum: a practice architectures perspective. J. Acad. Lang. Learn. **12**, 97–114 (2018)
5. Crawley, E., Malmqvist, J., Ostlund, S., Brodeur, D.: Rethinking Engineering Education: The CDIO Approach. Springer, Heidelberg (2007)
6. Bruner, J.S.: The Process of Education. Harvard University Press, Cambridge (1960)
7. Wiliam, D.: What is assessment for learning? Stud. Educ. Eval. **37**, 3–14 (2011)
8. O'Donovan, B., Rust, C., Price, M.: A scholarly approach to solving the feedback dilemma in practice. Assess. Eval. High. Educ. **41**, 938–949 (2016)
9. Nicol, D.J., Macfarlane-dick, D.: Formative assessment and self-regulated learning: a model and seven principles of good feedback practice. Stud. High. Educ. **31**, 199–218 (2006)
10. Boud, D., Molloy, E.: Rethinking models of feedback for learning: the challenge of design. Assess. Eval. High. Educ. **38**, 698–712 (2013)
11. Carless, D., Kam, K., Chan, H., To, J., Lo, M., Barrett, E.: Developing students' capacities for evaluative judgment through analysing exemplars. In: Boud, D., Ajjawi, R., Dawson, P., Tai, J. (eds.) Developing Evaluative Judgement in Higher Education: Assessment for Knowing and Producing Quality Work. Taylor & Francis (2018)
12. Liu, N.F., Carless, D.: Peer feedback: the learning element of peer assessment. Teach. High. Educ. **11**, 279–290 (2006)
13. Huisman, B., Saab, N., van Driel, J., van den Broek, P.: Peer feedback on academic writing: undergraduate students' peer feedback role, peer feedback perceptions and essay performance. Assess. Eval. High. Educ. **43**, 955–968 (2018)
14. Zhou, J., Zheng, Y., Tai, J.H.: Grudges and gratitude: the social-affective impacts of peer assessment. Assess. Eval. High. Educ. **45**, 345–358 (2019)
15. Liang, J., Tsai, C.: Learning through science writing via online peer assessment in a college biology course. Internet High. Educ. **13**, 242–247 (2010)
16. O'Mahony, T.: The impact of a constructivist approach to assessment and feedback on student satisfaction and learning: a case-study. AISHE **9**, 2871–2879 (2017)
17. McLoone, S.C., Maloco, J.: A cost-effective hardware-based laboratory solution for demonstrating PID control. In: 2016 UKACC International Conference on Control UKACC Control 2016, pp. 1–6 (2016)

Systematic Assessment of Interactive Instructional Technologies in Higher Engineering Education

Aurelia Ciupe(✉), Serban Meza, and Bogdan Orza

Multimedia Systems and Applications Laboratory, Technical University of Cluj-Napoca, 28 Memorandumului Street, 400114 Cluj-Napoca, Romania
aurelia.ciupe@com.utcluj.ro

Abstract. Given the diversity of interactive instructional systems targeting higher engineering education, a systematic assessment of such systems is desirable in order to offer an overview on the existing landscape. Following a Systematic Mapping protocol (search, extraction, selection, mapping/visualization and discussion), 294 studies have been selected from a publication interval between 2010–2019 and discussed according to venue characterization and field applicability distribution. The resulting analysis demonstrated the various forms of research under which studies are conducted (from Validation Research, Experience Papers to Opinion Papers and Solution Proposal), while offering a specific depiction on the interactive technologies (Augmented/Virtual/Mixed/Simulated Reality, Game-based strategies, Simulation environments, Conversational tools/other bots) used in educational processes (teaching, learning, training, tutoring, practicing, supporting). The current study demonstrates the overall applicability of the Systematic Assessment frameworks to in-depth characterization of such a landscape, providing the preliminaries for a further statistical analysis.

Keywords: Interactive assistive technology · Higher engineering education · Evaluation methodologies · Systematic Mapping Study

1 Introduction

The demanding need of efficiently shaping the learning and instructional processes becomes a visionary direction of the *Future of Education and Skills* 2030 [1] project, proposed by the Organization for Economic Co-operation and Development (OECD) in the context of redefining the instructional systems to effectively address the students' cognitive resources of knowledge, skills, attitudes. The challenge currently relies in building and assessing instructional designs on a framework basis derived from pedagogy practices, the OECD Learning Framework 2030, that would economically, socially and environmentally adapt a worldwide shared perspective on education, to the fast evolving world. Updated instructional paradigms are required to support the process of building and assessing procedural knowledge through practical problem-oriented learning. Quality in higher education through *engagement* and *interactivity* have been the key pillars since the Policy Level 3 articulation, of the OCED manifesto

M. E. Auer and T. Rüütmann (Eds.): ICL 2020, AISC 1328, pp. 797–804, 2021.
https://doi.org/10.1007/978-3-030-68198-2_75

on *IMHE Guide for Higher Education Institutions* [2]. Additionally, student engagement needs to be redefined as a catalyst of quality teaching, where ICT resources are to provide a meaningful, consistent environment that would foster instant feedback and satisfaction evaluation among both students and teachers [2].

Given the existing ICT support, such assistive interactive technologies should be mapped in a qualitative manner to provide an overview on possible gaps or strengths of the current instructional designs, practices, environments. Therefore, the current study aims at conducting a systematic assessment of the case-studied instructional tools that provide a more interactive educational environment, apart from the dominant learning platforms of Content and Learning Management Systems.

Two main premises define the current context and further, the research methodology: the premise of defining the interactive instructional technologies from a functionality and environmentally perspective and the premise of applying the framework of the Systematic Mapping Study (SM) to collect and analyze the existing body of evidence. The work is structured as follows: Section 2 offers an overview on secondary studies protocols and depicts several pioneering works conducted in the field of interactive assistive technologies, that are applicable to higher engineering education. Section 3 presents the applied research methodology, while Sect. 4 presents the mapping results in form of graphical representations. Section 5 assesses current implications and further directions for extending the current protocol.

2 Background and Preliminaries

The framework of systematic assessment in the form of Systematic Mapping (SM or MS), Systematic Literature Review (SLR) or Tertiary Review, has been translated from evidence-based medicine, towards summarizing the body of research in technology or business areas of research. The rigorous process of Systematic Reviewing has been documented since 2007 as a set of guidelines by Kitchenham et al. [3], with applicability to software engineering. Further notable literature interventions have been submitted by Petersen et al. [4], defining the framework of Systematic Mapping as an expanded SLR in terms of scope and specificity. An updated description of the SM process, focusing on the protocol phases has been published by Petersen et al. [5], having the same software engineering domain of applicability. [6] enhance the existing guidelines with a proposal of constructing the search strategy following the *mentefacto conceptual* method and tool, based on a graphic model of semantic thesaurus following 4 classes of concepts: 1) isoordinated, 2) superordinate, 3) excluded, and 4) infraordinated.

With respect to SLRs/MSs applicability, more recent investigations progress towards broader IC&T, engineering or educational fields. In terms of assistive technologies in education, MSs and SLRs have been conducted at all educational stages: gamification principles in higher education [7] (on 34 studies), Augmented Reality learning at both STEM [8] (on 28 studies) and higher education [9] (on 79 articles) levels, Virtual Reality implications in High School education [10] (on 21 studies). A model of further systematic expansion applied to Computer-Supported Collaborative Learning has been conducted in the form of meta-analysis by [11], where 143 studies between 2005–2014, have been assessed to statistically describe the effectiveness of CSCL in STEM education. The landscape of assistive instructional technologies has been assessed from an evaluation perspective, under a SLR protocol, by [12]. 365 studies, between 2015–2017 have been included in the analysis, where the results have been investigated under 5 themes: learning, technology, affective elements, presence and behaviour. A SLR focusing on immersive environments in educational processes has been conducted by [21], comprising both lessons learnt as well as forecasting on the existing trends and further perspectives.

A summary of prominent secondary studies in interactive educational environments with implications applicable to engineering education, according to the research scope and the number of selected studies in the protocol design, is presented in Table 1.

Table 1. PICO strategy applied to search strategy.

Authors	Type of study	Research scope	Number of studies
Diegmann [13]	SLR	Enhanced learning environments, AR	523/25 selected
(Bacca, 2014)	SLR	Inclusive learning, AR	NA/32 selected
Akçayır [15]	SLR	Advantages and challenges, AR	102/68 selected
(Martin, 2011)	SLR	Technology trends	NA/10 selected
Santos [17]	MA	Tech-enhanced curriculum, AR	553/87 selected
Sirakaya [18]	SLR	Avantages, AR	NA/105 selected
Broadbent [19]	MA	Self-regulated learning	1789/118 selected
Ali [20]	SLR	Collaborative environments, MR	NA/148 selected
Radianti [21]	SLR	Collaborative learning, AR, VR	3219/590 selected
(Faculdade, 2016)	MS	Increased interaction, 3D, VR, AR, MR	419/113 selected
Avellar [23]	MS	Teaching-learning support, AR, VR, MR	2249/46 selected
Dicheva [7]	MS	Trends, gamification/GBL	1647/34
Borges [24]	MS	Collaborative learning, gamification/GBL	357/26
Krassmann [25]	MS	Cognitive evaluation, gamification/GBL	2174/47
(Souza, 2018)	MS	Landscape mapping, gamification/GBL	2675/156
Subhash [27]	SLR	Advantages, gamification/GBL	602/41
Vlachopoulos [28]	SLR	Effects, simulations, gamification/GBL	8859/123
Cook [29]	MA	Effectiveness, simulations	10903/289
Battistella [30]	SLR	Teaching support and learning objectives, GBL	NA/107
Gu [31]	SLR	Opportunities, simulations	36612/90

3 Systematic Mapping – Protocol Design and Discussion

The applied SM protocol has been conducted according to the guidelines depicted by Kitchenham et al. [3] and Petersen et al. [4], comprising of 5 phases: 1) research scope definition, 2) search strategy definition, 3) research body extraction; 4) assessment and filtering and 5) visual mapping and discussion. The following Research Questions (RQ1, RQ2) define the scope of the study: RQ1. Which is the publication venue of the instructional technologies applied to higher engineering education? RQ2. To which higher education fields have the interactive instructional technologies been applied?

The search strategy has been constructed based on the PICO description (Population, Intervention, Comparison and Outcomes), where 2 dimensions have been extracted: the *Population dimension* that addresses the area of research, the *Intervention dimension* that exploits the interactive educational technologies, the *Outcome dimension* that characterizes the levels of educational actions and processes. Depiction of the 3 dimensions is presented together with the assembled search string based on Boolean operators are presented in Table 2.

Table 2. PICO strategy applied to search strategy.

PICO dimension	Dimension depiction
Population	Higher engineering education
Intervention	I1: Virtual Reality (VR); I2: Augmented Reality (AR); I3: gamified/Game-based; I4: code-/chat -bots; I5: Simulations or 3D; I6: Simulated Reality (SR)/Mixed Reality (MR)/eXtended Reality (XR)/Second Life
Outcome	Learning, teaching, tutoring, training, guiding, coaching, practicing, supporting, assisting, educating

Search strings have been adapted to be applied on 3 main representative academic databases to which institutional access is provided (Table 3): IEEE Xplore (https://ieeexplore.ieee.org/Xplore/home.jsp), Science Direct of Elsevier (https://www.sciencedirect.com/) and Springer Link (https://link.springer.com/). An interval between 2010–2019 has been considered representative, comprising most of the research conducted on the topic and grouped on 5 intervals: 2010–2011, 2012–2013, 2014–2015, 2016–2017, 2018–2019.

Table 3. Search strings construction and outputs

#	Search string	IEEE Xplore	ScienceDirect	SpringerLink
S1	<I1> AND <O> AND <P>	327	279	431
S2	<I2> AND <O> AND <P>	165	452	553
S3	<I3> AND <O> AND <P>	711	471	1308
S4	<I4> AND <O> AND <P>	660	177	921
S5	<I5> AND <O> AND <P>	1580	1820	2652
S6	<I6> AND <O> AND <P>	29	301	581

Selection of the primary studies to be included in the Systematic Mapping has been done through: Inclusion (IC), Exclusion (EC) and Relevance Criteria (RC), as presented in Table 4.

Table 4. .

Criterion	Examples
IC	Peer-reviewed manuscripts (conference or journal articles), English language
EC	Book chapters, technical reports, grey literature, work in progress, keynotes, book reviews, workshop papers
RC	Applicability in technical engineering education, research that refers to a specific technology, tool or framework, higher education level

4 Systematic Mapping – Mapping and Discussion

A total number of 294 resulting primary studies have been extracted and further analyzed. The contributions this paper proposes result from the mappings assessed to respond each RQ, as follows:

RQ1. Which is the publication venue of the instructional technologies applied to higher engineering education? One relevant distribution applies the Wieringa principle of study classification, as exemplified in (Petersen, 2015), where studies have been classified according to the scope of the intervention (Fig. 1): Validation Research (VR) – 7% papers, Evaluation Research (ER) – 13% papers, Solution Proposal (SP) – 28% papers, Philosophical Papers (PP) – 22% papers, Opinion Paper (OP) – 11%, Experience Papers (EP) - 18% papers. The most representative interval of publication, results to be 2018–2019 with 30% of the total number of studies.

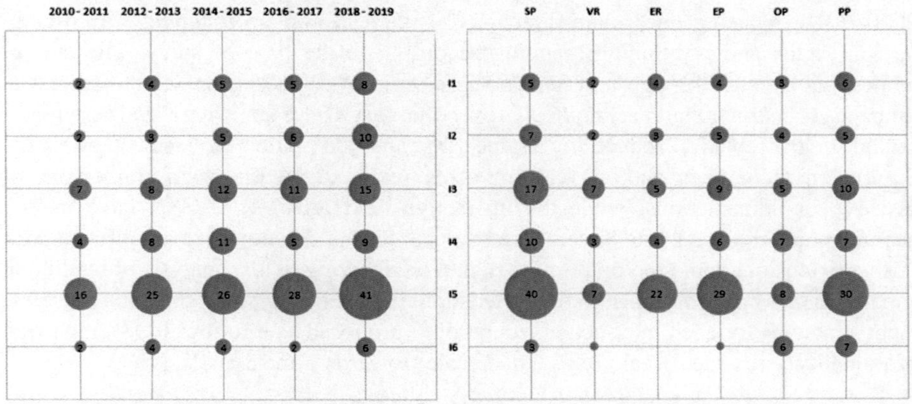

Fig. 1. Mapping distributions according to a) Years b) Type of research

RQ2. To which higher engineering education fields have the interactive instructional technologies been applied? The most relevant distribution maps the allocation of the predefined interactive instructional technologies to engineering fields according to the type of intervention (as defined in the PICO – I dimension). The best represented field results to be computer science engineering - CE (23%) and mechanical engineering as a field of Production Engineering – PE (20%), where the most documented interactive instructional techniques result to be simulative environments and tools (28%) (Fig. 2).

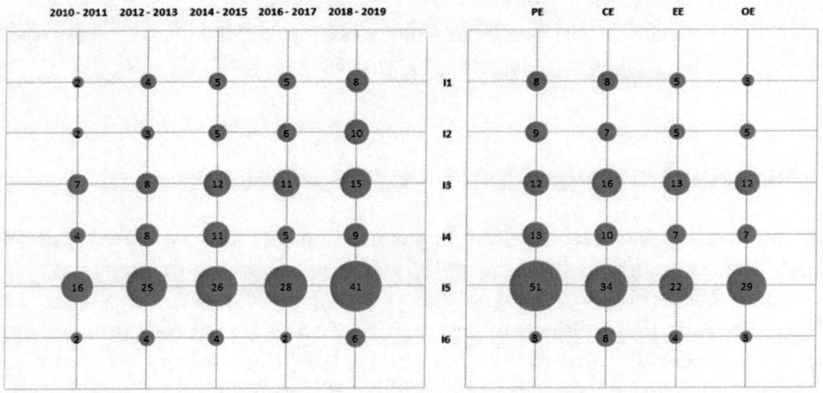

Fig. 2. Mapping distribution according to a) Years b) Main engineering field

5 Conclusion

The current study provides an overview of interactive assistive technologies applied to higher engineering education, through a Systematic Mapping Study conducted between 2010–2019, focusing on 2 main directions of interest: domain mapping within engineering fields and positioning mapping according to the type of study. The current work demonstrates through a systematic assessment the growing re-search interest in Interactive Instructional Technologies, over the past 10 years, where the best represented fields remain Production Engineering and Computer Science Engineering. Simulation techniques and 3D environments result to be the main framework of assistive implementations, while the interactive fields AR/VR/MR/XR have mainly become popular since 2016. Several directions of further development are proposed for further extension: depiction on primary higher engineering fields, analysis of intelligent enhancements applied to instructional technologies, specific focus on impact assessment strategies for each inter-active technology (proposed as a further RQ3), inclusion of a quality assessment scale (e.g. Jadad scale) towards a meta-analysis.

References

1. OECD: The future of education and skills: education 2030. OECD Education Working Paper, p. 23 (2018)
2. Hénard, F., Roseveare, D.: Fostering Quality Teaching in Higher Education : Policies and Practices, p. 54 (2012)

3. Kitchenham, B., Charters, S.: Guidelines for performing systematic literature reviews in software engineering. In: Engineering, vol. 2, p. 1051 (2007)
4. Petersen, K., Feldt, R., Mujtaba, S., Mattsson, M.: Systematic mapping studies in software engineering. In: Proceedings of the 12th International Conference on Evaluation and Assessment in Software Engineering, pp. 68–77 (2008)
5. Petersen, K., Vakkalanka, S., Kuzniarz, L.: Guidelines for conducting systematic mapping studies in software engineering: an update. Inf. Softw. Technol. **64**, 1–8 (2015)
6. Torres-Carrion, P.V., Gonzalez-Gonzalez, C.S., Aciar, S., Rodriguez-Morales, G.: Methodology for systematic literature review applied to engineering and education. In: IEEE Global Engineering Education Conference EDUCON, April 2018, pp. 1364–1373 (2018)
7. Dicheva, D., Dichev, C., Agre, G., Angelova, G.: Gamification in education: a systematic mapping study. J. Educ. Technol. Soc. **18**, 75–88 (2015)
8. Ibáñez, M.B., Delgado-Kloos, C.: Augmented reality for STEM learning: a systematic review. Comput. Educ. **123**, 109–123 (2018)
9. Arici, F., Yildirim, P., Caliklar, Ş, Yilmaz, R.M.: Research trends in the use of augmented reality in science education: content and bibliometric mapping analysis. Comput. Educ. **142**, 103647 (2019)
10. Díaz, S., Díaz, J., Ahumada, D.: Virtual reality in high school: a systematic mapping study. In: Proceedings - International Conference of the Chilean Computer Science Society SCCC, vol. 2018, pp. 1–7 (2019)
11. Jeong, H., Hmelo-Silver, C.E., Jo, K.: Ten years of computer-supported collaborative learning: a meta-analysis of CSCL in STEM education during 2005–2014. Educ. Res. Rev. **28**, 100284 (2019)
12. Lai, J.W., Bower, M.: How is the use of technology in education evaluated? A systematic review. Comput. Educ. **133**, 27–42 (2019)
13. Diegmann, P., Schmidt-Kraepelin, M., Eynden, S., Basten, D.: Benefits of augmented reality in educational environments-a systematic literature review. Benefits **3**(6), 1542–1556 (2015)
14. Acosta, B., Luis, J., Baldiris, S., Gesa, R., Graf, S.: Augmented reality trends in education: a systematic review of research and applications. J. Educ. Technol. Soc. **17**(4), 133–149 (2014)
15. Akçayır, M., Akçayır, G.: Advantages and challenges associated with augmented reality for education: a systematic review of the literature. Educ. Res. Rev. **20**, 1–11 (2017)
16. Martin, S., Diaz, G., Sancristobal, E., Gil, R., Castro, M., Peire, J.: New technology trends in education: seven years of forecasts and convergence. Comput. Educ. **57**(3), 1893–1906 (2017)
17. Santos, M.E.C., Chen, A., Taketomi, T., Yamamoto, G., Miyazaki, J., Kato, H.: Augmented reality learning experiences: Survey of prototype design and evaluation. IEEE Trans. Learn. Technol. **7**(1), 38–56 (2013)
18. Sirakaya, M., Alsancak Sirakaya, D.: Trends in educational augmented reality studies: a systematic review. Malays. Online J. Educ. Technol. **6**(2), 60–74 (2018)
19. Broadbent, J., Poon, W.L.: Self-regulated learning strategies & academic achievement in online higher education learning environments: a systematic review. Internet High. Educ. **27**, 1–3 (2015)
20. Ali, A.A., Dafoulas, G.A., Augusto, J.C.: Collaborative educational environments incorporating mixed reality technologies: a systematic mapping study. IEEE Trans. Learn. Technol. **12**(3), 321–332 (2019)
21. Radianti, J., Majchrzak, T.A., Fromm, J., Wohlgenannt, I.: A systematic review of immersive virtual reality applications for higher education: design elements, lessons learned, and research agenda. Comput. Educ. **147**, 103778 (2020)

22. Nunes, F.B., Herpich, F., Paschoal, L.N., De Lima, J.V., Tarouco, L.M.R.: Systematic review of virtual worlds applied in education. In: Brazilian Symposium on Computers in Education, vol. 27, no. 1, p. 657 (2016)
23. Avellar, G.M.N., Barbosa, E.F.: Virtual and augmented reality in the teaching and learning of programming: a systematic mapping study. In: Brazilian Symposium on Computers in Education (Simpósio Brasileiro de Informática na Educação-SBIE), vol. 30, no. 1, p. 664 (2019)
24. de Sousa Borges, S., Durelli, V.H., Reis, H.M., Isotani, S.: A systematic mapping on gamification applied to education. In: Proceedings of the 29th Annual ACM Symposium on Applied Computing, pp. 216–222 (2014)
25. Krassmann, A.L., Paschoal, L.N., Falcade, A., Medina, R.D.: Evaluation of game-based learning approaches through digital serious games in computer science higher education: a systematic mapping. In: 2015 14th Brazilian Symposium on Computer Games and Digital Entertainment, pp. 43–51 (2015)
26. Maurício, R.D.A., Veado, L., Moreira, R.T., Figueiredo, E., Costa, H.: A systematic mapping study on game-related methods for software engineering education. Inf. Softw. Technol. **95**, 201–218 (2018)
27. Subhash, S., Cudney, E.A.: Gamified learning in higher education: a systematic review of the literature. Comput. Hum. Behav. **87**, 192–206 (2018)
28. Vlachopoulos, D., Makri, A.: The effect of games and simulations on higher education: a systematic literature review. Int. J. Educ. Technol. High. Educ. **14**(1), 22 (2017)
29. Cook, D.A., Hamstra, S.J., Brydges, R., Zendejas, B., Szostek, J.H., Wang, A.T., Hatala, R.: Comparative effectiveness of instructional design features in simulation-based education: systematic review and meta-analysis. Med. Teach. **35**(1), 867–898 (2013)
30. Battistella, P., von Wangenheim, C.G.: Games for teaching computing in higher education–a systematic review. IEEE Technol. Eng. Educ. **9**(1), 8–30 (2016)
31. Gu, X., Blackmore, K.L.: A systematic review of agent-based modelling and simulation applications in the higher education domain. High. Educ. Res. Dev. **34**(5), 883–898 (2015)

Author Index

M. E. Auer and T. Rüütmann (Eds.): ICL 2020, AISC 1328, pp. 805–808, 2021.
https://doi.org/10.1007/978-3-030-68198-2